Foundations of Nuclear and Particle Physics

This textbook brings together nuclear and particle physics, presenting a balanced overview of both fields as well as the interplay between the two. The theoretical as well as the experimental foundations are covered, providing students with a deep understanding of the subject. In-chapter exercises ranging from basic experimental to sophisticated theoretical questions provide an important tool for students to solidify their knowledge. Suitable for upper undergraduate courses in nuclear and particle physics as well as more advanced courses, the book includes road maps guiding instructors on tailoring the content to their course. Online resources including color figures, tables, and a solutions manual complete the teaching package. This textbook will be essential for students preparing for further study or a career in the field who require a solid grasp of both nuclear and particle physics.

Key features

- Contains up-to-date coverage of both nuclear and particle physics, particularly the areas where the two overlap, equipping students for the real-world occasions where aspects of both fields are required for study
- Covers the theoretical as well as the experimental foundations, providing students with a deep understanding of the field
- Exercises ranging from basic experimental to sophisticated theoretical questions provide an important tool for readers to consolidate their knowledge

Thomas William Donnelly is a Senior Research Scientist at MIT. He received his PhD in Theoretical Nuclear Physics in 1967 from the University of British Columbia.

Joseph Angelo Formaggio is an Associate Professor of Physics at MIT. He received his PhD in Physics at Columbia University in 2001. He has been a member on a number of experiments including the Sudbury Neutrino Observatory and the KATRIN neutrino experiment.

Barry R Holstein is an Emeritus Professor Physics at the University of Massachusetts. He received his PhD in Physics from Carnegie Mellon University in 1969. He is Editor of *Annual Reviews of Nuclear and Particle Physics,* Consulting Editor of the *American Journal of Physics*, and Associate Editor of the *Journal of Physics G.*

Richard Gerard Milner is a Professor of Physics at MIT. He received his PhD from the California Institute of Technology in 1985. He has proposed and led experiments at SLAC, DESY, MIT-Bates, and Jefferson Laboratory.

Bernd Surrow is a Professor of Physics at Temple University. He gained his PhD in Physics at the University of Hamburg in 1998. He has been a member of a number of experiments including the STAR experiment at BNL, the CMS and OPAL experiments at CERN and the ZEUS experiment at DESY.

Foundations of Nuclear and Particle Physics

T. W. DONNELLY
Massachusetts Institute of Technology, Cambridge, MA

J. A. FORMAGGIO
Massachusetts Institute of Technology, Cambridge, MA

B. R. HOLSTEIN
University of Massachusetts, Amherst, MA

R. G. MILNER
Massachusetts Institute of Technology, Cambridge, MA

B. SURROW
Temple University, Philadelphia, PA

CAMBRIDGE
UNIVERSITY PRESS

CAMBRIDGE
UNIVERSITY PRESS

University Printing House, Cambridge CB2 8BS, United Kingdom

Cambridge University Press is part of the University of Cambridge.

It furthers the University's mission by disseminating knowledge in the pursuit of
education, learning, and research at the highest international levels of excellence.

www.cambridge.org
Information on this title: www.cambridge.org/9780521765114

© Cambridge University Press 2017

This publication is in copyright. Subject to statutory exception
and to the provisions of relevant collective licensing agreements,
no reproduction of any part may take place without the written
permission of Cambridge University Press.

First published 2017

Printed in the United States of America by Sheridan Books, Inc.

A catalogue record for this publication is available from the British Library.

Library of Congress Cataloguing in Publication Data
Names: Donnelly, T. W. (T. William), 1943– author. | Formaggio, Joseph A., 1974– author. |
Holstein, Barry R., 1943– author. | Milner, Richard Gerard, 1956– author. | Surrow, Bernd, 1998– author.
Title: Foundations of nuclear and particle physics / T. William Donnelly
(Massachusetts Institute of Technology), Joseph A. Formaggio (Massachusetts Institute
of Technology), Barry R. Holstein (University of Massachusetts, Amherst), Richard G. Milner
(Massachusetts Institute of Technology), Bernd Surrow (Temple University, Philadelphia).
Description: Cambridge, United Kingdom ; New York, NY : Cambridge
University Press, [2016] | Includes index.
Identifiers: LCCN 2016026959| ISBN 9780521765114 (hardback) |
ISBN 0521765110 (hardback)
Subjects: LCSH: Nuclear physics–Textbooks. | Particles (Nuclear physics)–Textbooks.
Classification: LCC QC776 .D66 2016 | DDC 539.7–dc23 LC record available
at https://lccn.loc.gov/2016026959

ISBN 978-0-521-76511-4 Hardback

Additional resources for this publication at www.cambridge.org/9780521765114.

Cambridge University Press has no responsibility for the persistence or accuracy
of URLs for external or third-party Internet Web sites referred to in this publication,
and does not guarantee that any content on such Web sites is, or will remain,
accurate or appropriate.

Bill $\Longleftrightarrow$ to Barbara

Joe $\Longleftrightarrow$ to Mike, Hamish, Janet, and John, for their unwavering wisdom;
to Jaymi, Coby, and Joshua, for their unquestioning love

Barry $\Longleftrightarrow$ to Jeremy and Jesse

Richard $\Longleftrightarrow$ to Liam Milner for inspiration
and to Eileen, Will, Sam, and David for love and support

Bernd $\Longleftrightarrow$ to Suzanne, Alec, Arianna, and Carl for their love and support

Contents

The first question one might ask about this book is: Why do we need another text on the subject of nuclear and particle physics when excellent texts already exist in both of these areas? Indeed, it is true that each sub-discipline has texts that range from elementary to very advanced and cover specific topics in varying degrees of depth that can be used for the appropriate types of courses. For instance, there are fine books on quantum field theory [Bjo64, Pes95, Wei05, Sch14], on the constituent quark model [Clo79], on high-energy physics [Gri08, Hal84], on hadron scattering [Col84], and on nuclear structure [Des74, Wal95, Won98, Pov08, Row10]. However, there are relatively few textbooks that cover several sub-disciplines in a coherent and balanced way, and those that do exist are either more elementary, e.g., Povh et al. [Pov08] than the present book, or are cast at a more theoretical level and are too advanced for the goals we as authors have set for ourselves. Having a book that stresses the interconnections between the two areas of subatomic physics is crucial, since increasingly one finds that the two fields overlap and that it is essential for a graduate student conducting frontier research and preparing for a career in the field to have an understanding of both. An example of this overlap occurs, for instance, in modern neutrino physics wherein experiments utilizing several-GeV neutrinos as probes almost always involve targets/detectors constructed from nuclei and specifics of nuclear structure are unavoidably required to properly interpret such data.

One specific decision we have made in designing this book is to assume that the reader is familiar with the basics of quantum field theory. More elementary texts typically do not make this assumption and thus much of the discussion, for instance, of lepton scattering from hadrons and nuclei, or of the foundations of chiral symmetry and effective field theory is limited and not at the frontier of the field. We realize that many students today do have at least an introductory course in quantum field theory, or are taking one simultaneously with a course that this book covers, and thus we have followed a somewhat more advanced approach than has been customary. We have included in Appendix B an overview of the essential aspects of quantum mechanics and quantum field theory that are needed for the book. Furthermore, the subject of many-body theory underlies much of nuclear physics and the presentation of this subject can also be rather elementary, as is usually the case in texts that cover the two fields, or too advanced for our purposes, focusing on Green's functions, diagrammatic techniques and nonperturbative approximations at a theoretical level. We have chosen a middle course: we have covered the basics of many-body theory, but also have introduced some of the important diagrammatic representations of the nonperturbative approximations employed very widely in quantum physics ranging from atomic and condensed matter physics to the present context of nuclear and hadronic physics.

The book's central focus is to describe the current understanding of the sub-atomic world within the framework of the Standard Model. The layout of the book is summarized as follows: In the first quarter of the book, the Standard Model is developed. The structure of the nucleon and few-body nuclei are discussed in the second quarter. In the third quarter, the structure and properties of atomic nuclei are described. Lepton scattering is the principal tool used in the central narrative of the book to understand hadrons. In the final quarter of the book we present extensions of the earlier focus on EM lepton scattering to include the weak interactions of leptons with nucleons and nuclei. This begins with a chapter on beta-decay and progresses to intermediate-to-high energy neutrino-induced reactions. These two chapters are followed by two more that build on what occurs earlier in the book, namely, on applications to nuclear and particle astrophysics and to studies of the hot, dense phase of matter formed in heavy-ion collisions The book closes with a brief perspective on physics beyond the Standard Model.

We should also emphasize that the use of word "foundations" in the title of the book is intentional, indicating that this text is not an encyclopedia where one might find material on all of the major topics in the field, albeit at a superficial level. Rather, we have consciously made choices in what and what not to present. We have, for instance, not developed the topic of intermediate-energy hadron scattering, emphasizing lepton scattering instead and have not attempted to cover the lattice approach to the solution of QCD. While the important areas of nuclear structure and the high-energy frontier are covered, we note that excellent, up-to-date, comprehensive textbooks on these important areas are available. Our intent has been to provide the reader with basic material upon which to build by subsequently employing the more advanced sources that exist when it becomes necessary for a more in-depth understanding of specific subjects. In this regard, we have included references to review articles, so that the interested reader can pursue material to a more advanced level. Just what to emphasize and what merely to refer to in passing is, of course, subjective; however, having five co-authors has allowed us to debate the choices we have made.

We view the approximately 120 exercises provided throughout the book and located at the end of each chapter as an important tool for the reader to consolidate their understanding of the material in the book. There exists significant variety in these exercises, ranging from basic experimental issues to sophisticated theoretical questions. Many owe their origins to other sources, but we have tried to tailor them to the material discussed here.

The authors have all taught courses of the type described above at various levels. Specifically, at MIT the book covers the scopes set out for the introductory first-year graduate course in nuclear and particle physics (8.701), together with the second-year graduate courses in nuclear (8.711) and particle (8.811) physics. All graduate students in experimental nuclear/particle physics at MIT are required to take the latter two, with the former being a prerequisite. Additionally, at MIT there is an advanced undergraduate course in nuclear/particle physics (8.276), as well as more advanced courses in many-body theory (8.361), nuclear theory (8.712) and electroweak interactions (8.841) – all taught by one of the authors (TWD) – for which at least some of this text is appropriate.

We acknowledge that the derivation of the QCD Lagrangian in Chapter 5 owes its origins to Professor Frank Wilczek. We acknowledge that Chapter 19 was shaped by the work of Professor Berndt Müller and his colleagues. We thank the Super-Kamiokande Collaboration for permission to use their image on the cover.

The book's evolution profited from its use in draft form as a resource for the MIT course 8.711 taught by one of us (RGM) and Dr. Stephen Steadman in the spring semesters of 2014, 2015, and 2016. We acknowledge the constructive feedback from the MIT graduate students in those classes. Further, we acknowledge careful and critical reading of drafts by Dr. Jan Bernauer, Charles Epstein, Dr. Douglas Hasell, Dr. Rebecca Russell, Dr. Axel Schmidt, Dr. Stephen Steadman, Reynier Cruz Torres and Constantin Weisser at MIT, Professor James Napolitano, Dr. Matt Posik, Devika Gunarathne, Amani Kraishan and Daniel Olvitt at Temple University, Rosi Reed at Lehigh University and Rosi Esha at UCLA. We are grateful to Dr. Brian Henderson for a careful reading of all of the exercises. We thank Connor Dorothy-Pachuta for his considerable expertise in creating many of the figures in the book. There are, of course, many others to thank who, over the years, have been our collaborators – we cannot list them all, but they will find their efforts reflected in many of our choices for what to present. We do, however, wish to acknowledge three who directly played roles in developing some of the figures in Chapters 16 and 18, namely, Professors Maria Barbaro and Juan Caballero, and Guillermo Megias.

In addition to being an integrated text, there are other aspects of this presentation that we feel are important. Specifically, we have attempted to make strong connections with contemporary experiments and have tried, whenever possible, to help the reader become aware of the relevant frontier experimental facilities available and planned worldwide. Doing so is, of course, time dependent; but we have tried to be as up to date as possible. We have also made liberal use of the Particle Data Group website [PDG14] as a resource with which we encourage all students to become familiar. Finally, in Appendix A we have collected information that we believe will be useful to readers.

Introduction

The past one hundred years has witnessed enormous advances in human understanding of the physical universe in which we have evolved. For the past fifty years or so, the Standard Model of the subatomic world has been systematically developed to provide the quantum mechanical description of electricity and magnetism, the weak interaction, and the strong force. Symmetry principles, expressed mathematically via group theory, serve as the backbone of the Standard Model. At this time, the Standard Model has passed all tests in the laboratory. Notwithstanding this success, most of the matter available to experimental physicists is in the form of atomic nuclei. The most successful description of nuclei is in terms of the observable protons, neutrons, and other hadronic constituents, and not the fundamental quarks and gluons of the Standard Model. Thus, the professional particle or nuclear physicist should be comfortable in applying the hadronic description of nuclei to understanding the structure and properties of nuclei. Experimentally, lepton scattering has proved to be the cleanest and most effective tool for unraveling the complicated structure of hadrons. Its application over different energies and kinematics to the nucleon, few-body nuclei, and medium- and heavy-mass nuclei has provided the solid body of precise experimental data on which the Standard Model is built.

In addition, the current understanding of the microcosm described in this book provides answers to many basic questions: How does the Sun shine? What is the origin of the elements? How old is the Earth? Further, it underscores many aspects of modern human civilization, e.g., MRI imaging uses the spin of the proton, nuclear isotopes are essential medical tools, nuclear reactions have powered the Voyager spacecraft since 1977 into interstellar space.

The purpose of the book is to allow the graduate student to understand the foundations and structure of the Standard Model, to apply the Standard Model to understanding the physical world with particular emphasis on nuclei, and to establish the frontiers of current research. There are many outstanding questions that the Standard Model cannot answer. In particular, astrophysical observation strongly supports the existence of dark matter, whose direct detection has thus far remained elusive.

Essential to making progress in understanding the subatomic world are the sophisticated accelerators that deliver beams of particles to experiments. Existing lepton scattering facilities include Jefferson Laboratory in the US, muon beams at CERN, and University of Mainz and University of Bonn in Germany. Intense photon beams are used at the $HI\gamma S$ facility at Duke University in the U.S., and in Japan at LEPS at SPring-8, and at Elphs at Tohoku University. Hadrons beams are used at the TRIUMF laboratory in Vancouver, Canada, using the COSY accelerator in Juelich, Germany, at the Paul Scherrer Institute (PSI) in Switzerland, and at the Joint Institute for Nuclear Research (JINR),

Dubna, Russia. Neutron beams are used for subatomic physics research at the Institut Laue-Langevin (ILL), Grenoble, France, at both the Los Alamos Neutron Science Center (LANSCE) and the Spallation Neutron Source (SNS) in the US, and at the future European Spallation Source (ESS) in Sweden. The hot, dense matter present in the early universe is studied using heavy-ion beams at the Relativistic Heavy Ion Collider (RHIC) in the US and at the Large Hadron Collider (LHC) at CERN. Of course, searches for new physics beyond the Standard Model are underway at the high-energy frontier of 13 TeV at CERN. Understanding the structure of nuclei, with particular emphasis on the limits of stability, is a major worldwide endeavor. The most powerful facility at present is the Rare Isotope Beam Facility (RIBF) at RIKEN in Japan. In the US, the frontier experiments at present are carried out at the National Superconducting Cyclotron Laboratory at Michigan State University (MSU) and at the ATLAS facility at Argonne National Laboratory. A future Facility for Rare Isotope Beams (FRIB) is under construction at MSU and is expected to have world-leading capabilities by 2022, as is a facility in South Korea, the Rare Isotope Science Project (RAON). Hadron beams for research are available at Los Alamos and the Spallation Neutron Source in the US, GSI in Germany, J-PARC in Japan, and NICA at Dubna, Russia. A major new facility FAIR is planned at GSI. Neutrino beams are generated at Fermilab, CERN, and J-PARC and directed at detectors located both at the Earth's surface and deep underground. A major new Deep Underground Neutrino Experiment (DUNE) is planned in the US using the Fermilab beam and the Sanford Underground Research Laboratory in South Dakota. Belle II, an experiment at the high luminosity e^+e^- collider SuperKEKB in Japan, will come online within the next several years and provide new stringent tests of flavor physics. Annihilation of electrons and positrons is used to probe the Standard Model at both the Double Annular ϕ Factory for Nice Experiments (DAFNE) collider in Frascati, Italy as well as the Beijing Electron Positron Collider (BEPC) in China. Finally, a high luminosity electron–ion collider has been widely identified by as the next machine to study the fundamental quark and gluon structure of nuclei and machine designs are under development in the US, Europe, and China.

To begin, let us remind the reader of the particles that comprise the Standard Model (see Fig. 1.1). As will be discussed in due course, the Standard Model starts with massless particles and then, through spontaneous symmetry breaking, these interacting particles acquire masses in almost all cases. The measured spectrum of masses is still a mystery; indeed, in the case of the neutrinos, intense effort is going into determining the actual pattern of masses in Nature. Note that at this microscopic level, but also at the hadronic/nuclear level, when one says that particles interact with one another what is meant is that some particle is exchanged between two other particles, thereby mediating the interaction. For instance, an electron can exchange a photon with a quark whereby the photon mediates the $e - q$ interaction. Or two nucleons (protons and neutrons) can exchange a pion and one has the long-range part of the NN interaction.

The organizational principle for this book centers on building from the underlying fundamental particles (leptons, quarks, and gauge bosons) to hadrons (mesons and baryons) built from $q\bar{q}$ and qqq, respectively, and on to many-body nuclei or hypernuclei built from these hadronic constituents. At very low energies and momenta the last are the relevant effective degrees of freedom, since, using the Heisenberg Uncertainty Principle, such kinematics

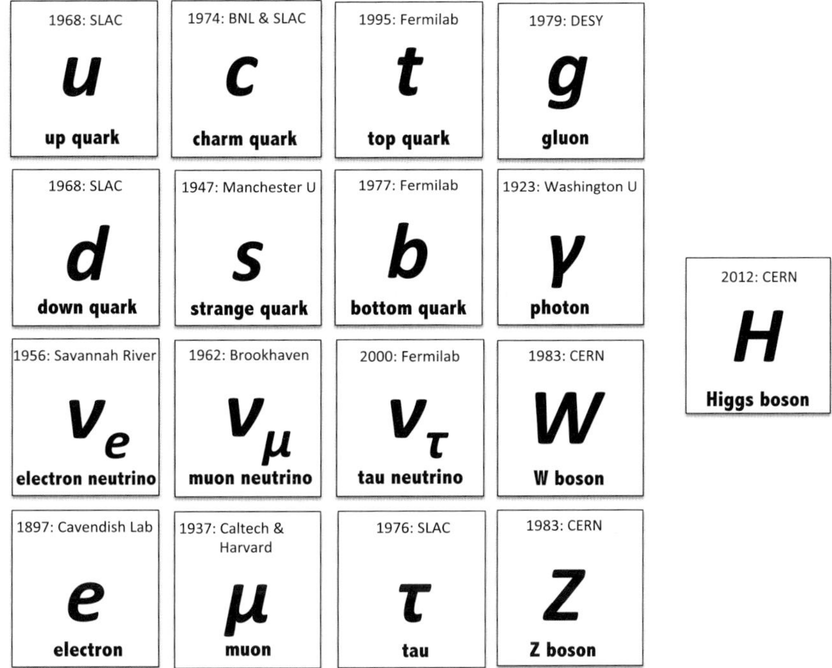

Fig. 1.1   The particles of the Standard Model.

translate into large distance scales where the microscopic ingredients are packaged into the macroscopic hadronic degrees of freedom. Then, as the energy/momentum is increased, more and more of the sub-structure becomes relevant, until at very high energy/momentum scales the QCD degrees of freedom must be used to represent what is observed.

Naturally, there can be a blending between the different degrees of freedom and, where they overlap, it may be possible to use one language or the other. And in some cases it turns out to be important to address both the "fundamental" physics issues and the larger-scale nuclear structure issues at the same time. This book attempts to present the foundations of the general field of nuclear/particle physics – sometimes called subatomic physics – in a single volume, trying to maintain a balance between the very microscopic QCD picture and the hadronic/nuclear picture.

The outline of the book is the following. After this introductory chapter, in Chapter 2 the basic ideas of symmetries are introduced. In general discussions of quantum physics it is often advantageous to exploit the exact (or at least approximate) symmetries in the problem, for then selection rules emerge where, for instance, matrix elements between specific initial and final states of certain operators can only take on a limited set of values. An example of what will be important in later discussions is the use of good angular momentum quantum numbers and the transformation properties of multipole operators (see Chapter 7) where conservation of angular momentum leads to a small set of allowed values for matrix elements of such operators taken between states that have known spins. Another example of an important (approximate) symmetry is provided by invariance under spatial inversion, namely, parity: to the extent that parity is a good symmetry again only specific transitions can occur. Other symmetries discussed in Chapter 2 include charge conjugation and time reversal, as well as discrete unitary flavor symmetries, the latter being important for classifying the hadrons built from constituent quarks, namely, the subject of Chapter 3.

After these introductory discussions the book proceeds to build up from particles to hadrons to many-body nuclei, starting in Chapter 4 with the Standard Model of particle physics. In this one begins with massless leptons, quarks, and gauge bosons together with the Higgs and then through spontaneous symmetry breaking generates the basic familiar building blocks with their measured masses. The recent successful discovery of the Higgs boson at the Large Hadron Collider (LHC) is summarized.

The Standard Model has proven to be extremely successful and, at the time of writing, there is as yet no clear evidence that effects beyond the Standard Model (BSM) are needed; in the final chapter of the book, Chapter 21 we return to summarize some of these BSM issues. For the present, following the path of increasing complexity, in Chapters 5 and 6 the ideas and models employed in descriptions of low-Q^2, strong coupling QCD are discussed in some detail, including what is not typically covered in a book at this level, namely, chiral symmetry.

Chapters 7 through 10 form a distinct section where the aim is to visualize the structure of the proton, neutron, and nuclei in terms of the fundamental quarks and gluons of QCD. At low and medium energies, this is carried out using lepton scattering where intense beams of high quality are available. Thus, snapshots of the nucleon charge and magnetism and quark momentum and spin distributions are directly obtainable in the form of structure functions and form factor distributions. Chapter 7 provides an introduction to lepton scattering, including both parity-conserving and parity-violating scattering. Since lepton scattering is being used as a common theme in much of the rest of the book, Chapter 7 is the first stop along the way where the multipole decomposition of the electromagnetic current is developed in some detail. This is followed in Chapter 8 by a discussion of elastic scattering from the nucleon. At this time, a direct connection between elastic scattering and QCD remains elusive and the most successful theoretical description is in terms of hadrons. Chapter 9 describes the current understanding of the structure of hadrons in terms of high-energy lepton scattering and this is directly interpretable in terms of perturbative QCD. Further, the gluon momentum and spin distributions are indirectly determined via the QCD evolution equations. The parton distributions are snapshots of nucleon structure over different spatial resolutions and with different shutter speeds. Lepton scattering constitutes a theme of the book at both high- and low-energy scales and with the full electroweak interaction. Due to the lack of suitable lepton beams, QCD is at present probed at the highest energies using hadron beams. This is the focus of Chapter 10 and the measurements extend and complement those with lepton beams in the previous chapters. For example, direct experimental information on the contribution of gluons to the spin of the proton has become possible only through polarized proton–proton collisions.

The above constitutes the first part of the book after which the building-up process moves from hadrons to nuclei. The next step is to deal with the simplest system that is not a single baryon, namely, the system of two nucleons, discussing NN scattering and the properties of the only bound state with baryons number two, the deuteron in

Chapter 11. For the latter the EM form factors and electrodisintegration are treated in some detail. After this, in Chapter 12 the so-called few-body nuclei, those with $A = 3$ and 4, constitute the focus.

For nuclei heavier than the $A = 2$, 3, and 4 cases, treating the many-body problem forms the basic issue, and accordingly in Chapter 13 an overview of the general nuclear "landscape" is presented, showing the typical characteristics of nuclei, including the regions where nuclei are stable (the "valley of stability") out to where they are just unstable (the "drip lines"), and their regions of especially tight binding (the "magic numbers"). Also in this chapter the concept of infinite nuclear matter and neutron matter is introduced and treated in some detail. This is followed in Chapter 14 by a discussion of a selection of typical nuclear models. As mentioned earlier, this book is not intended to be a theoretical text on nuclear many-body theory. That said, this chapter has sufficient detail that the basic issues in this area can be appreciated. Importantly, the tools used in this part of the field must be capable of dealing with nonperturbative interacting systems and accordingly this provides a theme in this chapter where discussions of the so-called Hartree–Fock (HF) and Random Phase Approximations (RPA) are provided together with an introduction to diagrammatic representations of the approximations. Also typical collective models are discussed as examples of how one may start with some classical oscillation or vibration of the nuclear fluid, make harmonic approximations to those movements, and then quantize the latter to arrive at semi-classical descriptions of nuclear excitations ("surfons," "rotons," etc.), as is done in many areas of physics where similar techniques are employed.

The above discussions are then followed by two chapters focused on electron scattering from nuclei, Chapter 15 where elastic scattering is treated in some detail, together with some applications of the models introduced in Chapter 14 for low-lying excited states. Chapter 16 continues this by treating higher-lying excitations where different modeling is required. Specifically, the Relativistic Fermi Gas (RFG) model is derived and used as a prototype for more sophisticated approaches. It is also the starting point for similar discussions of neutrino scattering from nuclei to follow in Chapter 18. Before those are presented, in Chapter 17 the weak interaction provides the focus and we see how precision beta-decay experiments can be used as a probe for beyond Standard Model physics. Chapter 18 deals with the subject of neutrinos and the fact that one flavor can oscillate into another, since neutrinos are known to have mass. At the time of writing, the detailed nature of the mass spectrum, whether or not CP violation is present in the leptonic sector and whether neutrinos are Dirac or Majorana particles are still under investigation and intensive efforts are being undertaken worldwide to shed light on these interesting questions.

In Chapter 19 the high-energy regime (essentially quark–quark scattering) is re-visited within the context of relativistic heavy-ion scattering. Here the nature of the modeling is somewhat different from that discussed in most of the rest of the book with statistical mechanics being called into play together with fluid dynamics. An informed practitioner in the general field of nuclear/particle physics should be familiar with this subject as well.

The book concludes with Chapter 20 on nuclear and particle astrophysics using many of the concepts treated in the rest of the book, and with Chapter 21 where the types of signatures of effects beyond the Standard Model are summarized, together with two appendices where some useful material is gathered.

While we strongly advocate using the book to explore both nuclear and particle physics in a coherent, balanced way, nevertheless it might be that it will also be used in a course that emphasizes one subfield or the other. Accordingly, we suggest the following "road maps" to help the reader negotiate the text for those purposes. When the emphasis is placed on particle physics we suggest paying the closest attention to Chapters 2 to 10 and 21, with some parts of Chapters 17, 18, and perhaps 19, and when the emphasis is on nuclear physics Chapters 2, 7, 11 to 18, 20 and perhaps 19.

We strongly recommend the following online resources as important tools for enhancing the material presented in this book.

1. The Review of Particle Physics, Particle Data Group
   http://pdg.lbl.gov includes a compilation and evaluation of measurements of the properties of the elementary particles. There is an extensive number of review articles on particle physics, experimental methods, and material properties as well as a summary of searches for new particles beyond the SM.
2. National Nuclear Data Center
   http://www.nndc.bnl.gov is a source of detailed information on the structure, properties, reactions, and decays of known nuclei. It contains an interactive chart of the nuclides as well as a listing of the properties for ground and isomeric states of all known nuclides.

We conclude this introductory chapter with some exercises designed to introduce some of the concepts which we hope our particle/nuclear students will be able to address.

Exercises

1.1 US Energy Production

In 2011, the United States of America required 3,856 billion kW-hours of electricity. About 20% of this power was generated by $\sim$100 nuclear fission reactors. About 67% was produced by the burning of fossil fuels, which accounted for about one-third of all greenhouse gas emissions in the US. The remaining 13% was generated using other renewable energy resources. Consider the scenario where all the fossil fuel power stations are replaced by new 1-GW nuclear fission reactors. How many such reactors would be needed?

1.2 Geothermal Heating

It is estimated that 20 TW of heating in the Earth is due to radioactive decay: 8 TW from ^{238}U decay, 8 TW from ^{232}Th decay, and 4 TW from ^{40}K decay. Estimate the total amount of ^{238}U, ^{232}Th, and ^{40}K present in the Earth in order to produce such heating.

1.3 Radioactive Thermoelectric Generators

A useful form of power for space missions which travel far from the Sun is a radioactive thermoelectric generator (RTG). Such devices were first suggested by the science fiction writer Arthur C. Clarke in 1945. An RTG uses a thermocouple to convert the heat released by the decay of a radioactive material into electricity by the Seebeck effect. The two Voyager spacecraft have been powered since 1977 by RTGs using ^{238}Pu. Assuming a mass of 5 kg of ^{238}Pu, estimate the heat produced and the electrical power delivered. (Do not forget to include the $\sim 5\%$ thermocouple efficiency.)

1.4   Fission versus Fusion

Energy can be produced by either nuclear fission or nuclear fusion.

a) Consider the fission of ^{235}U into ^{117}Sn and ^{118}Sn, respectively. Using the mass information from a table of isotopes, calculate (i) the energy released per fission and (ii) the energy released per atomic mass of fuel.

b) Consider the deuteron–triton fusion reaction

$$^{2}\mathrm{H} + {}^{3}\mathrm{H} \rightarrow {}^{4}\mathrm{He} + \mathrm{n} \, .$$

Using the mass information from the periodic table of the isotopes, calculate (i) the energy released per fusion and (ii) the energy released per atomic mass unit of fuel.

1.5   Absorption Lengths

A flux of particles is incident upon a thick layer of absorbing material. Find the absorption length, the distance after which the particle intensity is reduced by a factor of $1/e \sim 37\%$ (the absorption length) for each of the following cases:

a) When the particles are thermal neutrons (i.e., neutrons having thermal energies), the absorber is cadmium, and the cross section is 24,500 barns.

b) When the particles are 2 MeV photons, the absorber is lead, and the cross section is 15.7 barns per atom.

c) When the particles are anti-neutrinos from a reactor, the absorber is the Earth, and the cross section is 10^{-19} barns per atomic electron.

Symmetries

2.1 Introduction

When studying quantum systems, exploiting knowledge about the inherent symmetries is usually an important step to take before addressing issues of dynamics [Sch55, Sak94, Rom64, Gri08]. This motivates a discussion of group theory, and so we shall begin by summarizing some of the basic elements needed, particularly when discussing symmetries in particle and nuclear physics. More details can be found in specialized texts on the subject [Ham62, Clo79]. Noether's theorem states that if the Hamiltonian is invariant under a continuous group of transformations, then there exist corresponding conserved quantities and accordingly one wants to discuss various natural symmetries and the conservation laws that accompany them (see [Rom64] Chapter IV for a clear discussion of Noether's theorem, and also see Exercise 2.1). Specifically, in Table 2.1 are several important examples that are believed to be absolute symmetries and hence exact conservation laws. Some of these specific examples are discussed in more detail in what follows.

Furthermore, there are symmetries that are not completely respected in Nature, although characterizing the states used in terms of eigenstates of these approximate symmetries often proves fruitful; some examples are given in Table 2.2. We shall be using all of these concepts throughout the book. Next let us turn to a brief discussion of some of the basics needed when treating symmetries using group theory.

Representations

By an n-dimensional representation of a group G one means a mapping

$$G \to GL\{p\}(n) \tag{2.1}$$

$$g \to A(g) \tag{2.2}$$

which assigns to every element g a linear operator $A(g)$ in some n-dimensional complex vector space, the so-called carrier space of the representation $GL(n)$, such that the image of the identity e is the unit operator I and that group operations are preserved

$$A(gg') = A(g)A(g') . \tag{2.3}$$

Throughout the book we shall frequently encounter infinite-dimensional continuous groups (Lie groups) whose elements are labeled uniquely by a set of parameters which can change continuously (see [Rom64] for an introductory discussion). An example is provided by

	Table 2.1 Exact conservation laws
Symmetry	Conservation law
translation in time	energy
translation in space	linear momentum
rotation in space	angular momentum
local gauge invariance	charge
transformations in color space	color

	Table 2.2 Approximate conservation laws
Approximate symmetry	Conservation law
spatial inversion	parity, P
particle–antiparticle interchange	charge conjugation, C
temporal inversion	time-reversal invariance, T
transformations in isospace	isospin, I (or T)
transformations in flavor space	flavor

the rotation group, that is, the group of continuous rotations. For the Lie groups that are encountered frequently in this book it is sufficient to study the mapping from the Lie algebra into $GL(n)$,

$$L_\alpha \to T_\alpha \,, \tag{2.4}$$

where the $\{T_\alpha\}$ preserve the Lie-algebra commutation relations. If a subspace of the carrier space of some representation is left unchanged by all operators T_α, it is called an invariant subspace and the representation is reducible; otherwise it is irreducible. If the correspondence

$$g \to A(g) \tag{2.5}$$

defines a representation of the group G, then the correspondence

$$g \to \left(A^{-1}\right)^T (g) = A^T(g^{-1}) \tag{2.6}$$

also defines a representation of the group, the so-called conjugate representation. For a Lie group we find that the representation matrices for the conjugate representation are given by

$$(T_\alpha)_{\text{conjugate}} = -T_\alpha^* \,. \tag{2.7}$$

When discussing the implications of symmetries in particle and nuclear physics one frequently encounters the special unitary groups in N dimensions, $SU(N)$, which can be represented using $N \times N$ matrices U satisfying

$$U^{-1} = U^\dagger \quad \text{and} \quad \det U = 1 \,. \tag{2.8}$$

The importance of the continuous Lie group $SU(N)$ lies in the fact that these matrices describe transformations between N basis states $\{|e_\alpha\rangle, \alpha = 1, \ldots, N\}$ preserving orthonormality

$$\langle Ue_\alpha \,|\, Ue_\beta \rangle = \left\langle e_\alpha \,\middle|\, U^\dagger Ue_\beta \right\rangle = \langle e_\alpha \,|\, e_\beta \rangle = \delta_{\alpha\beta} \ . \tag{2.9}$$

We shall see several examples of physical states labeled using various symmetries, specifically by spin and by isospin ($SU(2)$), by flavor and by color ($SU(3)$), or by higher groups, e.g., $SU(6)$ for spin-flavor. Within the context of $SU(N)$, a representation is reducible if it is possible to choose a basis in which the matrices T_α take the block form

$$T_\alpha = \begin{pmatrix} A & 0 & 0 & 0 \\ 0 & B & 0 & 0 \\ 0 & 0 & C & 0 \\ 0 & 0 & 0 & \cdots \end{pmatrix}, \tag{2.10}$$

where $A, B, C, \ldots$ are lower-dimensional irreducible sub-matrices when the original matrix T_α is fully reduced. Given an irreducible representation $\{T_\alpha\}$, the only linear operators $\mathcal{O}$ which commute with every T_α are multiples of the identity and also the converse:

$$[T_\alpha, \mathcal{O}] = 0 \ \forall\, \alpha \implies \mathcal{O} = \lambda I \ . \tag{2.11}$$

Any unitary matrix can be written as

$$U = e^{iH} = 1 + iH - \frac{1}{2!}H^2 + \cdots, \tag{2.12}$$

where H is a traceless Hermitian matrix. For a Lie group the elements of the group are characterized by a finite number of real parameters $\{a_\alpha\}$ and for $SU(N)$ one finds that there are $n = N^2 - 1$ such parameters. Accordingly, one can write

$$H = \sum_{\alpha=1}^{n} a_\alpha L_\alpha \ , \tag{2.13}$$

where the $\{L_\alpha\}$ form a basis for the $N \times N$ Hermitian matrices known as the generators of the group $SU(N)$. To study the representations, it is sufficient to study the generators and their commutation relations,

$$\left[L_\alpha, L_\beta \right] = ic_{\alpha\beta\gamma} L_\gamma \ , \tag{2.14}$$

where the latter are characterized by the antisymmetric structure constants $\left\{ c_{\alpha\beta\gamma} \right\}$.

2.2 Angular Momentum and *SU*(2)

Let us begin by discussing the representations of $SU(2)$ in a systematic way. The basis space is three-dimensional and is spanned by $S = (S_1, S_2, S_3)$, that satisfy the commutation relations [Edm74]

$$\left[S_i, S_j \right] = i\epsilon_{ijk} S_k \ , \tag{2.15}$$

where ϵ_{ijk} is the antisymmetric tensor, $+1$ if ijk is an even permutation of 123, -1 if an odd permutation and zero otherwise. In the carrier space a Hermitian scalar product exists:

$$\langle a \,|\, b \rangle = \langle b \,|\, a \rangle^* \,. \tag{2.16}$$

Next we need to label the states in the carrier space using the Cartan subalgebra, namely, the maximal set of mutually commuting operators that span the space. For $SU(2)$ the subalgebra only contains a single operator, usually chosen to be S_z, where the z-axis is chosen by convention to point in some convenient direction; later in Section 2.4 we shall see that for $SU(N)$ with $N \geq 3$ the situation is more complicated. The importance of devising such a mutually commuting set is well-known from quantum mechanics: it is then possible to diagonalize all of the matrices in the set simultaneously and to label the states with the corresponding eigenvalues. From this set of generators there are special operators that can be constructed which commute with all generators of the group, namely, the so-called Casimir operators. Again for $SU(2)$ there is only one such operator (although more for $SU(N)$ with $N \geq 3$) namely the quadratic Casimir operator

$$C = \sum_i S_i S_i = S_x^2 + S_y^2 + S_z^2 = \mathbf{S}^2 \,. \tag{2.17}$$

As discussed above, such operators commute with all generators of the group,

$$\left[\mathbf{S}^2, S_i \right] = 0 \,, \tag{2.18}$$

and hence must be proportional to the unit matrix, i.e., their eigenvalues may be used to label the representations. Let us now proceed to construct explicit representations using the commutation relations. One labels the basis states or representations with λ, the eigenvalues of the Casimir operator, and with quantum numbers m, the eigenvalues belonging to the operators in the Cartan subalgebra,

$$\mathbf{S}^2 \,|\lambda, m\rangle \equiv \lambda \,|\lambda, m\rangle \tag{2.19}$$

$$S_z \,|\lambda, m\rangle \equiv m \,|\lambda, m\rangle \,. \tag{2.20}$$

Since $\mathbf{S}^2$ and S_z are Hermitian, λ and m are both real, and moreover, λ is positive and may be chosen by convention to be

$$\lambda \equiv j(j+1) \,, \tag{2.21}$$

where j then labels the representation. Correspondingly, we now have

$$\mathbf{S}^2 \,|j, m\rangle = j(j+1) \,|j, m\rangle \quad (j \geq 0, \text{ real}) \tag{2.22}$$

$$S_z \,|j, m\rangle = m \,|j, m\rangle \quad (m \text{ real}) \tag{2.23}$$

and, being eigenstates of Hermitian matrices, the states $|j, m\rangle$ are orthogonal and can be normalized. Defining raising and lowering operators

$$S_\pm \equiv S_x \pm iS_y \,, \tag{2.24}$$

it is straightforward to show that

$$[S_z, S_\pm] = \pm S_\pm \tag{2.25}$$

$$[S_+, S_-] = 2S_z \,. \tag{2.26}$$

Next using Eq. (2.25) one proves that, after operating on the states $|j, m\rangle$ with the raising or lowering operators to form new states, $S_\pm |j, m\rangle$, the latter are also eigenstates of S^2 and S_z,

$$S^2 \left(S_\pm |j, m\rangle \right) = S_\pm S^2 |j, m\rangle = S_\pm j(j+1) |j, m\rangle$$
$$= j(j+1) \left(S_\pm |j, m\rangle \right) \tag{2.27}$$
$$S_z \left(S_\pm |j, m\rangle \right) = S_\pm (S_z \pm 1) |j, m\rangle = (m \pm 1) \left(S_\pm |j, m\rangle \right) , \tag{2.28}$$

that is, with eigenvalues $j(j+1)$ and $m \pm 1$. Writing this result in the form

$$S_\pm |j, m\rangle \equiv C_\pm |j, m \pm 1\rangle \tag{2.29}$$

and using the fact that

$$\langle j, m | S_+^\dagger S_+ | j, m \rangle = \langle j, m | S_- S_+ | j, m \rangle \tag{2.30}$$
$$= \langle j, m | (S^2 - S_z^2 - S_z) | j, m \rangle \tag{2.31}$$
$$= j(j+1) - m(m+1) , \tag{2.32}$$

one then has that

$$|C_+|^2 = j(j+1) - m(m+1) . \tag{2.33}$$

Since $S^2 = S_x^2 + S_y^2 + S_z^2$ is made up from quadratic Hermitian operators, one has that

$$m^2 \leq j(j+1) < (j+1)^2 , \tag{2.34}$$

and, since the allowed m-values change only in steps of 1 with the highest value m_{max} occurring when

$$S_+ |j, m_{\mathrm{max}}\rangle = 0 , \tag{2.35}$$

one finds that $m_{\mathrm{max}} = j$, justifying the choice made in the definition in Eq. (2.21). Collecting these developments together, in summary we have basis states characterizing the representation $\{|j, m\rangle\}$ with non-negative Casimir quantum number j and quantum number m having values running in steps of unity from $-j$ to $+j$; thus the dimension of the representation is $2j + 1$. The choice of phase usually made [Con35] is such that the raising and lowering operators acting on states $|j, m\rangle$ yield real c-numbers times states with $m \pm 1$:

$$S_\pm |j, m\rangle \equiv \sqrt{j(j+1) - m(m \pm 1)} \, |j, m \pm 1\rangle . \tag{2.36}$$

Next let us focus on spin $SU(2)$, taking $j \to S$ with $m \to S_z$ and discuss the lowest-dimensional representations in somewhat more detail. The simplest is the one-dimensional, singlet representation ($S = 0$) with basis state $|0, 0\rangle$ and having $S_z |0, 0\rangle = S_+ |0, 0\rangle = S_- |0, 0\rangle = 0$. The first nontrivial representation is the so-called fundamental one, which for $SU(2)$ is two-dimensional ($S = 1/2$) with basis states $|S = 1/2, S_z = \pm 1/2\rangle$

$$|1/2, 1/2\rangle \equiv \begin{pmatrix} 1 \\ 0 \end{pmatrix} \quad |1/2, -1/2\rangle \equiv \begin{pmatrix} 0 \\ 1 \end{pmatrix} . \tag{2.37}$$

Letting $S_\pm$ and S_z act on the basis states, it is straightforward to obtain explicit expressions for the representation matrices:

$$S_z = \frac{1}{2}\begin{pmatrix} 1 & 0 \\ 0 & -1 \end{pmatrix} \quad S_+ = \begin{pmatrix} 0 & 1 \\ 0 & 0 \end{pmatrix} \quad S_- = \begin{pmatrix} 0 & 0 \\ 1 & 0 \end{pmatrix} \tag{2.38}$$

or equivalently

$$S_x = \frac{1}{2}\begin{pmatrix} 0 & 1 \\ 1 & 0 \end{pmatrix} \quad S_y = \frac{1}{2}\begin{pmatrix} 0 & -i \\ i & 0 \end{pmatrix} \quad S_z = \frac{1}{2}\begin{pmatrix} 1 & 0 \\ 0 & -1 \end{pmatrix}. \tag{2.39}$$

Conventionally one writes $S_i \equiv \sigma_i/2$, thereby defining the Pauli matrices σ_i, with $i = 1, 2, 3$, corresponding to x, y, z, respectively; we use the two types of notation interchangably. A more complicated case is the one for $S = 1$ (dimension three) with basis states labeled $|S, S_z\rangle$:

$$|1, 1\rangle = \begin{pmatrix} 1 \\ 0 \\ 0 \end{pmatrix} \quad |1, 0\rangle = \begin{pmatrix} 0 \\ 1 \\ 0 \end{pmatrix} \quad |1, -1\rangle = \begin{pmatrix} 0 \\ 0 \\ 1 \end{pmatrix}. \tag{2.40}$$

As above, letting $S_\pm$ and S_z act on these states, one obtains

$$S_z = \begin{pmatrix} 1 & 0 & 0 \\ 0 & 0 & 0 \\ 0 & 0 & -1 \end{pmatrix} \quad S_+ = \begin{pmatrix} 0 & \sqrt{2} & 0 \\ 0 & 0 & \sqrt{2} \\ 0 & 0 & 0 \end{pmatrix} \quad S_- = \begin{pmatrix} 0 & 0 & 0 \\ \sqrt{2} & 0 & 0 \\ 0 & \sqrt{2} & 0 \end{pmatrix} \tag{2.41}$$

or equivalently

$$S_x = \frac{1}{\sqrt{2}}\begin{pmatrix} 0 & 1 & 0 \\ 1 & 0 & 1 \\ 0 & 1 & 0 \end{pmatrix} \quad S_y = \frac{1}{\sqrt{2}}\begin{pmatrix} 0 & -i & 0 \\ i & 0 & -i \\ 0 & i & 0 \end{pmatrix} \quad S_z = \begin{pmatrix} 1 & 0 & 0 \\ 0 & 0 & 0 \\ 0 & 0 & -1 \end{pmatrix}. \tag{2.42}$$

This is the so-called adjoint or regular representation. This is an example of an $N^2 - 1$ dimensional representation of $SU(N)$ given by the mapping in Eqs. (2.1) and (2.2) with structure constants (see Eq. (2.14))

$$c_{\alpha\beta\gamma} = -i\,(T_\alpha)_{\gamma\beta}\;. \tag{2.43}$$

Later when building hadrons in Chapter 3 we shall find it convenient to use weight diagrams. Since the generator in the Cartan subalgebra can be used to label states of a representation, the corresponding eigenvalues can be plotted in a diagram of this type, which here for $SU(2)$ amounts to drawing a line with dots to indicate where the eigenvalues occur, as shown in Fig. 2.1. Below we shall see that in $SU(N)$ with $N \geq 3$ one has patterns in $(N - 1)$-dimensional space.

Coupling of Angular Momentum

By taking the direct product of two representations, we find a new representation which in general is reducible. For instance, as an example in $SU(2)$ let us consider the direct product of two $S = 1/2$ (two-dimensional) representations (see Eq. (2.37)), written in the

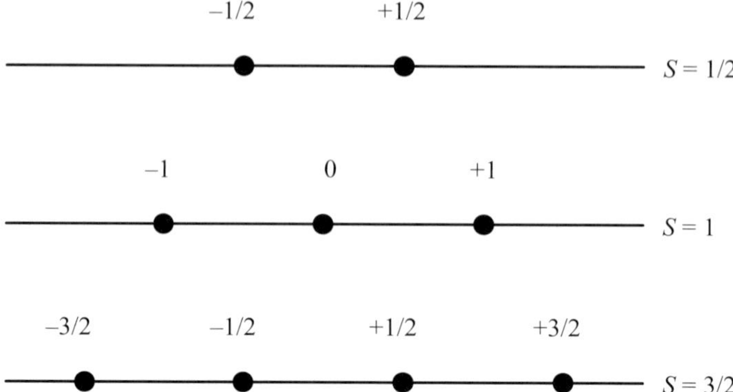

Fig. 2.1   Weight diagrams for $SU(2)$ for spins $S = 1/2, 1$, and $3/2$.

following way $|S(1) = 1/2, S_3(1) = \pm 1/2; S(2) = 1/2, S_3(2) = \pm 1/2\rangle$, now for brevity simply indicated $|\pm\pm\rangle$, yielding four states

$$|++\rangle, \quad |+-\rangle, \quad |-+\rangle, \quad |--\rangle, \tag{2.44}$$

with

$$|++\rangle = \begin{pmatrix} 1 \\ 0 \\ 0 \\ 0 \end{pmatrix}, \text{etc.} \tag{2.45}$$

The representation matrices in the direct product space are

$$S_x = \frac{1}{2}\begin{pmatrix} 0 & 1 & 1 & 0 \\ 1 & 0 & 0 & 1 \\ 1 & 0 & 0 & 1 \\ 0 & 1 & 1 & 0 \end{pmatrix} \quad S_y = \frac{1}{2}\begin{pmatrix} 0 & -i & -i & 0 \\ i & 0 & 0 & -i \\ i & 0 & 0 & -i \\ 0 & i & i & 0 \end{pmatrix}$$

$$S_z = \begin{pmatrix} 1 & 0 & 0 & 0 \\ 0 & 0 & 0 & 0 \\ 0 & 0 & 0 & 0 \\ 0 & 0 & 0 & -1 \end{pmatrix}. \tag{2.46}$$

If, instead of the basis in Eq. (2.44), one uses

$$|++\rangle, \quad \frac{1}{\sqrt{2}}\{|+-\rangle + |-+\rangle\}, \quad |--\rangle, \quad \frac{1}{\sqrt{2}}\{|+-\rangle - |-+\rangle\}, \tag{2.47}$$

it is straightforward to check that the matrices in Eq. (2.46) are now reduced to block form with a 3×3 block and a 1×1 block along the diagonal, the former being the representation found above for spin 1 (triplet) and the latter being for spin 0 (singlet). Using the quantum number S to label a representation one can write

$$[1/2] \otimes [1/2] = [1] \oplus [0] \tag{2.48}$$

or, equivalently, using the dimensions of the representations

$$\mathbf{2} \otimes \mathbf{2} = \mathbf{3} \oplus \mathbf{1} \; ; \tag{2.49}$$

both types of notation will be used and will be generalized for $SU(N)$ with $N \geq 3$.

Equations (2.48) and (2.49) are the most basic versions of what is called the Clebsch–Gordan series: the product of two states with given angular momenta can be rewritten (recoupled) and expressed as a sum of states with good angular momentum quantum numbers. For $SU(2)$, which is involved in recoupling angular momenta, only one state of each angular momentum occurs,

$$[j_1] \otimes [j_2] = [j_{\min}] \oplus \cdots \oplus [j_{\max}] \; , \tag{2.50}$$

where $j_{\min} = |j_1 - j_2|$ and $j_{\max} = j_1 + j_2$. Later we shall see that the situation for $SU(N)$ with $N \geq 3$ is more complicated. The Clebsch–Gordan or vector coupling coefficients, $\langle j_1 m_1 j_2 m_2 | j_1 j_2 j m \rangle$, enter when states in a coupled scheme $|(j_1 j_2) j m \rangle$ are written in terms of states in an uncoupled scheme $|j_1 m_1 ; j_2 m_2 \rangle$:

$$|(j_1 j_2) j m \rangle = \sum_{m_1 m_2} \langle j_1 m_1 j_2 m_2 | j_1 j_2 j m \rangle \, |j_1 m_1 ; j_2 m_2 \rangle \; ; \tag{2.51}$$

here the other quantum numbers characterizing the states are suppressed. These coefficients are related to the 3-j symbols which are generally more convenient to use (we employ the conventions of [Edm74] throughout the book):

$$\langle j_1 m_1 j_2 m_2 | j_1 j_2 j_3 m_3 \rangle = (-)^{j_1 - j_2 + m_3} [j_3] \begin{pmatrix} j_1 & j_2 & j_3 \\ m_1 & m_2 & -m_3 \end{pmatrix} , \tag{2.52}$$

where $[x] \equiv \sqrt{2x + 1}$ and in the 3-j symbol one has the following: angular momentum conservation requires that $(j_1 j_2 j_3)$ add vectorially, indicated $\Delta(j_1 j_2 j_3)$, while the projections add to zero algebraically, $m_1 + m_2 + m_3 = 0$; any even permutation of the columns leaves the symbol unchanged, while any odd permutation of the columns or an overall change $m_i \rightarrow -m_i$, $i = 1, 2, 3$, leads to a potential sign change of $(-)^{j_1 + j_2 + j_3}$. One special 3-j symbol that occurs frequently is the so-called parity coefficient,

$$\begin{pmatrix} j_1 & j_2 & j_3 \\ 0 & 0 & 0 \end{pmatrix}, \tag{2.53}$$

which by the symmetry relationships above is zero if $j_1 + j_2 + j_3 = $ odd. One has the orthogonality relation

$$[j_3]^2 \sum_{m_1 m_2} \begin{pmatrix} j_1 & j_2 & j_3 \\ m_1 & m_2 & m_3 \end{pmatrix} \begin{pmatrix} j_1 & j_2 & j_3' \\ m_1 & m_2 & m_3' \end{pmatrix} = \delta_{j_3 j_3'} \delta_{m_3 m_3'} \tag{2.54}$$

with inverse

$$\sum_{j_3 m_3} [j_3]^2 \begin{pmatrix} j_1 & j_2 & j_3 \\ m_1 & m_2 & m_3 \end{pmatrix} \begin{pmatrix} j_1 & j_2 & j_3 \\ m_1' & m_2' & m_3 \end{pmatrix} = \delta_{m_1 m_1'} \delta_{m_2 m_2'} \; . \tag{2.55}$$

The only irreducible tensor operators are those that can be expressed in a spherical basis, the so-called spherical tensors. These tensors transform under rotations as representations of $SU(2)$, namely with angular momentum quantum numbers k and projections $-k \leq m \leq k$. The tensors are combined with the following generalized multiplication rule:

$$T_k^m = \left[T_{k_1} \otimes T_{k_2}\right]_k^m = \sum_{m_1 m_2} \langle k_1 m_1 k_2 m_2 | k_1 k_2 k m \rangle \, T_{k_1}^{m_1} T_{k_2}^{m_2} \tag{2.56}$$

with adjoint

$$T_k^{m\dagger} = (-)^{k+m} T_k^{-m} \tag{2.57}$$

and inverse

$$T_{k_1}^{m_1} T_{k_2}^{m_2} = \sum_{km} \langle k_1 m_1 k_2 m_2 | k_1 k_2 k m \rangle \, T_k^m \,. \tag{2.58}$$

The tensors satisfy the following commutation relations with the angular momentum operators

$$\left[J_0, T_k^m\right] = m T_k^m \tag{2.59}$$

$$\left[J_\pm, T_k^m\right] = \sqrt{(k \mp m)(k \pm m + 1)} T_k^{m\pm 1} \,, \tag{2.60}$$

where $J_\pm = J_x \pm J_y \equiv \mp\sqrt{2} J_{\pm 1}$. From this a very useful result can be obtained (see [Edm74]), namely, the Wigner–Eckart theorem:

$$\langle j'm' | T_k^\mu | jm \rangle = (-)^{j'-m'} \begin{pmatrix} j' & k & j \\ -m' & \mu & m \end{pmatrix} \langle j' \, \| T_k \| \, j \rangle \,, \tag{2.61}$$

where the reduced matrix element $\langle j' \, \| T_k \| \, j \rangle$ has been introduced. What is captured by this result is that all of the matrix elements with the different projections allowed are in fact proportional to a single number, the reduced matrix element, weighted by well-known coefficients, the 3-j symbols.

2.3   *SU*(2) of Isospin

Because of their closeness in mass and the very similar roles they play, as will be discussed later in Chapter 8, one may consider the proton (mass $m_p = 938.27\,\text{MeV}$) and neutron (mass $m_n = 939.57\,\text{MeV}$) to be two states of a common particle, the nucleon N. In analogy to a system with spin $S = 1/2$ and spin projections $S_3 = \pm 1/2$ (see Section 2.2) one may assign a new quantum number to the nucleon, namely, isospin $T = 1/2$. The convention typically used in particle physics is to assign an isospin projection $T_3 = +1/2$ to the proton and $-1/2$ to the neutron, although in nuclear physics the opposite convention is sometimes employed. In this book we use the former convention and so one has the two basic states of the nucleon, written $|T = 1/2, T_3 = \pm 1/2\rangle$, respectively:

$$|1/2, +1/2\rangle = \begin{pmatrix} 1 \\ 0 \end{pmatrix} \equiv |p\rangle \text{ proton; and} \tag{2.62}$$

$$|1/2, -1/2\rangle = \begin{pmatrix} 0 \\ 1 \end{pmatrix} \equiv |n\rangle \text{ neutron.} \tag{2.63}$$

The analog of Eq. (2.13) is

$$H = \sum_{i=1}^{3} a_i T_i \; . \tag{2.64}$$

As with spin, for isospin one has raising and lowering operators $T_\pm = T_1 \pm i T_2$ that satisfy the analogs of Eqs. (2.25,2.26) and of Eq. (2.36)

$$T_\pm \, |T, T_3\rangle \equiv \sqrt{T(T+1) - T_3(T_3 \pm 1)} \, |T, T_3 \pm 1\rangle \; ; \tag{2.65}$$

which will be employed in Chapter 3 when moving across the isospin subspace of the baryon multiplets. We also have $T_i \equiv \tau_i/2$, yielding the analogs of the Pauli matrices (see Eqs. (2.39))

$$\tau_1 = \begin{pmatrix} 0 & 1 \\ 1 & 0 \end{pmatrix} \quad \tau_2 = \begin{pmatrix} 0 & -i \\ i & 0 \end{pmatrix} \quad \tau_3 = \begin{pmatrix} 1 & 0 \\ 0 & -1 \end{pmatrix}, \tag{2.66}$$

that obey the relations

$$\mathrm{Tr}\left(\tau_i \tau_j\right) = \delta_{ij} \tag{2.67}$$

$$\left[\tau_i, \tau_j\right] = 2i\epsilon_{ijk}\tau_k \tag{2.68}$$

$$\left\{\tau_i, \tau_j\right\} = 2\delta_{ij} \; . \tag{2.69}$$

We shall also see in Chapter 3 that isospin will be used in building hadrons from quarks. In this case the fundamental representation **2** (at the level of $SU(2)$) will be the doublet $\{|u\rangle, |d\rangle\}$, up and down, corresponding to isospin $1/2$ with projections $\pm 1/2$, respectively:

$$|u\rangle = \begin{pmatrix} 1 \\ 0 \end{pmatrix} \; ; \text{ and}$$

$$|d\rangle = \begin{pmatrix} 0 \\ 1 \end{pmatrix} . \tag{2.70}$$

Following the developments in Section 2.1 (see Eq. (2.7)), the conjugate representation that enters when discussing antiparticles, $\{|\bar{u}\rangle, |\bar{d}\rangle\}$, has representation matrices

$$-\tau_1^* = \begin{pmatrix} 0 & -1 \\ -1 & 0 \end{pmatrix} \quad -\tau_2^* = \begin{pmatrix} 0 & -i \\ i & 0 \end{pmatrix} \quad -\tau_3^* = \begin{pmatrix} -1 & 0 \\ 0 & 1 \end{pmatrix} . \tag{2.71}$$

If we change to the basis $\{-|\bar{d}\rangle, |\bar{u}\rangle\}$ then the representations in Eqs. (2.66) and (2.71) become identical. Note that for $SU(2)$ the conjugate representation is equivalent to the fundamental representation – we shall see that this is not the case for $SU(N)$ with $N \geq 3$.

As with ordinary spin, it is possible when considering many-particle systems to vector-couple the isospins. For example, when treating NN scattering in Chapter 11, one has an isosinglet (the analog of the spin singlet discussed above)

$$|T = 0, T_3 = 0\rangle \equiv \frac{1}{\sqrt{2}} \left(|pn\rangle - |np\rangle\right) \tag{2.72}$$

and the three states of the isotriplet

$$|T = 1, T_3 = 1\rangle \equiv |pp\rangle \tag{2.73}$$

$$|T = 1, T_3 = 0\rangle \equiv \frac{1}{\sqrt{2}} \left(|pn\rangle + |np\rangle \right) \tag{2.74}$$

$$|T = 1, T_3 = -1\rangle \equiv |nn\rangle \ . \tag{2.75}$$

In strong-interaction physics the isospin appears to be a good or nearly good symmetry and so, for example, the nuclear force is (approximately) invariant under isospin transformations. Note, however, that it is broken by the electromagnetic interaction – pp interactions differ from nn and pn interactions, for instance.

2.4   Extensions to Flavor $SU(3)$

In analogy to the developments in the previous section for the case of $SU(2)$ with up and down quarks an extended basis of three states, up, down, and strange, occurs when discussing flavor for low-lying mesons and baryons (see Chapter 3) and the special unitary group $SU(3)$ becomes relevant:

$$|u\rangle = \begin{pmatrix} 1 \\ 0 \\ 0 \end{pmatrix} \quad |d\rangle = \begin{pmatrix} 0 \\ 1 \\ 0 \end{pmatrix} \quad |s\rangle = \begin{pmatrix} 0 \\ 0 \\ 1 \end{pmatrix} . \tag{2.76}$$

As in Eq. (2.64) one has

$$H = \sum_{a=1}^{8} a_a F_a \ , \tag{2.77}$$

now with eight transformation matrices rather than three for $SU(2)$ making up the representation (the "eight-fold way"). When acting on the basis states in Eqs. (2.76) these matrices induce transformations within the basis, just as the Pauli matrices induce transformations within the doublet space of $SU(2)$ in Eqs. (2.70). Defining the Gell-Mann λ-matrices by $\lambda_a \equiv 2F_a$ these are conventionally given by

$$\lambda_1 = \begin{pmatrix} 0 & 1 & 0 \\ 1 & 0 & 0 \\ 0 & 0 & 0 \end{pmatrix} \quad \lambda_2 = \begin{pmatrix} 0 & -i & 0 \\ i & 0 & 0 \\ 0 & 0 & 0 \end{pmatrix} \quad \lambda_3 = \begin{pmatrix} 1 & 0 & 0 \\ 0 & -1 & 0 \\ 0 & 0 & 0 \end{pmatrix}$$

$$\lambda_4 = \begin{pmatrix} 0 & 0 & 1 \\ 0 & 0 & 0 \\ 1 & 0 & 0 \end{pmatrix} \quad \lambda_5 = \begin{pmatrix} 0 & 0 & -i \\ 0 & 0 & 0 \\ i & 0 & 0 \end{pmatrix} \quad \lambda_6 = \begin{pmatrix} 0 & 0 & 0 \\ 0 & 0 & 1 \\ 0 & 1 & 0 \end{pmatrix} \tag{2.78}$$

$$\lambda_7 = \begin{pmatrix} 0 & 0 & 0 \\ 0 & 0 & -i \\ 0 & i & 0 \end{pmatrix} \quad \lambda_8 = \frac{1}{\sqrt{3}} \begin{pmatrix} 1 & 0 & 0 \\ 0 & 1 & 0 \\ 0 & 0 & -2 \end{pmatrix}$$

together with the identity

$$I = \begin{pmatrix} 1 & 0 & 0 \\ 0 & 1 & 0 \\ 0 & 0 & 1 \end{pmatrix}. \tag{2.79}$$

The λ-matrices satisfy the relations (cf. Eqs. (2.67)–(2.69))

$$\mathrm{Tr}\,(\lambda_a \lambda_b) = 2\delta_{ab} \tag{2.80}$$

$$[\lambda_a, \lambda_b] = 2if_{abc}\lambda_c \tag{2.81}$$

$$\{\lambda_a, \lambda_b\} = \frac{4}{3}\delta_{ab} + 2d_{abc}\lambda_c\,, \tag{2.82}$$

where the ds are symmetric and the fs are antisymmetric under interchange of any two indices. One has the following nonzero $SU(3)$ structure constants

$$\begin{aligned}
f_{123} &= 1 \\
f_{147} &= f_{246} = f_{257} = f_{345} = f_{516} = f_{637} = 1/2 \\
f_{458} &= f_{678} = \sqrt{3}/2 \\
d_{118} &= d_{228} = d_{338} = -d_{888} = 1/\sqrt{3} \\
d_{146} &= d_{157} = d_{256} = d_{344} = d_{355} = 1/2 \\
d_{247} &= d_{366} = d_{377} = -1/2 \\
d_{448} &= d_{558} = d_{668} = -1/2\sqrt{3}\,,
\end{aligned} \tag{2.83}$$

together with all permutations. The rest are all zero, for instance, f_{168}.

2.5 Young Tableaux

The above discussions are focused on some of the basics of $SU(2)$ (spin and isospin) and $SU(3)$ (flavor) and can be generalized to higher-dimensional unitary symmetries. Most of the developments in later chapters will not require detailed treatments of the latter, although the ability to enumerate the dimensions of the various representations proves useful. The general techniques for doing this can be found in [Ham62]; for our present purposes we follow the discussions in [Clo79] which are adequate. These involve specifying the rules for determining the dimension of a specific representation of $SU(N)$ using the elegant constructions called Young tableaux. Following [Clo79] we state the rules without proof and give a few examples (more examples are found in the Exercises). The fundamental representation N in $SU(N)$ of dimension N is denoted by a box $\square$, and the conjugate representation $\overline{N}$ by a column of $N-1$ boxes. As noted above, in $SU(2)$ the fundamental $\mathbf{2}$ and conjugate $\overline{\mathbf{2}}$ representations are the same, namely, a single box, whereas for $SU(N)$ with $N \geq 3$ this is not the case; for instance, for $SU(3)$ the fundamental $\mathbf{3}$ and conjugate $\overline{\mathbf{3}}$ representations are those shown in Fig. 2.2.

The next set of rules is the following: Any row of boxes is totally symmetric under particle interchanges, while any column is totally antisymmetric. In $SU(N)$ one can have a maximum of N boxes in a column and a single column with N boxes in $SU(N)$ is a singlet.

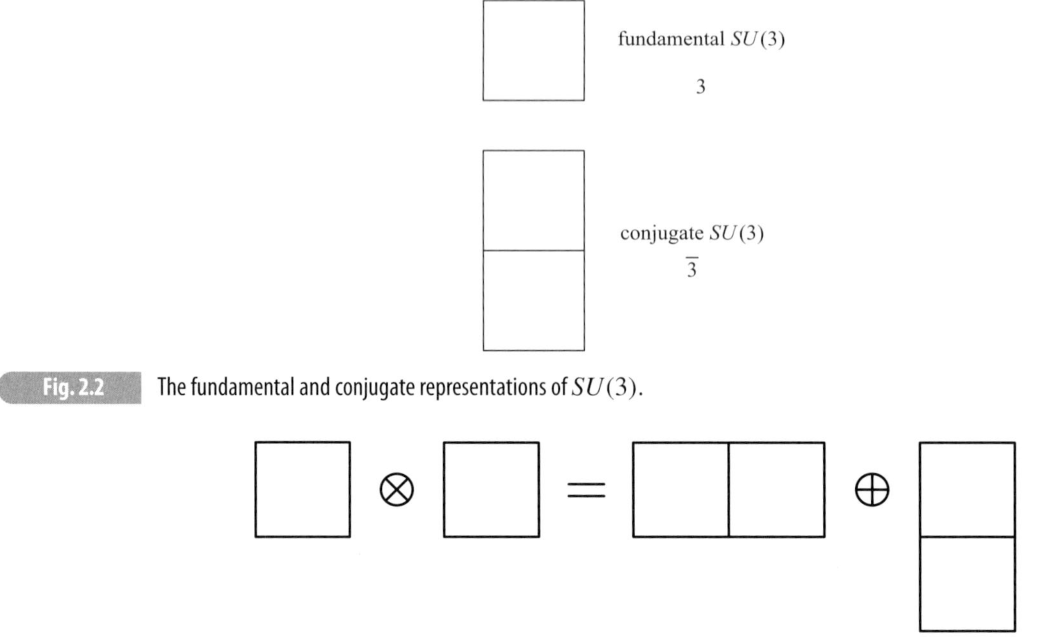

Fig. 2.2 The fundamental and conjugate representations of $SU(3)$.

Fig. 2.3 Multiplication of two fundamental representations.

For example, two boxes in a column in $SU(2)$ for spin denotes the singlet spin-0 state (likewise for the isosinglet state) and three boxes in a column in $SU(3)$ flavor denotes the totally antisymmetric flavor singlet state. As in the above discussions of coupling of angular momenta, one can take the direct product of two or more representations and decompose this into a direct sum of representations. For instance, suppose that two fundamental representations are multiplied together, then one obtains two-box tableaux of the type shown in Fig. 2.3.

As we shall see in Chapter 3, it is important to be able to determine the dimension of a specific tableau and hence the number of particles that fit into related multiplets. The rules for computing the dimension may be found in [Clo79]; these yield a ratio of two numbers which is the dimension of the representation. Working in $SU(N)$, for the numerator one starts in the upper left-hand corner and inserts the number N down the diagonal of the tableau, then $N + 1$ for the next box to the right of the left-hand corner and down the diagonal lying above the main diagonal, $N + 2$ for the next box to the right and down its diagonal, and so on, and then one does the same with $N - 1$ starting with the box below the upper left-hand corner and its diagonal, and so on. The structure obtained is made clearer with an example – see Fig. 2.4.

Note that only tableaux of this type, namely concave downwards and to the right, are allowed. Then the numerator of the ratio is given by the product of all of these numbers. For the denominator one uses the following rule: one draws a line entering the tableau from the right-hand side for each row, for each line one terminates the line in all possible ways, i.e., ending in all possible boxes it encounters, and for each choice the line turns downwards exiting the tableau via the particular column being considered.

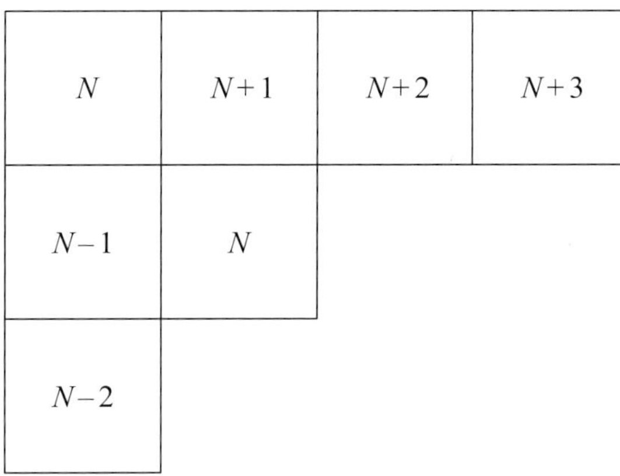

Fig. 2.4   Labeling of Young tableaux to determine the dimension of the representation.

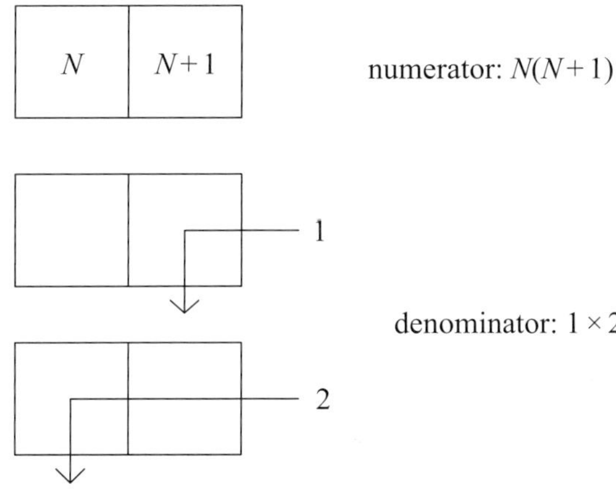

Fig. 2.5   Illustration of the hook rule for two boxes in a row.

All such constructions, which Close [Clo79] calls "hooks," are made, the number of boxes encountered enumerated and then finally the denominator is the product of all of those numbers. The detailed proofs of how to determine the dimension of a specific tableau are given in [Ham62]. Again an example should help to make this rule clear. Consider two boxes in a row in $SU(N)$ as in Fig. 2.5. The numerator is the product $N(N + 1)$ and the denominator is 2, yielding dimension $N(N+1)/2$. For two boxes in a column in $SU(N)$ as in Fig. 2.6, one finds a numerator of $N(N - 1)$ and a denominator of 2, yielding dimension $N(N - 1)/2$. For $SU(2)$ these two dimensions are 3 and 1, respectively, while for $SU(3)$ they are 6 and 3, namely one has found that (see also Eq. (2.49))

$$\mathbf{2} \otimes \mathbf{2} = \mathbf{3} \oplus \mathbf{1} \quad SU(2)$$
$$\mathbf{3} \otimes \mathbf{3} = \mathbf{6} \oplus \mathbf{\bar{3}} \quad SU(3).$$

Next let us take the direct product of three fundamental representations. Starting from the two results above with either two boxes in a row or two in a column, the procedure is to add a third box in all possible ways that yield diagrams that are concave downwards and to

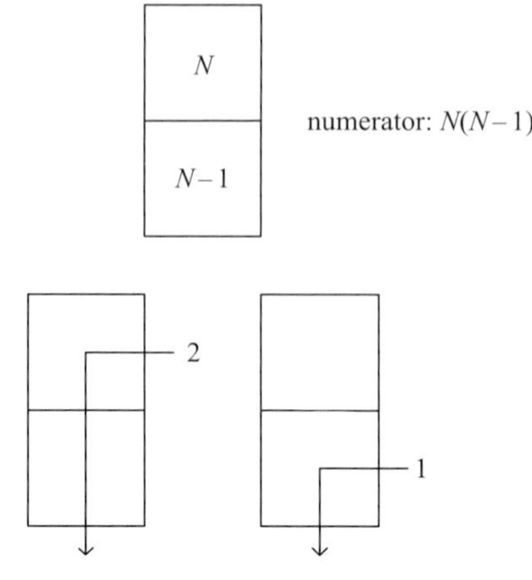

Fig. 2.6 Illustration of the hook rule for two boxes in a column.

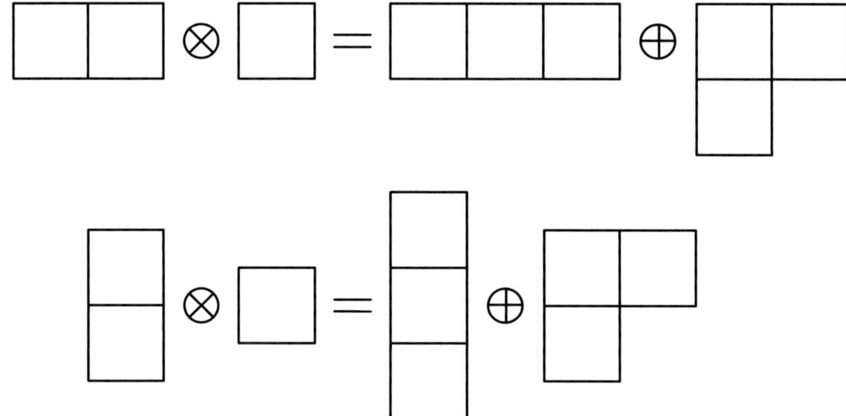

Fig. 2.7 Direct products of three fundamental representations (see Fig. 2.3).

the right, resulting in the tableaux in Fig. 2.7. Using the rules stated above the dimensions are immediately found to be $N(N+1)(N+2)/6$ and $(N-1)N(N+1)/3$ for the upper left- and right-hand tableaux, respectively, and $N(N-1)(N-2)/6$ and $(N-1)N(N+1)/3$ for the lower left- and right-hand tableaux, respectively. For $SU(2)$ the tableau with a column of three boxes cannot occur, whereas it can for $SU(3)$, and so we arrive at the answers

$$(\mathbf{2} \otimes \mathbf{2}) \otimes \mathbf{2} = (\mathbf{4} \oplus \mathbf{2}) \oplus \mathbf{2} \qquad SU(2)$$
$$(\mathbf{3} \otimes \mathbf{3}) \otimes \mathbf{3} = (\mathbf{10} \oplus \mathbf{8}) \oplus (\mathbf{8} \oplus \mathbf{1}) \quad SU(3)$$

namely, a quartet and two doublets for $SU(2)$ and a decuplet, two octets, and a singlet for $SU(3)$. In Chapter 12 few-body nuclei will be discussed and there one has both spin and isospin as $SU(2)$ properties of the three-body states obtained for ^{3}He and ^{3}H, and in Chapter 3 we shall see how to build low-lying baryons from triplets of u, d, and s (spin-1/2) quarks requiring both the $SU(2)$ characterization of the spin content together with the $SU(3)$ characterizations of their flavor and color. The tableau with three boxes

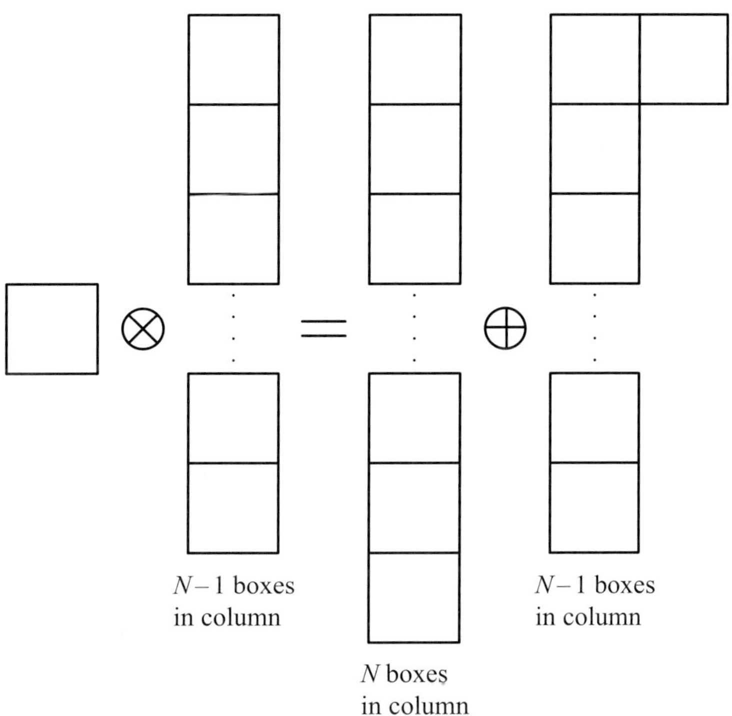

Fig. 2.8　Direct product of the fundamental and conjugate representations.

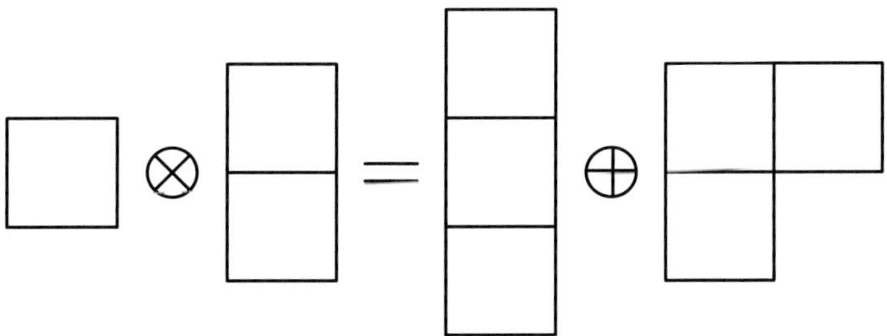

Fig. 2.9　Direct product of the fundamental and conjugate representations in $SU(3)$.

in the same row is completely symmetric, the one with three in a column is completely antisymmetric, whereas the others with both rows and columns is of mixed symmetry, i.e., for particles in the first row (column) it is symmetric (antisymmetric), but has no good permutation symmetry between the upper right-hand and lower boxes.

Finally, let us consider the direct product of the fundamental representation N (a single box) with the conjugate representation $\overline{N}$, namely, a column of $N - 1$ boxes as shown in Fig. 2.8; here $SU(2)$ is uninteresting, since a single box represents both, although for $SU(N)$ with $N \geq 3$ important new results are found. One has $N \otimes \overline{N} = \mathbf{1} \oplus (N^2-1)$, namely, a singlet and a representation of dimension $N^2 - 1$. For instance, in $SU(3)$ this yields the result in Fig. 2.9 and using the above rules one finds that $\mathbf{3} \otimes \overline{\mathbf{3}} = \mathbf{1} \oplus \mathbf{8}$, that is, one has a singlet and an octet. These will be used in Chapter 3 when discussing the flavor structure of the low-lying mesons built from $q\bar{q}$ pairs of u, d, and s quarks.

2.6 Discrete Symmetries: *P*, *C*, and *T*

A finite group is one which contains only a finite number of elements. In particle and nuclear physics we encounter very simple discrete groups with just two elements, namely the identity e and an element g satisfying $g^2 = e$. For such groups, the element g can be represented by a unitary (or antiunitary) operator, which we refer to as $U(g)$. If the dynamics of the system are invariant under operation by g, the $U(g)$ all satisfy

$$[U, H] = 0 \tag{2.84}$$

$$U^2 |p\rangle = p^2 |p\rangle \implies p = \pm 1 . \tag{2.85}$$

This means that if the system is an eigenstate of U then transitions can only occur between eigenstates with the same eigenvalue ± 1. As mentioned at the beginning of this chapter, even when the symmetry is not exact it may still be useful to characterize the true states in terms of eigenstates of the symmetry.

A first important example of such a bi-modal symmetry is spatial inversion

$$P\psi(\mathbf{r}) \to \psi(-\mathbf{r}) \tag{2.86}$$

with associated parity eigenvalues $P = \pm 1$ (even or odd, respectively). A few things to know about parity, some of which will be the subject of later discussions in the book, are the following: (1) Parity is a multiplicative quantum number; (2) Parity is believed to be conserved in strong and electromagnetic interactions, but is certainly violated by the weak interaction; (3) Fermion states have opposite parities for particles and antiparticles, whereas bosons have the same parities for particles and antiparticles; and (4) Fermions and bosons may be catalogued according to their spin-parity, J^P, half-integer for the former and integer for the latter. Bosons are denoted 0^+ (scalar), 0^- (pseudoscalar), 1^- (vector), 1^+ (axial-vector), and so on. For example, the pion is a 0^- pseudoscalar meson.

One example of using parity properties when discussing quantum systems is provided by the case of solutions in a spherically symmetric potential. There one has $H(\mathbf{r}) = H(-\mathbf{r}) = H(r)$, so that $[P, H] = 0$ and bound states in the potential have definite parity. A simple example involves a system moving in such a potential having no other internal degrees of freedom where the wavefunctions have the form

$$\psi(r, \theta, \phi) = R(r) Y_L^M(\theta, \phi) , \tag{2.87}$$

where $Y_L^M(\theta, \phi)$ is a spherical harmonic [Edm74]. It can be shown that making the transformation $\mathbf{r} \to -\mathbf{r}$, which implies that $\theta \to \pi - \theta$ and $\phi \to \pi + \phi$ yields $Y_L^M \to (-)^L Y_L^M$ and consequently such wavefunctions have parity $P = (-)^L$. Another example involves electromagnetic transitions between nuclear states which will be discussed in more detail in Chapter 7. There it will be shown that electric transitions of multipolarity J entail natural parity changes, $\Delta P = (-)^J$, whereas magnetic transitions are the opposite, having non-natural parity changes, $\Delta P = (-)^{J+1}$. Thus, given good parity quantum numbers for two nuclear states, only transitions of a given multipolarity of one type or the other can occur. For instance, if an excited state with spin-parity

1^- decays electromagnetically to a ground state with spin-parity 0^+, only electric dipole ($E1$) radiation is involved; see Chapter 7 for a discussion of multipole operators.

A second important example of bi-modal symmetry is provided by the operation of charge conjugation, C, which reverses the signs of the charge and magnetic moment of a particle. In classical electromagnetic theory this is seen through the invariance of Maxwell's equations under changes of the sign of the charge and current density or of electric and magnetic fields. In relativistic quantum mechanics one observes that C is manifested in the interchange of particles with antiparticles. Both the electromagnetic and strong interactions are observed to be invariant under charge conjugation, whereas the weak interaction is not. In fact, the weak interaction is not invariant under either P or C, although the product CP is a better symmetry than P and C are separately.

Finally, let us consider time reversal. Starting with the Schrödinger equation

$$i\frac{\partial \psi}{\partial t} = H\psi \tag{2.88}$$

and, using the fact that the Hamiltonian is Hermitian, the complex conjugate yields

$$-i\frac{\partial \psi^*}{\partial t} = H\psi^* \implies i\frac{\partial \psi^*}{\partial(-t)} = H\psi^* \,, \tag{2.89}$$

leading to the conclusion that if ψ is a solution to the original equation then ψ^* is a solution to the time-reversed equation. The time-reversal operator T is somewhat unusual, being anti-unitary: with K being complex conjugation and U_T being a unitary operator, one has

$$K\psi = \psi^* \tag{2.90}$$

and therefore

$$T\psi(t) = \psi^*(-t)\,. \tag{2.91}$$

An important principle of quantum field theory states that all interactions are invariant under the combined action of the three operators C, P, and T taken in any order, i.e., that CPT is an exact symmetry. Since CP is violated (weakly), this implies that time-reversal invariance is also not an exact symmetry, but must transform oppositely to maintain the product CPT. Implications of CPT invariance are that particles and antiparticles must have the same mass and lifetime, and must have magnetic moments that are equal in magnitude but opposite in sign. Later in the book both CP violation and time-reversal violation will be discussed in more detail (see Chapters 4, 17, 18, and 21).

Exercises

2.1   Noether's Theorem

The mathematician Emmy Noether proved a very important theorem to physics which states that for any invariance of the classical action under a continuous field transformation there exists a classical charge Q which is independent of time and which is connected to a conserved current, $\partial_\mu J^\mu = 0$. That is, the existence of a symmetry requires the validity of a corresponding conservation law. Well-known examples in classical physics are the invariance of the equations of motion under spatial translation, time translation, and rotation, which lead to conservation of momentum, energy, and angular momentum respectively. The purpose of this exercise is to prove this theorem.

We begin with an action

$$S = \int d^4x \, \mathcal{L}(\phi, \partial_\mu \phi)$$

and consider an infinitesimal transformation of the field

$$\phi \to \phi' = \phi + \epsilon f(\phi) \,,$$

where $f(\phi)$ is some function of the field.

a) Find the equation of motion for a constant value of the infinitesimal ϵ and show that one finds the Euler–Lagrange equation

$$\frac{\delta \mathcal{L}}{\delta \phi} - \partial_\mu \frac{\delta \mathcal{L}}{\delta \partial_\mu \phi} = 0 \,.$$

b) Calculate the change of the action under this field transformation in the case that $\epsilon = \epsilon(x)$ and show that

$$S \to S' = S + \int d^4x \, \partial_\mu \epsilon \, j^\mu \,,$$

where

$$j^\mu = \frac{\delta \mathcal{L}}{\delta \partial_\mu \phi} f(\phi) \,.$$

c) Now integrate by parts and show that if the action is invariant we require that $\partial_\mu j^\mu = 0$, up to a total derivative.

d) Integrate the equation

$$\frac{\partial j^0}{\partial t} = -\nabla \cdot \boldsymbol{j} = 0$$

over all space, and show that for a local "charge" distribution we have

$$\frac{dQ}{dt} = 0 \qquad \text{where} \quad Q = \int d\mathbf{x} \, j^0 \,,$$

which proves Noether's theorem.

2.2   Rotation Matrices and Finite Rotations

The effect on a spin eigenstate $|S, S_z\rangle$ under a rotation by angle χ about an axis $\hat{\boldsymbol{n}}$ is given by

$$|S, S_z\rangle' = R^S(\hat{\boldsymbol{n}}, \chi) \, |S, S_z\rangle \,,$$

where

$$R^S(\hat{\boldsymbol{n}}, \chi) = \exp(-i\chi \boldsymbol{S} \cdot \hat{\boldsymbol{n}}) \,,$$

and where S are the $(2S + 1) \times (2S + 1)$ component spin matrices constructed from the relations

$$S_z |S, m\rangle = m |S, m\rangle$$

$$(S_x \pm iS_y)|S, m\rangle = \sqrt{(S \mp m)(S \pm m + 1)} \, |S, m \pm 1\rangle \ .$$

a) Evaluate the rotation matrix for a spin-1/2 system and show that

$$R^{\frac{1}{2}}(\hat{\boldsymbol{n}}, \chi) = \exp(-i\chi S^{\frac{1}{2}} \cdot \hat{\boldsymbol{n}}) = \cos\frac{\chi}{2} - i\boldsymbol{\sigma} \cdot \hat{\boldsymbol{n}} \sin\frac{\chi}{2} \ ,$$

where $\boldsymbol{\sigma}$ are the Pauli matrices.
b) Calculate the rotated spin-1/2 state $|1/2, m\rangle'$ for the initial state $|1/2, m\rangle$ for $m = \pm\frac{1}{2}$ using $\hat{\boldsymbol{n}} = \hat{\boldsymbol{e}}_z$ and rotation angle χ and demonstrate that

$$|1/2, m\rangle' = \exp(-im\chi) |1/2, m\rangle \ .$$

c) Verify the commutation relations for the representations of the spin-1 operators in Eq. (2.41),

$$[S_z, S_\pm] = \pm S_\pm$$

$$[S_+, S_-] = 2S_z \ .$$

d) Now, evaluate the rotation matrix for a spin-1 system and show that

$$R^1(\hat{\boldsymbol{n}}, \chi) = \exp(-i\chi S^1 \cdot \hat{\boldsymbol{n}}) = 1 - (S^1 \cdot \hat{\boldsymbol{n}})^2(1 - \cos\chi) - iS^1 \cdot \hat{\boldsymbol{n}} \sin\chi \ ,$$

where S^1 are the 3×3 spin matrices constructed in the text.
e) Calculate the rotated spin state $|1, m\rangle'$ for the initial state $|1, m\rangle$ for the cases $m = 1, 0, -1$ using $\hat{\boldsymbol{n}} = \hat{\boldsymbol{e}}_z$ and rotation angle χ and demonstrate that

$$|1, m\rangle' = \exp(-im\chi) |1, m\rangle \ .$$

2.3   **Symmetry and Dipole Moments**

Since the only three-vector associated with an elementary particle in its rest frame is the spin, any dipole moment of the particle must be along the spin direction. For example, in the case of a spin-1/2 system such as the nucleon, one writes

$$\boldsymbol{m} = g_m \frac{e}{2m} \boldsymbol{S} \ ,$$

where $\boldsymbol{m}$ is the magnetic dipole moment, g_m is the gyromagnetic ratio, and $e/2m$ is called the Bohr magneton. When such a dipole is placed in a magnetic field $\boldsymbol{B}$, the corresponding interaction energy is

$$U_M = -\boldsymbol{m} \cdot \boldsymbol{B} \ .$$

a) Analyze this interaction from the point of view of parity and time reversal, and demonstrate that U_M is even under both.

Now suppose that we define an analogous electric dipole moment p in terms a gyroelectric ratio g_e

$$p = g_e \frac{e}{2m} S \,.$$

When placed in an external electric field E there will exist an interaction energy

$$U_E = -p \cdot E \,.$$

b) Analyze this interaction from the point of view of parity and time reversal and demonstrate that U_E is odd under both.

Thus, an elementary particle cannot possess an electric dipole moment if parity and/or time reversal are conserved.

c) We know that atoms and molecules can have large static electric dipole moments. Explain why this fact does this not indicate a violation of parity and time reversal invariance.

2.4   Spin Coupling

Consider a state with two spin-1/2 particles having spinors $\chi_{1/2}^{m_{s_1}}(1)$ and $\chi_{1/2}^{m_{s_2}}(2)$, where the particles are labeled 1 and 2 and the spin projections along some axis of quantization have the values $m_{s_i} = \pm\frac{1}{2}$. If the spin angular momenta of the two particles are coupled to total spin S with projection M_S one has the state

$$A_S^{M_S}(1,2) \equiv \left[\chi_{1/2}(1) \otimes \chi_{1/2}(2)\right]_S^{M_S} \,.$$

What values can S and M_S have? By explicit evaluation of the Clebsch–Gordan coefficients or 3-j symbols (see [Edm74]) show that the familiar answers are obtained (see [Sch55]). Using the properties of the 3-j symbols prove upon interchange of the coordinates of the two particles that one has

$$A_S^{M_S}(2,1) = (-)^{S+1} A_S^{M_S}(1,2) \,.$$

These results can trivially be extended to the case of two isospin-1/2 particles, for instance two nucleons, where one has isospinors $\xi_{1/2}^{m_{t_1}}(1)$ and $\xi_{1/2}^{m_{t_2}}(2)$ with $m_{t_i} = \pm\frac{1}{2}$, one has the coupled state

$$B_T^{M_T}(1,2) \equiv \left[\xi_{1/2}(1) \otimes \xi_{1/2}(2)\right]_T^{M_T} \,,$$

and obtains the symmetry under interchange

$$B_T^{M_T}(2,1) = (-)^{T+1} B_T^{M_T}(1,2) \,.$$

What happens upon coordinate interchange for a system of two nucleons (NN) where one has both spin and isospin?

2.5   L-S Coupling and Central Potentials

Consider a nonrelativistic system of two particles, which interact via a potential. In general such an interaction can depend on spin and isospin as well as the particle separation. This problem considers how to deal with such a potential.

If we represent the spatial coordinates of the two particles by r_1 and r_2, then one can change variables to a center-of-mass variable $R = (r_1 + r_2)/2$ and a relative coordinate $r = r_1 - r_2$. The two-particle spatial wavefunction may then be written as a product of a plane wave for the center-of-mass (CM) part times a function involving the relative coordinate, $\Phi(r)$. If the state has good orbital angular momentum L, with projection M_L, where L is an integer and $-L \leq M_L \leq L$, then the spatial wavefunction can be written

$$\Phi_L^{M_L}(r) = R_L(r) Y_L^{M_L}(\Omega_r) ,$$

where $R_L(r)$, the radial wavefunction, depends only on $r = |r|$ and the dependence on the polar and azimuthal angles $\Omega_r = (\theta, \phi)$ specifying the direction of r is captured in the spherical harmonic $Y_L^{M_L}(\Omega_r)$. For a general discussion of these basic ideas see one of the standard books on quantum mechanics, such as [Sch55]; see also [Edm74].

a) Prove that under a parity transformation, namely under inversion of coordinates $r \to -r$ one has $Y_L^{M_L}(\Omega_r) \to (-)^L Y_L^{M_L}(\Omega_r)$. One may now form the total wavefunction for an NN system in a given partial wave (i.e., with good orbital angular momentum),

$$\Psi_{(LS)J;T}^{M_J;M_T}(1, 2) = R_{(LS)J;T}(r) [Y_L(\Omega_r) \otimes A_S(1, 2)]_J^{M_J} B_T^{M_T}(1, 2) ,$$

where here for simplicity the CM plane wave has been omitted. The total state is presumed to have orbital and spin angular momenta coupled to total angular momentum J with projection M_J. What happens upon interchange of all of the coordinates of the two nucleons? Since the Pauli exclusion principle states that the total wavefunction must be antisymmetric, what quantum numbers are allowed for the NN system?

b) Suppose that one has a spin-isospin dependent central potential of the form

$$V_C(r) = V_1(r) + V_2(r)\sigma(1) \cdot \sigma(2) + V_3(r)\tau(1) \cdot \tau(2)$$
$$+ V_4(r)\sigma(1) \cdot \sigma(2)\tau(1) \cdot \tau(2) ,$$

where the factors $\sigma(i)$ and $\tau(i)$ are the familiar Pauli matrices discussed in this chapter. For a state of the type introduced in the previous exercise evaluate the spin and isospin matrix elements of the potential.

c) If a so-called spin–orbit potential of the form

$$V_{LS}(r) = \{V_5(r) + V_6(r) (\tau(1) \cdot \tau(2))\} (L \cdot S)$$

were to be added to the central potential in the previous Exercise, what would be the form of the orbital-spin-isospin matrix elements?

d) One might also want to add a so-called tensor potential of the form

$$V_T(r) = \{V_7(r) + V_8(r) (\tau(1) \cdot \tau(2))\} S_{12}(1, 2),$$

involving the tensor operator

$$S_{12}(1, 2) = 3(r \cdot \sigma(1))(r \cdot \sigma(2))/r^2 - \sigma(1) \cdot \sigma(2) .$$

Find the result of applying the tensor operator to a singlet spin state, i.e., a state having $S = 0$. For a potential that contains a tensor term one finds that the orbital angular momentum is not a good quantum number and that states with differing values of L must mix; which states are these? Do you know a relatively familiar example of this mixing? Can you think of other types of potentials not included in the types discussed here in Exercises 2.4 and 2.5?

2.6   Single-Particle Wavefunctions

As we shall see later in the book, it is useful to employ a basis of single-particle wavefunctions when discussing the nuclear many-body problem. One starts with some mean-field potential in which the individual nucleons move and, upon coupling the orbital and spin angular momenta to form the total angular momentum, forms the single-particle wavefunctions

$$\psi_{(\ell s)j;t}^{m_j;m_t}(r) = R_{(\ell s)j;t}(r)\,[Y_\ell(\Omega_r) \otimes \chi_s]_j^{m_j}\,\xi_t^{m_t}\,,$$

where $s = t = 1/2$ and here the spatial coordinate is with respect to the origin of the potential. What values are allowed for the orbital and total angular momentum quantum numbers ℓ and j? If the potential has a spin–orbit term such as in Exercise 2.5c, what does one expect for the splitting between states with the same orbital angular momentum, but different values of j?

2.7 $SU(6)$ Symmetry

Employing the rules given in the text for determining the dimensions of representations of $SU(N)$ groups using the Young tableaux, consider the direct product of three copies of the fundamental in $SU(6)$, namely **6**,

$$\mathbf{6} \otimes \mathbf{6} \otimes \mathbf{6}$$

(in the text the cases of $SU(2)$ and $SU(3)$ were both discussed). Determine the resulting direct sum of the representations that emerge and find their dimensions.

2.8   Spontaneous Symmetry Breaking in Classical Mechanics

Consider a frictionless bead of mass m free to slide on a hoop of radius R, which is rotating about a vertical axis with angular velocity ω.

a) Show that the potential energy is given by

$$V(\theta) = -\frac{1}{2}m\omega^2 R^2 \sin^2\theta - mgR\cos\theta\,,$$

where θ is the angle of the bead as measured from the bottom of the hoop.

b) Show that for $\omega^2 < \frac{g}{R}$ the shape of the potential has a single minimum, so that the position of stable equilibrium is at $\theta = 0$.

c) Show that for $\omega^2 > \frac{g}{R}$ the shape of the potential has two minima, so that there exist two positions of stable equilibrium at

$$\theta = \pm\cos^{-1}\frac{g}{\omega^2 R}\,.$$

Let us commence the discussions in this chapter by recalling the arguments made in the previous chapter. Here we use the concepts of symmetry group representations to explore the possible ways the quarks are combined to form hadrons. We begin with the minimal version of the problem, restricting our attention to u and d quarks, and later extend the model to include s quarks. Both mesons and baryons are considered below, but let us consider the latter with only u and d quarks active to get started.

Being a fermion, a baryon's wavefunction must be completely antisymmetric under quark exchange, i.e., under exchange of all quantum numbers (spatial, spin, isospin, and color coordinates – see Chapter 5). Let us assume that the lowest-lying baryons are made of three quarks, qqq, and have symmetric spatial wavefunctions, $(1s)^3$, which at least nonrelativistically is reasonable, since non-s-wave configurations have more curvature and hence higher energies. This implies that baryons must belong to a completely antisymmetric representation with respect to

$$\text{spin} \otimes \text{isospin} \otimes \text{color} , \tag{3.1}$$

where, as discussed in Chapter 2 the spin and isospin content involves $SU(2)$ subgroups, while the color subgroup involves $SU(3)$ (see Chapter 5). Namely, one has an antisymmetric representation within $SU(12)$, which can be decomposed into spin/isospin and color subgroups,

$$SU(12)_{SIC} = SU(4)_{SI} \otimes SU(3)_C . \tag{3.2}$$

In fact, we are only interested in antisymmetric color configurations (singlets), and so the spin/isospin part of the wavefunction must be symmetric. The spin/isospin parts of the wavefunction may be decomposed into the individual $SU(2)$ subgroups,

$$SU(4)_{SI} = SU(2)_S \otimes SU(2)_I . \tag{3.3}$$

There are two combinations possible, one with spin and isospin both equal to 3/2, viz., both symmetric, and one with spin and isospin both equal to 1/2, i.e., both mixed-symmetric, but put together such that the product wavefunction is symmetric. As we shall see later in this chapter, the former case is the multiplet that includes the prominent spin-3/2, isospin-3/2 resonance seen at 1232 MeV (the Δ), while the latter includes the nucleon.

Let us continue with the characterizations in terms of the spin/isospin $SU(2)$ subgroups. Recalling from the discussion in Chapter 2 what happens when we recouple the quantum numbers for a system of two spin-1/2, isospin-1/2 fermions we obtained the Clebsch–Gordan series

$$\mathbf{2} \otimes \mathbf{2} = \mathbf{3}_S \oplus \mathbf{1}_A . \tag{3.4}$$

On the right-hand side of this equation the first term is symmetric under interchange of the two particles, while the other is antisymmetric; we have added the labels "S" and "A" to emphasize the permutation symmetry. Now let us proceed to couple in a third spin-1/2, isospin-1/2 particle, obtaining

$$\mathbf{2} \otimes \mathbf{2} \otimes \mathbf{2} = \mathbf{4}_S \oplus \mathbf{2}_{M,S} \oplus \mathbf{2}_{M,A} \,, \tag{3.5}$$

again labeling the representation with their permutation symmetries, a quartet that is symmetric and two mixed-symmetry doublets, one symmetric under interchange of the first two labels, 12, and one antisymmetric under this interchange, but with no simple symmetry under 13 or 23 interchanges. For simplicity, letting "↑" and "↓" indicate the two projections of spin, and writing the product of the three single-particle wavefunctions in order $↑↓_1↑↓_2↑↓_3$, we have the following: for representation $\mathbf{4}_S$

$$\psi^S_{+3/2} = |↑↑↑\rangle \tag{3.6}$$

$$\psi^S_{+1/2} = [|↑↑↓\rangle + |↑↓↑\rangle + |↓↑↑\rangle]/\sqrt{3} \tag{3.7}$$

$$\psi^S_{-1/2} = [|↓↓↑\rangle + |↓↑↓\rangle + |↑↓↓\rangle]/\sqrt{3} \tag{3.8}$$

$$\psi^S_{-3/2} = |↓↓↓\rangle \,, \tag{3.9}$$

where the four states correspond to the $+3/2, +1/2, -1/2, -3/2$ projections in spin space and are normalized to unity. Likewise for the doublet $\mathbf{2}_{M,S}$ representation

$$\psi^{M\{12\}}_{+1/2} = [\{|↑↓↑\rangle + |↓↑↑\rangle\} - 2|↑↑↓\rangle]/\sqrt{6} \tag{3.10}$$

$$\psi^{M\{12\}}_{-1/2} = [\{|↓↑↓\rangle + |↑↓↓\rangle\} - 2|↓↓↑\rangle]/\sqrt{6} \tag{3.11}$$

and the doublet $\mathbf{2}_{M,A}$ representation

$$\psi^{M[12]}_{+1/2} = [|↑↓↑\rangle - |↓↑↑\rangle]/\sqrt{2} \tag{3.12}$$

$$\psi^{M[12]}_{-1/2} = [|↓↑↓\rangle - |↑↓↓\rangle]/\sqrt{2} \,. \tag{3.13}$$

Note that we have added the label {12} or [12] to remind ourselves that the states are either symmetric or antisymmetric with respect to interchange of particles 1 and 2. Later in constructing the baryons we shall use as basis states those from the $\mathbf{2}_{M,A}$ doublet which are antisymmetric in (12), (23), and (13). Let us now turn to specifics for light mesons in the next section and for baryons in the section to follow.

3.1   Light Mesons Built from *u*, *d*, and *s* Quarks

Given the introduction above where only *u* and *d* quarks are considered, let us now extend the model to include *s* quarks; we begin with a discussion of the light mesons. To obtain the flavor $SU(3)$ structure of light mesons ($q\bar{q}$ states made from *u*, *d*, and *s*) quarks and antiquarks one takes the product of $\mathbf{3}$ (q) with $\bar{\mathbf{3}}$ ($\bar{q}$):

$$\mathbf{3} \otimes \bar{\mathbf{3}} = \mathbf{1} \oplus \mathbf{8} \,, \tag{3.14}$$

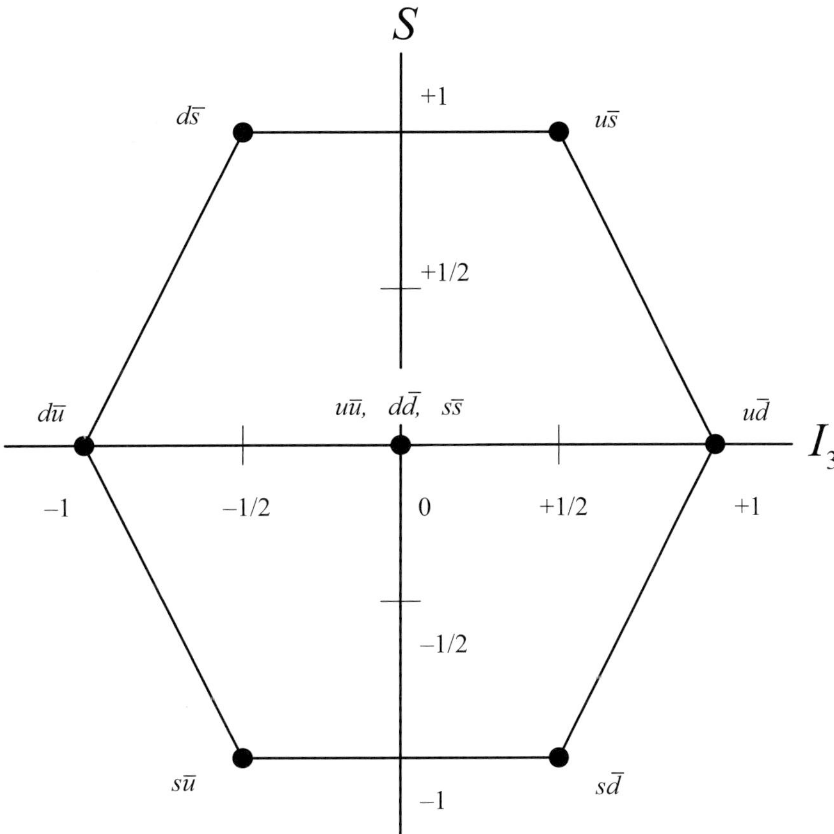

Fig. 3.1   Weight diagram for the meson nonet, where the vertical axis is labeled by strangeness S and the horizontal axis is labeled by the third component of isospin I_3. By convention, strange quarks carry strangeness $S = -1$, whereas strange antiquarks carry the opposite.

i.e., a singlet and an octet. In terms of the $SU(2)$ spin content one has Eq. (3.4),

$$\mathbf{2} \otimes \mathbf{2} = \mathbf{3} \oplus \mathbf{1} \, , \tag{3.15}$$

and so spin-0 (pseudoscalar, $J^P = 0^-$) and spin-1 (vector, $J^P = 1^-$) mesons. The weight diagram showing the $q\bar{q}$ structures of the meson nonet is shown in Fig. 3.1 while the low-lying physical pseudoscalar and vector mesons are indicated in Fig. 3.2.

The states at the centers of the nonets require a bit of discussion. Let us label the states in the nonet $|f; I, I_3\rangle$ by flavor symmetry f and isospin I with its projection I_3.[1] There are three possible combinations of $u\bar{u}$, $d\bar{d}$, and $s\bar{s}$ that can be written, one singlet

$$|\underline{1}; 0, 0\rangle = \left[|u\bar{u}\rangle + |d\bar{d}\rangle + |s\bar{s}\rangle \right] / \sqrt{3} \equiv \phi_1 \tag{3.16}$$

---

[1] Note that, whilst in most of the book we have used T and T_3 to denote isospin, as is more common in discussions of nuclear physics, in this chapter instead we have employed the notation that is more common in particle physics. namely, I and I_3.

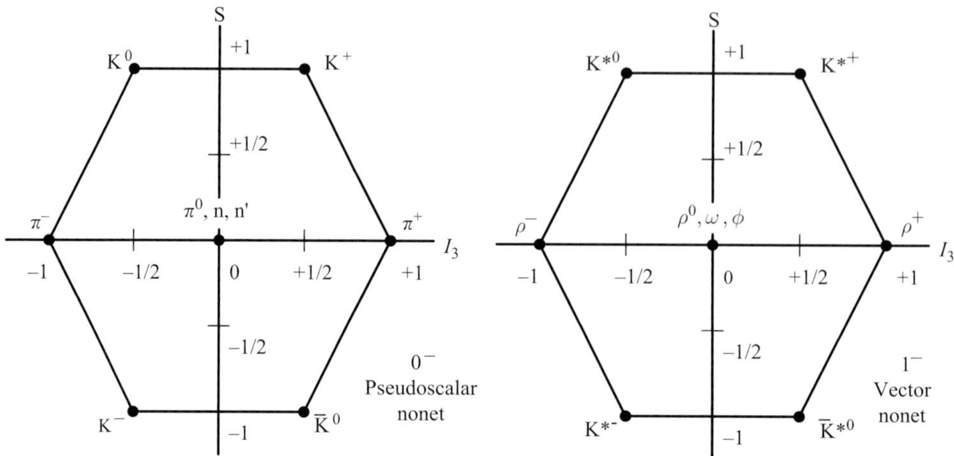

Fig. 3.2   Low-lying mesons of the pseudoscalar and vector nonets. The measured masses in MeV are $\pi^{\pm}$(140), π^0(135), K^0, $\overline{K}^0$(498), $K^{\pm}$(494), η (548), η'(958) and $\rho^{\pm,0}$(773), ω (783), ϕ (1019), K^{*0}, $\overline{K}^{*0}$(896), $K^{*\pm}$(892), respectively.

and two octet states

$$|\underline{8}; 1, 0\rangle = \left[|u\bar{u}\rangle - |d\bar{d}\rangle\right]/\sqrt{2} \tag{3.17}$$

$$|\underline{8}; 0, 0\rangle = \left[|u\bar{u}\rangle + |d\bar{d}\rangle - 2\,|s\bar{s}\rangle\right]/\sqrt{3} \equiv \phi_8\,, \tag{3.18}$$

where, as in the discussion of spin in the first section, the flavor labels refer to particles 1 and 2, respectively. In fact, since the strange quark is heavier than the up and down quarks one state is closer to being purely $s\bar{s}$, while the other two have almost no strange quark content. Thus the physical vector meson states are better represented by

$$\rho^0 = \left[|u\bar{u}\rangle - |d\bar{d}\rangle\right]/\sqrt{2} \tag{3.19}$$

$$\omega = \left[|u\bar{u}\rangle + |d\bar{d}\rangle\right]/\sqrt{2} = \sqrt{2/3}\phi_1 + \sqrt{1/3}\phi_8 \tag{3.20}$$

$$\phi = |s\bar{s}\rangle = \sqrt{1/3}\phi_1 - \sqrt{2/3}\phi_8\,. \tag{3.21}$$

That is, the ω and ϕ are mixtures of $SU(3)$ singlet and octet configurations. This is called ideal mixing and correlates well with the observation that these two mesons are nearly degenerate in mass, while the ϕ is much heavier.

A similar pattern roughly emerges for the pseudoscalar mesons at the center of the nonet, with the π^0 being the analog of the ρ^0 and the analogs of the states called $\phi_{1,8}$ above being labeled $\eta_{1,8}$. For ideal mixing one would have

$$\eta = \cos\theta_p\eta_1 + \sin\theta_p\eta_8 \tag{3.22}$$

$$\eta' = \sin\theta_p\eta_1 - \cos\theta_p\eta_8\,, \tag{3.23}$$

with $\cos\theta_p = \sqrt{2/3} \leftrightarrow \theta_p = 35.3^0$, whereas the measured masses and decay branching ratios actually favor $\theta_p \cong 11^0$. The reason that the pseudoscalar mesons do not ideally mix is thought to arise from possible couplings via two-gluon exchange (see Chapter 5), which affects only the flavor singlet combinations. The total Hamiltonian is thus not diagonal as far as the non-strange/strange basis is concerned (as happens for the vector mesons), but contains contributions that tend to drive the physical states towards being singlet and octet states.

Let us next discuss the concept of G-parity

$$G = C e^{i\pi I_2}, \tag{3.24}$$

where C is the charge conjugation operator, interchanging particles and antiparticles, and $I_2 = \tau_2/2$. Acting on the states

$$\psi_S \equiv \left[|u\bar{d}\rangle + |\bar{d}u\rangle \right] / \sqrt{2} \tag{3.25}$$

$$\psi_A \equiv \left[|u\bar{d}\rangle - |\bar{d}u\rangle \right] / \sqrt{2}\,, \tag{3.26}$$

where as usual the particles are 1 to the left and 2 to the right, with the G-parity operator then yields

$$G\psi_S = \left[- |\bar{d}u\rangle - |u\bar{d}\rangle \right] / \sqrt{2} = -\psi_S \tag{3.27}$$

$$G\psi_A = \left[- |\bar{d}u\rangle + |u\bar{d}\rangle \right] / \sqrt{2} = +\psi_A\,, \tag{3.28}$$

and so these states are eigenstates of the G-parity operator with eigenvalues $G = -1\ (+1)$ for $S\ (A)$ – the former is the π and the latter the ρ. For the charge-0 meson states we have

$$\pi^0 = \left[\left(|d\bar{d}\rangle - |u\bar{u}\rangle \right) + \left(|\bar{d}d\rangle - |\bar{u}u\rangle \right) \right] / 2 \tag{3.29}$$

$$\rho^0 = \left[\left(|d\bar{d}\rangle - |u\bar{u}\rangle \right) - \left(|\bar{d}d\rangle - |\bar{u}u\rangle \right) \right] / 2\,, \tag{3.30}$$

that is, states ψ_S^0 and ψ_A^0, respectively. The former has G-parity $+1$ and the latter has -1. Thus, one sees the utility of characterizing the meson states by their G-parity, since this correlates with the type of particle, as illustrated here for mesons built from u and d quarks. Extensions to include s quarks are straightforward [Clo79]. Labeling the symmetric spin-triplet and antisymmetric spin-singlet wavefunctions by $\chi_{S,A}$, respectively, we have the following symmetry properties for the $\psi\chi$ product wavefunctions: under interchange of particles 1 and 2 one has symmetric wavefunctions $\psi_S\chi_S$ and $\psi_A\chi_A$ and antisymmetric wavefunctions $\psi_S\chi_A$ and $\psi_A\chi_S$.

Given the structure of the light mesons as $q\bar{q}$ configurations in $SU(3)$, let us next explore a simple model for their masses. For this we start with the sum of the masses of the constituent quark and antiquark masses, m_1 and m_2,[2]

$$M_0^{\text{meson}} = m_1 + m_2 \tag{3.31}$$

and add a spin-spin interaction term of the type

$$M_{\text{int}}^{\text{meson}} \equiv \xi \, \frac{\mathbf{S}_1 \cdot \mathbf{S}_2}{m_1 m_2} \tag{3.32}$$

to form the total mass

$$M^{\text{meson}} = M_0^{\text{meson}} + M_{\text{int}}^{\text{meson}}\,, \tag{3.33}$$

---

[2] The distinction between "constituent" and "current" quarks will become clear when discussed in Chapters 4 and 5.

where the constituent quark masses and the constant ξ are parameters to be fit using the measured meson masses. The spin-spin interaction is motivated by recalling the hyperfine interaction in the hydrogen atom [Sch55, Mer98], the present case being the QCD analog due to gluon exchange rather than the photon exchange occurring in the atomic case (see Chapter 5 for more discussion of hadron masses). Since the total spin is given by $\mathbf{S} = \mathbf{S}_1 + \mathbf{S}_2$ with $\mathbf{S}_1 \cdot \mathbf{S}_2 = [S(S+1) - 3/2]/2$, it is straightforward to show that

$$M_{\text{int}} = \frac{\xi}{4m_1 m_2} \times \begin{cases} -3 & \text{pseudoscalar} \\ +1 & \text{vector} . \end{cases} \tag{3.34}$$

Upon fitting the physical masses one finds good agreement for the pseudoscalar π and K mesons and for all four vector mesons using as parameters $m_u = m_d \equiv m = 308\,\text{MeV}$, $m_s \equiv m' = 483\,\text{MeV}$ and $\xi/4m^2 = 159\,\text{MeV}$. Indeed, this simple model yields values that are typically in agreement to better than 1%, the largest deviation being for the kaon at 1.9%. On the other hand, the η and η' mesons, which as discussed above are not found to have ideal mixing, are more complicated [Clo79] (see the Exercises).

3.2 Baryons

Having summarized the familiar problem of recoupling angular momenta (or, equivalently, isospins) for three spin-1/2 fermions in the first section, we now turn to the extensions of similar ideas to representations of flavor $SU(3)$, i.e., using u, d, and s quarks in qqq configurations. Focusing on the low-lying baryons, the fundamental representation now has three states, rather than two $\uparrow\downarrow$ as in $SU(2)$, and, anticipating our identifying these with quark flavors, we will now use the labels u, d, and s. Recoupling two such fermions (the analog of Eq. (3.4); see also Fig. 2.3) one has for the nine states in the product six that are symmetric and three that are antisymmetric:

$$\mathbf{3} \otimes \mathbf{3} = \mathbf{6}_S \oplus \overline{\mathbf{3}}_A . \tag{3.35}$$

Continuing for the product of three fundamental representations for the flavor representations for baryons, in analogy with Eq. (3.5) one finds the following for the recoupling of the 27 states obtained:

$$\mathbf{3} \otimes \mathbf{3} \otimes \mathbf{3} = \mathbf{10}_S \oplus \mathbf{8}_{M,S} \oplus \mathbf{8}_{M,A} \oplus \mathbf{1}_A , \tag{3.36}$$

namely, a symmetric decuplet, two mixed-symmetry octets (labeled as above where $SU(2)$ was being discussed) and an antisymmetric singlet. Recalling the arguments made in the introduction to this chapter, we see that when building baryons from qqq configurations in terms of spin, one has a symmetric spin-3/2 quartet and two mixed-symmetry spin-1/2 doublets (see Eq. (3.5)), while in terms of the flavor content one has a singlet, two mixed-symmetry octets and a decuplet (see Eq. (3.36)). The flavor weight diagrams for the octets and the decuplet are shown in Fig. 3.3 (the singlet is not shown, since it is trivial).

Recalling that the Pauli exclusion principle requires the total wavefunction to be antisymmetric and that we have already argued that the color $SU(3)$ wavefunction must

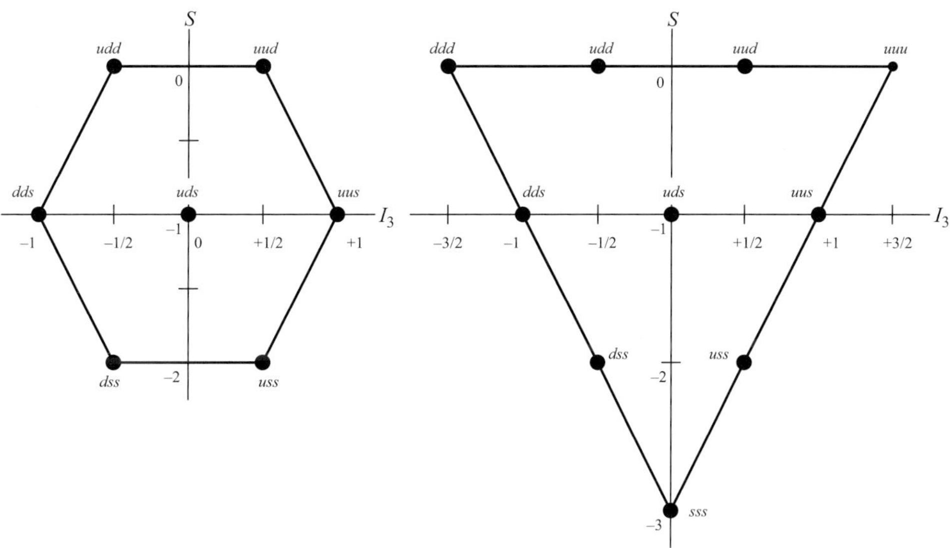

Fig. 3.3   Weight diagrams for the baryon octet and decuplet.

be antisymmetric, and specializing to the situation where the spatial wavefunctions of the three quarks are assumed all to be in symmetric s-states (s^3), we require the flavor/spin wavefunctions of the baryons to be symmetric, and hence only specific products of the $SU(2)$ spin representations and the flavor $SU(3)$ representations are allowed. In terms of the $SU(6)$ product representations we have

$$\mathbf{6} \otimes \mathbf{6} \otimes \mathbf{6} = \mathbf{56}_S \oplus \mathbf{70}_{M,S} \oplus \mathbf{70}_{M,A} \oplus \mathbf{20}_A \ . \tag{3.37}$$

In terms of the $(SU(3)_F, SU(2)_S)$ flavor/spin content one has

$$(\mathbf{10}, \mathbf{4}) \oplus (\mathbf{8}, \mathbf{2}) \quad \mathbf{56}_S \tag{3.38}$$

$$(\mathbf{10}, \mathbf{2}) \oplus (\mathbf{8}, \mathbf{4}) \oplus (\mathbf{8}, \mathbf{2}) \oplus (\mathbf{1}, \mathbf{2}) \quad \mathbf{70}_M \tag{3.39}$$

$$(\mathbf{8}, \mathbf{2}) \oplus (\mathbf{1}, \mathbf{4}) \quad \mathbf{20}_A \ . \tag{3.40}$$

Thus, the required symmetric $SU(6)$ representation is the 56-dimensional one with corresponding low-mass $J^P = \frac{1}{2}^+$ octet and $J^P = \frac{3}{2}^+$ decuplet baryon multiplets. Indeed, as we shall see below, the feature that the correct numbers of baryons are found to fit with this picture and to have approximately the correct ground-state properties (for instance, their magnetic moments are about right) is the basic reason for introducing the fully antisymmetric color state; were this to be absent the flavor/spin states here would have to be antisymmetric and would not fit the observed pattern.

Let us begin by discussing the easier of the two multiplets, that is, the spin-3/2 decuplet, whose spin structure is the $\mathbf{4}_S$ representation discussed earlier. This case will illustrate how to proceed with the somewhat more complicated spin-1/2 octet. We consider first the maximal spin projection of $+3/2$ in which case the spin wavefunction is simply $\psi^S_{+3/2} = |\uparrow\uparrow\uparrow\rangle$ and begin with the maximal charge case having three u quarks,

$$|\Delta^{++}\rangle = |uuu\rangle \ . \tag{3.41}$$

The full flavor-spin state is actually $|u\uparrow u\uparrow u\uparrow\rangle$; however, since the spin content is universal for the decuplet, for simplicity we suppress it here (compare with the discussion of the octet to follow). Recall that the action of the isospin lowering operator is to change a u quark into a d quark (see Eq. (2.65)) and to annihilate a d or s quark. Consider applying the single-particle isospin-lowering operator

$$I_- = \sum_i I_-(i) \ . \tag{3.42}$$

This is one of a set of useful single-particle operators (one-body operators in second quantization; see the discussion later in the book) which have the form

$$\mathcal{O}^{[1]} = \sum_i \mathcal{O}^{[1]}(i) \ , \tag{3.43}$$

namely symmetric sums over the quarks one at a time. We find that

$$|\Delta^+\rangle = \frac{1}{\sqrt{3}} I_- |\Delta^{++}\rangle = \frac{1}{\sqrt{3}} I_- |uuu\rangle \tag{3.44}$$

$$= \frac{1}{\sqrt{3}} (|duu\rangle + |udu\rangle + |uud\rangle) \ . \tag{3.45}$$

Applying the I-spin lowering operator two more times we find that

$$|\Delta^0\rangle = \frac{1}{2} |\Delta^+\rangle = \frac{1}{\sqrt{3}} (|ddu\rangle + |dud\rangle + |udd\rangle) \tag{3.46}$$

$$|\Delta^-\rangle = \frac{1}{\sqrt{3}} |\Delta^0\rangle = |ddd\rangle \ . \tag{3.47}$$

In each case one can check that the states are normalized to unity. The same idea can be followed in applying the so-called V-spin-lowering operator

$$V_- = \sum_i V_-(i)$$

which changes u quarks into s quarks, and annihilates d and s quarks:

$$|\Sigma^{*+}\rangle = \frac{1}{\sqrt{3}} V_- |\Delta^{++}\rangle = \frac{1}{\sqrt{3}} V_- |uuu\rangle \tag{3.48}$$

$$= \frac{1}{\sqrt{3}} (|suu\rangle + |usu\rangle + |uus\rangle) \tag{3.49}$$

$$\left|\Xi^{*0}\right\rangle = \frac{1}{2}\left|\Sigma^{*+}\right\rangle = \frac{1}{\sqrt{3}}\left(\left|ssu\right\rangle + \left|sus\right\rangle + \left|uss\right\rangle\right) \tag{3.50}$$

$$\left|\Omega^{-}\right\rangle = \frac{1}{\sqrt{3}}\left|\Xi^{*0}\right\rangle = \left|sss\right\rangle . \tag{3.51}$$

One could just as well use the other $SU(2)$ subgroup, the so-called U-spin, which changes d quarks into s quarks, and annihilates u and s quarks. Referring back to Fig. 3.3, we see that I_- moves horizontally to the left, V_- moves down and to the right and U_- moves down and to the right in the weight diagram. I-spin, V-spin, and U-spin form the three $SU(2)$ subgroups of $SU(3)$.

Then the rest of the multiplet may be constructed by applying the I-spin lowering operator to the states in Eqs. (3.49-3.51), as before:

$$\begin{aligned}\left|\Sigma^{*0}\right\rangle = \frac{1}{\sqrt{2}}I_-\left|\Sigma^{*+}\right\rangle = \frac{1}{\sqrt{6}}(&\left|dsu\right\rangle + \left|dus\right\rangle \\ &+ \left|sdu\right\rangle + \left|uds\right\rangle \\ &+ \left|sud\right\rangle + \left|usd\right\rangle)\end{aligned} \tag{3.52}$$

$$\left|\Sigma^{*-}\right\rangle = \frac{1}{\sqrt{2}}I_-\left|\Sigma^{*0}\right\rangle = \frac{1}{\sqrt{3}}\left(\left|sdd\right\rangle + \left|dsd\right\rangle + \left|dds\right\rangle\right) \tag{3.53}$$

$$\left|\Xi^{*-}\right\rangle = I_-\left|\Xi^{*0}\right\rangle = \frac{1}{\sqrt{3}}\left(\left|ssd\right\rangle + \left|sds\right\rangle + \left|dss\right\rangle\right) , \tag{3.54}$$

which nicely fits the pattern of the known baryon decuplet, as displayed on the right-hand side of Fig. 3.4. In particular, a great triumph for this multiplet model – the eightfold way [Gel64b] – was the prediction using the symmetry properties discussed here that the

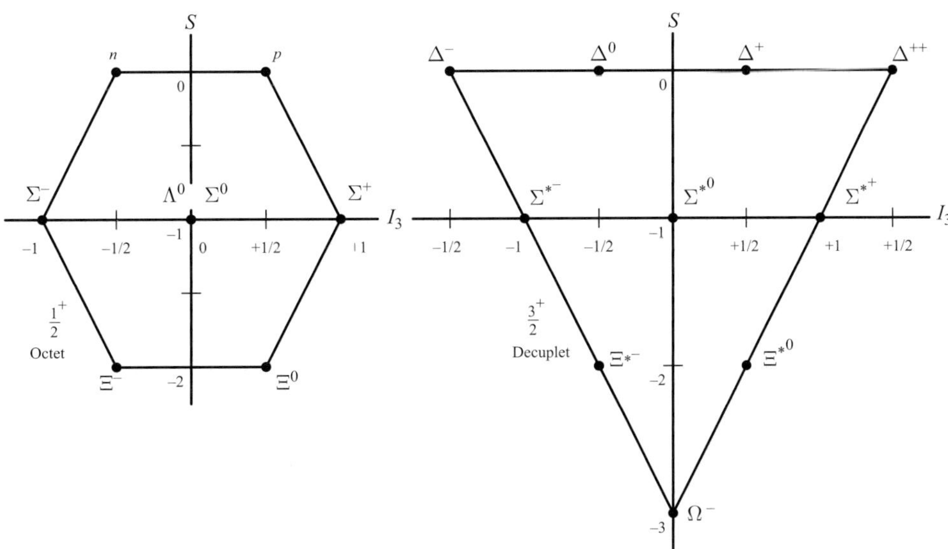

Fig. 3.4   Octet and decuplet of low-lying baryons. The masses in MeV are n (940), p (938), Λ^0 (1116), Σ^-(1197), Σ^0(1193), Σ^+(1189), Ξ^-(1322), Ξ^0(1315) for the octet, and Δ(1232), Σ^{*-}(1387), Σ^{*0}(1384), Σ^{*+}(1383), Ξ^{*-}(1535), Ξ^{*0}(1532), Ω^-(1672) for the decuplet.

particle with negative charge and strangeness -3, that is, the Ω^- should exist to complete the multiplet. As we shall see below, even its mass can be predicted. The particle was found with the appropriate quantum numbers and mass, lending weight to the constituent quark model.

The spin-1/2 baryon octet may be constructed in a similar manner, the complication being that now we must deal with mixed-symmetry representations such as for the $\mathbf{2}_{M,S}$ and $\mathbf{2}_{M,A}$ spin configurations discussed above. Starting with the mixed-symmetry spin states in Eqs. (3.12, 3.13) and specializing to the spin projection $+1/2$ case,

$$\psi^{M[12]}_{+1/2} = [|\uparrow\downarrow\uparrow\rangle - |\downarrow\uparrow\uparrow\rangle]/\sqrt{2}$$
$$\psi^{M[23]}_{+1/2} = [|\uparrow\uparrow\downarrow\rangle - |\uparrow\downarrow\uparrow\rangle]/\sqrt{2} \tag{3.55}$$
$$\psi^{M[13]}_{+1/2} = [|\uparrow\uparrow\downarrow\rangle - |\downarrow\uparrow\uparrow\rangle]/\sqrt{2}$$

and also specializing at first to states built from two u quarks and one d quark, namely the proton, the flavor analogs of Eqs. (3.12, 3.13) for projection $+1/2$ are the following:

$$\phi^{M[12]}_{+1/2} = [|udu\rangle - |duu\rangle]/\sqrt{2}$$
$$\phi^{M[23]}_{+1/2} = [|uud\rangle - |udu\rangle]/\sqrt{2} \tag{3.56}$$
$$\phi^{M[13]}_{+1/2} = [|uud\rangle - |duu\rangle]/\sqrt{2}\,.$$

Clearly the states as written are antisymmetric under exchange of particles 1 and 2, 2 and 3, or 1 and 3, as indicated. So, for example, the first of Eqs. (3.55) is spin singlet for particles 1 and 2 and the first of Eqs. (3.56) has particles 1 and 2 in an isospin singlet state. One can show that

$$\psi^{M,A}_{+1/2}(13) = \psi^{M,A}_{+1/2}(12) + \psi^{M,A}_{+1/2}(23) \tag{3.57}$$
$$\phi^{M,A}_{+1/2}(13) = \phi^{M,A}_{+1/2}(12) + \phi^{M,A}_{+1/2}(23)\,, \tag{3.58}$$

although we shall not need to use these facts. Since the total wavefunction must be antisymmetric and we have the color part antisymmetric together (in the simple model being constructed here) with a spatially symmetric $(1s)^3$ wavefunction, this means we need to have the spin-flavor parts of the wavefunction completely symmetric. We accomplish the by combining the states above in the following way:

$$|p\uparrow\rangle = \phi^{M[12]}_{+1/2}\phi^{M[12]}_{+1/2} + \phi^{M[23]}_{+1/2}\psi^{M[23]}_{+1/2} + \phi^{M[13]}_{+1/2}\psi^{M[13]}_{+1/2} \tag{3.59}$$
$$= (2\,|u\uparrow u\uparrow d\downarrow\rangle + 2\,|u\uparrow d\downarrow u\uparrow\rangle + 2\,|d\downarrow u\uparrow u\uparrow\rangle$$
$$- |u\uparrow u\downarrow d\uparrow\rangle - |u\uparrow d\uparrow u\downarrow\rangle - |d\uparrow u\uparrow u\downarrow\rangle$$
$$- |u\downarrow u\uparrow d\uparrow\rangle - |u\downarrow d\uparrow u\uparrow\rangle - |d\uparrow u\downarrow u\uparrow\rangle)/\sqrt{18}\,. \tag{3.60}$$

We can obtain the wavefunction for the neutron (again with spin projection $+1/2$) by applying the isospin lowering operator I_-. Applying this to Eq. (3.60) yields

$$|n\uparrow\rangle = (2\,|d\uparrow d\uparrow u\downarrow\rangle + 2\,|d\uparrow u\downarrow d\uparrow\rangle + 2\,|u\downarrow d\uparrow d\uparrow\rangle$$
$$- |d\uparrow d\downarrow u\uparrow\rangle - |d\uparrow u\uparrow d\downarrow\rangle - |u\uparrow d\uparrow d\downarrow\rangle \tag{3.61}$$
$$- |d\downarrow d\uparrow u\uparrow\rangle - |d\downarrow u\uparrow d\uparrow\rangle - |u\uparrow d\downarrow d\uparrow\rangle)/\sqrt{18}\,.$$

We can use the same strategy to generate some of the other states. In particular, we can apply U- and V-spin raising or lowering operators to move to other positions on the octet diagram shown in Fig. 3.4:

$$
\begin{aligned}
\left|\Sigma^+ \uparrow\right\rangle &= U_- \left|p \uparrow\right\rangle \\
\left|\Sigma^- \uparrow\right\rangle &= V_- \left|n \uparrow\right\rangle \\
\left|\Xi^- \uparrow\right\rangle &= U_- \left|\Sigma^- \uparrow\right\rangle \\
\left|\Xi^0 \uparrow\right\rangle &= V_- \left|\Sigma^+ \uparrow\right\rangle \\
\left|\Sigma^0 \uparrow\right\rangle &= \frac{1}{\sqrt{2}} I_- \left|\Sigma^+ \uparrow\right\rangle .
\end{aligned}
\tag{3.62}
$$

These states can be written out in detail; for instance, applying U_- to the proton to get the Σ^+ (making the replacements $d \to s$ and $u, s \to 0$ symmetrically for particles 1, 2, and 3), ignoring overall signs one has

$$
\begin{aligned}
\left|\Sigma^+ \uparrow\right\rangle = [\,&2\left|u \uparrow u \uparrow s \downarrow\right\rangle + 2\left|u \uparrow s \downarrow u \uparrow\right\rangle + 2\left|s \downarrow u \uparrow u \uparrow\right\rangle \\
&- \left|u \uparrow u \downarrow s \uparrow\right\rangle - \left|u \uparrow s \uparrow u \downarrow\right\rangle - \left|s \uparrow u \uparrow u \downarrow\right\rangle \\
&- \left|u \downarrow u \uparrow s \uparrow\right\rangle - \left|u \downarrow s \uparrow u \uparrow\right\rangle - \left|s \uparrow u \downarrow u \uparrow\right\rangle]/\sqrt{18} ,
\end{aligned}
\tag{3.63}
$$

leading to

$$
\begin{aligned}
\left|\Sigma^0 \uparrow\right\rangle = \frac{1}{6} [\,&2\left(\left|u \uparrow d \uparrow s \downarrow\right\rangle + \left|d \uparrow u \uparrow s \downarrow\right\rangle\right) + 2\left(\left|u \uparrow s \downarrow d \uparrow\right\rangle + \left|d \uparrow s \downarrow u \uparrow\right\rangle\right) \\
&+ 2\left(\left|s \downarrow u \uparrow d \uparrow\right\rangle + \left|s \downarrow d \uparrow u \uparrow\right\rangle\right) \\
&- \left(\left|u \uparrow d \downarrow s \uparrow\right\rangle + \left|d \downarrow u \uparrow s \uparrow\right\rangle\right) - \left(\left|u \downarrow d \uparrow s \uparrow\right\rangle + \left|d \uparrow u \downarrow s \uparrow\right\rangle\right) \\
&- \left(\left|u \uparrow s \uparrow d \downarrow\right\rangle + \left|d \downarrow s \uparrow u \uparrow\right\rangle\right) - \left(\left|u \downarrow s \uparrow d \uparrow\right\rangle + \left||d \uparrow s \uparrow u \downarrow\right\rangle\right) \\
&- \left(\left|s \uparrow u \uparrow d \downarrow\right\rangle + \left|s \uparrow d \downarrow u \uparrow\right\rangle\right) - \left(\left|s \uparrow u \downarrow d \uparrow\right\rangle + \left|s \uparrow d \uparrow u \downarrow\right\rangle\right)]
\end{aligned}
\tag{3.64}
$$

and so on for the other cases in Eqs. (3.62) (see the Exercises). Note that the isospin structure with respect to the u and d quark content in Eq. (3.64) has been grouped to bring out the fact that this pair forms an isotriplet, as it must, since there are three states Σ^-, Σ^0, and Σ^+.

At this point we have seven of the eight states in the ground-state baryon octet and all that remains is to obtain the final state. This is accomplished by noting that we already have used the isotriplet structure in Eq. (3.64) for Σ^0 and have another structure for the completely antisymmetric state

$$
\begin{aligned}
\phi_A = \,&\left|uds\right\rangle - \left|dus\right\rangle - \left|usd\right\rangle \\
&+ \left|dsu\right\rangle + \left|sud\right\rangle - \left|sdu\right\rangle .
\end{aligned}
\tag{3.65}
$$

However, we cannot combine this with a spin state to "cancel" the symmetry to leave a completely symmetric flavor-spin wavefunction, since no completely antisymmetric state exists. We can proceed by forming a flavor state built from all three flavors of quarks that is orthogonal both to ϕ_A and to the flavor structure in Eq. (3.64), specifically,

$$\widetilde{\phi}^{M[12]}_{+1/2} = [2\,(|uds\rangle - |dus\rangle) + (|usd\rangle - |dsu\rangle) - (|sud\rangle - |sdu\rangle)]\,/\sqrt{12}$$

$$\widetilde{\phi}^{M[23]}_{+1/2} = [2\,(|sud\rangle - |sdu\rangle) + (|usd\rangle - |dsu\rangle) - (|uds\rangle - |dus\rangle)]\,/\sqrt{12}$$

$$\widetilde{\phi}^{M[13]}_{+1/2} = [2\,(|usd\rangle - |dsu\rangle) + (|uds\rangle - |dus\rangle) - (|sud\rangle - |sdu\rangle)]\,/\sqrt{12}\,. \tag{3.66}$$

These three states are again antisymmetric under exchange of particles 1 and 2, 2 and 3, or 1 and 3, as indicated; also, in contrast to Eq. (3.64), the isospin structure with respect to the u and d quarks is antisymmetric, i.e., one has an isosinglet state in this case. Proceeding as in building the proton one now has for the remaining state at the center of the octet

$$\left|\Lambda^0\uparrow\right\rangle = \widetilde{\phi}^{M[12]}_{+1/2}\phi^{M[12]}_{+1/2} + \widetilde{\phi}^{M[23]}_{+1/2}\psi^{M[23]}_{+1/2} + \widetilde{\phi}^{M[13]}_{+1/2}\psi^{M[13]}_{+1/2} \tag{3.67}$$

$$\begin{aligned}
= \frac{1}{\sqrt{12}}\big[&(|u\uparrow d\downarrow s\uparrow\rangle - |d\uparrow u\downarrow s\uparrow\rangle) - (|u\downarrow d\uparrow s\uparrow\rangle - |d\downarrow u\uparrow s\uparrow\rangle) \\
&+ (|u\uparrow s\uparrow d\downarrow\rangle - |d\uparrow s\uparrow u\downarrow\rangle) - (|u\downarrow s\uparrow d\uparrow\rangle - |d\downarrow s\uparrow u\uparrow\rangle) \\
&+ (|s\uparrow u\uparrow d\downarrow\rangle - |s\uparrow d\uparrow u\downarrow\rangle) - (|s\uparrow u\downarrow d\uparrow\rangle - |s\uparrow d\downarrow u\uparrow\rangle)\big]\,.
\end{aligned}$$

$$\tag{3.68}$$

It is straightforward to see that acting on this state by $I_\pm$, $U_\pm$, or $V_\pm$ gives zero, that is, this state is a singlet with respect to all three of the $SU(2)$ subgroups.

3.3 Baryon Ground-State Properties

One is now in a position to compute matrix elements of various operators. Let us begin with the case of single-particle operators. For instance, we might want to obtain the baryon ground-state expectation value $\langle B\,|O^{[1]}|\,B\rangle$. If the operator acts only in spin-flavor-coordinate space (i.e., not in color space) then the active parts of the baryon wavefunctions are symmetric and hence we can permute the indices so that only particle number 1 occurs in the sum above:

$$\langle B\,|\mathcal{O}^{[1]}|\,B\rangle = 3\,\langle B\,|\mathcal{O}^{[1]}(1)|\,B\rangle\,, \tag{3.69}$$

which simplifies the calculation. Moreover, in the special cases where a given configuration $|q_a q_b q_c\rangle$ is an eigenstate of the operator, $\mathcal{O}^{[1]}(1)\,|q_a q_b q_c\rangle = K_a\,|q_a q_b q_c\rangle$ with quantum number K_a, the calculation is further simplified by using the orthogonality of the individual quark configurations. For instance, consider the charge operator

$$Q = \sum_i q(i)\,, \tag{3.70}$$

with eigenvalues $e_q = +2/3$ for u quarks and $-1/3$ for d and s quarks. For the $|\Delta^{++}\rangle = |uuu\rangle$ clearly one has charge $+2$, while for the proton with spin up one has charge $3\times\left(\frac{2}{3}\times 12 - \frac{1}{3}\times 6\right)/18 = +1$ from the individual contributions in Eq. (3.60). Likewise for the neutron with spin up in Eq. (3.61) one has charge $3\times\left(\frac{2}{3}\times 6 - \frac{1}{3}\times 12\right)/18 = 0$. These cases are very straightforward; a little more complicated, but still straightforward, is the computation of the magnetic dipole moments of the baryons in the octet. Assuming that the quarks are all in $1s$ configurations, viz., $(1s)^3$, in which case there is no contribution from their orbital motion, then the net magnetic moment is simply given by the vector sum of the magnetic moments of the three quarks:

$$\mu = \sum_i \mu(i) \,, \tag{3.71}$$

where the quarks have magnetic moments

$$\mu_q = e_q \frac{e}{2m_q} \tag{3.72}$$

with m_q being the corresponding quark mass. By definition the baryon magnetic moment is the ground-state expectation value of the z-component of this operator in the state with maximal spin projection:

$$\begin{aligned} \mu_B &= \langle B; S, M_S = S | \mu_z | B; S, M_S = S \rangle \\ &= 3 \langle B; S, M_S = S | \mu(1)_z | B; S, M_S = S \rangle \,. \end{aligned} \tag{3.73}$$

For example, consider the case of the proton where using the wavefunction in Eq. (3.60) one has

$$\mu_p = 3 \times (8\mu_u - 2\mu_d)/18 = \frac{4}{3}\mu_u - \frac{1}{3}\mu_d \tag{3.74}$$

and for the neutron using Eq. (3.61)

$$\mu_n = 3 \times (8\mu_d - 2\mu_u)/18 = \frac{4}{3}\mu_d - \frac{1}{3}\mu_u \,. \tag{3.75}$$

Note that if $m_u = m_d$ one has $\mu_u/\mu_d = -2$ and therefore $\mu_n/\mu_p = -2/3$, in reasonably good accord with the measured ratio of -0.684979 (see also the discussion in Chapter 8). The other cases in the baryon octet can be treated in the same way (see the Exercises). One finds that the constituent quark model does a good job of accounting for the magnetic moments of the lowest-lying baryon octet, and this is assuming only valence u, d, and s quarks with no explicit contributions from sea quarks, gluons, or orbital angular momentum.

Next, let us employ the same type of model for the baryon masses that was introduced in discussing the meson spectrum. Namely, starting with

$$M_0^{\text{baryon}} = m_1 + m_2 + m_3 \,, \tag{3.76}$$

the sum of the masses of the three constituent quarks, and adding an interaction term of the form in Eq. (3.32),

$$M_{\text{int}}^{\text{baryon}} \equiv \xi' \left[\frac{\mathbf{S}_1 \cdot \mathbf{S}_2}{m_1 m_2} + \frac{\mathbf{S}_1 \cdot \mathbf{S}_3}{m_1 m_3} + \frac{\mathbf{S}_2 \cdot \mathbf{S}_3}{m_2 m_3} \right] , \tag{3.77}$$

one can write the total mass as

$$M^{\text{baryon}} = M_0^{\text{baryon}} + M_{\text{int}}^{\text{baryon}} . \tag{3.78}$$

For the baryon decuplet one finds that all quark pairs are in spin-triplet states (see Eqs. (3.9)) and so each pair has $\mathbf{S}_i \cdot \mathbf{S}_j = 1/4$, yielding

$$M_{\text{int}}^{\text{baryon[10]}} = \frac{1}{4}\xi' \left[\frac{1}{m_1 m_2} + \frac{1}{m_1 m_3} + \frac{1}{m_2 m_3} \right] \tag{3.79}$$

and hence

$$M^{\text{baryon[10]}} = M_0^{\text{baryon}} \left[1 + \frac{\xi'}{4 m_1 m_2 m_3} \right]. \tag{3.80}$$

An example is the Ω^- which was discussed above. With $m = 363\,\text{MeV}$, $m' = 538\,\text{MeV}$ and $\xi'/4m^2 = 50\,\text{MeV}$ (see below) and here with $m_1 = m_2 = m_3 = m'$ one obtains $M_{\Omega^-} = 1682\,\text{MeV}$, less than 1% larger than its physical mass.

The baryon octet is more complicated. Simplifying things with the assumption that one has only quarks of mass m and m' as was done above for the mesons, and using the fact that $\mathbf{S}_1 \cdot \mathbf{S}_2 + \mathbf{S}_1 \cdot \mathbf{S}_3 + \mathbf{S}_2 \cdot \mathbf{S}_3 = [S(S+1) - 9/4]/2$ the nucleon can be shown to have mass

$$M_N = 3m - \frac{3\xi'}{4m^2}. \tag{3.81}$$

For the light-quark components of the Σ and Λ one can prove that the pairs that enter are spin-triplet and spin-singlet configurations, respectively (see, for instance, Eqs. (3.63), (3.64), and (3.68)), and accordingly the factor $\mathbf{S}(m) \cdot \mathbf{S}(m) = +1/4\,(\Sigma)$, $-3/4\,(\Lambda)$, where $\mathbf{S}(m)$ denotes the spin of a u or d quark. Then one can write

$$M_{\Sigma,\Lambda} = 2m + m' + \xi' \left[\frac{\mathbf{S}(m) \cdot \mathbf{S}(m)}{m^2} + \frac{\mathbf{S}_1 \cdot \mathbf{S}_2 + \mathbf{S}_1 \cdot \mathbf{S}_3 + \mathbf{S}_2 \cdot \mathbf{S}_3 - \mathbf{S}(m) \cdot \mathbf{S}(m)}{mm'} \right] \tag{3.82}$$

$$= 2m + m' + \xi' \left[\frac{\mathbf{S}(m) \cdot \mathbf{S}(m)}{m^2} + \frac{[S(S+1) - 9/4]/2 - \mathbf{S}(m) \cdot \mathbf{S}(m)}{mm'} \right], \tag{3.83}$$

that is,

$$M_\Sigma = 2m + m' + \frac{\xi'}{4} \left[\frac{1}{m^2} - \frac{4}{mm'} \right] \tag{3.84}$$

$$M_\Lambda = 2m + m' - \frac{3\xi'}{4m^2}. \tag{3.85}$$

Finally, for the cascade, Ξ, one notes that the two s quarks are in a spin triplet and, using the same ideas as for the $S = -1$ octet baryons, one can show that

$$M_\Xi = m + 2m' + \frac{\xi'}{4} \left[\frac{1}{m'^2} - \frac{4}{mm'} \right]. \tag{3.86}$$

Upon fitting the physical masses of the baryons, one finds the masses and interaction strength parameter given above are not the same values that worked for the mesons. This should not necessarily be surprising, since the $q\bar{q}$ interaction for the latter need not be the same as the interactions within the qqq system (see also Chapters 5 and 6).

Note that if $SU(N)$ flavor symmetry were to be exact then all members of a given $SU_F(N)$ multiplet would have the same mass, which is not the case. While the masses are usually similar, they are not exactly degenerate and we see some violation of the symmetry (see Figs. 3.2 and 3.4 where the physical masses are listed). In fact $SU_F(2)$ (u, d) is broken at the level of about 5 MeV, $SU_F(3)$ (u, d, s) at the level of about 100 MeV and $SU_F(4)$ (u, d, s, c) by more than 1 GeV. If the hadron's mass were to be attributed simply to the quark masses then one would have $m_u \cong m_d \cong 0.4$ GeV, $m_s \cong 0.5$ GeV and $m_c \cong 1.6$ GeV. Such effective quark masses are called *constituent quark* masses and are what results when all other degrees of freedom except the valence quarks themselves are suppressed – that is, they are the effective degrees of freedom of dressed quarks being treated as quasi-particles. These ideas are to be contrasted with the concept of *current quarks*, those that enter in the basic QCD Lagrangian, as discussed in the next two chapters.

Exercises

3.1 **Isospin Symmetry and $K \to 2\pi$ Decay**

The dominant decay modes of the K mesons are the two-pion channels

$$K^0 \to \pi^+\pi^-, \quad K^0 \to \pi^0\pi^0, \quad K^\pm \to \pi^\pm\pi^0 .$$

a) Using the fact that pions are spinless and are therefore bosons, show that the final two-pion state resulting from the decay of a kaon must have either isospin zero or two.

b) Since $\bar{u}d$ carries $I = 1$ while $\bar{s}u$ carries $I = \frac{1}{2}$ the strangeness-changing non-leptonic Hamiltonian

$$\mathcal{H}_w \sim \frac{G}{\sqrt{2}} \bar{s}u\bar{u}d$$

involves a linear combination of $I = \frac{1}{2}$ and $I = \frac{3}{2}$ components, leading to $K \to 2\pi$ decay amplitudes a_1 and a_3. Show that the decay amplitude can be written as

$$A(K^+ \to \pi^+\pi^0) = \frac{3}{2\sqrt{2}} a_3$$

$$A(K^0 \to \pi^+\pi^-) = -a_1 + \frac{1}{2} a_3$$

$$A(K^0 \to \pi^0\pi^0) = a_1 + a_3 .$$

c) Compare this parametrization with the experimental lifetimes for the following channels

$$\tau_{K^0 \to \pi^+\pi^-} \simeq \tau_{K^0 \to \pi^0\pi^0} \ll \tau_{K^\pm \to \pi^\pm\pi^0} .$$

What do you conclude about the amplitudes a_1 and a_3?

3.2 **Isospin Invariance**

Just as in the case of angular momentum invariance where the Hamiltonian is required to be a scalar under arbitrary rotations, isospin invariance requires that the

Hamiltonian be a scalar under rotations in isospin space. Spin and isospin invariance have another similarity in that, since the commutation relations of spin and isospin operators have the same structure

$$[S_i, S_j] = i\epsilon_{ijk}S_k \,, \qquad [I_i, I_j] = i\epsilon_{ijk}I_k \,,$$

the irreducible representations must be the same – particles must lie in representations labeled by $|I, I_3\rangle$ with $I = 0, \frac{1}{2}, 1, \frac{3}{2}, \ldots$ and $-I \leq I_3 \leq I$.

In the case of the pion-nucleon interaction, the interaction Lagrangian must be of the form

$$\mathcal{L}_{\text{int}} = g_{\pi NN}\bar{N}\boldsymbol{\tau} \cdot \boldsymbol{\pi}\gamma_5 N \,,$$

where $\boldsymbol{\tau}$ are the Pauli isospin matrices, $\boldsymbol{\pi}$ is the isovector pion field and the presence of γ_5 is required by the feature that the pion is a pseudoscalar particle.

a) Since the nucleon has isospin $\frac{1}{2}$ while the pion has isospin 1, demonstrate that under an infinitesimal isospin rotation by angle $\delta\boldsymbol{\chi}$

$$N \to N + \frac{i}{2}\delta\boldsymbol{\chi} \cdot \boldsymbol{\tau}N$$

$$\boldsymbol{\pi} \to \boldsymbol{\pi} - \delta\boldsymbol{\chi} \times \boldsymbol{\pi} \,.$$

b) From these transformation properties show that the interaction Lagrangian given above is invariant:

$$\mathcal{L}_{\text{int}} \xrightarrow{\text{rot}} \mathcal{L}_{\text{int}} \,.$$

c) Using the representations

$$\pi^+ = -\sqrt{\frac{1}{2}}(\pi_1 + i\pi_2), \quad \pi^0 = \pi_3, \quad \pi^- = \sqrt{\frac{1}{2}}(\pi_1 - i\pi_2)$$

of the pions in terms of their Cartesian components demonstrate that the isospin invariance of $\mathcal{L}_{\text{int}}$ requires the pion couplings to nucleons to have the form

$$\sqrt{\frac{1}{2}}g(np\pi^-) = -\sqrt{\frac{1}{2}}g(pn\pi^+) = g(pp\pi^0) = -g(nn\pi^0) = g_{\pi NN} \,.$$

d) Compare the results of c) with the requirements of the Wigner–Eckart theorem

$$g(N_aN_b\pi_c) = \left\langle \frac{1}{2}m_a 1 m_c \,\middle|\, \frac{1}{2} 1\frac{1}{2}m_b \right\rangle \langle N||g||\pi N\rangle \,,$$

where $\left\langle \frac{1}{2}m_a 1 m_c \,\middle|\, \frac{1}{2} 1\frac{1}{2}m_b \right\rangle$ is a Clebsch–Gordan coefficient (see Chapter 2).

Of course, unlike angular momentum invariance which is exact, these isospin predictions are expected to be broken at the percent level due to the small differences between the n, p and π^+, π^0, π^- masses.

3.3 $SU(3)$ Invariance

The near degeneracy of the u, d quarks as well as the n, p and π^+, π^0, π^- masses indicate that isospin should be a quite good symmetry, with breaking at the percent level. In the case of $SU(3)$ the symmetry is still useful, with breaking, however, at the $\sim 20\%$ level, as suggested by the difference between the N, Δ, and ρ, K^* masses. Since the lightest pseudoscalar mesons and spin-$\frac{1}{2}$ baryons are both members of $SU(3)$ octet representations, it is interesting to see what the implications of $SU(3)$ symmetry are both for the masses and the couplings of these particles. In order to study these predictions it is useful to represent these octets in the form

$$M = \sum_{j=1}^{8} \lambda_j \phi_j = \begin{pmatrix} \sqrt{\frac{1}{2}}\pi^0 + \sqrt{\frac{1}{6}}\eta_8 & \pi^+ & K^+ \\ \pi^- & -\sqrt{\frac{1}{2}}\pi^0 + \sqrt{\frac{1}{6}}\eta_8 & K^0 \\ K^- & \bar{K}^0 & -2\sqrt{\frac{1}{6}}\eta_8 \end{pmatrix}$$

$$B = \sum_{j=1}^{8} \lambda_j B_j = \begin{pmatrix} \sqrt{\frac{1}{2}}\Sigma^0 + \sqrt{\frac{1}{6}}\Lambda^0 & \Sigma^+ & p \\ \Sigma^- & -\sqrt{\frac{1}{2}}\Sigma^0 + \sqrt{\frac{1}{6}}\Lambda^0 & n \\ \Xi^- & \Xi^0 & -2\sqrt{\frac{1}{6}}\Lambda^0 \end{pmatrix}$$

where λ_j are the Gell-Mann matrices introduced in Chapter 2. Then an $SU(3)$ rotation is given by

$$M' = UMU^\dagger, \quad B' = UBU^\dagger \quad \text{with} \quad U = \exp\left(\frac{i}{2}\sum_{j=1}^{8}\alpha_j\lambda_j\right),$$

and the simplest possible $SU(3)$ invariants are constructed using the trace

$$\mathrm{Tr}(\bar{B}B), \quad \frac{1}{2}\mathrm{Tr}(\bar{B}\{B,M\}), \quad \frac{1}{2}\mathrm{Tr}(\bar{B}[B,M]) .$$

That such forms are invariant is clear from the properties of the trace

$$\mathrm{Tr}(ABC) \to \mathrm{Tr}(UAU^\dagger UBU^\dagger UCU^\dagger) = \mathrm{Tr}(ABC) .$$

Thus the Lagrangian describing the pseudoscalar meson–baryon couplings should have the form

$$\mathcal{L} = \sum_{abc} g_{abc} M_a \bar{B}_b \gamma_5 B_c$$

with

$$g_{abc} = \frac{1}{4}D\mathrm{Tr}(\lambda_a\lambda_b, \lambda_c) + \frac{1}{4}F\mathrm{Tr}(\lambda_a[\lambda_b, \lambda_c]) = Dd_{abc} + iFf_{abc},$$

where d_{abc} and f_{abc} are the $SU(3)$ structure constants and D and F are the corresponding couplings.

a) Use these results to write out the various possible meson–baryon couplings in terms of their $SU(3)$ decompositions, e.g.,

$$g(p\bar{p}\pi^0) = \frac{1}{2}(D + F)$$

$$g(p\bar{p}\eta^0) = -\frac{1}{2\sqrt{3}}(D - 3F) .$$

b) From the experimental results

$$g_{p\bar{p}\pi^0} \cong 13.5 \quad \text{and} \quad g_{p\bar{p}\eta^0} \cong -3.9$$

determine the $SU(3)$ couplings D, F.

To the extent that the mass operator can be written in terms of an $SU(3)$ singlet plus octet forms, we can then fit the pseudoscalar and $\frac{1}{2}^+$ baryon masses.

c) Show that the masses can then be written in the form

$$M_{ij}^2 = M_0^2 \delta_{ij} + M_D^2 d_{ij8} + i M_F^2 f_{ij8} \quad \text{mesons}$$
$$B_{ij} = B_0 \delta_{ij} + B_D d_{ij8} + i B_F f_{ij8} \quad \text{baryons}.$$

d) Determine the constants M_0^2, M_D^2, M_F^2 for mesons and B_0, B_D, B_F for baryons from experimental data.

e) Show that the mass predictions obey the Gell-Mann–Okubo sum rules

$$\frac{1}{2}(m_N + m_\Xi) = \frac{1}{4}(3m_\Lambda + m_\Sigma)$$

$$\frac{1}{2}(m_{K^+}^2 + m_{K^0}^2) = \frac{1}{4}(3m_\eta^2 + m_\pi^2)$$

and compare with experiment.

3.4   *SU*(4) **Representations**

Just as in the case of $SU(3)$ symmetry, where one looks at rotations among $u, d,$ and s quarks, one can look at $SU(4)$ symmetry which involves rotations among u, d, s and c (charm) quarks. Of course, since the c quark is very much heavier than the others, $SU(4)$ should be strongly broken as a dynamical symmetry, but nevertheless its irreducible representations in the meson and baryon sectors should be useful in understanding the shape of the particle spectrum, and that is the goal of this exercise.

In the case of $SU(2)$ we know that the irreducible representations can be given in terms of a line along which the particles having the same total isospin but different values of I_3 are given. In the case of $SU(3)$ we require a two dimensional picture in which I_3 is given along the x-axis and hypercharge Y, defined through the relationship $Q = I_3 + Y/2$, is plotted along the y-axis, leading to the familiar octet and decuplet representations. For $SU(4)$ we will require a three-dimensional plot with charm plotted along the z-axis and a stacking of $SU(3)$ representaions

a) In the case of the pseudoscalar mesons, show that we expect a 15-dimensional representation, and plot the result in terms of the quark content in a three-dimensional plot along I_3, Y, C axes.

b) Analyze this 15-dimensional representation in terms of its $SU(3)$ content and show that we have $\{15\}_{SU(4)} = (8 + 3 + \bar{3} + 1)_{SU(3)}$.

c) In the case of the spin-1/2 baryons show that we expect a 20-dimensional representation, and plot the results in terms of the quark content in a three-dimensional plot along I_3, Y, C axes.

d) Analyze this 20-dimensional representation in terms of its $SU(3)$ content and show that we have $\{20\}_{SU(4)} = \left(8 + 6 + 3 + \overline{3}\right)_{SU(3)}$.

e) Compare these representations with what has been observed experimentally.

3.5   **Symmetry and Weak Nonleptonic Λ Decay**

The $\Lambda(1115)$ has spin-1/2 and a lifetime of about 200 ps. Its primary decay modes are into a proton and a negatively charged pion or to a neutron and a neutral pion. The $\sim 10^{-10}$ second lifetime and the fact that strangeness is changed by one unit indicates that this decay is due to the weak interaction. Hence the decay amplitude will possess a parity-violating as well as a parity-conserving component.

a) Show that angular momentum conservation requires that the transition amplitude must be either S-wave or P-wave so that the decay amplitude must have the form

$$\mathcal{M} = A_S \chi_p^\dagger \chi_\Lambda + A_P \chi_p^\dagger \boldsymbol{\sigma} \cdot \hat{\boldsymbol{p}}_\pi \chi_\Lambda \, ,$$

where $\boldsymbol{\sigma}$ is a Pauli spin matrix and $\hat{\boldsymbol{p}}_\pi$ is a unit vector in the direction of the outgoing pion.

b) Evaluate the decay distribution from an unpolarized Λ and show that the distribution is isotropic.

c) Evaluate the decay distribution for the case of decay from a Λ with polarization $P\hat{\boldsymbol{n}}$, and show that it has the form

$$\frac{d\Gamma}{d\Omega} \sim 1 + A_1 P\boldsymbol{n} \cdot \hat{\boldsymbol{p}}_\pi$$

with

$$A_1 = \frac{2\mathrm{Re}A_S^* A_P}{|A_S|^2 + |A_P|^2} \, .$$

d) Show that a nonzero value of A_1 implies that pions tend to be emitted either parallel or antiparallel to the Λ spin direction according to whether A_1 is positive or negative and show, using mirror arguments, that the existence of such an asymmetry requires a violation of parity invariance. Compare with the explicit form for the asymmetry in terms of A_S and A_P and comment.

e) Evaluate the polarization of the final-state nucleon in the direction $\hat{\boldsymbol{n}} \times \hat{\boldsymbol{p}}_\pi$ for the case of decay from a Λ with polarization $P\hat{\boldsymbol{n}}$ and show that it is of the form

$$\frac{\boldsymbol{P}_N \cdot \hat{\boldsymbol{n}} \times \hat{\boldsymbol{p}}_\pi}{|\hat{\boldsymbol{n}} \times \hat{\boldsymbol{p}}_\pi|} = P\frac{2\mathrm{Im}A_S^* A_P}{|A_S|^2 + |A_P|^2 + 2P\mathrm{Re}A_S^* A_P \hat{\boldsymbol{n}} \cdot \hat{\boldsymbol{p}}_\pi}|\hat{\boldsymbol{n}} \times \hat{\boldsymbol{p}}_\pi| \, .$$

f) Analyze this result from the point of view of simple time reversal invariance arguments and show that one might have expected this polarization to vanish.

g) What does the fact that the experimental value for this polarization is nonzero say about the weak decay amplitudes?

3.6   **Mesonic States**

Consider the simple mesonic states which can be constructed from binding a quark and antiquark.

a) Write down all of the $q\bar{q}$ states in the pseudoscalar and vector nonets having good G-parity.

b) Using Eq. (3.34), obtain the masses of the pseudoscalar and vector nonets shown in Fig. 3.2, and compare with the physical values (note: the source for the masses is [PDG14], with which the reader should become familiar).

3.7 Baryonic States

Consider the simple baryonic states which can be constructed from binding three quarks.

a) Using the proton and neutron states given in Eqs. (3.60) and (3.61) together with the charge operator, verify that the correct charges are obtained.

b) Using the magnetic dipole operator given in Eq. (3.71) with the baryon octet states given in the text, obtain predictions for the magnetic moments. Compare these results with the physical values given in [PDG14].

c) Using the mass formulas given in the text, obtain predictions for the masses of both the baryon octet and decuplet (see Fig. 3.4). Compare these results with the physical values given in [PDG14].

4

The Standard Model

During the early part of the 19th century physicists believed that electric and magnetic effects were components of different interactions. However, with Maxwell's synthesis, electric and magnetic effects were unified into a single electromagnetic interaction. Similarly, during the first part of the twentieth century physicists believed that electromagnetic and weak interactions represented separate phenomena. Nevertheless, with the development of the Weinberg–Salam model in 1967 it was realized that both are components of a single electroweak interaction, wherein such effects are mediated by the exchange of gauge bosons [Wei67, Sal69, Gla61]. Soon thereafter developed the idea that strong interactions could be treated in a parallel fashion by introducing color and the exchange of colored gluons. Although some have attempted to go even further by unifying the electroweak and strong interactions via so-called grand unified theories (GUTs), there is as yet no convincing evidence for the correctness of this unification. Still, the combination of color-gluon-mediated strong interactions together with gauge-boson-mediated electromagnetic and weak interactions has been extraordinarily successful in explaining the entire range of elementary particle and nuclear interactions, so much so that it is now called the "Standard Model" (as discussed in Chapter 1). In this chapter we lay out the basic features of this picture.

4.1 Electroweak Interaction: The Weinberg–Salam Model

In order to motivate the form of the electroweak interaction, we begin by reviewing the electromagnetic interaction and its connection to local gauge invariance. The Lagrangian which describes the free Dirac equation for a spin-$1/2$ particle of mass m is:

$$\mathcal{L}_0 = \bar{\psi}(i \not{\nabla} - m)\psi. \tag{4.1}$$

It is clear that this Lagrangian is invariant under a global phase change

$$\psi(x) \rightarrow e^{ie\chi}\psi(x)$$
$$\bar{\psi}(x) \rightarrow \bar{\psi}(x)e^{-ie\chi}, \tag{4.2}$$

where e is the fermion charge and χ is a position-independent constant. This property is called global $U(1)$ invariance. It is global in that the phase $e\chi$ is a constant and is the same whether on earth or in some distant galaxy. It is a $U(1)$ invariance since it is unitary, i.e., $(e^{ie\chi})^{\dagger}e^{ie\chi} = 1$, and involves a single parameter χ. Now suppose we wish to make the theory invariant under a *local* phase change, i.e., a phase change $\exp(ie\chi(x))$ using a function $\chi(x)$ which varies from point to point in spacetime. This goal can be achieved provided that one is willing to introduce an extra vector field $A_\mu(x)$ into the problem. If we define the covariant derivative

$$iD_\mu = i\partial_\mu - eA_\mu$$

the Lagrangian becomes

$$\mathcal{L}(x) = \bar{\psi}(x)(i\,\slashed{D} - m)\psi(x), \tag{4.3}$$

which is seen to be invariant under the local gauge transformation

$$\psi(x) \rightarrow e^{ie\chi(x)}\psi(x) \tag{4.4}$$

provided that the field $A_\mu(x)$ transforms as

$$A_\mu(x) \rightarrow A_\mu(x) - \partial_\mu\chi. \tag{4.5}$$

Indeed the covariant derivative $iD_\mu\psi(x)$ transforms covariantly in that

$$iD_\mu\psi(x) \rightarrow (i\partial_\mu - e(A_\mu(x) - \partial_\mu\chi))e^{ie\chi(x)}\psi(x) = e^{ie\chi(x)}iD_\mu\psi(x) \tag{4.6}$$

so that the Lagrangian is unchanged.

$$\begin{aligned}
\mathcal{L}(x) &= \bar{\psi}(x)(i\,\slashed{D} - m)\psi(x) \\
&\longrightarrow \bar{\psi}(x)e^{-ie\chi(x)}(i\,\slashed{D} - m)e^{ie\chi(x)}\psi(x) \\
&= \bar{\psi}(x)e^{-ie\chi(x)}e^{ie\chi(x)}(i\,\slashed{D} - m)\psi(x) \\
&= \bar{\psi}(x)(i\,\slashed{D} - m)\psi(x) = \mathcal{L}(x).
\end{aligned} \tag{4.7}$$

Since the field tensor $F_{\mu\nu}$ defined via

$$F_{\mu\nu} = \frac{-i}{e}[D_\mu, D_\nu] = \partial_\mu A_\nu - \partial_\nu A_\mu \tag{4.8}$$

is itself invariant,

$$F_{\mu\nu} \rightarrow \partial_\mu A_\nu - \partial_\nu A_\mu - e(\partial_\mu\partial_\nu - \partial_\nu\partial_\mu)\chi(x) = F_{\mu\nu}, \tag{4.9}$$

the full QED Lagrangian

$$\mathcal{L}_{\text{QED}}(x) = -\frac{1}{4}F_{\mu\nu}F^{\mu\nu} + \bar{\psi}(x)(i\,\slashed{D} - m)\psi(x) \tag{4.10}$$

is seen to be unchanged under local $U(1)$ transformations, where $A_\mu(x)$ is the photon field. Since inclusion of a mass term $\frac{1}{2}m_\gamma^2 A_\mu(x)A^\mu(x)$ would spoil this local gauge invariance, the photon is predicted to be a massless particle. Indeed, experimentally the upper bound on the photon mass is extremely tight, $m_\gamma < 10^{-28}$ eV, representing strong evidence for the existence of local gauge invariance.

One can make a simple generalization of these ideas in order to produce a proper electroweak interaction. Since the leptons e and ν_e emitted in beta-decay processes are found experimentally to be (predominantly) left-handed, and since there are two such particles we shall invoke the gauge group $SU(2)_L$ and place the left-handed components of these fields into a doublet representation

$$L \equiv \begin{pmatrix} \psi_{\nu_e} \\ \psi_e \end{pmatrix}_L, \tag{4.11}$$

where the subscript L means we have projected out the left-handed component of the field via the chirality operator

$$\psi_L \equiv \frac{1}{2}(1 - \gamma_5)\psi. \tag{4.12}$$

A general $SU(2)_L$ transformation can be written in the form

$$\psi_L \rightarrow \exp\left(\frac{i}{2}\boldsymbol{\tau} \cdot \boldsymbol{\beta}\right)\psi_L, \tag{4.13}$$

where $\boldsymbol{\beta}$ is an arbitrary three-vector and $\boldsymbol{\tau}$ are the Pauli isospin matrices as defined in Chapter 2. Calling the $SU(2)_L$ coupling constant g, we can then define the covariant derivative via

$$iD_\mu = i\partial_\mu - \frac{g}{2}\boldsymbol{\tau} \cdot \boldsymbol{W}_\mu, \tag{4.14}$$

where $\boldsymbol{W}_\mu$ represents a triplet of vector fields. Since we wish to combine such $SU(2)_L$ rotations with corresponding $U(1)$ transformations in order that the weak and electromagnetic interactions be unified, we also define a $U(1)$ coupling constant g' with a corresponding vector hypercharge field B_μ. Here the relation between the electric charge Q and hypercharge Y is by convention

$$Q = I_3 + \frac{1}{2}Y \tag{4.15}$$

so that the hypercharge of the (e, ν_e) doublet is $Y_L = -1$. The full $SU(2)_L \times U(1)$ covariant derivative then becomes

$$iD_\mu = i\partial_\mu - \frac{g}{2}\boldsymbol{\tau} \cdot \boldsymbol{W}_\mu - \frac{g'}{2}YB_\mu \tag{4.16}$$

and, defining the field tensors $\boldsymbol{F}_{\mu\nu}$, $G_{\mu\nu}$ via

$$[D_\mu, D_\nu] = \frac{g}{2}\boldsymbol{\tau} \cdot \boldsymbol{F}_{\mu\nu} + \frac{g'}{2}YG_{\mu\nu}, \tag{4.17}$$

we identify

$$\boldsymbol{F}_{\mu\nu} = \partial_\mu\boldsymbol{W}_\nu - \partial_\nu\boldsymbol{W}_\mu - g\boldsymbol{W}_\mu \times \boldsymbol{W}_\nu \tag{4.18}$$

$$G_{\mu\nu} = \partial_\mu B_\nu - \partial_\nu B_\mu. \tag{4.19}$$

The final addition which we must make to the ingredients of the model is to include, in the case of the electron, a right-handed $SU(2)$-singlet field,

$$R \equiv \psi_{eR}, \tag{4.20}$$

where the subscript R indicates that the right-handed component has been projected out via

$$\psi_{eR} = \frac{1}{2}(1 + \gamma_5)\psi_e. \tag{4.21}$$

The point here is that, since the electron has a nonzero mass, which can be represented via

$$m_e \bar{\psi}\psi = m_e(\bar{\psi}_{eL}\psi_{eR} + \bar{\psi}_{eR}\psi_{eL}),\tag{4.22}$$

both left- and right-handed components of the field must be present. (In the simplest version of the Standard Model the neutrino is taken to be massless, so that no right-handed neutrino component is required; more about that later.) From Eq. (4.15) we see that the hypercharge carried by the right-handed field ψ_{eR} must be $Y_R = -2$. Thus far, we have included only the electron and electron neutrino in our theory, but in the real world, of course, there exist three such lepton families. Summing over the three lepton flavors e, μ, τ we have the leptonic component of the Standard Model Lagrangian

$$\mathcal{L}_{\text{lep}} = \sum_{j=1}^{3} \bar{L}_j \left(i\,\partial\!\!\!/ - \frac{g}{2}\boldsymbol{\tau}\cdot\boldsymbol{W} - \frac{g'}{2}Y_L\,B\!\!\!/ \right) L_j$$

$$+ \sum_{j=1}^{3} \bar{R}_j \left(i\,\partial\!\!\!/ - \frac{g'}{2}Y_R\,B\!\!\!/ \right) R_j - \frac{1}{4}\boldsymbol{F}_{\mu\nu}\cdot\boldsymbol{F}^{\mu\nu} - \frac{1}{4}G_{\mu\nu}G^{\mu\nu}\tag{4.23}$$

and it is straightforward to demonstrate invariance under both $SU(2)_L$ rotations

$$L(x) \;\rightarrow\; \exp\left(\frac{i}{2}\boldsymbol{\tau}\cdot\boldsymbol{\beta}(x)\right) L(x)$$

$$R(x) \;\rightarrow\; R(x)$$

$$\frac{1}{2}\boldsymbol{\tau}\cdot\boldsymbol{W}_\mu \;\rightarrow\; \exp\left(\frac{i}{2}\boldsymbol{\tau}\cdot\boldsymbol{\beta}(x)\right)\frac{1}{2}\boldsymbol{\tau}\cdot\boldsymbol{W}_\mu \exp\left(-\frac{i}{2}\boldsymbol{\tau}\cdot\boldsymbol{\beta}(x)\right)$$

$$-\frac{i}{g}\exp\left(\frac{i}{2}\boldsymbol{\tau}\cdot\boldsymbol{\beta}(x)\right)\partial_\mu \exp\left(-\frac{i}{2}\boldsymbol{\tau}\cdot\boldsymbol{\beta}(x)\right)$$

$$B_\mu(x) \;\rightarrow\; B_\mu(x)$$

$$\frac{1}{2}\boldsymbol{\tau}\cdot\boldsymbol{F}_{\mu\nu} \;\rightarrow\; \exp\left(\frac{i}{2}\boldsymbol{\tau}\cdot\boldsymbol{\beta}(x)\right)\frac{1}{2}\boldsymbol{\tau}\cdot\boldsymbol{F}_{\mu\nu}\exp\left(-\frac{i}{2}\boldsymbol{\tau}\cdot\boldsymbol{\beta}(x)\right)$$

$$G_{\mu\nu} \;\rightarrow\; G_{\mu\nu}\tag{4.24}$$

and a $U(1)$ gauge change

$$L(x) \;\rightarrow\; \exp\left(\frac{i}{2}Y_L\gamma(x)\right) L(x)$$

$$R(x) \;\rightarrow\; \exp\left(\frac{i}{2}Y_R\gamma(x)\right) R(x)$$

$$\boldsymbol{F}_{\mu\nu} \;\rightarrow\; \boldsymbol{F}_{\mu\nu}$$

$$G_{\mu\nu} \;\rightarrow\; G_{\mu\nu},\tag{4.25}$$

so that we now have a coupling of the charged W bosons, $W_\mu^\pm = \sqrt{\frac{1}{2}}(W_\mu^1 \pm iW_\mu^2)$ to the lepton currents

$$\sum_{j=1}^{3} \bar{L}_j \frac{1}{2}\tau^\mp \gamma_\mu L_j\tag{4.26}$$

as desired. However, these W bosons are, so far, massless, since addition of a term $\frac{1}{2}m_W^2 \boldsymbol{W}_\mu \cdot \boldsymbol{W}^\mu$ to the Lagrangian would break the $SU(2)_L \times U(1)$ invariance. Likewise, the leptons are massless, since addition of a term $m_e \bar{\psi}_e \psi_e = m_e(\bar{\psi}_{eR}\psi_{eL} + \bar{\psi}_{eR}\psi_{eL})$ to the Lagrangian would also violate the gauge symmetry. The solution to this conundrum is to invoke the phenomenon of spontaneous symmetry breaking, also commonly referred to as the Higgs mechanism, to restore the desired Lagrangian invariance, as described in the following section.

4.2 The Higgs Mechanism

In the phenomenon of spontaneous symmetry breaking, the vacuum state of a system breaks the symmetry of a system even though the Lagrangian remains invariant. A classic example of this is a permanent ferromagnet, wherein the atom–atom interaction is rotationally invariant, but the magnetization selects a particular direction. In the Standard Model the spontaneous symmetry breaking is realized via the so-called Higgs mechanism [Eng64, Hig64], wherein we introduce an $SU(2)_L$ doublet of scalar fields

$$\phi(x) = \begin{pmatrix} \phi^+(x) \\ \phi^0(x) \end{pmatrix} \tag{4.27}$$

for which we see that the hypercharge is $Y_\phi = 1$. Using the covariant derivative

$$iD_\mu \phi = \left(i\partial_\mu - \frac{g}{2}\boldsymbol{\tau} \cdot \boldsymbol{W}_\mu - \frac{g'}{2}YB_\mu \right) \phi \tag{4.28}$$

we can define a gauge-invariant contribution

$$\mathcal{L}_\phi = D_\mu \phi^\dagger D^\mu \phi - V(\phi^\dagger \phi), \tag{4.29}$$

where the potential V is quartic and is chosen to have a double well-shape

$$V(\phi^\dagger \phi) = -\mu^2 \phi^\dagger \phi + \lambda(\phi^\dagger \phi)^2. \tag{4.30}$$

In the case that $\mu^2 < 0$, we see that the minimum of the potential occurs when $\phi^\dagger \phi = 0$ and there is nothing unusual. However, if $\mu^2 > 0$, the minimum occurs when

$$\phi^\dagger \phi = \frac{\mu^2}{\lambda}. \tag{4.31}$$

We now assume that the symmetry is broken spontaneously, with the neutral scalar field developing a vacuum expectation value

$$< 0|\phi|0 >= \begin{pmatrix} 0 \\ v \end{pmatrix} \text{ with } \quad v = \sqrt{\frac{\mu^2}{\lambda}}. \tag{4.32}$$

Note that the Lagrangian remains invariant in this scenario. We see then that the kinetic energy piece of the scalar Lagrangian assumes the form

$$(D_\mu \phi)^\dagger D^\mu \phi = \frac{1}{2}v^2 \left(\frac{g^2}{4} \boldsymbol{W}_\mu \cdot \boldsymbol{W}^\mu + \frac{g'^2}{4} B_\mu B^\mu - \frac{gg'}{2} W_\mu^0 B^\mu \right). \tag{4.33}$$

Equation (4.33) can be diagonalized via the definitions

$$W_\mu^\pm = \sqrt{\frac{1}{2}}(W_\mu^1 \pm iW_\mu^2)$$
$$Z_\mu = \cos\theta_W W_\mu^0 - \sin\theta_W B_\mu$$
$$A_\mu = \sin\theta_W W_\mu^0 + \cos\theta_W B_\mu, \tag{4.34}$$

where the weak mixing angle θ_W is defined via

$$\tan\theta_W = \frac{g'}{g}. \tag{4.35}$$

This component of the Lagrangian then gives mass to the vector bosons $W_\mu^\pm$, Z_μ while the field A_μ remains massless

$$(D_\mu\phi)^\dagger D^\mu\phi = \frac{1}{2}M_W^2(W_\mu^+ W^{\mu-} + W_\mu^- W^{\mu+}) + \frac{1}{2}M_Z^2 Z_\mu Z^\mu \tag{4.36}$$

with

$$M_Z = \frac{v}{2}\sqrt{g^2 + g'^2}, \quad M_W = \frac{v}{2}g = \cos\theta_W M_Z, \quad M_A = 0. \tag{4.37}$$

The electroweak coupling assumes the form

$$\begin{aligned}
\mathcal{L}_{int} = &-\frac{g}{\sqrt{2}}\left[(\bar{\nu}_{eL}\gamma^\mu e_L W_\mu^- + \text{h.c.}) + (e \to \mu) + (e \to \tau)\right] \\
&+ g\sin\theta_W(\bar{e}\gamma^\mu e + \bar{\mu}\gamma^\mu\mu + \bar{\tau}\gamma^\mu\tau)A_\mu \\
&- \frac{g}{\cos\theta_W}\left(\sum_{j=1}^{3}\bar{\psi}_{jL}\frac{1}{2}\tau_3\gamma^\mu\psi_{jL} + \sin^2\theta_W(\bar{e}\gamma^\mu e + \bar{\mu}\gamma^\mu\mu + \bar{\tau}\gamma^\mu\tau)\right)Z_\mu.
\end{aligned} \tag{4.38}$$

We then can identify the electric charge e as

$$e = g\sin\theta_W \tag{4.39}$$

and the effective weak interaction as

$$\mathcal{H}_w = \frac{g^2}{2(m_W^2 - q^2)}J_\mu^{W\dagger}J^{W\mu} + \frac{g^2}{\cos^2\theta_W(M_Z^2 - q^2)}J_\mu^{Z\dagger}J^{Z\mu}, \tag{4.40}$$

where q^μ is the four-momentum of the exchanged boson and

$$J_\mu^W = \bar{\nu}_e\gamma_\mu e + \bar{\nu}_\mu\gamma_\mu\mu + \bar{\nu}_\tau\gamma_\mu\tau \tag{4.41}$$

is the charged weak current and

$$J_\mu^Z = \sum_{j=1}^{3}\bar{\psi}_{jL}\frac{1}{2}\tau_3\gamma_\mu\psi_{jL} + \sin^2\theta_W(\bar{e}\gamma_\mu e + \bar{\mu}\gamma_\mu\mu + \bar{\tau}\gamma_\mu\tau) \tag{4.42}$$

is its neutral analog. In the low-energy limit $|q^2| << M_Z^2,\, M_W^2$ we can write the effective weak Hamiltonian in terms of a contact interaction

$$\mathcal{H}_w^{\text{eff}} = \frac{G}{\sqrt{2}}\left(J_\mu^{W\dagger}J^{W\mu} + 2J_\mu^{Z\dagger}J^{Z\mu}\right), \tag{4.43}$$

where

$$G_F = \frac{g^2}{\sqrt{2}M_W^2} \tag{4.44}$$

is called the Fermi constant. Thus, the spontaneously broken symmetry scenario has provided masses for the vector bosons, while keeping the photon massless. These masses have been determined experimentally with great precision. In the case of the W, the mass is found to be [PDG14]:

$$M_W = 80.385 \pm 0.015 \text{ GeV} \tag{4.45}$$

while that of the Z boson has measured as

$$M_Z = 91.1876 \pm 0.0021 \text{ GeV}. \tag{4.46}$$

Independently the weak mixing angle has been measured to be

$$\sin^2\theta_W = 0.23136 \pm 0.00005. \tag{4.47}$$

We find then

$$\left(\frac{M_Z \cos\theta_W}{M_W}\right) = 0.9945 \tag{4.48}$$

and the less than 1% deviation from unity is on account of small radiative correction effects. We see then that spontaneously broken symmetry leads to a very successful picture of vector boson mass generation. However, the leptons remain massless. This problem can be addressed by the addition of a gauge invariant Yukawa interaction

$$\mathcal{L}_m = \sum_{j=1}^{3} \frac{1}{2}K_j\left(\bar{\psi}_{jL}\phi\psi_{jR} + \bar{\psi}_{jR}\phi^\dagger\psi_{jL}\right) \tag{4.49}$$

which becomes, when spontaneous symmetry breaking takes place,

$$\mathcal{L}_m = \frac{v}{2\sqrt{2}}\left[K_e(\bar{e}_L e_R + \bar{e}_R e_L) + K_\mu(\bar{\mu}_L\mu_R + \bar{\mu}_R\mu_L)\right.$$
$$\left. + K_\tau(\bar{\tau}_L\tau_R + \tau_R\bar{\tau}_L)\right] = \frac{v}{\sqrt{2}}\left(K_e\bar{e}e + K_\mu\bar{\mu}\mu + K_\tau\bar{\tau}\tau\right). \tag{4.50}$$

We see then that the neutrinos remain massless, while the charged leptons pick up mass values

$$m_e = \frac{v}{\sqrt{2}}K_e, \quad m_\mu = \frac{v}{\sqrt{2}}K_\mu, \quad m_\tau = \frac{v}{\sqrt{2}}K_\tau. \tag{4.51}$$

The Higgs mechanism then has produced an entirely satisfactory picture of both leptonic weak and electromagnetic interactions. Inclusion of quarks can be accomplished in a parallel fashion by inclusion of left-handed quark doublets

$$\begin{pmatrix} u \\ d \end{pmatrix}, \begin{pmatrix} c \\ s \end{pmatrix}, \begin{pmatrix} t \\ b \end{pmatrix} \quad \text{with} \quad Y_{qL} = \frac{1}{3} \tag{4.52}$$

together with right-handed singlets

$$u_R, c_R, t_R \quad \text{with} \quad Y_{qR}^u = \frac{4}{3} \tag{4.53}$$

and

$$d_R, s_R, b_R \quad \text{with} \quad Y_{qR}^d = -\frac{2}{3}. \tag{4.54}$$

An $SU(2)_L \times U(1)$-invariant Lagrangian which couples quarks and vector bosons can then be constructed

$$\begin{aligned} \mathcal{L}_q &= \sum_{j=1}^{3} \bar{\psi}_{jL} \left(i\,\not{\partial} - \frac{g}{2}\boldsymbol{\tau}\cdot\boldsymbol{W} - \frac{g'}{6}\,B \right) \psi_{jL} \\ &+ \sum_{j=1}^{3} \left[\bar{\psi}_{jR}^u \left(i\,\not{\partial} - \frac{2}{3}g'\,B \right) \psi_{jR}^u + \bar{\psi}_{jR}^d \left(i\,\not{\partial} + \frac{1}{3}g'\,B \right) \psi_{jR}^d \right]. \end{aligned} \tag{4.55}$$

The generation of quark masses can be achieved by noting that if the Higgs spinor

$$\phi = \begin{pmatrix} \phi^+ \\ \phi^0 \end{pmatrix}$$

transforms as a doublet with $Y_\phi = 1$ then the conjugate spinor

$$\bar{\phi} = i\sigma_2\phi^\dagger = \begin{pmatrix} \bar{\phi}^0 \\ -\phi^- \end{pmatrix}$$

also transforms as an $SU(2)_L$ doublet, but with $Y_{\bar\phi} = -1$. We can then write down a general $SU(2)_L \times U(1)$ invariant Yukawa coupling

$$\mathcal{L}_{m_q} = \sum_{i,j=1}^{3} \left[\Gamma_{ij}^u \bar{\psi}_{iL}^u \phi \psi_{jR}^u + \Gamma_{ij}^d \bar{\psi}_{iL}^d \phi \psi_{jR}^d \right] + \text{h.c.} \tag{4.56}$$

Under spontaneous symmetry breaking.

$$\phi \to \begin{pmatrix} 0 \\ \frac{v}{\sqrt{2}} \end{pmatrix} \quad \text{and} \quad \bar{\phi} \to \begin{pmatrix} \frac{v}{\sqrt{2}} \\ 0 \end{pmatrix} \tag{4.57}$$

the quark mass Lagrangian Eq. (4.56) becomes

$$\mathcal{L}_{m_q} = \sum_{i,j=1}^{3} \frac{v}{\sqrt{2}} \left[\Gamma_{ij}^u \bar{\psi}_{iL}^u \begin{pmatrix} 0 \\ 1 \end{pmatrix} \psi_{jR}^u + \Gamma_{ij}^d \bar{\psi}_{iL}^d \begin{pmatrix} 0 \\ 1 \end{pmatrix} \psi_{jR}^d \right] + \text{h.c.}, \tag{4.58}$$

which has the general form of a quark mass term, but is not diagonal. The diagonalization can be achieved by independent unitary transformations, which leads to another unique and experimentally testable aspect of the Standard Model, quark mixing, which is discussed in Section 4.4.

4.3 The Higgs Boson

So far, we have treated the scalar field ϕ as being a simple numerical constant. However, in the real world there also exist quantum fluctuations around this value, so that we should write

$$\phi \to \phi(x) = \begin{pmatrix} 0 \\ \frac{v}{\sqrt{2}} \end{pmatrix} + \begin{pmatrix} \phi^+(x) \\ \phi^0(x) \end{pmatrix}. \tag{4.59}$$

It might appear that there exist then four scalar degrees of freedom, since there are two complex fields $\phi^+(x)$ and $\phi^0(x)$. However, this is not the case, as can be seen by writing the doublet instead as

$$\phi(x) = \exp(\frac{i}{v}\boldsymbol{\tau} \cdot \boldsymbol{\zeta}(x))\sqrt{\frac{1}{2}} \begin{pmatrix} 0 \\ v + h(x) \end{pmatrix}, \tag{4.60}$$

so that the four degrees of freedom are now represented by the vector field $\boldsymbol{\zeta}(x)$ together with the scalar field $h(x)$. However, the vector field can be eliminated via the gauge transformation

$$\phi(x) \to \psi'(x) = \exp\left(-\frac{i}{v}\boldsymbol{\tau} \cdot \boldsymbol{\zeta}(x)\right)\phi(x), \tag{4.61}$$

so that the only physical degree of freedom is the scalar field $h(x)$, which is called the (neutral) Higgs boson. The mass of the Higgs is undetermined, since it depends on the detailed form of the potential function $V(\phi^\dagger \phi)$.

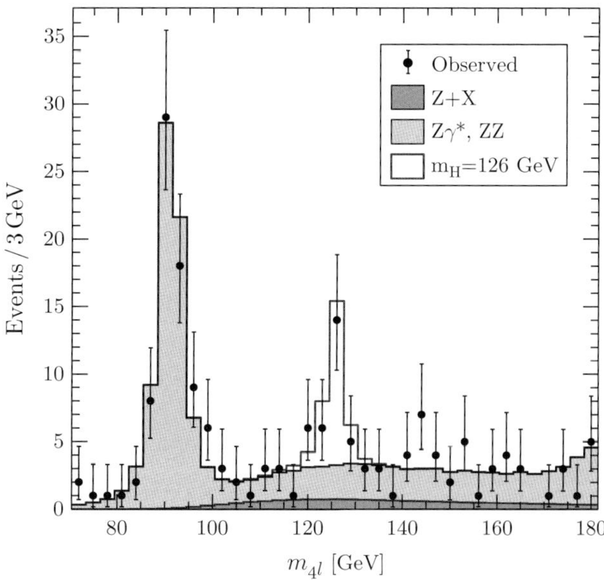

Fig. 4.1   Evidence for the Higgs boson from the Compact Muon Solenoid (CMS) experiment at CERN [Cha14].

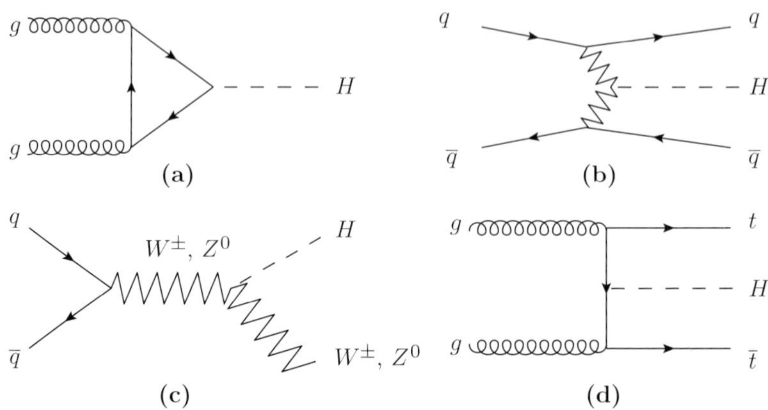

Fig. 4.2 — Higgs production Feynman diagrams [PDG14].

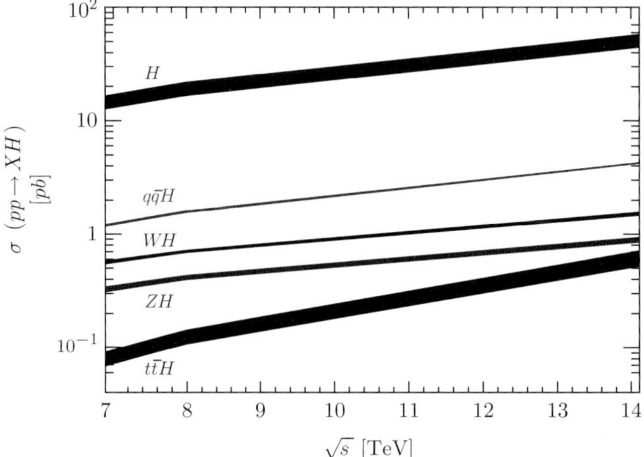

Fig. 4.3 — The SM Higgs boson production cross sections as a function of the center-of-mass energy for pp collisions [PDG14].

The Higgs boson was discovered at the LHC in pp collisions at CERN (see Fig.4.1) by the ATLAS and CMS experiments with a mass of approximately 125 GeV decaying into $\gamma\gamma$, WW, and ZZ bosons [Atl12, Cms12]. This discovery, which was awarded with a Nobel prize in physics to Higgs and Engler in 2013, marks a major triumph for particle physics in confirming electroweak symmetry breaking, as described in Chapter 4. The main production and decay mechanisms are discussed below followed by the highlights of the actual discovery.

The main production mechanisms in pp collisions are gg fusion, heavy-gauge boson fusion, associated production with a gauge boson and associated production with top quarks. Figure 4.2 shows the leading-order Feynman diagrams for those production mechanisms. The size of production cross section for a Higgs mass of 125 GeV is shown in Fig. 4.3 along with each individual process contribution. The decay modes are shown in Fig. 4.4 as a function of the Higgs mass.

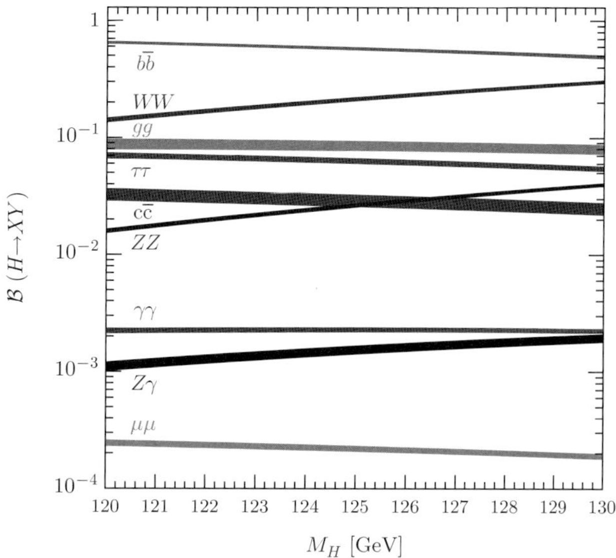

Fig. 4.4   The branching ratios for the main decays of the SM Higgs boson. The theoretical uncertainties are indicated by the bands [PDG14].

4.4 Quark Mixing

The Standard Model accommodates flavor mixing among quarks and among leptons. This general mixing among flavors is one of the signature features of the weak force and allows for great insight into the nature of the Standard Model. Just as a yardstick of how important a feature mixing is, the presence of a third generation in the Standard Model was first hinted at by quark mixing in rare processes. Likewise, in the lepton sector, flavor mixing among neutrinos provided the first positive evidence of nonzero neutrino mass. The mixing matrix that governs the quark sector is known as the Cabibbo–Kobayashi–Maskawa (CKM) matrix [Cab63, Kob73], while for the lepton sector it is the Pontecorvo–Maki–Nagawa–Sakata (PMNS) matrix [Pon57, Mak62]. Both, of course, share very similar properties and therefore can be discussed collectively.

We begin by considering a general $N \times N$ unitary matrix. In general, such a matrix can possess N^2 independent parameters. These parameters can be subdivided into $N(N-1)/2$ mixing angles and $N(N+1)/2$ phases. In the case of the CKM and PMNS matrices, we always consider the product of two unitary $N \times N$ matricies:

$$V = V_L^{u\dagger} V_L^d. \tag{4.62}$$

So, naively one might think that there exist $2N^2$ parameters that describe the above mixing. Fortunately, one can quickly constrain the number of degrees of freedom by looking at the structure of the weak current. In the case of the weak charged-current involving

$$J_\mu^W = \bar{q}_L^u \gamma_\mu V q_L^d, \tag{4.63}$$

by applying a global phase transformation to the quark fields of the type

$$q_j^u \rightarrow e^{i\phi_j^u} q_j^u, \qquad q_k^d \rightarrow e^{i\phi_k^d} q_k^d \tag{4.64}$$

the Lagrangian should remain unchanged. All terms in the Standard Model Lagrangian are impervious to this transformation (including the neutral current terms, thanks to the Glashow–Iliopoulos–Maiani (GIM) mechanism [Gla70][1]) except for the charged weak current. Here, the change can be absorbed by the mixing matrix itself. Rewriting the weak current under the new transformation rules, we find

$$J_\mu^W = \sum_j^N \sum_k^N \bar{q}_L^u \gamma^\mu e^{-i\phi_j^u} V e^{i\phi_k^d} q_L^d. \tag{4.65}$$

We can then express the weak current via

$$j_\mu^W = e^{-i(\phi_l^u - \phi_m^d)} \sum_j^N \sum_k^N \bar{q}_L^u \gamma_\mu e^{-i(\phi_j^u - \phi_l^u)} V e^{i(\phi_k^d - \phi_m^d)} q_L^d. \tag{4.66}$$

The $e^{-i(\phi_j^u - \phi_l^u)}$ and $e^{i(\phi_k^d - \phi_m^d)}$ terms each remove $N - 1$ arbitrary phases from the unitary matrix, while the global term $e^{-i(\phi_l^u - \phi_m^d)}$ removes one such term (provided that the baryon number is conserved). As a result, the number of true independent parameters present in the mixing matrix is reduced to $N(N - 1)/2$ mixing angles and $(N - 1)(N - 2)/2$ phases.

Let us illustrate this feature for the case of simple two-quark mixing ($N = 2$). The most general complex unitary matrix is of the form

$$V = \begin{pmatrix} \cos\theta \; e^{i\delta_1} & \sin\theta \; e^{i(\delta_2 + \eta)} \\ -\sin\theta \; e^{i(\delta_1 - \eta)} & \cos\theta \; e^{i\delta_2} \end{pmatrix}$$

and the quarks would involve only the up, down, charm and strange flavors, i.e.

$$q_L^u = \begin{pmatrix} u_L \\ c_L \end{pmatrix}, \; q_L^d = \begin{pmatrix} d_L \\ s_L \end{pmatrix}. \tag{4.67}$$

Again, one can add phases to the quarks without altering the Lagrangian in any physically observable manner. In this particular example, one can rewrite the quark flavors as

$$u_L \rightarrow e^{i(\delta_1 + \eta)} u_L, \; d_L \rightarrow e^{i\eta} d_L, \; c_L \rightarrow e^{i\delta_2} c_L. \tag{4.68}$$

in which case, the mixing matrix simply becomes

$$V = \begin{pmatrix} \cos\theta_C & \sin\theta_C \\ -\sin\theta_C & \cos\theta_C \end{pmatrix}.$$

---

[1] In the days before the discovery of the charm quark there existed only three known flavors (u, d, s) and it was not understood why flavor-changing neutral currents did not exist, i.e., if there were $\bar{d}d$ and $\bar{s}s$ currents why not $\bar{s}d$? In addition, there would be a large quadratically divergent $\Delta S = 2$ interaction having the quark content $\bar{s}d\bar{d}s$ which would contribute to $K^0 - \bar{K}^0$ mixing. Both problems were solved by Glashow, Iliopoulis, and Maiani who proposed the existence of what we now call the charm quark which, together with the strange quark, was a member of an $SU(2)$ doublet. In this picture the weak neutral current was required to be quark-flavor diagonal and the $\Delta S = 2$ operator became logarithmically divergent and small. Soon thereafter the charm quark was discovered experimentally.

As expected, the number of physical angles is $\frac{2(2-1)}{2} = 1$ and the number of phases is $\frac{(2-1)(2-2)}{2} = 0$. The above is not an arbitrary exercise, but rather an experimental observable. The angle θ_C is known as the Cabbibo angle and represents a two-quark model simplification of the more accurate 3×3 quark representation, but is a fair representation of what happens when energies are low enough that there is no significant contribution from bottom and top quark loops.

We can now expand our discussion to the full three-quark model. Having expanded to a 3×3 matrix allows for $\frac{3(3-1)}{2} = 3$ physical angles and $\frac{(3-1)(3-2)}{2} = 1$ phase angle. The possible representation space of how such a matrix can be written is quite large, but the physics community has generally settled on a particular convention

$$
V = \begin{pmatrix} V_{ud} & V_{us} & V_{ub} \\ V_{cd} & V_{cs} & V_{cb} \\ V_{td} & V_{ts} & V_{tb} \end{pmatrix}
$$

$$
= \begin{pmatrix} c_{12}c_{13} & s_{12}c_{13} & s_{13}e^{-i\delta} \\ -s_{12}c_{23} - c_{12}s_{23}s_{13}e^{i\delta} & c_{12}c_{23} - s_{12}s_{23}s_{13}e^{i\delta} & s_{23}c_{13} \\ s_{12}s_{23} - c_{12}c_{23}s_{13}e^{i\delta} & -c_{12}s_{23} - s_{12}c_{23}s_{13}e^{i\delta} & c_{23}c_{13} \end{pmatrix} \tag{4.69}
$$

where $s_{ij} = \sin\theta_{ij}$ and $c_{ij} = \cos\theta_{ij}$, θ_{12}, θ_{23}, and θ_{13} are the three physical angles of the system and δ is the phase angle.

As long as the CKM matrix is diagonally dominant with a clear hierarchy (in this case $s_{13} \ll s_{23} \ll s_{12} \ll 1$), it is possible to parametrize the individual entries in a manner that simplifies experimental comparison. The parametrization, known as the Wolfenstein parametrization, is valid up to fourth order in the expansion [Wol83]:

$$
V = \begin{pmatrix} 1 - \lambda^2/2 & \lambda & A\lambda^3(\rho - i\eta) \\ -\lambda & 1 - \lambda^2/2 & A\lambda^2 \\ A\lambda^3(1 - \rho - i\eta) & -A\lambda^2 & 1 \end{pmatrix} + \mathcal{O}(\lambda^4).
$$

It is easy to show that the relations between the Wolfenstein parametrization and the various matrix elements are as follows:

$$
s_{12} = \lambda = \frac{|V_{us}|}{\sqrt{|V_{ud}|^2 + |V_{us}|^2}}, \qquad s_{23} = A\lambda^2 = \lambda \left| \frac{V_{cb}}{V_{us}} \right|,
$$

$$
s_{13}e^{i\delta} = V_{ub}^* = A\lambda^3(\rho + i\eta). \tag{4.70}
$$

Again, the entire 3×3 unitary matrix can be characterized by four distinct parameters (λ, A, η, and ρ). However, given that the elements of the CKM matrix can be accessed through a number of different channels, the emphasis in the experimental program has been in looking at *over-constraining* the matrix, possibly to reveal channels of new physics.

Indeed historically, experimental probes of the CKM physics have helped reveal previously hidden aspects of the Standard Model itself. For example, in 1970 Glashow, Iliopoulos, and Maiani used the unitary quark-mixing ansatz to postulate the existence of a fourth quark, the charm quark [Gla70]. The subsequent discovery of bottom and top quarks, and even a third lepton generation, as well as the observation of direct *CP* violation in the kaon system, were all correctly predicted from the general framework posed by the CKM mixing.

The Unitarity Triangle

The unitarity of the CKM matrix invites a more elegant representation of its elements. The unitarity condition imposes the following condition on any row or column in the matrix:

$$\sum_i V_{ij} V_{ik}^* = \delta_{jk};$$

$$\sum_j V_{ij} V_{kj}^* = \delta_{ik}. \tag{4.71}$$

Within the 3×3 formalism, the six combinations can be more readily represented as triangles in the complex plane. Take, for example, one vanishing combination:

$$V_{ud} V_{ub}^* + V_{cd} V_{cb}^* + V_{td} V_{tb}^* = 0 . \tag{4.72}$$

Dividing Eq. (4.72) by $V_{cd} V_{cb}^*$ yields

$$\frac{V_{ud} V_{ub}^*}{V_{cd} V_{cb}^*} + 1 + \frac{V_{td} V_{tb}^*}{V_{cd} V_{cb}^*} = 0. \tag{4.73}$$

In the complex plane, one side of the triangle, see Fig.4.5, is constrained to have a length of 1 (with vertex (0,0) and (1,0)), while the free vertex will be set by the value of $\frac{V_{ud} V_{ub}^*}{V_{cd} V_{cb}^*}$, with value (ρ, η). The absolute value $\left| \frac{V_{ud} V_{ub}^*}{V_{cd} V_{cb}^*} \right|$ yields the length of one of the sides of the triangle, while $\left| \frac{V_{td} V_{tb}^*}{V_{cd} V_{cb}^*} \right|$ yields the length of the remaining side. Likewise, the angles of the triangle yield equally valuable information:

$$\alpha = \arg\left(-\frac{V_{td} V_{tb}^*}{V_{ud} V_{ub}^*} \right), \quad \beta = \arg\left(-\frac{V_{cd} V_{cb}^*}{V_{td} V_{tb}^*} \right), \quad \gamma = \arg\left(-\frac{V_{ud} V_{ub}^*}{V_{cd} V_{cb}^*} \right). \tag{4.74}$$

Other triangles are possible to construct, although all triangles will have the same area. Various consistency checks have been carried out on the elements of the unitarity triangles. As an example, the sum of all the angles (α, β, γ) should add up to 180 degrees. To date, the constraint on the sum of the angles yields $(178 \pm 11)°$. Various rows and columns in the matrix can also be checked to satisfy unitarity, and currently they all do, within their known uncertainties.

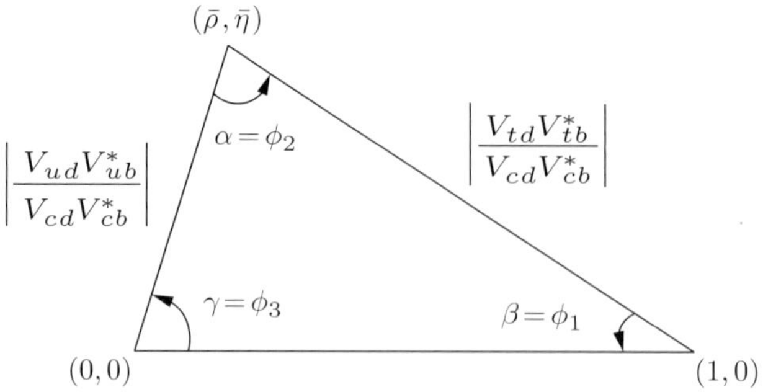

 The quark unitarity triangle [PDG14].

A final convention-free parametrization of the unitarity triangle commonly used is the Jarlskog invariant, J [Jar85]:

$$J = \text{Im}(V_{us}V_{cb}V_{ub}^{*}V_{cs}^{*}). \tag{4.75}$$

In addition to representing twice the area of the unitarity triangle, the Jarlskog variable is also nonzero only if CP is not conserved, and this provides a measure of the CP-violating phase predicted from the CKM mechanism.

The CKM Matrix and CP Violation

As discussed in Chapter 2, the majority of the forces associated with particle physics respect certain symmetry principles. Both electromagnetism, gravity, and the strong force respect charge (C) and parity (P) symmetry. The notorious odd man out is the weak force, which violates C and P maximally. That said, the combination of C and P (CP) appears to be nearly an exact symmetry.

Violations of CP imply that there is a measurable difference between particles and antiparticles, in either their decay rates or interaction probabilities, and the existence of the CKM matrix implies the possibility of such a violation existing in Nature. Although the 2×2 mixing matrix does not contain any CP-violating component, the extension to three families introduces a single CP-violating phase in the Standard Model. Indeed, it was the observation of indirect CP violation (via the suppression of certain decays in the neutral kaon sector) that led people to believe the existence of a third generation prior to its direct discovery.

CP-violation manifests itself in slight differences in amplitudes between particles and antiparticles, either via decay or via neutral meson mixing, all of which are essentially interference phenomena. Consider the case of a given observable state (say M) and its CP conjugate state $\bar{M}$ which can decay to a different state f and its CP conjugate $\bar{f}$. Under the Standard Model Hamiltonian, the amplitude connecting the states f and M can be represented as follows:

$$A_f = <f|H|M>, \quad \bar{A}_{\bar{f}} = <\bar{f}|H|\bar{M}> . \tag{4.76}$$

For the case that $\left| \frac{\bar{A}_{\bar{f}}}{A_f} \right| \neq 1$, CP is no longer a preserved symmetry of the system. In describing CP violation in decay processes, it is conventional to describe each of the above quantities as a sum of different amplitudes with distinct CP phases:

$$A_f = \sum_k |A_k| e^{i(\delta_k + \phi_k)}, \quad \bar{A}_{\bar{f}} = \sum_k |A_k| e^{i(\delta_k - \phi_k)}. \tag{4.77}$$

Here, ϕ_k (δ_k) represent relative phases that are odd (even) under CP. One can therefore construct an asymmetry that explicitly extracts $|\bar{A}_{\bar{f}}/A_f|$:

$$A_{\text{asym}} = \frac{\Gamma(M^- \to f^-) - \Gamma(M^+ \to f^+)}{\Gamma(M^- \to f^-) + \Gamma(M^+ \to f^+)}. \tag{4.78}$$

Assuming just two possible amplitude contributions and using the convention mentioned above, the asymmetry thus reduces to:

$$A_{\text{asym}} = \frac{2|A_1||A_2|\sin(\phi_1 - \phi_2)\sin(\delta_1 - \delta_2)}{|A_1|^2 + |A_2|^2 + 2|A_1||A_2|\cos(\phi_1 - \phi_2)\cos(\delta_1 - \delta_2)}. \tag{4.79}$$

If the amplitude ratio $|A_1/A_2|$ is known, then the phase difference $\phi_1 - \phi_2$ can be directly extracted. In any case, a nonzero value in this difference points to CP violation. Such charged meson decays have been readily observed in the kaon sector, as well as in the decay of $B^0 \to K^+\pi^-$.

Alternatively, one could study the time evolution of mixtures of different flavor eigenstates. For example, suppose that two physical mass states, M_L and M_H are really mixtures of different flavor states (in this case, M^0 and $\bar{M}^0$):

$$|M_H> = p|M> +q|\bar{M}>, \quad |M_L> = p|M> -q|\bar{M}> . \tag{4.80}$$

If $|M_H>$ and $|M_L>$ are not orthogonal to one another, then $|p|^2 - |q|^2 \neq 0$, implying that the quantity $|q/p| \neq 1$. Under such conditions, CP symmetry is violated and, as the system evolves in time, it will exhibit different decay or branching ratios which depend either on the mass differences or decay differences.

Historically, CP violation was first observed in $K \to \pi\pi$ via the mechanism described above. The mass eigenstates of the neutral kaon have distinct lifetimes (K_L and K_S, long and short) and also are distinct CP states, the former being predominantly odd and the latter being predominantly even. Without CP violation, the decay of the CP-odd state to a CP-even state (such as $\pi\pi$) should be forbidden. But such is not the case for K_L, where decays to the two-pion final state have been observed (for both charged and neutral pions).

One final channel for observing CP violation occurs when both flavor states can decay to the same final state. If the mass difference between the two states is sufficiently small, there will exist a time-dependent asymmetry that arises from CP violation:

$$\begin{aligned}
A_{\text{asym}} &= \frac{\Gamma(\bar{M}^0 \to f) - \Gamma(M^0 \to f)}{\Gamma(\bar{M}^0 \to f) + \Gamma(M^0 \to f)}, \\
&= \frac{2\,\text{Im}\lambda_f \sin(\Delta mt) + (|\lambda_f|^2 - 1)\cos(\Delta mt)}{1 + |\lambda_f|^2},
\end{aligned} \tag{4.81}$$

where $\lambda_f = \frac{q}{p}\frac{\bar{A}_{\bar{f}}}{A_f}$ and Δm is the difference in mass eigenstates.

But perhaps the most obvious observation for CP violation appears to be the most puzzling of all. Indeed, our observable Universe most readily exhibits the asymmetry between matter and antimatter. Although the CKM mechanism describes beautifully CP violation as observed in meson decay, it fails spectacularly in using the same mechanism to describe our observed matter-dominated universe. Another CP-violation mechanism must be at play in order to account for this discrepancy. A variety of searches (electron dipole moment, CP violation in the lepton sector, etc.), are currently underway to help resolve this mystery.

4.5 Majorana Mass

Before looking at the analogous mixing which occurs in the lepton sector, it is important to point out a wrinkle that occurs only in the neutrino sector, i.e., the possibility of a Majorana mass. Above we saw that the mass term of a Dirac particle, e.g., an electron, appears in the Lagrangian as

$$m_e \bar{\psi}_e \psi_e = m_e(\bar{\psi}_{eL}\psi_{eR} + \bar{\psi}_{eR}\psi_{eL}) \tag{4.82}$$

and if the particle in question carries a nonzero quantum number such as charge, hypercharge, etc., then the form of this (Dirac) mass term is unique. Here $L, R = \frac{1}{2}(1 \mp \gamma_5)$ are projection operators which identify the chirality of the Dirac field and the full field is the sum of its positive- and negative-chirality components, $\psi = \psi_L + \psi_R$. However, there exists another possible form for a spin-1/2 mass term if the particle being described is neutral and carries no quantum numbers so that it is its own antiparticle, as pointed out by Majorana [Maj37]. In order to construct such terms we use the charge conjugate fields

$$\psi^c \equiv C\gamma^0 \psi, \qquad (\psi_{L,R})^c = \left(\frac{1}{2}(1 \mp \gamma_5)\psi\right)^c, \tag{4.83}$$

where $C = i\gamma^0\gamma^2$ is the charge conjugation operator. The possibility of a Majorana mass term arises because the combination $\psi^T C \psi$ is a Lorentz invariant, meaning that a mass term

$$\begin{aligned}
\mathcal{L}_m &= -\frac{m_{L,R}}{2}\left[\bar{\psi}^c_{L,R}\psi_{L,R} + \bar{\psi}_{L,R}\psi^c_{L,R}\right] \\
&= -\frac{m_{L,R}}{2}\left[(\psi_{L,R})^T C\psi_{L,R} - (\psi^*_{L,R})^T C\psi^*_{L,R}\right]
\end{aligned} \tag{4.84}$$

is permitted in the Dirac Lagrangian. Since $(\psi_R)^T C\psi_L = 0$, a Majorana mass term involves the coupling of two left-handed or two right-handed chiral fields and the left- and right-handed masses m_L and m_R are independent. Treating ψ and ψ^* as independent fields we find the equations of motion

$$i\,\slashed{\partial}\psi_{L,R} = m_{L,R}\psi^*_{L,R} \quad \text{and} \quad i\,\slashed{\partial}\psi^*_{L,R} = m_{L,R}\psi_{L,R}. \tag{4.85}$$

A Majorana mass term mixes particle and antiparticle so that if we use the self-conjugate field

$$\psi^M_{L,R} = \sqrt{\frac{1}{2}}(\psi_{L,R} + \psi^c_{L,R}) \tag{4.86}$$

the Dirac Lagrangian assumes the familiar form

$$\begin{aligned}
\mathcal{L}_D &= \bar{\psi}_{L,R}i\,\slashed{\partial}\psi_{L,R} - \frac{m_{L,R}}{2}\left[\bar{\psi}^c_{L,R}\psi_{L,R} + \bar{\psi}_{L,R}(\psi^c_{L,R})\right] \\
&= \bar{\psi}^M_{L,R}i\,\slashed{\partial}\psi^M_{L,R} - m_{L,R}\bar{\psi}^M_{L,R}\psi_{L,R}.
\end{aligned} \tag{4.87}$$

The only known particle which might possibly have a Majorana mass component is the neutrino, but determining whether or not this is the case is a challenge which requires the search for neutrinoless double-beta decay as will be described in detail in Chapter 18. As of yet there is no evidence for the existence of such a Majorana mass, but the search continues.

4.6 Lepton Mixing

We end this chapter with a brief discussion of the leptonic side of the mixing picture. Although hints for mixing in the leptonic sector made themselves visible as early as the 1960s, this mixing (and by this, we mean neutrino mixing) was not truly experimentally confirmed until the end of the twentieth century. Neutrino masses and mixings will be discussed more thoroughly in Chapter 18; here, we will cover the basics.

Paralleling the quark sector, the terms in the Standard Model Lagrangian allow the weak eigenstates to be distinct from the mass eigenstates of the system. Again, we can write down the weak current that arises from the leptonic sector in the Standard Model Lagrangian as follows:

$$J_\mu^W = \bar{\nu}_L U^\dagger \gamma_\mu l_L \,, \tag{4.88}$$

where $\bar{\nu}_L$ is the neutrino field and l_L is the charged lepton field. Here again, the matrix U represents a unitary mixing matrix relating the weak states to the mass eigenstates. In this regard, the neutrino mixing (PMNS) matrix, mirrors much of the same structure as discussed for its CKM counterpart. However, there potentially exists some strong differences between the two. In the case of the quarks, the nature of the kind of particles involved is known, i.e. they are Dirac particles. Since the Dirac fields are invariant under $U(1)$ global gauge transformations, one can apply three independent phases to each quark field without any change to the physics of the system. In such cases, one is able to remove some of the free parameters of the matrix elements (in the 3×3 case, two terms are removed, leaving three physical angles and one CP-violating phase). Such a condition, however, does not necessarily apply to the neutrino sector. As discussed previously, neutrinos could either be Dirac or Majorana fields. The Majorana fields are *not* invariant under $U(1)$ transformations and removal of those phases is not possible.

Not to worry, though. If neutrinos are indeed Majorana in Nature, one can still write the PMNS matrix as a combination of both the Dirac case, plus an additional diagonal term:

$$U = U^d U^M,$$
$$U^M = (e^{i\alpha_1}, e^{i\alpha_2}, e^{i\alpha_3})_{\text{diag}}. \tag{4.89}$$

Since a purely global phase has no impact on the Lagrangian, the above diagonal matrix can be re-written as

$$U^M = (1, e^{i(\alpha_2-\alpha_1)}, e^{i(\alpha_3-\alpha_1)})_{\text{diag}}. \tag{4.90}$$

The diagonal matrix U^M has no impact on neutrino oscillation experiments.

The CKM matrix exhibited very diagonal dominance, with $s_{13} \ll s_{23} \ll s_{12} \ll 1$. However, such is not the case with neutrino mixing, where some of the mixing angles are very near maximal. Although nothing in the Standard Model dictates what these angles should be, there certainly existed early theoretical bias that the leptonic mixing matrix should closely mirror its Dirac counterpart in terms of its values as well as its structure. Experimentally, however, this was shown to be far from the truth.

Lepton mixing is only manifested in neutrino mixing,[2] as neutrinos initially created in one flavor state transmute to another flavor state over time. The observation of neutrino oscillations can be treated as a coherent evolution of different mass states over time. The amplitude of the oscillation probability can be written in terms of the PMNS matrix introduced above:

$$A_{\nu_\alpha \to \nu_\beta} = \sum_i U^*_{\alpha i} \Phi(\nu_i) U_{i\beta}, \tag{4.91}$$

where $\Phi(\nu_i)$ represents the coherent sum of the propagation of the neutrino mass eigenstates as the neutrino travels between interaction points. The form of the propagation term can be derived if we consider the Hamiltonian acting on the neutrino mass state over time. In the rest frame of the neutrino with mass m_i and proper time τ_i, we have:

$$i\frac{\partial}{\partial \tau_i}|\nu_i(\tau_i)\rangle = m_i|\nu_i(\tau_i)\rangle$$

$$\implies |\nu_i(\tau_i)\rangle = e^{-im_i\tau_i}|\nu_i(0)\rangle. \tag{4.92}$$

Hence

$$\Phi(\nu_i) = \langle \nu_i(\tau_i)|\nu_i(0)\rangle = e^{-im_i\tau_i}. \tag{4.93}$$

Note that the quantity $m_i\tau_i$ arising in the phase of the propagating system is invariant. Boosting from the rest frame to the laboratory frame (where the neutrino has energy E_i and momentum p_i), the phase term becomes

$$m_i\tau_i = E_i t - p_i L \simeq Et - (E - m_i^2/2E)L = E(t - L) + \frac{m_i^2}{2E}, \tag{4.94}$$

where L is the travel distance. Here we have made the assumption that the neutrino mass eigenstates share a common energy E in order to act coherently, in which case $p_i = \sqrt{E^2 - m_i^2} \simeq E - \frac{m_i^2}{2E}$. The first term in the phase is an overall phase shift which has no observable effect. If the masses are not degenerate, then an interference term will arise between the different mass eigenstates. The corresponding amplitude for oscillation (up to a constant phase) is thus given as:

$$A_{\nu_\alpha \to \nu_\beta} = \sum_i U^*_{\alpha i} e^{-im_i^2 \frac{L}{2E}} U_{i\beta}. \tag{4.95}$$

The corresponding probability of oscillation is thus given as

$$P(\nu_\alpha \to \nu_\beta) = |A_{\nu_\alpha \to \nu_\beta}|^2$$

$$= \delta_{\alpha\beta} - 4 \sum_{i>j} \mathrm{Re}(U^*_{\alpha i} U_{\beta i} U^*_{\alpha j} U_{\beta j}) \sin^2\left(\delta m^2_{ij}\frac{L}{4E}\right)$$

$$\pm 2 \sum_{i>j} \mathrm{Im}(U^*_{\alpha i} U_{\beta i} U^*_{\alpha j} U_{\beta j}) \sin\left(\delta m^2_{ij}\frac{L}{2E}\right). \tag{4.96}$$

---

[2] If you ask "Why can't charged leptons mix?," recall that mixing implies that one can write the observed (in this case, the weak eigenstate) in terms of the mass eigenstate. In the case of the charged leptons, these particles *are* the mass states. In fact, the only distinguishing feature amongst the e, μ, and τ particles are that they possess different masses, and nothing else. This is different from the neutrino case, where a particular state is tagged indirectly via its weak isodoublet counterpart and, therefore, can be considered (and are) superpositions of different mass eigenstates.

Much like in the quark sector, the PMNS matrix accommodates three real angles and one *CP*-violating phase (and, possibly, two Majorana phases). The sign notation on the imaginary terms indicates whether one deals with neutrino or antineutrino oscillations. Note that the *CP*-violating term can again be related to the Jarlskog invariant as we described in quark mixing (up to a sign):

$$|J| = |\text{Im}(U_{\alpha i}^* U_{\beta i} U_{\alpha j}^* U_{\beta j})|. \tag{4.97}$$

A value of $J \neq 0$ would imply *CP*-violation in the neutrino sector. However, unlike the case of quarks, neutrino *CP*-violation has yet to be observed experimentally.

We close our discussion of electroweak theory by studying what happened to the three scalar degrees of freedom which were eliminated via the gauge transformation. The answer can be found by simple counting arguments. We began our analysis with:

i) Four massless vector boson (gauge) fields with two degrees of freedom each.
ii) Four scalar fields with one degree of freedom each before spontaneous symmetry breaking.

There are thus total 12 degrees of freedom. After spontaneous symmetry breaking, we have:

i) Three massive gauge bosons with three degrees of freedom each.
ii) One massless gauge boson with two degrees of freedom.
iii) One scalar boson with one degree of freedom.

Again the total is 12 degrees of freedom, but now it is clear what has happened. The three originally massless gauge bosons are said to have *eaten* the three scalar degrees of freedom, becoming massive in the process.

In conclusion, we have constructed a very successful model which unifies the electromagnetic and weak interactions. However, this is only one part of the Standard Model. Our next task is to look at the strong interaction component, which is described by quantum chromodynamics in the following chapter.

Exercises

4.1 Z^0 **Decays**

The Z^0 boson decays via the weak neutral current to a pair of fundamental fermions: $\bar{u}u, \bar{d}d, \ldots, \bar{\nu}_\tau \nu_\tau$

a) Show that the decay amplitude of the Z^0 boson to a pair of fundamental spin-$1/2$ fermions is given by

$$\text{Amp} = i\frac{g_2}{\cos\theta_W}\epsilon_Z^\mu \bar{u}(p_1)\gamma_\mu \left[\left(I_3^f + 2\sin^2\theta_W Q_f\right) + I_3^f \gamma_5\right] v(p_2),$$

where I_3^f is the $SU(2)_L$ eigenvalue of the fermion, Q_f is the fermion charge, and θ_W is the weak mixing angle.

b) Calculate the decay width for each quark and lepton in the approximation that $m_f \ll M_{Z^0}$.

c) Compare your results with the results from the particle data book [PDG14],

$$\Gamma(\bar{e}e) = \Gamma(\bar{\mu}\mu) = \Gamma(\bar{\tau}\tau) \simeq 84 \,\mathrm{MeV}$$

$$\Gamma(\bar{u}u) = \Gamma(\bar{c}c) \simeq 300 \,\mathrm{MeV}$$

$$\Gamma(\bar{d}d) = \Gamma(\bar{s}s)) = \Gamma(\bar{b}b) \simeq 380 \,\mathrm{MeV}.$$

d) Calculate the width due to (unseen) neutrino modes.

4.2 Yang–Mills Theory

The Dirac Lagrangian for a spin-$1/2$ particle of mass m

$$\mathcal{L}[\psi(x)] = \bar{\psi}(x)\,[i\,\slashed{\partial} - m]\,\psi(x)$$

is invariant under the $U(1)$ *global* phase transformation

$$\psi(x) \to \psi'(x) = \exp(i\beta)\psi(x)\,.$$

a) Demonstrate this invariance.

It is also well known that the Dirac Lagrangian for a spin-$1/2$ particle of mass m and charge e can be made invariant under a *local* $U(1)$ phase transformation

$$\psi(x) \to \psi'(x) = \exp(ie\beta(x))\psi(x) \equiv U\psi(x).$$

b) Demonstrate that the Lagrangian $\mathcal{L}[\psi(x)]$ is *not* invariant under a local $U(1)$ transformation.

The invariance can be accomplished if we introduce an additional dynamical vector field $A_\mu(x)$ with the transformation property

$$A_\mu(x) \to U A_\mu(x) U^{-1} - \frac{i}{e} U \partial_\mu U^{-1} = A_\mu(x) - \partial_\mu \beta(x)$$

and modify the Lagrangian to become

$$\mathcal{L}[\psi(x), A_\mu(x)] = \bar{\psi}(x)\,[i\,\slashed{D} - m]\,\psi(x) - \frac{1}{4}F_{\mu\nu}(x)F^{\mu\nu}(x),$$

where

$$D_\mu = \partial_\mu + ie A_\mu(x)$$

is the covariant derivative and has the property

$$D_\mu \psi(x) \to D'_\mu \psi'(x) = U D_\mu \psi(x).$$

c) Verify this transformation property of the covariant derivative.

d) Demonstrate the invariance of $\mathcal{L}[\psi(x), A_\mu(x)]$ under a local $U(1)$ transformation.

Yang and Mills showed how to generalize this local invariance to the case of non-Abelian transformations. We consider the case of $SU(2)$ with a doublet of spin-1/2 fields

$$N(x) = \begin{pmatrix} \psi_1(x) \\ \psi_2(x) \end{pmatrix}$$

which transforms as

$$N(x) \rightarrow N'(x) = \exp\left(ig\frac{1}{2}\boldsymbol{\tau}\cdot\boldsymbol{\beta}(x)\right)N(x) \equiv UN(x)$$

under a local $SU(2)$ gauge transformation, where $\boldsymbol{\tau}$ are the Pauli isospin matrices and satisfy the commutation relations

$$[\tau_i, \tau_j] = 2i\epsilon_{ijk}\tau_k.$$

e) Demonstrate that the Lagrangian $\mathcal{L}[N(x)]$ is *not* invariant under a local $SU(2)$ gauge transformation.

　　Now define a triplet of vector fields $\boldsymbol{A}_\mu(x)$ which transform as

$$\frac{1}{2}\boldsymbol{\tau}\cdot\boldsymbol{A}_\mu(x) \rightarrow U\frac{1}{2}\boldsymbol{\tau}\cdot\boldsymbol{A}_\mu(x)U^{-1} - iU\partial_\mu U^{-1}$$

and a covariant derivative

$$D_\mu = \partial_\mu + ig\frac{1}{2}\boldsymbol{\tau}\cdot\boldsymbol{A}(x).$$

f) Demonstrate that the covariant derivative has the property

$$D_\mu N(x) \rightarrow D'_\mu N'(x) = UD_\mu\psi(x).$$

where $U = \exp\left(i\frac{1}{2}\boldsymbol{\tau}\cdot\boldsymbol{\beta}(x)\right)$.

g) For an infinitesimal $\delta\boldsymbol{\beta}$ show that

$$\boldsymbol{A}_\mu(x) \rightarrow \boldsymbol{A}_\mu(x) + \partial_\mu\delta\boldsymbol{\beta}(x) - g\delta\boldsymbol{\beta}(x) \times \boldsymbol{A}_\mu(x).$$

h) Demonstrate that the Lagrangian

$$\mathcal{L}[N(x), \boldsymbol{A}(x)](x) = \bar{N}(x)\left[i\not{D} - m\right]N(x) - \frac{1}{4}\boldsymbol{F}_{\mu\nu}(x)\cdot\boldsymbol{F}^{\mu\nu}(x)$$

is invariant under a local $SU(2)$ gauge transformation provided that $\boldsymbol{F}_{\mu\nu}(x)$ is defined via

$$[D_\mu, D_\nu]N(x) = -ig\frac{1}{2}\boldsymbol{\tau}\cdot\boldsymbol{F}_{\mu\nu}(x)N(x)$$

i.e.,

$$\boldsymbol{F}_{\mu\nu}(x) = \partial_\mu\boldsymbol{A}_\nu(x) - \partial_\nu\boldsymbol{A}_\mu(x) - g\boldsymbol{A}_\mu(x) \times \boldsymbol{A}_\nu(x)$$

and has the transformation property

$$\frac{1}{2}\boldsymbol{\tau}\cdot\boldsymbol{F}_{\mu\nu} \to U\frac{1}{2}\boldsymbol{\tau}\cdot\boldsymbol{F}_{\mu\nu}U^{-1}.$$

4.3   Neutrino Oscillations with Two Flavors

Consider the standard analysis of neutrino oscillations supposing that there are only two types of neutrinos. Production of these neutrinos in a weak interaction emphasizes the *flavor* states, ν_a and ν_b, while the neutrino states with definite *mass* are ν_1 and ν_2. These two pairs of states are related by a 2×2 unitary matrix with a single parameter, the mixing angle θ_{12}, via

$$\begin{pmatrix} \psi_a \\ \psi_b \end{pmatrix} = \begin{pmatrix} \cos\theta_{12} & \sin\theta_{12} \\ -\sin\theta_{12} & \cos\theta_{12} \end{pmatrix} \begin{pmatrix} \psi_1 \\ \psi_2 \end{pmatrix},$$

which implies the states a and b can transform back and forth between each other. Consider plane-wave states of neutrinos 1 and 2 that have well-defined energies E_i and momenta P_i large compared with their rest masses m_i, which propagate essentially at the speed of light.

a) For propagation along the x-axis, show that the momenta P_i are

$$P_i \approx \frac{E_i}{c}\left(1 - \frac{m_i^2 c^4}{2E_i^2}\right),$$

    and therefore

$$\psi_i(x,t) = \psi_{i,0}e^{i(P_i x - E_i t)/\hbar} \approx \psi_{i,0}e^{E_i(x/c-t)/\hbar}e^{-im_i^2 c^3 x/2E_i\hbar}.$$

b) The first phase factor can be ignored to write

$$\psi_i(x,t) \approx \psi_{i,0}e^{-im_i^2 c^3 x/2E_i\hbar},$$

    where $x \approx ct$. This exponential phase factor is slightly different for mass states 1 and 2, which leads to an oscillatory interference term in the spatial/time dependence of an initial single-flavor state. Consider a neutrino created at time $t = 0$ in a pure flavor state a with $\psi_{a,0} = 1$, $\psi_{b,0} = 0$. Find the initial mass states $\psi_{1,0}$ and $\psi_{2,0}$.

c) Determine the evolution of flavor states $\psi_a(x)$ and $\psi_b(x)$.

d) Show that the probability that the initial state flavor a is still a after the neutrino has traveled a distance x is given by

$$P_{a \to a}(x) = 1 - \sin^2 2\theta_{12} \sin^2 \frac{\Delta m_{12}^2 x}{4E},$$

    where $E = \frac{1}{2}(E_1 + E_2)$ is the average energy of the two neutrinos and $\Delta m_{12}^2 = \left(\frac{m_1^2}{E_1} - \frac{m_2^2}{E_2}\right)E.$

e) Explain how neutrino oscillation between different mass eigenstates is consistent with conservation of energy.

4.4   Neutrino Oscillations Using QFT

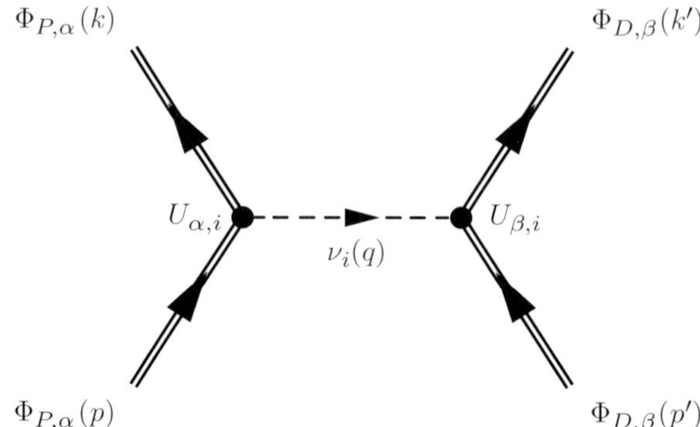

In this exercise, we consider neutrino mixing from a diagrammatic perspective. Consider the case that a nucleus, represented by the wavefunction $\Phi_{P,\alpha}(p)$ makes a transition to a final state $\Phi_{P,\alpha}(k)$ by emitting a virtual massive neutrino ν_i. The neutrino propagates and then interacts with a second nucleus, located at some distance L after some time T such that $\Phi_{D,\beta}(p') + \nu_i \rightarrow \Phi_{D,\beta}(k')$. Thus $p(p')\,k(k')$ represents the production (detection) momentum, and we consider the coupling term at each vertex to have the form $U_{\alpha,i}$ and $U_{\beta,i}$ respectively.

a) Using the function Φ to represent the initial and final wavefunction for the production and detection nucleus, write out the transition amplitude for the reaction assuming that there exist n neutrinos, each with a distinct mass m_i, which participate in the interaction. Assume the momentum transfer of the interaction is given by the four-vector q. One can take into account the fact that the production and detection locations are spatially separated by explicitly considering the plane-wave term of the wavefunction $e^{-iq\cdot x}$, where $x = (t_P - t_D, x_P - x_D) = (T, L)$ represents the spacetime separation of the two interaction points. Show, by rewriting the energy–momentum conservation delta function in terms of its Fourier transform, how the $e^{-iq\cdot x}$ emerges.

b) Using complex integration methods show that

$$\int \frac{d\mathbf{p}}{(2\pi)^3} \frac{\phi(q)e^{i\mathbf{p}\cdot L}}{A - \mathbf{p}^2} = -\frac{1}{4\pi|L|}\phi\left(\frac{\sqrt{A}L}{|L|}\right)e^{i\sqrt{A}|L|}.$$

c) Using the integral from b) and the completeness of neutrino spinors

$$\sum_s u_j^s \bar{u}_j^s = \slashed{p} + m_j$$

extract the amplitude and probability for the exchange to occur. Show that the phase term, when expanded in terms of the mass, is consistent with the usual oscillation form.[3]

---

[3] You may find that your expression remains a function of the nuclear wavefunctions. These functions disappear once the proper normalization is applied. You are free to treat them as a constant for this problem.

4.5   **Spontaneous Symmetry Breaking: Meissner Effect for Scalar Electrodynamics**

The Meissner effect refers to the exclusion of a magnetic field from the interior of a superconductor, and will be explored in this exercise.

The Lagrangian density for the classical electrodynamics of charged spinless particles is given by

$$\mathcal{L} = -\frac{1}{4}F_{\mu\nu}F^{\mu\nu} + (D_\mu\phi)^*(D^\mu\phi) - V(\phi)$$

with

$$D_\mu = \partial_\mu - ieA_\mu$$

being the covariant derivative and

$$V(\phi) = \frac{1}{2}m^2(\phi^*\phi) + \frac{\lambda}{4}(\phi^*\phi)^2$$

representing the potential energy. Here ϕ is a complex scalar field.

a) Using Noether's theorem, identity the electromagnetic current of the ϕ field.
b) Find the Hamiltonian density.
c) If $m^2 > 0$ show that the ground state of the system has $\phi = 0, A_\mu = 0$. (In this case the theory is that of normal electrodynamics.)
d) There can exist another phase of the theory if the parameter $m^2 = \mu^2 < 0$. In this case show that the lowest energy state has $A_\mu = 0, \phi = $ constand Calculate ϕ. What is the mass of the photon? This phase has electromagnetic fields being screened. To see this calculate the potential between two static point charges $- \rho(x) = Q\delta^3(x) -$ using Green's function techniques (as would be done in Jackson [Jac75]), but including the photon mass in the Green's function equation. What sets the scale of the screening distance?
e) Now add an external field to the problem. One way to do this is to add an external current to generate the field by including a new term in the Lagrangian

$$\mathcal{L} \rightarrow \mathcal{L}_0 - A_\mu J^\mu_{\text{ext}}$$

with

$$J^\mu_{\text{ext}} = \partial_\mu F^{\nu\mu}_{\text{ext}}.$$

If the fields vanish at infinity this is equivalent to

$$\mathcal{L}_0 + \frac{1}{2}F_{\mu\nu}F^{\mu\nu}_{\text{ext}}.$$

In order to see that this does act like an applied field, show that if one disregards the scalar field the equations of motion require $F^{\mu\nu} = F^{\mu\nu}_{\text{ext}}$.

f) Show that there exist two simple solutions to the equations of motion in the presence of a constant applied field

i) $\phi = 0, \quad F^{\mu\nu} = F^{\mu\nu}_{\text{ext}}$
ii) $\phi = $ const., $\quad F^{\mu\nu} = 0$

where const. is the value of ϕ calculated in part d) above. Again, these correspond to unscreened and screened phases for the electromagnetic field.

g) Calculate the energy of the two phases when $F^{\mu\nu}_{\text{ext}}$ describes a constant magnetic field. Show that solution ii) is preferred for $B < B_{\text{critical}}$ and that i) has the lower energy for $B > B_{\text{critical}}$. Then, there exists a phase transition from the screened to the unscreened phase as one increases the external magnetic field – this is the simple analogy of the Meissner effect.

Suggestions:

In a) and b) remember that ϕ is a complex scalar field, i.e., $\phi \neq \phi^*$, and that it has two degrees of freedom. These can be chosen as $\phi = \phi_1 + i\phi_2$, with ϕ_1, ϕ_2 being two real fields. However, it is probably easier (and more common) to treat ϕ and ϕ^* in applying Lagrangian techniques.

The photon mass can be obtained from the equation of motion when $\phi = \text{const}$.

In e), f), and g) continue to assume that $m^2 = -\mu^2 < 0$.

It is best to work with the second form of the source term, $\frac{1}{2}F_{\mu\nu}F^{\mu\nu}_{\text{ext}}$.

In g), do not forget to include the source term in the Hamiltonian density.

QCD and Confinement

5.1 Introduction

The modern theory of the strong interaction began with the introduction in 1963 of the concept of quarks, purely from theoretical considerations [Gel64a, Zwe64]. As discussed in Chapter 3, the observed spectrum of mesons and baryons could be readily understood as bound states of quark–antiquark and three quarks. The existence of three different species or "flavors" (u, d, s) of spin-1/2 quarks with different quantum numbers (electric charge, isospin, strangeness), but approximately the same strong interaction rationalized the immensely successful $SU(3)$ "eight-fold way" symmetry introduced by Gell-Mann earlier [Gel64b]. The idea that structureless quarks could form a fundamental basis for the description of hadrons became plausible when relations derived from the quantum field theory of quarks, namely the algebra of quark currents and their divergences, were successfully applied.

A second major concept, that each flavor of quark should come in three colors, arose from several, separate considerations. The basic idea that quarks carry a new internal quantum number was introduced by Han and Nambu in 1965 [Han65]. Hints also came from the fact that $q\bar{q}$ and qqq states have lower masses, which can be understood if there is an $SU(3)$ symmetry among the different colors of quarks and for some reason color singlets are much lighter than non-singlets. Quark models indicated that the wavefunctions of baryons qqq should be symmetric in the interchange of the spatial, spin, and flavor quantum numbers of the quarks. Bose symmetry for spin-1/2 particles contradicts the principles of relativistic field theory. Antisymmetry in color quantum numbers can restore the expected Fermi statistics. Later, calculations of the decay rate for $\pi^0 \to \gamma\gamma$ and of the cross section for $e^+e^- \to$ hadrons further strengthened the evidence for the color degree of freedom.

A third major development leading to QCD was the experimental discovery [Tay91] that the proton was composed of pointlike constituents, as demonstrated in the scaling of deep inelastic electron–nucleon data, first taken at the Stanford Linear Accelerator Center (SLAC). The parton model [Bjo67] was developed to explain the high-energy, inelastic data. The cornerstone of the parton model was that hadrons consisted of pointlike constituents with simple properties. Some of these constituents were soon identified as quarks. The parton model, though strikingly successful as a description of data, was an intuitive model and did not have a rigorous microscopic basis. In the 1970s a microscopic basis for analyzing the processes described by the parton model in quantum field theory was developed. Important elements included showing that non-Abelian gauge theories possess the property of asymptotic freedom and the work of 't Hooft on renormalization of gauge theories. Thus, a synthesis of the previous ideas concerning quarks, color and partons was at hand. The color degree of freedom of the quarks could be gauged to yield an asymptotically free theory – using the degrees of freedom determined spectroscopically to justify the dynamical hypotheses of the parton model. Finally, a mathematically well-defined Lagrangian field theory with a microscopic basis was available. In the Standard Model, the strong interaction is described by quantum chromodynamics (QCD) [Wil82], a theory which describes hadrons in terms of a color interaction between massive spin-1/2 fermions, called quarks, via exchange of massless bosons, called gluons. QCD is a non-Abelian $SU(3)_{color}$ gauge theory with N_f flavors of quarks, three of them light (u, d, s) and the other three heavy (c, b, t). See Chapter 2 for a review of $SU(3)$ group symmetry. Here, light and heavy refer to a typical hadron mass scale, e.g., the proton mass. Unlike in the atom or in the nucleus, which are well described in terms of fermions moving nonrelativistically in a mean field, QCD describes a world where the virtual particles, i.e., quark–antiquark pairs and gluons, are dominant. Further, neither quarks nor gluons can be isolated experimentally, i.e., the color forces become infinite at long distance scales. Thus, QCD possesses two unique properties:

Asymptotic freedom, which means that at high energies quarks and gluons interact weakly. This aspect of QCD was discovered by Politzer [Pol73] and by Wilczek and Gross [Gro73] in the early 1970s. For this work they were awarded the 2004 Nobel Prize in Physics.

Confinement, which means that the attractive force between two quarks increases as their separation increases. Consequently, it would require an infinite amount of energy to separate two quarks, and thus they are always bound in color neutral hadrons, e.g., protons and neutrons. While confinement is unproven analytically, it is consistent with the nonobservation of free quarks.

Since essentially all of the visible matter in the universe arises from QCD interactions, understanding the structure and properties of hadrons in terms of the pointlike quarks and gluons of QCD is a major research thrust worldwide. A large body of high-energy, experimental data confirms the validity of QCD, but exact calculations in this theory are in general not possible at present because of the theory's complexity. Data are interpreted using QCD-inspired models, using approximate numerical solutions of QCD that rely on advanced computer simulations, and by using effective filed theory (EFT) techniques which are valid in a given energy range (cf. Chapter 6). Major open questions include: How does the spin-1/2 of the nucleon arise from the spin-1/2 quark and spin-1 gluon constituents? At high energies do the self-interacting gluons in atomic nuclei constitute a universal, saturated form of gluonic matter?

5.2 Renormalization

Before proceeding with our QCD derivation, it is useful to examine the concept of renormalization. Perhaps one of the simplest examples of this idea can be found in the analysis of the energy difference between the $2S_{\frac{1}{2}}$ and $2P_{\frac{1}{2}}$ states of the hydrogen atom, which are predicted to be degenerate, even when the spin–orbit interaction is included in the analysis or the static Coulomb interaction is included to all orders in α. Nevertheless, in 1947 this $\sim 1050\,$MHz splitting was found to be nonzero by Lamb and Retherford and is generally called the *Lamb shift* [Lam47]. Its origin is the inclusion of coupling to the electromagnetic radiation field A, which in second order perturbation theory leads to an energy shift for a hydrogen atom state $|B, 0 >$ of [1]

$$\Delta E_B = \Lambda + P \sum_{I,\rho} \frac{\left| < I, \rho | \hat{V} | B, 0 > \right|^2}{E_B^{(0)} - E_I^{(0)} - \omega_\rho} \, , \tag{5.1}$$

where P denotes the principal value and the notation $|B, 0 >$ indicates a hydrogen atom in the state $|B >$ with no photons present. The term Λ, arising from the quadratic interaction $e^2 A \cdot A$, is independent of state and hence may be neglected if only energy differences *between* states are considered. (This is equivalent to changing the reference point from which energies are measured.) We shall therefore confine our attention to the principal value integral and note first that the sum on directions and polarizations of the virtual photon ρ may easily be performed. Making the long wavelength approximation, we have

$$\begin{aligned}
\Delta E_B &= \sum_I \int \frac{k^2 dk}{(2\pi)^3} d\Omega_{\hat{k}} \sum_\lambda \left(\frac{e}{m}\right)^2 \frac{\left| \hat{\epsilon}_{k,\lambda} \cdot \boldsymbol{p}_{IB} \right|^2}{E_B^{(0)} - E_I^{(0)} - k} \frac{1}{2k} \\
&= \sum_I \int \frac{k dk}{(2\pi)^3} d\Omega_{\hat{k}} \frac{e^2}{2m^2} \frac{\boldsymbol{p}_{BI} \cdot \boldsymbol{p}_{IB} - \boldsymbol{p}_{BI} \cdot \hat{\boldsymbol{k}} \boldsymbol{p}_{IB} \cdot \hat{\boldsymbol{k}}}{E_B^{(0)} - E_I^{(0)} - k} \\
&= \sum_I \int \frac{k dk}{(2\pi)^3} \frac{e^2}{2m^2} \cdot 4\pi \cdot \frac{2}{3} \frac{\left| \boldsymbol{p}_{BI} \right|^2}{E_B^{(0)} - E_I^{(0)} - k} \\
&= -\frac{2}{3\pi} \frac{\alpha}{m^2} \int_0^\infty dk\, k \sum_I \frac{\left| \boldsymbol{p}_{BI} \right|^2}{k - E_B^{(0)} + E_I^{(0)}} \, . \tag{5.2}
\end{aligned}$$

Equation (5.2) is a divergent integral – $\Delta E_B = -\infty$. The divergence is linear in $k - \int_0^\infty dk$ for large k. (In the relativistic theory, using the Dirac equation and taking account of the existence of positrons as well as electrons, the divergence is logarithmic – $\int^\infty \frac{dk}{k}$ for large k.) This divergent contribution to the energy caused a good deal of consternation but was largely ignored until the experimental measurement by Lamb and Retherford compelled attention to this problem.

The basic suggestion for the removal of the divergence was made by Kramers, who noticed that for a free electron of momentum $\boldsymbol{p}$ there is also an infinite energy shift – ΔE_p – obtainable from the previous expression if we replace atomic states by plane waves. Since plane waves are eigenstates of the momentum operator, off-diagonal matrix elements vanish and the sum over I reduces to a single term with $\boldsymbol{p}_{BB}$ representing the free electron momentum $\boldsymbol{p}$. Then

$$\Delta E_p = -\frac{2}{3\pi} \frac{\alpha}{m^2} \boldsymbol{p}^2 \int_0^\infty dk \, . \tag{5.3}$$

---

[1] A more complete discussion is given in [Hol14a].

The theory is, of course, incorrect for very large photon momenta. However, we may suppose that due to additional effects the integral over k has an effective cutoff $K \sim m$, since for $k \gg K$, e.g., relativistic effects become important. Then

$$\frac{\Delta E_p}{E_p} \approx -\frac{4}{3\pi}\alpha <<< 1 \tag{5.4}$$

so that this represents only a tiny correction to the electron kinetic energy.

Since $\Delta E_p \propto \boldsymbol{p}^2$ we may consider the energy shift to be associated with a change δm in the electron rest mass

$$\Delta E_p = \frac{\boldsymbol{p}^2}{2(m+\delta m)} - \frac{\boldsymbol{p}^2}{2m} \approx -\frac{\boldsymbol{p}^2}{2m^2}\delta m \tag{5.5}$$

due to the interaction with the radiation field. We thus identify

$$\delta m = \frac{4\alpha}{3\pi}\int_0^K dk \ . \tag{5.6}$$

If this suggestion is correct, then the quantity m which, up to now, we have employed in the Hamiltonian is *not* the *experimental* electron mass, but is rather a *fictitious* mass which the electron would possess if somehow interaction with the radiation field could be turned off. The mass is said to be *renormalized*. The physical – experimental – electron mass is given by

$$m_{\text{exp}} = m + \delta m \ , \tag{5.7}$$

which suggests that we should rewrite our Hamiltonian in terms of this measurable quantity

$$\begin{aligned}
\hat{H} &= \frac{\boldsymbol{p}^2}{2m} + e\phi(\boldsymbol{r}) + \hat{H}_{\text{RAD}}^{(r)} + \hat{V} \\
&= \frac{\boldsymbol{p}^2}{2m_{\text{exp}}} + e\phi(\boldsymbol{r}) + \hat{H}_{\text{RAD}}^{(r)} + \hat{V} + \left(\frac{\boldsymbol{p}^2}{2m} - \frac{\boldsymbol{p}^2}{2(m+\delta m)}\right) \\
&\equiv \hat{H}_0' + \hat{V}' \ .
\end{aligned} \tag{5.8}$$

Here $\hat{H}_0'$ is simply the usual Hamiltonian $\hat{H}_0$ but with m_{exp} substituted for the electron mass. However, $\hat{V}'$ now consists of two pieces

$$\hat{V}' = \hat{V} + \frac{\boldsymbol{p}^2}{2m^2}\delta m \equiv \hat{V}_1 + \hat{V}_2 \tag{5.9}$$

and *both* must be included in the energy shift calculation, as first done by Bethe [Bet47]. From $\hat{V}_1$ we find as before

$$\Delta E_1 = -\frac{2}{3\pi}\frac{\alpha}{m^2}\int_0^K dk\, k \sum_I \frac{<B\,|\boldsymbol{p}|\,I> \cdot <I\,|\boldsymbol{p}|\,B>}{k + E_I^{(0)} - E_B^{(0)}} \tag{5.10}$$

while for the piece $\hat{V}_2$

$$\Delta E_2 = \frac{1}{2m^2}\delta m < B\,|\boldsymbol{p}\cdot\boldsymbol{p}|\,B >$$

$$= \text{(using completeness of atomic states)}$$

$$\frac{4\alpha}{3\pi}\int_0^K dk\,\frac{1}{2m^2}\sum_I < B\,|\boldsymbol{p}|\,I > \cdot < I\,|\boldsymbol{p}|\,B >$$

$$= \frac{2\alpha}{3\pi m^2}\int_0^K dk\,\sum_I \left(k + E_I^{(0)} - E_B^{(0)}\right)\frac{< B\,|\boldsymbol{p}|\,I > \cdot < I\,|\boldsymbol{p}|\,B >}{k + E_I^{(0)} - E_B^{(0)}}. \tag{5.11}$$

Adding the two contributions, we find

$$\Delta E_{\text{tot}} = \Delta E_1 + \Delta E_2$$

$$= \frac{2\alpha}{3\pi m^2}\int_0^K dk\,\sum_I \frac{\left(E_I^{(0)} - E_B^{(0)}\right) < B\,|\boldsymbol{p}|\,I > \cdot < I\,|\boldsymbol{p}|\,B >}{k + E_I^{(0)} - E_B^{(0)}}. \tag{5.12}$$

The integral now diverges only logarithmically for large k – $\int \frac{dk}{k}$ – and we may be encouraged to hope that in a proper relativistic treatment there will exist an effective cutoff for wavenumbers larger than some $K \approx m$. This does in fact occur and the corresponding integral in the relativistic theory is finite. (Numerical details can be found in [Hol14].) The relativistic calculation was independently performed by Feynman, Schwinger, and Tomonaga, for which they were awarded the 1965 Nobel Prize in physics.

What is important here is that by use of renormalization, a calculation which contained a divergence was made finite via identification of the proper physical quantity m_{phys}, and the use of this procedure is just as important relativistically. Consider, for example, QED. Evaluating the self-energy diagram shown in Fig. 5.1, wherein a particle of mass m and charge e emits and absorbs a virtual photon, we find the result

$$-i\Sigma(p) \equiv (-ie)^2 \int \frac{d^4k}{(2\pi)^4}\frac{-i\eta^{\mu\nu}}{k^2 + i\epsilon}\gamma_\mu\frac{i}{\not{p}- \not{k} - m + i\epsilon}\gamma_\nu$$

$$= (-ie)^2 \int \frac{d^4k}{(2\pi)^4}\frac{\eta^{\mu\nu}}{k^2 + i\epsilon}\frac{\gamma_\mu(\not{p}- \not{k} + m)\gamma_\nu}{k^2 - 2p\cdot k + p^2 - m^2 + i\epsilon}. \tag{5.13}$$

From the fact that $\Sigma(p)$ must be a Lorentz as well as a Dirac scalar, it is clear that we must be able to represent

$$\Sigma(p) = A(p^2) + B(p^2)(\not{p} - m) . \tag{5.14}$$

For electrons which are near the mass shell – $p^2 \approx m^2$ – we can write

$$\Sigma(p) = A(m^2) + \left(B(m^2) + 2mA'(m^2)\right)(\not{p} - m) + \cdots \tag{5.15}$$

where we have used

$$p^2 - m^2 = (\not{p} + m)(\not{p} - m) \approx 2m(\not{p} - m) . \tag{5.16}$$

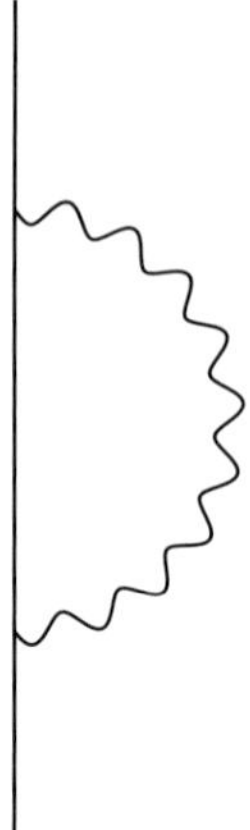

Fig. 5.1 The self-energy diagram. Here the solid line represents an electron and the wiggly line designates a photon.

In order to begin this calculation, it is useful to examine the radiative correction to the electron propagator which, to lowest order in e, is given by the simple expression

$$iS_F^{(0)}(p) = \frac{i}{\not{p} - m + i\epsilon} \tag{5.17}$$

and is represented by a straight line in a Feynman diagram. On the other hand, due to the effects of the electromagnetic interactions, to first order in e^2 we have

$$iS_F(p) = iS_F^{(0)}(p) + iS_F^{(0)}(p) \times -i\Sigma(p) \times iS_F^{(0)}(p) + \cdots . \tag{5.18}$$

To order α then

$$iS_F(p) = \frac{i}{\not{p} - m - \Sigma(p)} \cong \frac{i}{\not{p} - m - A(m^2) - (\not{p} - m)\left(B(m^2) + 2mA'(m^2)\right)}$$

$$\approx \frac{i}{\left(1 - B(m^2) - 2mA'(m^2)\right)\left(\not{p} - m - A(m^2)\right)} \approx \frac{iZ_2}{\not{p} - m_{\text{phys}}} , \tag{5.19}$$

where we have defined

$$m_{\text{phys}} = m + A(m^2) \equiv m + \delta m$$

$$Z_2 \cong 1 + B(m^2) + 2mA'(m^2) \equiv 1 + \tilde{B}(m^2) . \tag{5.20}$$

Since the propagator has a pole at the physical mass of the particle, we see that $A(m^2)$ performs the role of a mass shift due to interaction with the electromagnetic field. The factor Z_2 is known as the wave function renormalization constant for the electron and represents the probability to find the "bare" electron state (i.e., an electron unencumbered by electromagnetic effects) in the physical electron wavefunction. This effect is a familiar one from ordinary time independent perturbation theory, where we represent the normalized eigenstate of the full Hamiltonian $H - |\psi_n> -$ in terms of "bare" eigenstates of the free Hamiltonian $H_0 - |\phi_n> -$ as

$$|\psi_n> = \sqrt{Z_n}\,|\phi_n> + \sum_{m\neq n} \frac{<\phi_m\left|\hat{V}\right|\phi_n>}{E_n^{(0)} - E_m^{(0)}}|\phi_m> + \cdots. \qquad (5.21)$$

Obviously,

$$Z_n = 1 - \sum_{m\neq n} \frac{\left|<\phi_m|\hat{V}|\phi_n>\right|^2}{\left(E_n^{(0)} - E_m^{(0)}\right)^2} + \cdots \qquad (5.22)$$

represents the "wave function renormalization" – the probability that the unperturbed eigenstate $|\phi_n>$ is to be found in the corresponding full eigenstate $|\psi_n>$ –

$$Z_n = |<\phi_n|\psi_n>|^2. \qquad (5.23)$$

So far, so good. However, we encounter problems when we attempt to actually calculate $A(p^2)$, $B(p^2)$ – they are divergent! This is easily seen from Eq. (5.13), which in the large k regime behaves as

$$\Sigma(p) \sim e^2 \int^\Lambda \frac{d^4k}{k^4} m \sim \alpha m \ln\left(\frac{\Lambda}{m}\right). \qquad (5.24)$$

Note: One might naively have expected a linear divergence. However,

$$\int \frac{d^4k}{k^4} k_\mu = 0 \qquad (5.25)$$

because the integrand is an odd function of k. Both the mass shift and wave function renormalization effects diverge logarithmically.

This result should not be unexpected. Indeed the corresponding nonrelativistic calculation discussed above involves a *linear* divergence! The logarithmic relativistic form is much more manageable. Even if the cutoff Λ were as large as 1 TeV, the self-energy correction would represent only a small fraction – $<<<10\%$ – of the electron mass. Still a divergence *is* a divergence and appears to be unsatisfactory.

Renormalization comes to our rescue. As we shall see, such divergences do *not* enter into physically measurable predictions of the theory and cancel against one another when expressed in terms of experimentally accessible quantities such as m_{phys} instead of mathematical artifacts such as m which appear in the zeroth order Lagrangian. Nevertheless, the use of some sort "regularization" procedure for divergent integrals is a useful intermediate step, since subtracting infinities from one another is a notoriously dangerous procedure. The most common is to use the method of 't Hooft and Veltman, involving working in a space with other than four dimensions. This technique is called *dimensional regularization*, wherein we perform the integration in not four but rather d dimensions, where $d < 4$. Defining the quantity $\epsilon = \frac{1}{2}(4 - d)$ the integration can be done using the results given in Appendix A, and divergences appear in the form

$$\frac{2}{\epsilon} \qquad \text{where} \qquad \epsilon = 4 - d \qquad (5.26)$$

occur. While the presence of such poles might seem strange, since certainly we live in a universe where d is identically four, there is no real problem. These poles are merely a calculational artifact. When all diagrams are added together any ϵ^{-1} dependence disappears and the limit $\epsilon \to 0$ can be taken.

Now let us see how such regularization is actually carried out. We begin by simplifying the form of the self-energy $\Sigma(p)$, by use of the identify

$$\gamma_\mu \, \slashed{A} \gamma^\mu = \gamma_\mu \left(\{ \slashed{A}, \gamma^\mu \} - \gamma^\mu \slashed{A} \right)$$
$$= 2\gamma_\mu A^\mu - \left(\gamma_\mu \gamma^\mu \right) \slashed{A} \ . \tag{5.27}$$

Since $\gamma_\mu \gamma^\mu = d$, where d is the number of dimensions, we find

$$\gamma_\mu \, \slashed{A} \gamma^\mu = (2 - d) \slashed{A} = (-2 + 2\epsilon) \slashed{A} \ . \tag{5.28}$$

Then we may evaluate the self-energy in the dimensional regularization scheme

$$\Sigma(p) = \frac{e^2}{(4\pi)^{d/2}} \left(\mu^2 \right)^{2 - \frac{d}{2}} \frac{\Gamma(2 - \frac{d}{2})}{\Gamma(1)\Gamma(1)} \int_0^1 dy \left[ym^2 + (1-y)\lambda^2 - p^2 y(1-y) \right]^{\frac{d}{2} - 2}$$
$$\times \left(4m - 2 \, \slashed{p}(1-y) + \frac{\epsilon}{2} \left(\slashed{p}(1-y) - m \right) \right) . \tag{5.29}$$

Here, as mentioned above, μ is an arbitrary (but necessary) parameter having dimensions of mass inserted in order that the coupling e^2 remain dimensionless when we make the change from four to d dimensions. Physical results must be independent of μ, but *some* value of μ must be chosen in order to *define* the theory.

Writing

$$\Gamma\left(2 - \frac{d}{2} \right) = \Gamma\left(\frac{\epsilon}{2} \right) = \frac{2}{\epsilon} \Gamma\left(1 + \frac{\epsilon}{2} \right) = \frac{2}{\epsilon} - \gamma + \mathcal{O}(\epsilon) \tag{5.30}$$

where $\gamma = -0.574216\ldots$ is Euler's constant, and using the identity

$$a^\epsilon = 1 + \epsilon \ln a + \mathcal{O}(\epsilon^2) \ , \tag{5.31}$$

we determine

$$\Sigma(p) = \frac{e^2}{(4\pi)^2} \left[(3m - (\slashed{p} - m)) \left(\frac{2}{\epsilon} - \gamma + \ln 4\pi + \ln \mu^2 \right) + \slashed{p} - 2m \right.$$
$$- \int_0^1 dy \, (2m(1+y) - 2(\slashed{p} - m)(1-y))$$
$$\left. \log \left(m^2 y + (1-y)\lambda^2 - y(1-y)p^2 \right) \right] \tag{5.32}$$

i.e.,

$$A(m^2) = m\frac{e^2}{(4\pi)^2} \left\{ \frac{12}{\epsilon} - 1 - 3\gamma - 2 \int_0^1 dy(1+y) \ln \frac{y^2 m^2}{4\pi \mu^2} \right\}$$
$$= m\frac{\alpha}{4\pi} \left\{ \frac{6}{\epsilon} - 3\gamma - 3 \ln \frac{m^2}{4\pi \mu^2} + 4 \right\}$$

$$\tilde{B}(m^2) = \frac{e^2}{(4\pi)^2}\left\{-\frac{2}{\epsilon} + 1 + \gamma + 2\int_0^1 dy(1-y)\ln\frac{y^2 m^2}{4\pi\mu^2}\right.$$
$$\left. + 4m^2\int_0^1 dy\frac{y(1-y^2)}{m^2 y^2 + (1-y)\lambda^2}\right\}$$
$$= -\frac{\alpha}{4\pi}\left\{\frac{2}{\epsilon} - \gamma - \ln\frac{m^2}{4\pi\mu^2} - 2\ln\frac{m^2}{\lambda^2} + 4\right\} \quad . \tag{5.33}$$

Both δm and Z_2 are divergent as $\epsilon \to 0$ and cannot contribute to physical processes. The mass correction can be eliminated in the same fashion as in the nonrelativistic case by rewriting the Dirac equation in terms of $m_{\text{phys}} = m + \delta m$

$$\left(i\,\slashed{\nabla} - m_{\text{phys}}\right)\psi(x) = (e\,\slashed{A}(x) - \delta m)\,\psi(x) \quad . \tag{5.34}$$

We observe that δm must be employed as a perturbation to the electron propagator to $\mathcal{O}(\alpha)$, as shown graphically in Fig. 5.2 where the cross indicates the perturbation introduced by δm.

Mathematically we have

$$iS_F(p) = iS_F^{(0)}(p) + iS_F^{(0)}(p) \times -i\Sigma(p) \times iS_F^{(0)}(p) + iS_F^{(0)}(p) \times i\delta m \times iS_F^{(0)}(p) + \cdots$$
$$= \frac{i}{\slashed{k}p - m_{\text{phys}} - \Sigma(p) + \delta m} + \mathcal{O}(\alpha^2) \approx \frac{iZ_2}{\slashed{p} - m_{\text{phys}}} \tag{5.35}$$

so that the divergent mass shift correction has been eliminated. The divergent wavefunction renormalization Z_2 remains and can be shown to be canceled by the vertex diagram shown in Fig. 5.3. Details can be found in any field theory book, such as [Bjo64].

There remains one term contributing to the $\mathcal{O}(\alpha)$ corrections to the electron–photon vertex which we have not considered. As shown in Fig. 5.4, this diagram accounts for the feature that the photon can dissociate into a virtual electron–positron pair. This virtual dissociation process is called "vacuum polarization" and is best described in terms of its modification of the photon propagator, as indicated in Fig. 5.5.

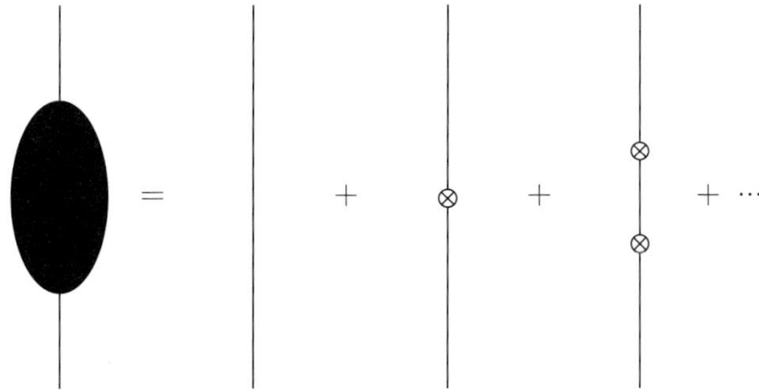

Fig. 5.2 The blob represents the electron propagator as modified by the mass correction δm, which is represented by the circled x.

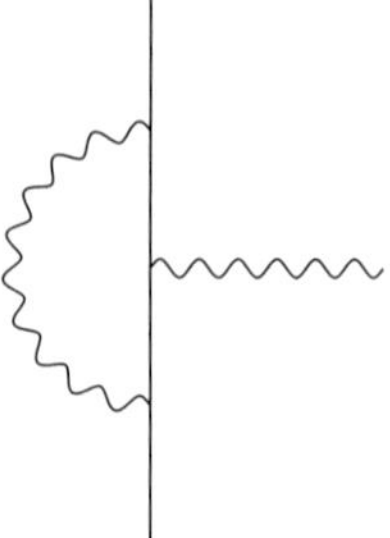

The vertex diagram, in which a photon, indicated by the wiggly line, is exchanged around a photon vertex. Here the solid line represents an electron.

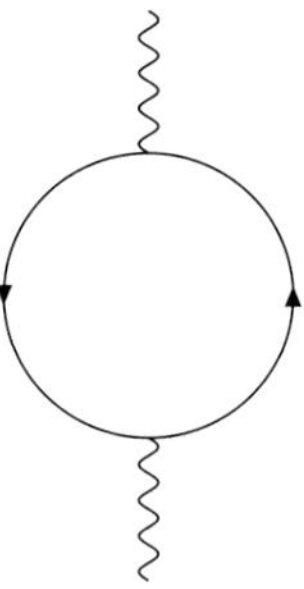

Shown is the vacuum polarization diagram, wherein a photon dissociates into an electron–positron pair and then recombines.

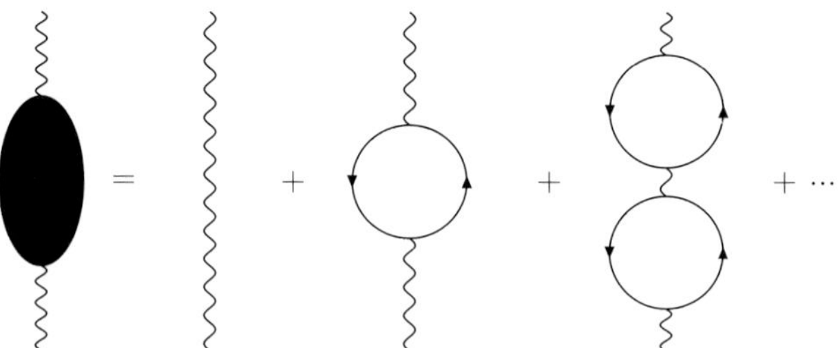

The photon propagator, as modified by vacuum polarization. Here the solid (wiggly) line represents an electron (photon).

That is,

$$
\begin{aligned}
iD_F^{\mu\nu}(q) &= -\frac{i\eta^{\mu\nu}}{q^2 + i\epsilon} - \frac{i\eta^{\mu\rho}}{q^2 + i\epsilon} \times (-)(-ie)^2 \int \frac{d^4k}{(2\pi)^4} \left(\gamma_\rho\right)_{ai} \left(\frac{i}{\not{k} - \not{q} - m + i\epsilon}\right)_{ij} \\
&\quad \times (\gamma_\lambda)_{j\ell} \left(\frac{i}{\not{k} - m + i\epsilon}\right)_{\ell a} \cdot -\frac{i\eta^{\lambda\nu}}{q^2 + i\epsilon} + \cdots \\
&= -\frac{i\eta^{\mu\nu}}{q^2 + i\epsilon} + \left(-\frac{i\eta^{\mu\rho}}{q^2 + i\epsilon} \times i\Pi^{\lambda\rho}(q) \times -\frac{i\eta^{\lambda\nu}}{q^2 + i\epsilon}\right) + \cdots
\end{aligned}
\tag{5.36}
$$

where

$$i\Pi_{\lambda\rho}(1) = -(-ie)^2 \int \frac{d^4k}{(2\pi)^4} \mathrm{Tr}\gamma_\rho \frac{i}{\not{k}-\not{q}-m+i\epsilon}\gamma_\lambda \frac{i}{\not{k}-m+i\epsilon} \tag{5.37}$$

is called the vacuum polarization tensor. By simple Lorentz invariance, the structure of $\Pi_{\lambda\rho}(q)$ must be

$$\Pi_{\lambda\rho}(q) = -\eta_{\lambda\rho}\left(K + q^2\Pi_1(q^2)\right) + q_\lambda q_\rho \Pi_2(q^2) \ . \tag{5.38}$$

However, something is clearly wrong here. If we temporarily keep only the term in K and iterate the vacuum polarization terms we would find an effective photon propagator of the form

$$\begin{aligned}
iD_F^{\mu\nu}(q) &\cong -i\frac{\eta^{\mu\nu}}{q^2+i\epsilon} + \left(-i\frac{\eta^{\mu\rho}}{q^2+i\epsilon}\right) \times (i\eta_{\rho\lambda}K) \times \left(-i\frac{\eta^{\lambda\nu}}{q^2+i\epsilon}\right) + \cdots \\
&= -i\frac{\eta^{\mu\nu}}{q^2+i\epsilon}\left(1 + K\frac{1}{q^2+i\epsilon} + K^2\left(\frac{1}{q^2+i\epsilon}\right)^2 + \cdots\right) \\
&= -i\frac{\eta^{\mu\nu}}{q^2-K+i\epsilon} \ .
\end{aligned} \tag{5.39}$$

$K \neq 0$ then would signify a nonzero photon mass, $m_\gamma^2 = K$, in clear contradiction to experiment. The condition $K = 0$ would seem to be a requirement on the form of the vacuum polarization tensor.

A further stricture may be found by considering Fig. 5.6, which represents the exchange of a virtual photon between the electron current and a current induced in the vacuum due to the presence of the photon field $A^\rho(q)$. Since $\Pi_{\lambda\rho}(q)$ characterizes the proportionality between the induced current and external potential it is called the polarization tensor. However, we know that this amplitude must be invariant under the gauge change

$$A^\rho(q) \to A^\rho(q) + \lambda q^\rho \tag{5.40}$$

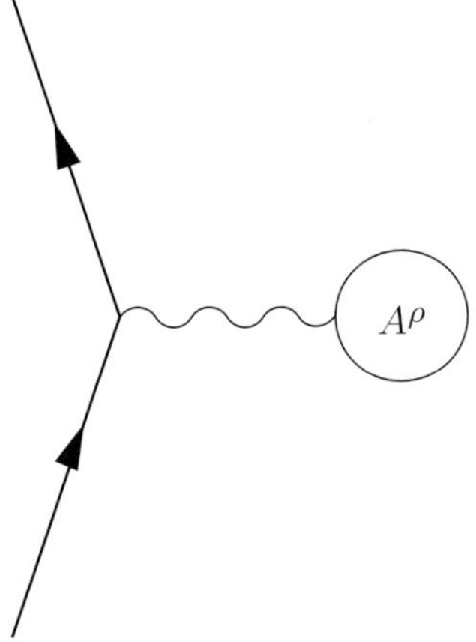

Fig. 5.6 The figure represents photon exchange between an electron, indicated by the solid line, and an external electromagnetic field, indicated by the vector potential A^ρ.

so we require

$$q^\rho \Pi_{\rho\lambda}(q) = 0 \ , \quad \text{i.e.,} \quad \Pi_1(q^2) = \Pi_2(q^2) \ . \tag{5.41}$$

Thus, we may write

$$\Pi_{\rho\lambda}(q) = \Pi(q^2)\left(-\eta_{\rho\lambda}q^2 + q_\rho q_\lambda\right) \tag{5.42}$$

as the most general form of the polarization tensor.

This is as far as one can go with arguments based upon physics alone – the explicit form of $\Pi(q^2)$ must be found from calculation. Using dimensional regularization, we find

$$\Pi_{\lambda\rho}(q) = ie^2\left(\mu^2\right)^{2-\frac{d}{2}}\int\frac{d^dk}{(2\pi)^d}\mathrm{Tr}\gamma_\rho(\not{k}-\not{q}+m)\gamma_\lambda(not k + m)$$
$$\times\frac{1}{\left((k-q)^2 - m^2\right)\left(k^2 - m^2\right)}$$
$$= i4e^2\left(\mu^2\right)^{2-\frac{d}{2}}\int\frac{d^dk}{(2\pi)^d}\frac{1}{\left((k-q)^2 - m^2\right)\left(k^2 - m^2\right)}$$
$$\times\left[(k-q)_\rho k_\lambda + (k-q)_\lambda k_\rho - \eta_{\rho\lambda}\left(k\cdot(k-q) - m^2\right)\right] \ . \tag{5.43}$$

The integration may be performed directly, yielding

$$\Pi_{\lambda\rho}(q) = \left(-q^2\eta_{\rho\lambda} + q_\rho q_\lambda\right)4e^2\left(\mu^2\right)^{2-\frac{d}{2}}\Gamma\left(2-\frac{d}{2}\right)\frac{1}{(4\pi)^{d/2}}$$
$$\times\int_0^1 dy\, 2y(1-y)\left(m^2 - y(1-y)q^2\right)^{\frac{d}{2}-2}$$
$$= \left(-q^2\eta_{\rho\lambda} + q_\rho q_\lambda\right)\frac{\alpha}{3\pi}\left(\frac{2}{\epsilon} - \gamma - \ln\frac{m^2}{4\pi\mu^2} + \frac{q^2}{5m^2} + \cdots\right) \tag{5.44}$$

which clearly is of the proper form.

We now can write the complete form for the radiatively modified electron–photon vertex

$$\mathrm{Amp} = -ieA^\lambda(q)\bar{u}(p_2)\left[\gamma_\lambda\left(1 + \frac{\alpha}{3\pi}\left(\frac{2}{\epsilon} - \gamma - \ln\frac{m^2}{4\pi\mu^2}\right)\right.\right.$$
$$\left.\left. + \mathcal{O}\left(\alpha\frac{q^2}{m^2}\right)\right) + \cdots\right]u(p_1) \tag{5.45}$$

where we have eliminated the vacuum polarization term involving $q_\lambda q_\rho$ since

$$\bar{u}(p_2)\,\not{q}u(p_1) = \bar{u}(p_2)\left(\not{p}_1 - \not{p}_2\right)u(p_1) = 0 \ . \tag{5.46}$$

Although Eq. (5.45) appears unsatisfactory since a divergence has crept back into the scattering amplitude, this is only illusory. Thus consider scattering of our electron probe by a Coulombic "potential"

$$A^\lambda(q) = -\frac{Ze}{q^2}2\pi\,\delta(q^0)\delta^{\lambda 0} \tag{5.47}$$

for which the radiatively corrected scattering amplitude becomes, in the $q \to 0$ limit

$$\text{Amp} \xrightarrow{q \to 0} i\frac{Ze^2}{\boldsymbol{q}^2} 2\pi \delta(q^0) \bar{u}(p_2)\gamma_0 u(p_1) \left(1 - \frac{\alpha}{3\pi}\left(\frac{2}{\epsilon} - \gamma - \ln\frac{m^2}{4\pi\mu^2}\right)\right). \tag{5.48}$$

The solution to our problem is now clear. Just as in the case of mass, where the quantity m which appears in the Dirac equation is *not* the physical or renormalized mass m_{phys}

$$m_{\text{phys}} = m + \delta m \tag{5.49}$$

the same is true of charge! The quantity e which appears in the Dirac equation is *not* the physical or renormalized quantity e_R measured in Coulomb scattering, defined by

$$\text{Amp} \xrightarrow{q \to 0} i\frac{Ze_R^2}{\boldsymbol{q}^2} 2\pi \delta(q^0) \bar{u}(p_2)\gamma^0 u(p_1) \ . \tag{5.50}$$

Instead we have

$$e_R^2 \equiv e^2 \left(1 - \frac{\alpha}{3\pi}\left(\frac{2}{\epsilon} - \gamma - \ln\frac{m^2}{4\pi\mu^2}\right)\right) \tag{5.51}$$

and it is this quantity which is given experimentally by the relation

$$\alpha_R = \frac{1}{4\pi}e_R^2 \approx \frac{1}{137} \ . \tag{5.52}$$

When the scattering amplitude is written in terms of the *renormalized* charge α_R everything is finite – the proper form of the radiatively corrected vertex becomes

$$\text{Amp} = -ie_R A_R^\lambda(q)\bar{u}(p_2)\left[\gamma_\lambda\left(1 + \mathcal{O}\left(\alpha_R\frac{q^2}{m^2}\right) + \cdots\right)\right]u(p_1). \tag{5.53}$$

Before seeing how this vertex can be confronted with an experimental test, it is important to realize that this form has been calculated in the limit $q^2 << m^2$. It is also of interest to examine the case $-q^2 >> m^2$ in which case

$$\text{Amp} \xrightarrow{q^2 >> m^2} -ieA^\lambda(q)\bar{u}(p_2)\gamma_\lambda\left(1 - \frac{\alpha}{3\pi}\left(\frac{2}{\epsilon} - \gamma - \ln\frac{-q^2}{4\pi\mu^2}\right) + \cdots\right)u(p_1). \tag{5.54}$$

The effective vertex retains the same form, but the coupling constant becomes a function of q^2

$$\alpha(q^2) = \alpha\left(1 - \frac{\alpha}{3\pi}\left(\frac{2}{\epsilon} - \gamma - \ln\frac{-q^2}{4\pi\mu^2}\right)\right). \tag{5.55}$$

In fact this relation can be used in order to *define* the renormalized coupling constant. If at some momentum transfer q_1^2 one *measures* a value $\alpha(q_1^2)$

$$\alpha(q_1^2) \equiv \alpha\left(1 - \frac{\alpha}{3\pi}\left(\frac{2}{\epsilon} - \gamma - \ln\frac{-q_1^2}{4\pi\mu^2}\right)\right) \tag{5.56}$$

then the value of the effective coupling at a different momentum transfer q_2^2 can be written as

$$
\begin{aligned}
\alpha(q_2^2) &= \alpha \left(1 - \frac{\alpha}{3\pi} \left(\frac{2}{\epsilon} - \gamma - \ln \frac{-q_2^2}{4\pi\mu^2} \right) \right) \\
&= \alpha(q_1^2) \left(1 + \frac{\alpha}{3\pi} \ln \frac{-q_2^2}{4\pi\mu^2} - \frac{\alpha}{3\pi} \ln \frac{-q_1^2}{4\pi\mu^2} \right) \\
&= \alpha(q_1^2) \left(1 + \frac{\alpha}{3\pi} \ln \frac{q_2^2}{q_1^2} \right) \ .
\end{aligned}
\tag{5.57}
$$

Thus, the effective coupling strength increases as the momentum transfer rises and $\alpha(q^2)$ is often called the "running" coupling constant. This result is consistent with the picture that at high q^2 the scattering electron is able to penetrate the cloud of virtual e^+e^- pairs which surround a given charge and thus get closer to its "bare" value. Often this running is expressed in terms of the QED *beta function*, defined as

$$
\beta(e^2) = \mu^2 \frac{\partial e^2(\mu^2)}{\partial \mu^2} = \frac{e^4}{12\pi^2}
\tag{5.58}
$$

or, equivalently, as

$$
\beta(e) = \mu \frac{\partial e}{\partial \mu} = \frac{e^3}{12\pi^2} \ .
\tag{5.59}
$$

In either case, the beta function is *positive*, which indicates a *stronger* coupling strength with increasing momentum transfer. In order to gauge the physical size of this effect, we compare the value of the fine structure constant at very low energies $-\alpha_R(m_e) \sim 1/137-$ with the corresponding value at the mass of the Z^0 boson

$$
\alpha_R(m_Z) = \alpha_R(m_e) \left[1 + \left(3 \times (-1)^2 + 2 \times \left(\frac{2}{3} \right)^2 + 3 \times \left(\frac{-1}{3} \right)^2 \right) \alpha_R(m_e) 3\pi \ln \frac{m_Z^2}{m_e^2} \right]
$$
$$
\simeq 1/127
\tag{5.60}
$$

where the factor $3 \times (-1)^2 + 2 \times (\frac{2}{3})^2 + 3 \times (\frac{-1}{3})^2$ comes from inclusion of all quark and gluon degrees of freedom in the vacuum polarization, except for the top quark, which is much more massive than the Z^0.

The importance of this change in the fine structure constant is clear too if the running coupling constant relation is written in a full "leading logarithm" expansion [Bjo64]

$$
\alpha(q_2^2) = \frac{\alpha(q_1^2)}{1 - \frac{\alpha(q_1^2)}{3\pi} \ln \frac{q_2^2}{q_1^2}}
\tag{5.61}
$$

which has a singularity at

$$
q_2^2 = m_e^2 \exp \left(\frac{3\pi}{\alpha(m_e^2)} \right) \sim 10^{560} m_e^2 \ .
\tag{5.62}
$$

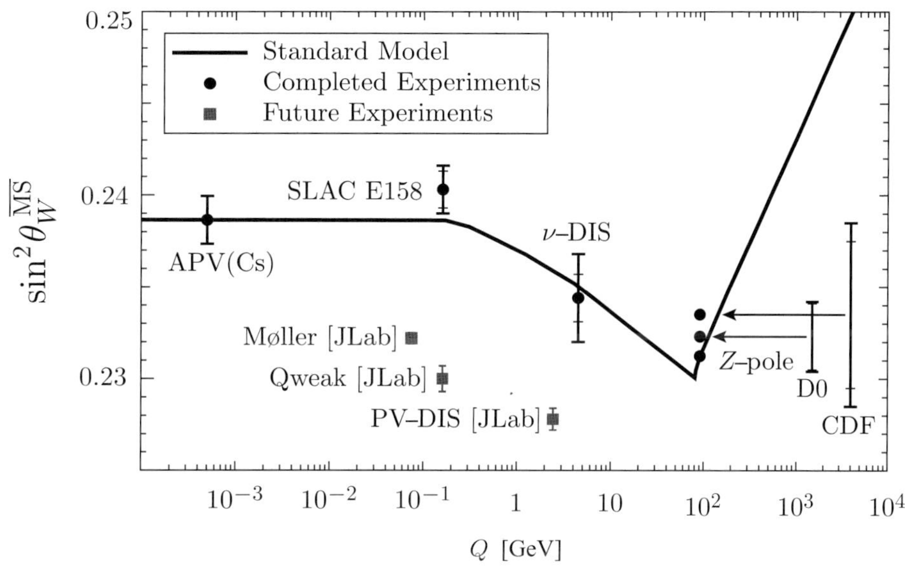

Fig. 5.7   The curve represents the running of $\sin^2\theta_W$ in the $\overline{MS}$ renormalization scheme [Ben10]. The Z-pole point represents the combined results of six LEP and SLC experiments.

which is called the Landau pole. However, such speculation is taking the expression for the effective charge far beyond the simple perturbative regime wherein its validity can be trusted.

Another quantity which displays running is the weak mixing angle $\sin^2\theta_W$, which is not surprising, since its definition is in terms of the ratio of the weak isospin and weak hypercharge coupling constants, both of which run. The running of $\sin^2\theta_W$ has been a focus of experimental programs at various accelerators since verification of the predicted shape, by measuring $\sin^2\theta_W$ at different values of q^2, as shown in Fig. 5.7, provides a strong test for the validity of QED.

The basic lesson of this section is that by writing amplitudes in terms of physical (*renormalized*) quantities, divergences can be removed and the result is completely well defined. It is the feature that the divergences can be put into the form of these quantities appearing in the lowest order Lagrangian which makes a theory such as QED renormalizable. In Chapter 6 we shall see that this is not always the case and will show how to deal with a nonrenormalizable Lagrangian. However, for the present we shall see how another renormalizable theory, quantum chromodynamics, can be developed in analogy to QED.

5.3   Formulation of the QCD Lagrangian

Quantum electrodynamics (QED) and quantum chromodynamics (QCD) can both be derived from such general principles as relativistic invariance and renormalizability plus the principle of local gauge invariance. Thus, QCD is a generalization of QED. Because of their importance, we revisit these theories a bit more deeply than done in Chapter 4 and derive their forms in parallel, following closely the foundational discussion of [Wil82].

In quantum mechanics, the conservation of charge is expressed as the commutation of the charge operator and the time-development operator (Hamiltonian) $[H, Q] = 0$. The unitary transformations $e^{iQ\theta}$ therefore leave the equations of motion unchanged. This is an example of the general connection between symmetries and conservation laws in quantum mechanics (see Chapter 2). The symmetry associated with the conservation of charge is called a gauge symmetry. Applied to a field $\psi(x)$, which creates quanta of charge q (and destroys quanta of charge $-q$), the gauge symmetry acts as a phase:

$$\psi'(x) = e^{iQ\theta}\psi(x)e^{-iQ\theta} = e^{iq\theta}\psi(x) \tag{5.63}$$

or

$$[Q, \psi(x)] = q\psi(x) . \tag{5.64}$$

Gauge symmetry is equivalent to conservation of charge, $dQ/dt = 0$.

To implement color symmetry we let $\psi(x)$ be a three-component vector and generalize the above so that arbitrary unitary transformations in this internal space are allowed. For every unitary transformation Ω in color space we have then a corresponding Hermitian operator ω so that

$$\psi'(x) = e^{i\omega}\psi(x)e^{-i\omega} = \Omega\psi(x) . \tag{5.65}$$

The space of $SU(3)$ unitary transformations in color is generated by eight infinitesimal transformations – a conventional choice of basis is the set $\frac{\lambda^a}{2}, a = 1,\ldots,8$ introduced in Chapter 2. Figure 5.8 shows how quark i can change color to quark j. Then we can write

$$\Omega = e^{ig\frac{\lambda^a}{2}\theta^a} \text{ and } \omega = e^{iT^a\theta^a} . \tag{5.66}$$

We then have $[T^a, \psi(x)] = g\frac{\lambda^a}{2}\psi(x)$. The passage to a local symmetry is made by postulating that θ above in the QED discussion or ω and Ω can depend on the spacetime location. Then the derivatives transform as follows:

QED:

$$\partial_\mu\psi'(x) = e^{iq\theta(x)}[\partial_\mu\psi(x) + iq\partial_\mu\theta(x)\psi(x)] \tag{5.67}$$

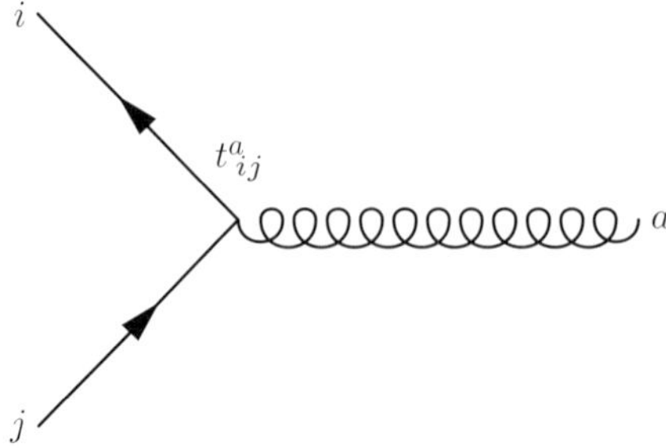

Fig. 5.8 The color of a quark can change from i to j by a gluon of color a coupled through the $SU(3)$ generator $t^a_{ij} = \frac{1}{2}\lambda^a_{ij}$.

QCD:

$$\partial_\mu \psi'(x) = \Omega(x)[\partial_\mu \psi(x) + \Omega^{-1}(x)\partial_\mu \Omega(x)\psi(x)] . \tag{5.68}$$

Derivatives are required to construct a reasonable Lagrangian. Otherwise, the equations of motion will give only constraints, not interesting dynamics. The inhomogeneous transformation law, however, makes it very awkward to construct Lagrangians invariant under the local symmetry. To remedy this problem, the derivative is modified by a correction term so that $D'\psi'$ and ψ' have the same homogeneous transformation law:

QED:

$$D'_\mu \psi'(x) = e^{iq\theta(x)} D_\mu \psi(x) \tag{5.69}$$

$$D_\mu \equiv \partial_\mu + iqA_\mu(x) \tag{5.70}$$

$$A'_\mu(x) = A_\mu(x) + \partial_\mu \theta(x) \tag{5.71}$$

QCD:

$$D'_\mu \psi'(x) = \Omega(x)D_\mu \psi(x)$$

$$D_\mu \equiv \partial_\mu + igB_\mu(x)$$

$$B'_\mu(x) = \Omega(x)B_\mu(x)\Omega^{-1}(x) + \frac{1}{ig}[\partial_\mu \Omega(x)]\Omega(x)^{-1}. \tag{5.72}$$

Where D_μ is called the covariant derivative. In the case of QED, $A_\mu(x)$ is a vector field of real numbers – the usual four-vector potential; in the case of QCD, $B^\mu(x) = B_a^\mu \lambda^a/2$, $a = 1,\ldots,8$ is a vector field of 3×3, traceless, Hermitian matrices. The trace part of $B_\mu(x)$ corresponds to an overall phase transformation of ψ.

We can now construct the invariant kinetic energy for the matter field ψ; if it is a Dirac spinor field then

$$L_{\text{kin}} = \overline{\psi}\, \overleftrightarrow{D}^{\,\mu} \gamma_\mu \psi . \tag{5.73}$$

Here $\overleftrightarrow{D}_\mu \equiv \frac{1}{2}(\overrightarrow{D}_\mu - \overleftarrow{D}_\mu)$, the symmetrized derivative, with $\overrightarrow{D}^\mu = \overrightarrow{\partial}^\mu + igA^\mu$ and $\overleftarrow{D}^\mu = \overleftarrow{\partial}^\mu - igA^\mu$. We are left with the problem of constructing a kinetic energy term for A_μ. In electromagnetism, the field strength $F_{\mu\nu} = \frac{-e}{e}[D_\mu, D_\nu] = \partial_\mu A_\nu - \partial_\nu A_\mu$ and the inhomogeneous term vanishes so that

$$\mathcal{L}_{\text{field}} = -\frac{1}{4}F^{\mu\nu}F_{\mu\nu} \tag{5.74}$$

is a suitable kinetic energy. In QCD a similar construction works. The commutator of two covariant derivatives contains derivatives of B_μ, and is guaranteed to transform homogeneously. We then have

$$[D_\mu, D_\nu]\psi = ig[B_\mu, B_\nu] \tag{5.75}$$

$$G_{\mu\nu} = \partial_\mu B_\nu - \partial_\nu B_\mu + ig[B_\mu, B_\nu] . \tag{5.76}$$

The transformation law for $G'_{\mu\nu}$ can then be written: $G'_{\mu\nu}(x) = \Omega(x)G_{\mu\nu}(x)\Omega(x)^{-1}$ so that the invariant kinetic energy for B_μ is

$$\mathcal{L}_{\text{field}} = -\frac{1}{4}\text{Tr}G^{\mu\nu}G_{\mu\nu}. \tag{5.77}$$

Here $Tr G^{\mu\nu}G_{\mu\nu} \equiv G^{\mu\nu a}G_{\mu\nu a}$, where the sum is over the gluon indices $a = 1,\ldots,8$. We can now construct Lagrangians for the interactions of colored quarks with local color symmetry. In fact, given any Lagrangian with a global color symmetry we need merely change ordinary into covariant derivatives to obtain a locally symmetric Lagrangian. However, the possibilities are much more limited if we insist that the theory be *renormalizable*, as discussed earlier.

In natural units for action and velocity, the action

$$S = \int d^4x \mathcal{L}(x) \tag{5.78}$$

is dimensionless so $\mathcal{L}(x)$ has dimensions of $(\text{mass})^4$. The form of the kinetic energy in Eq. (5.73) therefore indicates that the fermion field ψ has dimensions $(\text{mass})^{\frac{3}{2}}$; the kinetic energies expressed in Eqs. (5.74) and (5.77) indicate that the fields A_μ and B_μ have dimensions of $(\text{mass})^1$; finally the couplings e and g are dimensionless. A term $\mu\bar{\psi}\psi$ in $\mathcal{L}$ appears with a coefficient μ with dimensions $(\text{mass})^1$ (this term represents a mass for the fermion), a term $KTr(G^{\mu\nu}G_{\mu\nu})^2$ would require the dimensions of K to be $(\text{mass})^{-4}$, etc. If we compute any physical process in a power series in K we find the answer in the form

$$a_0 + a_1 K\Lambda^4 + a_2 K^2\Lambda^8 + \cdots , \tag{5.79}$$

where $a_0, a_1, a_2, \ldots$ have a common dimension and where Λ, with dimensions of $(\text{mass})^1$ represents a large momentum or energy scale at which we cut off the integrals. Couplings with dimensions of mass to a negative power, i.e., operators in the Lagrangian with mass dimension greater than four, are called *unrenormalizable*. If we suppose that unrenormalizable couplings do not appear in the Lagrangian, the possible gauge-invariant terms are limited. Let us suppose that we have n flavors, labeled by $j = 1,\ldots,n$ of color triplet spin-1/2 quark fields ψ_j in addition to the gauge field B_μ. Then the allowed terms are as follows:

$$Z\left(-\frac{1}{4}\text{Tr}G^{\mu\nu}G_{\mu\nu}\right) \tag{5.80}$$

$$Z_{jk}^L \bar{\psi}_j (iD^\mu\gamma_\mu)\frac{(1-\gamma_5)}{2}\psi_k = Z_{jk}^L \bar{\psi}_j^L (iD^\mu\gamma_\mu)\psi_k^L \tag{5.81}$$

$$Z_{jk}^R \bar{\psi}_j (iD^\mu\gamma_\mu)\frac{(1+\gamma_5)}{2}\psi_k = Z_{jk}^R \bar{\psi}_j^R (iD^\mu\gamma_\mu)\psi_k^R \tag{5.82}$$

$$M_{jk}\bar{\psi}_j\psi_k = M_{jk}(\bar{\psi}_j^L\psi_k^R + \bar{\psi}_j^R\psi_k^L) \tag{5.83}$$

$$M_{jk}^5\bar{\psi}_j(i\gamma_5)\psi_k = M_{jk}^5(\bar{\psi}_j^L\psi_k^R - \bar{\psi}_j^R\psi_k^L) \tag{5.84}$$

$$\frac{\theta g^2}{32\pi^2}\varepsilon_{\mu\nu\rho\sigma}G^{\mu\nu}G^{\rho\sigma} . \tag{5.85}$$

Equation (5.80) represents the kinetic energy of the gluons; Eqs. (5.81) and (5.82) represent kinetic energies for the left- and right-handed quarks, respectively; Eqs. (5.83) and (5.84) are generalized mass terms for the quarks; and Eq. (5.85) is the θ-term to be discussed below.

The Lagrangian containing terms in Eqs. (5.80) to (5.85) can be brought into a simple canonical form by the following steps:

a) The gauge field B_μ^{old} is replaced by $B_\mu^{new} \equiv Z^{\frac{1}{2}} B_\mu^{old}$, and similarly $g^{new} \equiv Z^{-\frac{1}{2}} g^{old}$. In terms of the new variables, the kinetic energy Eq. (5.80) appears with coefficient with $Z = 1$.
b) New left- and right-handed quark fields can be introduced so that Z_{ij}^L and Z_{ij}^R become unit matrices.
c) With the redefinition $\psi_j^{L,new} = U_{jk}\psi_k^{L,old}$, $\psi_j^{R,new} = U_{jk}\psi_k^{R,old}$, M_{jk}^5 can be eliminated and M_{jk} can be diagonalized.

With these redefinitions, the Lagrangian assumes the canonical form

$$\mathcal{L}_{QCD} = -\frac{1}{4}\mathrm{Tr}G^{\mu\nu}G_{\mu\nu} + \sum_{j=1}^{n_f} \overline{\psi}_j(iD_\mu\gamma^\mu - M_j)\psi_j + \theta - \text{term} . \tag{5.86}$$

Here $j_{\max} = n_f = 6$ is the number of quark flavors with mass M_j. With $g = 0$, $\mathcal{L}_{QCD}$ represents a free field theory of spin-1/2 fermions, quarks and vector bosons. On expanding the traceless Hermitian matrix $B_\mu(x)$ in terms of the λ matrices (see Chapter 2)

$$B_\mu(x) = B_\mu^a(x)\frac{\lambda^a}{\sqrt{2}} , \quad a = 1,\ldots,8, \quad \text{and} \quad \frac{1}{2}\mathrm{Tr}\lambda^a\lambda^b = \delta^{ab} , \tag{5.87}$$

we find that the gluon kinetic energy can be expanded in the λ matrices, implying that there are eight independent vector gluon degrees of freedom. To first order in g, we find gluon couplings which mix quarks of different colors, changing each color of quark to the other ones. Six gluons mediate this type of interaction; the other two, like the photon in QED, couple to but do not change colors. In addition, because of the commutator term in $G_{\mu\nu}$ (see Eq. (5.76)), there exist trilinear couplings of the gluons to each other at order g. This self-coupling of the gauge fields, as shown in Fig. 5.9, which is absent for QED, leads to profound dynamical consequences.

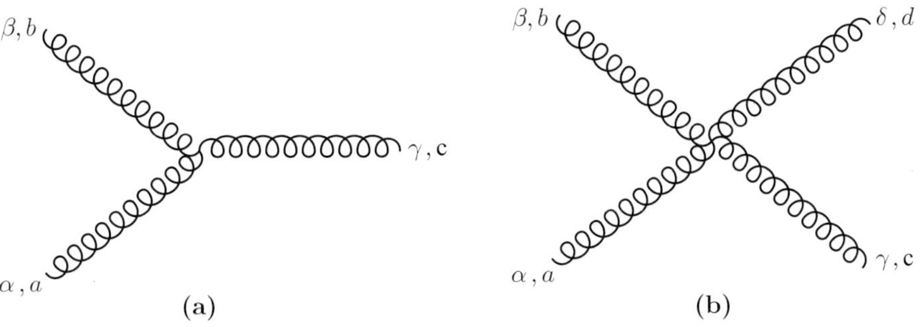

Fig. 5.9   Self-interactions of gluons which are responsible for many important features of QCD, such as asymptotic freedom, color confinement and chiral symmetry breaking which have no QED analog.

Symmetry Properties

The Lagrangian $\mathcal{L}_{\text{QCD}}$ of Eq. (5.86) has many important symmetry properties.

a) *Discrete symmetries P, C, and T*: In the absence of the θ-term, $\mathcal{L}_{\text{QCD}}$ is manifestly invariant under the discrete spacetime symmetries of parity and time reversal, and under charge conjugation. (The θ-term destroys P and T symmetry; see below.)

b) *Flavor conservation*: $\mathcal{L}_{\text{QCD}}$ is invariant under the phase rotations $\psi_j \rightarrow e^{i\theta_j}\psi_j$, which is equivalent to conservation of the additive quantum number that counts the number of quarks of flavor type j. These conservation laws correspond to the conservation of strangeness, baryon number, I_3 the third component of isospin, electric charge, charm, etc., by the strong interaction.

c) *Approximate flavor symmetry*: If quark masses are neglected, $\mathcal{L}_{\text{QCD}}$ is invariant under transformations of the type $\psi_j = U_{jk}\psi_k$ where U_{jk} is a unitary matrix. (cf. Chapter 3). The existence of two very light quarks u and d, and a third fairly light one, s, accounts for the accurate isospin and approximate $SU(3)$ invariance of the strong interaction. It should be noted that the coupling constant g is the same for every kind of quark. This arises because the transformation law in Eq. (5.72) for the gauge potential itself involves g. It would be mathematically inconsistent, therefore, to have different couplings for different quarks. This is to be contrasted with QED where, in principle, A_μ may couple to fields of arbitrary charge. QED in itself neither requires nor explains quantization of charge.

d) *Approximate chiral symmetry*: Ignoring the masses of the u, d, and s quarks, $\mathcal{L}_{\text{QCD}}$ is invariant under separate unitary rotations

$$\psi_j^{L'} = U_{jk}^L \psi_k^L \;\;, \;\; \psi_j^{R'} = U_{jk}^R \psi_k^R \tag{5.88}$$

among left- and right-handed light quarks, generating the group $U(3)_L \times U(3)_R$ of the chiral symmetry group. The hypothesis that these symmetries are spontaneously broken is discussed below.

In summary, there is almost a perfect match between the known symmetries of the strong interaction and the symmetry properties of the Lagrangian $\mathcal{L}_{\text{QCD}}$.

$U(1)$ Problem, Anomalies, θ-term

One apparent, approximate symmetry of $\mathcal{L}_{\text{QCD}}$ is not observed in the strong interactions. This is the chiral $U(1)$ symmetry under which the light left-handed quark fields u, d, and s are multiplied by a common phase $e^{-i\rho}$. This symmetry is associated with approximate conservation of the axial-vector "light baryon number" current

$$j_\mu^5 = \bar{u}\gamma_\mu\gamma_5 u + \bar{d}\gamma_\mu\gamma_5 d + \bar{s}\gamma_\mu\gamma_5 s \tag{5.89}$$

$$\partial^\mu j_\mu^5 \propto \text{light quark masses} . \tag{5.90}$$

This symmetry appears superficially to be of the same general nature as the other chiral symmetries. If this were the case, however, we would predict, in addition to the eight observed light pseudoscalars, a ninth light would-be Goldstone boson, an $SU(3)$ singlet pseudoscalar particle associated with the spontaneous breakdown of axial-vector baryon number symmetry. There is a particle, the η' meson, with the correct quantum numbers, but, with a mass of 958 MeV, it is much too heavy. The absence of the ninth light pseudoscalar is called the $U(1)$ problem.

The reason for the failure of Eq. (5.90) is quite deep. Equation (5.90) is correct at the classical level, but fails quantum mechanically, a phenomenon which is called anomalous symmetry breaking. Perhaps the most direct way to see the reason for the failure is to consider the amplitude for the axial-vector current j_μ^5 to produce two gluons by the triangle graph. This amplitude is linearly divergent at large internal momentum and this infinity must be regulated. The regularization procedure preserves P, C, T, Lorentz, gauge, and flavor invariances, but the regularized axial-vector current is not conserved:

$$\partial^\mu j_\mu^5 = \frac{g^2}{32\pi^2} \varepsilon_{\mu\nu\rho\sigma} \mathrm{tr} G^{\mu\nu} G^{\rho\sigma} + \text{(quark mass terms)} . \tag{5.91}$$

It can be argued that the anomaly Eq. (5.91) holds independently of the details of the regularization procedure and demonstrates that the unwanted extra $U(1)$ symmetry is not present in QCD.

A similar anomaly is crucial to the analysis of $\pi^0 \to \gamma\gamma$ decay. Soft-pion theorems allow one to express amplitudes for processes with pions at small momentum in terms of amplitudes involving divergences of the axial-vector currents with the same quantum numbers. The amplitude for $\pi^0 \to \gamma\gamma$ is controlled by the anomalous contribution to the divergence of the isovector axial-vector current A_μ:

$$\partial_\mu (\overline{u}\gamma_\mu\gamma_5 u - \overline{d}\gamma_\mu\gamma_5 d) = \frac{\alpha}{4\pi}\varepsilon_{\mu\nu\rho\sigma} F^{\mu\nu} F^{\rho\sigma} , \tag{5.92}$$

where $F^{\mu\nu}$ is the electromagnetic field strength. The agreement of the calculated rate for $\pi^0 \to \gamma\gamma$ with experiment is a remarkable triumph of quantum field theory. It should be remarked that the color degrees of freedom are crucial here – one gets a factor of three in amplitude because of three different colors circulating in the anomalous triangle graph; without color the calculated rate would be too small by a factor of nine.

The other apparent mismatch between the observed symmetries of the strong interaction and the symmetries of QCD regards the θ-term, which arises from an anomalous symmetry breaking of QCD and violates P and T invariance. This term cannot be neglected, and the observed P and T invariance of the strong interaction, (e.g., the smallness of the neutron electric dipole moment) force us to conclude that $\theta \approx 0$ to high accuracy. Experimentally from neutron EDM limits, $\theta < 10^{-10}$. QCD by itself cannot explain the smallness of the θ-term, which is known as the *strong CP problem*. In 1977, Peccei and Quinn postulated [Pec77] a solution where $\overline{\theta}$ was assumed to be a field. This was accomplished by adding a new global symmetry that becomes spontaneously broken. This results in a new particle, called the *axion*, which results in the value of $\overline{\theta}$ being zero. Non-trivial QCD vacuum effects give a small mass to the axion, which can be viewed as a Nambu–Goldstone boson. Axions are actively being sought in experimental searches.

The interested reader is directed to [Don78] for early discussions of the spontaneous symmetry breaking of the $U(1)$ symmetry involved and the subsequent prediction that the axion may exist (one of the authors of that paper who was instrumental in recognizing the potential for the existence of such a Nambu–Goldstone boson – Peccei – in fact advocated calling it a "higglet"; however, the name axion stuck). One of the themes in this book has been the interplay of particle and nuclear physics, and that paper underlines this theme. For once the particle physics aspects of the problem lead one to studies of how such a particle might interact with standard matter, the procedures of developing currents, multipole operators, etc., just as in our treatments of the electroweak interaction, are to be followed and cross sections for potentially interesting experiments obtained. In the cited reference this was, in fact, done and led to several early attempts to detect axions.

The QCD Coupling Constant

The definition of g involves picking some renormalization scale μ and this scale is at our disposal. The mathematical theory showing how the same theory may be expressed with different values of μ is called the renormalization group equation. It can be shown that $g(\mu)$ becomes small for large μ, so that the large-momentum behavior of amplitudes is calculable and approximates that of a free ($g = 0$) theory. This is the phenomenon called *asymptotic freedom*.

The μ-dependence of the QCD coupling constant g is described by the β-function

$$\beta = \frac{\partial}{\partial \ln \mu} g(\mu) , \tag{5.93}$$

which in perturbation theory, is calculated as

$$\beta(g) = -\frac{1}{16\pi^2}\left(11 - \frac{2}{3}n_f\right)g^3 - \left(\frac{1}{16\pi^2}\right)^2\left(102 - \frac{46}{3}n_f\right)g^5 + \cdots \equiv -b_0 g^3 - b_1 g^5 + \cdots , \tag{5.94}$$

where n_f is the number of quark flavors. For small g and $n_f < 17$, the β function is negative, i.e., from Eq. (5.93) the effective coupling decreases as we increase the momentum scale. Equations (5.94) and (5.93) for the coupling can be solved at large μ to yield

$$\frac{1}{g^2(\mu)} \rightarrow \frac{1}{2b_0}\ln \mu + \frac{b_1}{b_0}\ln \ln \mu \tag{5.95}$$

with corrections that are finite as $\mu \rightarrow \infty$. The effective coupling approaches zero essentially logarithmically with the scale μ for large momenta.

Asymptotic freedom is a rare phenomenon in quantum field theories and is unique to non-Abelian gauge theories. It means that the vacuum antiscreens charge unlike in QED where the effective charge seen at large distances (small momenta) becomes small. Thus, QCD acts like a dielectric medium with $\varepsilon < 1$.

The strong coupling α_s is related to the QCD coupling constant g via $\alpha_s = g^2/4\pi$. Experimentally, the QCD coupling constant is determined independently from measurement of the following processes: Z decay and e^+e^- total rates; deep inelastic scattering (DIS); fragmentation functions; event shapes and jet counting; π decay; lattice gauge theory; heavy quark systems; hadron–hadron scattering.

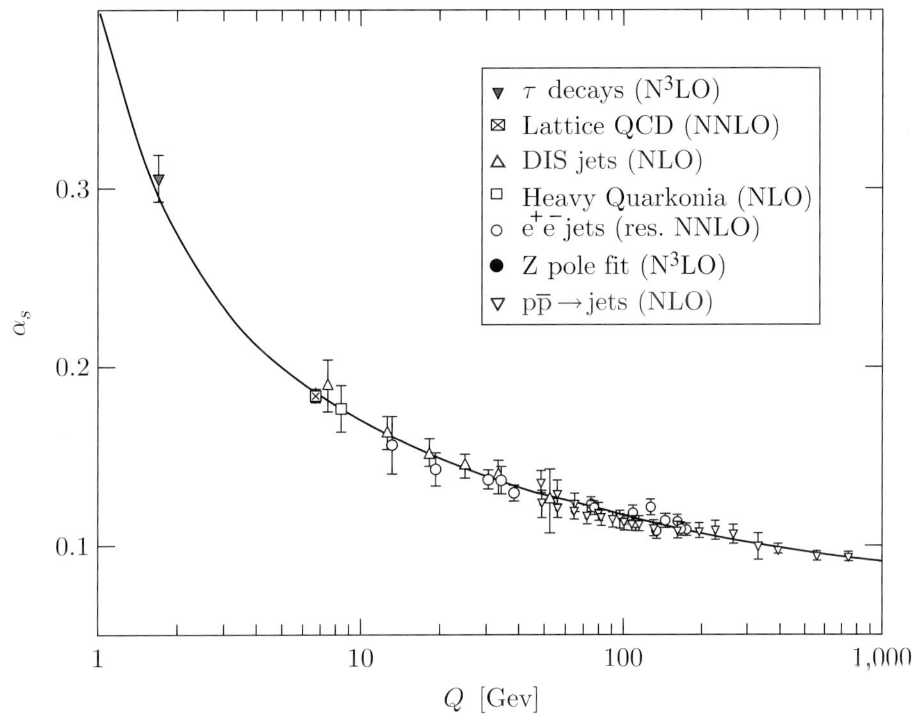

Fig. 5.10   The running of α_s from experiment; figure adapted from [PDG14].

One finds the following expression in leading order for the strong coupling constant $\alpha_s(Q^2)$, see Figs.5.10 and 5.11:

$$\alpha_s(Q^2) = \frac{4\pi}{(11 - 2n_f/3)\ln(\frac{Q^2}{\Lambda^2})}, \tag{5.96}$$

where n_f is the number of quark flavors. The free scale parameter, Λ, discussed in Sec. 5.3.4, follows, e.g., from a measurement of $\alpha_s(Q^2)$ which requires a well-defined description of $\alpha_s(Q^2)$ to obtain a unique value for Λ. At higher orders in perturbation theory, $\alpha_s(Q^2)$ depends on the renormalization procedure used. This sets the scale at which the coupling constant $\alpha_s(Q^2)$ becomes large. For large values of Q^2, α_s decreases logarithmically and perturbative calculations can be performed. At small values of Q^2, where α_s gets large, perturbative calculations can no longer be trusted and one enters the region of nonperturbative QCD.

The QCD Scaling Parameter Λ

The parameter Λ is one of the fundamental quantities of QCD. It sets the scale for the running constant $\alpha_s(\mu)$, and it is the only parameter of the theory in the chiral limit of vanishing quark mass. Usually Λ is defined by writing $\alpha_s(\mu)$ as an expansion in inverse powers of $\ln(\mu^2/\Lambda^2)$. For such a relationship to remain valid for all values of μ, Λ must

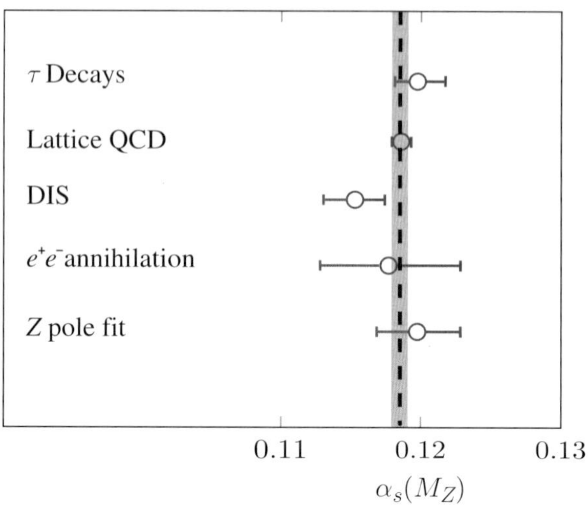

Fig. 5.11 Summary of the determination of $\alpha_s(M_Z)$ at the mass of the Z boson; figure adapted from [PDG14]. Note that lattice QCD provides the most precise determination.

change as flavor thresholds are crossed: $\Lambda \to \Lambda_{n_f}$, where n_f indicates the effective number of light (with respect to the scale μ) quarks. A specific renormalization scheme must be used to calculate Λ.

Λ has been determined [Goc06] by a lattice QCD calculation in the $\overline{MS}$ (modified minimal subtraction) renormalization scheme for $n_f = 2$ flavors

$$\Lambda^{\overline{MS}}_{n_f=2} = 261(17)(26)\,\text{MeV} .$$

Λ sets the scale at which the coupling constant becomes large and the physics becomes nonperturbative, i.e., it sets the scale for strong interaction physics. All dimensionful QCD results without external momentum scale with Λ when quark masses are negligible. For instance, Λ is largely responsible for the mass scale of protons and neutrons, and hence the scale of the baryonic mass in the universe. On the other hand, pure QCD predictions with massless quarks are independent of Λ and hence are pure numbers.

Quark Masses

The quark masses M_j appearing in the QCD Lagrangian are the "current quark" masses. The masses and charges of the six different quark flavors are tabulated in Table 5.1. These current quark masses are determined from current algebra, chiral symmetry breaking effects and QCD sum rules. As described in Chapter 4, the quark masses are generated through a symmetry-breaking phase transition of the electroweak interactions via the Higgs mechanism. Note that the $M_u - M_d$ mass difference is a source of isospin breaking, together with EM effects. The current quark masses are to be distinguished from the constituent quark masses (≈ 300 MeV, see Chapter 3) which include the QCD interactions in the hadron.

Table 5.1 Quark masses in the $\overline{MS}$ renormalization scheme at a scale of $\mu = 2$ GeV

Quark flavor	up	down	strange	charm	bottom	top
Mass (MeV)	1.5–4	4–8	150	1250	4250	175,000
Charge (e)	$\frac{2}{3}$	$-\frac{1}{3}$	$-\frac{1}{3}$	$\frac{2}{3}$	$-\frac{1}{3}$	$\frac{2}{3}$

Quark and Gluon Color

To make calculations of physical processes using QCD, it is necessary to understand how the Feynman rules are applied. Denoting the three colors as *red* (R), *blue* (B), and *green* (G), each external quark wavefunction must have a color vector:

$$c_R = \begin{pmatrix} 1 \\ 0 \\ 0 \end{pmatrix} \quad c_B = \begin{pmatrix} 0 \\ 1 \\ 0 \end{pmatrix} \quad c_G = \begin{pmatrix} 0 \\ 0 \\ 1 \end{pmatrix}.$$

When a gluon interacts with a quark, the quark color can change. This means that a gluon carries one unit of color and one unit of anti-color. From $SU(3)$ color symmetry, there exist eight gluons. Explicitly, $3 \otimes 3 = 8 \oplus 1$ means that we have both a color octet and a color singlet. The color singlet is $|9> = \frac{1}{\sqrt{3}}(|R\overline{R}> +|B\overline{B}> +|G\overline{G}>)$ and is invariant with respect to a rotation in color space. Thus, it has no effect in color space and cannot be exchanged between color charges. The octet states form a basis from which all other color states can be constructed. The way in which these eight states are constructed from colors and anti-colors is a matter of convention. One possible choice is

$$|1> = |R\overline{G}>, |2> = |R\overline{B}>, |3> = |B\overline{B}>,$$

$$|4> = |G\overline{R}>, |5> = |B\overline{R}>, |6> = |B\overline{G}>, \tag{5.97}$$

$$|7> = \frac{1}{\sqrt{2}}(|R\overline{R}> -|G\overline{G}>), \tag{5.98}$$

$$|8> = \frac{1}{\sqrt{6}}(|R\overline{R}> +|G\overline{G}> - 2|B\overline{B}> . \tag{5.99}$$

The gluon wavefunctions will consist of a polarization vector (ϵ^μ) and an 8-component color vector B^α: the basic QCD vertex, as shown in Fig. 5.8, contributes a factor $-\frac{ig_s}{2}\lambda^a\gamma^\mu$.

Each internal gluon connects two vertices of $\lambda^a\gamma^\mu$ and $\lambda^b\gamma^\nu$ so the gluon propagator has the form $\frac{-ig_{\nu\mu}\delta^{ab}}{q^2}$. The quark propagator has the familiar form $\frac{i(q^\mu\gamma_\mu+M)}{(q^2-M^2)}$. The sign of q matters – we take it to be in the same direction as the fermion arrow.

5.4 Lattice QCD

Lattice QCD is a formulation of QCD on a discretized spacetime lattice, which has the important consequence that the lattice theory can be simulated numerically on a computer. At present, this is a well established technique for carrying out nonperturbative calculations of QCD. Lattice QCD is based on the path integral formulation of quantum mechanics. In classical physics, Hamilton's principle states that the path followed by a particle from time t_i to time t_f is that which minimizes the action

$$\delta \int_{t_i}^{t_f} dt L_{\text{classical}}\left(x, \frac{dx}{dt}\right) = 0 \tag{5.100}$$

from which Lagrange's equation of motion can be derived. In quantum mechanics, all possible paths between t_i and t_f must play a role, including those which do not bear any resemblance to the classical path. Feynman's path integral formulation says that

$$< x_f t_f | x_i t_i > = \int_{x_i}^{x_f} D[x(t)] \exp\left[i \int_{t_i}^{t_f} dt \frac{L_{\text{classical}}\left(x, \frac{dx}{dt}\right)}{h}\right], \tag{5.101}$$

where $D[x(t)]$ indicates a sum over all possible paths. This Lagrangian formulation of quantum mechanics can be shown to be completely equivalent to that of Schrödinger's Hamiltonian-based wave mechanics.

In lattice QCD, spacetime is represented not as being continuous, but as a crystalline lattice $L^3 \otimes N_t$ with lattice spacing a. The vertices are connected by lines, while quarks (fermion fields) reside on the vertices and the gluons (gauge fields) travel along the lines. As the spacing between the vertices $a \rightarrow 0$ the theory will approach continuum QCD. The QCD action depends on the number of quark flavors, the quark masses and the gauge coupling. The volume $V = (aL)^3$ and the total time is aN_t.

The lattice spacing depends on the gauge coupling and is set by some physical (zero-kelvin) quantity such as hadron mass. In general, calculations are affected by finite-size and cutoff effects, the latter depending on the particular discretization chosen. The task then is to compute observables for various volumes and lattice schemes and extrapolate to thermodynamic ($V \rightarrow \infty$) and continuum limits ($a \rightarrow 0, N_t \rightarrow \infty$), while keeping aN_T finite.

In practice, calculations are severely limited by computational resources. The evaluation of the fermionic content is the most computationally intensive part. For this reason, the *quenched approximation*, which neglects all quark loop contributions, has often been invoked. In dynamical simulations at finite temperature, the procedures called the "Wilson" and "staggered-fermion" actions are predominantly used because they are computationally cheapest to simulate, although they violate the chiral symmetry of QCD, and another choice with better chiral properties uses "domain-wall" fermions. The lattice formulation of QCD naturally introduces a momentum cut-off of order $1/a$ which regularizes the theory and, as a result, lattice QCD is mathematically well-defined. Lattice QCD calculations often involve analysis at different spacings to determine the lattice spacing dependence, which can then be extrapolated to the continuum.

Important limitations to the lattice technique in achieving more realistic QCD calculations include: a dramatic increase in simulation costs with more realistic light quark masses due to the large correlation lengths of the light states and the use of large lattice volumes to avoid finite-size effects.

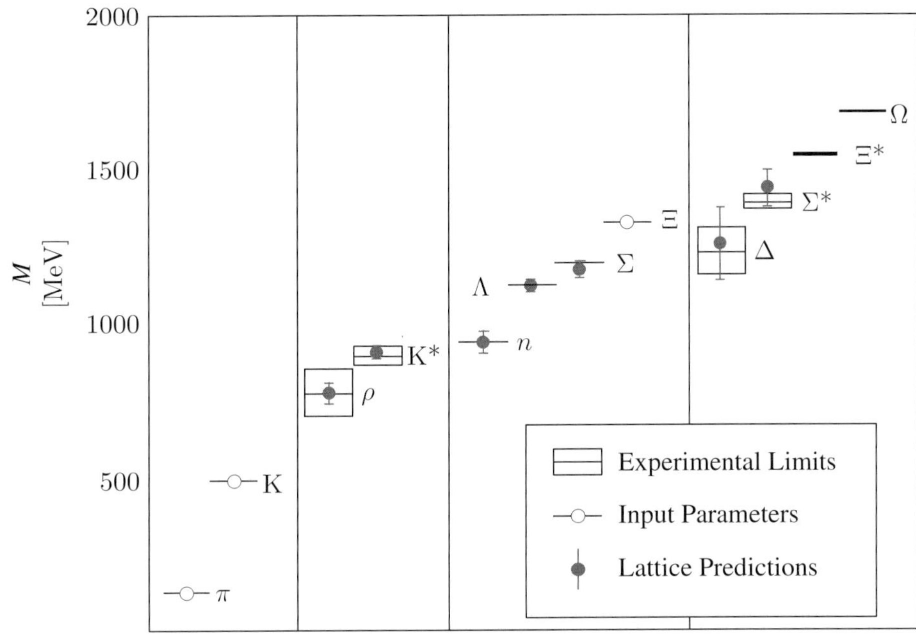

Fig. 5.12 The hadron spectrum from lattice QCD [Fod12]. The boxes denote experimentally measured masses with uncertainties.

Figure 5.12 shows an overview of lattice QCD calculations of the hadron spectrum compared with experimental measurement. The agreement is in general very good. Understanding the thermodynamics of QCD is a subject of great current interest. Experimentally, hot, dense partonic matter is created in the laboratory by colliding relativistic heavy nuclei; this is discussed in detail in Chapter 19. Lattice QCD predicts a phase transition to a deconfined plasma of quarks and gluons at $T \sim 150\,\mathrm{MeV}$. Such temperatures were realized in the early Universe, which during its cooling expansion passed through the plasma phase and the quark-hadron transition on its way to its present state. Hot, dense partonic matter was discovered at RHIC and this has been confirmed by the LHC.

5.5 Nucleon Models

Models have historically played an important role in the development of hadronic physics as stepping stones to more fundamental theories. The spin–orbit interaction discovered by Goeppert-Mayer was the underpinning for the nuclear shell model – see Chapter 13. The Veneziano model for hadron–hadron scattering amplitudes spearheaded the development of string theory. The nonrelativistic quark model and the bag model embody important features of hadron spectra and structure. Model building is a valuable tool for developing intuition about complex systems and for connecting results of *ab initio* computations and experimental data. Here three important models of the nucleon are introduced.

Dyson–Schwinger Equations and the Nucleon Mass

A nonperturbative method based upon a continuum description can be built on the infinite tower of Dyson–Schwinger equations (DSEs). The DSEs are coupled integral equations which relate the Green's functions of a field theory to each other. Solving these equations provides a solution of the theory. For studies based on DSEs it is unavoidable that the infinite tower of coupled equations must be truncated at some point. The omitted functions are approximated, in some way guided by the need to maintain certain properties of the theory, including various local and global symmetries. In addition, there is hope that future lattice gauge theory simulations will be able to provide additional insight into the form of the higher-order functions.

The DSE calculations allow us to understand how the *constituent* light quarks (mass $\approx 300\,\mathrm{MeV}$) of the parton model can be built from the *current* light quarks (mass $\approx 5\,\mathrm{MeV}$) of the fundamental Lagrangian. Figure 5.13 shows the interplay between DSE results and lattice QCD, and confirms that almost all of the constituent quark mass in a proton arises from the gluons that propagate with the quark: the $938\,\mathrm{MeV}$ mass of the proton arises from binding and kinetic energy of massless gluons. One observes the current quark of perturbative QCD evolving into a constituent quark as its momentum becomes smaller. The constituent quark mass arises from a cloud of low-momentum gluons attaching themselves to the current quark. This is an example of *dynamical chiral symmetry breaking*: an

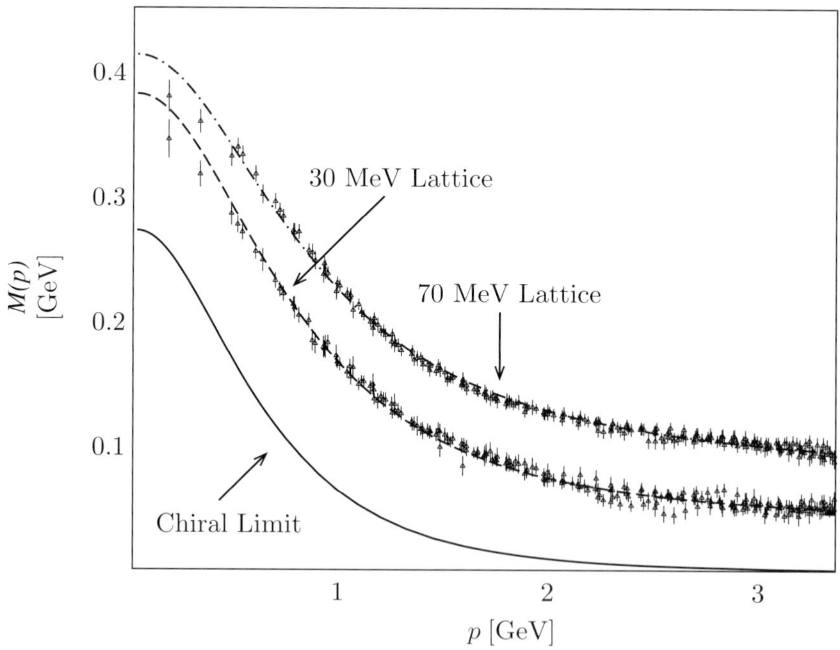

Fig. 5.13   Lattice QCD calculations confirm model predictions (solid curves) that the vast bulk of the constituent mass of a light quark comes from a cloud of gluons that are dragged along by the quark as it propagates; figure adapted from [Hol10].

essentially nonperturbative effect that generates a constituent quark mass from quark–gluon interactions [Hol10].

Ji [Ji95] has carried out a QCD analysis of the mass structure of the nucleon. The nucleon derives its mass of 939 MeV from the quark–gluon dynamics of its underlying structure. We have seen that lattice QCD is successful in reproducing the measured mass from the QCD Lagrangian, but the approach provides little insight on its origins. Ji's analysis provides a description within QCD using a deep inelastic momentum sum rule and the trace anomaly. The energy–momentum tensor of QCD is

$$T^{\mu\nu} = \frac{1}{2}\overline{\psi}\, i \overleftrightarrow{D}{}^{(\mu}\gamma^{\nu)}\psi + \frac{1}{4}g^{\mu\nu}G^2 - G^{\mu\alpha}G_\alpha^\nu\,, \tag{5.102}$$

where ψ is the quark field with color, flavor, and Dirac indices; $G^{\mu\nu}$ is the gluon field strength with color indices and $G^2 \equiv G^{\alpha\beta}G_{\alpha\beta}$. The covariant derivative is defined in the discussion of Eq. (5.73). The symmetrization of the indices μ and ν in the first term is indicated by $(\mu\nu)$. $T^{\mu\nu}$ is a symmetric and conserved tensor whose 00-component defines the Hamiltonian of QCD

$$H_{\text{QCD}} = \int d\mathbf{x}\, T^{00}(0, \mathbf{x})\,, \tag{5.103}$$

which is finite and scale-independent. Further, the matrix element of the tensor operator in the nucleon state is

$$< P|T^{\mu\nu}|P > = P^\mu P^\nu / m_N\,, \tag{5.104}$$

where $|P>$ is the nucleon state with momentum P^μ and m_N is the mass of the nucleon. Finally, the trace of the tensor can be expressed as

$$\text{Tr}T^{\mu\nu} = \frac{1}{4}g^{\mu\nu}\left[(1 + \gamma_m)\overline{\psi}m\psi + \frac{\beta(g)}{2g}F^2\right]\,, \tag{5.105}$$

where m is a quark mass matrix, γ_m is the anomalous dimension of the quark operator, and $\beta(g)$ is the β-function of QCD, discussed above. The second term in the last equation is known as the *trace anomaly* and is generated in the process of renormalization, as discussed below. The mass of the nucleon is expressed as

$$m_N = \frac{< P|\int d^3 x T^{00}(0, \mathbf{x})|P >}{< P|P >} \equiv < T^{00} >\,, \tag{5.106}$$

in the nucleon rest frame.

Ji's analysis produces a separation of the nucleon mass into contributions from the quark kinetic and potential energy, the gluon energy, and the trace anomaly. The largest uncertainty is from the strange quark matrix element $< P|m_s\overline{s}s|P >$. Table 5.2 summarizes the results. The quark kinetic and potential energies contribute in total about one-third of the nucleon mass, with a sizable cancellation in the sum. The quark masses contribute in total about 5% of the nucleon mass. The normal gluon energy is about one-third of the nucleon mass and the trace anomaly contribution is about one-quarter. This allows a determination of the color-electric and color-magnetic fields in the nucleon at $\alpha_s(m_N) \approx 0.4$,

$$< P|\mathbf{E}^2|P > = +1700 \text{ MeV} \tag{5.107}$$

$$< P|\mathbf{B}^2|P > = -1050 \text{ MeV}\,. \tag{5.108}$$

Table 5.2 A separation of the nucleon mass into different contributions at the scale of $\mu^2 = m_N^2$ from [Ji95]

Mass type	$m_s \to 0$	$m_s \to \infty$
Quark energy	270	300
Quark mass	160	110
Gluon energy	320	320
Trace anomaly	190	210

This interesting behavior of the color fields in the presence of quarks may shed light on the structure of the QCD vacuum.

The MIT Bag Model

To understand the long-distance behavior of QCD, various phenomenological models have been proposed. The MIT bag model [DeT83] is an example of such a model. It describes hadrons as extended objects of quarks and gluons immersed in a vacuum that exerts an inward pressure B. The long-distance effects of QCD are represented by this bag pressure, while short-distance effects are treated in QCD perturbation theory confined to a finite region of space. The Lagrangian density for massless quarks is

$$\mathcal{L}_{\text{bag}} = (i\overline{\psi}\gamma^\mu \partial_\mu \psi - B)\theta_V - \frac{1}{2}\overline{\psi}\psi \delta_S , \qquad (5.109)$$

where θ_V is zero outside the spacetime volume occupied by the quark and gluon fields and is unity inside and where δ_S is a δ-function at the bag surface. The constant B is an energy per unit volume required to create a region of space where the vacuum is perturbative – as opposed to the region outside the bag, where there are no free quarks and the vacuum is nonperturbative. The use of the step function θ_V to define the bag surface renders the theory difficult to quantize, but leads naturally to the simple, semi-classical cavity approximation. This makes possible an extensive phenomenological analysis. The MIT bag model embodies some of the essential features of light hadrons, namely that they are bound systems of interacting quarks and gluons moving at relativistic speeds, confined to a region of space with size about 1 fm. The theory also respects color gauge invariance. It amounts to a relativistic shell model for hadronic structure.

We assume the static cavity approximation, i.e., in which the surface of the bag is frozen. Consider the bag with only quarks present. Demanding that the action associated with $\mathcal{L}_{\text{bag}}$ be stationary with respect to variations in the field and the bag surface, S, leads to three equations. The first is the free Dirac equation for a massless quark inside the bag

$$i\gamma^\mu \partial_\mu \psi = 0 .$$

The other two equations are boundary conditions for ψ on S, a linear boundary condition

$$i\gamma^\mu \eta_\mu \psi(x) = \psi(x), \qquad x \text{ on } S ,$$

and a nonlinear boundary condition

$$B = -\frac{1}{2}\eta_\mu \partial^\mu (\overline{\psi}\psi(x)), \ x \text{ on } S ,$$

where η^μ is the covariant normal to the surface. The first two equations describe a free Dirac particle moving in a cavity with a boundary condition that is tantamount to requiring that the normal component of the vector current $J_\mu = \overline{q}\gamma_\mu q$ vanish at the surface. Therefore, the vector current is conserved. The third condition, the nonlinear boundary condition, requires that the outward pressure of the quark field balance the bag pressure B. The cavity approximation freezes the bag volume V at a shape and location that agrees with the nonlinear boundary condition in a time-averaged sense. For static bag configurations, this is equivalent to requiring that the energy be a minimum as a function of shape and size.

The static cavity Hamiltonian is easily derived from the action above:

$$H = \int_V d\mathbf{x} \left[\psi^\dagger (-i\boldsymbol{\alpha}\cdot\nabla)\psi + \psi^\dagger \beta M \psi + B \right] . \tag{5.110}$$

Thus, the bag pressure B is equivalent to a constant energy density everywhere inside the bag. This feature and the confinement of color electric flux enforce quark confinement, since an infinite quark separation requires an infinite bag volume, i.e., an infinite energy.

The quark fields are usually expressed in terms of an expansion in cavity normal modes that satisfy the equations above. For a spherical cavity of radius R, they are easily expressed in terms of spinor spherical harmonics. For example, the lowest zero-mass cavity eigenfunction is a $J^P = \frac{1}{2}^+$ state

$$\psi_s = N \left\{ \begin{array}{c} j_0(\omega r)U \\ i\boldsymbol{\sigma}\cdot\widehat{\mathbf{r}}j_1(\omega r)U \end{array} \right\} e^{-i\omega t} , \tag{5.111}$$

where j_0 and j_1 are spherical Bessel functions, $\omega R = 2.043\ldots$, is the lowest root of the linear boundary condition $j_0(\omega R) = j_1(\omega R)$, U is a two-component spinor, and N is fixed by the normalization condition $\int d\mathbf{x}\psi_s^\dagger \psi_s = 1$. The next lowest orbitals are of odd parity: $p_{\frac{3}{2}}$ at $\omega = \frac{3.204}{R}$ and $p_{\frac{1}{2}}$ at $\omega = \frac{3.812}{R}$ for $M = 0$, respectively. For nonspherical cavities, a variational approach can be used.

Consider a spherical baryon made of n massless quarks or antiquarks in the $s_{\frac{1}{2}}$ orbital. The expectation value of the Hamiltonian on such a state is

$$E = <H> = \frac{2.043n}{R} + \frac{4\pi}{3}BR^3 - \sum_{njlm} \omega_{njlm} . \tag{5.112}$$

The first term is the kinetic energy of n quarks, the second is the bag volume energy, and the third is the energy of the Dirac sea, i.e., the zero-point energy of the fermion modes. The calculation of the zero-point energy is problematic and this is usually approximated as $-\frac{Z_0}{R}$ where Z_0 is an adjustable parameter. Then, minimizing E with respect to R yields

$$R^4 = \frac{(2.043n - Z_0)}{4\pi B} \tag{5.113}$$

$$E = \frac{4}{3}(2.043n - Z_0)^{\frac{3}{4}}(4\pi B)^{\frac{1}{4}} . \tag{5.114}$$

This example illustrates the basic features of spectroscopic calculations in the static cavity approximation. To complete the physical description of the spectrum, it remains to include the level splitting induced by the color interactions among the quarks and to remove from E the contribution due to the kinetic energy of the motion of the center-of-mass.

The following sub-sections describe some bag states.

Strings

The ground-state particles have been assumed to correspond to a spherical bag. However, the bag states of high angular momentum J are likely to deform into rotating tubes with quarks and/or antiquarks at the ends and a flux of color tubes connecting them. Such a structure resembles a *string* with a constant energy per unit length or string tension T_0. The bag string tension is $T_0 = (32\pi\alpha_s B/3)^{\frac{1}{2}}$. With values of α_s and B required for light hadron spectroscopy, one obtains $T_0 = 856$ MeV fm^{-1} in good agreement with experiment.

Heavy quark–antiquark bound states

The spectrum of the J/ψ and Υ particles and their excitations have been successfully explained using an interquark potential being a Coulombic one for short distances and a linear potential for long distances. Using the adiabatic Born–Oppenheimer approximation, the static energy of the $q\bar{q}$ system is calculated for fixed quark sources in the bag model. This energy is then used as the potential in the Schrödinger equation and the short-range Coulombic/long-range linear potential form is obtained.

Glueballs

In addition to bound states of quarks, it is expected that in QCD bound states of gluons, called *glueballs*, should exist. In the bag model, the gluon fields are massless gauge-invariant Lagrangian fields. The lowest-lying gluon modes in a spherical cavity are: $J^P = 1^+$ at $\omega = \frac{2.744}{R}$, $J^P = 2^-$ at $\omega = \frac{3.87}{R}$, and $J^P = 1^-$ at $\omega = \frac{4.49}{R}$. Glueball states can be formed from these cavity modes. The lowest-energy two-gluon states have $J^{PC} = 0^{++}, 2^{++}$, while the first excited states have $J^{PC} = 0^{-+}, 2^{-+}$. The lowest three-gluon states have $J^{PC} = 0^{++}, 1^{+-}, 3^{+-}$.

Exotics

In addition to glueballs, there are other bag states that do not occur in the simplest quark models. One can form color singlet combinations of constituents from $q\bar{q}q\bar{q}$, $q\bar{q}G$, $qqqq\bar{q}$, and $qqqqqq$. Some of them have quantum numbers that do not occur for the combinations of $q\bar{q}$ or qqq. They are called *exotics*.

Recent theoretical progress indicates the existence of a family of exotic particles, *hybrid mesons*, in which the role of the glue may be observed more readily. Theory predicts that the gluons not only hold the quarks together, but can also move collectively and contribute more than just mass. The simplest such motion is a rotation of the glue. The masses of the hybrid mesons are related to the energy in the rotation; thus information on their masses will provide information on the confining gluon field.

The Lee–Friedberg Soliton Model

An alternative approach (*soliton bag model*) is to use an elementary scalar field to define the bag geometry [Fri78]. In the soliton picture, the effective mass of the quark can be very light inside, but extremely heavy outside, thus giving rise to the quark confinement mechanism. It is possible to obtain confined quark states from soliton solutions of an effective Lagrangian which contains quarks and a scalar (σ) field. We now describe this approach.

We assume a phenomenological Lagrangian of the form

$$L = -\frac{1}{4}\kappa(\sigma)F^{a\mu\nu}F_{a\mu\nu}$$

$$+\overline{\psi}(x)\left[\frac{i}{2}\gamma_{\mu}\partial^{\mu} - \frac{1}{2}gA^{a}_{\mu}\lambda^{a}\gamma^{\mu} - M + f\sigma(x)\right]\psi(x) \tag{5.115}$$

$$-\frac{1}{2}\left[\partial_{\mu}\sigma\right]^{2} - U(\sigma), \tag{5.116}$$

where $\sigma(x)$ is a scalar field and $U(\sigma)$ is a phenomenological potential function nonlinear in σ. Although there is no obvious relation to QCD in this Lagrangian, one can think of the σ field as describing the long-range, nonperturbative QCD effects, the nonlinear features of which are being provided by $U(\sigma)$. Compared with the fundamental QCD Lagrangian above, the Lagrangian has the following new features: a dynamical scalar field σ with self-interaction given by $U(\sigma)$; a dielectric constant κ that depends on σ; and a Yukawa coupling with strength f between the scalar field and the quark field. The scalar field self-interaction ordinarily causes it to develop a nonzero vacuum expectation value. However, in the presence of quarks or gluons, a soliton-like or bag-like solution for the classical field equations emerges, i.e., the quark, gluon, and scalar field energy densities are nonzero over a localized region of space. Outside the bag the fermion has a large effective mass determined by the nonzero vacuum-expectation value of the scalar field. Inside the bag the fermion has essentially only its bare mass M. There is also an effective bag pressure coming from the self-interaction of the scalar field. The dielectric constant κ depends on σ in such a way that inside the bag $\kappa = 1$ and outside $\kappa = 0$, leading to confinement of color electric flux. The scalar field is to be regarded as an effective composite field and so is usually treated semi-classically. This approximation is analogous to the static cavity approximation of the MIT bag model.

In conclusion, we have developed the Standard Model theory of the nuclear force in this chapter. Based on a color force between quarks and gluons that cannot be observed directly, the consequences are profound, as we shall discuss in the coming chapters. In particular, we shall see that the most effective theory to describe nuclei is based on hadrons. Connecting the hadrons to the quarks and gluons of QCD is a central thrust of modern subatomic physics research. It is still very much a work in progress.

Exercises

5.1 Quark–Antiquark Interaction in QCD

Consider a quark and antiquark interacting via gluon exchange, as shown in the figure.

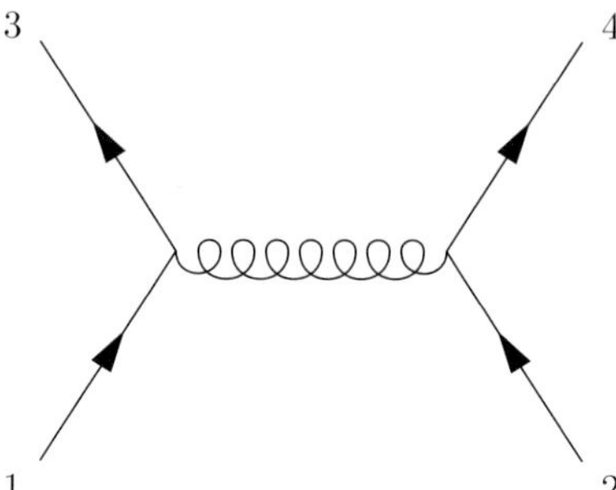

a) Calculate the interaction amplitude $\mathcal{M}$ and show that it has the form

$$\mathcal{M} = -\frac{g_s^2 f_C}{q^2}\, \bar{u}_3 \gamma^\mu u_1 \bar{v}_4 \gamma_\mu v_2 \,,$$

where

$$f_C = \frac{1}{4} c_3^\dagger \lambda^a c_1 c_2^\dagger \lambda^a c_4$$

is a color factor.

Note that the amplitude has the same form as the QED amplitude for $e^+ e^- \to \mu^+ \mu^-$ scattering except for the color factor f_C.

b) Show that if the quark–antiquark are in an octet color configuration $f_C = -\frac{1}{6}$, while if they are in a color singlet state $f_C = \frac{4}{3}$.

c) If the quark–antiquark are close enough to justify the use of perturbation theory show that we have the potential

$$V_{q\bar{q}}(r) = \left\{ \begin{array}{ll} -\frac{4\alpha_s}{3r} & \text{color singlet} \\ +\frac{\alpha_s}{6r} & \text{color triplet} \end{array} \right\} .$$

We see then that the quark–antiquark interaction is attractive (repulsive) in the color singlet (triplet) configuration, which explains why only color singlet mesons are found.

However, this form of the potential is only valid at very short distances. At larger separations it is necessary to take gluon–gluon interactions into account. In QED the interaction between a pair of particles, one with charge $+e$ and one with charge $-e$ has the familiar dipole form, as shown in the left-hand side of the figure overleaf. On the other hand, in QCD the strong gluon–gluon interaction squeezes the lines of force together into a *flux tube* as shown in the right-hand side of the figure overleaf.

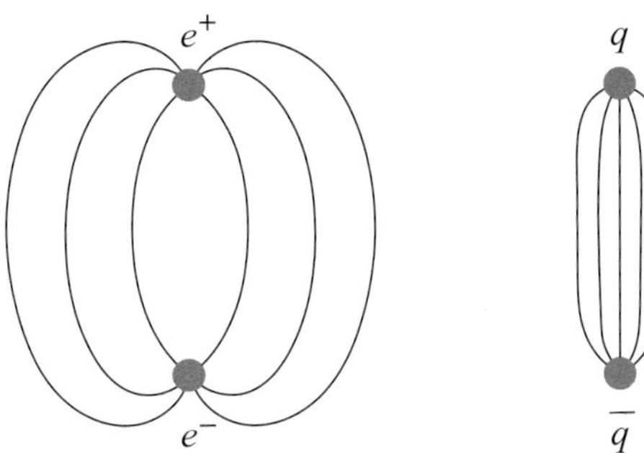

d) Show that the existence of a flux tube implies the quark-antiquark potential of the form

$$V(r) = -\frac{4}{3}\frac{\alpha_s}{r} + \lambda r\,,$$

as shown in the following figure.

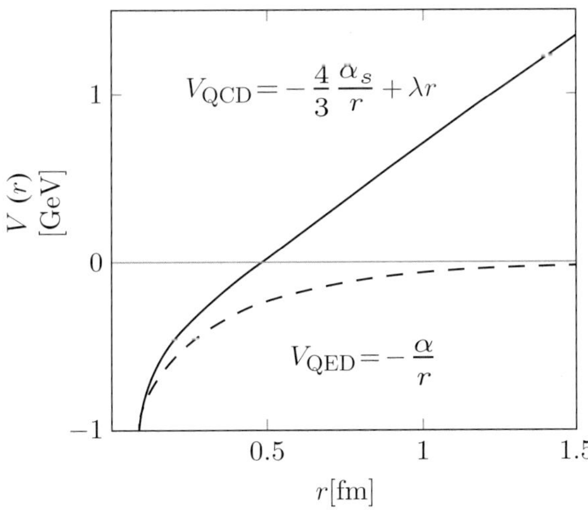

It is this flux tube which leads to confinement. The linear potential implies that it requires an infinite amount of energy to completely separate the quark and antiquark, i.e., they are confined. Rather, as the quark and antiquark separate they produce additional $q\bar{q}$ pairs.

5.2   Anomalous Symmetry Breaking

Anomalous symmetry breaking is said to occur when the process of quantization destroys a symmetry which was present in the classical theory. This phenomenon generally occurs in quantum field theory, where the trace anomaly, chiral anomaly, etc., are well-known manifestations. However, there exist two examples of anomalous symmetry breaking in ordinary quantum mechanics: the two-dimensional delta function potential and the $1/r^2$ potential. In this exercise we explore the former.

Note that this is *not* the familiar delta function potential bound state which is solved in many quantum mechanics courses and is not anomalous. Rather, we deal here with a *two-dimensional* delta function. It is in two dimensions that classical scale invariance is valid and for which a quantum mechanical anomaly arises, as studied below.

Consider a particle of mass m moving in two dimensions under the influence of a potential

$$V(r) = -\lambda \delta^2(r) .$$

The time-independent Schrödinger equation then reads

$$\left(-\frac{1}{2m} \nabla^2 - \lambda \delta^2(r) \right) \psi(r) = E\psi(r) .$$

a) Show that the Schrödinger equation is invariant under a scale transformation

$$r \to r/\zeta \quad \text{and} \quad E \to \zeta^2 E .$$

Such an invariance means that there can exist no bound state, since any solution $\psi_E(r)$ with energy E must have a sister solution $\psi_{\zeta^2 E}(r/\zeta)$, implying that if there exists one negative energy bound state solution then there must exist a continuum of solutions with energies ranging from zero to negative infinity. Furthermore, because the system is not bounded from below, a particle which finds itself in a negative-energy state would cascade wildly to negative infinity, releasing unlimited amounts of energy in the process, so that there can exist no such bound (negative-energy) state. However, this exercise will show that such a state *does* exist. We begin by finding a positive energy (scattering) solution of the Schrödinger equation $\psi_k^+(r)$.

b) Demonstrate that in momentum space the Schrödinger equation becomes

$$\frac{p^2}{2m} \phi_E^+(p) - \lambda \psi_E^+(0) = \frac{k^2}{2m} \phi_E^+(p) ,$$

where

$$\phi_k^+(p) = \int dr\, e^{-ip \cdot r} \psi_k^+(r)$$

is the momentum space wavefunction.

c) Solve this equation and show that

$$\phi_k^+(p) = (2\pi)^2 \delta^2(p - k) + \frac{2m\lambda \psi_k^+(0)}{p^2 - k^2 - i\epsilon} .$$

d) Demonstrate that

$$\psi_k^+(0) = \frac{1}{1 - \lambda G^+(0)} ,$$

where

$$G^+(0) = 2m \int \frac{dp}{(2\pi)^2} \frac{\theta(\Lambda^2 - p^2)}{p^2 - k^2 - i\epsilon} ,$$

and we have been forced to regularize the Green's function by use of a cutoff Λ, since otherwise there would exist a divergence at large p.

e) Solve for the Green's function and show that

$$G^{+}_{\text{reg}}(\mathbf{0}) = \frac{m}{\pi} \log \frac{\Lambda^2}{-k^2} .$$

f) Using

$$2m \int \frac{d\mathbf{p}}{(2\pi)^2} \frac{e^{i\mathbf{p}\cdot\mathbf{r}}}{p^2 - k^2 - i\epsilon} = \frac{im}{2} H^1_0(kr)$$

show that the scattering solution can be written as

$$\psi_{\mathbf{k}}(\mathbf{r}) = e^{i\mathbf{k}\cdot\mathbf{r}} + \left[\frac{1}{2m\lambda} - \frac{1}{4\pi} \log \left(\frac{\Lambda^2}{-k^2} \right) \right]^{-1} \frac{i}{4} H^1_0(kr) .$$

Here $J_0(kr)$, $N_0(kr)$ are the regular, irregular Bessel functions of order zero, while $H^{(1)}_0(kr) = J_0(kr) + iN_0(kr)$ is the related circular Hankel function.

g) Find the scattering amplitude and show that

$$f(\theta) = \sqrt{\frac{1}{2\pi k}} \left[\frac{1}{m\lambda} - \frac{1}{2\pi} \log \frac{\Lambda^2}{k^2} - \frac{i}{2} \right]^{-1} .$$

h) Making the analytic continuation $k \to i\kappa$, show that $f(\theta)$ has a pole at energy

$$E_{b.s.} = -\frac{\Lambda^2}{2m} \exp \left(-\frac{2\pi}{m\lambda} \right) ,$$

which corresponds to a bound state.

The existence of this bound state shows that the scale invariance has been broken – there exists an anomaly. The reason for the anomaly is seen to be the breaking of scale invariance due to introduction of a cutoff in order to define the Green's function, and is necessary in order to regularize the short-distance behavior of the theory.

5.3   Probing Color Experimentally

Consider the process of e^+e^- annihilation to either quark–antiquark pairs of flavor f or $\mu^+\mu^-$ pairs: $e^+ + e^- \to f + \bar{f}$. From the Feynman rules for QED, we expect that the cross section for this process should be proportional to Q^2_f, where Q_f is the charge of the fermion.

a) Show that

$$R \equiv \frac{\sigma(e^+e^- \to \text{hadrons})}{\sigma(e^+e^- \to \mu^+\mu^-)} = N_c \sum_f Q^2_f ,$$

where N_c is the number of colors of quark and the sum over f is over the number of active flavors (see also Section 10.1).

b) Calculate R for (i) $f = u, d, s$, that is, below charm threshold; (ii) for $f = u, d, s, c$, that is, below bottom threshold; and (iii) for $f = u, d, s, c, b$, that is, below top threshold. Compare your results with data from the Particle Data Group [PDG14].

The agreement here between theory and experiment constitutes strong evidence for the existence of color degrees of freedom.

5.4 Color Factors for Quarks

The color factor for quark i–quark j interaction $\rightarrow$ quark k–quark l via single-gluon exchange is

$$\sum_a <i|t_{ij}^a|j><k|t_{kl}^a|l> = \sum_a t_{ij}^a t_{kl}^a = \frac{1}{2}\delta_{il}\delta_{jk} - \frac{1}{6}\delta_{ij}\delta_{kl} \equiv f(ijkl) .$$

The color factor $f(ijkl)$ depends on whether pairs of colors are the same or not.

a) Show that the color factor for quark–quark scattering with quarks of the same color $(i=j=k=l)$ is $f = \frac{1}{3}$.

b) Show that the color factor for quark–quark scattering with $i=j$ and $k=l$ is $f = -\frac{1}{6}$.

c) Show that the color factor for quark–quark scattering with $i=l$ and $j=k$ is $f = \frac{1}{2}$.

d) Show that f is otherwise zero.

5.5 Color Factors C_F, C_A, and T_F for Gluons

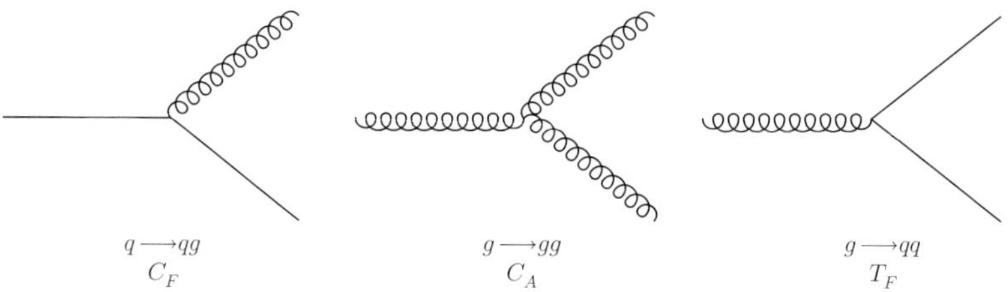

Consider the three gluon processes in the figure.

a) Show that the three color factors are

$$\sum_{a=1}^{8}\sum_{j=1}^{3} t_{ij}^a t_{jk}^a = \delta_{ik} C_F$$

$$\sum_{a,b=1}^{8} f^{abc} f^{abd} = \delta^{cd} C_A$$

$$\mathrm{Tr}(t^a t^b) = \delta^{ab} T_F ,$$

where t^a are the generators of $SU(3)$ ($\frac{1}{2}\lambda^a$, where λ^a are the Gell-Mann matrices) and f^{abc} are the structure constants of $SU(3)$.

b) Show that $C_F = \frac{4}{3}$, $C_A = 3$, and $T_F = \frac{1}{2}$.

The color factor ratios C_A/C_F and T_F/C_F have been determined experimentally in measurements of angular correlations, of event shapes, and of the differences between gluon and quark multi-jet events in e^+e^- collisions. The data are in very good agreement with predictions and directly validate the $SU(3)$ gauge group structure of QCD.

5.6 Positronium and Quarkonium

The energy levels of the hydrogen atom are

$$E_n = -\left[\frac{m_e\alpha^2}{2}\right]\frac{1}{n^2} = -13.6\frac{1}{n^2}\,\mathrm{eV}\,.$$

For positronium, we can change to the reduced mass of the e^+e^- system, i.e., $\mu_r = \frac{1}{2}m_e$ to find the energy levels $E_n = -6.8/n^2$ eV. The lowest energy level of positronium ($n = 1$) is -6.8 eV. The next highest energy level ($n = 2$) is -1.7 eV. Both $n = 1$ and $n = 2$ levels have singlet and triplet level splittings. The lower-energy $1^1\mathrm{S}_0$ (singlet) state and $1^3\mathrm{S}_1$ (triplet) states are split by 8.4×10^{-4} eV. For $n = 2$, there is also a splitting of the $2^1\mathrm{S}_0$ and $2^3\mathrm{S}_1$ states as well as of the $2\mathrm{P}_{0,1,2}$ states.

Take the bound state of the quark–antiquark system where the quarks have the same flavor (quarkonium) and (i) using the positronium level structure described above and (ii) scaling from the Coulomb potential of positronium to the QCD potential of Exercise 5.1, determine the level structure of quarkonium. To a good approximation, this procedure works well for charmonium and bottomonium. Note that toponium does not exist, since the top quark decays through the electroweak interaction before a bound state can form.

Chiral Symmetry and QCD

6.1 Introduction to Chiral Symmetry

Although it is a very successful renormalizable theory of strong, weak, and electromagnetic interactions, the strong interaction sector of the Standard Model cannnot be solved at low energies except by numerical methods such as lattice gauge theory. The reasons for this problem are at least four-fold:

i) The Standard Model is nonlinear, containing triple-gluon and quartic-gluon couplings, unlike QED where there exist no photon self-couplings.

ii) The size of the strong interaction coupling $\alpha_{st} = g_c^2/4\pi$ is order unity at low energies unlike QED where the low-energy fine structure constant $\alpha_{em} = e^2/4\pi$ is $\sim 1/137$ so that perturbative techniques are possible, enabling high precision electromagnetic calculations.

iii) The Standard Model is written in terms of quark and gluon fields, while at low energies the appropriate strong interaction degrees of freedom are hadronic: $\pi, K, \eta, N, \Delta, \Lambda, \cdots$.

iv) A real understanding of the confinement mechanism, whereby the quarks and gluons are replaced by hadrons has not been achieved.

Nevertheless, it *is* possible to generate a rigorous approach to hadronic interactions at low energy by using the methods of effective field theory (EFT). Specifically one can calculate such low energy processes using what has become to be called chiral perturbation theory (χpt) [Gas84], which exploits the spontaneously broken chiral symmetry of QCD. In this chapter we explore this technique, whereby we simply accept confinement as a fact. That is, we deal not with quarks and gluons but rather with confined states, specifically pions and nucleons, and avoid a detailed dynamical discussion of how such particles are constructed. We generate an effective Lagrangian written in terms of these hadronic degrees of freedom, but which accurately encodes the predictions of the Standard Model. Such techniques have become a staple of contemporary nuclear and particle physics and it is the purpose of this chapter to develop how the ways in which χpt can be used to describe the low energy manifestation of QCD.[1]

In order to understand the relevance of spontaneous symmetry breaking within QCD, we must introduce the idea of "chirality," introduced in Chapter 4. We consider by the operators

---

[1] Note that much of this discussion is based on lectures given at the Hadron Physics 2000 meeting [Nav01].

$$\Gamma_{R,L} = \frac{1}{2}(1 \pm \gamma_5) = \frac{1}{2}\begin{pmatrix} 1 & \mp 1 \\ \mp 1 & 1 \end{pmatrix} \tag{6.1}$$

which project left- and right-handed components of the Dirac wavefunction via

$$\psi_L = \Gamma_L \psi \qquad \psi_R = \Gamma_R \psi \quad \text{with} \quad \psi = \psi_L + \psi_R . \tag{6.2}$$

In terms of these chirality states the quark component of the QCD Lagrangian can be written as

$$\bar{q}(i \not{D} - m)q = \bar{q}_L i \not{D} q_L + \bar{q}_R i \not{D} q_R - \bar{q}_L m q_R - \bar{q}_R m q_L. \tag{6.3}$$

The reason that these chirality states are called left- and right-handed can be seen by examining helicity eigenstates of the free Dirac equation.[2] In the high-energy (or massless) limit we note that

$$u(p) = \sqrt{\frac{E+m}{2E}}\begin{pmatrix} \chi \\ \frac{\sigma \cdot p}{E+m}\chi \end{pmatrix} \underset{E \gg m}{\sim} \sqrt{\frac{1}{2}}\begin{pmatrix} \chi \\ \sigma \cdot \hat{p}\chi \end{pmatrix}. \tag{6.4}$$

Left- and right-handed helicity eigenstates then can be identified as

$$u_L(p) \sim \sqrt{\frac{1}{2}}\begin{pmatrix} \chi \\ -\chi \end{pmatrix}, \qquad u_R(p) \sim \sqrt{\frac{1}{2}}\begin{pmatrix} \chi \\ \chi \end{pmatrix}. \tag{6.5}$$

But

$$\Gamma_L u_L = u_L \quad \Gamma_R u_L = 0$$
$$\Gamma_R u_R = u_R \quad \Gamma_L u_R = 0 \tag{6.6}$$

so that in this limit chirality is identical with helicity, i.e., a right-(left-)handed chirality eigenstae is a right-(left-)handed helicty state.

With this background, we now return to QCD and observe that in the limit as $m \to 0$

$$\mathcal{L}_{\text{QCD}} \to \bar{q}_L i \not{D} q_L + \bar{q}_R i \not{D} q_R \tag{6.7}$$

would be invariant under *independent* global left- and right-handed rotations

$$q_L \to \exp\left(i \sum_j \lambda_j \alpha_j\right) q_L, \qquad q_R \to \exp\left(i \sum_j \lambda_j \beta_j\right) q_R \tag{6.8}$$

where $\lambda_i i$ are the eight Gell-Mann matrices introduced in Chapter 2, while α_i, β_i, $i = 1, 2, \ldots, 8$ are 16 arbitrary constants. (Of course, in this massless limit the heavy quark component, involving c, b, t quarks, is also invariant, but since $m_{c,b,t} \gg \Lambda_{\text{QCD}}$, where $\Lambda_{\text{QCD}} \simeq 250\,\text{MeV}$ is the scale defined by the running of the strong gluon coupling constant discussed in Chapter 5, it would be silly to consider this as even an approximate symmetry in the real world.) This invariance is called $SU(3)_L \otimes SU(3)_R$ or chiral $SU(3) \otimes SU(3)$. Continuing to neglect the light quark masses, we see that in a chiral symmetric world

---

[2] The value of the helicity h is found by quantizing along the direction of a particle's three-momentum $\hat{p}$. Then, for a particle of spin J, we have $-J \leq m_J \leq J$ and the helicity is defined to be $h \equiv m_J$.

one would expect to have sixteen – eight left-handed and eight right-handed – conserved Noether currents

$$\bar{q}_L\gamma_\mu\frac{1}{2}\lambda_i q_L\,, \qquad \bar{q}_R\gamma_\mu\frac{1}{2}\lambda_i q_R\,. \tag{6.9}$$

Equivalently, by taking the sum and difference we would have eight conserved vector and eight conserved axial-vector currents

$$V^i_\mu = \bar{q}\gamma_\mu\frac{1}{2}\lambda_i q, \qquad A^i_\mu = \bar{q}\gamma_\mu\gamma_5\frac{1}{2}\lambda_i q\,. \tag{6.10}$$

In the vector case, this is just a simple generalization of isospin ($SU(2)$) invariance to the case of $SU(3)$. There exist *eight* ($3^2 - 1$) time-independent charges

$$Q_i = \int d\mathbf{x} V^i_0(\mathbf{x}, t) \tag{6.11}$$

and there exist various supermultiplets of particles having identical spin-parity and (approximately) the same mass in the configurations, singlet, octet, decuplet, etc. demanded by $SU(3)$ invariance.

If chiral symmetry were realized in the conventional (Wigner–Weyl) fashion one would expect there also to exist corresponding nearly degenerate but *opposite* parity states generated by the action of the time-independent axial-vector charges $Q_i^5 = \int d\mathbf{x} A^i_0(\mathbf{x}, t)$ on these states. Indeed, since

$$H|P\rangle = E_P|P\rangle$$
$$H(Q_5|P\rangle) = Q_5(H|P\rangle) = E_P(Q_5|P\rangle)\,, \tag{6.12}$$

we see that $Q_5|P\rangle$ must also be an eigenstate of the Hamiltonian with the same eigenvalue as $|P >$, which would seem to require the existence of parity doublets. However, experimentally this does not appear to be the case. Indeed, although the $J^\pi = \frac{1}{2}^+$ nucleon has a mass of about 1 GeV, the nearest $\frac{1}{2}^-$ resonance lies nearly 600 MeV higher in energy. Likewise in the case of the 0^- pion which has a mass of about 140 MeV, the nearest corresponding 0^+ state (if it exists at all) is nearly 700 MeV or so higher in energy.

Goldstone's Theorem

One can resolve this apparent paradox by postulating that parity-doubling is avoided because the axial-vector symmetry is *spontaneously broken.* Then according to a theorem due to Goldstone, when a continuous symmetry is broken in this fashion there must also be generated a *massless* boson having the quantum numbers of the broken generator, in this case a pseudoscalar. When an axial-vector charge Q_5 acts on a single-particle eigenstate $|P >$, one does *not* produce a nearly degenerate single-particle eigenstate of opposite parity in return [Gol61], as parity doubling would suggest. Rather, one generates one or more of these massless pseudoscalar bosons

$$Q_5|P\rangle \sim |Pa\rangle + \cdots \tag{6.13}$$

and the interactions of such "Goldstone bosons" to each other and to other particles is found to vanish as the four-momentum involved in any interaction of these particles goes to zero.

In order to see how the corresponding situation develops in QCD, it is useful to study a simple pedagogical example (toy model): a scalar field theory [Bur98]

$$\mathcal{L} = \partial_\mu \phi^* \partial^\mu \phi - V(\phi^* \phi) \quad \text{with} \quad V(x) = \frac{\lambda}{4}\left(x - \frac{\mu^2}{\lambda}\right)^2 , \tag{6.14}$$

which is obviously invariant under the global $U(1)$ (phase) transformation $\phi \to e^{i\alpha}\phi$. The vacuum (lowest-energy) state of the system can be found by minimizing the Hamiltonian density

$$\mathcal{H} = \dot{\phi}^* \dot{\phi} + \boldsymbol{\nabla}\phi^* \cdot \boldsymbol{\nabla}\phi + V(\phi^* \phi) . \tag{6.15}$$

Since $\mathcal{H}$ is the sum of positive definite terms, the vacuum (lowest energy) state is easily found by minimizing the potential energy $V(\phi^* \phi)$ and choosing ϕ to be real, which is permitted by the $U(1)$ symmetry of the Hamiltonian. The resulting vacuum solution is then $\phi_0 = v = \mu/\sqrt{\lambda}$. Of course, once this solution has been chosen, the $U(1)$ symmetry is broken (spontaneous symmetry breaking has taken place) and Goldstone's theorem is applicable.

In order to see how Goldstone's prediction of a massless boson comes about we select the real and imaginary components of ϕ as independent fields, where one has $\rho \equiv \sqrt{2}(\text{Re}\phi - v)$ and $\chi \equiv \sqrt{2}\text{Im}\phi$, in terms of which the Lagrangian density becomes

$$\mathcal{L} = \frac{1}{2}\partial_\alpha \rho \partial^\alpha \rho + \frac{1}{2}\partial_\alpha \chi \partial^\alpha \chi - \frac{1}{2}\mu^2 \rho^2 - \frac{\lambda}{16}(\rho^2 + \chi^2)^2 - \frac{\mu\sqrt{\lambda}}{2\sqrt{2}}\rho(\rho^2 + \chi^2) . \tag{6.16}$$

We observe that the field χ is massless (χ is the Goldstone boson) while the field ρ has a mass μ. The Noether current

$$j_\alpha = -i(\phi^* \partial_\alpha \phi - \partial_\alpha \phi^* \phi) = \sqrt{2}v\partial_\mu \chi + \rho\partial_\alpha \chi - \chi\partial_\alpha \rho \tag{6.17}$$

possesses a nonzero matrix element between χ and the vacuum

$$< \chi(p)|j_\alpha|0 > = \sqrt{2}vp_\alpha e^{-ip\cdot x} \tag{6.18}$$

provided that $v \neq 0$. Also there exist complicated self-interactions as well as mutual interactions between ρ and χ. However, if we calculate the tree-level amplitude for $\rho\chi$ scattering, using the diagrams illustrated in Fig. 6.1 we find

$$\text{Amp}(\rho(q) + \chi(p) \to \rho(q') + \chi(p')) = \frac{\lambda}{2} + \frac{3}{2}\lambda^2 v^2 \frac{1}{(p+p')^2 - \mu^2}$$
$$+ \frac{1}{2}\lambda^2 v^2 \left(\frac{1}{(p+q)^2} + \frac{1}{(p-q')^2}\right), \tag{6.19}$$

and in the soft momentum limit for the Goldstone bosons, $p, p' \to 0$, we find that

$$\lim_{p,p' \to 0} \text{Amp} = \frac{\lambda}{2} - \frac{3}{2}\frac{\lambda^2 v^2}{\mu^2} + \frac{\lambda^2 v^2}{\mu^2} = 0 , \tag{6.20}$$

i.e., the amplitude vanishes, as asserted above.

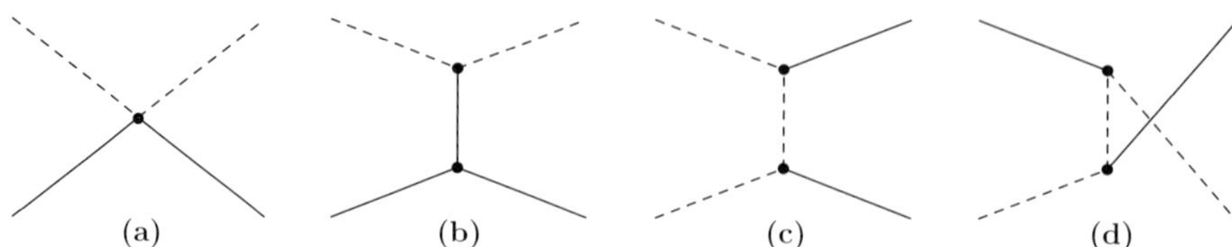

Fig. 6.1   Toy model $\rho\chi$ scattering diagrams. Here ρ (χ) is designated by a solid (dashed) line.

Thus our toy model certainly has all the right stuff, but our representation of the fields is not the optimal one in order to display the Goldstone properties. Instead of using the real and imaginary parts of the field ϕ, it is advantageous to exploit a polar coordinate representation in which the Goldstone boson appears in the guise of a local $U(1)$ transformation – $\phi = \left(v + \sqrt{\tfrac{1}{2}}\xi \right) \exp i\theta/v\sqrt{2}$ – whereby the Lagrangian density assumes the form

$$\mathcal{L} = \frac{1}{2}\partial_\alpha\xi\partial^\alpha\xi + \frac{1}{2}\left(1 + \frac{1}{\sqrt{2}}\frac{\xi}{v}\right)^2 \partial_\alpha\theta\partial^\alpha\theta - \frac{1}{2}\mu^2\xi^2 - \frac{\mu\lambda}{2\sqrt{2}}\xi^3 - \frac{\lambda}{16}\xi^4 . \tag{6.21}$$

We see in this form that θ is the massless Goldstone field, while the field ξ has mass μ. The Noether current

$$j_\mu = \sqrt{2}v\left(1 + \sqrt{\frac{1}{2}}\frac{\xi}{v}\right)^2 \partial_\mu\theta \tag{6.22}$$

clearly has a nonzero vacuum-Goldstone matrix element, $< 0|j_\mu|0 >$, which agrees with Eq. (6.18). What is particularly useful about this representation is the feature that the Goldstone modes couple only through interactions containing derivatives. Since a derivative brings in an energy–momentum, the feature that any such interactions must vanish in the soft-momentum limit is displayed explicitly, making Eq. (6.20) trivial.

Now back to QCD: According to Goldstone's argument, one would expect there to exist eight *massless* pseudoscalar states, one for each spontaneously broken $SU(3)$ axial-vector generator, which would be the Goldstone bosons of QCD. However, no such massless 0^- particles exist. There do exist eight 0^- particles, $\pi^\pm, \pi^0, K^\pm, K^0, \bar{K}^0, \eta$, which are much lighter than their baryonic siblings, cf. Chapter 3. However, these states are certainly not massless and this causes us to ask what has gone wrong with what appears to be rigorous reasoning. The answer is found in the feature that our discussion thus far has neglected quark mass, i.e., the piece

$$\mathcal{L}^m_{\mathrm{QCD}} = -(\bar{u}_L u_R + \bar{u}_R u_L)m_u - (\bar{d}_L d_R + \bar{d}_R d_L)m_d \tag{6.23}$$

of the QCD Lagrangian. Since clearly this term breaks the chiral symmetry,

$$\bar{q}_L q_R \rightarrow \bar{q}_L \exp\left(-i\sum_j \lambda_j\alpha_j\right) \times \exp\left(i\sum_j \lambda_j\beta_j\right) q_R$$

$$\neq \bar{q}_L q_R , \tag{6.24}$$

and we have a violation of the conditions under which Goldstone's theorem applies. The associated pseudoscalar bosons are *not* required to be massless, $m_G^2 \neq 0$, but since their mass arises only from the breaking of the symmetry, the various "would-be" Goldstone masses are expected to be proportional to the symmetry breaking parameters,

$$m_G^2 \propto m_u, m_d, m_s \ .$$

To the extent that such quark masses are small, the eight pseudoscalar particles are not required to be massless, but rather simply much lighter than other hadronic masses in the spectrum.

Effective Chiral Lagrangian

The existence of a set of particles, the pseudoscalar mesons, which are notably less massive than other hadrons suggests the possibility of generating an effective field theory [Man96, Don14] which correctly incorporates the (broken) chiral symmetry of the underlying QCD Lagrangian in describing the low-energy interactions of these would-be Goldstone particles. As found in our pedagogical example, and in Exercise 6.1, this can be formulated in a variety of ways, but the most transparent is done by including the Goldstone modes in terms of the argument of an exponential $U = \exp(i\tau \cdot \phi/v)$ such that under the chiral transformations

$$\begin{aligned}
\psi_L &\to L\psi_L \\
\psi_R &\to R\psi_R
\end{aligned} \tag{6.25}$$

we have

$$U \to LUR^\dagger \ . \tag{6.26}$$

Then a form such as

$$\mathrm{Tr}(\partial^\mu U \partial_\mu U^\dagger) \to \mathrm{Tr}(L\partial^\mu U R^\dagger R \partial_\mu U^\dagger L^\dagger) = \mathrm{Tr}(\partial^\mu U \partial_\mu U^\dagger) \tag{6.27}$$

is invariant under chiral rotations and can be used as part of the effective Lagrangian. However, this form is also not one that we can use in order to realistically describe Goldstone interactions in Nature, since according to Goldstone's theorem, a completely invariant Lagrangian must also have zero pion mass, in contradiction to experiment. We *must* include a term which uses the quark masses to generate chiral symmetry breaking and thereby a nonzero pion mass.

We infer then that the *lowest-order* effective chiral Lagrangian can be written as

$$\mathcal{L}_2 = \frac{v^2}{4}\mathrm{Tr}(\partial_\mu U \partial^\mu U^\dagger) + \frac{m_\pi^2}{4}v^2\mathrm{Tr}(U + U^\dagger) \ , \tag{6.28}$$

where the subscript 2 indicates that we are working at two-derivative order or one power of chiral symmetry breaking, i.e., m_π^2. This Lagrangian is also *unique*, i.e., if we expand to lowest-order in ϕ

$$\mathrm{Tr}(\partial_\mu U \partial^\mu U^\dagger) = \mathrm{Tr}\left(\frac{i}{v}\tau \cdot \partial_\mu \phi \times \frac{-i}{v}\tau \cdot \partial^\mu \phi\right) = \frac{2}{v^2}\partial_\mu \phi \cdot \partial^\mu \phi \ , \tag{6.29}$$

where τ is a Pauli isospin matrix, we reproduce the free pion Lagrangian, as required,

$$\mathcal{L}_2 = \frac{1}{2}\partial_\mu\boldsymbol{\phi}\cdot\partial^\mu\boldsymbol{\phi} - \frac{1}{2}m_\pi^2\boldsymbol{\phi}\cdot\boldsymbol{\phi} + \mathcal{O}(\phi^4) \; . \tag{6.30}$$

At the $SU(3)$ level, including a generalized chiral symmetry breaking term, there is even predictive power: one has

$$\frac{v^2}{4}\mathrm{Tr}\partial_\mu U\partial^\mu U^\dagger = \frac{1}{2}\sum_{j=1}^{8}\partial_\mu\phi_j\partial^\mu\phi_j + \cdots \tag{6.31}$$

and

$$\frac{v^2}{4}\mathrm{Tr}2B_0 m(U + U^\dagger) = \mathrm{const.} - \frac{1}{2}(m_u + m_d)B_0\sum_{j=1}^{3}\phi_j^2$$

$$-\frac{1}{4}(m_u + m_d + 2m_s)B_0\sum_{j=4}^{7}\phi_j^2$$

$$-\frac{1}{6}(m_u + m_d + 4m_s)B_0\phi_8^2 + \cdots , \tag{6.32}$$

where B_0 is a constant and $m = (m_u, m_d, m_s)_{\mathrm{diag}}$ is the quark mass matrix. We can then identify the meson masses as

$$m_\pi^2 = (m_u + m_d)B_0$$
$$m_K^2 = \left(\frac{1}{2}(m_u + m_d) + m_s\right)B_0$$
$$m_\eta^2 = \frac{1}{3}(m_u + m_d + 4m_s)B_0. \tag{6.33}$$

This system of three equations is *overdetermined*, and we find by simple algebra

$$3m_\eta^2 + m_\pi^2 - 4m_K^2 = 0 \; , \tag{6.34}$$

which is known as the Gell-Mann–Okubo mass relation and is well-satisfied experimentally [Gel61].

Currents

Since under a vector (axial vector) transformation $\alpha_L = \pm\alpha_R$ we have

$$U \to LUR^\dagger \begin{cases} \overset{V}{\simeq} U + i\left[\sum_j \alpha_j\lambda_j, U\right] \\ \overset{A}{\simeq} U + i\left\{\sum_j \alpha_j\lambda_j, U\right\} \end{cases} , \tag{6.35}$$

which leads to the vector and axial-vector currents

$$\{V, A\}_\mu^k = -i\frac{v^2}{4}\mathrm{Tr}\lambda^k(U^\dagger\partial_\mu U \pm U\partial_\mu U^\dagger) \; . \tag{6.36}$$

At this point the constant v can be identified by use of the axial-vector current. In $SU(2)$ we find

$$U^\dagger \partial_\mu U - U \partial_\mu U^\dagger = 2i \frac{1}{v} \boldsymbol{\tau} \cdot \partial_\mu \boldsymbol{\phi} + \cdots \tag{6.37}$$

so that

$$A_\mu^k = i \frac{v^2}{4} \mathrm{Tr} \tau^k 2i \frac{1}{v} \boldsymbol{\tau} \cdot \partial_\mu \boldsymbol{\phi} + \cdots = -v \partial_\mu \phi^k + \cdots . \tag{6.38}$$

If we set $k = 1 - i2$, then this represents the axial-vector component of the $\Delta S = 0$ charged weak current and

$$A_\mu^{1-i2} = -v \partial_\mu \phi^{1-i2} = -\sqrt{2} v \partial_\mu \phi^- . \tag{6.39}$$

Comparing with the conventional definition

$$\langle 0 | A_\mu^{1-i2}(0) | \pi^+(p) \rangle = i \sqrt{2} F_\pi p_\mu , \tag{6.40}$$

we find that, to lowest-order in chiral symmetry, $v = F_\pi$, where $F_\pi = 92.2\,\mathrm{MeV}$ is the pion decay constant [Hol90a].

Likewise in $SU(2)$, we note that

$$U^\dagger \partial_\mu U + U \partial_\mu U^\dagger = \frac{2i}{v^2} \boldsymbol{\tau} \cdot \boldsymbol{\phi} \times \partial_\mu \boldsymbol{\phi} + \cdots , \tag{6.41}$$

so that the *vector current* is

$$V_\mu^k = -i \frac{v^2}{4} \mathrm{Tr} \tau^k \frac{2i}{v^2} \boldsymbol{\tau} \cdot \boldsymbol{\phi} \times \partial_\mu \boldsymbol{\phi} + \cdots$$

$$= (\boldsymbol{\phi} \times \partial_\mu \boldsymbol{\phi})^k + \cdots . \tag{6.42}$$

We can identify V_μ^k as the (isovector) electromagnetic current by setting $k = 3$ so that

$$V_\mu^{\mathrm{em}} = \phi^+ \partial_\mu \phi^- - \phi^- \partial_\mu \phi^+ + \cdots . \tag{6.43}$$

Comparing with the conventional definition

$$\langle \pi^+(p_2) | V_\mu^{\mathrm{em}}(0) | \pi^+(p_1) \rangle = F_1(q^2)(p_1 + p_2)_\mu , \tag{6.44}$$

we identify the pion form factor: $F_1(q^2) = 1$. Thus to lowest-order in chiral symmetry the pion has unit charge, but is pointlike and structureless. We shall see below how to insert structure.

$\pi\pi$ Scattering

At two-derivative level we can generate additional predictions by extending our analysis to the case of $\pi\pi$ scattering. Expanding $\mathcal{L}_2$ to order $\boldsymbol{\phi}^4$ we find

$$\mathcal{L}_2 : \phi^4 = \frac{1}{6v^2} \boldsymbol{\phi}^2 \boldsymbol{\phi} \cdot \Box \boldsymbol{\phi} + \frac{1}{2v^2} (\boldsymbol{\phi} \cdot \partial_\mu \boldsymbol{\phi})^2 + \frac{m_\pi^2}{24v^2} \boldsymbol{\phi}^4, \tag{6.45}$$

which yields for the $\pi\pi$ T matrix

$$T(q_a, q_b; q_c, q_d) = \frac{1}{F_\pi^2}\left[\delta^{ab}\delta^{cd}(s - m_\pi^2) + \delta^{ab}\delta^{bd}(t - m_\pi^2) + \delta^{ad}\delta^{bc}(u - m_\pi^2)\right]$$
$$-\frac{1}{3F_\pi^2}(\delta^{ab}\delta^{cd} + \delta^{ac}\delta^{bd} + \delta^{ad}\delta^{bc})(q_a^2 + q_b^2 + q_c^2 + q_d^2 - 4m_\pi^2),$$

$$(6.46)$$

where $s = (q_a + q_b)^2$, $t = (q_a - q_c)^2$ and $u = (q_a - q_d)^2$ are the usual Mandlestam variables. Working on the mass shell, so that $q_a^2 = q_b^2 = q_c^2 = q_d^2 = m_\pi^2$ and defining more generally

$$T_{ab;cd}(s, t, u) \equiv A(s, t, u)\delta_{ab}\delta_{cd} + A(t, s, u)\delta_{ac}\delta_{bd} + A(u, t, s)\delta_{ad}\delta_{bc}, \qquad (6.47)$$

we can write the chiral prediction in terms of the more conventional isospin language by taking appropriate linear combinations [Don14]

$$\begin{aligned}
T^{I=0}(s, t, u) &= 3A(s, t, u) + A(t, s, u) + A(u, t, s)\\
T^{I=1}(s, t, u) &= A(t, s, u) - A(u, t, s)\\
T^{I=2}(s, t, u) &= A(t, s, u) + A(u, t, s).
\end{aligned} \qquad (6.48)$$

Partial-wave amplitudes, projected out via (cf. Appendix B)

$$T_\ell^I(s) = \frac{1}{64\pi}\int_{-1}^{1} d(\cos\theta)P_\ell(\cos\theta)T^I(s, t, u), \qquad (6.49)$$

can be used to identify the associated scattering phase shifts $\delta_\ell^I(s)$ via

$$T_\ell^I(s) = \left(\frac{1}{k(s)}\right)^{\frac{1}{2}} e^{i\delta_\ell^I(s)} \sin\delta_\ell^I(s), \qquad (6.50)$$

where $k(s) = \sqrt{\frac{s - 4m_\pi^2}{s}}$. Then from the lowest-order chiral form

$$A_0(s, t, u) = \frac{s - m_\pi^2}{F_\pi^2} \qquad (6.51)$$

we determine the pion scattering lengths and effective ranges, defined via

$$T_\ell^I(s) = k^{2\ell}(s)\left[a_\ell^I + 2b\frac{k^2(s)}{m_\pi^2} + \mathcal{O}(k^4(s))\right] \qquad (6.52)$$

to be

$$a_0^0 = \frac{7m_\pi^2}{32\pi F_\pi^2}, \qquad a_0^2 = -\frac{m_\pi^2}{16\pi F_\pi^2}, \qquad a_1^1 = \frac{m_\pi^2}{24\pi F_\pi^2},$$

$$b_0^0 = \frac{m_\pi^2}{4\pi F_\pi^2}, \qquad b_0^2 = -\frac{m_\pi^2}{8\pi F_\pi^2}, \qquad (6.53)$$

comparison of which with experimental numbers is shown in Table 6.1.

Table 6.1 The pion scattering lengths and slopes compared with predictions of chiral symmetry

	Experimental	Lowest Order[3]	First Two Orders[3]
a_0^0	0.220 ± 0.005	0.16	0.20
b_0^0	0.250 ± 0.030	0.18	0.26
a_0^2	-0.044 ± 0.001	-0.045	-0.041
b_0^2	-0.082 ± 0.008	-0.089	-0.070
a_1^1	0.038 ± 0.002	0.030	0.036
b_1^1		0	0.043
a_2^0	$(17 \pm 3) \times 10^{-4}$	0	20×10^{-4}
a_2^2	$(1.3 \pm 3) \times 10^{-4}$	0	3.5×10^{-4}

Despite the obvious success of this and other such predictions [Gas69] it is clear that we do not really have at this point a satisfactory theory, since the strictures of unitarity are violated. Indeed, since we are working at tree level, i.e., to leading order, with no loop corrections included, all our amplitudes are real. However, unitarity of the S-matrix requires transition amplitudes to contain an imaginary component, since

$$0 = S^\dagger S - 1 = i\left(<f|T^\dagger|i> - <f|T|i> \right) + <f|T^\dagger T|i>$$

$$\text{so} \quad 2\text{Im}\left(<f|T|i> \right) = \sum_n <f|T^\dagger|n><n|T|i> \neq 0 . \tag{6.54}$$

The solution of such problems with unitarity are well known, i.e., the inclusion of loop corrections to these simple tree-level calculations. Insertion of such loop terms removes the unitarity violations, but comes with a high price in that numerous divergences are introduced and this difficulty prevented progress in this field for nearly a decade until a paper by Weinberg suggested the solution [Wei79]. One can deal with such divergences, just as in QED, by introducing phenomenologically determined counterterms into the Lagrangian in order to absorb the infinities. We see in the next section how this can be accomplished.

6.2 Renormalization

Effective Chiral Lagrangian

We can now apply Weinberg's solution to the effective chiral Lagrangian, Eq. (6.28). As noted earlier, when loop corrections are made to tree level (lowest-order) amplitudes in order to enforce unitarity, divergences inevitably arise. However, there is an important difference from the familiar case of QED discussed in Chapter 5 in that the form of the divergences is *different* from their lower-order counterparts, i.e., the theory is no longer renormalizable. In QED the form of any loop corrections was found to be identical to structures appearing in the original Lagrangian. They could therefore be absorbed and made to disappear by redefining the meaning of the bare couplings, the charge e_0 and mass m_0, via

$$m_{\text{exp}} \equiv m_0 + \delta m_{\text{loop}}, \quad e_{exp} \equiv e_0 + \delta e_{\text{loop}} . \tag{6.55}$$

The reason that the effective chiral Lagrangian is different can be seen from a simple example. Consider $\pi\pi$ scattering. In lowest-order there exists a tree-level contribution from $\mathcal{L}_2$ which is $\mathcal{O}(p^2/F_\pi^2)$, where p represents the energy–momentum of a particle in the process being considered. The fact that p appears to the second power is due to the feature that its origin is the *two*-derivative Lagrangian $\mathcal{L}_2$. Now suppose that $\pi\pi$ scattering is examined at one-loop order. Since the scattering amplitude must still be dimensionless, but now the amplitude involves a factor $1/F_\pi^4$, the numerator must involve *four* powers of energy–momentum. Thus any quantity, often called a *counterterm*, which is introduced into the Lagrangian in order to absorb this divergence must be *four*-derivative in character and cannot be present in the original two-derivative Lagrangian. Gasser and Leutwyler [Gas84] have studied this problem and have written the most general form of such an order-four counterterm in chiral $SU(3)$ as

$$\begin{aligned}
\mathcal{L}_4 = \sum_{i=1}^{10} L_i \mathcal{O}_i = {} & L_1 \left[\text{Tr}(D_\mu U D^\mu U^\dagger) \right] + L_2 \text{Tr}(D_\mu U D_\nu U^\dagger) \cdot \text{Tr}(D^\mu U D^\nu U^\dagger) \\
& + L_3 \text{Tr}(D_\mu U D^\mu U^\dagger D_\nu U D^\nu U^\dagger) + L_4 \text{Tr}(D_\mu U D^\mu U^\dagger) \text{Tr}\left(\chi U^\dagger + U \chi^\dagger \right) \\
& + L_5 \text{Tr}\left(D_\mu U D^\mu U^\dagger \left(\chi U^\dagger + U \chi^\dagger \right) \right) + L_6 \left[\text{Tr}\left(\chi U^\dagger + U \chi^\dagger \right) \right]^2 \\
& + L_7 \left[\text{Tr}\left(\chi^\dagger U - U \chi^\dagger \right) \right]^2 + L_8 \text{Tr}\left(\chi U^\dagger \chi U^\dagger + U \chi^\dagger U \chi^\dagger \right) \\
& + i L_9 \text{Tr}\left(F_{\mu\nu}^L D^\mu U D^\nu U^\dagger + F_{\mu\nu}^R D^\mu U^\dagger D^\nu U \right) + L_{10} \text{Tr}\left(F_{\mu\nu}^L U F^{R\mu\nu} U^\dagger \right),
\end{aligned} \tag{6.56}$$

where the covariant derivative is defined via

$$D_\mu U = \partial_\mu U + \{A_\mu, U\} + [V_\mu, U] , \tag{6.57}$$

the constants $L_i, i = 1, 2, \ldots, 10$ are arbitrary (not determined from chiral symmetry alone) and $F_{\mu\nu}^L, F_{\mu\nu}^R$ are external field strength tensors defined via

$$F_{\mu\nu}^{L,R} = \partial_\mu F_\nu^{L,R} - \partial_\nu F_\mu^{L,R} - i[F_\mu^{L,R}, F_\nu^{L,R}], \quad F_\mu^{L,R} = V_\mu \pm A_\mu . \tag{6.58}$$

Now just as in the case of QED the bare parameters L_i which appear in this Lagrangian are not physical quantities. Instead the experimentally relevant (renormalized) values of these parameters are obtained by appending to these bare values the divergent one-loop contributions having the form

$$L_i^r = L_i - \frac{c_i}{32\pi^2} \left[\frac{-2}{\epsilon} - \ln(4\pi) + \gamma_E - 1 \right], \tag{6.59}$$

Table 6.2 Gasser–Leutwyler counterterms and the means by which they are determined

Coefficient	Value ($\times 10^{-3}$)	Origin
L_1^r	0.53 ± 0.06	$\pi\pi$ scattering
L_2^r	0.81 ± 0.04	and
L_3^r	-3.07 ± 0.20	$K_{\ell 4}$ decay
L_5^r	1.01 ± 0.06	F_K/F_π
L_9^r	6.90 ± 0.70	π charge radius
L_{10}^r	-5.22 ± 0.06	$\pi \to e\nu\gamma$

where γ_E is Euler's constant, while c_i, $i+1, 2, \ldots, 10$ are constants which are determined by the loop integration. By comparing with experiment, Gasser and Leutwyler were able to determine empirical values for each of these ten parameters. While ten sounds like a rather large number, we shall see below that this picture is actually quite *predictive*. Typical values for the parameters are shown in Table 6.2 [Bij14].

The important question to ask at this point is: Why stop at order-four derivatives? Clearly if two-loop amplitudes from $\mathcal{L}_2$ or one-loop corrections from $\mathcal{L}_4$ are calculated, divergences will arise which are of six-derivative character. Why not include these? The answer is that the chiral procedure represents an expansion in energy–momentum. Corrections to the lowest-order (tree-level) predictions from one-loop corrections from $\mathcal{L}_2$ or tree-level contributions from $\mathcal{L}_4$ are $\mathcal{O}(E^2/\Lambda_\chi^2)$, where $\Lambda_\chi \sim 4\pi F_\pi \sim 1\,\text{GeV}$ is the chiral scale [Man84]. Thus, chiral perturbation theory is a *low-energy* procedure. It is only to the extent that the energy is small compared to the chiral scale that it makes sense to truncate the expansion at the four-derivative level. Realistically this means that we deal with processes involving $E < 500\,\text{MeV}$, and, as we shall describe below, for such reactions the procedure is found to work very well.

Now Gasser and Leutwyler, besides giving the form of the $\mathcal{O}(p^4)$ chiral Lagrangian, have also performed the one-loop integration and have written the result in a simple algebraic form. Users merely need to look up the result in their paper. However, in order to really understand what they have done, it is useful to study a simple example of a chiral perturbation theory calculation in order to see how it is performed and in order to understand how the experimental counterterm values are actually determined. We consider the pion electromagnetic form factor, which by Lorentz- and gauge-invariance has the structure

$$\langle \pi^+(p_2)|J_{\text{em}}^\mu|\pi^+(p_1)\rangle = F_1(q^2)(p_1 + p_2)^\mu. \tag{6.60}$$

We begin by identifying the electromagnetic current as

$$J_{\text{em}}^\mu = -\frac{\partial \mathcal{L}}{\partial(eA_\mu)} = (\boldsymbol{\phi} \times \partial^\mu \boldsymbol{\phi})_3 \left[1 - \frac{1}{3F^2}\boldsymbol{\phi} \cdot \boldsymbol{\phi} + \mathcal{O}(\boldsymbol{\phi}^4) \right]$$

$$+ (\boldsymbol{\phi} \times \partial^\mu \boldsymbol{\phi})_3 \left[16L_4 + 8L_5 \right] \frac{m_\pi^2}{F^2} + \frac{4L_9}{F^2} \partial^\nu (\partial^\mu \boldsymbol{\phi} \times \partial_\nu \boldsymbol{\phi})_3 + \cdots, \tag{6.61}$$

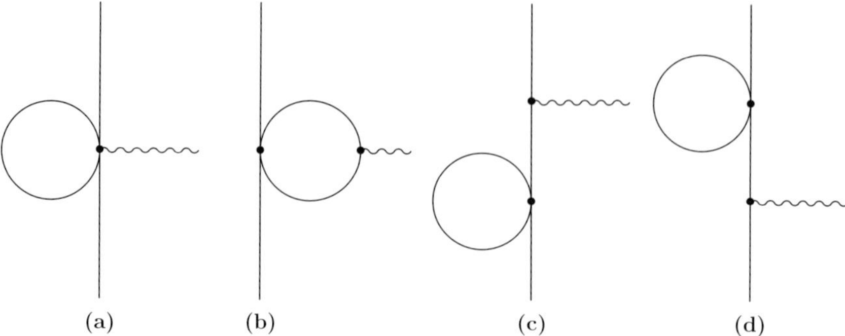

(a)   (b)   (c)   (d)

Fig. 6.2   Loop corrections to the charged pion electromagnetic form factor. Here the charged pion (photon) is designated by a solid (wiggly) line.

where we have expanded to *fourth*-order in the pseudoscalar fields. Defining

$$\delta_{jk} I(m^2) = i\Delta_{Fjk}(0) = \langle 0|\phi_j(x)\phi_k(x)|0\rangle,$$

$$\text{with} \quad I(m^2) = \mu^{4-d} \int \frac{d^d k}{(2\pi)^d} \frac{i}{k^2 - m^2} = \frac{\mu^{4-d}}{(4\pi)^{d/2}} \Gamma\left(1 - \frac{d}{2}\right) (m^2)^{\frac{d}{2}-1},$$

$$\delta_{jk} I_{\mu\nu}(m^2) = -\partial_\mu \partial_\nu i\Delta_{Fjk}(0) = \langle 0|\partial_\mu \phi_j(x)\partial_\nu \phi_k(x)|0\rangle,$$

$$\text{with} \quad I_{\mu\nu}(m^2) = \mu^{4-d} \int \frac{d^d k}{(2\pi)^d} k_\mu k_\nu \frac{i}{k^2 - m^2} = g_{\mu\nu}\frac{m^2}{d} I(m^2), \tag{6.62}$$

we calculate the one-loop correction shown in Fig. 6.2a to be

$$J^\mu_{\text{em}}|_{(a)} = -\frac{5}{3F_\pi^2} (\boldsymbol{\phi} \times \partial^\mu \boldsymbol{\phi})_3 I(m_\pi^2). \tag{6.63}$$

We also need the one-loop correction shown in Fig. 6.2b. For this piece we require the form of the $\pi\pi$ scattering amplitude which arises from $\mathcal{L}_2$

$$\langle \pi^+(k_1)\pi^-(k_2)|\pi^+(p_1)\pi^-(p_2)\rangle = \frac{i}{3F_0^2}\left(2m_0^2 + p_1^2 + p_2^2 + k_1^2 + k_2^2 - 3(p_1 - k_1)^2\right)$$

$$\tag{6.64}$$

and we shall perform the loop integration using the method of dimensional regularization, which yields

$$\langle J^\mu_{\text{em}}\rangle_{(b)} = \frac{1}{(4\pi F_\pi)^2}(p_1 + p_2)^\mu \int_0^1 dx (m_\pi^2 - q^2 x(1-x))$$

$$\times \left[\left(-\frac{2}{\epsilon} + \gamma - 1 - \ln 4\pi\right) + \ln \frac{m_\pi^2 - q^2 x(1-x)}{\mu^2}\right]. \tag{6.65}$$

Performing the x-integration we find, finally

$$\langle J^\mu_{\text{em}}\rangle_{(b)} = \frac{1}{(4\pi F_\pi)^2}(p_1 + p_2)^\mu \left\{\left(m_\pi^2 - \frac{1}{6}q^2\right)\left[-\frac{2}{\epsilon} + \gamma - 1 - \ln 4\pi + \ln \frac{m_\pi^2}{\mu^2}\right]\right.$$

$$\left. + \frac{1}{6}(q^2 - 4m_\pi^2)H\left(\frac{q^2}{m_\pi^2}\right) - \frac{1}{18}q^2\right\}, \tag{6.66}$$

where $\epsilon = 4 - d$ and the function $H(a)$ is given by

$$
H(a) \equiv \int_0^1 dx \ln(1 - ax(1-x))
$$

$$
= \begin{cases}
2 - 2\sqrt{\frac{4}{a} - 1}\,\cot^{-1}\sqrt{\frac{4}{a} - 1} & (0 < a < 4) \\[2ex]
2 + \sqrt{1 - \frac{4}{a}}\left[\ln\frac{\sqrt{1-\frac{4}{a}}-1}{\sqrt{1-\frac{4}{a}}+1} + i\pi\theta(a-4)\right] & \text{(otherwise)}
\end{cases} \tag{6.67}
$$

and contains the imaginary component required by unitarity.

We are not done yet, however, since we must also include mass and wavefunction effects, cf. Figs. 6.2c,d. In order to do so, we expand $\mathcal{L}_2$ to fourth-order in $\phi(x)$, and $\mathcal{L}_4$ to second-order:

$$
\mathcal{L}_2 = \frac{1}{2}[\partial^\mu\boldsymbol{\phi} \cdot \partial_\mu\boldsymbol{\phi} - m_0^2\boldsymbol{\phi} \cdot \boldsymbol{\phi}] + \frac{m_0^2}{24F_0^2}(\boldsymbol{\phi} \cdot \boldsymbol{\phi})^2
$$

$$
+ \frac{1}{6F_0^2}[(\boldsymbol{\phi} \cdot \partial^\mu\boldsymbol{\phi})(\boldsymbol{\phi} \cdot \partial_\mu\boldsymbol{\phi}) - (\boldsymbol{\phi} \cdot \boldsymbol{\phi})(\partial^\mu\boldsymbol{\phi} \cdot \partial_\mu\boldsymbol{\phi})] + \mathcal{O}(\phi^6),
$$

$$
\mathcal{L}_4 = \frac{m_0^2}{F_0^2}[16L_4 + 8L_5]\frac{1}{2}\partial_\mu\boldsymbol{\phi} \cdot \partial^\mu\boldsymbol{\phi}
$$

$$
- \frac{m_0^2}{F_0^2}[32L_6 + 16L_8]\frac{1}{2}m_0^2\boldsymbol{\phi} \cdot \boldsymbol{\phi} + \mathcal{O}(\phi^4) . \tag{6.68}
$$

Performing the loop integrations on the $\phi^4(x)$ component of Eq. 6.68 yields

$$
\mathcal{L}_{\text{eff}} = \frac{1}{2}\partial^\mu\boldsymbol{\phi}\partial_\mu\boldsymbol{\phi} - \frac{1}{2}m_0^2\boldsymbol{\phi} \cdot \boldsymbol{\phi} + \frac{5m_\pi^2}{12F_\pi^2}I(m_\pi^2)\boldsymbol{\phi} \cdot \boldsymbol{\phi}
$$

$$
+ \frac{1}{6F_\pi^2}(\delta_{ik}\delta_{jl} - \delta_{ij}\delta_{kl})I(m_\pi^2)(\delta_{ij}\partial^\mu\phi_k\partial_\mu\phi_l + \delta_{kl}m_\pi^2\phi_i\phi_j)
$$

$$
+ \frac{1}{2}\partial_\mu\boldsymbol{\phi} \cdot \partial^\mu\boldsymbol{\phi}\frac{m_\pi^2}{F_\pi^2}[16L_4 + 8L_5)] - \frac{1}{2}m_\pi^2\boldsymbol{\phi} \cdot \boldsymbol{\phi}\frac{m_\pi^2}{F_\pi^2}[32L_6 + 16L_8]
$$

$$
= \frac{1}{2}\partial^\mu\boldsymbol{\phi} \cdot \partial_\mu\boldsymbol{\phi}\left[1 + (16L_4 + 8L_5)\frac{m_\pi^2}{F_\pi^2} - \frac{2}{3F_\pi^2}I(m_\pi^2)\right]
$$

$$
- \frac{1}{2}m_0^2\boldsymbol{\phi} \cdot \boldsymbol{\phi}\left[1 + (32L_6 + 16L_8)\frac{m_\pi^2}{F_\pi^2} - \frac{1}{6F_\pi^2}I(m_\pi^2)\right] \tag{6.69}
$$

from which we can now read off the wavefunction renormalization term Z_π.

When this is done we find

$$
Z_\pi F_1^{\text{(tree)}}(q^2) = \left[1 - \frac{8m_\pi^2}{F_\pi^2}(2L_4 + L_5\right.
$$

$$
\left. + \frac{m_\pi^2}{24\pi^2F_\pi^2}\left\{-\frac{2}{\epsilon} + \gamma - 1 - \ln 4\pi + \ln\frac{m_\pi^2}{\mu^2}\right\}\right]
$$

$$\times \left[1 + \frac{8m_\pi^2}{F_\pi^2}(2L_4 + L_5) + 2q^2 \frac{L_9}{F_\pi^2} \right]$$

$$= \left[1 + \frac{m_\pi^2}{24\pi^2 F_\pi^2}\left(-\frac{2}{\epsilon} + \gamma - 1 - \ln 4\pi + \ln \frac{m_\pi^2}{\pi^2} \right) + \frac{2L_9}{F_\pi^2}q^2 \right], \quad (6.70)$$

while from the loop diagrams given in Fig. 6.2

$$F_1(q^2)\bigg|_{(a)} = -\frac{5m_\pi^2}{48\pi^2 F_\pi^2}\left\{ -\frac{2}{\epsilon} + \gamma - 1 - \ln 4\pi + \ln \frac{m_\pi^2}{\mu^2} \right\}$$

$$F_1(q^2)\bigg|_{(b)} = \frac{1}{16\pi^2 F_\pi^2}\left\{ \left(m_\pi^2 - \frac{1}{6}q^2 \right)\left[-\frac{2}{\epsilon} + \gamma - 1 - \ln 4\pi + \ln \frac{m_\pi^2}{\mu^2} \right] \right.$$

$$\left. + \frac{1}{6}(q^2 - 4m_\pi^2)H\left(\frac{q^2}{m_\pi^2} \right) - \frac{1}{18}q^2 \right\}. \quad (6.71)$$

Adding everything together we have the final result, which when written in terms of the renormalized value $L_9^{(r)} = 6.90 \pm 0.70 \times 10^{-3}$ as shown in Table 6.2 is *finite*!

$$F_1(q^2) = 1 + \frac{2L_9^{(r)}}{F_\pi^2}q^2 + \frac{1}{96\pi^2 F_\pi^2}\left[(q^2 - 4m_\pi^2)H\left(\frac{q^2}{m_\pi^2} \right) - q^2 \ln \frac{m_\pi^2}{\mu^2} - \frac{q^2}{3} \right].$$

$$(6.72)$$

Expanding to lowest-order in q^2 we find

$$F_1(q^2) = 1 + q^2\left[\frac{2L_9^{(r)}}{F_\pi^2} - \frac{1}{96\pi^2 F_\pi^2}\left(\ln \frac{m_\pi^2}{\mu^2} + 1 \right) \right] + \cdots \quad (6.73)$$

which can be compared with the phenomenological description in terms of the pion charge radius

$$F_1(q^2) = 1 + \frac{1}{6}\langle r_\pi^2 \rangle q^2 + \cdots . \quad (6.74)$$

By equating these two expressions and using the experimental value of the pion charge radius, $\langle r_\pi^2 \rangle_{\exp} = (0.44 \pm 0.01)$ fm^2 [Dal82], we determine the value of the counterterm $L_9^{(r)}$ shown in Table 6.2.

As seen in Fig. 6.3 this form gives a reasonable representation of the experimental pion form factor near threshold, but deviates substantially from the empirical result as the ρ resonance is approached. This is not surprising as any perturbative approach will be unable to reproduce resonant behavior. This failure should not be considered a failure of chiral perturbative techniques *per se*, just that as one approaches higher energy the importance of two-loop ($\mathcal{O}(p^6)$) and higher terms become important. Although for simple processes such two-loop studies have been performed, the number of p^6 counterterms is well over a hundred and a general chiral analysis at two-loop level is not feasible. Nevertheless, things are certainly not hopeless, and at the end of this chapter we present some approaches to extend the validity of chiral predictions to higher energy.

More relevant at this point is to stay near threshold and ask if chiral pertubation methods are predictive. Can they be used as a *test* of QCD, for example? The answer is definitely yes! We do not have space to present a detailed presentation of the status of such

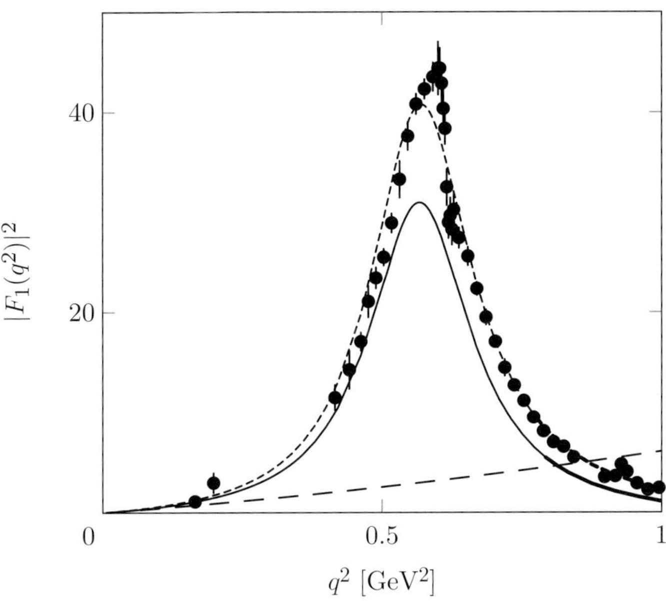

Fig. 6.3 Calculations of the modulus of the pion form factor squared compared with experimental results. The figure is adapted from [Tru88]. Here the solid line gives the result of the inverse amplitude method, as discussed at the end of this chapter, while the dashed line gives the one-loop chiral perturbation theory prediction. The dotted line shows an empirical simulation of the inelastic $\omega\pi$ contribution obtained by multiplying the inverse amplitude result by the factor $1 + 0.15 s/s_\omega$ [Tru88].

tests; a simple example will have to suffice [Don89a]. We have seen above how the pion charge radius enables the determination of one of the chiral parameters, L_9^r. A second, L_{10}^r, can be found from measurement of the axial-vector structure constant h_A in radiative pion decay, $\pi^+ \to e^+ \nu_e \gamma$ or $\pi^+ \to e^+ \nu_e e^+ e^-$, for which the decay amplitudes can be written

$$\mathcal{M}_{\pi^+ \to e^+ \nu_e \gamma} = -\frac{eG_F}{\sqrt{2}} \cos\theta_1 M_{\mu\nu}(p,q)\varepsilon^{\mu*}(q)\bar{u}(p_\nu)\gamma^\nu(1+\gamma_5)v(p_e)$$

$$\mathcal{M}_{\pi^+ \to e^+ \nu_e e^+ e^-} = -\frac{e^2 G_F}{\sqrt{2}} \cos\theta_1 M_{\mu\nu}(p,q)\frac{1}{q^2}$$
$$\times \bar{u}(p_2)\gamma^\mu v(p_1)\bar{u}(p_\nu)\gamma^\nu(1+\gamma_5)v(p_e), \tag{6.75}$$

and the hadronic component of $M_{\mu\nu}$ has the structure

$$M_{\mu\nu}(p,q) = \int d^4x\, e^{iq\cdot x} < 0|T(J_\mu^{\text{em}}(x)J_\nu^{1-i2}(0)|\pi(p) >= \text{Born terms}$$
$$- h_A((p-q)_\mu q_\nu - g_{\mu\nu}q\cdot(p-q)) - r_A(q_\mu q_\nu - g_{\mu\nu}q^2)$$
$$+ ih_V \epsilon_{\mu\nu\alpha\beta}q^\alpha p^\beta, \tag{6.76}$$

where h_A, r_A, h_V are unknown structure functions. (Note that r_A can be measured only via the rare Dalitz decay $\pi^+ \to e^+ \nu_e e^+ e^-$.)

We also note that the related amplitude for Compton scattering can be written in the form

$$- iT_{\mu\nu}(p, p', q) = -i \int d^4 x e^{iq_1 \cdot x} < \pi^+(\boldsymbol{p}')|T(J_\mu^{\text{em}}(x)J_\nu^{\text{em}}(0)|\pi^+(\boldsymbol{p}) >$$

$$= \text{Born terms} + \sigma (q_{2\mu} q_{1\nu} - g_{\mu\nu} q_1 \cdot q_2) + \cdots . \tag{6.77}$$

The $\gamma \pi^+ \to \gamma \pi^+$ reaction is often analyzed in terms of the pion electric and magnetic polarizabilities α_E and β_M which describe the response of the pion to external electric and magnetizing fields [Hol90b]. In the static limit such fields induce electric and magnetic dipole moments

$$\boldsymbol{p} = 4\pi \alpha_E \boldsymbol{E}, \quad \boldsymbol{\mu} = 4\pi \beta_M \boldsymbol{H}, \tag{6.78}$$

which correspond to an interaction energy

$$U = -\frac{1}{2} \left(4\pi \alpha_E \boldsymbol{E}^2 + 4\pi \beta_M \boldsymbol{H}^2 \right) . \tag{6.79}$$

Use of chiral perturbation theory yields the results

$$h_V = \frac{N_c}{12\sqrt{2}\pi^2 F_\pi} = 0.027 m_\pi^{-1}, \quad \frac{h_A}{h_V} = 32\pi^2 (L_9^r + L_{10}^r)$$

$$\frac{r_A}{h_V} = 32\pi^2 \left[L_9^r - \frac{1}{192\pi^2} \left(\ln \frac{m_\pi^2}{\mu^2} + 1 \right) \right], \quad \alpha_E + \beta_M = 0$$

$$\alpha_E = \frac{\alpha}{2m_\pi} \sigma = \frac{4\alpha}{m_\pi F_\pi^2} \left(L_9^r + L_{10}^r \right) . \tag{6.80}$$

Use of the experimental result [Poc14]

$$\frac{h_A}{h_V} = 0.45 \pm 0.07 \quad \text{gives} \quad L_{10}^r(\mu = m_\eta) = -0.0056(3) \tag{6.81}$$

and once this is determined chiral symmetry makes *four* predictions among these parameters! As shown in Table 6.3, three of the four are found to be in good agreement with experiment. The possible exception involves a relation between the charged pion polarizability and the axial-vector structure constant h_A measured in radiative pion decay. In this case there exist three conflicting experimental results, two of which agree and one of which does not agree with the theoretical prediction. It is important to resolve this potential discrepancy, since such chiral predictions are firm ones. There is no way (other than introducing perversely large higher-order effects) to bring things into agreement were some large violation of a chiral prediction to be verified, since the only ingredient which goes into such predictions is the (broken) chiral symmetry of QCD itself! In this regard a recent report by the COMPASS collaboration, using the Primakoff effect, has reported [Fri12]

$$\alpha_E = -\beta_M = (1.9 \pm 0.7 \pm 0.8) \times 10^{-4} \text{ fm}^3 , \tag{6.82}$$

in agreement with the chiral prediction.

Table 6.3 Chiral predictions and data in radiative pion processes

Reaction	Quantity	Theory	Experiment
$\pi^+ \to e^+ \nu_e \gamma$	$h_V(m_\pi^{-1})$	0.027	0.029 ± 0.017 [PDG14]
$\pi^+ \to e^+ \nu_e e^+ e^-$	r_V/h_V	2.6	2.3 ± 0.6 [PDG14]
$\gamma \pi^+ \to \gamma \pi^+$	$(\alpha_E + \beta_M)\,(10^{-4}\,\mathrm{fm}^3)$	0	1.4 ± 3.1 [Ant85]
	$\alpha_E\,(10^{-4}\,\mathrm{fm}^3)$	2.8	6.8 ± 1.4 [Ant83]
			12 ± 20 [Ahr05]
			2.1 ± 1.1 [Bab92]
			1.9 ± 1.0 [Fri12]

6.3 Baryon Chiral Perturbation Theory

Our discussion of chiral methods given above was limited to the study of the interactions of the pseudoscalar mesons (would-be Goldstone bosons) with leptons and with each other. In the real world, of course, interactions with baryons also take place and it is important to develop a useful predictive scheme based on chiral invariance for such processes. Again much work has been done in this regard [Gas88], but there remain important problems, such as the covergence of the chiral expansion and the extension of the $SU(2)$ results to $SU(3)$. [Ber95]. Writing down the lowest-order chiral Lagrangian at the $SU(2)$ level is straightforward:

$$\mathcal{L}_{\pi N} = \bar{N}(i\,\slashed{D} - m_N + \frac{g_A}{2}\slashed{u}\gamma_5)N \, , \tag{6.83}$$

where g_A is the usual nucleon axial-vector coupling in the chiral limit, the covariant derivative $D_\mu = \partial_\mu + \Gamma_\mu$ is given by

$$\Gamma_\mu = \frac{1}{2}[u^\dagger, \partial_\mu u] - \frac{i}{2}u^\dagger(V_\mu + A_\mu)u - \frac{i}{2}u(V_\mu - A_\mu)u^\dagger \, , \tag{6.84}$$

and u_μ represents the axial-vector structure

$$u_\mu = iu^\dagger \nabla_\mu U u^\dagger \, , \tag{6.85}$$

where $U \equiv u^2$. Expanding to lowest-order we find

$$\mathcal{L}_{\pi N} = \bar{N}(i\slashed{\partial} - m_N)N + g_A \bar{N}\gamma^\mu \gamma_5 \frac{1}{2}\tau N \cdot \left(\frac{i}{F_\pi}\partial_\mu \boldsymbol{\pi} + 2\boldsymbol{A}_\mu\right)$$
$$- \frac{1}{4F_\pi^2}\bar{N}\gamma^\mu \boldsymbol{\tau} N \cdot \boldsymbol{\pi} \times \partial_\mu \boldsymbol{\pi} + \cdots \, , \tag{6.86}$$

which yields the Goldberger–Treiman relation, connecting strong and axial-vector couplings of the nucleon system [Gol58]

$$F_\pi \, g_{\pi NN} = m_N \, g_A. \tag{6.87}$$

Using the present best values for these quantities, we find

$$92.4\,\mathrm{MeV} \times 13.05 = 1206\,\mathrm{MeV} \quad \text{versus} \quad 1193\,\mathrm{MeV} = 939\,\mathrm{MeV} \times 1.270 \tag{6.88}$$

and the agreement to better than two percent strongly confirms the validity of chiral symmetry in the nucleon sector. Actually the Goldberger–Treiman relation is only strictly true at the unphysical point $g_{\pi NN}(q^2 = 0)$ and one expects about a 1% discrepancy to exist. An interesting "wrinkle" in this regard is the use of the so-called Dashen–Weinstein relation which uses simple $SU(3)$ symmetry breaking to predict this discrepancy in terms of corresponding numbers in the strangeness changing sector [Das69].

A second prediction of the lowest-order chiral Lagrangian deals with charged pion photoproduction. As emphasized previously, chiral symmetry requires any pion coupling to be in terms of a (covariant) derivative. Hence there exists a $\bar{N}N\pi^{\pm}\gamma$ contact interaction (the Kroll–Ruderman term) [Kro54] which contributes to threshold charged pion photoproduction. Here what is measured is the s-wave or E_{0+} multipole, defined via

$$\text{Amp} = 4\pi(1 + \mu)E_{0+}\boldsymbol{\sigma} \cdot \hat{\boldsymbol{\epsilon}} + \cdots, \tag{6.89}$$

where $\mu = m_\pi/m_N$. In addition to the Kroll–Ruderman piece there exists, at the two derivative level, a second contact term which arises from

$$\mathcal{L}^{(2)}_{\pi\gamma NN} = \frac{eg_A}{8m_N F_\pi}v \cdot qP_+[(1 + \tau_3)\,A^\perp, \gamma_5\tau^a]P_+ = \frac{eg_A}{2m_N F_\pi}S \cdot \epsilon v \cdot q(\tau^a + \delta^{a3}). \tag{6.90}$$

Adding these two contributions yields the result [Deb70]

$$\begin{aligned} E_{0+} &= \pm\frac{1}{4\pi(1 + \mu)}\frac{eg_A}{\sqrt{2}F_\pi}(1 \mp \frac{\mu}{2}) = \frac{eg_A}{4\sqrt{2}F_\pi}\begin{pmatrix} 1 - \frac{3}{2}\mu & \pi^+ \\ -1 + \frac{1}{2}\mu & \pi^- \end{pmatrix} \\ &= \begin{cases} +26.3 \times 10^{-3}/m_\pi & \pi^+ n \\ -31.3 \times 10^{-3}/m_\pi & \pi^- p \end{cases}, \end{aligned} \tag{6.91}$$

and the numerical predictions are found to be in excellent agreement with the present experimental results,

$$E_{0+}^{\text{exp}} = \begin{cases} (+27.9 \pm 0.5) \times 10^{-3}/m_\pi & [\text{Bur65}] & \pi^+ n \\ (+28.8 \pm 0.7) \times 10^{-3}/m_\pi & [\text{Ada66}] & \\ (+27.6 \pm 0.3) \times 10^{-3}/m_\pi & [\text{Ber98}] & \\ (-31.4 \pm 1.3) \times 10^{-3}/m_\pi & [\text{Gol66}] & \pi^- p \\ (-32.2 \pm 1.2) \times 10^{-3}/m_\pi & [\text{Kov97}] & \\ (-31.5 \pm 0.8) \times 10^{-3}/m_\pi & [\text{Ber98}]. & \end{cases} \tag{6.92}$$

Heavy-Baryon Methods

Extension to $SU(3)$ gives additional successful predictions, e.g., the linear Gell-Mann–Okubo relation as well as the generalized Goldbeger–Treiman relation. However, difficulties arise when one attempts to include higher-order corrections to this formalism. The difference from the Goldstone case is that there now exist *two* dimensionful parameters, m_N and F_π, in the problem rather than *one*, F_π. Thus, loop effects can be of order $(m_N/4\pi F_\pi)^2 \sim 1$ and we no longer have a reliable perturbative scheme. A consistent power-counting mechanism can be constructed provided that we eliminate the nucleon mass from the leading-order Lagrangian. This is done by considering the nucleon to be very heavy. Then we can write its four-momentum as [Jen92]

$$p_\mu = m_N v_\mu + k_\mu \,, \tag{6.93}$$

where v_μ is the four-velocity and satisfies $v^2 = 1$, while k_μ is a small off-shell momentum, with $v \cdot k << m_N$. One can construct eigenstates of the projection operators $P_\pm = \frac{1}{2}(1 \pm \not v)$, which in the rest frame project out upper, lower components of the Dirac wavefunction, so that [Ber92]

$$\psi = e^{-i m_N v \cdot x}(H_v + h_v) \,, \tag{6.94}$$

where

$$H_v = P_+ \psi \,, \qquad h_v = P_- \psi \,. \tag{6.95}$$

The effective Lagrangian can then be written in terms of N, h as

$$\mathcal{L}_{\pi N} = \bar{H}_v \mathcal{A} H_v + \bar{h}_v \mathcal{B} H_v + \bar{H}_v \gamma_0 \mathcal{B}^\dagger \gamma_0 h_v - \bar{h}_v \mathcal{C} h_v \,, \tag{6.96}$$

where the operators $\mathcal{A}, \mathcal{B}, \mathcal{C}$ have the low-energy expansions

$$\begin{aligned}
\mathcal{A} &= i v \cdot D + g_A u \cdot S + \cdots \\
\mathcal{B} &= i \not{D}^\perp - \frac{1}{2} g_A v \cdot u \gamma_5 + \cdots \\
\mathcal{C} &= 2 m_N + i v \cdot D + g_A u \cdot S + \cdots \,.
\end{aligned} \tag{6.97}$$

Here $D_\mu^\perp = (g_{\mu\nu} - v_\mu v_\nu) D^\nu$ is the transverse component of the covariant derivative and $S_\mu = \frac{i}{2}\gamma_5 \sigma_{\mu\nu} v^\nu$ is the Pauli–Lubanski spin vector and satisfies the relations

$$S \cdot v = 0, \quad S^2 = -\frac{3}{4}, \quad \{S_\mu, S_\nu\} = \frac{1}{2}(v_\mu v_\nu - g_{\mu\nu}), \quad [S_\mu, S_\nu] = i\epsilon_{\mu\nu\alpha\beta} v^\alpha S^\beta. \tag{6.98}$$

We see that the two components H, h are coupled in this expression for the effective action. However, the system may be diagonalized by use of the field transformation

$$h' = h - \mathcal{C}^{-1} \mathcal{B} H \,, \tag{6.99}$$

in which case the Lagrangian becomes

$$\mathcal{L}_{\pi N} = \bar{H}_v (\mathcal{A} + (\gamma_0 \mathcal{B}^\dagger \gamma_0) \mathcal{C}^{-1} \mathcal{B}) H_v - \bar{h}'_v \mathcal{C} h'_v \,. \tag{6.100}$$

The piece of the Lagrangian involving H only contains the mass in the operator $\mathcal{C}^{-1}$ and is the effective Lagrangian that we desire. The remaining piece involving h'_v can be thrown away, as it does not couple to the H_v physics. (In path-integral language we simply integrate out this component yielding an uninteresting overall constant.) Of course, when loops are calculated, a set of counterterms will be required and these are given at leading (two-derivative) order by

$$\mathcal{A}^{(2)} = \frac{m_N}{F_\pi^2}(c_1 \mathrm{Tr} \chi_+ + c_2 (v \cdot u)^2 + c_3 u \cdot u + c_4 [S^\mu, s^\nu] u_\mu u_\nu$$

$$+ c_5 (\chi_+ - \mathrm{Tr}\chi_+) - \frac{i}{4 m_N}[S^\mu, S^\nu]((1 + c_6) F_{\mu\nu}^+ + c_7 \mathrm{Tr} f_{\mu\nu}^+))$$

$$\mathcal{B}^{(2)} = \frac{m_N}{F_\pi^2}\left(\left(-\frac{c_2}{4}i[u^\mu, u^\nu] + c_6 f_+^{\mu\nu} + c_7 Tr f_+^{\mu\nu}\right)\sigma_{\mu\nu} - \frac{c_4}{2}v_\mu \gamma_\nu Tr u^\mu u^\nu\right)$$

$$\mathcal{C}^{(2)} = -\frac{m_N}{F_\pi^2}\left(c_1 Tr\chi_+ + \left(-\frac{c_2}{4}i[u^\mu, u^\nu] + c_6 f_+^{\mu\nu} + c_7 tr F_+^{\mu\nu}\right)\sigma_{\mu\nu}\right.$$

$$\left. -\frac{c_3}{4}Tr u^\mu u_\nu - \left(\frac{c_4}{2} + m_N c_5\right)v_\mu v_\nu Tr u^\mu u^\nu\right). \tag{6.101}$$

Expanding $\mathcal{C}^{-1}$ and the other terms in terms of a power series in $1/m_N$ leads to an effective heavy-nucleon Lagrangian of the form (to $\mathcal{O}(q^3)$)

$$\mathcal{L}_{\pi N} = \bar{H}_v\{\mathcal{A}^{(1)} + \mathcal{A}^{(2)} + \mathcal{A}^{(3)} + (\gamma_0 \mathcal{B}^{(1)\dagger}\gamma_0)\frac{1}{2m_N}\mathcal{B}^{(1)}$$

$$+ \frac{(\gamma_0 \mathcal{B}^{(1)\dagger}\gamma_0)\mathcal{B}^{(2)} + (\gamma_0 \mathcal{B}^{(2)\dagger}\gamma_0)\mathcal{B}^{(1)}}{2m_N}$$

$$- (\gamma_0 \mathcal{B}^{(1)\dagger}\gamma_0)\frac{i(v\cdot D) + g_A(u\cdot S)}{(2m_N)^2}\mathcal{B}^{(1)}\}H_v + \mathcal{O}(q^4). \tag{6.102}$$

A set of Feynman rules can now be written down and a consistent power counting scheme developed, as shown by Meissner and his collaborators [Ber95].

Applications

As an example of the use of this formalism, called heavy-baryon chiral perturbation theory (HBχpt), consider the nucleon–photon interaction. To lowest (one-derivative) order we have from $\mathcal{A}^{(1)}$

$$\mathcal{L}_{\gamma NN}^{(1)} = ie\bar{N}\frac{1}{2}(1 + \tau_3)\epsilon \cdot vN, \tag{6.103}$$

while at two-derivative level we find

$$\mathcal{L}_{\gamma NN}^{(2)} = \bar{N}\left\{\frac{e}{4m_N}(1 + \tau_3)\epsilon \cdot (p_1 + p_2) + \frac{ie}{2m_N}[S\cdot\epsilon, S\cdot k](1 + \kappa_S + \tau_3(1 + \kappa_V))\right\}N, \tag{6.104}$$

where we have made the identifications $c_6 = \kappa_V$, $c_7 = \frac{1}{2}(\kappa_S - \kappa_V)$. We can now reproduce the low-energy theorems for Compton scattering. Consider the case of the proton. At the two-derivative level, we have the tree-level prediction

$$(\gamma_0 \mathcal{B}^{(1)\dagger}\gamma_0)\frac{1}{2M}\mathcal{B}^{(1)}|_{\gamma pp} = \frac{e^2}{2m_N}A_\perp^2, \tag{6.105}$$

which yields the familiar Thomson amplitude

$$\mathrm{Amp}_{\gamma\gamma pp} = -\frac{e^2}{m_N}\hat{\epsilon}' \cdot \hat{\epsilon}. \tag{6.106}$$

On the other hand, at order q^3 we find a contribution from Born diagrams with two-derivative terms at each vertex, yielding

$$\mathrm{Amp}_{\gamma\gamma pp} = \left(\tfrac{e}{m_N}^2\right)\tfrac{1}{\omega}\,[(\hat{\boldsymbol{\epsilon}}' \cdot \boldsymbol{k}\boldsymbol{S} \cdot \hat{\boldsymbol{\epsilon}} \times \boldsymbol{k} - \hat{\boldsymbol{\epsilon}} \cdot \boldsymbol{k}'\boldsymbol{S} \cdot \hat{\boldsymbol{\epsilon}}' \times \boldsymbol{k}')(1 + \kappa_p)$$

$$+ i\boldsymbol{S} \cdot (\hat{\boldsymbol{\epsilon}} \times \boldsymbol{k}) \times (\hat{\boldsymbol{\epsilon}}' \times \boldsymbol{k}')(1 + \kappa_p)^2]\,. \tag{6.107}$$

The full result must also include contact terms at order q^3 from the last piece of Eq. (6.102)

$$- eP_+ \not{A}^\perp \frac{iv \cdot D}{(2m_N)^2} e\,\not{A}^\perp P_+ = -\frac{e^2}{2m_N^2}\boldsymbol{S} \cdot \boldsymbol{A} \times \dot{\boldsymbol{A}} \tag{6.108}$$

and from the third

$$\frac{1}{2m_N}P_+ \left\{ e\,\not{A}^\perp, \kappa_p\sigma_{\mu\nu}F^{\mu\nu} \right\} P_+ = \kappa_p \frac{e^2}{m_N^2}\boldsymbol{S} \cdot \boldsymbol{A} \times \dot{\boldsymbol{A}}\,. \tag{6.109}$$

When added to the Born contributions the result can be expressed in the general form [Ber95]

$$\mathrm{Amp} = \hat{\boldsymbol{\epsilon}} \cdot \hat{\boldsymbol{\epsilon}}'A_1 + \hat{\boldsymbol{\epsilon}}' \cdot \boldsymbol{k}\hat{\boldsymbol{\epsilon}} \cdot \boldsymbol{k}'A_2 + i\boldsymbol{\sigma} \cdot (\hat{\boldsymbol{\epsilon}}' \times \hat{\boldsymbol{\epsilon}})A_3$$

$$+ i\boldsymbol{\sigma} \cdot (\boldsymbol{k}' \times \boldsymbol{k})\hat{\boldsymbol{\epsilon}}' \cdot \hat{\boldsymbol{\epsilon}}A_4 + i\boldsymbol{\sigma} \cdot [(\hat{\boldsymbol{\epsilon}}' \times \boldsymbol{k})\hat{\boldsymbol{\epsilon}} \cdot \boldsymbol{k}' - (\hat{\boldsymbol{\epsilon}} \times \boldsymbol{k}')\hat{\boldsymbol{\epsilon}}' \cdot \boldsymbol{k}]A_5$$

$$+ i\boldsymbol{\sigma} \cdot [(\hat{\boldsymbol{\epsilon}}' \times \boldsymbol{k}')\hat{\boldsymbol{\epsilon}} \cdot \boldsymbol{k}' - (\hat{\boldsymbol{\epsilon}} \times \boldsymbol{k})\hat{\boldsymbol{\epsilon}}' \cdot \boldsymbol{k}]A_6$$

$$\tag{6.110}$$

with[3]

$$A_1 = -\frac{e^2}{m_N},\quad A_2 = \frac{1}{m_N^2\omega},\quad A_3 = \frac{e^2\omega}{2M^2}(1 + 2\kappa - (1 + \kappa)^2\hat{\boldsymbol{k}} \cdot \hat{\boldsymbol{k}}')$$

$$A_4 = -A_5 = -\frac{e^2(1 + \kappa)^2}{2m_N^2\omega},\quad A_6 = -\frac{e^2(1 + \kappa)}{2m_N^2\omega} \tag{6.111}$$

which agrees with the usual result derived in this order via Low's theorem [Low54].

A full calculation at order q^3 must also, of course, include loop contributions. Using the lowest-order (one-derivative) pion-nucleon interactions

$$\mathcal{L}_{\pi NN} = \frac{g_A}{F_\pi}\bar{N}\tau^a \boldsymbol{S} \cdot \boldsymbol{q}N$$

$$\mathcal{L}_{\pi\pi NN} = \frac{1}{4F_\pi^2}v \cdot (q_1 + q_2)\epsilon^{abc}\bar{N}\tau_c N$$

$$\mathcal{L}_{\gamma\pi NN} = \frac{ieg_A}{F_\pi}\epsilon^{a3b}\bar{N}\boldsymbol{\epsilon} \cdot \boldsymbol{S}\tau_b N\,, \tag{6.112}$$

these can be calculated using the diagrams shown in Fig. 6.4. Of course, from Eq. (6.102) the propagator for the nucleon must have the form $1/iv \cdot k$ where k is the off-shell momentum. Thus, for example, the seagull diagram, Fig. 6.4a, is of the form

$$\mathrm{Amp} = 4e^2 \left(\frac{g_A}{F_\pi}\right)^2 \hat{\boldsymbol{\epsilon}} \cdot \hat{\boldsymbol{\epsilon}}' \int \frac{d^4k}{(2\pi)^4}\frac{\boldsymbol{S} \cdot \boldsymbol{k}\boldsymbol{S} \cdot \boldsymbol{k}}{v \cdot k(k^2 - m_\pi^2)((k + q_1 - q_2)^2 - m_\pi^2)}\,. \tag{6.113}$$

---

[3] Here we have used the identity

$$\boldsymbol{\sigma} \cdot (\hat{\boldsymbol{\epsilon}}' \times \boldsymbol{k}') \times (\hat{\boldsymbol{\epsilon}} \times \boldsymbol{k}) = \boldsymbol{\sigma} \cdot (\boldsymbol{k}' \times \boldsymbol{k})\hat{\boldsymbol{\epsilon}} \cdot \hat{\boldsymbol{\epsilon}}' + \boldsymbol{\sigma} \cdot (\hat{\boldsymbol{\epsilon}}' \times \hat{\boldsymbol{\epsilon}})\boldsymbol{k}' \cdot \boldsymbol{k} + \boldsymbol{\sigma} \cdot (\hat{\boldsymbol{\epsilon}} \times \boldsymbol{k}')\hat{\boldsymbol{\epsilon}}' \cdot \boldsymbol{k} - \boldsymbol{\sigma} \cdot (\hat{\boldsymbol{\epsilon}}' \times \boldsymbol{k})\hat{\boldsymbol{\epsilon}} \cdot \boldsymbol{k}'.$$

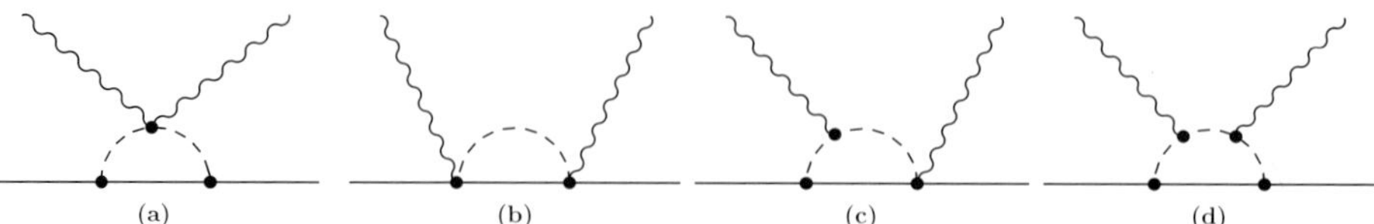

Fig. 6.4 Heavy-baryon loop diagrams for nucleon Compton scattering. Here the nucleon, pion, and photon are represented by solid, dashed, and wiggly lines respectively. Note that the diagrams (b), (c), and (d) must be augmented by appropriate cross diagrams.

Since there are no additional counterterms at this order q^3, the sum of loop diagrams must be finite and yields, to lowest-order in energy and after considerable calculation

$$A_1^{\text{loop}} = \xi \left(\frac{11\omega^2}{24m_\pi} + \frac{t}{48m_\pi} \right), \quad A_2^{\text{loop}} = \xi \left(\frac{1}{24m_\pi} \right), \quad A_3^{\text{loop}} = \xi \left(\frac{\omega t}{\pi m_\pi^2} + \frac{\omega^3}{3\pi m_\pi^2} \right)$$

$$A_4^{\text{loop}} = \xi \left(\frac{\omega}{6\pi m_\pi^2} \right), \quad A_5^{\text{loop}} = -A_6^{\text{loop}} = -\xi \left(\frac{13\omega}{12\pi m_\pi^2} \right) \tag{6.114}$$

and $\xi = g_A^2/8\pi F_\pi^2$.

The experimental implications of these results may be seen by first considering the case of an unpolarized proton target. Writing

$$\text{Amp}_{\text{unpol}} = \left\{ \hat{\boldsymbol{\epsilon}} \cdot \hat{\boldsymbol{\epsilon}}' \left(-\frac{e^2}{m_N} + 4\pi\alpha_E\omega^2 \right) + (\hat{\boldsymbol{\epsilon}} \times \boldsymbol{k}) \cdot (\hat{\boldsymbol{\epsilon}}' \times \boldsymbol{k}')4\pi\beta_M \right\}, \tag{6.115}$$

where α_E, and β_M are the proton electric and magnetic polarizabilities, we identify the one-loop chiral predictions [Ber91a]

$$\alpha_E^{\text{theo}} = 10\beta_M^{\text{theo}} = \frac{5e^2 g_A^2}{384\pi^2 F_\pi^2 m_\pi} = 12.2 \times 10^{-4}\ \text{fm}^3, \tag{6.116}$$

which are in reasonable agreement with the recently measured values [Fed91]

$$\alpha_E^{\text{exp}} = (12.0 \pm 0.6) \times 10^{-4}\ \text{fm}^3, \qquad \beta_M^{\text{exp}} = (1.9 \pm 0.5) \times 10^{-4}\ \text{fm}^3. \tag{6.117}$$

For the case of spin-dependent forward scattering, we find, in general

$$\frac{1}{4\pi}\text{Amp} = f_1(\omega^2)\hat{\boldsymbol{\epsilon}} \cdot \hat{\boldsymbol{\epsilon}}' + i\omega f_2(\omega^2)\boldsymbol{\sigma} \cdot \hat{\boldsymbol{\epsilon}}' \times \hat{\boldsymbol{\epsilon}} \tag{6.118}$$

with

$$f_1(\omega^2) = -\frac{e^2}{4\pi m_N} + (\alpha_E + \beta_M)\omega^2 + \mathcal{O}(\omega^4)$$

$$f_2(\omega^2) = -\frac{e^2 \kappa_p^2}{8\pi^2 m_N^2} + \gamma_S\omega^2 + \mathcal{O}(\omega^4), \tag{6.119}$$

where γ_S is a sort of "spin-polarizability" and is related to the classical Faraday effect. Assuming that the amplitudes f_1, f_2 obey once-subtracted and unsubtracted dispersion relations respectively we find the sum rules

$$\alpha_E + \beta_M = \frac{1}{4\pi^2} \int_{\omega_0}^{\infty} \frac{d\omega}{\omega^2} (\sigma_+(\omega) + \sigma_-(\omega))$$

$$\frac{\pi e^2 \kappa_p^2}{2m_N^2} = \int_{\omega_0}^{\infty} \frac{d\omega}{\omega} [\sigma_+(\omega) - \sigma_-(\omega)]$$

$$\gamma_S = \frac{1}{4\pi^2} \int_{\omega_0}^{\infty} \frac{d\omega}{\omega^3} [\sigma_+(\omega) - \sigma_-(\omega)] , \tag{6.120}$$

where here $\sigma_\pm(\omega)$ denote the photoabsorption cross sections for scattering circularly polarized photons on polarized nucleons involving total γN helicity 3/2 and 1/2 respectively. Here the first is the well-known Baldin sum rule for the sum of the electric and magnetic polarizabilities, while the second is the equally familiar Drell–Hearn–Gerasimov (DHG) sum rule [Dre66]. The third is less well known, but follows from that of DHG and offers a new check of the chiral predictions.

Another venue where there has been a great deal of work is that of neutral pion photoproduction, for which the one-derivative contribution vanishes. In this case the leading contribution arises from the two derivative term given in Eq. (6.102), augmented by the three derivative contribution from Born diagrams. The net result is

$$\text{Amp}^{(2)} = \frac{e g_A}{2 F_\pi} \mu \boldsymbol{\sigma} \cdot \hat{\boldsymbol{\epsilon}} \times \begin{cases} 1 & \pi^0 p \\ 0 & \pi^0 n \end{cases} \tag{6.121}$$

for the contact term and

$$\text{Amp}^{(3)} = -\frac{e}{2m_N}[S \cdot \epsilon, S \cdot k](1 + \kappa_p)\frac{1}{v \cdot q} \frac{g_A}{2m_N F_\pi} S \cdot (2p - q)m_\pi = -\frac{e g_A}{4 F_\pi} \mu^2 (1 + \kappa_p)\boldsymbol{\sigma} \cdot \hat{\boldsymbol{\epsilon}} \tag{6.122}$$

for the pole terms. (Note: only the cross term is nonvanishing at threshold.)

We must append the loop contributions which arise from the graphs shown in Fig. 6.5

$$\text{Amp}^{\text{loop}} = -\frac{e g_A m_N}{64\pi F_\pi^2} \mu^2 \boldsymbol{\sigma} \cdot \hat{\boldsymbol{\epsilon}} . \tag{6.123}$$

The result is the prediction [Ber91b]

$$E_{0+} = \frac{e g_A}{8\pi m_N} \mu \left\{ 1 - \left[\frac{1}{2}(3 + \kappa_p) + \left(\frac{m_N}{4 F_\pi} \right)^2 \right] \mu + \mathcal{O}(\mu^2) \right\} . \tag{6.124}$$

However, comparison with experiment is tricky because of the existence of isospin breaking in the pion and nucleon masses, so that there are *two* thresholds, one for $\pi^0 p$

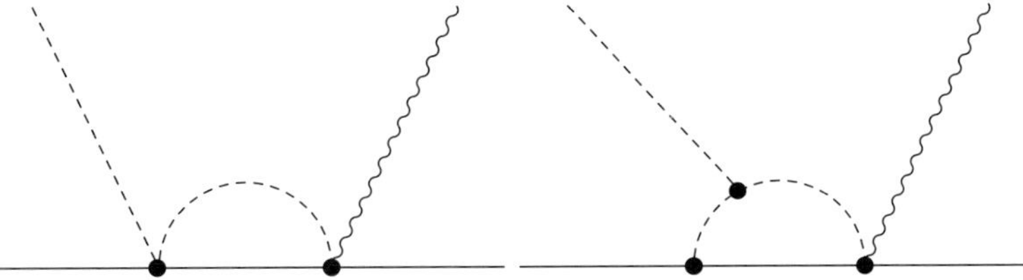

Fig. 6.5 Heavy-baryon diagrams for neutral pion photoproduction. Here the nucleon, pion, and photon are represented by solid, dashed, and wiggly lines, respectively. Both diagrams should be accompanied by appropriate cross diagrams.

	theory	expt.
Table 6.4 Threshold parameters for neutral pion photoproduction		
$E_{0+}(\pi^0 p)(\times 10^{-3}/m_\pi)$	-1.2	-1.31 ± 0.08 [Fuc96]
		-1.32 ± 0.11 [Ber96]
$E_{0+}(\pi^0 n)(\times 10^{-3}/m_\pi)$	2.1	1.9 ± 0.3 [Arg88]
$P_1/|\boldsymbol{q}|(\pi p)(\times \mathrm{GeV}^{-2})$	0.48	0.47 ± 0.01 [Fuc96]
		0.41 ± 0.03 [Ber96]

and the second for $\pi^+ n$, only 7 MeV apart. When the physical masses of the pions are used, recent data from both Mainz and from Saskatoon agree with the chiral prediction. However, there are concerns about the convergence of the chiral expansion, which reads $E_{0+} = C(1 - 1.26 + 0.59 + \cdots)$. There also exist chiral predictions for threshold p-wave amplitudes, which are in good agreement with experiment, as shown in Table 6.4, and for which the convergence is calculated to be rapid.

Finally, there exists a chiral symmetry prediction for the reaction $\gamma n \to \pi^0 n$

$$E_{0+} = -\frac{e g_A}{8\pi m_N}\mu^2 \left\{ \frac{1}{2}\kappa_n + \left(\frac{m_N}{4F_\pi}\right)^2 \right\} + \cdots = 2.13 \times 10^{-3}/m_\pi \, . \qquad (6.125)$$

However, the experimental measurement of such an amplitude involves considerable challenge, and must be accomplished either by use of a deuterium target with the difficult subtraction of the proton contribution and of meson-exchange contributions or by use of a ^{3}He target. Neither of these are straightforward, although some limited data already exist [Arg88].

Other areas wherein chiral predictions can be confronted with experiment include the electric dipole amplitude in electroproduction as well as in weak interactions such as muon capture. However, we do not have the space here to cover this work. We can succinctly summarize the situation by stating that at the present time there exist no significant disagreements with experimental findings, but the predictive power in the baryon sector is only in the near-threshold region, due among other things, to the feature that the expansion is in terms of p/m rather than p^2/m^2 which occurs in the meson case.

6.4 On to Higher Energy: Dispersion Relations

Above we have seen the power of chiral perturbation theory in addressing near-threshold phenomenology. However, we have also observed its associated weakness, i.e., loss of predictive ability in the meson sector once $E, p \geq\sim m_\rho/2$ and in the baryon sector even sooner. In this closing section, we try to address the question of whether one can somehow extend the success of χpt to higher energy without losing its model-independent connections to QCD. Strictly speaking the answer is no, since as the energy–momentum increases one *must* go to higher and higher order in the chiral expansion, which means increasing the number of loops and of unknown counterterms. Any other approach must be model-dependent at some level and hence use more than simply the (broken) chiral symmetry of QCD as an input. Nevertheless, there are some techniques which keep such model dependence at a minimum. One such procedure is to marry the results of χpt with the strictures of dispersion relations [Don97], the validity of which, given that they rely only on the causality properties of the theory, are not in question. In the case of the pion form factor studied above in one-loop χpt the causality condition asserts that the function $F(q^2)$ is analytic in the entire complex q^2-plane except for a cut along the real axis which extends from $4m_\pi^2 < q^2 < \infty$. Along this cut $F(q^2)$ has an imaginary part which is given by the unitarity condition

$$(p_2 - p_1)_\mu \operatorname{Im}(F_1(q^2)) = \frac{1}{2} \int \frac{d^3k_1}{(2\pi)^3 2E_1} \frac{d^3k_2}{(2\pi)^3 2E_2} (2\pi)^4 \delta^4(p_1 + p_2 - k_1 - k_2)$$
$$\times (< \pi\pi |T| \pi\pi >)^* < \pi\pi |j_\mu| 0 > . \tag{6.126}$$

Use of Cauchy's theorem involving the contour as shown in Fig. 6.6 and the assumption that $F(s) \to 0$ as $s \to \infty$ then leads to the dispersion relation

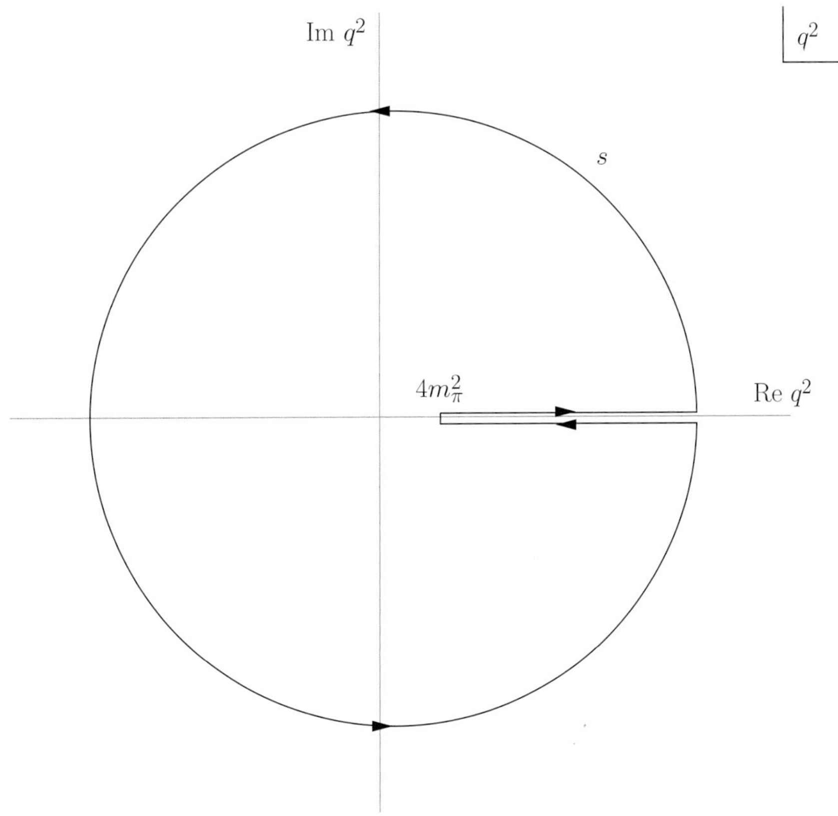

 Shown is the s-plane contour for the charged pion electromagnetic form factor dispersion integral, Eq. (6.127). Here Im $F_1(q^2)$ is assumed to vanish as the radius becomes large and there exists a discontinuity along the real axis in the region $4m_\pi^2 < s < \infty$.

$$F_1(q^2) = \frac{1}{\pi} \int_{4m_\pi^2}^{\infty} \frac{\operatorname{Im} F_1(s) ds}{s - q^2 - i\epsilon} .$$

(6.127)

If the asymptotic condition is *not* satisfied one can use various numbers of subtractions in order to guarantee convergence. For example, provided that $F_1(s)/s \to 0$ as $s \to \infty$ we have

$$
\begin{aligned}
F_1(q^2) - F_1(0) &= \frac{1}{\pi} \int_{4m_\pi^2}^{\infty} \operatorname{Im} F_1(s) ds \left(\frac{1}{s - q^2 - i\epsilon} - \frac{1}{s - i\epsilon} \right) \\
&= \frac{q^2}{\pi} \int_{4m_\pi^2}^{\infty} \frac{\operatorname{Im} F_1(s) ds}{s(s - q^2 - i\epsilon)}
\end{aligned}
$$

(6.128)

and further subtractions may be performed if necessary. Even if such subtractions are not required it may be advantageous to employ them. The point is that in order to get a prediction one requires the value of the imaginary component of the form factor at *all* values of s. However, the more subtractions the less the sensitivity to input from large s, where in general the form factor is not as well determined.

In the case of the pion form factor it is convenient to perform two subtractions, whereby

$$F_1(q^2) = F_1(0) + q^2 F_1'(0) + \frac{q^4}{\pi} \int_{4m_\pi^2}^{\infty} \frac{\operatorname{Im} F_1(s) ds}{s^2(s - q^2 - i\epsilon)} .$$

(6.129)

Here by current conservation we may set $F_1(0) = 1$ while from experiment we have $F_1'(0) = < r_\pi^2 > /6 = (0.073 \pm 0.002)$ fm^2. In order to find $\operatorname{Im} F(s)$ we begin by projecting onto the p-wave $\pi\pi$ channel via

$$T_1^1(s) = \frac{1}{64\pi} \int_{-1}^{1} d(\cos\theta) P_1(\cos\theta) < \pi\pi|T|\pi\pi > = \sqrt{\frac{s}{s - 4m_\pi^2}} e^{i\delta_1^1} \sin\delta_1^1 ,$$

(6.130)

whereby the unitarity condtion Eq. (6.126) reads

$$\operatorname{Im} F_1(s) = e^{-i\delta_1^1} \sin\delta_1^1 F_1(s) .$$

(6.131)

Of course, the solution of this equation is simply the Fermi–Watson theorem

$$F_1(s) = |F_1(s)| \exp i\delta_1^1(s)$$

(6.132)

which states that the phase of the form factor is the p-wave $\pi\pi$ phase shift. Now at lowest-order we have the chiral prediction

$$T_1^1(s) = \frac{s - 4m_\pi^2}{96\pi F_\pi^2}$$

(6.133)

and if this is substituted into Eq. (6.131) along with the lowest-order result $F_1(s) \approx F_1(0) = 1$ we find

$$F_1(q^2) = 1 + \frac{< r_\pi^2 >}{6} q^2 + \frac{1}{96\pi^2 F_\pi^2} \left((q^2 - 4m_\pi^2) H(q^2) + \frac{2}{3} q^2 \right) ,$$

(6.134)

where we have used the integral

$$\int_{4m_\pi^2}^{\infty} \frac{ds}{s^2} \sqrt{\frac{s - 4m_\pi^2}{s}} \left(\frac{a + bs}{s - q^2 - i\epsilon} \right) = \frac{a + bq^2}{q^4} H(q^2) - \frac{a}{6m^2 q^2} .$$

(6.135)

Comparison with the chiral perturbative result, Eq. (6.72), reveals that the results are identical provided that one identifies

$$\frac{< r_\pi^2 >}{6} = \frac{2L_9^r}{F_\pi^2} + \frac{1}{96\pi^2 F_\pi^2}\left(1 + \ln\frac{m_\pi^2}{\mu^2}\right) . \tag{6.136}$$

In this form then we see that the chiral perturbative predictions are simply a result of use of the lowest-order chiral forms in the input to the dispersion integral, while a real world calculation would use experimental data. In this context it is clear that the renormalized counterterms simply play the role of subtraction constants.

So far our dispersive calculation has simply reproduced the chiral perturbative results. How can one do better? Clearly by using experimental data rather then the simple lowest-order chiral forms [Don97]. For example, if we divide Eq. (6.129) by q^2 and then take the limit as $q^2 \to \infty$ we find a sum rule for $L_9^r(\mu)$

$$L_9^r(\mu) = \frac{F_\pi^2}{6\pi}\int_{4m_\pi^2}^\infty \frac{ds}{s^2}\operatorname{Im}F_1(s) - \frac{1}{192\pi^2}\left(1 + \ln\frac{m_\pi^2}{\mu^2}\right) , \tag{6.137}$$

which can be evaluated from experiment. Actual data on $\operatorname{Im}F(s)$ are available only from the $\pi\pi$ threshold up to the middle of the resonance region so other methods must be found in order to generate the high-energy component required in order to perform the dispersive integral. For very high energies we can use the prediction of perturbative QCD

$$F_1(s) \longrightarrow -\frac{64\pi^2}{9}\frac{F_\pi^2}{s\ln\frac{-s}{\Lambda^2}} , \tag{6.138}$$

which corresponds to the asymptotic form

$$\operatorname{Im}F_1(s) \longrightarrow -\frac{64\pi^3}{9}\frac{F_\pi^2}{s\ln^2\frac{-s}{\Lambda^2}} . \tag{6.139}$$

Knowing the low- and high-energy forms of the function, we can generate a smooth matching which joins them in the intermediate-energy region, as shown in Fig. 6.7. Obviously the most striking part of this function is the strong ρ peak, but note also the long negative tail at high energy. That the high-energy component must be negative is easy to see from the fact that Eq. (6.139) guarantees that the dispersion integral converges even without any subtractions. Then, taking the $q^2 \to \infty$ limit produces two additional sum rules

$$0 = \frac{1}{\pi}\int_{4m_\pi^2}^\infty ds\operatorname{Im}F_1(s) \tag{6.140a}$$

$$1 = \frac{1}{\pi}\int_{4m_\pi^2}^\infty \frac{ds}{s}\operatorname{Im}F_1(s) . \tag{6.140b}$$

In particular the former sum rule requires that the large positive contribution generated by integrating over the ρ peak be cancelled by a significant negative result from higher energy. Now in fact both sum rules are satisfied by the spectral function shown in Fig. 6.7,

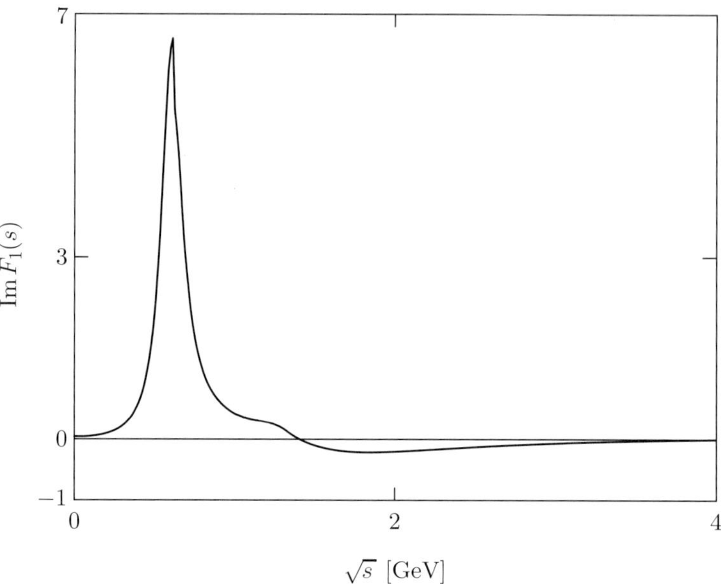

Fig. 6.7 Fit to input data for the charged pion electromagnetic form factor dispersion relation [Don97]. The large peak at $\sqrt{s} \simeq 0.75$ GeV is due to the ρ meson.

as they were used in its construction and are responsible for the small bump to the right of the ρ tail. Having generated a consistent form for Im $F(s)$ we can now use Eq. 6.137 in order to generate a prediction for L_9^r. We find

$$a)\ L_9^r(m_\eta)|^{\text{disp.rel.}} = 0.0074 \ [\text{Don97}]$$

$$\text{compared with} \qquad b)\ L_9^r(m_\eta)|^{\text{expt.}} = 0.0071(3) \tag{6.141}$$

obtained from analysis of low-energy phenomenology (specifically the pion charge radius). Obviously there is good agreement here, confirming the validity of the dispersive approach (and therefore causality!).

However, while we have gained a new understanding of the chiral perturbative results we have not yet succeeded in extending their validity to higher energy. Indeed, Fig. 6.3 shows that, while the form factor $F(q^2)$ matches onto the experimental result at low q^2, it is strongly at variance by the time $q^2 \approx 400\,\text{MeV}^2$ due to the strong rise of the ρ. In fact the prominence of the ρ peak suggests an oft-used approximation, i.e., replacement of the spectral density by a simple delta function

$$\text{Im}\,F_1(s) = \pi m_\rho^2 \delta(s - m_\rho^2)\ . \tag{6.142}$$

This drastic approximation, called "vector dominance," does not satisfy the barely convergent sum rule Eq. (6.140a), but explicitly satisfies the better damped Eq. (6.140b). With respect to that for L_9 we find

$$< r_\pi^2 > = 6m_\rho^2 \simeq 0.40\ \text{fm}^2 \tag{6.143}$$

in reasonable agreement with experiment.

An approach which does allow us to move to higher energy is the "inverse amplitude method," wherein one looks not at the form factor but rather at its inverse [Tru88, Han99]. The unitarity condition in this case reads

$$\text{Im}\,\frac{1}{F_1(s)} = -\frac{e^{i\delta_1^1(s)}\sin\delta_1^1(s)}{F_1(s)} = -\sqrt{\frac{s-4m_\pi^2}{s}}\,\frac{T_1^1(s)}{F_1(s)} \tag{6.144}$$

and a doubly-subtracted dispersion relation reads

$$\frac{1}{F_1(q^2)} = 1 - \frac{<r_\pi^2>}{6}q^2 - \frac{1}{\pi}\int_{4m_\pi^2}^\infty \frac{ds}{s^2(s-q^2-i\epsilon)}\sqrt{\frac{s-4m_\pi^2}{s}}\,\frac{T_1^1(s)}{F_1(s)}\,. \tag{6.145}$$

Using the leading chiral forms, the integration can be performed using Eq. (6.135) and yields

$$F_1(q^2) = \frac{1}{1 - \frac{<r_\pi^2>}{6}q^2 - \frac{1}{96\pi^2 F_\pi^2}\left((q^2-4m_\pi^2)H(q^2) + \frac{2}{3}q^2\right)}\,. \tag{6.146}$$

Expanding in q^2 we observe that this form is consistent with the $\mathcal{O}(p^4)$ chiral result and represents summation of the Lippman–Schwinger equation in terms of a geometric series. (Sometimes this is also called the Pade (1,1) approximant form [Tru88].) Phenomenologically it has the right stuff. Indeed, as shown in Fig. 6.3, the ρ resonance arises very naturally as a pole as $s \to m_\rho^2$ and the phase shift, given by $\delta_1^1 = \arg F(s)$ is in reasonable agreement with experiment. We stress that it is *not* a simple result of chiral symmetry; rather it is a chirally *motivated* form which reduces to the chiral result at low energy.

It is interesting to note that there is an exact solution to the problem of finding a function whose phase is δ_1^1 along the real axis from $4m_\pi^2 < s < \infty$. This is the Omnes solution and has the general form [Omn58]

$$F_1^{\text{Omnes}}(q^2) = P(q^2)\exp\frac{q^2}{\pi}\int_{4m_\pi^2}^\infty \frac{ds}{s}\,\frac{\delta_1^1(s)}{s-q^2-i\epsilon}\,. \tag{6.147}$$

Here $P(s)$ is an arbitrary polynomial which for our case is set equal to unity. Using *experimental* $\pi\pi$ phase shifts one can construct the function $F_0(q^2)$ and compare with experiment and with the inverse amplitude form. Agreement with experiment is basically quite good. Note that the Omnes solution assumes the validity of the Fermi–Watson theorem along the entire real axis and hence ignores the substantial inelasticity which arises above $\sim 1.2\,\text{GeV}$. Thus this form should presumably only be trusted up to energies where such inelasticity takes over.

One can in similar fashion construct a form for the partial-wave scattering amplitude which is unitary

$$\text{Im}\,\frac{1}{T_1^1(s)} = -i\sqrt{\frac{s-4m_\pi^2}{s}} \tag{6.148}$$

and agrees with the $\pi\pi$ phase shifts found from the form factor as

$$T_1^1(s) = \frac{1}{96\pi^2 F_\pi^2} \frac{(s - 4m_\pi^2)}{1 - \frac{\langle r_\pi^2 \rangle}{6} s - \frac{1}{96\pi^2 F_\pi^2}\left((s - 4m_\pi^2)H(s) + \frac{2}{3}s\right)} . \tag{6.149}$$

In fact, such a form was written down long before the development of chiral perturbative techniques as a simple unitary generalization of the lowest-order (Weinberg) scattering amplitude [Bro68]. It is sometimes called the N/D form, where D is the inverse form factor and satisfies the Fermi–Watson theorem along the right-hand cut, while N is the lowest-order chiral amplitude and is analytic along this discontinuity. Of course, Eq. (6.149) violates crossing symmetry in that it does not have the proper discontinuity along the left-hand (u-channel) cut, but we may hope that this does xnot matter along the physical (right-hand) cut, as it is far away.

We see then that by the use of chiral-based methods, one *can* construct analytic forms for observables which satisfy the strictures of unitarity exactly (not perturbatively as in the case of chiral perturbation theory itself) and which reduce to the forms demanded by chiral symmetry at low energy. They are admittedly no longer model-independent, nor do they satisfy all the strictures of field theory (such as crossing symmetry). However, any reasonable model must assume a similar form and they can be used with some confidence to construct a successful phenomenology.

Exercises

6.1 $\pi\pi$ Scattering Lengths

Consider the lowest-order chiral Lagrangian

$$\mathcal{L} = \frac{F_\pi^2}{4}\mathrm{Tr}(D_\mu U D^\mu U^\dagger) + \frac{F_\pi^2}{4}\mathrm{Tr}(2B_0 m(U + U^\dagger)),$$

where m is the quark mass matrix and $B_0 = m_\pi^2/(m_u + m_d)$.

a) Expand to $\mathcal{O}(\phi^2)$ and demonstrate that

$$\mathcal{L}_2 = \sum_{j=1}^{3}\left(\frac{1}{2}\partial^\mu \phi_j \partial_\mu \phi_j - \frac{m_\pi^2}{2}\phi_j^2\right).$$

b) Expand to $\mathcal{O}(\phi^4)$ and show that in $SU(2)$

$$\mathcal{L}_4 = \frac{1}{6F_\pi^2}\boldsymbol{\phi}\cdot\boldsymbol{\phi}\boldsymbol{\phi}\cdot\Box\boldsymbol{\phi} + \frac{1}{2F_\pi^2}(\boldsymbol{\phi}\cdot\partial_\mu\boldsymbol{\phi})^2 + \frac{m_\pi^2}{24F_\pi^2}(\boldsymbol{\phi}\cdot\boldsymbol{\phi})^2 .$$

c) Evaluate s-wave scattering and show that the I = 0, 2 scattering lengths are given by

$$a_0^0 = \frac{7m_\pi^2}{32\pi F_\pi^2} \qquad a_0^2 = -\frac{m_\pi^2}{16\pi F_\pi^2} .$$

d) Evaluate p-wave scattering and show that the $I = 1$ scattering length is given by

$$a_1^1 = \frac{m_\pi^2}{24\pi F_\pi^2}.$$

These are the so-called Weinberg scattering lengths.

6.2　Pion Form Factor and Unitarity

Verify that the pion form factor as calculated in chiral perturbation theory

$$F_{1\pi}(q^2) = 1 + \frac{2L_9^r(\mu)}{F_\pi^2}q^2 + \frac{1}{96\pi^2 F_\pi^2}\left[(q^2 - 4m_\pi^2)H\left(\frac{q^2}{m_\pi^2}\right) - q^2\log\frac{m_\pi^2}{\mu^2} - \frac{q^2}{3}\right],$$

where

$$H(x) = 2 + \sqrt{\frac{x-4}{4}}\left[\log\frac{-1+\sqrt{\frac{x-4}{x}}}{1+\sqrt{\frac{x-4}{x}}} + i\pi\theta(x-4)\right],$$

satisfies the unitarity requirement

$$2\mathrm{Im}\,F_{1\pi}(q^2)(p_1 - p_2)_\mu = -\int \frac{d^3q_1 d^3q_2}{(2\pi)^6 2q_1^0 2q_2^0}(2\pi)^4\delta^4(p_1 + p_2 - q_1 - q_2)$$
$$\times (q_1 - q_2)_\mu < \pi^+(q_1)\pi^-(q_2)|T|\pi^+(p_1)\pi^-(p_2) >,$$

where T is the two-derivative (tree-level) pion–pion scattering amplitude.

6.3　Haag's Theorem

A popular model which is chiral symmetric and renormalizable is the linear sigma model, which involves the interaction of pions with a scalar meson σ and can be written in the "standard" form

$$\mathcal{L}_{\mathrm{std}} = \frac{1}{2}\partial_\mu\boldsymbol{\pi}\cdot\partial^\mu\boldsymbol{\pi} + \frac{1}{2}[\partial_\mu\sigma\partial^\mu\sigma - m_\sigma^2\sigma^2]$$
$$- \lambda F_\pi\sigma(\sigma^2 + \boldsymbol{\pi}^2) - \frac{1}{4}\lambda(\sigma^2 + \boldsymbol{\pi}^2)^2,$$

where $m_\sigma = \sqrt{2\lambda}F_\pi$. After integrating out the σ this model can be written as an effective Lagrangian involving only the pions

$$\mathcal{L}_{\mathrm{eff}} = \frac{F_\pi^2}{4}\mathrm{Tr}(\partial_\mu U\partial^\mu U^\dagger),$$

where

$$U = \exp\frac{i}{F_\pi}\boldsymbol{\tau}\cdot\boldsymbol{\pi}.$$

Calculate $\pi^+\pi^0$ scattering at lowest (tree) order using two very different forms of the chiral effective Lagrangian.

a) In the standard form of the linear σ-model show that

$$\mathcal{L}_{\mathrm{std}} = -\frac{\lambda}{4}(\boldsymbol{\pi}\cdot\boldsymbol{\pi})^2 - \lambda F_\pi\sigma\boldsymbol{\pi}\cdot\boldsymbol{\pi} + \cdots$$

and

$$\mathcal{M}_{\pi^+\pi^0 \to \pi^+\pi^0} = -2i\lambda + (-2i\lambda F_\pi)^2 \frac{i}{q^2 - m_\sigma^2}$$

$$= -2i\lambda \left[1 + \frac{2\lambda F_\pi^2}{q^2 - 2\lambda F_\pi^2} \right] = i\frac{q^2}{F_\pi^2} + \cdots,$$

where $q = p_{\pi_f^+} - p_{\pi_i^+}$ is the momentum transfer.

b) In the exponential representation, demonstrate that

$$\mathcal{L}_{eff} = \frac{1}{6F_\pi^2} \left[(\boldsymbol{\pi} \cdot \partial_\mu \boldsymbol{\pi})^2 - \boldsymbol{\pi} \cdot \boldsymbol{\pi} \partial_\mu \boldsymbol{\pi} \cdot \partial^\mu \boldsymbol{\pi} \right] + \cdots$$

and

$$\mathcal{M}_{\pi^+\pi^0 \to \pi^+\pi^0} = i\frac{q^2}{F_\pi^2} + \cdots .$$

Note that despite very different looking forms, the physics is the same. This is guaranteed by a deep result in quantum field theory called Haag's theorem.

6.4   Compton Scattering in Effective Field Theory

The lowest order Hamiltonian for Compton scattering is given in quantum mechanics by

$$H_0 = \frac{e^2 \mathbf{A}^2}{2m} \phi^* \phi$$

where ϕ is the wavefunction of a spinless particle, where e, m are the particle's charge, mass respectively.

a) Calculate the laboratory frame cross section for unpolarized photons from this spin zero system and show that it takes the form

$$\frac{d\sigma}{d\Omega} = \frac{\alpha^2}{2m^2} \left(\frac{\omega'}{\omega}\right)^2 (1 + \cos^2\theta)$$

where $\alpha = e^2/4\pi$ is the fine structure constant and ω, ω' are the frequencies of incoming and outgoing photons, respectively.

Now, include the first order corrections due to electric and magnetic polarizabilities α_E and β_M, which are described by the Hamiltonian

$$H_1 = -\frac{1}{2}[4\pi \alpha_E \mathbf{E}^2 + 4\pi \beta_M \mathbf{H}^2] \phi^* \phi .$$

b) Calculate the cross section, including the polarizability effects to first order, and show that

$$\frac{d\sigma}{d\Omega} = \frac{\alpha^2}{m^2} \left(\frac{\omega'}{\omega}\right)^2 \left\{ \frac{1}{2}(1 + \cos^2\theta) - \frac{m\omega\omega'}{2\alpha} \cdot [(\alpha_E + \beta_M)(1 + \cos\theta)^2 \right.$$

$$\left. + (\alpha_E - \beta_M)(1 - \cos\theta)^2] \right\} .$$

Thus, polarizabilities may be measured via experimental study of the Compton cross section, as discussed in the text.

6.5   **Applied Compton Scattering: Why the Sky is Blue**

Quantum mechanics should be able to give a simple account of two very obvious facts:

a) The sky is blue (and the sunset red).
b) The skylight is highly polarized.

(If you haven't had direct experience with the polarization, borrow a pair of polaroid sunglasses and study photons coming from the sun scattered nearly perpendicularly.) For the classical explanation of these phenomena read the *Feynman Lectures On Physics*, Vol. I, pp. 32–6 to 32–9 [Fey63].

a) Frequency dependence of the scattering cross section: As a model for the sky, consider a gas of one-electron atoms in the ground state $|A\rangle$. Consider elastic (Rayleigh) scattering of photons by one of the atoms in the case that the photon energy is much smaller than the energy separation between $|A\rangle$ and any excited states $|I\rangle$. (Note: $\omega \approx \frac{1}{2}$ eV for visible light compared to $(E_I - E_A)_{\min} = 10.4$ eV for hydrogen.) Show that, in the long wavelength approximation, the cross section is given by

$$\frac{d\sigma}{d\Omega} = m^2 r_0^2 \omega^4 \left| \sum_I \left[\left(\hat{\epsilon}_f^* \cdot \boldsymbol{r} \right)_{AI} \left(\hat{\epsilon}_i \cdot \boldsymbol{r} \right)_{IA} + \left(\hat{\epsilon}_i \cdot \boldsymbol{r} \right)_{AI} \left(\hat{\epsilon}_f^* \cdot \boldsymbol{r} \right)_{IA} \right] \frac{1}{\omega_{IA}} \right|^2 .$$

The strong increase in cross section with frequency ($\sim \omega^4$) gives the blue color of the sky and the red color of the sunset. Rayleigh was the first to understand this many years ago.

Hint: Start from the Kramers–Heisenberg formula. Take $|B\rangle = |A\rangle$ and $\omega_f = \omega_i = \omega$. Expand the energy denominators in powers of ω/ω_{IA}

$$-\frac{1}{E_A - E_I - \omega} = \frac{1}{\omega_{IA}} \left(1 + \frac{\omega}{\omega_{IA}} + \frac{\omega^2}{\omega_{IA}^2} + \cdots \right)$$

keeping terms up to second order. Show that the lowest order terms combine to give $-\hat{\epsilon}_f^* \cdot \hat{\epsilon}_i$, and exactly cancel the "seagull term" arising from $\boldsymbol{A} \cdot \boldsymbol{A}$. Eliminate the $\frac{1}{\omega_{IA}}$ by noting that

$$<\boldsymbol{p}>_{AI} = -im\omega_{IA} <\boldsymbol{r}>_{AI} , \qquad <\boldsymbol{p}>_{IA} = im\omega_{IA} <\boldsymbol{r}>_{IA} .$$

Why?

Use completeness to sum over states, $|I\rangle$ and note the commutation role of operators $\boldsymbol{r}$ and $\boldsymbol{p}$. Show that the first order terms give zero (by repeating the previous trick). Finally, demonstrate that the second order terms combine to give the desired result, neglecting higher powers of ω.

b) Polarization of scattered light: Study the case $\hat{k}_i = \hat{e}_z, \hat{k}_f = \hat{e}_y$. If $\hat{e}_i = \hat{e}_x$ show that

$$\text{Amp}(\hat{\epsilon}_f = \hat{e}_z) = 0 \qquad \text{Amp}(\hat{\epsilon}_f = \hat{e}_x) \neq 0 \ .$$

If $\hat{\epsilon}_i = \hat{e}_y$ show that there is no scattering in the y direction.

c) What can you say about the polarization of photons scattered in the y-direction if the incident photon beam is completely unpolarized? What is the relevance of this result to the effectiveness of polaroid sunglasses?

d) Why isn't the sky violet, since even more violet light is scattered than blue?

We now progress from focusing primarily on issues of the Standard Model and QCD. Since throughout much of the rest of the book a focus will be placed on lepton scattering, particularly on electron scattering, we collect in this chapter some of the basic ideas behind this subject. The emphasis will be placed on the general aspects of the leptonic and hadronic tensors which, when contracted together, yield the basic quantity that determines the semi-leptonic cross section. Furthermore, the detailed discussions will be limited to inclusive scattering, namely, to reactions where the scattered lepton is assumed to be detected, but where any outgoing particles other than this lepton are not detected; the latter reaction where the lepton and at least one other particle are detected is called semi-inclusive and will be treated in later chapters (Chapter 9 for particle physics and Chapter 16 for nuclear physics). In addition to the general formalism for parity-conserving and parity-violating electron scattering, an introduction to the development of electromagnetic (EM) multipole operators is presented, as this will be important for later chapters where discrete states in nuclei that have good angular momentum and parity quantum numbers are considered.

After the formal developments in the present chapter we progress in Chapter 8 to *elastic* electron scattering from the proton and the insights that this gives us into the structure of the nucleon and its relevance for low-Q^2, strong-coupling QCD. Chapter 9 goes on to treat *inelastic ep* scattering with its underlying partonic modeling. Then, after an insertion (Chapter 10) where high-energy hadronic reactions are discussed as the way one studies quark–quark, quark–gluon and gluon–gluon interactions, we continue with the succeeding chapters where the developments here are employed – in few-body nuclei in Chapters 11 and 12, and in many-body nuclei in Chapters 13–16.

The lepton scattering process from a nucleon target provides a detailed "snapshot" of the target structure at a particular level of resolution. The highest resolution is provided by energetic, charged leptons, which interact with individual quarks and antiquarks inside a proton or neutron. These interactions, being sensitive to the motion of the struck quarks, can map the probability for finding the various constituents as a function of the fraction they carry of the nucleon's overall momentum. Such detailed maps provide crucial tests for theoretical calculations of nucleon structure. At the same time, less energetic electron beams must be used to obtain a lower-resolution, but more global view of the nucleon's properties. These include its overall size and shape, distributions of charge and magnetism, and its deformability when subjected to external electric and magnetic fields.

Using electron scattering to study hadron structure has many advantages. Both the weakness and precision understanding of the probe provide powerful capabilities. Multiple scattering effects are negligible and perturbation of the initial state of the nucleus is minimal. The ability to vary independently the momentum and energy transferred to

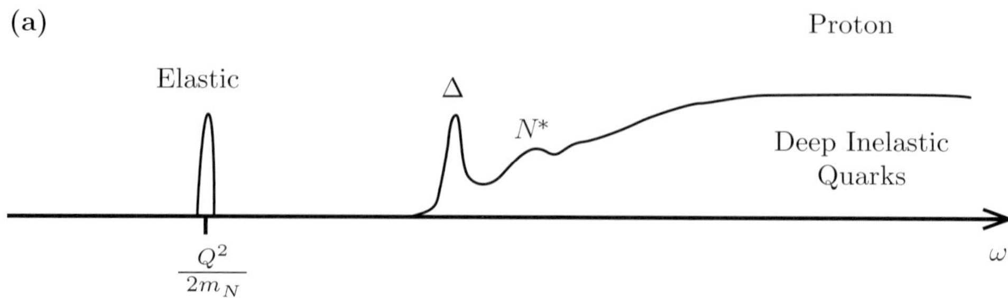

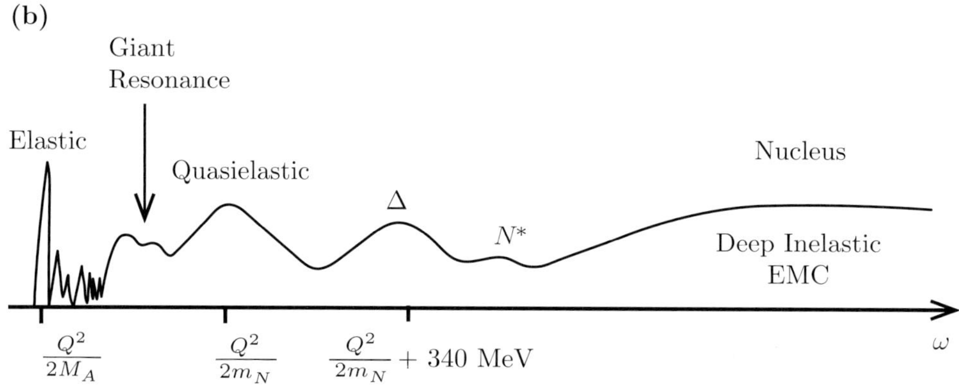

Fig. 7.1 Schematic cross section versus incident electron energy loss for inclusive electron scattering from a proton (a) and from a nucleus (b); the figure is adapted from [Fro87].

the nucleon or nucleus allows the mapping of the spatial distributions of the constituent particles. Because electrons are point particles, they provide superb spatial resolution that can be adjusted to the scale of processes that need to be studied. The spatial scale probed, Δx, is related to the momentum transfer q from the scattered electron. They are approximately related via the Heisenberg uncertainty equation as $\Delta x \approx 200\,\text{MeV}/c\cdot\text{fm}\cdot/q$. We will see in Chapter 9 that electron energies of order $10\,\text{GeV}$ have a spatial resolution sufficient to probe quark distributions, while incident electron energies of about $0.5\,\text{GeV}$ probe a spatial resolution of order $0.5\,\text{fm}$, which is ideally suited for the study of nucleon distributions in nuclei. In Fig. 7.1, the cross section is shown schematically as a function of electron energy loss for a proton (Fig. 7.1a) and for a nucleus (Fig. 7.1b). With increasing energy loss in electron scattering from the proton, there is the elastic final-state, next the resonance states of the proton dominated by the $\Delta(1232)$, and finally, at sufficiently high energies, the deep inelastic regime as described in detail in Chapter 9. With increasing energy loss in electron scattering from a nucleus, there is again the elastic final-state, then the low-lying excited states of the nucleus, next the giant resonance region, followed by the quasielastic region, then the region of the nucleon resonances, and finally the deep inelastic scattering regime, this time for the nucleus.

The realization of powerful beams of electrons for study of hadron structure had its origins in the development of radar technology in the middle of the twentieth

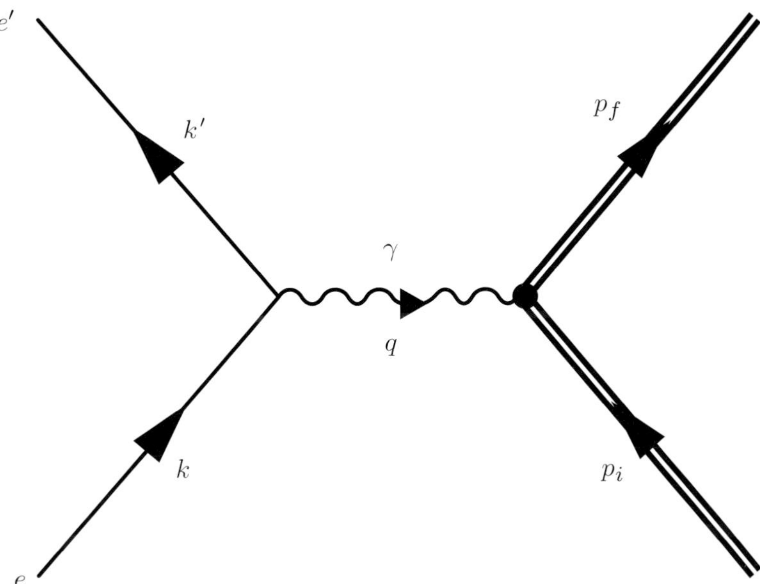

Fig. 7.2 The basic Feynman diagram for electron scattering from a nucleus or from an individual nucleon. The kinematic variables indicated in the figure are discussed in more detail in the text.

century. Devices, especially the magnetron and klystron, were developed, which allowed acceleration of beams of electrons to relativistic energies. In the succeeding decades, the linear accelerator technology was pushed to ever increasing energies, e.g., at the Stanford Linear Accelerator Center (SLAC). Over the past decade, new technical developments have made available intense continuous electron beams, e.g., at the continuous electron beam accelerator facility (CEBAF) at the Thomas Jefferson national accelerator facility (which we will refer to as Jefferson Laboratory).

Let us begin with some of the essentials of charged lepton (electron or muon) scattering. Referring to Fig. 7.2, an electron with four-momentum[1] $k^\mu = (\epsilon, \mathbf{k})$ and spin projection λ is scattered through an angle θ_e to four-momentum $k'^\mu = (\epsilon', \mathbf{k}')$. The virtual photon exchanged in the process carries four-momentum transfer $q^\mu = (\omega, \mathbf{q})$ and, in interacting with the target, causes it to proceed from state $|i\rangle$ with total four-momentum p_i^μ and rest mass M_i to state $|f\rangle$ with total four-momentum p_f^μ and rest mass M_f. Note that in high-energy physics discussions it is more conventional to use ν instead of ω to denote the energy transfer in the scattering. Conservation of four-momentum tells us that $q^\mu = k^\mu - k'^\mu = p_f^\mu - p_i^\mu$. Furthermore, we employ the so-called SLAC convention and define $Q^2 \equiv -q^\mu q_\mu = q^2 - \omega^2 > 0$ for lepton scattering corresponding to the exchange

---

[1] Four-vectors are generally indicated by $k^\mu = (k^0, k^1, k^2, k^3) = (k^0, \mathbf{k})$, where μ runs over the Lorentz indices 0, 1, 2, and 3; also one has $k_\mu = (k^0, -k^1, -k^2, -k^3)$. Boldface $\mathbf{k}$ is used to indicate a three-vector and $k = |\mathbf{k}|$ its magnitude. A dot-product of two four-vectors a^μ and b^μ is given by $a \cdot b = a^\mu b_\mu = a^0 b^0 - (a^1 b^1 + a^2 b^2 + a^3 b^3) = a^0 b^0 - \mathbf{a} \cdot \mathbf{b}$, where the summation convention occurs when one has repeated indices, as here. For other conventions see the Appendices.

of spacelike ($Q^2 > 0$) virtual photons. The real-photon point involved in photoexcitation or gamma decay corresponds to the limit $Q^2 \to 0$.

For completeness, and because they are used throughout much of the book, we collect some of the basic kinematical relationships here. The energy transfer is given by

$$\omega = \nu = \epsilon - \epsilon' \,, \tag{7.1}$$

the square of the three-momentum transfer is given by

$$q^2 = k^2 + k'^2 - 2kk' \cos\theta_e \,, \tag{7.2}$$

and therefore the square of the four-momentum transfer is

$$Q^2 = -q_\mu q^\mu = q^2 - \omega^2 = q^2 - \nu^2 \tag{7.3}$$

$$= 2\left[\epsilon\epsilon' - kk' \cos\theta_e - m_e^2\right] \tag{7.4}$$

$$= 4kk' \sin^2\theta_e/2 + 2\left[\epsilon\epsilon' - kk' - m_e^2\right] \geq 0 \,. \tag{7.5}$$

Note that for inelastic scattering one has $Q^2 > 0$ and that for elastic scattering, while Q^2 can be zero, this happens only when $q = \omega = 0$. See Exercise 7.7 for more on what happens when the scattering angle approaches very close to zero. In the long wavelength limit (LWL) where $\epsilon \gg m_e$ and $k' \gg m_e$, implying that $k \cong \epsilon$ and $k' \cong \epsilon'$, one has

$$Q^2 \cong 4\epsilon\epsilon' \sin^2\theta_e/2 \,. \tag{7.6}$$

Another variable often used in high-energy physics is the so-called Bjorken-x variable,

$$x \equiv \frac{Q^2}{2p_i \cdot q} = \frac{Q^2}{2M_i\nu}\,\text{(Lab system)} \,, \tag{7.7}$$

where $0 \leq x \leq 1$. Finally, the invariant mass of the final state, M_f, is often denoted by W, where one has

$$W = \sqrt{(M_i + \nu)^2 - q^2} = \sqrt{M_i^2 + 2M_i\nu - Q^2} \geq M_i \,. \tag{7.8}$$

Referring again to Fig. 7.1, we see that the schematic behavior shown there for electron scattering from the proton and from nuclei can be plotted versus ω as shown (or $\nu = \omega$), versus W using Eq. (7.8), or versus x using Eq. (7.7) – any choice captures the same information and is simply a matter of what is the commonly adopted convention. For example, the Bjorken-x variable is often used when discussing electron scattering from the proton. In that case, elastic scattering occurs at $x = 1$, while inelastic scattering (production of the Δ, N^*s, DIS, etc., as discussed in later chapters) occurs when $x < 1$.

The cross section involves the square of the absolute value of the invariant matrix element, $\mathcal{M}_{fi}$, which is in turn made up as a product of three factors – the electron current j^μ, the photon propagator $g_{\mu\mu'}/Q^2$ and the hadronic current matrix element, $J_{fi}^{\mu'}$:

$$d\sigma \sim |\mathcal{M}_{fi}|^2 \sim \left| j_\mu \frac{1}{Q^2} J_{fi}^\mu \right|^2 = \frac{1}{Q^4} j_\mu^* j_\nu \cdot J_{fi}^{\mu*} J_{fi}^\nu \,. \tag{7.9}$$

We must perform the appropriate average-over-initial and sum-over-final states (indicated by $\overline{\sum}$) in obtaining the cross section pertaining to the actual experimental conditions (e.g., charged lepton polarized or not, target polarized or not, inclusive or exclusive scattering, etc.). This yields the leptonic and hadronic tensors

$$L_{\mu\nu} \equiv \overline{\sum}_{\text{leptons}} j_\mu^* j_\nu \tag{7.10}$$

$$W^{\mu\nu} \equiv \overline{\sum}_{\text{hadrons}} J_{fi}^{\mu*} J_{fi}^\nu , \tag{7.11}$$

whose contraction is involved in forming the cross section:

$$d\sigma \sim \frac{1}{Q^4} L_{\mu\nu} W^{\mu\nu} . \tag{7.12}$$

It proves useful to decompose both leptonic and hadronic tensors into pieces which are symmetric (s) or antisymmetric (a) under the interchange $\mu \leftrightarrow \nu$:

$$W^{\mu\nu} = W_s^{\mu\nu} + W_a^{\mu\nu} \tag{7.13}$$

$$L_{\mu\nu} = L_{\mu\nu}^s + L_{\mu\nu}^a . \tag{7.14}$$

In contracting leptonic and hadronic tensors no cross-terms are allowed. Furthermore, we shall make use of the fact that both the leptonic and hadronic electromagnetic currents are conserved:

$$q_\mu W_s^{\mu\nu} = q_\mu W_a^{\mu\nu} = 0 \tag{7.15}$$

$$q^\mu L_{\mu\nu}^s = q^\mu L_{\mu\nu}^a = 0 . \tag{7.16}$$

Here we focus on the parity-conserving (PC) leptonic tensor $L_{\mu\nu}$. Substituting for the electron current

$$j_\mu(k', \lambda'; k, \lambda) \sim \bar{u}(k', \lambda') \gamma_\mu u(k, \lambda) \tag{7.17}$$

in Eq. (7.10), we have

$$L_{\mu\nu} = \overline{\sum} \left[\bar{u}(k, \lambda) \gamma_\mu u(k', \lambda') \right] \left[\bar{u}(k', \lambda') \gamma_\nu u(k, \lambda) \right] , \tag{7.18}$$

where properties [Bjo64] of the spinors u and gamma matrices γ have been used to obtain the complex conjugate of the lepton current as the first group of three factors in Eq. (7.18). We can guarantee that electrons and not positrons occur by inserting projection operators[2] $(\not{k}' + m_e)/2m_e$ and $(\not{k} + m_e)/2m_e$ in the appropriate places:

$$L_{\mu\nu} = \frac{1}{4m_e^2} \overline{\sum} \bar{u}(k, \lambda) \gamma_\mu (\not{k}' + m_e) u(k', \lambda') \bar{u}(k', \lambda') \gamma_\nu (\not{k} + m_e) u(k, \lambda) . \tag{7.19}$$

7.1 Unpolarized Electron Scattering

Let us now consider unpolarized electron scattering, in which case $\overline{\sum}$ corresponds to an unrestricted sum over all four components in the spinors divided by two for the

---

[2]   As is common practice, we use a slash through a four-vector to indicate the scalar product of the four-vector with a gamma-matrix.

initial-state spin average. This is in fact simply a trace which can be evaluated using standard techniques [Bjo64]:

$$L_{\mu\nu}^{\text{unpol}} = \frac{1}{8m_e^2} \text{Tr} \left\{ \gamma_\mu (\slashed{k}' + m_e) \gamma_\nu (\slashed{k} + m_e) \right\} \tag{7.20}$$

$$= \frac{1}{2m_e^2} \left[k_\mu k_\nu' + k_\mu' k_\nu - g_{\mu\nu} \left(k \cdot k' - m_e^2 \right) \right] . \tag{7.21}$$

This tensor is symmetric under the interchange $\mu \leftrightarrow \nu$ and satisfies the current conservation condition, $q^\mu L_{\mu\nu}^{\text{unpol}} = 0$. Various classes of response will arise when particular combinations of the Lorentz indices μ and ν are selected. For example, unpolarized single-arm electron scattering (detection only of the scattered electron) involves the combinations

$$\ell_L \equiv 4m_e^2 \times [L_{00} - 2L_{03} + L_{33}]$$
$$= 4m_e^2 \times [L_{00}(Q^2/q^2)^2] \tag{7.22}$$

$$\ell_T \equiv 4m_e^2 \times \frac{1}{2}[L_{11} + L_{22}] , \tag{7.23}$$

where here the labels L and T refer to projections of the current matrix elements longitudinal and transverse to the virtual photon direction, respectively; q is taken to be along the z-axis. In the extreme relativistic limit (ERL), where the electron's mass may be neglected with respect to its energy, one obtains the expressions, $\ell_L \to v_0 v_L$ and $\ell_T \to v_0 v_T$, involving the longitudinal and transverse electron kinematic factors

$$v_L = \rho^2 \tag{7.24}$$

$$v_T = \frac{1}{2}\rho + \tan^2 \theta_e/2 , \tag{7.25}$$

where $v_0 \equiv (\epsilon + \epsilon')^2 - q^2$ and $\rho \equiv Q^2/q^2 = 1 - (\omega/q)^2$ which implies that $0 \leq \rho \leq 1$.

We next construct the parity-conserving hadronic tensor. The kinematic situation is shown in Fig. 7.2. The strategy is the following: we must build the hadronic tensors from q^μ, p_i^μ, and p_f^μ. In fact, we can use momentum conservation to eliminate one, say $p_f^\mu = p_i^\mu + q^\mu$ leaving two independent four-momenta: $\{q^\mu, p_i^\mu\}$. The possible Lorentz scalars in the problem are Q^2, p_i^2, and $q \cdot p_i$. Since we presumably know what the target is, and since $p_i^2 = M_i^2$, we are left with two independent scalars to vary: $\{Q^2, q \cdot p_i\}$. Moreover, since $Q^2 = q^2 - \omega^2$ and $q \cdot p_i = \omega M_i$ in the laboratory system, we can regard our hadronic tensors to be functions of $\{Q^2, q \cdot p_i\}$, $\{q, \omega\}$, $\{q, v\}$ or $\{Q^2, x\}$, where $x \equiv Q^2/2m_N v$, Bjorken-x (later used in Chapter 9), with m_N being the nucleon mass.

Now we wish to write $W_s^{\mu\nu}$ and $W_a^{\mu\nu}$ in terms of the two independent four-vectors q^μ and p_i^μ. Alternatively, instead of p_i^μ it turns out to be useful to employ the four-vector

$$V_i^\mu \equiv \frac{1}{M_i} \left\{ p_i^\mu + \left(\frac{q \cdot p_i}{Q^2} \right) q^\mu \right\} , \tag{7.26}$$

which is especially convenient because $q \cdot V_i = 0$, by construction. The tensor $W^{\mu\nu}$ must be of second-rank and so we can write the following general expansions:

$$W_s^{\mu\nu} = X_1 g^{\mu\nu} + X_2 q^\mu q^\nu + X_3 V_i^\mu V_i^\nu + X_4 \left(q^\mu V_i^\nu + V_i^\mu q^\nu \right) \tag{7.27}$$

$$W_a^{\mu\nu} = iY_1 \left(q^\mu V_i^\nu - V_i^\mu q^\nu \right) + iY_2 \epsilon^{\mu\nu\alpha\beta} q_\alpha V_{i\beta} , \tag{7.28}$$

where the scalar response functions (the Xs and Ys) depend on the scalars discussed above:

$$X_\ell = X_\ell\left(Q^2, q \cdot p_i\right) \qquad \ell = 1,\ldots,4$$

$$Y_\ell = Y_\ell\left(Q^2, q \cdot p_i\right) \qquad \ell = 1,2 \,. \tag{7.29}$$

In the absence of parity-violating (PV) effects from the weak interaction (see below), the hadronic electromagnetic current matrix elements are polar-vectors and so the tensors here must have specific properties under spatial inversion. In particular, the ϵ-terms in Eq. (7.28) have the wrong behavior and so Y_2 must vanish. Finally, we must make use of the current conservation conditions in Eqs. (7.15, 7.16).

These lead us to the following expressions (recall that $q \cdot V_i = 0$)

$$q_\mu W_s^{\mu\nu} = 0 = (X_1 - X_2 Q^2)q^\nu - (X_4 Q^2)V_i^\nu \tag{7.30}$$

$$q_\mu W_a^{\mu\nu} = 0 = -(Y_1 Q^2)V_i^\nu \tag{7.31}$$

and so $X_1 - X_2 Q^2 = 0$, $X_4 = 0$, and $Y_1 = 0$, using the linear independence of q^ν and V_i^ν. Using more conventional nomenclature, defining $W_1 \equiv -X_1$ and $W_2 \equiv X_3$, we have then rederived the results [Von60]

$$W_s^{\mu\nu} = -W_1\left(g^{\mu\nu} + \frac{q^\mu q^\nu}{Q^2}\right) + W_2 V_i^\mu V_i^\nu \tag{7.32}$$

$$W_a^{\mu\nu} = 0 \,. \tag{7.33}$$

Contracting these hadronic tensors with the leptonic tensors obtained above, we obtain for unpolarized electron scattering

$$\chi_{\mu\nu}^{\text{unpol}} W_s^{\mu\nu} \sim W_2 + 2W_1 \tan^2 \theta_e/2 \,, \tag{7.34}$$

where only the symmetric responses enter. This form for inclusive unpolarized electron scattering (see [Von60]) contains only two independent response functions W_1 and W_2 which may be separated at fixed q and ω (or, equivalently, fixed Q^2 and $q \cdot P_i$) by varying the electron scattering angle to make a so-called Rosenbluth decomposition.

On the other hand, for polarized PC electron scattering

$$\chi_{\mu\nu}^{\text{pol}} W_a^{\mu\nu} = 0 \,, \tag{7.35}$$

where now only the antisymmetric responses enter. Thus, if only PC interactions are considered and if only inclusive scattering with no hadronic polarizations is discussed, then no differences will be seen when the electron's helicity is flipped from $+1$ to -1; however, this statement does not obtain for the case of PV electron scattering, as we shall see in the last section of this chapter.

When we contract the unpolarized leptonic tensor developed above with the general hadronic tensor for inclusive electron scattering from unpolarized targets, we obtain a sum involving projections of the current matrix elements. It is convenient to choose these to be transverse (T) or longitudinal (L) with respect to the direction $\mathbf{q}$. Thus, we obtain a structure of the form $\ell_L W^L + \ell_T W^T$ (see Eqs. 7.22, 7.23)) and can write the inclusive cross section in the laboratory system as

$$\left(\frac{d\sigma}{d\Omega_e d\omega}\right)^{\text{unpol}} = \frac{1}{M_i}\sigma_{\text{Mott}}\left\{v_L W^L + v_T W^T\right\} \tag{7.36}$$

$$= \frac{\rho}{2M_i\mathcal{E}}\sigma_{\text{Mott}}\left\{2\rho\mathcal{E}W^L + W^T\right\}, \tag{7.37}$$

using the ERL and hence Eqs. (7.24, 7.25). Here the Mott cross section is given by

$$\sigma_{\text{Mott}} = \left\{\frac{\alpha\cos\theta_e/2}{2\epsilon\sin^2\frac{\theta_e}{2}}\right\}^2 \tag{7.38}$$

and we have introduced the quantity

$$\mathcal{E} \equiv \left[1 + \frac{2}{\rho}\tan^2\theta_e/2\right]^{-1}, \tag{7.39}$$

referred to as the photon's degree of longitudinal polarization to represent the "virtualness" of the exchanged photon ($\mathcal{E} \to 1(0)$ when $\theta_e \to 0°(180°)$). The responses W^L and W^T are related to W_1 and W_2 by

$$W^L = (W_2 - \rho W_1)/\rho^2$$
$$W^T = 2W_1 . \tag{7.40}$$

If discrete states are involved, then a single-differential cross section can be obtained (see the detailed discussions below as well as in [Don86, Def66]):

$$\left(\frac{d\sigma}{d\Omega_e}\right)^{\text{unpol}} = 4\pi\sigma_{\text{Mott}}f_{\text{rec}}^{-1}\left\{v_L F_L^2(q) + v_T F_T^2(q)\right\}$$

$$= 4\pi\sigma_{\text{Mott}}f_{\text{rec}}^{-1}F^2(q,\theta_e), \tag{7.41}$$

where the recoil factor is given by $f_{\text{rec}} = 1 + \frac{2\epsilon}{M_i}\sin^2\theta_e/2$ and where the form factors F_L^2 and F_T^2 result from integrating the responses W^L and W^T (the form factors are only functions of q, where ω is fixed by the excitation energy and the momentum transfer). For given angular momentum and parity quantum numbers only a finite set of multipole form factors can in general occur:

$$F_L^2(q) = \sum_{J\geq 0} F_{CJ}^2(q) \tag{7.42}$$

$$F_T^2(q) = \sum_{J\geq 1}\left\{F_{EJ}^2(q) + F_{MJ}^2(q)\right\}. \tag{7.43}$$

As discussed in [Don86], when polarized targets are involved or when final-state nuclear polarizations are determined, a much richer variety of polarization observables becomes accessible. In general these contain interferences between the various form factors and consequently a complete decomposition into the underlying electromagnetic matrix elements can, in principle, be achieved (up to a simple phase ambiguity in the arbitrary spin case – see [Don86]). For such cases the helicity asymmetry is generally nonzero *even for strictly PC interactions.* Some cases where polarization observables are considered are those in Chapters 8, 11, and 12.

Moreover, for situations where no hadronic polarizations are specified, but where coincidence reactions are considered, there can also be nonzero helicity asymmetries *again, even for strictly PC interactions* – an example is the so-called "5th response" [Don85, Don86a]. Further discussion of coincidence reactions will occur in Chapter 16.

7.2 Spin-Dependent Lepton–Nucleon Scattering

With polarized electrons, things are somewhat more complicated. Let us assume that the scattered electron's polarization is not measured, but that the incident electron beam is prepared with its spin pointing in some direction characterized by the four-vector S^μ, which must satisfy [Bjo64] $S^2 = -1$ and $k \cdot S = 0$. The more general situation with both incident and scattered electrons polarized is discussed in [Don86]. We may then insert into Eq. (7.19) just after the factor γ_ν the spin projection operator $(1 + \gamma_5 \not{S})/2$ and once again now have a sum over all four spinor components and hence a trace:

$$\begin{aligned} L_{\mu\nu} &= \frac{1}{16m_e^2} \operatorname{Tr}\left\{\gamma_\mu \left(\not{k} + m_e\right) \gamma_\nu \left(1 + \gamma_5 \not{S}\right) \left(\not{k} + m_e\right)\right\} \\ &\equiv \frac{1}{2}\left[L_{\mu\nu}^{\text{unpol}} + L_{\mu\nu}^{\text{pol}}\right], \end{aligned} \tag{7.44}$$

where $L_{\mu\nu}^{\text{unpol}}$ is the leptonic tensor in Eq. (7.21).

Upon evaluation of the trace, a new piece occurs which contains all reference to the electron spin and is given by

$$L_{\mu\nu}^{\text{pol}} = \frac{1}{2m_e^2}\left(i\epsilon_{\mu\nu\alpha\beta}\, m_e S^\alpha q^\beta\right). \tag{7.45}$$

Note that this tensor is antisymmetric under the interchange $\mu \leftrightarrow \nu$ and satisfies the current conservation condition, $q^\mu L_{\mu\nu}^{\text{pol}} = 0$.

In the case of purely longitudinally polarized electrons we have

$$m_e S^\mu = h\epsilon\left(\beta, \boldsymbol{u}_L\right), \tag{7.46}$$

while in the case of purely transversely-polarized electrons we have

$$m_e S^\mu = \frac{1}{\gamma} \cdot h\epsilon\left(0, \boldsymbol{u}_\perp\right), \tag{7.47}$$

where $\beta = k/\epsilon = \sqrt{1 - (m_e/\epsilon)^2}$ and $\gamma = 1/\sqrt{1 - \beta^2} = \epsilon/m_e$ are the relativistic factors. Transverse polarization effects are suppressed relative to longitudinal effects by a factor $1/\gamma$.

Considering only purely longitudinally polarized electrons and the ERL where $\beta \to 1$, $\gamma \to \infty$ one has:

$$K^\mu \longrightarrow \epsilon(1, \boldsymbol{u}_L) \tag{7.48}$$

$$(m_e S^\mu)_L \longrightarrow h\epsilon(1, \boldsymbol{u}_L) = hK^\mu ; \tag{7.49}$$

that is, $h = \pm 1$ becomes the electron helicity. The leptonic tensors under these ERL conditions become

$$2m_e^2 L_{\mu\nu}^{\text{unpol}} \to k_\mu k_\nu' + k_\mu' k_\nu - g_{\mu\nu} k \cdot k' \equiv \chi_{\mu\nu}^{\text{unpol}} \tag{7.50}$$

$$2m_e^2 L_{\mu\nu}^{\text{pol}} \to -ih\epsilon_{\mu\nu\alpha\beta} k^\alpha k'^\beta \equiv \chi_{\mu\nu}^{\text{pol}} , \tag{7.51}$$

and we see that they are in general comparable in magnitude, both being characterized by $\epsilon\epsilon'$. Note that in the ERL one has $k \cdot k' \to Q^2/2$.

With these developments, we are in a position to proceed with discussions of PC electron scattering with or without a polarized beam. One important consequence of the arguments summarized in this section is that only *longitudinally polarized electrons* are relevant for most studies in nuclear or particle physics.

7.3 Electron–Nucleus Scattering

Electron scattering is the unique experimental technique that has been used to great effect to determine nuclear structure since the mid twentieth century. The availability of intense beams from electron accelerators, polarized sources, and targets and specialized detectors led to a period of intense experimental activity at laboratories worldwide. Coupled with tremendous advances in theory, this "golden age" yielded our current understanding of how the electroweak theory of the Standard Model applies to nuclei. Electron accelerators at Illinois, Stanford, MIT-Bates, Saclay, NIKHEF, and Mainz pioneered the study of nuclear structure using electron beams starting about 1955. More recently, the development of continuous wave (CW) beams enhanced the ability to carry out coincidence experiments. At present, this research is carried out at Jefferson Laboratory in the US and at the universities of Bonn and Mainz in Germany. In the chapters which follow, the interaction of electrons with nuclei is developed for different kinematic conditions and energy scales.

In Chapters 15 and 16, the interaction of electron beams with nuclei is discussed in the context of the different kinematic regimes of Fig. 7.1. The deep-inelastic regime for electron scattering from nuclei will also be discussed in Chapter 9. In particular, in the present chapter the multipole analysis of the electromagnetic current provides the focus in Section 7.4. When dealing with the discrete states of a nucleus having angular momentum and parity J^π as good quantum numbers, it is advantageous to expand the general current in operators that transform appropriately, i.e., are irreducible tensors under rotations and have specific parity, for then one can exploit the symmetry properties discussed in Chapter 2.

We begin by considering the situation shown in Fig. 7.3 involving electroexcitation from a ground state labeled $|i\rangle$ having parity π_i and angular momentum quantum numbers J_i, M_i to some excited state having π_f and J_f, M_f. Assuming that parity is a good quantum number one has $\pi = \pi_i \pi_f$ and in general (see below), assuming conservation

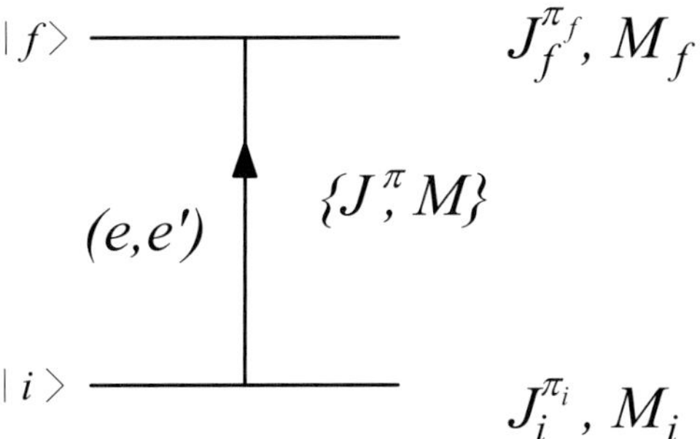

Fig. 7.3 Electron scattering excitation of discrete states; see text for notation.

of angular momentum, all multipolarities J, where $|J_i - J_f| \le J \le J_i + J_f$. Let us begin to rewrite the matrix elements of the four-vector EM current operator $\hat{J}^{\mu} = (\hat{\rho}, \hat{\boldsymbol{j}})$ in terms of irreducible tensor operators, viz., operators that transform under rotations with good angular momentum quantum numbers. In discussing the matrix elements of the current operators and their multipole projections, it is usual to work in the nuclear Hilbert space where the states are represented in terms of configurations of the A nucleons in the nucleus. The details of how this works for typical nuclear modeling is the central theme of Chapters 13 and 14. For the present, we note only that any operators working in this space can be characterized as one-body, two-body, three-body operators, etc., that is, as operators that act on one, two, three, etc., nucleons at time respectively. Or, stated more concretely, they go as $\sum_{\alpha,\beta} \langle \alpha | \mathcal{O}^{[1]} | \beta \rangle a_{\alpha}^{\dagger} a_{\beta}$, $\sum_{\alpha\alpha'\beta\beta'} \langle \alpha'\beta' | \mathcal{O}^{[2]} | \beta\alpha \rangle a_{\alpha'}^{\dagger} a_{\beta'}^{\dagger} a_{\alpha} a_{\beta}$, etc., for one-body, two-body, etc., operators labeled [1], [2], $\dots$, respectively. Here $\langle \alpha | \mathcal{O}^{[1]} | \beta \rangle$ is the single-particle matrix element, $< \alpha'\beta' | \mathcal{O}^{[2]} | \beta\alpha >$ the two-particle matrix element, and the expansions are over complete sets of single-particle quantum numbers, and involve the creation and destruction operators $a_{\alpha}^{\dagger}$ and a_{β}. An example of what this implies is shown in Figs. 7.4 and 7.5 where in the former a virtual photon interacts with a single nucleon (one-body current operator), taking it from a state with quantum numbers p, Λ (four-momentum p^{μ} and helicity Λ) to a state having p', Λ'. In the latter case, the example involves the so-called two-body meson-exchange current (MEC) contributions: the various pieces of the current typically used are shown in the figure and listed in the caption. As noted, the longest-range terms are those arising from single-pion exchange, to be discussed in Chapter 11. In the present chapter we proceed assuming a general form for the current and its matrix elements, and only in the following chapters do we invoke particular models for the one- two- and possibly three-body contributions.

From the developments above, we know that the electron tensor for unpolarized electron scattering is given by

$$2m_e^2 L_{\mu\nu} = k_{\mu} k_{\nu}' + k_{\mu}' k_{\nu} - g_{\mu\nu}(k \cdot k' - m_e^2) \tag{7.52}$$

$$= 2R_{\mu} R_{\nu} - \frac{1}{2}\left[q_{\mu} q_{\nu} - Q^2 g_{\mu\nu} \right], \tag{7.53}$$

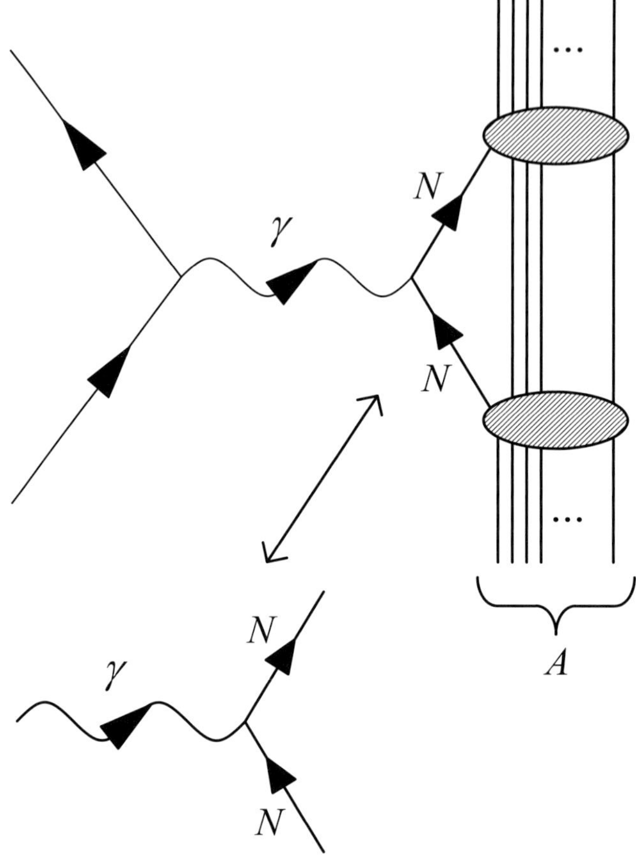

Fig. 7.4 The interaction of a virtual photon with a single nucleon, leading to one-body contributions to the EM current. Shown here is the diagram in free space where one has eN scattering, as discussed in later chapters, together with what happens in a nucleus, where one typically imbeds the EM interaction with a single nucleon in the nuclear many-body problem as a one-body operator; this constitutes what is often called the impulse approximation (IA). Note that the IA does *not* imply that either the initial or final nuclear states are especially simple, for instance plane waves for the final outgoing nucleon.

where, as usual, $q_\mu = k_\mu - k'_\mu$ and now we have also defined $R_\mu \equiv (k_\mu + k'_\mu)/2$. Note that the continuity equation is manifestly satisfied, namely, contracting the four-momentum transfer with $L_{\mu\nu}$ yields zero, since $q \cdot R = 0$. Using Eq. (7.53) and contracting with a general product involving the hadronic current matrix elements, $J_{fi}^{\mu*}(\mathbf{q})J_{fi}^{\nu}(\mathbf{q})$, we are led to consider the quantity [Def66]

$$\mathcal{R}_{fi} \equiv \left|2R \cdot J_{fi}(\mathbf{q})\right|^2 - Q^2 \left|J_{fi}(\mathbf{q})\right|^2 , \tag{7.54}$$

where current conservation, $q \cdot J_{fi}(\mathbf{q}) = 0$, has been used, and one has $-q_\mu q^\mu \equiv Q^2 = q^2 - \omega^2$.[3] For spacelike processes such as electron scattering, one has $Q^2 > 0$. The hadronic

---

[3] Note that in numerous publications by the first author a different convention is used for Q^2, where the sign is taken to be opposite to the one used here.

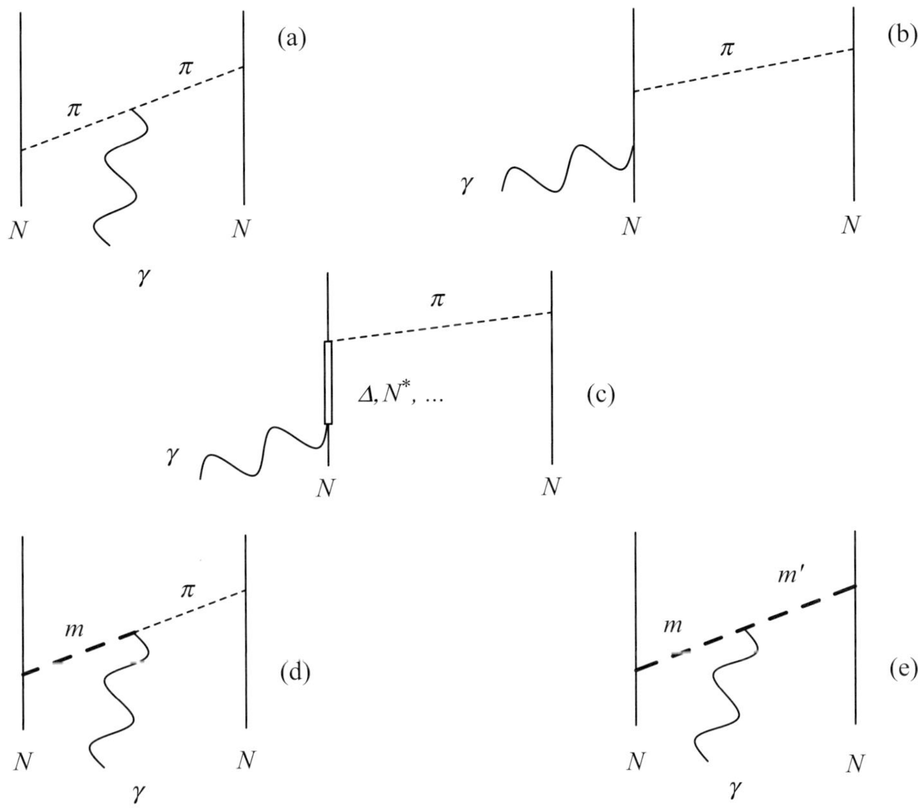

Fig. 7.5 Typical two-body meson-exchange current contributions to the EM current arising from a) the so-called pion-in-flight diagram, b) a contribution with an intermediate nucleon propagator plus exchange of a pion, c) contributions from intermediate Δ and N^* propagators together with the exchange of a pion, d) diagrams like the pion-in-flight diagram, but now with one pion and one heavier meson, and e) contributions involving the exchange of two heavier mesons. Typically the first three diagrams involving the exchange of the low-mass pion are dominant, as they provide the longest-range MEC contributions.

matrix elements are in turn the three-dimensional Fourier transforms of general transition matrix elements $|i\rangle \to |f\rangle$ of the electromagnetic current operator (see later discussions):

$$J^\mu_{fi}(\mathbf{q}) = \int d\mathbf{x}\, e^{i\mathbf{q}\cdot\mathbf{x}} J^\mu_{fi}(\mathbf{x})$$

$$= \int d\mathbf{x}\, e^{i\mathbf{q}\cdot\mathbf{x}} \langle f\, |\widehat{J}^\mu(\mathbf{x})|\, i\rangle. \tag{7.55}$$

Let us examine $J^\mu_{fi}(\mathbf{q})$ more closely. We have for the $\mu = 0$ component $\rho_{fi}(\mathbf{q}) =$ Fourier transform of the transition charge density, while for the spatial components $\mu = 1, 2, 3$ we have $\mathbf{J}_{fi}(\mathbf{q}) =$ Fourier transform of the transition three-vector current density, where the latter has contributions from convection and magnetization currents. Throughout this book we use a coordinate system having its z-axis along $\mathbf{q}$, the virtual photon direction, $\mathbf{u}_z \equiv \mathbf{q}/q$, with the y-axis normal to the electron scattering plane, $\mathbf{u}_y \equiv (\mathbf{k} \times \mathbf{k}')\,/\,|\mathbf{k} \times \mathbf{k}'|$ and with the x-axis in the scattering plane and orthogonal to these two unit vectors, $\mathbf{u}_x \equiv \mathbf{u}_y \times \mathbf{u}_z$ [Don86]. Since we shall be making use of tensors, etc., that transform appropriately under rotations, it proves convenient to introduce unit spherical vectors, $\mathbf{e}(\mathbf{q}, m)$, $m = 0, \pm 1$ [Edm74] and then any three-vector can be expanded in this basis. In particular, for the current matrix elements we have

$$\mathbf{J}_{fi}(\mathbf{q}) \equiv \sum_{m=0,\pm 1} J_{fi}(\mathbf{q}; m)\mathbf{e}(\mathbf{q}, m)^*, \tag{7.56}$$

where

$$J_{fi}(\mathbf{q}; m) = \mathbf{e}(\mathbf{q}, m) \cdot \mathbf{J}_{fi}(\mathbf{q}) \tag{7.57}$$

and hence

$$\begin{aligned}\mathbf{J}_{fi}(\mathbf{q}) &= J_{fi}(\mathbf{q}; +1)\mathbf{e}(\mathbf{q}, +1)^* + J_{fi}(\mathbf{q}; -1)\mathbf{e}(\mathbf{q}, -1)^* \\ &\quad + J_{fi}(\mathbf{q}; 0)\mathbf{e}(\mathbf{q}, 0)^*.\end{aligned} \tag{7.58}$$

Now current conservation implies that

$$\begin{aligned} q \cdot J_{fi}(\mathbf{q}) = 0 &= Q^0 J_{fi}^0(\mathbf{q}) - \mathbf{q} \cdot \mathbf{J}_{fi}(\mathbf{q}) \\ &= \omega \rho_{fi}(\mathbf{q}) - q J_{fi}(\mathbf{q}; 0) \,, \end{aligned} \tag{7.59}$$

that is, there is a relationship between the longitudinal and charge matrix elements

$$J_{fi}(\mathbf{q}; 0) = \left(\frac{\omega}{q}\right) \rho_{fi}(\mathbf{q}) \tag{7.60}$$

and we shall usually eliminate the former in terms of the latter, leaving three independent projections of the four-vector current matrix elements to deal with:

$$\begin{aligned} \rho_{fi}(\mathbf{q}) &\quad \text{charge (1)} \\ J_{fi}(\mathbf{q}; \pm 1) &\quad \text{transverse current (2)} \,. \end{aligned}$$

Let us next rewrite some of the terms that enter in Eq. (7.54): one can show that

$$\begin{aligned} (\mathbf{k} + \mathbf{k}') \cdot \mathbf{J}_{fi}(\mathbf{q}) &= (\epsilon + \epsilon')\left(\frac{\omega}{q}\right)^2 \rho_{fi}(\mathbf{q}) \\ &\quad - \sqrt{2}\frac{1}{q} kk' \sin\theta_e \left[J_{fi}(\mathbf{q}; +1) - J_{fi}(\mathbf{q}; -1)\right], \end{aligned} \tag{7.61}$$

using Eq. (7.60) to eliminate the longitudinal projection of the three-vector current matrix elements. The quantity $\mathcal{R}_{fi}$ in Eq. (7.54) then becomes

$$\begin{aligned} \mathcal{R}_{fi} &= \left[\left((\epsilon + \epsilon')^2 - q^2\right)\left(\frac{Q^2}{q^2}\right)^2\right] |\rho_{fi}(\mathbf{q})|^2 \\ &\quad + \left[2\left(\frac{1}{q}kk'\sin\theta_e\right)^2 + Q^2\right]\left\{|J_{fi}(\mathbf{q}; +1)|^2 + |J_{fi}(\mathbf{q}; -1)|^2\right\} \end{aligned}$$

$$+\left[\sqrt{2}\left(\epsilon+\epsilon'\right)\left(\frac{Q^2}{q^2}\right)\frac{1}{q}kk'\sin\theta_e\right]$$
$$\times\left\{2\mathrm{Re}\left\{\rho_{fi}(\mathbf{q})^*\left(J_{fi}(\mathbf{q};+1)-J_{fi}(\mathbf{q};-1)\right)\right\}\right\}$$
$$+\left[-2\left(\frac{1}{q}kk'\sin\theta_e\right)^2\right]\left\{2\mathrm{Re}\left\{J_{fi}(\mathbf{q};+1)^*J_{fi}(\mathbf{q};-1)\right\}\right\} \tag{7.62}$$

$$\equiv v_0\left\{v_L W_{fi}^L+v_T W_{fi}^T+v_{TL}W_{fi}^{TL}+v_{TT}W_{fi}^{TT}\right\}. \tag{7.63}$$

By convention one removes the factor $v_0\equiv\left(\epsilon+\epsilon'\right)^2-q^2=4\epsilon\epsilon'\cos^2\theta_e/2$, where the second equality arises when, as is typical, the electron mass can be neglected with respect to its energy. In that high-energy limit one also has $Q^2=4\epsilon\epsilon'\sin^2\theta_e/2$. Then one is in a position to define the various combinations of bilinear products of current matrix elements as

$$W_{fi}^L\equiv\left|\rho_{fi}(\mathbf{q})\right|^2 \tag{7.64}$$

$$W_{fi}^T\equiv\left|J_{fi}(\mathbf{q};+1)\right|^2+\left|J_{fi}(\mathbf{q};-1)\right|^2 \tag{7.65}$$

$$W_{fi}^{TL}\equiv-2\mathrm{Re}\left\{\rho_{fi}(\mathbf{q})^*\left(J_{fi}(\mathbf{q};+1)-J_{fi}(\mathbf{q};-1)\right)\right\} \tag{7.66}$$

$$W_{fi}^{TT}\equiv 2\mathrm{Re}\left\{J_{fi}(\mathbf{q};+1)^*J_{fi}(\mathbf{q};-1)\right\}, \tag{7.67}$$

leaving the remaining kinematic factors ("generalized Rosenbluth factors"). Again, assuming the ERL for the electron, one finds the following:

$$v_L\equiv\rho^2 \tag{7.68}$$

$$v_T\equiv\frac{1}{2}\rho+\tan^2\theta_e/2 \tag{7.69}$$

$$v_{TL}\equiv-\frac{1}{\sqrt{2}}\rho\sqrt{\rho+\tan^2\theta_e/2} \tag{7.70}$$

$$v_{TT}\equiv-\frac{1}{2}\rho, \tag{7.71}$$

where ρ is defined in Section 7.1. The labels L, T, TL, and TT refer to longitudinal (and charge, see above), transverse unpolarized, transverse-longitudinal interference and transverse-transverse interference responses; in fact, the last can be shown to carry information on linearly-polarized transverse virtual photons. One can extend the analysis given here to include polarized electron scattering [Don86] and then one finds two more classes of response denoted TL' and T', although these go beyond the scope of the present discussions. For real-photon processes, the same results obtain except that only purely transverse responses can occur.

7.4 Electromagnetic Multipole Operators

Next let us consider the responses in Eqs. (7.64)–(7.67) in more detail. We begin with the terms that involve only the matrix elements of the charge density. From Eq. (7.55) we have

$$\rho_{fi}(\mathbf{q})=\int d\mathbf{x}\,e^{i\mathbf{q}\cdot\mathbf{x}}\,\langle f\,|\widehat{\rho}(\mathbf{x})|\,i\rangle\,, \tag{7.72}$$

into which we substitute the expansion of a plane wave [Edm74]

$$e^{i\mathbf{q}\cdot\mathbf{x}} = 4\pi \sum_{J\geq 0, M} i^J j_J(qx) Y_J^M(\Omega_x) Y_J^M(\Omega_q)^* ,$$

(7.73)

yielding

$$\rho_{fi}(\mathbf{q}) = 4\pi \sum_{J\geq 0, M} i^J Y_J^M(\Omega_q)^* \langle f | \widehat{M}_{JM}(q) | i \rangle,$$

(7.74)

where we have defined the *Coulomb multipole operator*

$$\widehat{M}_{JM}(q) \equiv \int d\mathbf{x}\, j_J(qx) Y_J^M(\Omega_x) \widehat{\rho}(\mathbf{x}) .$$

(7.75)

The Coulomb multipoles are denoted $C0, C1, C2 \ldots$ (corresponding to $J = 0, 1, 2 \ldots$) and have natural parity, $\pi = (-)^J$, as discussed below. The response in Eq. (7.64) then reads

$$\begin{aligned} W_{fi}^L &\equiv \left| \rho_{fi}(\mathbf{q}) \right|^2 \\ &= (4\pi)^2 \sum_{J\geq 0, M} \sum_{J'\geq 0, M'} (-i)^J i^{J'} Y_J^M(\Omega_q) Y_{J'}^{M'}(\Omega_q)^* \\ &\quad \times \langle f | \widehat{M}_{JM}(q) | i \rangle^* \langle f | \widehat{M}_{J'M'}(q) | i \rangle . \end{aligned}$$

(7.76)

Since in discussing discrete states that are almost always labeled with good angular momentum and parity quantum numbers, $J_i^{\pi_i}$ and $J_f^{\pi_f}$ (see Fig. 7.3), we are now able to employ the Wigner–Eckart theorem to write the multipole matrix elements in Eq. (7.76) in terms of 3-j symbols and reduced matrix elements (see Chapter 2). We presume that the states and operator all have the same axis of quantization to which the z-projections of angular momentum are referred. The 3-j symbol embodies the angular momentum conservation conditions $|J_f - J_i| \leq J \leq J_f + J_i$ and $M = M_f - M_i$. Now, specifically for unpolarized scattering, where no hadronic polarizations are specified, we have to perform the average-over-initial and sum-over-final, meaning summing over M_i and M_f and dividing by $2J_i + 1$, and therefore have

$$\overline{\sum_{if}} W_{fi}^L = \overline{\sum_{if}} \left| \rho_{fi}(\mathbf{q}) \right|^2$$

(7.77)

$$\begin{aligned} &= \frac{(4\pi)^2}{2J_i + 1} \sum_{M_i} \sum_{M_f} \sum_{J\geq 0, M} \sum_{J'\geq 0, M'} (-i)^J i^{J'} Y_J^M(\Omega_q) Y_{J'}^{M'}(\Omega_q)^* \\ &\quad \times \begin{pmatrix} J_f & J & J_i \\ -M_f & M & M_i \end{pmatrix} \begin{pmatrix} J_f & J' & J_i \\ -M_f & M' & M_i \end{pmatrix} \\ &\quad \times \langle J_f \| \widehat{M}_J \| J_i \rangle^* \langle J_f \| \widehat{M}_{J'} \| J_i \rangle . \end{aligned}$$

(7.78)

One can then use the orthogonality of the 3-j symbols (see Eq. (2.54)) together with an identity for the spherical harmonics [Edm74] to obtain the standard result for the charge contributions to *unpolarized inclusive scattering*:

$$\overline{\sum_{if} W_{fi}^L} \equiv \left(W^L\right)^{\mathrm{unpol}} \tag{7.79}$$

$$= \frac{4\pi}{2J_i + 1} \sum_{J \geq 0} \left|\langle J_f \,\|\widehat{M}_J\| \, J_i\rangle\right|^2 , \tag{7.80}$$

namely, proportional to the incoherent sum of the squares of the reduced matrix elements of the allowed Coulomb multipole operators.

A similar analysis can be performed for the other responses in Eqs. (7.65)–(7.67), now with the three-vector projections of the current matrix elements. We use the spherical vector notation introduced above and so consider

$$J_{fi}(\mathbf{q}; m) = \mathbf{e}(\mathbf{q}, m) \cdot \mathbf{J}_{fi}(\mathbf{q})$$

$$= \int d\mathbf{x} \, \langle f \,|\, e^{i\mathbf{q}\cdot\mathbf{x}} \mathbf{e}(\mathbf{q}, m) \cdot \widehat{\mathbf{J}}_{fi}(\mathbf{x}) \,|\, i\rangle , \tag{7.81}$$

where, using the fact that

$$e^{i\mathbf{q}\cdot\mathbf{x}} \mathbf{e}(\mathbf{q}, m) = \tag{7.82}$$

$$-\sqrt{2\pi} \times \begin{cases} \sum_{J \geq 0} i^J [J] \sqrt{2} \frac{i}{q} \nabla M_J^0(q\mathbf{x}) & m = 0 \\ \sum_{J \geq 1} i^J [J] \left(m\mathbf{M}_{JJ}^m(q\mathbf{x}) + \frac{1}{q} \nabla \times \mathbf{M}_{JJ}^m(q\mathbf{x}) \right), & m = \pm 1 , \end{cases}$$

with

$$M_J^M(q\mathbf{x}) \equiv j_J(qx) Y_J^M(\Omega_x) \tag{7.83}$$

$$\mathbf{M}_{JL}^M(q\mathbf{x}) \equiv j_L(qx) \mathbf{Y}_{JL1}^M(\Omega_x) , \tag{7.84}$$

these developments then lead us to define the *transverse electric and magnetic multipole operators*

$$\widehat{T}_{JM}^{\mathrm{el}}(q) \equiv \int d\mathbf{x} \left[\frac{1}{q} \nabla \times \mathbf{M}_{JJ}^M(q\mathbf{x}) \right] \cdot \widehat{\mathbf{J}}(\mathbf{x}) \tag{7.85}$$

$$\widehat{T}_{JM}^{\mathrm{mag}}(q) \equiv \int d\mathbf{x} \, \mathbf{M}_{JJ}^M(q\mathbf{x}) \cdot \widehat{\mathbf{J}}(\mathbf{x}) , \tag{7.86}$$

with $J = 1, 2 \ldots$ and corresponding multipoles $E1, E2, \ldots$ or $M1, M2, \ldots$ respectively. The analogous expression for the *Coulomb multipole operator* defined above is

$$\widehat{M}_{JM}(q) \equiv \int d\mathbf{x} \, M_J^M(\Omega_x) \widehat{\rho}(\mathbf{x}) , \tag{7.87}$$

and, to complete the set, we may also define a *longitudinal multipole operator*

$$\widehat{L}_{JM}(q) \equiv \int d\mathbf{x} \left[\frac{i}{q} \nabla M_J^M(q\mathbf{x}) \right] \cdot \widehat{\mathbf{J}}(\mathbf{x}) , \tag{7.88}$$

with $J = 0, 1, 2, \ldots$ and multipoles $L0, L1, L2, \ldots$ The current conservation condition allows us to relate matrix elements of the last two types of operators:

$$\langle f \,|\widehat{L}_{JM}(q)|\, i\rangle = -\left(\frac{\omega}{q}\right) \langle f \,|\widehat{M}_{JM}(q)|\, i\rangle . \tag{7.89}$$

Together with the Coulomb multipoles, the electric and longitudinal multipole operators have natural parity, $\pi = (-)^J$, whereas the magnetic multipole operators have non-natural parity, $\pi = (-)^{J+1}$. The results in Eq. (7.81) may now be written in the form

$$
J_{fi}(\mathbf{q}; m) = \begin{cases}
-\sqrt{4\pi}\, \sum_{J \geq 0} i^J [J] \langle f \,|\widehat{L}_{J0}(q)|\, i \rangle, & m = 0 \\[4pt]
-\sqrt{2\pi}\, \sum_{J \geq 1} i^J [J] \{ \langle f\, |\widehat{T}^{el}_{Jm}(q)|\, i \rangle & \\[2pt]
\qquad +m \langle f\, |\widehat{T}^{mag}_{Jm}(q)|\, i \rangle \}, & m = \pm 1 \,.
\end{cases}
\tag{7.90}
$$

The cross section involves bilinear products of the above current matrix elements, $J(\mathbf{q}; m)^* J(\mathbf{q}; m')$ (see Eqs. (7.65))–(7.67)), and again we can employ the Wigner–Eckart theorem and the orthogonality of the 3-j symbols when performing the average-over-initial and sum-over-final to obtain a factor, $(2J+1)^{-1} \delta_{JJ'} \delta_{mm'}$, as above. Since the longitudinal multipoles have $m = 0$, and are proportional to the Coulomb multipoles by Eq. (7.89), whereas the transverse electric and magnetic multipoles have $m = \pm 1$, the fact that m' must equal m implies that

$$
\overline{\sum_{if}} W^{TL}_{fi} \equiv \left(W^{TL}\right)^{\mathrm{unpol}} = \overline{\sum_{if}} W^{TT}_{fi} \equiv \left(W^{TT}\right)^{\mathrm{unpol}} = 0 \,.
\tag{7.91}
$$

Note that for situations with nuclear polarizations, these interference responses are nonzero. We then have for the purely transverse terms in Eq. (7.65), that is, those having $m = \pm 1$,

$$
\overline{\sum_{if}} J(\mathbf{q}; m)^* J(\mathbf{q}; m') = \delta_{mm'} \frac{4\pi}{2J_i + 1} \times \frac{1}{2} \sum_{J \geq 1} \left| \langle J_f \,\|\widehat{T}^{el}_{J}(q)\|\, J_i \rangle \right.
$$
$$
\left. +m \langle J_f \,\|\widehat{T}^{mag}_{J}(q)\|\, J_i \rangle \right|^2 \,.
\tag{7.92}
$$

Furthermore, for the potential interferences that might occur here between electric and magnetic multipoles, note that, given a specific parity $\pi = \pi_i \pi_f$, we have under the sum over multipolarity J only an electric contribution when $\pi = (-)^J$ or a magnetic contribution when $\pi = (-)^{J+1}$. That is, one has only one or the other, but not both. In the present completely unpolarized situation we see that there can be no interferences at all and have obtained in addition to the result in Eqs. (7.79) and (7.80) simply the following:

$$
\overline{\sum_{if}} W^{T}_{fi} \equiv \left(W^{T}\right)^{\mathrm{unpol}}
\tag{7.93}
$$

$$
= \frac{4\pi}{2J_i + 1} \sum_{J \geq 1} \left(\left| \langle J_f \,\|\widehat{T}^{el}_{J}(q)\|\, J_i \rangle \right|^2 + \left| \langle J_f \,\|\widehat{T}^{mag}_{J}(q)\|\, J_i \rangle \right|^2 \right),
\tag{7.94}
$$

namely, the standard result for the transverse parts of the *unpolarized inclusive scattering* cross section.

We are now in a position to summarize the basic results for single-arm, inclusive scattering of unpolarized electrons from unpolarized targets. The cross section in the laboratory system then takes on its standard form

$$
\frac{d\sigma}{d\Omega_e} \equiv \sigma_M f_{rec}^{-1} F^2(q, \theta_e)
\tag{7.95}
$$

for the excitation of a discrete state with $\omega = E_f - E_i$ and we have for the square of the total form factor

$$F^2(q, \theta_e) = v_L F_L^2(q) + v_T F_T^2(q) \tag{7.96}$$

with longitudinal and transverse contributions

$$F_L^2(q) = \sum_{J \geq 0} F_{CJ}^2(q) \tag{7.97}$$

$$F_T^2(q) = \sum_{J \geq 1} \left\{ F_{EJ}^2(q) + F_{MJ}^2(q) \right\} , \tag{7.98}$$

involving *Coulomb, electric, and magnetic form factors*

$$F_{CJ}(q) \equiv \frac{\sqrt{4\pi}}{[J_i]} \langle J_f \| \widehat{M}_J \| J_i \rangle \tag{7.99}$$

$$F_{EJ}(q) \equiv \frac{\sqrt{4\pi}}{[J_i]} \langle J_f \| \widehat{T}_J^{\mathrm{el}}(q) \| J_i \rangle \tag{7.100}$$

$$F_{MJ}(q) \equiv \frac{\sqrt{4\pi}}{[J_i]} \langle J_f \| i\widehat{T}_J^{\mathrm{mag}}(q) \| J_i \rangle, \tag{7.101}$$

respectively. We shall see below that the form factors as defined here can all be chosen to be real quantities with a specific phase convention. The parity and time-reversal properties of the multipoles are discussed in more detail below. Here we note only that all multipolarities occur except for those where $J = 0$ and the parity changes $\pi = -$ (i.e., in the one-photon approximation) and, that for the allowed monopole transitions, those with $\pi = +$, only Coulomb terms $C0$ appear, since the lowest multipoles in the transverse case are $E1$ and $M1$. Thus, to the order considered one may have, for example, $0^{\pm} \leftrightarrow 0^{\pm}$ transitions, but no $0^{\pm} \leftrightarrow 0^{\mp}$ cases.

Finally, it is useful to record the formulas required in studying *real-γ processes*. For γ-decay it may be shown that the rate is given by

$$\omega_\gamma (J'; T'M_T \to J; TM_T) = \left[\frac{2J + 1}{2J' + 1} 8\pi \alpha q F_T^2(q) \right]_{q=\omega}, \tag{7.102}$$

where the transverse form factor for electroexcitation $J; TM_T \to J'; T'M_T$ is to be evaluated on the lightcone, $q = \omega$. The photoabsorption cross section integrated over an absorption line is given through a similar expression:

$$\Sigma_\gamma (J; TM_T \to J'; T'M_T) = \left[\frac{(2\pi)^3 \alpha}{q} F_T^2(q) \right]_{q=\omega}. \tag{7.103}$$

Note: only $F_T^2(q)$ enters here and hence no monopole transitions are allowed. Specifically, in electron scattering one has $0^+ \to 0^+$ and $0^- \to 0^-$, but no $0^+ \to 0^-$ or $0^- \to 0^+$ transitions, although the last two do occur for the weak interaction (see Chapter 17) where the axial-vector current plays a role. For real photons none of these occurs.

Let us examine the nature of the hadronic EM current in a little more detail. Specifically, the spatial components of the current density operator may be decomposed into convection and magnetization contributions, $\widehat{\mathbf{j}}(\mathbf{x})$ and $\nabla \times \widehat{\mu}(\mathbf{x})$, respectively,

$$\widehat{\mathbf{J}}(\mathbf{x}) = \widehat{\mathbf{j}}(\mathbf{x}) + \nabla \times \widehat{\mu}(\mathbf{x}) \,. \tag{7.104}$$

The EM interaction Hamiltonian in an external EM field $A^{\mu} = (\Phi, \mathbf{A})$ has the form

$$\begin{aligned}
\widehat{H}' &\sim \int d\mathbf{x}\, \widehat{J} \cdot A \\
&= \int d\mathbf{x}\, \left\{ \widehat{\rho}(\mathbf{x})\Phi(\mathbf{x}) - \widehat{\mathbf{J}}(\mathbf{x}) \cdot \mathbf{A}(\mathbf{x}) \right\} \\
&= \int d\mathbf{x}\, \left\{ \widehat{\rho}(\mathbf{x})\Phi(\mathbf{x}) - \widehat{\mathbf{j}}(\mathbf{x}) \cdot \mathbf{A}(\mathbf{x}) - \widehat{\mu}(\mathbf{x}) \cdot \mathbf{B}(\mathbf{x}) \right\} ,
\end{aligned} \tag{7.105}$$

using the fact that $\mathbf{B} = \nabla \times \mathbf{A}$. Inserting Eq. (7.104) into the expressions above for the multipole operators and using identities involving the multipole projectors [Edm74], yields the following for the transverse multipole operators

$$\widehat{T}^{\mathrm{el}}_{JM}(q) \equiv \int d\mathbf{x} \left\{ \left[\frac{1}{q}\nabla \times \mathbf{M}^{M}_{JJ}(q\mathbf{x}) \right] \cdot \widehat{\mathbf{j}}(\mathbf{x}) + q\mathbf{M}^{M}_{JJ}(q\mathbf{x}) \cdot \widehat{\mu}(\mathbf{x}) \right\} \tag{7.106}$$

$$\widehat{T}^{\mathrm{mag}}_{JM}(q) \equiv \int d\mathbf{x} \left\{ \mathbf{M}^{M}_{JJ}(q\mathbf{x}) \cdot \widehat{\mathbf{J}}(\mathbf{x}) + q \left[\frac{1}{q}\nabla \times \mathbf{M}^{M}_{JJ}(q\mathbf{x}) \right] \cdot \widehat{\mu}(\mathbf{x}) \right\} . \tag{7.107}$$

Let us conclude this section by discussing the properties of the states and multipole operators introduced above under spatial and temporal inversion. First, under spatial inversion $\widehat{\Pi}$, the properties of the spherical harmonics immediately tell us that

$$\widehat{M}_{JM}(q) \rightarrow \widehat{\Pi}\widehat{M}_{JM}(q)\widehat{\Pi}^{-1} = (-)^{J}\widehat{M}_{J,M}(q) \tag{7.108}$$

$$\widehat{T}^{\mathrm{el}}_{JM}(q) \rightarrow \widehat{\Pi}\widehat{T}^{\mathrm{el}}_{JM}(q)\widehat{\Pi}^{-1} = (-)^{J}\widehat{T}^{\mathrm{el}}_{J,M}(q) \tag{7.109}$$

$$\widehat{T}^{\mathrm{mag}}_{JM}(q) \rightarrow \widehat{\Pi}\widehat{T}^{\mathrm{mag}}_{JM}(q)\widehat{\Pi}^{-1} = (-)^{J+1}\widehat{T}^{\mathrm{mag}}_{J,M}(q) \,. \tag{7.110}$$

Using, for example, the Coulomb operator in reduced matrix elements such as occur in Eq. (7.99), one finds that

$$\langle J_f \| \widehat{M}_J \| J_i \rangle = \left\langle J_f \left\| \widehat{\Pi}^{-1}\widehat{\Pi}\widehat{M}_J\widehat{\Pi}^{-1}\widehat{\Pi} \right\| J_i \right\rangle \tag{7.111}$$

$$= \pi_f\pi_i(-)^{J} \langle J_f \| \widehat{M}_J \| J_i \rangle, \tag{7.112}$$

where the states are assumed to eigenstates of parity: $\widehat{\Pi}\,|i\rangle = \pi_i\,|i\rangle$ and $\widehat{\Pi}\,|f\rangle = \pi_f\,|f\rangle$. Thus, parity conservation implies that $(-)^{J} = \pi = \pi_i\pi_f$ for natural parity multipoles as these Coulomb cases. A similar treatment of the transverse cases yields the already-stated results, namely, the electric (magnetic) multipoles have natural (non-natural) parity.

Second, one can invoke time-reversal invariance to derive additional properties. For this, we follow [Def66] and take the following phase convention for the states

$$\widehat{T}\,|J_kM_k\rangle = (-)^{J_k+M_k}\,|J_k, -M_k\rangle \,, \tag{7.113}$$

where $k = i$ or f, and $\widehat{T}$ is the time-reversal operator with its antiunitary behavior (see Chapter 2)

$$\left\langle a \left| \widehat{T}^{-1} \right| b \right\rangle = \langle \widehat{T}a | b \rangle \tag{7.114}$$

$$\widehat{T}c\widehat{T}^{-1} = c^{*}, \qquad c = \text{complex number.} \tag{7.115}$$

The pieces of the current transform in the following way under time-reversal:

$$\widehat{\rho}(\mathbf{x}) \rightarrow \widehat{T}\widehat{\rho}(\mathbf{x})\widehat{T}^{-1} = \widehat{\rho}(\mathbf{x}) \tag{7.116}$$

$$\widehat{\mathbf{j}}(\mathbf{x}) \rightarrow \widehat{T}\,\widehat{\mathbf{j}}(\mathbf{x})\widehat{T}^{-1} = -\widehat{\mathbf{j}}(\mathbf{x}) \tag{7.117}$$

$$\widehat{\mu}(\mathbf{x}) \rightarrow \widehat{T}\widehat{\mu}(\mathbf{x})\widehat{T}^{-1} = -\widehat{\mu}(\mathbf{x}). \tag{7.118}$$

In discussing the hermiticity and time-reversal properties of the multipole operators, it is convenient to define new operators that transform in simple ways (this is analogous to the definition of [Edm74])

$$\widehat{\mathcal{O}}'_{JM} \equiv i^J \widehat{\mathcal{O}}_{JM}\,, \tag{7.119}$$

where $\widehat{\mathcal{O}}_{JM}$ is any one of the multipole operators defined above; clearly the primed operators can be used everywhere by making simple adjustments of factors of i. The hermiticity properties are

$$\left(\widehat{\mathcal{O}}'_{JM}\right)^{\dagger} = (-)^{J+M+\sigma}\,\widehat{\mathcal{O}}'_{J,-M}\,, \tag{7.120}$$

where $\sigma = 0$ for Coulomb multipoles and $\sigma = 1$ for transverse electric and magnetic multipoles. Under time-reversal the primed operators transform very simply:

$$\widehat{\mathcal{O}}'_{JM} \rightarrow \widehat{T}\widehat{\mathcal{O}}'_{JM}\widehat{T}^{-1} = (-)^{J+M}\widehat{\mathcal{O}}'_{J,-M}\,. \tag{7.121}$$

Using the Wigner–Eckart theorem we then find that,

$$\left\langle J_f \left\| \widehat{\mathcal{O}}'_J \right\| J_i \right\rangle^{*} = (-)^{J_i - J_f + J + \sigma} \left\langle J_i \left\| \widehat{\mathcal{O}}'_J \right\| J_f \right\rangle\,, \tag{7.122}$$

and, upon inserting $\widehat{T}^{-1}\widehat{T}$ as in the spatial-inversion case discussed above, we find that the reduced matrix elements of the primed multipole operators are real,

$$\left\langle J_f \left\| \widehat{\mathcal{O}}'_J \right\| J_i \right\rangle^{*} = \left\langle J_f \left\| \widehat{\mathcal{O}}'_J \right\| J_i \right\rangle\,, \tag{7.123}$$

for the phase convention chosen here. Using the last two equations we may also write

$$\left\langle J_f \left\| \widehat{\mathcal{O}}'_J \right\| J_i \right\rangle = (-)^{J_i - J_f + J + \sigma} \left\langle J_i \left\| \widehat{\mathcal{O}}'_J \right\| J_f \right\rangle\,, \tag{7.124}$$

which will have implications in the discussions of elastic scattering to follow in Chapter 15. With this brief treatment of EM multipole operators in hand, one could go on to generalize the problem to the full electroweak interaction with an extended set of multipoles. We shall not do so, but only refer the reader to other references where the general problem is presented, for instance in [Wal75, Don75, Don79].

7.5 Parity-Violating Lepton Scattering

In addition to photon exchange, in the Standard Model the lepton can exchange an intermediate vector boson Z^0 with a hadronic target. This is indicated in Fig. 7.6. While the contribution of Z^0-exchange to the cross section is extremely small, by polarizing the electron beam and measuring the scattering spin asymmetry effects of order 10^{-6} can be measured with small uncertainties, as discussed below. This was first used in a SLAC

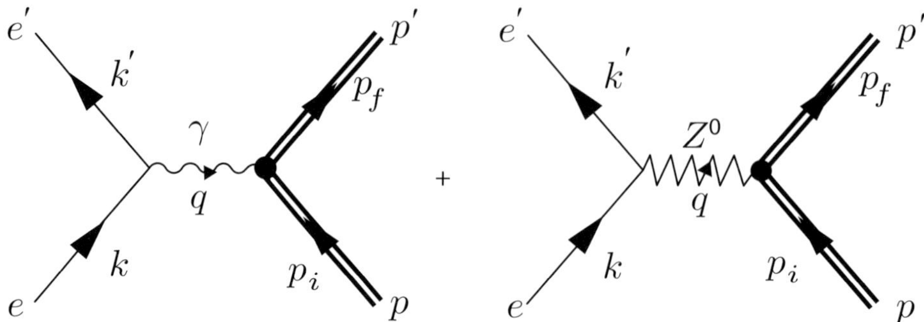

Fig. 7.6   The elastic electron–proton scattering process to leading-order in the electroweak theory, including both single-γ and single-Z^0 exchanges.

experiment in the 1970s to verify the Standard Model. In the last two decades, it has been used as a tool to explore hadron structure. It can also be used as a means to look for new physics beyond the Standard Model. Here we restrict our attention to three quark flavors (*up*, *down*, and *strange*) and the three types of form factors which are relevant (E = electric, M = magnetic, A = axial-vector). There are nine form factors to be separated, and as we shall see, this cannot be accomplished using elastic scattering from the proton alone. Since neutrons must also be used, it is natural to study elastic and inelastic scattering from nuclei. Upon combining cross sections and asymmetries for electron scattering from the proton and from nuclei, useful constraints on this set of form factors can emerge.

The single-photon exchange amplitude is the one introduced above and is proportional to $(e^2/Q^2)j_\mu^{\rm EM}J_{\rm EM}^\mu$. The Z^0-exchange amplitude, on the other hand, is proportional to $(g^2/M_Z^2)j_\mu^{\rm WNC}J_{\rm WNC}^\mu$, where g is the weak current coupling constant, M_Z is the Z^0 mass and the label "WNC" denotes the weak neutral current (to be distinguished from "EM" for the electromagnetic current). When the sum of these two amplitudes is squared one obtains three classes of terms: (1) the EM-terms squared – this is the PC problem summarized above and occurs with typical strength α; (2) the WNC-terms squared – this occurs with typical strength G^2 (G = the Fermi constant) and hence is very small – we shall ignore such contributions in the following;[4] and (3) interference terms EM×WNC and WNC×EM which enter with typical strength αG.

These last are the ones of interest for PV electron scattering, since now parity-violating effects can enter through the weak neutral current. Thus, in addition to the PC contributions introduced above which involved $|j_\mu^{\rm EM}J_{\rm EM}^\mu|^2$ and hence $L_{\mu\nu}^{\rm PC}W_{\rm PC}^{\mu\nu}$ (see Eqs. (7.9–7.12), where now the labels "EM" and "PC" have been included), we have additional terms to consider $\sim 2{\rm Re}(L_{\mu\nu}^{\rm PV}W_{\rm PV}^{\mu\nu})$, where

$$L_{\mu\nu}^{\rm PV} \equiv \overline{\sum}_{\rm leptons} j_\mu^{\rm EM*} j_\nu^{\rm WNC} \tag{7.125}$$

$$W_{\rm PV}^{\mu\nu} \equiv \overline{\sum}_{\rm hadrons} \left[J_{\rm EM}^{\mu*}\right]_{fi} \left[J_{\rm WNC}^{\nu}\right]_{fi}. \tag{7.126}$$

---

[4] In this work we consider only modest values of the momentum transfer ($Q^2 \ll M_Z^2$) and hence the Z^0 propagator is very small, in contrast to the massless photon case, resulting in the small effective coupling G^2.

The form of the EM electron current was given in Eq. (7.9),

$$j_\mu^{\mathrm{EM}}(k',\lambda';k,\lambda) \sim \bar{u}(k',\lambda')\gamma_\mu u(k,\lambda)\,, \tag{7.127}$$

while the WNC electron current may be written [Don79]

$$j_\mu^{\mathrm{WNC}}(k',\lambda';k,\lambda) \sim \bar{u}(k',\lambda')\left[a_V\gamma_\mu + a_A\gamma_\mu\gamma_5\right]u(k,\lambda)\,, \tag{7.128}$$

where in the standard electroweak model [Wei67, Sal69, Gla70] the vector (V) and axial-vector (A) tree-level couplings are given by

$$a_V \equiv 4\sin^2\theta_W - 1 \cong -0.08 \tag{7.129}$$

$$a_A \equiv -1\,; \tag{7.130}$$

see also Chapter 4 and [Don79].

In the product of the two currents there are terms that are parity violating. We arrive at the following expression for the EM×WNC part of the PV leptonic tensor

$$\left[L_{\mu\nu}^{\mathrm{PV}}\right]_{\mathrm{EM}\times\mathrm{WNC}} = \frac{1}{16m_e^2}\left(a_V\,\mathrm{Tr}\left\{\gamma_\mu\left(\slashed{k}'+m_e\right)\gamma_\nu\left(1+\gamma_5\slashed{S}\right)\left(\slashed{k}+m_e\right)\right\}\right.$$
$$\left.+\,a_A\,\mathrm{Tr}\left\{\gamma_\mu\left(\slashed{k}'+m_e\right)\gamma_\nu\gamma_5\left(1+\gamma_5\slashed{S}\right)\left(\slashed{k}+m_e\right)\right\}\right) \tag{7.131}$$

to which one must add the reverse, WNC×EM. The sum may be written

$$L_{\mu\nu}^{\mathrm{PV}} \equiv \frac{1}{2}\left[a_V L_{\mu\nu} + a_A \tilde{L}_{\mu\nu}\right]\,. \tag{7.132}$$

The first term here is just the EM case again, given in Eq. (7.44), containing the individual unpolarized and polarized leptonic tensors in Eqs. (7.21) and (7.45), respectively. The second term that results is an interference between vector (EM) and axial-vector (WNC) contributions; by evaluating the trace in Eq. (7.131), it is straightforward to show that this term can be written

$$\tilde{L}_{\mu\nu} = \frac{1}{2}\left[\tilde{L}_{\mu\nu}^{\mathrm{unpol}} + \tilde{L}_{\mu\nu}^{\mathrm{pol}}\right] \tag{7.133}$$

with

$$\tilde{L}_{\mu\nu}^{\mathrm{unpol}} \equiv -\frac{1}{2m_e^2}\left(i\epsilon_{\mu\nu\alpha\beta}k^\alpha k'^\beta\right) \tag{7.134}$$

$$\tilde{L}_{\mu\nu}^{\mathrm{pol}} \equiv \frac{1}{2m_e^2}\left[k'_\mu(m_e S_\nu) + k'_\nu(m_e S_\mu) - g_{\mu\nu}k'\cdot(m_e S)\right]\,. \tag{7.135}$$

Here the term containing the electron spin four-vector (pol) is symmetric under $\mu \leftrightarrow \nu$, whereas the non-spin-dependent term (unpol) is antisymmetric. The pattern should be clear: in Eqs. (7.21, 7.134, 7.135), terms of the form VV or AA (L^{unpol} and $\tilde{L}^{\mathrm{pol}}$) are symmetric and real, whereas terms of the form VA or AV (L^{pol} and $\tilde{L}^{\mathrm{unpol}}$) are antisymmetric and imaginary. The occurrences of axial-vectors come from two sources, the primordial weak interaction axial-vector currents and the electron's spin. As in

Eqs. (7.50, 7.51), we usually want the ERL expressions for the leptonic tensors. The new terms in Eqs. (7.134, 7.135) yield

$$2m_e^2 \tilde{L}_{\mu\nu}^{\text{unpol}} \rightarrow -i\epsilon_{\mu\nu\alpha\beta} k^\alpha k'^\beta \equiv \tilde{\chi}_{\mu\nu}^{\text{unpol}} \tag{7.136}$$

$$2m_e^2 \tilde{L}_{\mu\nu}^{\text{pol}} \rightarrow h\left[k_\mu k'_\nu + k'_\mu k_\nu - g_{\mu\nu} k \cdot k'\right] \equiv \tilde{\chi}_{\mu\nu}^{\text{pol}} . \tag{7.137}$$

Thus, in the ERL a very simple pattern emerges: defining symmetric and anti-symmetric tensors

$$\chi_{\mu\nu}^{(1)} \equiv k_\mu k'_\nu + k'_\mu k_\nu - g_{\mu\nu} k \cdot k' \tag{7.138}$$

$$\chi_{\mu\nu}^{(2)} \equiv -i\epsilon_{\mu\nu\alpha\beta} k^\alpha k'^\beta , \tag{7.139}$$

we find that

$$\begin{aligned}
\chi_{\mu\nu}^{\text{unpol}} &= \chi_{\mu\nu}^{(1)} \\
\chi_{\mu\nu}^{\text{pol}} &= h\chi_{\mu\nu}^{(2)} \\
\tilde{\chi}_{\mu\nu}^{\text{unpol}} &= \chi_{\mu\nu}^{(2)} \\
\tilde{\chi}_{\mu\nu}^{\text{pol}} &= h\chi_{\mu\nu}^{(1)} .
\end{aligned} \tag{7.140}$$

Reassembling the ERL limit of the full leptonic tensor, we now have

$$2m_e^2 L_{\mu\nu}^{\text{PV}} \rightarrow \chi_{\mu\nu}^{\text{PV}} = \frac{1}{2}(a_V + ha_A)\left[\chi_{\mu\nu}^{(1)} + h\chi_{\mu\nu}^{(2)}\right] , \tag{7.141}$$

or, calculating the helicity difference (i.e., the leptonic tensor to be employed when using longitudinally polarized electrons) $\{h = +1\} - \{h = -1\}$, we obtain the result

$$\chi_{\mu\nu}^{\text{PV,helicity}-\text{difference}} = a_A \chi_{\mu\nu}^{(1)} + a_V \chi_{\mu\nu}^{(2)} . \tag{7.142}$$

The treatment of the PV hadronic tensor proceeds strictly in parallel with the developments above. We begin with the same structure as in Eqs. (7.27, 7.28), except now with scalar response functions, $\tilde{X}_\ell$ and $\tilde{Y}_\ell$, that are constructed from products of the hadronic current matrix elements in the form EM×WNC (see Eq. (8.3)); we temporarily use a tilde over the response functions to indicate their PV nature. The vector part of the WNC is assumed to be conserved, and thus $\tilde{X}_1 - \tilde{X}_2 Q^2 = 0$ and $\tilde{X}_4 = 0$. Moreover, the $\tilde{Y}_1$ term in the antisymmetric PV hadronic tensor transforms as the product of two polar-vectors, both of which are conserved, and thus as before $\tilde{Y}_1 = 0$.

The term that differs from our previous analysis above is the one containing $\tilde{Y}_2$ and involving the tensor $\epsilon^{\mu\nu\alpha\beta}$, the analog of the last term in Eq. (7.28). Defining $\tilde{W}_1 \equiv -\tilde{X}_1$ and $\tilde{W}_2 \equiv \tilde{X}_3$, and moreover introducing $\tilde{W}_3 \equiv -\tilde{Y}_2$, we have

$$\tilde{W}_s^{\mu\nu} = -\tilde{W}_1\left(g^{\mu\nu} + \frac{q^\mu q^\nu}{Q^2}\right) + \tilde{W}_2 V_i^\mu V_i^\nu \tag{7.143}$$

$$\tilde{W}_a^{\mu\nu} = i\tilde{W}_3 \epsilon^{\mu\nu\alpha\beta} Q_\alpha V_{i\beta} . \tag{7.144}$$

As before, contracting the leptonic and hadronic tensors we obtain

$$\chi_{\mu\nu}^{(1)} \tilde{W}_s^{\mu\nu} \sim \tilde{W}_2 + 2\tilde{W}_1 \tan^2 \theta_e/2 \tag{7.145}$$

$$\chi_{\mu\nu}^{(2)} \tilde{W}_a^{\mu\nu} \sim 2\tilde{W}_3(\epsilon + \epsilon') \tan^2 \theta_e/2 . \tag{7.146}$$

We are now in a position to put these various pieces together to express the results for PV electron scattering in standard form. The reactions of interest here involve the measurement of the inclusive (single-arm) helicity-asymmetry, i.e., the difference between electron scattering of right- and left-handed electrons divided by the sum. The former is parity violating, while the latter is proportional to the usual PC electromagnetic cross section, as we have seen above.

Using the expressions obtained above, the helicity-difference asymmetry may be written

$$\mathcal{A} = \frac{d\sigma^{+1} - d\sigma^{-1}}{d\sigma^{+1} + d\sigma^{-1}} = \mathcal{A}_0 \times \frac{W^{\text{PV}}}{W^{\text{PC}}}\,, \tag{7.147}$$

where the characteristic size of the asymmetry is set by

$$\mathcal{A}_0 \equiv \frac{G_F \kappa}{2\pi\alpha\sqrt{2}}\, Q^2 \cong 7.00 \times 10^{-6}\, Q^2 (\text{fm}^{-2}) \tag{7.148}$$

$$\cong 1.80 \times 10^{-4}\, Q^2 ((\text{GeV}/\text{c})^2)\,. \tag{7.149}$$

Here G_F is the Fermi coupling and κ (usually equal to unity) is an overall constant involved in general discussions of weak neutral current effects in nuclei [Don79].

The hadronic content in the problem is contained in the ratio of the PV response W^{PV} to the PC electromagnetic response W^{PC}. As we found above (see Eq. (7.36)), the latter is given by

$$W^{\text{PC}} = v_L W^L + v_T W^T\,, \tag{7.150}$$

where the θ_e-dependence is contained in the electron kinematical factors v_L and v_T given in Eqs. (7.24, 7.25). The PV response may be cast in a similar form using the results in the previous section:

$$W^{\text{PV}} = v_L W^L_{AV} + v_T W^T_{AV} + v_{T'} W^{T'}_{VA}\,. \tag{7.151}$$

Here the subscripts denote interferences involving axial-vector leptonic currents with hadronic vector currents when written AV and the reverse when written VA. A third electron kinematic factor is now required

$$v_{T'} \equiv \tan\theta_e/2\sqrt{\rho + \tan^2\theta_e/2}\,, \tag{7.152}$$

where ρ is defined in Section 7.1.

The final response functions introduced above are simply related to the previous quantities: $W^L = (W_2 - \rho W_1)/\rho^2$, $W^T = 2W_1$ (see Eq. (7.40)) for the PC (EM) cases and $W^L_{AV} = (\widetilde{W}_2 - \rho\widetilde{W}_1)/\rho^2$, $W^T_{AV} = 2\widetilde{W}_1$ together with $W^{T'}_{VA} = 2q\widetilde{W}_3$ for the PV cases. All of these response functions are functions of q and ω (or equivalently Q^2 and $Q \cdot P_i$), but not of θ_e. Hence, by considering both PC and PV electron scattering, and by choosing at least three values for θ_e and so varying the electron kinematic factors, in principle it is possible to separate the five response functions.

This concludes the formal development of the basics of lepton scattering from nucleons and nuclei. In the following chapter we go on to apply these ideas to the problem of elastic electron scattering from the nucleon. Both parity-conserving and parity-violating scatterings are treated and some special aspects of using hadronic polarizations, together with polarized electron scattering, will be introduced. There we will see that electron scattering (and some other special problems such as the electronic and muonic Lamb shifts in hydrogen-atom-like systems) provide us with the best determinations of the ground-state structure of the nucleon. The modeling discussed in Chapter 5 and lattice QCD simulations can then be brought to bear through comparisons with the EM, strangeness and axial-vector form factors of the nucleon.

Exercises

7.1   Mott Scattering

The Mott cross section for a electron of incident energy E_0 scattered through an angle θ_e is written as

$$\frac{d\sigma}{d\Omega} = \frac{(\alpha \cos \theta_e/2)^2}{(2E_0 \sin^2 \theta_e/2)^2} \,,$$

where α is the fine structure constant. Plot the Mott cross section for $E_0 = 1$ GeV in units of barns/sr versus θ_e. Note in particular the sharp rise at small angles, which can be very useful in luminosity measurements.

7.2   Electron Scattering from a Spinless Target

Consider scattering of unpolarized electrons of energy E from a static, spinless charge distribution $Ze\rho(\mathbf{r})$ with

$$\nabla^2 \phi(\mathbf{r}) = -Ze\rho(\mathbf{r})$$

normalized so that $\int \rho(\mathbf{r})d\mathbf{r} = 1$.

a) From first-order perturbation theory, show that the amplitude for scattering from the initial state to the final state $T_{fi} = -i \int j_\mu^{fi} A^\mu d^4x$, where A^μ is the electromagnetic field and $j_\mu^{fi} = -e\bar{u}_f \gamma_\mu u_i e^{i(p_f-p_i)\cdot x}$, is given by

$$T_{fi} = -i2\pi\delta(E_f - E_i)(-e\bar{u}_f \gamma^0 u_i) \int e^{i\mathbf{q}\cdot\mathbf{r}} \phi(\mathbf{r})d\mathbf{r} \,.$$

b) Show that

$$\int e^{i\mathbf{q}\cdot\mathbf{r}} \phi(\mathbf{r})d\mathbf{r} = \frac{ZeF(\mathbf{q})}{|\mathbf{q}|^2} \,,$$

where $F(\mathbf{q})$ is the Fourier transform of the charge distribution.

c) By summing over final and averaging over initial electron spins show that

$$\frac{1}{2}\Sigma_{s_f,s_i}|\bar{u}_f \gamma^0 u_i|^2 = 4E^2(1 - v^2 \sin^2 \theta_e/2) \,,$$

where θ_e is the electron scattering angle.

d) Show that

$$\frac{d\sigma}{d\Omega} = \left(\frac{d\sigma}{d\Omega}\right)_{\text{Mott}} |F(\mathbf{q})|^2 \,.$$

e) Finally, show that if an electron beam is replaced by a beam of pointlike, spinless particles, the only change is that the factor $(1-v^2 \sin^2 \theta_e/2)$ in c) above is replaced by unity. Why does the electron spin make no difference in the nonrelativistic limit $v \to 0$? By considering the electron helicity, explain why you would anticipate the $\cos^2 \theta_e/2$ behavior of c) in the extreme relativistic limit (ERL).

7.3 Longitudinal-Transverse Decomposition of Electron Scattering

It is useful to decompose the electron scattering cross section into its longitudinal and transverse components, since the various contributions often depend on rather different aspects of the projections of the EM current.

a) Verify the basic expression for the electromagnetic response given in terms of the matrix elements of the EM current, namely, Eqs. (7.62) and (7.63):

$$
\begin{aligned}
\mathcal{R}_{fi} = & \left[\left((\epsilon + \epsilon')^2 - q^2 \right) \left(\frac{Q^2}{q^2} \right)^2 \right] |\rho_{fi}(\mathbf{q})|^2 \\
& + \left[2 \left(\frac{1}{q} kk' \sin \theta_e \right)^2 + Q^2 \right] \left\{ |J_{fi}(\mathbf{q}; +1)|^2 + |J_{fi}(\mathbf{q}; -1)|^2 \right\} \\
& + \left[\sqrt{2} \left(\epsilon + \epsilon' \right) \left(\frac{Q^2}{q^2} \right) \frac{1}{q} kk' \sin \theta_e \right] \\
& \times \left\{ 2\mathrm{Re} \left\{ \rho_{fi}(\mathbf{q})^* \left(J_{fi}(\mathbf{q}; +1) - J_{fi}(\mathbf{q}; -1) \right) \right\} \right\} \\
& + \left[-2 \left(\frac{1}{q} kk' \sin \theta_e \right)^2 \right] \left\{ 2\mathrm{Re} \left\{ J_{fi}(\mathbf{q}; +1)^* J_{fi}(\mathbf{q}; -1) \right\} \right\} \\
\equiv & \ v_0 \left\{ v_L W_{fi}^L + v_T W_{fi}^T + v_{TL} W_{fi}^{TL} + v_{TT} W_{fi}^{TT} \right\} .
\end{aligned}
$$

b) Prove the identity

$$
\langle f \,|\widehat{L}_{JM}(q)|\, i \rangle = - \left(\frac{\omega}{q} \right) \langle f \,|\widehat{M}_{JM}(q)|\, i \rangle
$$

that results from invoking the continuity equation for the EM current and its matrix elements.

c) Using angular momentum properties of the matrix elements of the current, provide the details in the proofs that yield the representations

$$
\overline{\sum_{if}} W_{fi}^L \equiv \left(W^L \right)^{\mathrm{unpol}} = \frac{4\pi}{2J_i + 1} \sum_{J \geq 0} |\langle J_f \,\|\widehat{M}_J\|\, J_i \rangle|^2
$$

and

$$
\overline{\sum_{if}} W_{fi}^T \equiv \left(W^T \right)^{\mathrm{unpol}} = \frac{4\pi}{2J_i + 1} \sum_{J \geq 1} \left(|\langle J_f \,\|\widehat{T}_J^{\mathrm{el}}(q)\|\, J_i \rangle|^2 + |\langle J_f \,\|\widehat{T}_J^{\mathrm{mag}}(q)\|\, J_i \rangle|^2 \right) ,
$$

(see Eqs. (7.80) and (7.94)).

7.4 Elastic Scattering of Fast Electrons by an Atom

Apply the Born approximation to the scattering of a fast electron by an atom, using as a perturbation, the interaction

$$V(\mathbf{r}) = -\frac{Z\alpha}{r} + \sum_{i=1}^{Z} \frac{\alpha}{|\mathbf{r} - \mathbf{r}_i|} \, ,$$

which represents the Coulomb potential between the incident electron, and the nucleon of charge $Z|e|$, and the individual electrons which make up the atom. For the initial-state wavefunction take

$$\langle \mathbf{r}, \mathbf{r}_1, \ldots, \mathbf{r}_Z | i \rangle = e^{i\mathbf{k}_i \cdot \mathbf{r}} \psi_0(\mathbf{r}_1, \ldots, \mathbf{r}_Z)$$

and for the final state

$$\langle \mathbf{r}, \mathbf{r}_1, \ldots, \mathbf{r}_Z | f \rangle = e^{i\mathbf{k}_f \cdot \mathbf{r}} \psi_0(\mathbf{r}_1, \ldots, \mathbf{r}_Z) \, ,$$

where $\psi_0(\mathbf{r}_1, \ldots, \mathbf{r}_Z)$ is the wavefunction of the ground state of the atom, assumed to be the same before and after the collision. (Note: In principle, the electron wavefunction should be properly antisymmetrized. However, at high energies this antisymmetrization can be shown to be relatively unimportant.)

a) Demonstrate that the differential cross section takes the form

$$\frac{d\sigma}{d\Omega} = |1 - F(\mathbf{q})|^2 \frac{d\sigma}{d\Omega}\bigg|_{\text{point}} \, ,$$

where

$$\frac{d\sigma}{d\Omega}\bigg|_{\text{point}} = \frac{Z^2 \alpha^2}{4m^2 v^4 \sin^4 \theta_e/2}$$

is the cross section for an electron of charge $|e|$ and velocity v to scatter from a point-particle target of charge $Z|e|$; here, $F(\mathbf{q})$ is the "form factor" for the atomic ground state $|\psi_0 >$

$$F(\mathbf{q}) = \frac{1}{Z} \left\langle \psi_0 \left| \sum_{j=1}^{Z} e^{i\mathbf{q} \cdot \mathbf{r}_j} \right| \psi_0 \right\rangle$$

$$= \frac{1}{Z} \prod_{i=1}^{Z} \int d\mathbf{r}_i \sum_{j=1}^{Z} e^{i\mathbf{q} \cdot \mathbf{r}_j} |\psi_0(\mathbf{r}_1, \ldots, \mathbf{r}_Z)|^2 \, .$$

b) For the specific case of electron scattering from the ground state of a hydrogen atom, show that the differential scattering cross section has the form

$$\frac{d\sigma}{d\Omega} = \frac{\alpha^2}{4m^2 v^4 \sin^4 \theta_e/2} \left(1 - \frac{1}{\left(1 + \frac{q^2 a_0^2}{4}\right)^2} \right)^2 \, ,$$

where a_0 is the Bohr radius.

In general, a form factor arises in problems in which scattering is due to a composite system made up of identical scatterers (electrons in our case), which are distributed in space. The form factor expresses the feature that the scattering amplitude generated by the scattering centers differ from one another because of purely *geometric* considerations – variation of the phase of the incident wave at the scattering center and of the scattered wave at the detector. Measurement of the form factor thus yields information about the distribution of scattering centers within the composite system, i.e., the Fourier transform of the spatial distribution. For a given momentum transfer $\boldsymbol{q}$, the form factor is sensitive to variations in the geometric distribution over distances of order d such that

$$qd \sim 1, \quad \text{i.e.,} \quad d \sim \frac{1}{q},$$

and this is the motivation behind the use of scattering experiments to map the distribution of constituent scatterers.

c) More quantitatively, consider the situation with a source S and a detector D with scattering centers at O, O'. The full scattering amplitude is then

$$\mathcal{M} = \frac{1}{r_i} e^{ikr_i} A(\boldsymbol{k}_i \to \boldsymbol{k}_f) \frac{1}{r_f} e^{ikr_f} + \frac{1}{r_i'} e^{ikr_i'} A(\boldsymbol{k}_i' \to \boldsymbol{k}_f') \frac{1}{r_f'} e^{ikr_f'} ,$$

where r_i (r_i') is the distance from the source to scattering center O (O') and r_f (r_f') is the distance from scattering center to the detector. Show that, if r_i, r_f are much larger than the distance ρ between the scattering centers, then it is reasonable to take the intrinsic scattering amplitude $A(\boldsymbol{k}_i \to \boldsymbol{k}_f)$ as being the same at both centers, so that

$$\mathcal{M} \simeq \frac{1}{r_i} e^{ikr_i} A(\boldsymbol{k}_i \to \boldsymbol{k}_f)(1 + e^{i\boldsymbol{q}\cdot\boldsymbol{\rho}}) \frac{1}{r_f} e^{ikr_f} ,$$

where $\boldsymbol{q} = \boldsymbol{k}_i - \boldsymbol{k}_f$ is the momentum transfer. Explain your result for atomic scattering obtained in a) of this exercise in light of this form.

d) Large momentum transfer: Show that if q is much larger than the atomic radius, then $|F(\boldsymbol{q})| \ll 1$ and $\frac{d\sigma}{d\Omega}$ is essentially the Rutherford cross section for scattering by the Coulomb field.

e) Small momentum transfer: Assume for simplicity that the atomic ground-state wavefunction ψ_0 has total angular momentum zero. Expand $F(\boldsymbol{q})$ in powers of $\boldsymbol{q}$ and show that

$$Z F(\boldsymbol{q}) = Z - \frac{1}{6} q^2 \left\langle \psi_0 \left| \sum_{i=1}^{Z} r_i^2 \right| \psi_0 \right\rangle + \cdots .$$

Is the differential scattering cross section finite or infinite at $\theta_e = 0$? Explain this result.

f) Analyze the hydrogen atom cross section calculated in part b) in terms of the large- and small-q limits discussed in d) and e).

7.5   Relativistic Coulomb Scattering

In this exercise, we calculate the differential cross section for the scattering of two non-identical spinless particles of charge and mass, $-e, m_1$ and Ze, m_2, respectively. In the case of electrodynamics, the equation-of-motion of the photon field $A_\mu(x)$ is

$$\Box A_\mu(x) = j_\mu(x) \,,$$

where $j_\mu(x)$ is the current density, and the term which couples $A_\mu(x)$ and $j^\mu(x)$ is

$$\mathcal{L}_{int} = - \int d^4 x j^\mu(x) A_\mu(x) \,.$$

For a charged scalar field we have

$$\langle p_1' | j_\mu(x) | p_1 \rangle = e^{i(p_1' - p_1)\cdot x} e(p_1 + p_1')_\mu \,.$$

a) Show that the interaction of the two particles is given by calculating the scattering amplitude via

$$S = \int d^4 x \mathcal{L}_{int}(x)$$

$$\times \int d^4 x \int d^4 y \int \frac{d^4 q}{(2\pi)^4} e^{i(p_2'-p_2)\cdot x} Ze(p_2 + p_2')_\mu e^{iq\cdot(x-y)} e^{i(p_2-p_1)\cdot y} e(p_1 + p_1')^\mu$$

$$= (2\pi)^4 \delta^4(p_1 + p_2 - p_1' - p_2') Ze(p_1 + p_1') \cdot (p_2 + p_2') \frac{1}{(p_1 - p_1')^2}$$

$$\equiv (2\pi)^4 \delta^4(p_1 + p_2 - p_1' - p_2') T(q) \,,$$

where

$$T(q) = 4\pi Z\alpha \frac{(p_1 + p_1') \cdot (p_2 + p_2')}{(p_1 - p_1')^2} \,.$$

b) Square to find the transition probability: interpret this result via

$$[(2\pi)^4 \delta^4(p_1 + p_2 - p_1' - p_2')]^2$$

$$= (2\pi)^4 \delta^4(p_1 + p_2 - p_1' - p_2') \int dt \int d\mathbf{x} \exp(-i(p_1 + p_2 - p_1' - p_2')\cdot x)$$

$$= (2\pi)^4 \delta^4(p_1 + p_2 - p_1' - p_2') TV \,,$$

where T is the interaction time and V is the volume. Since the cross section is the transition rate per unit volume divided by the incident flux, show that

$$d\sigma = \frac{1}{\text{Flux}} \int \frac{d\mathbf{p}_1'}{2E_1'} \frac{d\mathbf{p}_2'}{2E_2'} (2\pi)^4 \delta^4(p_1 + p_2 - p_1' - p_2') |T(q)|^2 \,,$$

where

$$\text{Flux} = 2E_1 \cdot 2E_2 \cdot |\mathbf{v}_1 - \mathbf{v}_2|$$

is the flux, with $2E$ representing relativistic normalization factors, due to the normalization condition

$$\int d^3 x \phi_k^*(x) i\partial_0 \phi_{k'}(x) - i\partial_0 \phi_k^*(x) \phi_{k'}(x) = 2E_k \delta^3(\mathbf{k} - \mathbf{k}')$$

for the Klein–Gordon states

c) Calculate the cross section in the laboratory frame (i.e., the frame in which the particle having mass m_2 and four-momentum p_2 is at rest) and show that

$$\frac{d\sigma}{d\Omega} = \frac{1}{(2\pi)^2} \frac{1}{16 m_2 p_1} \frac{p_1'^2}{E_2' p_1' + E_1'(p_1' - p_1 \cos\theta)} |\mathcal{T}(q)|^2$$

with

$$\mathcal{T}(q) = 4\pi\alpha Z \frac{m_2(E_1 + E_1')}{m_1^2 - E_1 E_1' + p_1 p_1' \cos\theta} \ .$$

d) Now transform to the frame where the target particle is very heavy,

$$m_2 \to \infty \ ,$$

and the incident particle is moving slowly

$$\frac{v_1}{c} \ll 1 \ ,$$

and compare your result with the Rutherford cross section

$$\frac{d\sigma_{\text{Ruth}}}{d\Omega} = \frac{Z^2 \alpha^2}{(2 m_1 v_1^2)^2 \sin^4 \theta_e/2} \ .$$

7.6   Gravitational Scattering

We can analyze the gravitational scattering of two massive particles in a fashion analogous to that used for Coulomb scattering. In the case of electrodynamics the equation-of-motion of the photon field $A_\mu(x)$ is determined by

$$\Box A_\mu(x) = j_\mu(x) \ ,$$

where $j_\mu(x)$ is the current density and the term which couples $A_\mu(x)$ and $j^\mu(x)$ is

$$\mathcal{L}_{int} = -\int d^4x\, j^\mu(x) A_\mu(x) \ .$$

For a charged scalar field we have

$$\langle p_1' | j_\mu(x) | p_1 \rangle = e^{i(p_1'-p_1)\cdot x} e(p_1 + p_1')_\mu \ .$$

Then the interaction of a two particles, one with charge e_1 and mass m_1 and a second with charge e_2 and mass m_2, is given by calculating the scattering amplitude via

$$S = -\int d^4x\, \mathcal{L}_{int}(x) = \frac{1}{\sqrt{4 E_1 E_2}} \frac{1}{\sqrt{4 E_1' E_2'}}$$

$$\times \int d^4x \int d^4y \int \frac{d^4q}{(2\pi)^4} e^{i(p_1'-p_1)\cdot x} e_1(p_1 + p_1') e^{iq\cdot(x-y)} e^{i(p_2'-p_2)\cdot y} e_2(p_2' + p_2)^\mu$$

$$= -(2\pi)^4 \delta^4(p_1 + p_2 - p_1' - p_2') e_1(p_1 + p_1') e_2 \cdot (p_2 + p_2') \frac{1}{(p_1 - p_1')^2}$$

$$\equiv (2\pi)^4 \delta^4(p_1 + p_2 - p_1' - p_2') \mathcal{T}_{em}(q) \ .$$

Then the potential energy is given by the Fourier transform of the transition amplitude

$$V_{em}(r) = \int \frac{d\mathbf{q}}{(2\pi)^3} e^{-i\mathbf{q}\cdot\mathbf{r}} T_{em}(\mathbf{q}) = \frac{e_1 e_2}{4\pi r} \,,$$

which is the familiar Coulomb potential.

A parallel formalism can be used to describe the lowest-order gravitational interaction. In this case, the graviton field, which carries spin 2, is described by a symmetric second-rank tensor, $h_{\mu\nu}(x)$, which obeys the equation of motion

$$\Box h_{\mu\nu}(x) = -16\pi G \left(T_{\mu\nu}(x) - \frac{1}{2}\eta_{\mu\nu}\mathrm{Tr}(T(x)) \right) ,$$

where here G is Newton's constant, $T_{\mu\nu}(x)$ is the energy–momentum tensor, and $\mathrm{Tr}\,(T(x)) = \eta^{\mu\nu} T_{\mu\nu}(x)$ is its trace. The fields $h_{\mu\nu}(x)$ and $T_{\mu\nu}(x)$ are coupled by the interaction term

$$\mathcal{L}_{\mathrm{int}} = -\frac{1}{2} \int d^4x\, T_{\mu\nu}(x) h^{\mu\nu}(x) ,$$

where for a scalar field we have

$$\langle p_1' | T_{\mu\nu}(x) | p_1 \rangle = e^{i(p_1' - p_1)\cdot x} \left[p_{1\mu} p_{1\nu}' + p_{1\mu}' p_{1\nu} - \eta_{\mu\nu}(p_1' \cdot p_1 - m^2) \right] \frac{1}{\sqrt{4E_1 E_1'}} \,.$$

The relation between the graviton field $h_{\mu\nu}(x)$ and the metric tensor $g_{\mu\nu}(x)$ is

$$g_{\mu\nu}(x) = \eta_{\mu\nu} + h_{\mu\nu}(x) ,$$

where

$$\eta_{\mu\nu} = \begin{pmatrix} 1 & 0 & 0 & 0 \\ 0 & -1 & 0 & 0 \\ 0 & 0 & -1 & 0 \\ 0 & 0 & 0 & -1 \end{pmatrix}$$

is the usual Minkowski metric.

a) Consider a very massive particle M (which could be the sun, for example) at rest at the origin so that

$$\langle p_2' | T_{\mu\nu}(x) | p_2 \rangle = e^{i(p_2' - p_2)\cdot x} M \delta_{\mu 0} \delta_{\nu 0} \,.$$

Using Fourier transform methods, solve the equation of motion for $h_{\mu\nu}(x)$ and show that

$$h_{00}(x) = -\frac{2GM}{|\mathbf{x}|}, \quad h_{ij}(x) = -\frac{2GM}{|\mathbf{x}|}\delta_{ij} \quad h_{i0}(x) = 0$$

when the mass is placed at the origin of coordinates. Compare this result with the Schwarzschild metric in the limit $GM/|x| << 1$ [Wei72].

b) Consider the scattering by M of a much lighter scalar particle with mass m. Show that the scattering amplitude is given by

$$S = \int d^4x \mathcal{L}_{int}(x) = \frac{-1}{\sqrt{4E_1 E_1'}}$$

$$\times \int d^4x \int d^4y \int \frac{d^4q}{(2\pi)^4} e^{i(p_1'-p_1)\cdot x} M(2\delta_{\mu 0}\delta_{\nu 0} - \eta_{\mu\nu}) e^{iq\cdot(x-y)} \frac{4\pi G}{q^2}$$

$$\times e^{i(p_2'-p_2)\cdot y} \left[p_{2\mu} p_{2\nu}' + p_{2\mu}' p_{2\nu} - \eta_{\mu\nu}(p_2 \cdot p_2' - m^2) \right]$$

$$= -(2\pi)^4 \delta^4(p_1 + p_2 - p_1' - p_2') 8\pi GM^2 \frac{1}{(p_2 - p_2')^2}(4E_1 E_1' - 2m^2) \frac{1}{2M} \frac{1}{\sqrt{4E_1 E_1'}}$$

$$\equiv (2\pi)^4 \delta^4(p_1 + p_2 - p_1' - p_2') \mathcal{T}_{gr}(q) \ .$$

c) Evaluate the Fourier transform of the static limit of the transition amplitude, and show that the mutual interaction of m and M is described by the potential energy

$$V_{gr}(r) = \int \frac{d\mathbf{q}}{(2\pi)^3} e^{-iq\cdot r} \mathcal{T}_{gr}(q) = -\frac{GMm}{r} \ ,$$

which is the conventional Newtonian result.

d) Calculate the differential scattering cross section using the transition amplitude $\mathcal{T}_{gr}(q)$ and show that

$$d\sigma = \frac{E_1}{p_1} \int \frac{d\mathbf{p}_2}{(2\pi)^3} 2\pi (E_1' - E_1) |\mathcal{T}_{gr}(q)|^2 = (2GM)^2 \frac{1}{q^2}(2E_1^2 - m^2)^2 d\Omega$$

$$= (2GM)^2 \frac{1}{16p_1^4 \sin^4 \theta/2}(2E_1^2 - m^2)^2 d\Omega \ .$$

e) We can connect with the corresponding classical result by using the relation between the impact parameter b and scattering angle θ

$$2\pi b\, db = d\sigma \ .$$

Show that in the massless ($m = 0$) limit this relation can be integrated to become

$$\theta = \frac{4GM}{b}$$

in the small angle scattering limit.

This is the relation derived by Einstein, and has been tested via the measurement of the deflection of starlight during an eclipse. (These experiments involved photons with spin-one rather than spin-zero particles as considered above. However, general relativity requires that all massless particles must follow the same trajectory, a geodesic, regardless of spin.)

7.7 Small-Angle Electron Scattering

Most frequently, the kinematics and leptonic factors such as in Eqs. (7.68–7.71) are given in the extreme relativistic limit, where the electron mass is neglected with respect to its energy or three-momentum. However, this simplification is not necessary, and under some conditions, it is essential to retain the mass.

a) Using the standard expressions for the square of the three-momentum transfer

$$q^2 = k^2 + k'^2 - 2kk' \cos \theta_e$$

and for the energy transfer

$$\omega = \epsilon - \epsilon',$$

and keeping the electron mass finite, obtain an expression for the four-momentum transfer squared. Here $q = |\mathbf{q}|$, $k = |\mathbf{k}|$ and $k' = |\mathbf{k}'|$.

b) By adding and subtracting a term $4kk' \sin^2 \theta_e/2$ to and from the expression obtained in part a), write the square of the four-momentum transfer in the form $Q^2 = 4kk' \sin^2 \theta_e/2 + \Delta$, and obtain an expression for Δ. When considering electron scattering at very small angles, the first term here goes to zero, but, in general, the second term does not. Find an approximate expression for the latter in the small-angle limit for a situation involving the excitation of a nucleus (or any system in general) from its ground state to an excited state at excitation energy E^*. In fact, one can see that the four-momentum transfer cannot ever be zero in such a situation and that one always has $Q^2 > 0$ (spacelike).

c) Using the results in parts a) and b), find the critical angle θ_e^c where, for $\theta_e \leq \theta_e^c$, one must retain the electron mass, while for $\theta_e > \theta_e^c$ it is safe to employ the ERL.

d) Obtain expressions for the analogs of the kinematic factors $v_{L,T,TL,TT}$ (usually denoted $V_{L,T,TL,TT}$) when the electron mass is kept finite.

7.8 Weisskopf Units

In the measurement of an electromagnetic transition probability within a nucleus, it is useful to compare the measured magnitude of a given quantity with a "natural" size, i.e., with what one might expect from a single nucleon making a transition from one orbital to another within the nucleus – the "single-particle value." For electric transitions, the $B(E\lambda)$ value is the square of the transition matrix element for a single proton, which can be written as

$$\langle J_f M_f | er^\lambda Y_{\lambda\mu} | J_i M_i \rangle = e(\text{radial integral}) \times (\text{angular integral}),$$

where the radial integral is

$$\langle r^\lambda \rangle = \int_0^\infty dr r^2 R^*_{n_f \ell_f}(r) r^\lambda R_{n_i \ell_i}(e)$$

and the angular integral yields 4π.

a) Assume the nucleus to be a uniformly charged sphere with radius $R_0 A^{\frac{1}{3}}$. Show that

$$\langle r^\lambda \rangle = \frac{3}{3 + \lambda} R_0^\lambda A^{\frac{\lambda}{3}}.$$

b) For the present purpose, take

$$B_{est}(E\lambda) = \frac{e^2}{4\pi}\langle r^\lambda\rangle^2$$

and show that the single-particle estimate becomes

$$B_W(E\lambda) = \frac{e^2}{4\pi}\left(\frac{3}{3+\lambda}\right)^2 R_0^{2\lambda} A^{\frac{2\lambda}{3}}\,,$$

which is called the *Weisskopf unit.*

7.9 Effective Field Theory and the EM Multipole Expansion

The application of effective field theoretic ideas to the electromagnetic interaction is manifested in a multipole expansion, which represents the interaction in terms of a few constants (multipoles).

a) Show that the electromagnetic interaction of a nucleus of spin J can be represented in terms of an expansion

$$H_{em} = Ze\phi_0 - \boldsymbol{M}\cdot\boldsymbol{B} - \frac{1}{6}\sum_{ij} Q_{ij}\frac{\partial E_i}{\partial x_j} + \cdots$$

with

$$\boldsymbol{M} = \mu\frac{e}{2m_N}\frac{\boldsymbol{J}}{J}$$

$$Q_{ij} = \frac{eQ}{J(2J-1)}\left[\frac{3}{2}(J_iJ_j + J_jJ_i) - \delta_{ij}\boldsymbol{J}^2\right].$$

Here μ is the magnetic dipole moment in units of nucleon magnetons and Q is the electric quadrupole moment of the particle having charge Ze.

b) Show that the matrix element of the electromagnetic current can be written at low momentum transfer as

$$\langle J, M_f, p_f | V_\mu^{em} | J, M_i, p_i\rangle = \delta_{M_i,M_f}\frac{P_\mu}{2M}F_1(q^2)$$

$$- iC_{J1;J}^{M_f k;M_i}\epsilon_{ijk}\eta_{\mu j}\frac{q^i}{2M}F_2(q^2)$$

$$+ C_{J2;J}M_f k; M_i\frac{P_\mu}{(2M)^3}\sqrt{\frac{4\pi}{5}}Y_2^k(\hat{q})q^2F_3(q^2) + \cdots\,,$$

where $P_\mu = (p_f + p_i)_\mu$ is the total momentum and $M = \frac{1}{2}(M_1 + M_2)$ is the average mass.

c) Verify that the current is conserved, in that

$$q^\mu\langle J, M_f, p_f | V_\mu^{em} | J, M_i, p_i\rangle = 0\,.$$

d) Show that the electromagnetic form factors $F_1(q^2)$, $F_2(q^2)$, $F_3(q^2)$ are connected to the moments via

$$F_1(0) = Ze$$

$$F_2(0) = \sqrt{\frac{J+1}{J}}\,\mu A$$

$$F_3(0) = -\sqrt{\frac{(J+1)(2J+3)}{J(2J-1)}}\,\frac{2M^2}{3}\,\mathcal{Q}\,,$$

where A is the atomic number.

7.10 The Anapole Moment

The electromagnetic interaction becomes more complex when the weak interaction, and its associated parity violation, is included. In addition to the charge, magnetic dipole, electric quadrupole, etc. moments, new effects arise. One such effect, an interaction between the nuclear spin and the electron momentum called the *anapole* moment, is explored in this exercise.

Consider the matrix element of the electromagnetic current between a pair of nucleons, which can be written as

$$\langle N(p_2)|V_\mu^{em}(0)|N(p_1)\rangle = \bar{u}_N(p_2)\left(F_1(q^2)\gamma_\mu - F_2(q^2)\frac{i\sigma_{\mu\nu}q^\nu}{2m_N}\right)u_N(p_1)\,,$$

where $q = p_1 - p_2$ is the four-momentum transfer.

a) Verify that the current matrix element is conserved, in that

$$q^\mu\langle N(p_2)|V_\mu^{em}(0)|N(p_1)\rangle = 0.$$

b) By going to the nonrelativistic limit,($q_0 \simeq 0$, $|\boldsymbol{q}| << m_N$), show that the matrix element becomes

$$\langle N(p_2)|V_0^{em}(0)|N(p_1)\rangle \xrightarrow{NR} F_1(0)\chi_2^\dagger\chi_1 + \cdots$$

$$\langle N(p_2)|V_j^{em}(0)|N(p_1)\rangle \xrightarrow{NR} F_1(0)\frac{p_{2j}+p_{1j}}{2m_N} - \frac{(F_1(0)+F_2(0))}{2m_N}\chi_2^\dagger i(\boldsymbol{\sigma}\times\boldsymbol{q})_j\chi_1 + \cdots\,.$$

c) Equating $\int d\mathbf{x}\,V_0^{em}$ with the electric charge, show that we require $F_1(0) = Q_N$, where Q_N is the nucleon charge.

d) Show that the nucleon magnetic moment is given by

$$\mu_N = \frac{e}{2m_N}(F_1(0)+F_2(0))\boldsymbol{\sigma}\,,$$

so that $F_2(0)$ is identified with the anomalous magnetic moment κ_N.

In the first part of this exercise, it was assumed that parity is conserved. However, in the presence of the weak interaction, parity is violated, and the general form of the nucleon matrix element of the electromagnetic current requires the addition of two additional form factors.

e) Show that in the presence of parity violation, the general form of the nucleon electromagnetic matrix element is

$$\langle N(p_2)|V_\mu^{em}(0)|N(p_1)\rangle = \bar{u}_N(p_2)\left(F_1(q^2)\gamma_\mu - F_2(q^2)\frac{i\sigma_{\mu\nu}q^\nu}{2m_N}\right.$$
$$\left. +\frac{F_3(q^2)}{2m_N}(\gamma_\mu q^2 - q_\mu \not{q})\gamma_5 - F_4(q^2)\frac{i\sigma_{\mu\nu}q^\nu\gamma_5}{2m_N}\right)u_N(p_1).$$

f) Verify that the result in a) still pertains.

g) Extend b) and show that

$$\langle N(p_2)|V_0^{em}(0)|N(p_1)\rangle \xrightarrow{NR} F_1(0)\chi_2^\dagger\chi_1 - \frac{F_4(0)}{2m_N}\chi_f^\dagger\boldsymbol{\sigma}\cdot\boldsymbol{q}\chi_i + \cdots$$

$$\langle N(p_2)|V_j^{em}(0)|N(p_1)\rangle \xrightarrow{NR} F_1(0)\frac{\boldsymbol{p}_2+\boldsymbol{p}_1}{2m_N} - \frac{(F_1(0)+F_2(0))}{2m_N}\chi_2^\dagger i(\boldsymbol{\sigma}\times\boldsymbol{q})_j\chi_1$$

$$+F_3(0)(q^2\chi_2\boldsymbol{\sigma}\,\chi_1 - q_j\chi_2^\dagger\boldsymbol{\sigma}\,\chi_1\cdot(\boldsymbol{p}_1+\boldsymbol{p}_2)) + \cdots$$

h) Show that a nonzero value of $F_4(0)$ indicates a violation of both parity and time-reversal invariance and corresponds to the existence of an electric dipole moment of size

$$d_E = F_4(0)\frac{e}{2m_N}\boldsymbol{\sigma}.$$

i) Suppose the nucleon interacts with an electron. Show that a nonzero value of $F_3(0)$ indicates a violation of parity, but not time reversal invariance, and signals the existence of a local interaction

$$H = \frac{1}{2m_N}F_3(0)\boldsymbol{\sigma}_N\cdot\gamma_0\boldsymbol{\gamma}_{elec}\,,$$

where $\boldsymbol{\sigma}_N$ is the nucleon spin and the subscript *elec* indicates the electron matrix element. Because of the local nature of the interaction, this sort of term is not part of the typical electromagnetic multipole expansion. This form factor $F_3(0)$ was called the *anapole* moment by Zeldovich.

8   Elastic Electron Scattering from the Nucleon

Having introduced the basics of lepton scattering in the previous chapter, we now go on to discuss applications of these ideas. The three Chapters 8, 9, and 10, form a well-defined section of the book where the primary goal is to understand the structure of hadrons in terms of the quarks and gluons of QCD. The electroweak interaction is the tool of choice to accomplish this. Up to energies of about 30 GeV, intense beams of polarized electrons have been developed and scattered from both unpolarized and polarized targets. At low energies, the dominant process is elastic electron–nucleon scattering and this is the primary focus of the present chapter. Elastic form factors describe the distribution of charge and magnetization resulting from the quarks of QCD. At higher energies, where Bjorken scaling is applicable, the cross section can be used to extract distributions of the partons, and this is discussed extensively in Chapter 9. With present accelerator limitations, the highest-energy studies of QCD can only be carried out with hadron beams, and this is the focus of Chapter 10.

Most of the empirical basis for the present understanding of the structure of hadrons has been obtained using the scattering of beams of charged leptons (electrons, positrons, and muons) from nucleon and nuclear targets via the electroweak interaction, the best understood theory in physics. In addition, the relatively weak strength of the interaction makes lepton scattering the tool of choice for understanding the structure and dynamics of hadrons, since the probe does not excessively perturb the target system. Most of the observable mass in the physical world around us is understood to reside in the form of strongly interacting matter, i.e., in the form of protons, neutrons, and nuclei. Over the last 50 years, tremendous progress has been made in understanding hadron structure, particularly by experimentalists using lepton beams. In this chapter, the study of the structure of the nucleon using lepton scattering will be introduced.

8.1 The Elastic Form Factors of the Nucleon

Consider elastic electron scattering from the nucleon. Here

$$W_1 = \tau G_M^2 \quad \text{and} \quad W_2 = \frac{1}{1+\tau}[G_E^2 + \tau G_M^2], \tag{8.1}$$

where the Sachs form factors $G_M(Q^2)$, the magnetic form factor, and $G_E(Q^2)$, the electric form factor, are introduced for both the proton and the neutron. Factors G_M and G_E are transverse and longitudinal form factors, respectively. The Pauli $F_1^N(Q^2)$ and Dirac $F_2^N(Q^2)$ form factors are defined for $N = p, n$ as linear combinations of the Sachs form factors by

$$F_1^N(Q^2) = \frac{1}{1+\tau}\left[G_E^N(Q^2) + \tau G_M^N(Q^2)\right] \tag{8.2}$$

$$F_2^N(Q^2) = \frac{1}{1+\tau}\left[G_M^N(Q^2) - G_E^N(Q^2)\right]. \tag{8.3}$$

The elastic form factors are defined in single-photon exchange and they contain detailed information on the distribution of charge and magnetism in the nucleon. Experimentally, it has been found that the proton electric and magnetic form factors and the neutron magnetic form factor have approximately the following Q^2 dependence (the so-called *dipole* form)

$$G_E^p, G_M^p/\mu_p, G_M^n/\mu_n = \frac{1}{[1 + \frac{Q^2}{0.71}]^2}, \tag{8.4}$$

where μ_p and μ_n are the proton and neutron static magnetic moments, respectively, and Q^2 is in units of $(\text{GeV}/c)^2$. It is useful to define the ratio

$$R_p \equiv \frac{\mu_p G_E^p(Q^2)}{G_M^p(Q^2)}. \tag{8.5}$$

The total neutron electric charge is zero, with the result that the neutron electric form factor $G_E^n(Q^2)$ is proportional to Q^2 at low Q^2, and, accordingly, is significantly smaller than its proton analog in that region.

The elastic cross section may be written as

$$\frac{\frac{d\sigma}{d\Omega}}{\left(\frac{d\sigma}{d\Omega}\right)_{\text{Mott}}} = \frac{G_E^2(Q^2) + \tau G_M^2(Q^2)}{1+\tau} + 2\tau G_M^2(Q^2)\tan^2\frac{\theta}{2}. \tag{8.6}$$

A reduced cross section $(\frac{d\sigma}{d\Omega})_{\text{red}}$ can then be defined as

$$\left(\frac{d\sigma}{d\Omega}\right)_{\text{red}} = \frac{\frac{d\sigma}{d\Omega}}{\left(\frac{d\sigma}{d\Omega}\right)_{\text{Mott}}}[\mathcal{E}(1+\tau)] \tag{8.7}$$

$$= \tau G_M^2(Q^2) + \mathcal{E}G_E^2(Q^2), \tag{8.8}$$

where $\mathcal{E}$ was defined in Chapter 7. In this way, over many decades the elastic form factors of the nucleon have been determined using unpolarized cross section measurements, which are typically plotted for fixed Q^2 as a reduced cross section versus $\mathcal{E}$, as shown in Fig. 8.1. A straight line is fitted to the data. The intercept yields the magnetic form factor, $G_M(Q^2)$, while the slope yields the electric form factor, $G_E(Q^2)$. The proton and neutron form factors have been determined in this way from scattering from hydrogen and deuterium targets, respectively. The use of this so-called Rosenbluth technique is limited by systematic uncertainties. For example, the determination in this way of the neutron electric form factor $G_E^n(Q^2)$ and the high-Q^2 extraction of $G_E^p(Q^2)$ is limited by uncertainties associated with cross section measurements. Figure 8.2 shows a compilation of the world data for the proton elastic form factors, as determined using the Rosenbluth technique.

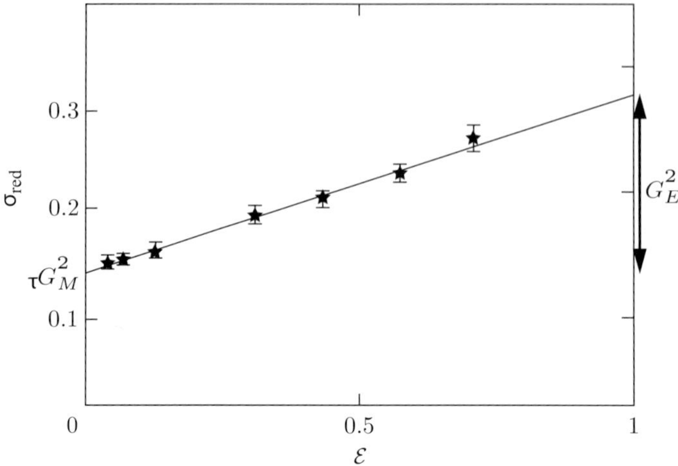

Fig. 8.1 The Rosenbluth method: the reduced cross section for elastic electron–proton scattering versus $\mathcal{E}$.

With the advent of high duty factor polarized electron beams in the GeV energy region at the end of the twentieth century, the major focus in the determination of the nucleon elastic form factors has been on the use of spin-polarization techniques. The spin-dependent cross section can be written as a sum of unpolarized (Σ) and polarized (Δ) contributions,

$$\frac{d\sigma}{d\Omega dE'} = \Sigma + p_e \cdot p_T \Delta \,, \tag{8.9}$$

where p_e and p_T are the beam and target polarizations, respectively. The cross sections can be written as

$$\Sigma = v_L G_E^2 + v_T G_M^2 \text{ and } \Delta = v_{TL'} G_E G_M \sin\theta^* \cos\phi^* + v_{T'} G_M^2 \cos\theta^* \,, \tag{8.10}$$

where the kinematic factors v_L, v_T, and $v_{T'}$ were defined in Eqs. (7.24), (7.25), and (7.152), respectively, and where now one also has

$$v_{TL'} = -\frac{1}{\sqrt{2}}\rho\sqrt{\rho + \tan^2\theta_e/2} \,. \tag{8.11}$$

Here θ^* and ϕ^* characterize the target spin direction – see Chapter 7 for a definition of the standard coordinate system. With a longitudinally polarized electron beam and a polarized target, by sorting the scattering events into the different polarization states the ratio of polarized to unpolarized cross sections can be extracted

$$\frac{\Delta}{\Sigma} = A_{meas}/(p_e \cdot p_T) \tag{8.12}$$

once the product of beam and target polarizations, $p_e \cdot p_T$, is known. By varying the direction of the target spin angle, the electric and magnetic form factors can be separated. Employing a polarimeter to measure the recoil polarization of the elastically scattered nucleon can be used instead of a polarized target. The use of the spin degree of freedom as a "knob" in this way is known as the "super-Rosenbluth" method, and for an optimally designed experiment produces substantially smaller systematic uncertainties than the standard Rosenbluth

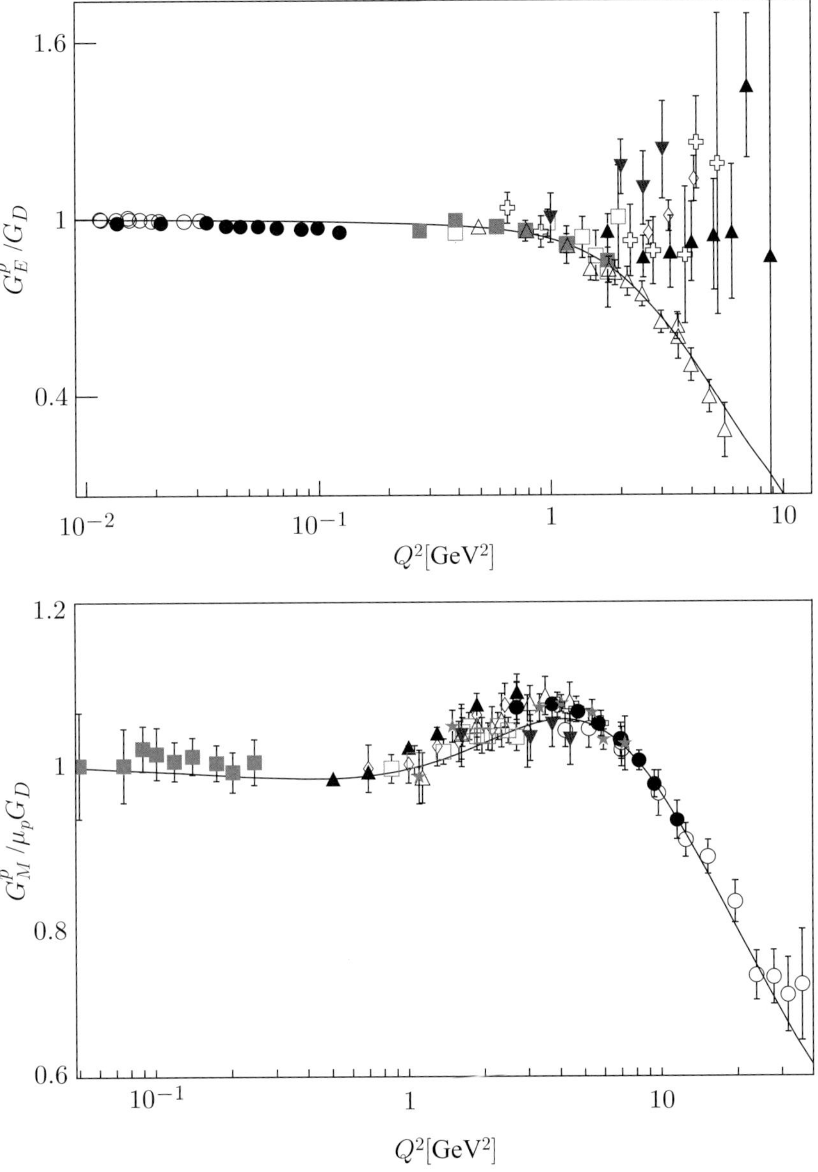

Fig. 8.2   Selection of world data for $G_E^p(Q^2)$ and $G_M^p(Q^2)$; figure adapted from [Cra10]. In the upper panel note that at high Q^2 the cross section data (solid triangles) are approximately flat in Q^2 but the polarization data (open triangles) decrease linearly with Q^2. This discrepancy is discussed in Section 8.3.

technique. A suite of carefully designed experiments employing polarization techniques has been used to great effect over the last decade to produce precision data on the nucleon elastic form factors for both the proton and neutron.

Figure 8.3 shows low-Q^2 data on $G_E^n(Q^2)$ as determined by spin polarization techniques [Gei08]. The shaded band shows the fit to the data; this fit is also constrained by the slope of $G_E^n(Q^2)$ at $Q^2 = 0$, which has been determined using scattering of low-energy neutrons from the atomic electrons.

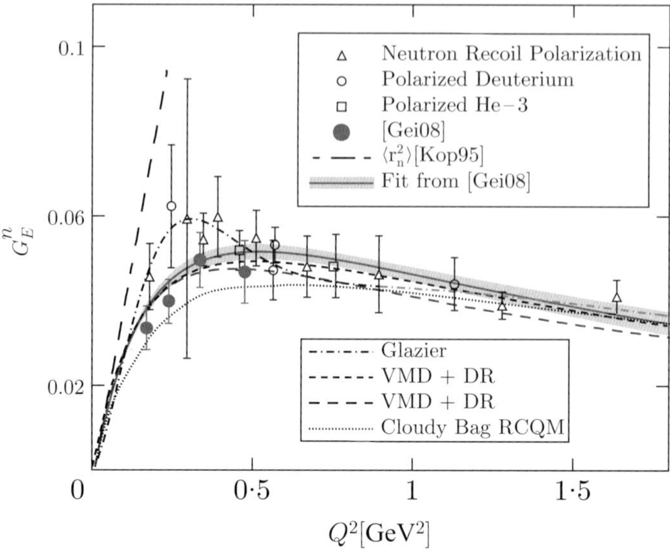

Fig. 8.3 The elastic electric form factor of the neutron; figure adapted from [Gei08]. Here the polarization data of fig. 8.2 are used.

Consider the expansion of the Fourier transform of the charge distribution at small Q^2:

$$\int \rho(r)e^{i\mathbf{q}\cdot\mathbf{r}}d\mathbf{r} = \int \left(1 - i\mathbf{q}\cdot\mathbf{r} - \frac{1}{2}(\mathbf{q}\cdot\mathbf{r})^2 + \cdots \right)\rho(r)d\mathbf{r} \tag{8.13}$$

$$= 1 - \frac{1}{6}q^2 \int r^2 \rho(r)d\mathbf{r} + \cdots \tag{8.14}$$

$$= 1 - \frac{1}{6}q^2 <r^2> + \cdots \tag{8.15}$$

where in the second line we assume a radially symmetric change distribution. Then $G_E(Q^2) = 1 - \frac{1}{6}Q^2 <r^2> + \cdots$ and we can define the charge radius of the nucleon as

$$<r^2> = -6\frac{dG_E(Q^2)}{dQ^2}\bigg|_{Q^2=0}. \tag{8.16}$$

The charge radius of the proton has been determined from electron–proton elastic scattering data, as well as from the Lamb shift in both electronic and muonic hydrogen. The electronic determinations (both scattering and Lamb shift) are in agreement and produce a value of 0.878 ± 0.005 fm for the proton charge radius. The muonic hydrogen determination is 0.841 ± 0.003 fm, which is in disagreement. There is no accepted explanation of this disagreement and it is the focus of considerable experimental and theoretical effort.

8.2 The Role of Mesons

Whether one uses hadronic language involving some set of baryons and mesons or QCD language with quarks and gluons, the nucleon is not a point Dirac particle, but has

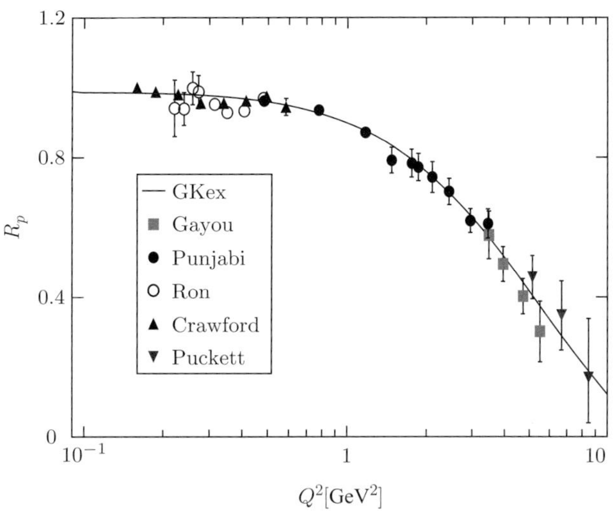

Fig. 8.4 The proton form factor ratio R_p showing the GKex universal fit together with recent data; figure adapted from [Cra10].

spatial extension. Its properties may be described, in large part, in terms of the elastic electric and magnetic form factors, G_E^p, G_E^n, G_M^p, and G_M^n. It is interesting to take the measured Sachs form factors for both the proton and neutron and try to gain insight into the distribution of charge and magnetism in the nucleon [Cra10]. Electrons couple through photons to the electromagnetic current provided by the charged constituents within the nucleons, yielding the form factors. Because the photon is a vector particle, at any parity-conserving vertex where it couples with hadrons it must connect to these particles with unit total angular momentum and negative parity. The photon does not conserve isospin and so these systems of hadrons may be isoscalar and isovector. The simplest such vertex connects the photon to a single vector meson (ρ, ω, ϕ, ...). It can also couple to systems of two or three pions or $K\bar{K}$ in a 1^- state, which in turn may couple to a ρ, ω, or ϕ meson. Since the latter are resonances of the multi-meson systems, the strength of the interaction is largest close to the masses of the vector mesons (i.e., near their poles). In leading-order, this vector meson dominance (VMD) limit of the photon-hadron interaction gives a good representation of the elastic proton and neutron form factor data over most of the present range of momentum transfers. The earliest reasonable fit was the VMD model of Iachello, Jackson, and Lande [Iac73] which incorporated a single vertex form factor, whereas Höhler and collaborators used [Hoh76] dispersion relations to obtain the contribution of the $\pi\pi$ continuum, giving the ρ meson its width. Gari and Krümpelmann [Gar86] developed a model in which VMD at low momentum transfers was extended by pQCD at high momentum transfers. Lomon further extended this work (the so-called GKex model, described in [Cra10]) to yield a theoretical framework which has successfully predicted the most recent data at high Q^2. Figures 8.4 and 8.5 show data for the form factor ratios R_p and R_n, which are well described by the GKex universal fit.

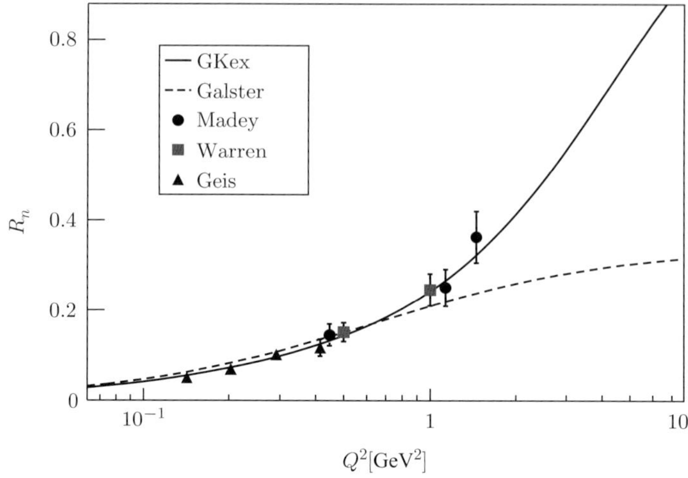

Fig. 8.5   The neutron form factor ratio R_n with the GKex universal fit and the so-called Galster parametrization together with recent data; figure adapted from [Cra10].

Coordinate Space Representation

Consider now the experimental data on nucleon elastic form factors and their interpretation using the GKex model in coordinate space. There are several motivations for doing this:

- We hope to obtain some insights into how charge is distributed in the nucleon;
- We are interested in how the various ingredients of the GKex approach are manifested differently in coordinate space than they are in momentum space;
- In particular, we wish to explore the role played by the coupling to the continuum and thereby to gain some insights into, for instance, what roles pions play in determining the nucleon's form factors;
- When characterizing the structure in coordinate space in terms of some set of basis functions, the correlations which occur are different from those that enter when doing the characterization in momentum space.

When choosing to represent the nucleon's properties one may choose any frame of reference, for instance, the initial-state rest frame, the final-state rest frame, choices in between or frames boosted to the lightcone. Inevitably, however, the initial state, the final state or both states must be moving and therefore boosts are required when attempting to relate to the properties in the nucleon rest frame. This makes the problem a relativistic one. Indeed, at high momentum transfers this makes the interpretation in terms of coordinate-space structure of the nucleon notoriously difficult, although at low enough momentum transfers it may be possible to make some connections between momentum and coordinate space. Problems occur in various guises, depending on the approach taken; for instance, rest-frame models may be very difficult to boost and lightcone models can have troubles when boosting from the infinite momentum frame back to physical frames of reference.

Note that a compromise is sometimes employed, that of Fourier transforming to coordinate space only with respect to the transverse directions (orthogonal to the boost), but leaving the third dimension in momentum space, thereby having a mixed representation [Mil07a]. While avoiding some of the inevitable problems discussed below, the nucleon's properties are harder to envision in that approach.

Clearly it is important to choose the least relativistic frame of reference to optimize one's chances. This choice is the so-called Breit frame, as may be seen simply by minimizing the product of the boost factors

$$\gamma_i = E_i/m_N$$
$$\gamma_f = E_f/m_N \tag{8.17}$$

for the boosts involved in relating the moving initial and final nucleon states to their rest frames. One has

$$p_f = -p_i = q/2 \tag{8.18}$$

$$\omega = 0 \ \leftrightarrow \ \sqrt{|Q^2|} = |\mathbf{q}| \tag{8.19}$$

$$\gamma_f = \gamma_i \equiv \gamma_{\mathrm{Breit}} = \sqrt{1+\tau}\,, \tag{8.20}$$

that is, the resulting Breit frame has the initial- and final-state nucleons moving with $\mp\mathbf{q}/2$, where $\mathbf{q}$ is the three-momentum of the virtual photon involved in the electron scattering process. The energy transfer that results is zero and hence $Q^2 = |\mathbf{q}|^2 = q^2$. One may then define the Breit-frame electric distributions as the Fourier transforms

$$4\pi r^2 \rho_{\mathrm{Breit}}^{p,n}(r) \equiv \frac{2}{\pi} \int_0^\infty dq \; qr \sin qr \; G_E^{p,n}(Q^2)\Big|_{\mathrm{Breit}}. \tag{8.21}$$

Note that this is only a definition. For the reasons mentioned above, the resulting functions are not generally to be interpreted as the proton and neutron charge distributions, although they are perfectly well-defined quantities.

To obtain some feeling for where the interpretations as charge distributions clearly should be invalid (and therefore for where they may be reasonable) it helps to compare the Compton wavelength $\lambda_C = \hbar c/m_N c^2 \cong 0.21$ fm, where m_N is the mass of the nucleon, with the characteristic scale probed at a given momentum transfer $\lambda(q) \sim \hbar c/q$. These become equal when $q \sim 1$ GeV/c, and thus one must expect functional dependence at even higher momentum transfers or, corresponding, smaller distance scales to lie beyond simplistic nonrelativistic intuition. At lower momentum transfers – corresponding to distance scales significantly larger than the nucleon's Compton wavelength – there may be some validity to the interpretation of the coordinate-space distributions as charge or spin distributions. An insightful discussion of what toy models have to offer in this long-wavelength regime is contained in [Isg99].

In Fig. 8.6 are shown the Breit-frame Fourier transforms of the charge (electric) form factors of the proton and neutron, respectively, together with the individual contributions from the vector mesons and pQCD. For the totals (the entire GKex model form factors) one has results which integrate to 1 (0) for the proton (neutron), since what is plotted is $4\pi r^2$ times the Breit-frame Fourier transforms. For the neutron one sees a positive contribution at

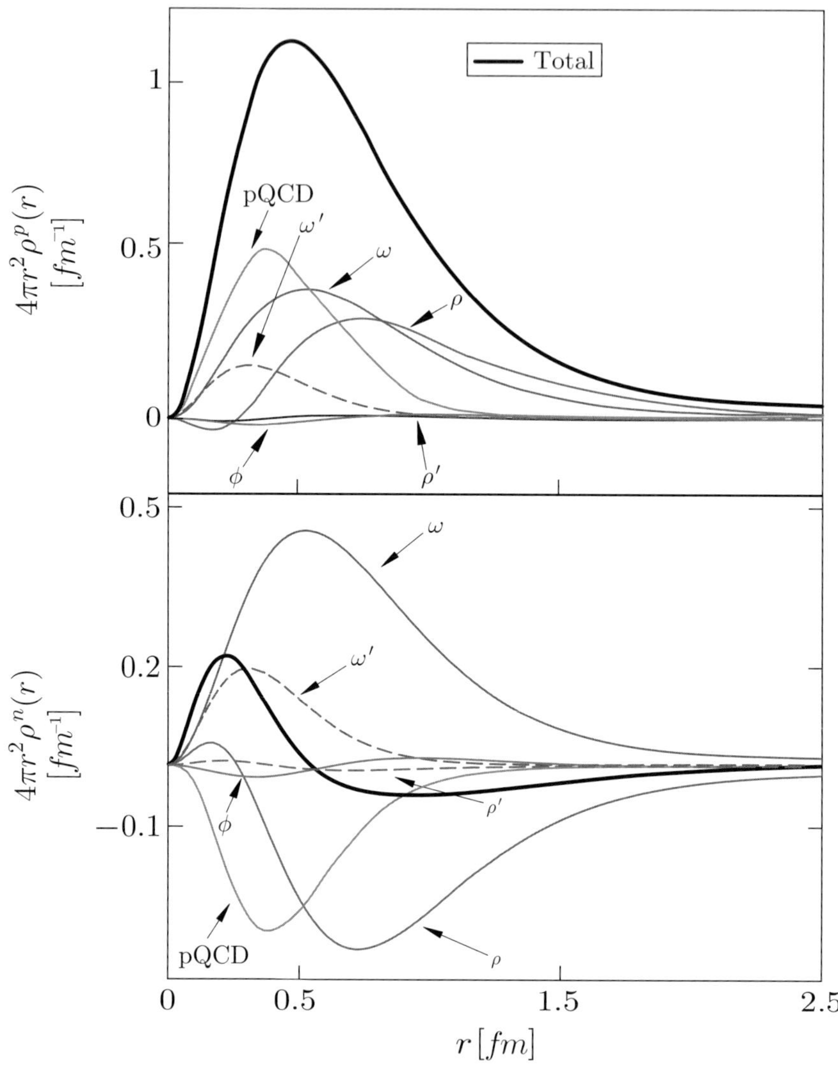

Fig. 8.6 The relative contributions [Cra10] to the Breit-frame distribution for the proton (top panel) and neutron (bottom panel) of the various vector mesons from the GKex model together with the pQCD contribution.

small distances and a negative one at large distances, which is consistent with the fact that the mean-square radius for the neutron is $\langle r^2 \rangle_{En} = -0.115 \pm 0.0035 \ \text{fm}^2$ [Kop95]. This is also consistent with a simple picture where isovector mesons such as the π and ρ extend to large distances and form the "meson cloud." For example, although unrealistically simple, a model where a neutron spends part of its time as a "proton + negative pion" would yield just such a charge polarization, and not the reverse with a negative "core" and a positive "cloud." Again, one is cautioned not to interpret these distributions as charge or spin distributions, except perhaps for their large-distance behavior. The issue of interpreting the RMS charge radius of the neutron is discussed in [Isg99].

Let us now discuss the individual contributions in somewhat more detail. As before, the ρ' and ϕ contributions are seen to be very small, while the rest play important roles. As seen in Fig. 8.6, for the Breit-frame Fourier transform of G_E^p these mostly add together to form the total, whereas for the Breit-frame Fourier transform of G_E^n the isoscalar mesons "fight" against the isovector mesons and the pQCD term to yield a relatively small net result. In both cases the longest-range effects arise from the ρ and next from the ω, while the ω' and pQCD contributions lie at small distances. Indeed, beyond about 0.7 fm most of the Breit-frame Fourier transform of G_E^p is contained in the ρ and ω alone (the neutron case is more complicated, due to the delicate cancellations seen in the figure).

With the caveats discussed above, the world data for $G_E^{p,n}$ may be Fourier-transformed using Eq. (8.21). In order to obtain Fourier transforms, the world data of G_E^p and G_E^n were fit to various parametrizations, which were then transformed numerically. Figures 8.7 and 8.8 show the Fourier transforms of these fits. The error bands in the figures were obtained by combining the variation from each fit parameter with the full covariance matrix. The calculated error bands, shown with dotted lines, have large oscillations in width, even dropping to $\delta\rho_{\mathrm{Breit}} \sim 0$ around $r = 0.37$ fm for the proton and $r = 0.75$ fm for the neutron. The calculated uncertainty for the proton also gets significantly smaller around $r = 0.75$ fm. This is clearly model dependence: the Fourier transform of this particular model has no flexibility at that point to respond to variations in the data. The shaded error bands in Figs. 8.7 and 8.8, were smoothed out to account for the model dependence.

This surprising behavior illustrates an interesting point, that a family of curves which fit the data well in momentum space may contain very little information or coverage of coordinate space. In choosing an appropriate model, one typically searches for the smoothest family of curves that fit the data with a reasonable χ^2. In contrast, the Fourier transform inherently includes information on all frequencies, not just smooth low frequencies. For example, the fit to a constant function $f(k) = a$ only determines a single point at the origin of the Fourier transform $\tilde{f}(x) = a\delta(x)$. Even arbitrary fit functions in

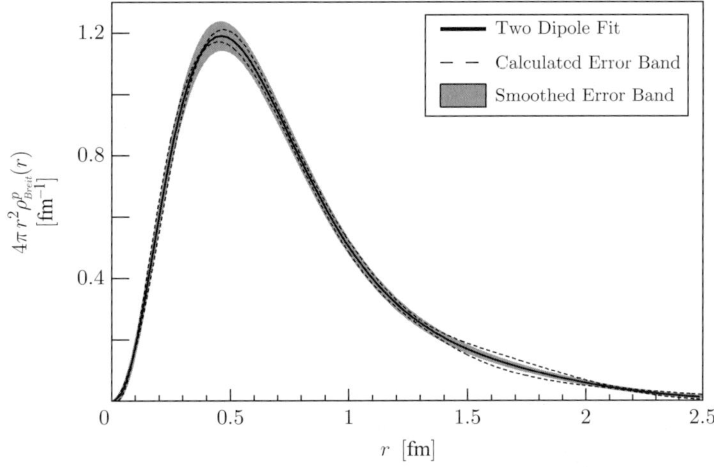

Fig. 8.7 Breit-frame Fourier transform of G_E^p from [Cra10], with both the calculated and smoothed error bands.

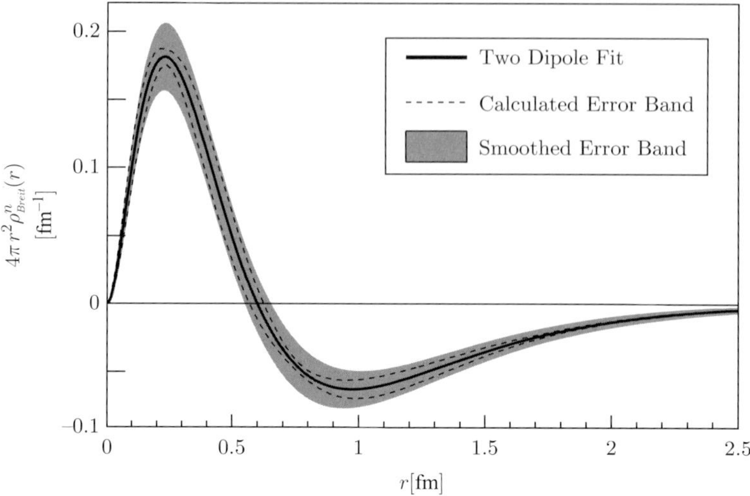

Fig. 8.8   Breit-frame Fourier transforms of G^n_E from [Cra10], with both the calculated and smoothed error bands.

one parameter can often be approximated by $f(k) = g(k) + a$ for a fixed function $g(k)$. In momentum space, that function will have a uniform error-band over the entire domain, but that error is completely correlated along the entire function. The Fourier transform has nonzero error bars only at the origin in position space.

The worldwide program over the past two decades to determine the elastic nucleon form factors using high duty factor electron accelerators to measure precisely polarization observables has been highly successful. It has yielded a data set of unprecedented precision and consistency for the nucleon elastic form factors at low and medium Q^2. Although the low-Q^2 polarized data constitute a very small part of the whole data set, they have cast doubt on indications seen in earlier results of structure at this low momentum transfer. These were attributed to a "pion cloud." Such structure is not present in the GKex representation, and indeed the coupling to explicit continuum pions is a relatively minor effect in this model. Future experimental work in this area is focused on pushing precision measurements to higher Q^2.

8.3 Beyond Single-Photon Exchange

The scattering of charged leptons from targets invariably involves processes where photons are radiated. As shown in Fig. 8.9, photons can arise from bremsstrahlung from the lepton legs as well as from the proton legs at high energies. There are vertex corrections, vacuum loops, and multiple-photon exchanges, which must be considered. To extract the cross sections and asymmetries arising from simple one-photon exchange, radiative corrections must be applied to the measurements. These correction procedures have been refined during over half a century of conducting experiments to the point where associated

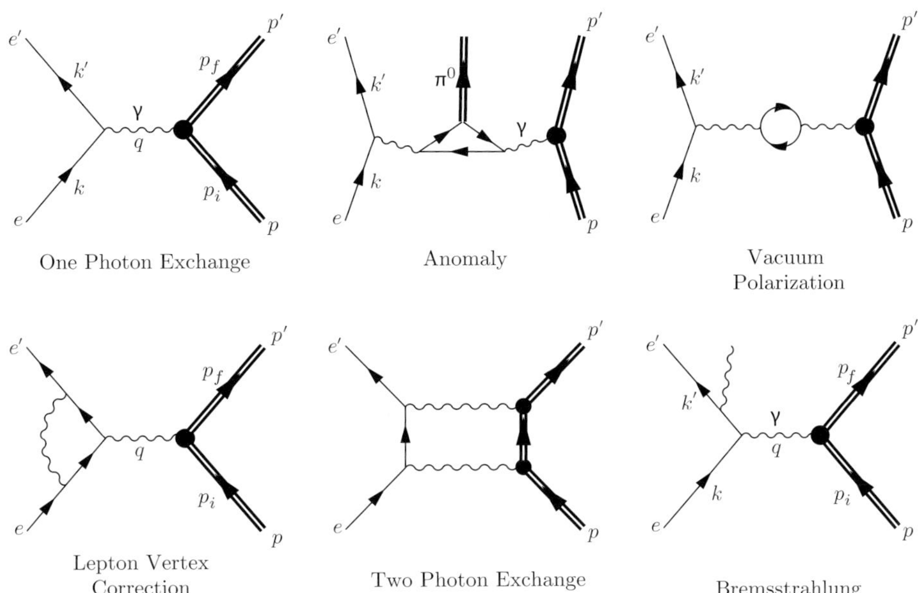

Fig. 8.9 The Feynman diagrams beyond leading order for elastic electron–proton scattering.

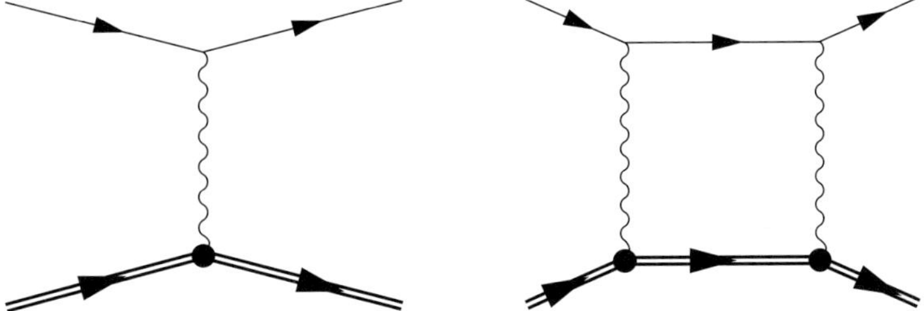

Fig. 8.10 The QED expansion of the scattering amplitude for elastic electron–proton scattering to second order.

uncertainties are typically not significant. The analysis of electron scattering data has been carried out essentially exclusively in the single-photon exchange approximation. We expect in the QED perturbative expansion, shown in Fig. 8.10, that the next term would be an interference between the first and second terms and would be of order α^3, down by a factor of α from the leading-order term. When the leading single photon exchange term is greatly suppressed, e.g., in the diffraction minimum in elastic scattering, two-photon effects are observed. In the late 1990s, measurement of the ratio R_p using the recoil polarization technique showed a dramatic linear decrease as a function of Q^2, as shown in Fig. 8.4. This differed dramatically from measurements at several laboratories using the Rosenbluth technique, which indicated a much flatter Q^2 dependence. Based on theoretical calculations with sizable uncertainties, it is widely believed that this dramatic discrepancy is due to the effects of multiple-photon exchange.

This can be directly probed in a precision comparison of e^+p and e^-p elastic scattering. As explained above, the first contribution beyond single-photon exchange will be an

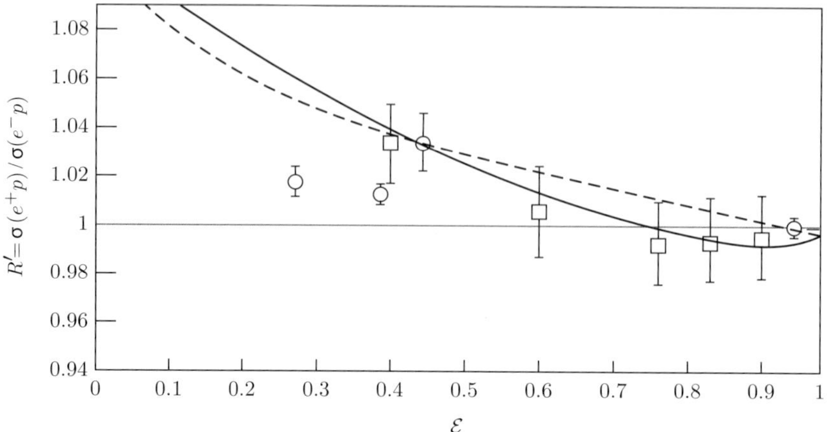

Fig. 8.11 Some recent determinations experimentally of the hard photon contribution to the ratio of positron–proton to electron–proton elastic scattering versus $\mathcal{E}$ from VEPP [Rad15] (open circles) and CLAS [Adi15] (open boxes). Note that the data are taken at different Q^2. The curves are expectations from Blunden, *et al.* [Blu05] (dashed) and from the Mainz fit of Bernauer, *et al.* [Ber14] (solid).

interference between one-photon and two-photon amplitudes. The interference term changes sign when the lepton sign is reversed. Thus, the ratio can be written as

$$\frac{\sigma_{e^+p}}{\sigma_{e^-p}} \approx 1 + 4\frac{Re(\mathcal{M}_{1\gamma}^{\dagger}\mathcal{M}_{2\gamma})}{\mathcal{M}_{1\gamma}^2} \, . \tag{8.22}$$

Experiments are underway at Jefferson Laboratory, Novosibirsk, and DESY [Mil14, Hen16] to precisely measure the ratio of positron-proton to electron-proton elastic scattering in the energy range 1–2 GeV.

8.4 PV Electron Scattering and Strange-Quark Content in the Nucleon

The strange-quark contribution to the nucleon mass can be addressed by studying the *sigma term* in pion–nucleon scattering. One first determines the value of the isospin-even πN scattering amplitude extrapolated to the pole at momentum transfer $Q^2 = 2m_\pi^2$ (the Cheng–Dashen point), $\Sigma_{\pi N}$. The most recent analyses indicate that a value of $\Sigma_{\pi N}(2m_\pi^2) \approx 60\,\text{MeV}$ is obtained from the data. This value is then extrapolated using theoretical input to $Q^2 = 0$ to give $\Sigma_{\pi N}(0) \approx 45\,\text{MeV}$. Furthermore, one may use hyperon mass relations to arrive at a prediction for the quantity

$$\sigma \equiv \frac{1}{2m_N} < p|\widehat{m}(\bar{u}u + \bar{d}d)|p > \simeq 25\,\text{MeV} \, , \tag{8.23}$$

where $\widehat{m} \equiv \frac{1}{2}(m_u + m_d)$. Flavor $SU(3)$ corrections to σ are expected to be about 10 MeV, yielding a predicted value of $\sigma \simeq 35$ MeV. In the absence of contributions from $s\bar{s}$ pairs in the nucleon, one would expect that $\Sigma_{\pi N} = \sigma$. The most recent analyses conclude that the difference between $\Sigma_{\pi N}(0)$ and σ implies a finite contribution of $s\bar{s}$ pairs to the nucleon mass

$$m_s < p|s\bar{s}|p >\sim 130 \text{ MeV} . \tag{8.24}$$

This result must be treated with considerable caution. Recent studies of this problem with some input from lattice QCD indicate that the strangeness contribution to the nucleon mass may vary from about twice as large as that quoted above or almost zero. Furthermore, other analyses of experimental work suggest that $\Sigma_{\pi N}$ may be significantly larger than previously thought.

One way to probe the strange quark content of the nucleon is via PV electron scattering. As we saw in Chapter 7, in addition to photon exchange, in the Standard Model an electron can exchange an intermediate vector boson Z^0 with a hadron target and, through the interference of this amplitude with the usual parity-conserving photon-exchange amplitude, introduce parity-violating effects in the helicity-difference asymmetry. The parity violating asymmetry for elastic ep scattering is given as follows

$$A = \left[\frac{-G_F Q^2}{4\sqrt{2}\pi\alpha} \right] \cdot \frac{\mathcal{E} G_E^\gamma G_E^Z + \tau G_M^\gamma G_M^Z - (1 - 4\sin^2\theta_W)\mathcal{E}' G_M^\gamma G_A^e}{\mathcal{E}(G_E^\gamma)^2 + \tau(G_M^\gamma)^2}, \tag{8.25}$$

where $\mathcal{E}' = \sqrt{\tau(1+\tau)(1 - \mathcal{E}^{\in})}$. The quantities $G_E^\gamma, G_M^\gamma, G_E^Z, G_M^Z$ are the (vector) form factors of the nucleon associated with γ and Z exchange. The neutral weak interaction also involves the axial-vector coupling G_A^e which appears in the third term of the numerator in Eq. (8.25). Note that this axial-vector term is suppressed by the factor $(1 - 4\sin^2\theta_W) \approx 0.075$.

We recall that the electromagnetic current has the form

$$V_\gamma^\mu = \frac{2}{3}\bar{u}\gamma^\mu u - \frac{1}{3}\bar{d}\gamma^\mu d - \frac{1}{3}\bar{s}\gamma^\mu s . \tag{8.26}$$

The WNC is given by an analogous expression

$$V_Z^\mu = \left(1 - \frac{8}{3}\sin^2\theta_W\right)\bar{u}\gamma^\mu u + \left(-1 + \frac{4}{3}\sin^2\theta_W\right)\bar{d}\gamma^\mu d + \left(-1 + \frac{4}{3}\sin^2\theta_W\right)\bar{s}\gamma^\mu s , \tag{8.27}$$

where the coefficients now depend on the weak mixing angle. The flavor structure here motivates measurement of the flavor decomposition of the form factors. The quark flavor structure of these form factors can be obtained by writing the matrix elements of individual quark currents in terms of form factors. Thus, we can write

$$G_{E,M}^\gamma = \frac{2}{3}G_{E,M}^u - \frac{1}{3}G_{E,M}^d - \frac{1}{3}G_{E,M}^s . \tag{8.28}$$

Similarly, we can express the WNC form factors in terms of the different flavor components as

$$G_{E,M}^Z = \left(1 - \frac{8}{3}\sin^2\theta_W\right)G_{E,M}^u + \left(-1 + \frac{4}{3}\sin^2\theta_W\right)G_{E,M}^d + \left(-1 + \frac{4}{3}\sin^2\theta_W\right)G_{E,M}^s . \tag{8.29}$$

It is essential to note that the $G_{E,M}^{u,d,s}$ appearing in both Eqs. (8.28) and (8.29) are identical. Assuming isospin symmetry and ignoring electroweak radiative corrections, one can eliminate the u and d quark contributions to the WNC form factors in terms of the proton and neutron electromagnetic form factors and obtain the expressions

$$G_{E,M}^{Z,p} = (1 - 4\sin^2\theta_W)G_{E,M}^{\gamma,p} - G_{E,M}^{\gamma,n} - G_{E,M}^s .$$
(8.30)

Equation (8.30) shows that the neutral weak form factors are related to the electromagnetic form factors of the proton and neutron plus a contribution from the strange (electric or magnetic) form factor. Thus, measurement of the WNC form factor together with sufficiently precise knowledge of the proton and neutron EM form factors, provides direct information on the strange form factors.

When studying parity-violating electron scattering from the nucleon it is often convenient to introduce parametrizations of the electric and magnetic strangeness form factors $G_E^s(Q^2)$ and $G_M^s(Q^2)$. In particular, in [Mus92, Mus94] the following forms were assumed:

$$G_E^s \equiv \mu_s G_D^V$$
$$G_M^s \equiv \tau\rho_s G_D^V ,$$

where a standard vector dipole form factor (the same as the EM dipole form factor) is assumed

$$G_D^V = [1 + 4.97\tau]^{-2},$$
(8.31)

with $\tau \equiv |Q^2|/4m_N^2$ as usual. The properties of the strange form factors near $Q^2 = 0$ are of particular interest in that they represent static properties of the nucleon. Thus, it is customary to define the quantity

$$\mu_s \equiv G_M^s(Q^2 = 0)$$
(8.32)

to be the strange magnetic moment of the nucleon. Since the nucleon has no net strangeness, $G_E^s(Q^2 = 0) = 0$, as guaranteed by the above parametrization, thus for electric strangeness we can use the slope of G_E^s at $Q^2 = 0$ and thereby characterize it in terms of a *strangeness radius* r_s

$$r_s^2 \equiv -6\left[\frac{dG_E^s}{dQ^2}\right]_{Q^2=0} = -\frac{3\rho_s}{2m_N^2}.$$
(8.33)

Theoretical estimates of the strange form factors $G_{E,M}^s(Q^2)$ have been made. Figure 8.12 shows examples of two physical processes that may contribute. These are generally known as loop effects and pole effects. The loop effects correspond to the fluctuation of the nucleon into a K meson and hyperon. The physical separation of the s and $\bar{s}$ in such processes (or the production of $s\bar{s}$ in a spin-singlet) leads to nonzero values of $G_{E,M}^s(Q^2)$. The pole processes are associated with the fluctuation of the virtual boson (photon or Z^0) into a ϕ meson, which is predominantly an $s\bar{s}$ pair (see Chapter 3 and the discussion of VMD above). Some attempts have been made to combine the two approaches using dispersion theoretical analyses. A general review of all theoretical calculations indicates that r_s^2 is small, but there is no general agreement on the sign; μ_s is expected to be negative, generally in the range of -0.8 to 0 nucleon magnetons.

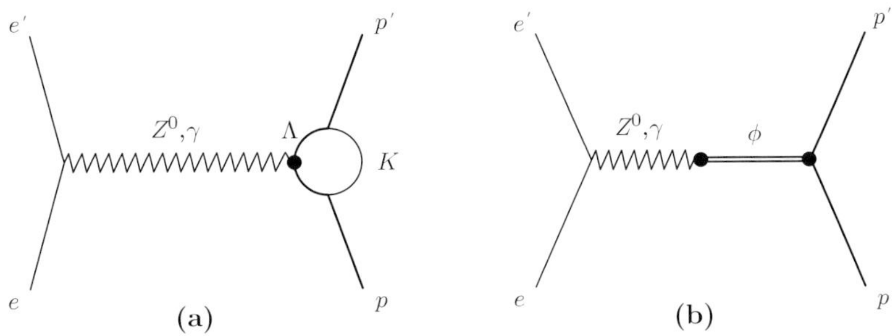

Fig. 8.12 Examples of (a) loop and (b) pole diagrams used to compute strangeness effects in the nucleon.

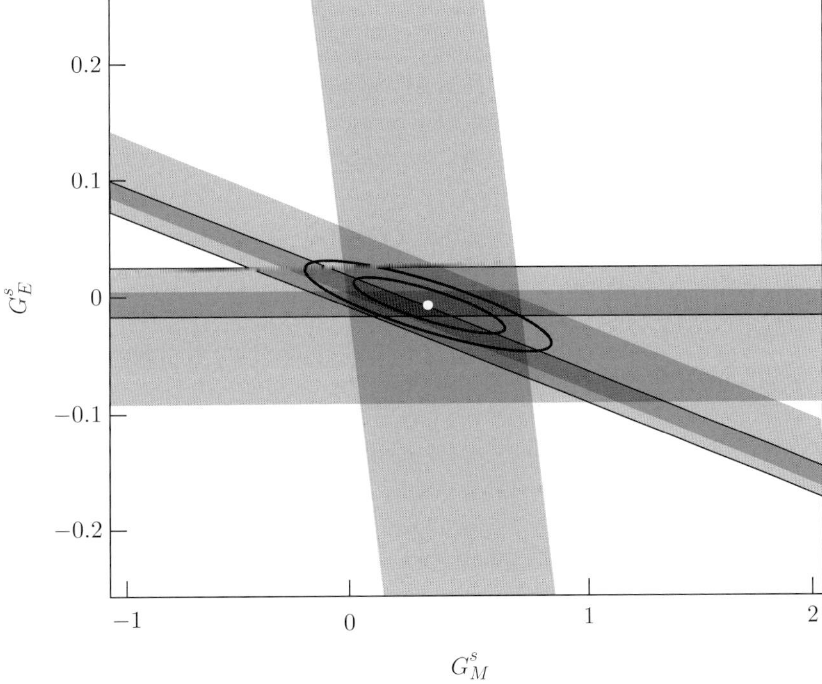

Fig. 8.13 The world data constraints on the strange quark form factors of the proton at $Q^2 = 0.1\,(\text{GeV/c})^2$ from [Arm12]. The inner and outer ellipses represent 68.27% and 95% confidence contours around the point of maximum likelihood.

Over the past two decades, the determination of the strange electric and magnetic form factors of the proton by elastic electron scattering has been a major focus of experimental nuclear physics. Typically, intense, highly polarized electron beams of energies from 100 to 1000 MeV have been directed on liquid hydrogen targets and the parity-violating asymmetry ($\approx 10^{-6}$) measured by reversal of the beam helicity. This program was initiated by the SAMPLE experiment at MIT-Bates, which was the first experiment to constrain the weak neutral form factor. Subsequent experiments at Jefferson Laboratory and Mainz have also placed limits on the electric form factor and have extended the kinematic range. Figure 8.13 shows the world data on the strange form factors of the nucleon at

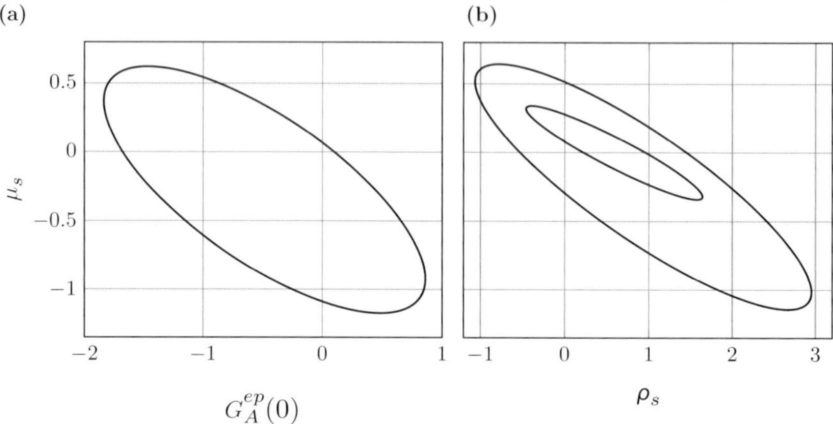

Fig. 8.14   Analysis of the world data on the nucleon's strange and neutral current axial-vector form factors. (a) 95% μ_s versus $G_A^{ep}(0)$ contour from a five-parameter fit; and (b) outer ellipse: 95% μ_s versus ρ_s contour from a five-parameter fit, and inner ellipse: likewise, but from a two-parameter fit (see the text for details and citations).

$Q^2 = 0.1$ (GeV/c)^2. The data indicate that the contribution of strange quarks to the proton electric and magnetic form factors is less than $\sim$5%. In the figure, results are shown not only for PV ep scattering, but in addition, there are results for PV quasielastic scattering from deuterium and for PV elastic scattering from ^{4}He – the latter process is discussed in Chapter 15 using the formalism of multipole matrix elements covered in Chapter 7 and so we postpone detailed treatment of that problem until later.

An alternative approach is to attempt to use all of the world PV elastic ep scattering data together to determine the allowed sizes of the strangeness form factors. This requires the use of some parametrization such as introduced in Eq. (8.31), as well as for the WNC axial-vector form factor for which one may take

$$G_A^{ep} \equiv G_A^{ep}(0)G_D^A \tag{8.34}$$

with axial-vector dipole form factor

$$G_D^A = [1 + 3.29\tau]^{-2}, \tag{8.35}$$

corresponding to an axial-vector mass $M_A = 1.036$ GeV/c^2. Of course, the EM form factors of the nucleon are also part of the PV asymmetry and state-of-the-art values for them should be used (see the discussion of parity-conserving elastic–electron nucleon scattering given above). In Fig. 8.14 we show an analysis of this type [Gon13, Gon14, Mor14]. Two strategies were adopted in this study. In [Gon13] everything except μ_s and ρ_s was fixed ($\equiv$ two-parameter fit), while in [Gon14] five parameters were allowed to vary, μ_s, ρ_s, $G_A^{ep}(0)$ together with the hadronic WNC couplings ξ_V^p and ξ_V^n. Clearly, the degree to which one can constrain the strangeness content of the nucleon is dependent on the assumptions made concerning the other issues that enter in PV ep scattering.

8.5 The Shape of the Proton

The possibility that hadrons would have non-spherical shapes was suggested by Glashow in 1979 on the basis of non-central, tensor interactions between quarks [Gla79]. This conjecture was based on the premise that there is a color spin-spin interaction between the quarks, as discussed in Chapter 3. There is experimental evidence to support this hypothesis, including the $E2$ transition in $\Delta \to N\gamma$, as well as quark-model calculations which show that the D-state admixture caused by the color hyperfine interaction corresponds to a nonzero neutron charge distribution and RMS charge radius. Physically, this is similar to the D-state component in the deuteron giving rise to a quadrupole moment – see the discussion in Chapter 11.

One cannot observe a static quadrupole moment for the spin-1/2 proton, and this has made the determination of the its shape quite challenging. On the other hand, the Δ has spin-3/2, and hence can have a quadrupole moment. The $\gamma^* N \to \Delta$ reaction has been studied for nonzero quadrupole amplitudes in the N and Δ. Due to spin and parity conservation in the

$$\gamma^* N(J^\pi = 1/2^+) \to \Delta(I^\pi = 3/2^+) \tag{8.36}$$

reaction, only three multipoles can contribute to the transition: the dominant magnetic dipole ($M1$), the electric quadrupole ($E2$), and the Coulomb quadrupole ($C2$) virtual photon amplitudes (see also Chapter 7). It is useful to define the Coulomb-to-magnetic ratio (CMR $\equiv C2/M1$) and electric-to-magnetic ratio (EMR $\equiv E2/M1$). QCD-inspired models predict values of CMR in the range -0.1% to 8% at low Q^2.

The notion of shape can be discussed in analogy to nuclear physics (see Chapter 14) where deformed nuclei with rotational bands are well known. Somewhat closer to the present consideration would be spin-1/2 nuclei with vibrational and rotational degrees of freedom, and as in the case of the proton, their static quadrupole moments in their ground states are zero. Their shape is derived through the study of their excitation spectrum. For hadrons like the proton and Δ, the situation is far more complex, since the constituents fluctuate (e.g., there are virtual $q\bar{q}$ pairs or mesons), relativity is important, and the system cannot be approximated as a rigid rotor. For the proton, the static quadrupole moment vanishes identically on account of its spin, even if there is a probability of D-states in the quark model and P-states for virtual pion emission.

The isolation of the $C2$ and $E2$ resonant amplitudes is complicated by the presence of nonresonant background processes which are coherent with the resonant excitations of the $\Delta(1232)$. The experimental difficulty is that the EMR and CMR ratios are small (typically -2 to -8% at low Q^2). The nonresonant background and resonant quadrupole amplitudes are the same order of magnitude and thus, the experiments have to be designed with great care to attain the required precision to separate the signal and background contributions.

The cross section for the H$(e, e'p)\pi^0$ reaction can be expressed as a sum of independent partial cross sections (σ_L, σ_T, σ_{TL}, σ_{TT}, and σ'_{TL}) which are proportional to the corresponding response functions

$$\frac{d^5\sigma}{d\omega \, d\Omega_e \, d\Omega_{pq}^{cm}} = \Gamma(\sigma_T + \mathcal{E}\sigma_L - v_{TL}\sigma_{TL}\cos\phi_{pq} + \mathcal{E}\sigma_{TT}\cos 2\phi_{pq} + hv_{TL'}\sigma_{LT'}) \,, \quad (8.37)$$

where $\mathcal{E}$ is the longitudinal polarization of the virtual photon, h is the helicity of the electron beam, Γ is the virtual photon flux, and ϕ_{pq} is the proton azimuthal angle with respect to the momentum transfer direction (see Chapter 16 for more detailed discussion of these so-called semi-inclusive reactions). Measurements at different azimuthal proton angles ϕ_{pq} and with polarized beam allow the extraction of the above partial cross sections, except for the first two terms which requires a Rosenbluth decomposition.

The $E2$ and $C2$ amplitudes manifest themselves most prominently through interference with the dominant magnetic dipole ($M1$) amplitude. The interference of the $C2$ amplitude with the $M1$ appears in the transverse-longitudinal (TL) response while the interference of the $E2$ amplitude with the $M1$ manifests itself in the transverse-transverse (TT) response.

At MIT-Bates, a dedicated out-of-plane spectrometer system (so-called OOPS) was explicitly developed to take advantage of the azimuthal angular dependence in the cross section, which was used as a lever arm for isolating the interference responses. Measurements of both the TL and TT responses were made and a global analysis performed. Figure 8.15 shows the resulting σ_{TL} versus θ_{pq} with the clear conclusion that the $N - \Delta$ system is deformed. The OOPS results are consistent with model results suggesting a prolate nucleon and an oblate Δ. The proton is slightly cigar-shaped, deviating from sphericity at the level of a few percent.

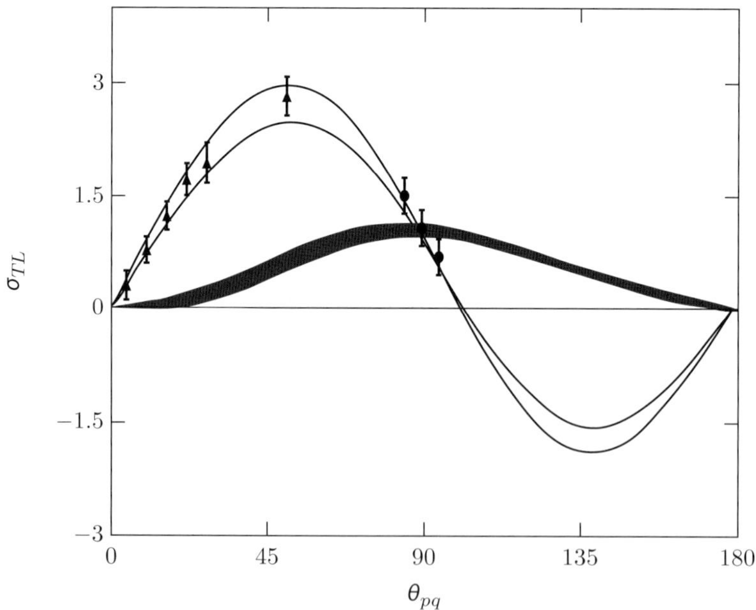

Fig. 8.15   The $N-\Delta$ system is deformed: the σ_{TL} response measured by the OOPS Collaboration at $W = 1232$ MeV [Pap03]. The shaded band represents the allowed uncertainty for a spherical nucleon. The region between the two curves represents the uncertainty allowed by all the data analyzed simultaneously.

8.6 Electromagnetic Form Factors in QCD

The asymptotic properties of EM form factors can be predicted in pQCD. The application of dimensional counting to the minimum quark-field components of a hadron can be shown to account for many of the experimental consequences of its compositeness. An elastic form factor can be considered as a measure of the probability for the hadron to stay intact after one of its quark constituents has undergone a hard scattering by a virtual photon. Thus, it decreases rapidly as a function of the momentum transfer Q^2 from the virtual photon. Ignoring logarithmic factors, one obtains a dimensional counting rule for any hadronic or nuclear form factor at large Q^2

$$F(Q^2) \sim \left(\frac{1}{Q^2}\right)^{n-1}, \tag{8.38}$$

which conserves baryon helicity, where n is the minimum number of fields in the hadron. Higher Fock-states lead to form factor contributions of successively higher-order in $1/Q^2$. Helicity non-conserving form factors should fall as an additional power of $1/Q^2$.

Consider the case of the nucleon where $n = 3$, i.e., there are three valence quarks. Then $F_1^N(Q^2) \sim 1/Q^2$, in agreement with the dipole form at large Q^2. The hard scattering picks out spatial configurations of the nucleon wavefunction, where the three valence quarks are close together. Using the Heisenberg uncertainty principle, the probability to find two pairs of quarks close together is $\frac{1}{Q} \cdot \frac{1}{Q} \sim G_M(Q^2)$. The Dirac form factor dominates for the nucleon, as the Pauli form factor $F_2(Q^2)$ is suppressed by an extra power of Q^2. Similarly, one has $F_\pi \sim 1/Q^2$ for the pion form factor, and $F_d(Q^2) \sim 1/Q^{10}$ for the deuteron elastic form factor. These predicted scaling behaviors are in reasonable agreement with the observed experimental data at high Q^2. However, this does not mean that pQCD is directly applicable at these relatively modest values of momentum transfer. It is generally accepted that pQCD becomes applicable only at Q^2 well beyond those currently attainable experimentally.

The conservation of helicity in exclusive scattering predicted by pQCD leads to the expectation that spin asymmetries in elastic scattering at high energies would be small. However, experiments have found large, unexpected asymmetries at high transverse momentum in pp spin-dependent elastic scattering. This observation lacks a well-accepted explanation.

A crucial check of the applicability of the pQCD is the verification of the color transparency idea (see Chapter 9). When the proton is in a nucleus and suffers a hard scattering by a virtual photon (so-called *quasielastic* scattering described in detail in Chapter 16), the final-state interactions of the recoiling proton in the surrounding nucleus are expected to be sensitive to the spatial size of the proton. This effect has been sought experimentally but has not been observed.

With an understanding of elastic lepton scattering from hadrons now completed, we are ready to proceed to consider inelastic final states. In the next chapter, we focus on deep inelastic scattering, accessed at higher lepton energies and a process which provides a crucial underpinning of the Standard Model.

Exercises

8.1 **Elastic Electron–Proton Scattering**

Consider the elastic scattering of electrons from a proton target.

a) Using the expressions for the virtual photoabsorption cross section, show that the *ep* elastic cross section may be written as

$$R = \frac{d^2\sigma}{d\Omega dE'}/\sigma_{\text{Mott}} = \left(\frac{Q^2}{q^2}\right)^2 R_L + \frac{Q^2}{2q^2}\frac{1}{\varepsilon}R_T \,,$$

where

$$q^2 = Q^2 + \nu^2$$

$$\varepsilon = \left[1 + 2\frac{q^2}{Q^2}\tan^2\frac{\theta_e}{2}\right]^{-1} \,.$$

b) Further, with $\tau = \frac{Q^2}{4m_N^2}$ show that

$$R_L = (1+\tau)G_E^2$$
$$R_T = 2\tau G_M^2 \,.$$

8.2 **Elastic Electron–Proton Scattering Kinematics**

Consider elastic electron–proton scattering. The electron beam of energy E is incident on a fixed proton target, while the final-state electron has energy E' and is scattered through an angle θ_e with respect to the incident beam direction.

a) Show that the final-state electron energy in elastic kinematics is given by

$$E' = \frac{E}{1 + \frac{2E}{M}\sin^2(\theta_e/2)} \,.$$

b) Show that the angle θ_q, between the direction of the three-momentum transfer $\mathbf{q}$ and the incident beam, is given by

$$\sin\theta_q = \frac{E'}{\sqrt{Q^2(1+\tau)}}\sin\theta_e \,,$$

where Q^2 is the four-momentum transfer squared and $\tau \equiv \frac{Q^2}{4M^2}$.

c) Consider an experiment to measure the elastic scattering cross section at an incident energy of 2 GeV as a function of

$$\varepsilon \equiv \left[1 + 2(1+\tau)\tan^2\theta_e/2\right]^{-1} \,,$$

the photon's degree of longitudinal polarization. It is desired to measure the elastically scattered electron and proton in coincidence over ε values of 0.3 to 1.0. Over what range in θ_e and θ_q must the electron and proton be detected, respectively?

8.3   Dipole Form Factor

Consider elastic electron scattering from the proton.

a) If the proton charge distribution $\rho(r)$ is assumed to have an exponential form

$$\rho_0 e^{-r/\alpha} ,$$

show that the form factor has the dipole form

$$F(|\mathbf{q}|) = \frac{1}{(1 + \alpha^2 q^2)^2} ,$$

with $q^2 = -|\mathbf{q}|^2$.

b) Show that the RMS proton charge radius is $\sqrt{12}\alpha$.

c) The Compton wavelength of a particle of mass m is defined as

$$\lambda_C = \frac{1}{m} .$$

It is the approximate length scale at which relativistic quantum field theory becomes necessary for an accurate description. First, write the above equation with appropriate factors $\hbar$ and c to make it dimensionally correct. Second, calculate the Compton wavelength of the proton. In electron scattering, the characteristic length scale probed at a given momentum transfer q is

$$\lambda \sim \frac{1}{q} .$$

Again, include factors $\hbar$ and c to make the expression dimensionally correct. Explain why the charge distribution of the proton is not well determined from experiment.

8.4   Charged Particle Scattering

The Compton wavelength of a particle of mass m is defined in the previous exercise.

a) Calculate the Compton wavelength for the nucleus ^{40}Ca.

b) As shown in Exercise 8.3, in electron scattering the characteristic scale probed at a given momentum transfer q is $\lambda \sim \hbar c/q$. Compare elastic electron scattering from the proton with ^{40}Ca. Explain why the charge distribution of ^{40}Ca is well determined from experiment but the charge distribution of the proton is not.

8.5   Continuous Beam

Consider an experiment to measure the reaction $(e, e'p)$ on a nucleus A. Assume that the beam comes in pulses of width T with f beam pulses per second. The electrons are detected with a time-averaged rate R_e in an electron spectrometer. The protons are detected with a time-averaged rate R_p in a proton spectrometer. Coincidence events are detected at a time-averaged rate R_{coinc} within a coincidence resolving time τ.

(a) Show that the fraction of the time the beam is on target, called the *duty factor d*, is given by $d = Tf$.

(b) Prove that the ratio of true events to accidental events can be approximated by

$$\frac{\text{trues}}{\text{accidentals}} = \frac{R_{coinc}d}{R_e R_p \tau} .$$

This shows that the quality of coincidence experiments is greatly improved by using continuous rather than pulsed electron beams. This has driven the technical development of continuous electron beams and has greatly improved the scientific scope of electron scattering experiments.

8.6   Measurement of Spin Asymmetries

Consider scattering of longitudinally polarized electrons from a polarized proton target. In a single-arm experiment, the final-state scattered electron is detected and its four-momentum is measured. The experimental spin asymmetry is measured by reversing the direction of either the incident electron spin or the target proton spin. The numbers of events for parallel spins (N_+) and antiparallel spin-states (N_-) are measured and their relative luminosity determined. Then, one can form the experimental asymmetry

$$A_{\mathrm{exp}} = \frac{N_+ - N_-}{N_+ + N_-} \,.$$

a) For small asymmetries $A \ll 1$, show that the statistical uncertainty in the experimental asymmetry is $\delta A = \frac{1}{\sqrt{N}}$, where $N \equiv N_+ + N_-$ is the total number of events.

b) In a real experiment, the electron beam polarization p_B and the target polarization p_T are less than unity. Further, there can be extraneous unpolarized material in the polarized target which dilutes the measured asymmetry by a dilution factor f. Show that

$$\delta A \sim (\sqrt{N} \cdot f \cdot p_B \cdot p_T)^{-1} \,.$$

A common target material used is ammonia (NH_3) where $f = 3/17$.

9

Hadron Structure via Lepton–Nucleon Scattering

In this chapter, the formalism developed in the previous chapter is used to study the fundamental quark and gluon structure of the nucleon. The discovery of deep inelastic scattering in the late 1960s at SLAC provided the basis for understanding the structure and properties of the nucleon and led to the development of quantum chromodynamics (QCD), the theory of the strong interaction in the Standard Model of Physics. While much progress in experiment and theory has been made over the last four decades towards a complete understanding of hadron structure, many open questions remain.

9.1 Deep Inelastic Scattering

The direct evidence that the proton and neutron contain pointlike quarks comes from experiments in the deep inelastic scattering (DIS) regime. This was first discovered at SLAC in the late 1960s and has been successfully pursued in succeeding decades with fixed-target experiments at CERN and FNAL and with the HERA electron–proton collider at DESY. At Jefferson Laboratory, for the last two decades the parton model has been successively and comprehensively applied to DIS at energies down to $4\,\text{GeV}$. Here we establish the basic formalism for DIS.

In general, when an electron scatters from a proton, in the final state the proton can be excited or broken apart, i.e., $e + p \rightarrow e' + $ hadrons. Denoting the four-momentum of the final hadronic state by W^μ, and applying momentum conservation at the photon–proton vertex, we find that

$$W^2 = W^\mu W_\mu = M^2 + Q^2 - 2M\nu. \tag{9.1}$$

Elastic scattering occurs when $W^2 = M^2$ and inelastic scattering arises when $W^2 > M^2$. Consider now the cross section for inelastic lepton scattering from the proton. Using the formalism developed in Chapter 7, we can write in lowest-order approximation

$$\frac{d^2\sigma}{d\Omega dE'} = \frac{\alpha^2}{Q^4}\frac{E'}{E}|A|^2 = \frac{\alpha^2}{Q^4}\frac{E'}{E}L_{\mu\nu}W^{\mu\nu}, \tag{9.2}$$

where $L_{\mu\nu}$ and $W^{\mu\nu}$ are the leptonic and hadronic tensors, respectively. If the initial proton is unpolarized, the most general form for $W^{\mu\nu}$ is

$$W^{\mu\nu} = W_1(\nu, Q^2)\left(-g^{\mu\nu} + \frac{q^\mu q^\nu}{q^2}\right) + W_2(\nu, Q^2)\left[\left(p^\mu - \frac{p.q}{q^2}q^\mu\right)\left(p^\nu - \frac{p.q}{q^2}q^\nu\right)\right]. \tag{9.3}$$

Contracting the lepton and hadron tensors yields

$$\eta_{\mu\nu} W^{\mu\nu} = 4W_1 k \cdot k' + \frac{W_2}{M^2} \left\{ 4(p \cdot k)(p \cdot k') - 2(k \cdot k')M^2 \right\}. \tag{9.4}$$

In the laboratory frame where the initial proton is stationary one finds

$$\frac{d^2\sigma}{d\Omega dE'} = \frac{4\alpha^2 (E')^2}{Q^4} \cos^2 \frac{\theta}{2} \left\{ W_2(\nu, Q^2) + 2W_1(\nu, Q^2) \tan^2 \frac{\theta}{2} \right\}. \tag{9.5}$$

Electron beams of energy greater than 20 GeV first became available at SLAC in the late 1960s. Pioneering measurements there discovered that in inclusive high energy electron scattering from the proton for sufficiently high $Q^2 \geq 1 \,(\text{GeV/c})^2$ and sufficiently deeply inelastic $W^2 \geq 4\,\text{GeV}$ that the structure functions $MW_1(\nu, Q^2)$ and $\nu W_2(\nu, Q^2)$ become functions of a single variable, $x \equiv -q^2/2p \cdot q = Q^2/2M\nu$. This is direct evidence (cf. electron–muon scattering) that the electron is scattering from pointlike, charged constituents in the proton. This is known as *Bjorken scaling*, after the physicist who first interpreted this and conventionally the structure functions in the scaling region are denoted by $F_{1,2}(x)$, i.e.,

$$MW_1(\nu, Q^2) \rightarrow F_1(x) \tag{9.6}$$
$$\nu W_2(\nu, Q^2) \rightarrow F_2(x). \tag{9.7}$$

In addition, it was found in the DIS regime that $\sigma_L \ll \sigma_T$ which implies that $F_2(x) = 2xF_1(x)$. This is known as *the Callan–Gross relation* and supports the conclusion that the pointlike constituents of the proton have spin $\frac{1}{2}$. More recently, the HERA electron–proton collider provided unique data on the deep inelastic scattering of 27 GeV electrons from 900 GeV protons from 1994 to 2007. The accessible Q^2, x, and $y = \nu/E$ are related to the square of the center-of-mass energy squared s by the relation

$$Q^2 = s \cdot x \cdot y. \tag{9.8}$$

Thus, to access high Q^2 and low x, requires high center-of-mass energies, which are most easily achieved using an electron–proton collider. HERA reached down to $x \sim 10^{-4}$ and as high as $Q^2 \sim 10^4 \,(\text{GeV/c})^2$. Figure 9.1 shows the world data for $F_2(x, Q^2)$ of the proton.

9.2 The Parton Model

The pointlike constituents of the proton were first called *partons* and a phenomenological model called *the parton model* was developed. It has been superseded by the development of QCD, but it is instructive to consider this model as it offers substantial physical insight into the structure of the nucleon. A cornerstone of the model is that the pointlike partons are identified as the fractionally charged quarks of Zweig and Gell-Mann. Thus, Bjorken scaling arises naturally if deep inelastic scattering arises from incoherent elastic scattering from the pointlike quarks. It is convenient to consider the DIS process in a particular frame of reference, *the infinite momentum frame*. Parton motion is slowed by time dilation and the hadron distribution is Lorentz-contracted. In the parton model, partons are defined

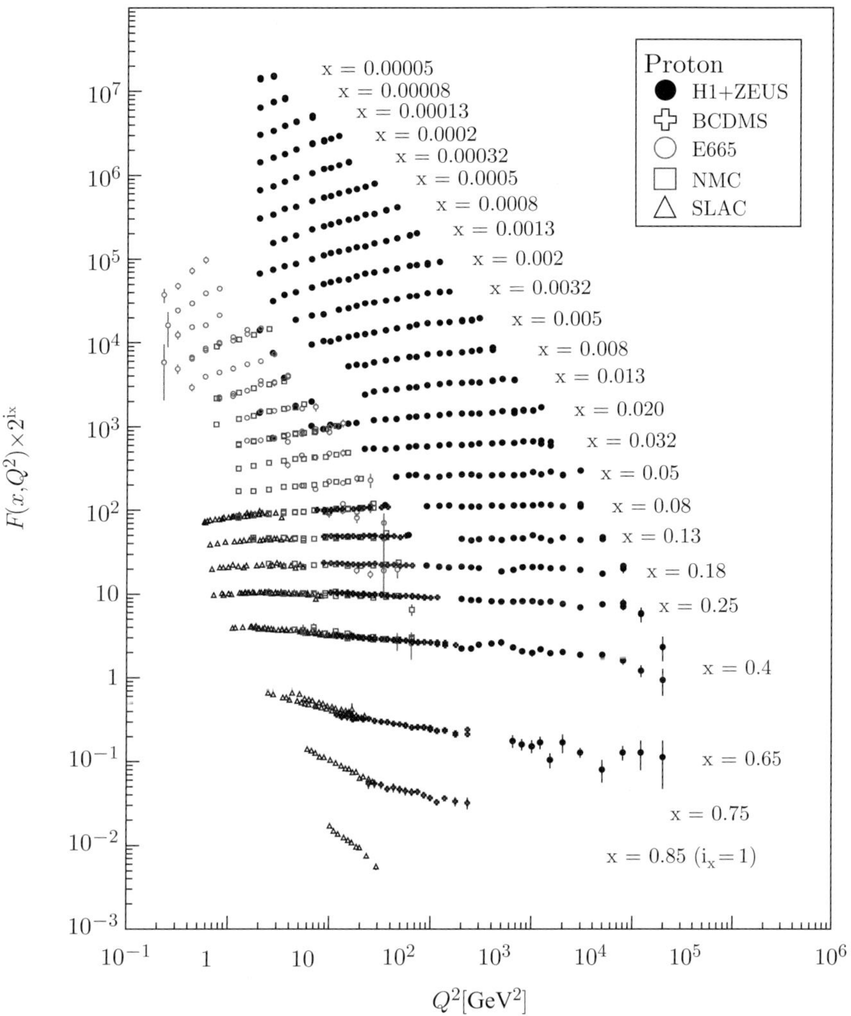

Fig. 9.1 A selection of world data on $F_2(x, Q^2)$ from [PDG14]. The data are plotted as a function of Q^2 in bins of x. For the purpose of plotting, F_2^p has been multiplied by 2^{i_x}, where i_x is the number of the x-bin, ranging from $i_x = 1$ ($x = 0.85$) to $i_x = 24$ ($x = 0.00005$). Note that the discovery of pointlike constituents in the proton originated in the flatness of F_2 at low Q^2 at moderate $Q^2 \sim 0.2$, and the presence of gluons causes F_2 to rise at low x and high Q^2.

with respect to a physical scale, namely the inverse of the momentum transfer. At low energies (long distance scales) a proton contains three valence quarks. At high energies, the virtual photon will couple to the $q\bar{q}$ pairs generated by the strong color force. In the infinite momentum frame, a parton carries a fraction x of the nucleon's four-momentum and if the parton's mass and transverse momentum are negligible, then

$$\nu W_2(x, Q^2) \longrightarrow F_2(x) = \sum_i e_i^2 x f_i(x), \tag{9.9}$$

where the sum i is over the various species of partons, i.e., over the different flavors of quarks $u, d, s, c, \ldots$ of charges e_i and $f_i(x)$ is the probability that the parton has momentum in the interval $(x, x + dx)$. In this way, the measurement of the structure function $F_2(x)$ can provide direct information on the distribution of the quark constituents of the proton.

The parton model can be considered as the subnucleonic analog of the impulse approximation of nuclear physics. It is assumed in this model that one can neglect the interactions between the partons during the time of the current interaction and that any interactions among partons in the final state can be ignored. Intuitively, one parton has been struck so violently that it has recoiled from its fellow partons and so can be regarded as quasifree, independent of their influence. In the nucleon, QCD tells us that the quarks are permanently confined. Thus, we assume that the final-state interactions which confine the quarks act at large spacetime distances of the order of the proton size, much larger than the parton size and the time scale of the current parton interaction. A physical picture of the quark structure of the nucleon can now be developed. In the case of the proton, the charge arises from three valence quarks: two *up* quarks and one *down* quark. In addition to these valence quarks, there are an infinite number of quark–antiquark pairs, the so-called *sea quarks*. The sea quarks arise through gluon emission by a valence quark. In analogy to photon emission in QED, the bremsstrahlung probability to produce a gluon of momentum k would be proportional to $dk/k \sim dx/x$. This means that gluon emission, and hence the quark–antiquark structure or sea, tends to be enhanced at small values of x. We can now develop an intuitive picture of the shape of $F_2(x)$. If the proton contained only three valence quarks, we would expect that $F_2(x) = \delta(x - \frac{1}{3})$. The presence of gluons would tend to broaden this distribution because it would allow for quarks of different momenta. Finally, the gluon bremsstrahlung and gluon internal conversion processes enhance the sea quark distribution at low x.

High-energy lepton scattering from nucleon targets can be analyzed to extract precise information on the quark structure of the nucleon. Using the parton model we can write

$$F_2(x) = 2xF_1(x) = \sum_{i=u,d,s,c} e^2 x f_i(x). \tag{9.10}$$

Expanding, we have respectively for the proton and neutron

$$\frac{1}{x}F_2^{ep}(x) = \frac{4}{9}[u^p(x) + \overline{u}^p(x)] + \frac{1}{9}[d^p(x) + \overline{d}^p(x)]$$

$$+ \frac{1}{9}[s^p(x) + \overline{s}^p(x)] + \cdots \tag{9.11}$$

$$\frac{1}{x}F_2^{en}(x) = \frac{4}{9}[u^n(x) + \overline{u}^n(x)] + \frac{1}{9}[d^n(x) + \overline{d}^n(x)]$$

$$+ \frac{1}{9}[s^n(x) + \overline{s}^n(x)] + \cdots . \tag{9.12}$$

Since the u, d quarks and p, n both form isospin doublets we have

$$u^p \equiv d^n \equiv u \tag{9.13}$$

$$d^p \equiv u^n \equiv d \tag{9.14}$$

$$s^p \equiv s^n \equiv s. \tag{9.15}$$

We can then rewrite

$$\frac{1}{x}F_2^{ep}(x) = \frac{4}{9}(u + \bar{u}) + \frac{1}{9}(d + \bar{d} + s + \bar{s}) + \cdots \tag{9.16}$$

$$\frac{1}{x}F_2^{en}(x) = \frac{4}{9}(d + \bar{d}) + \frac{1}{9}(u + \bar{u} + s + \bar{s}) + \cdots . \tag{9.17}$$

The ratio of neutron to proton structure functions is

$$\frac{F_2^{en}}{F_2^{ep}}(x) = \frac{u + \bar{u} + 4(d + \bar{d} + s + \bar{s})}{4(u + \bar{u}) + d + \bar{d} + s + \bar{s}}. \tag{9.18}$$

It can be shown that

$$\frac{1}{4} \leq \frac{F_2^{en}}{F_2^{ep}}(x) \leq 1$$

by considering the decomposition of the quark distribution functions in terms of a sum of valence and sea quarks, i.e., $q(x) = q_V(x) + s(x)$ where the sea momentum distribution $s(x)$ is taken to be flavor symmetric. $F_2^{en}/F_2^{ep}(x) \to 1$ as $x \to 0$, i.e., in the regime where sea quarks dominate. $F_2^{en}/F_2^{ep}(x) \to \frac{1}{4}$ for large $x \geq 0.2$ where valence quarks dominate. Figure 9.2 shows the distribution of momentum for the valence $u(x)$ and $d(x)$ quarks as well as the other flavors. These momentum distributions are determined by fits to lepton–nucleon scattering data at different energies. Note the dominant role of the gluons at low x. It is interesting to determine the total momentum carried by the quarks, i.e.,

$$\int_0^1 x[u + \bar{u} + d + \bar{d} + s + \bar{s}] \cdot dx = 1 - \epsilon_g, \tag{9.19}$$

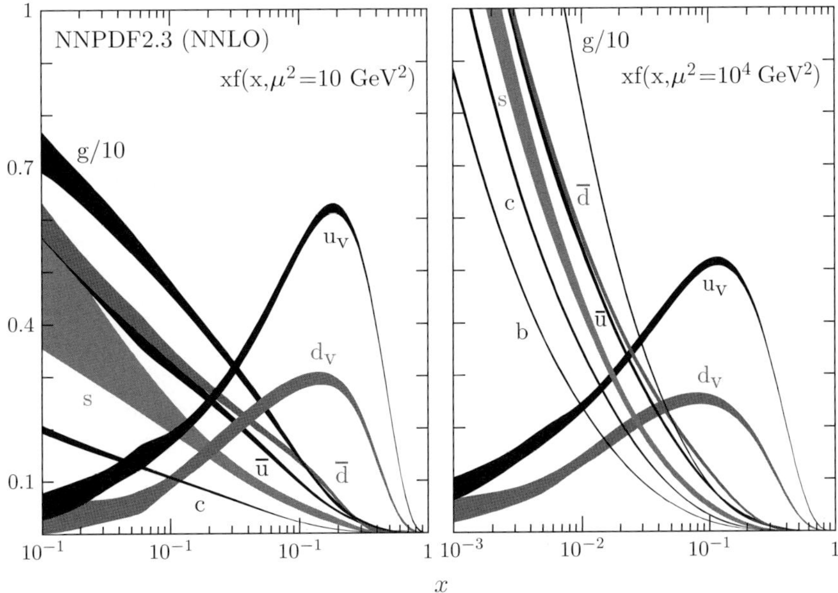

Fig. 9.2 The quark distribution functions extracted from DIS data from [PDG14].

where ϵ_g is the momentum fraction of the nucleon carried by the gluons. From the data, we find $\epsilon_g \sim 0.5$.

9.3 Evolution Equations

We now understand that the quark–parton model is the asymptotic limit $Q^2 \to \infty$ of QCD. QCD modifies the conclusions drawn from the discussion above where the modifications result from the dynamics among the quarks via exchange of the force-carrying gluons. Gluon radiation leads to transverse momentum of the quarks and therefore to the possibility to couple to the longitudinally polarized virtual photon which is probing the proton. For example, the longitudinal structure function F_L will no longer vanish and the Callan–Gross relation is modified.

The calculation of the inclusive ep cross section in QCD requires the knowledge of the cross section for the interaction between the virtual photon and a quark of momentum fraction x inside the proton as well as the knowledge of the momentum distribution which defines the probability to find a particular quark having a momentum fraction between x and $x + dx$. This approach emerges from the parton model. The first short-distance process can be calculated using perturbative calculations, whereas the second long-range process is not calculable within the framework of perturbative QCD (pQCD). This is the physical concept of the QCD factorization theorem. Factorization introduces an additional scale, the factorization scale μ_F, to denote the separation into short- and long-range interactions. By employing the QCD factorization theorem, the structure functions F_1 and F_2 can be expressed in the following form [Ste95]:

$$
F_1^{\gamma^* p} = \sum_{f,\bar{f},g} \int_x^1 \frac{dz}{z} C_1^{\gamma^* i}\left(\frac{x}{z}; \frac{Q^2}{\mu_R^2}; \frac{\mu_F^2}{\mu_R^2}; \alpha_S(\mu_R^2)\right) f_{i/p}(z; \mu_F; \mu_R^2) + \mathcal{O}\left(\frac{\Lambda^2}{Q^2}\right) \quad (9.20)
$$

$$
F_2^{\gamma^* p} = \sum_{f,\bar{f},g} \int_x^1 dz\, C_2^{\gamma^* i}\left(\frac{x}{z}; \frac{Q^2}{\mu_R^2}; \frac{\mu_F^2}{\mu_R^2}; \alpha_S(\mu_R^2)\right) f_{i/p}(z; \mu_F; \mu_R^2) + \mathcal{O}\left(\frac{\Lambda^2}{Q^2}\right), \quad (9.21)
$$

where the so called higher-twist terms $\mathcal{O}(\frac{\Lambda^2}{Q^2})$ which arise from parton–parton interactions have been neglected. The above expressions therefore consider only the leading-twist case which is the dominant contribution at high Q^2. The process-dependent coefficient functions $C_j^{\gamma^* i}$ ($j = 1, 2$) denote the short-range interactions and are calculable using perturbative QCD (pQCD). The $f_{i/p}$ are the universal, process independent parton distribution functions of the hadron under consideration. And μ_F denotes the factorization scale, whereas μ_R is the renormalization scale. Using these expressions for F_1 and F_2, one can determine the inclusive ep cross section. There exist several renormalization schemes in order to compute F_1 and F_2. In the DIS scheme, the structure function $F_2(x, Q^2)$ is given by

$$
F_2(x, Q^2) = x \sum_i e_i^2 f_{i/p}(x, Q^2). \quad (9.22)
$$

The parton distribution functions $f_{i/p}(x, Q^2)$ cannot be calculated within the framework of pQCD and therefore have to be determined experimentally. However, knowing $f_{i/p}(x, Q_0^2)$ at a particular value Q_0^2 within the range of applicability of pQCD, allows the determination of $f_{i/p}(x, Q^2)$ and therefore F_2 at any other value of Q^2. This increases the predictive power of pQCD enormously. This procedure is called the evolution of structure functions. The evolution in Q^2 follows from a set of integro-differential equations known as the DGLAP (Dokshiteer–Gribov–Lipatov–Altarelli–Parisi) equations [Alt77, Dok77, Gri72]. Both F_1 and F_2 are measurable quantities and therefore cannot depend on the choice of the factorization scale μ_F. Therefore, requiring that $\mu_F^2 (dF_i(x, Q^2)/d\mu_F^2) = 0$ $(i = 1, 2)$, one obtains the DGLAP equations.

The DGLAP equations can be written in their most general form as follows:

$$\frac{df_i(x, Q^2)}{d \ln Q^2} = \frac{\alpha_s(Q^2)}{2\pi} \sum_{j=-n_f}^{n_f} \int_x^1 \frac{dz}{z} P_{ij}\left(\frac{x}{z}\right) f_j(z, Q^2), \tag{9.23}$$

where $f_j(x, Q^2)$ are the parton density distributions and $P_{ij}(x/z)$ are the QCD splitting functions. The splitting functions can be interpreted as the probability for finding a parton of type i with momentum fraction x which originated from a parton j having a momentum fraction z. n_f is the number of flavors. Here, $f_{j>0}$ are the quark distribution functions $q_j(x, Q^2)$, $f_{j<0}$ are the antiquark distribution functions $\bar{q}_j(x, Q^2)$ and $f_{j=0}$ is the gluon distribution function $g(x, Q^2)$.

The above equation represents a set of $2n_f + 1$ coupled integro-differential equations. These equations can be simplified for the evolution of flavor singlet and flavor non-singlet quark distributions. The singlet distribution is given by: $\Sigma(x, Q^2) = \sum_{i=1}^{n_f} [q_i(x, Q^2) + \bar{q}_i(x, Q^2)]$ which gives rise to the following evolution equations:

$$\frac{d\Sigma(x, Q^2)}{d \ln Q^2} = \frac{\alpha_s(Q^2)}{2\pi} \int_x^1 \frac{dz}{z} \left[P_{qq}\left(\frac{x}{z}\right) \Sigma(z, Q^2) + P_{qg}\left(\frac{x}{z}\right) g(z, Q^2) \right] \tag{9.24}$$

$$\frac{dg(x, Q^2)}{d \ln Q^2} = \frac{\alpha_s(Q^2)}{2\pi} \int_x^1 \frac{dz}{z} \left[P_{gq}\left(\frac{x}{z}\right) \Sigma(z, Q^2) + P_{gg}\left(\frac{x}{z}\right) g(z, Q^2) \right]. \tag{9.25}$$

The non-singlet distributions, $q_i^-(x, Q^2) = q_i(x, Q^2) - \bar{q}_i(x, Q^2)$ and $q_i^+(x, Q^2) = q_i(x, Q^2) + \bar{q}_i(x, Q^2) - (1/n_f)\Sigma(x, Q^2)$ evolve as follows, independently of the gluon distribution:

$$\frac{dq_i^\pm(x, Q^2)}{d \ln Q^2} = \frac{\alpha_s(Q^2)}{2\pi} \int_x^1 \frac{dz}{z} \left[P_\pm\left(\frac{x}{z}\right) q_i^\pm(z, Q^2) \right]. \tag{9.26}$$

The splitting functions are presently calculated up to next-to-next leading order. Expressions for the splitting functions P_{ij} and $P_\pm$ in next-to-leading-order (NLO) can be found in [Fur80]. The quark and gluon distributions are expressed in terms of singlet and non-singlet distributions which are then evolved using the above equations.

The range of validity for the DGLAP equations is given by:

$$\alpha_s(Q^2) \ln(Q^2) \sim \mathcal{O}(1) \qquad \alpha_s(Q^2) \ln \frac{1}{x} \ll 1. \tag{9.27}$$

The Q^2-dependence of the proton structure function F_2 can be written in the following form:

$$\frac{dF_2(x, Q^2)}{d \ln Q^2} = \frac{\alpha_s(Q^2)}{2\pi} \left[\int_x^1 \frac{dz}{z} \frac{x}{z} P_{qq}(x/z) F_2(z, Q^2) \right.$$

$$\left. + 2 \sum_q e_q^2 \int_x^1 \frac{dz}{z} \frac{x}{z} P_{qg}(x/z) zg(z, Q^2) \right]. \tag{9.28}$$

The Prytz LO method [Pry93] allows one to extract the gluon density directly from a measurement of F_2 which follows from the last equation by ignoring the contributions of quarks, since the gluon density is expected to be the dominant contribution at low x:

$$\frac{dF_2(x, Q^2)}{d \ln Q^2} \approx \frac{5\alpha_s(Q^2)}{9\pi} \frac{2}{3} xg(2x, Q^2). \tag{9.29}$$

In the derivation of the DGLAP equations, only terms of the form $\alpha_s^n \cdot (\ln Q^2)^n$ are kept and summed to all orders which give the dominant contribution at large x and large Q^2. This approximation is known as the *leading logarithmic approximation (LLA)*. At small values of x, terms containing $\ln \frac{1}{x}$ are no longer negligible. The double leading logarithmic approximation (DLLA) provides for moderate values in x, i.e., for $\alpha_s(Q^2) \ln Q^2 \ll 1$, $\alpha_s(Q^2) \ln \frac{1}{x} \ll 1$ and $\alpha_s(Q^2) \ln Q^2 \ln \frac{1}{x} \sim \mathcal{O}(1)$, a scheme to include leading terms $\ln \frac{1}{x}$ accompanied by leading terms $\ln Q^2$.

Within the DLLA approximation, one can determine the small x behavior of the gluon distribution function:

$$xg(x, Q^2) \sim \exp \sqrt{\left[\frac{48}{11 - \frac{2}{3} n_f} \ln \left(\frac{\ln \frac{Q^2}{\Lambda^2}}{\ln \frac{Q_0^2}{\Lambda^2}} \right) \ln \frac{1}{x} \right]}, \tag{9.30}$$

where Q_0^2 is the starting scale for the evolution in Q^2. This behavior is numerically compatible with a power-like behavior of $x^{-0.4}$ [Lev97].

9.4 Hadronization/Fragmentation

The process that leads from the partons produced in the elementary interaction to the hadrons observed experimentally is commonly referred to as *hadronization* or *fragmentation*. Theoretical estimates indicate that the hadronization process occurs over length scales varying from less than 1 fm to several tens of fms. At these length scales the magnitude of the strong coupling constant is such that perturbative techniques cannot be applied. Hence, hadronization is intrinsically a nonperturbative QCD process, for which only approximate theoretical approaches are presently available. Experimental data are vital for supporting these theoretical developments, since they can be used to gauge or guide the calculations.

The narrow cone of hadrons and other particles produced by the hadronization of a quark or gluon is known as a *jet*. Jets are produced in QCD hard scattering processes, creating quarks with high transverse momentum p_t. For example, at the LHC the physics of top quark jets is a major area of study, both experimentally and theoretically. At lower energies, the hadronization products in lepton scattering are often characterized by the leading hadron, usually a meson. As the parton produced in a hard scatter exits the interaction, the strong coupling constant will increase with its separation. The probability for QCD radiation increases, giving rise to radiated gluons at small angles to the originating parton's momentum direction. These gluons can subsequently produce $q\bar{q}$ pairs which can further give rise to QCD radiation. Each new parton is nearly collinear with its parent. Parton showering produces fragments of successively lower energy and eventually perturbative QCD is not applicable. Phenomenological models are then used to describe the showering. One example is the Lund string model [And83], which is widely used in many simulations. It treats all but the highest energy gluons as field lines, which are attracted to each other due to the gluon self-interaction and so form a narrow tube (or string) of strong colored fields. String fragmentation describes many features of hadronization well. In particular, the model predicts that in addition to the particle jets found along the original paths of the two separating quarks, there will be a spray of hadrons produced between the jets by the string itself – which is precisely what is observed.

Consider a quark q and antiquark $\bar{q}$ emitted back-to-back in the CM frame, i.e., the typical situation in an e^+e^- event. Because of the three-gluon coupling, the color flux lines will not spread out over all space, as the electromagnetic field lines do, but rather are constrained to a thin, tubelike region. Within this tube, new $q\bar{q}$ pairs can be created from the available field energy. Thus, the original system breaks into smaller and smaller pieces, until hadrons remain. Some of these primary hadrons may be unstable and decay further into stable hadrons, leptons, and photons that can be detected. In the end, these observable particles come out essentially aligned along the original $q\bar{q}$ axis, with transverse momenta of around 400 MeV/c. The total fragmentation function for hadrons of type h in e^+e^- annihilation at a center-of-mass energy $\sqrt{s}$ is defined as:

$$F^h(z, s) = \frac{1}{\sigma_{tot}} \frac{d\sigma}{dx}(e^+e^- \to hX). \tag{9.31}$$

Note, that x here refers to the scaled hadron energy quantified as $x = 2E_h/\sqrt{s}$. The full integral over all x values yields the average multiplicity of hadrons of type h:

$$< n_h(s) > = \int_0^1 dx F^h(x, s). \tag{9.32}$$

The definition of $F^h(x, s)$ includes contributions for all types of partons. The representation in terms of different types of partons is given as follows:

$$F^h(z, s) = \sum_i \int_x^1 \frac{dz}{z} C_i(s, z, \alpha_S) D_i^h(x/z, s). \tag{9.33}$$

Here C_i stands for the coefficient function for parton i and $D_i^h(x/z, s)$ refers to the fragmentation function of hadron h relative to parton i.

Fragmentation functions cannot be calculated within a perturbative QCD framework. However, similar evolution equations as for parton distribution functions can be defined and thus their scale dependence can be evaluated, which provides enormous predictive power. In a fragmentation function x represents the fraction of the parton's momentum carried away by a hadron of type h, whereas in a parton distribution function, this scaling variable represents the momentum fraction of a parton relative to the hadron's momentum. The scale dependence of a fragmentation $D_i(x, Q)$ is given as follows:

$$Q\frac{\partial}{\partial t}D_i(x, t) = \sum_j \int_x^1 \frac{dz}{z}\frac{\alpha_S}{2\pi}P_{ij}(z, \alpha_S)D_j(x/z, Q).\qquad(9.34)$$

Here P_{ij} refers to the perturbative calculable splitting functions. Average multiplicity distributions for identified hadrons h play an important role as an experimental constrain for determining fragmentation functions in global analyses [Flo07a, Flo07b].

The hadronization process can be studied by means of semi-inclusive deep inelastic scattering (SIDIS) of electrons or positrons from the nucleon. A new variable is introduced, namely $z \equiv E_H/\nu$, which is the ratio of the hadron energy E_H to the energy lost by the incoming lepton. The lepton beam interacts with a q (or $\bar{q}$), which subsequently produces the detected hadron. As $z \to 1$, the most probable occurrence is that the hadron contains the struck $q(\bar{q})$ in its valence Fock state. In practice, the limit $z \to 1$ is not experimentally accessible, but SIDIS where the leading hadron is detected in coincidence with the scattered lepton is a powerful tool for studying nucleon structure.

Consider the process $l + N \to l' + H(z)$ where the target four-momentum is P, the momentum transfer is q, the hadron H momentum is p, and $z \equiv p \cdot P/q \cdot P = E_{lab}/q_0$. Then

$$2F_1^{p/H}(x, z) = \sum_f e_f^2[q_f(x)D_f^H(z) + \bar{q}_f D_{\bar{f}}^H(z)]\qquad(9.35)$$

$$2g_1^{p/H}(x, z) = \sum_f e_f^2[\Delta q_f(x)D_f^H(z) + \Delta \bar{q}_f D_{\bar{f}}^H(z)]\text{ and}\qquad(9.36)$$

$$A_1^{p/H}(x, z) = \frac{g_1^{p/H}(x, z)}{F_1^{p/H}(x, z)},\qquad(9.37)$$

where $D_f^H(z)$ is the fragmentation function for the parton of flavor f to produce hadron H and k_T of the produced parton is integrated over. The fragmentation functions are of two classes: those "favored" as $z \to 1$ (e.g., $D_u^{\pi^+}$ and those related by isospin or charge conjugation) and those "unfavored" (e.g., $D_u^{\pi^-}$). These have been measured [Aub85] and may parametrized as

$$D_u^{\pi^+}(z) = \eta(z)D_u^{\pi^-}(z) = \frac{0.7(1-z)^{1.75}}{z},\qquad(9.38)$$

where $\eta(z) \equiv \frac{1+z}{1-z}$. Factorization theorems exist for calculating high-energy cross sections involving hadrons described by QCD. These theorems allow the derivation of predictions by separating (factorizing) universal long-distance from calculable short-distance behavior in a systematic fashion. Some examples of processes for which one expects a factorization

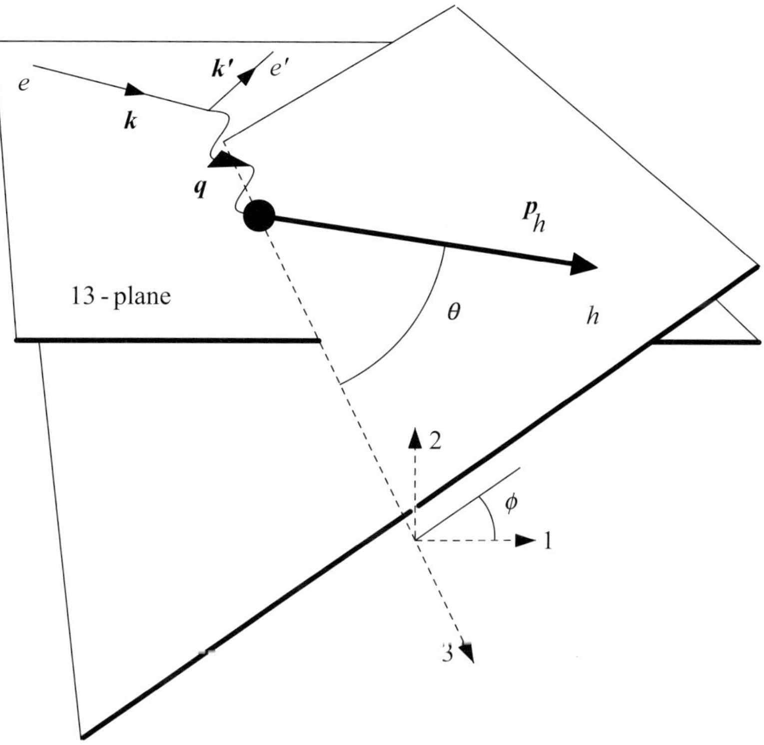

Fig. 9.3   Kinematic planes for hadron production in semi-inclusive deep inelastic scattering and definitions of the relevant lepton and hadron variables.

theorem to hold include (denoting hadrons by A, B, C, ...): deep inelastic scattering $l + N \rightarrow l' + X$; $e^+ + e^- \rightarrow A + X$; the Drell–Yan process $A + B \rightarrow X + (\mu^+ + \mu^-)/(e^+ + e^-)/W/Z$; $A + B \rightarrow \mathrm{jet} + X$; $A + B \rightarrow$ heavy quark $+ X$. The universal long-distance functions include the parton distributions, fragmentation functions, multi-parton correlation functions, generalized parton distributions and form factors. These universal long-distance functions are measured by experiment. Figure 9.3 shows the scattering and hadron planes with associated angles used in the analysis of semi-inclusive data.

There are a number of phenomena observed in SIDIS. The *Cahn effect* arises from the presence of nonzero intrinsic transverse momentum of the quarks in the nucleon. It shows up as a $\cos\phi$ azimuthal distribution of the final-state hadrons. When the proton is polarized, new important hadronization effects come into play. They include several effects. The *Sivers effect* is the correlation between the transverse momentum of the struck quark $\mathbf{k_t}$ and the spin $\mathbf{S}$ and momentum $\mathbf{p}$ of its parent nucleon, $\sim \mathbf{S} \cdot (\mathbf{k_t} \times \mathbf{p})$. The k_t-dependence means that the Sivers distribution is sensitive to nonzero orbital angular momentum in the nucleon, albeit with significant model dependence. The Sivers distribution is odd under time reversal so that either initial- or final-state interactions are required for it to be nonzero. It has been realized that initial-state color interactions in the case of Drell–Yan and final-state interactions in the case of SIDIS would lead to a process-dependent sign difference in the Sivers distribution.

The *Collins effect* arises from a spin-momentum correlation in the hadronization process $\mathbf{s_q} \cdot (\mathbf{k_q} \times \mathbf{p_t})$ with a hadron produced in the fragmentation process having some transverse momentum $\mathbf{p_t}$ with respect to the momentum direction $\mathbf{k_q}$ of a transversely polarized fragmenting quark with a spin $\mathbf{s_q}$. The Collins fragmentation function has been studied in semi-inclusive lepton–nucleon scattering and e^+e^- annihilation. It arises from a dependence on the transverse spin of the struck quark as it hadronizes, giving rise to an azimuthal dependence of the distribution of hadrons in SIDIS.

9.5 The Spin Structure of the Nucleon: Lepton Scattering

One of the central research thrusts of modern hadron physics is to understand how the spin $1/2$ of the nucleon arises from the quark and gluon constituents. We expect that

$$\frac{1}{2} = \frac{1}{2} \sum_q \Delta q + \Delta g + L_q + L_g, \tag{9.39}$$

where $\frac{1}{2} \sum_q \Delta q$ is the contribution of the quarks, Δg is the contribution of the gluons, and L_q and L_g denote the orbital angular momentum contributions. High-energy scattering experiments using polarization observables probe the quarks and gluons in the nucleon and aim to determine their contribution to the total spin [Hug99, Aid13]. A new generation of polarized lepton beams and targets has been employed in fixed-target experiments at CERN, DESY, Jefferson Laboratory, and SLAC over the last two decades to provide substantial new insight and to raise important, new questions concerning our understanding of the spin structure of the nucleon. At RHIC, polarized proton beams are collided at high energy and the contribution of the gluons to the proton's spin has been directly determined for the first time.

In analogy to the unpolarized structure function $F_1(x)$

$$F_1(x) = \frac{1}{2} \sum_f e_f^2 [q_f^\uparrow(x) + q_f^\downarrow(x)], \tag{9.40}$$

we define the spin-dependent structure function $g_1(x)$:

$$g_1(x) = \frac{1}{2} \sum_f e_f^2 [q_f^\uparrow(x) - q_f^\downarrow(x)]. \tag{9.41}$$

Extensive measurements of the asymmetry in scattering of longitudinally polarized leptons from longitudinally polarized targets for the proton, deuteron, and ^{3}He (effective polarized neutron target) have been perfomed. The scattering asymmetry is

$$A = \frac{d\sigma(\uparrow\uparrow) - d\sigma(\uparrow\downarrow)}{d\sigma(\uparrow\uparrow) + d\sigma(\uparrow\downarrow)} = \frac{g_1(x)}{F_1(x)} \tag{9.42}$$

and can be considered in a simple quark model of the nucleon. Invoking the Pauli exclusion principle for three valence quarks in an s-state, the identical flavors (such as u in the proton and d in the neutron) couple symmetrically to spin 1. Coupling the $S = 1$ to the $S = 1/2$ of the remaining valence quark to form overall $J = 1/2$ yields

$$p^\uparrow = \sqrt{\frac{2}{3}}(u^\uparrow u^\uparrow)d^\downarrow + \sqrt{\frac{1}{3}}(u^\uparrow u^\downarrow)d^\uparrow, \tag{9.43}$$

where the arrow refers to the component of quark spin along the nucleon (see Chapter 3). We can derive the spin-flavor weights as:

$$u^\uparrow = \frac{5}{3}; \quad u^\downarrow = \frac{1}{3}; \quad d^\uparrow = \frac{1}{3}; \quad d^\downarrow = \frac{2}{3}. \tag{9.44}$$

Then we can form the unpolarized and spin-dependent momentum distributions for the u and d flavors as follows

$$u \equiv u^\uparrow + u^\downarrow \equiv \int u(x)dx = 2 \tag{9.45}$$

$$d \equiv d^\uparrow + d^\downarrow \equiv \int d(x)dx = 1 \tag{9.46}$$

$$\Delta u \equiv u^\uparrow - u^\downarrow \equiv \int \Delta u(x)dx = \frac{4}{3} \tag{9.47}$$

$$\Delta d \equiv d^\uparrow - d^\downarrow \equiv \int \Delta d(x)dx = -\frac{1}{3}. \tag{9.48}$$

The total contribution to the spin of the proton of the up and down quarks is $\Delta u + \Delta d = 1$, i.e., all of the proton's spin is carried by the valence quarks in this simplistic model. The ratio of the proton to neutron magnetic moments is given by

$$\frac{\mu_p}{\mu_n} = \frac{\frac{2}{3}\Delta u - \frac{1}{3}\Delta d}{\frac{2}{3}\Delta d - \frac{1}{3}\Delta u} = \frac{2\left(\frac{\Delta u}{\Delta d}\right) - 1}{2 - \left(\frac{\Delta u}{\Delta d}\right)}, \tag{9.49}$$

where $\Delta d^n = \Delta u^p = \Delta u$ and $\Delta u^n = \Delta d^p = \Delta d$. Thus, one finds $\mu_p/\mu_n = -3/2$, in reasonable agreement with experimental data.

Consider next the charge squared weighted Δq_i. This is closely related to what is probed in spin-dependent DIS for $x \geq 0.2$, where valence quarks dominate. The asymmetry is

$$A = \frac{\sum e_i^2 \Delta q_i}{\sum e_i^2 q_i}. \tag{9.50}$$

For the neutron

$$A^n = \frac{1}{9}\Delta u + \frac{4}{9}\Delta d = 0, \tag{9.51}$$

while for the proton

$$A^p = \frac{4}{9}\Delta u + \frac{4}{9}\Delta d = \frac{5}{9}. \tag{9.52}$$

These values are very consistent with the experimentally measured asymmetries in the valence quark region.

More sophisticated models of the nucleon include relativistic effects and indicate that some of the proton's spin originates in the orbital angular momentum of the quarks. The lower component of the four-component Dirac spinor for the quarks is a p-wave with intrinsic spin primarily pointing in the opposite direction to the spin of the nucleon. In the MIT bag model [Jaf90], 35% of the proton's spin is shifted into orbital angular momentum through the confinement potential.

The QCD predictions exist for the x-dependence of both unpolarized and spin-dependent structure functions. Definite predictions do exist for first moments, e.g., $\Gamma_1 = \int_0^1 g_1(x)dx$. The most fundamental of these is the Bjorken sum rule [Bjo66]

$$\Gamma_1^p - \Gamma_1^n = \int_0^1 [g_1^p(x) - g_1^n(x)] \cdot dx = \frac{1}{6}\frac{g_A}{g_V} \tag{9.53}$$

where g_A and g_V are the axial and vector weak coupling constants of neutron β-decay. The sum rule was originally derived by Bjorken in 1966 from lightcone current algebra assuming isospin invariance.

In considering spin-dependent inclusive DIS, sum rules, i.e., integrals in x over the structure functions, are found to be particularly important. We assume that only the lightest three flavors, u, d, and s, make significant contributions. Using the moments of the structure functions $\Delta q_i = \int_0^1 [q_i^\uparrow(x) - q_i^\downarrow(x)] \cdot dx$, three linear combinations related to the singlet, triplet, and octet weak axial-vector couplings are used:

$$a_0 = \Delta u + \Delta d + \Delta s \equiv \Delta\Sigma \tag{9.54}$$

$$a_3 = \Delta u - \Delta d = \frac{g_A}{g_V} \tag{9.55}$$

$$a_8 = \Delta u + \Delta d - 2\Delta s. \tag{9.56}$$

Where $\Delta\Sigma$ is the fraction of the nucleon spin carried by the quarks. Equation (9.50) is the fundamental Bjorken sum rule, written down first in 1966, which can be used to provide a test of QCD. In QCD, moments of structure functions can be analyzed in the operator product expansion (OPE). In this framework, the sum rule can be derived in the limit $Q^2 \longrightarrow \infty$. At finite values of Q^2, the sum rule is subject to QCD radiative corrections

$$\Gamma_1^p(Q^2) - \Gamma_1^n(Q^2) = \frac{1}{6}\frac{g_A}{g_V}\left[1 - \frac{\alpha_s(Q^2)}{\pi} + \cdots\right], \tag{9.57}$$

where α_s is the strong coupling constant. The QCD corrections have been calculated up to $O(\alpha_s^3)$ and estimated to $O(\alpha_s^4)$. Separate sum rules for the proton and neutron were derived by Ellis and Jaffe [Ell74] by invoking additional assumptions. Assuming only three quark flavors (u, d, s) integration of the integral equation yields

$$\Gamma_1^p = \frac{1}{2}\left(\frac{4}{9}\Delta u + \frac{1}{9}\Delta d + \frac{1}{9}\Delta s\right) \tag{9.58}$$

and, invoking isospin invariance,

$$\Gamma_1^n = \frac{1}{2}\left(\frac{1}{9}\Delta u + \frac{4}{9}\Delta d + \frac{1}{9}\Delta s\right), \tag{9.59}$$

where the Δq_i are the moments of the spin-dependent quark distributions in the proton. We can then rewrite in terms of the axial-vector couplings above as

$$\Gamma_1^{p(n)} = \frac{1}{12}\left(\pm a_3 + \frac{1}{3}a_8\right) + \frac{1}{9}a_0. \tag{9.60}$$

If $SU(3)$ flavor symmetry is further assumed, then a_3 and a_8 are related to the symmetric and antisymmetric weak $SU(3)_f$ couplings F and D of the baryon octet via

$$a_3 = \frac{g_A}{g_V} = F + D \tag{9.61}$$

$$a_8 = 3F - D, \tag{9.62}$$

as introduced in Exercise 3.3.

Assuming that $SU(3)_f$ symmetry is valid, which is an approximation (see Chapter 3), measurements of F/D in hyperon decays can be used to predict a_8. There are no predictions for a_0. With the additional assumption that the strange sea in the nucleon is unpolarized ($\Delta s = 0$), $a_0 = a_8$ and we have the Ellis–Jaffe sum rules for the proton and neutron [Ell74]

$$\Gamma_1^{p(n)} = \pm\frac{1}{12}(F + D) + \frac{5}{36}(3F - D), \tag{9.63}$$

which are valid in the scaling limit. Like the Bjorken sum rule, they can be derived using the OPE and they are subject to QCD radiative corrections. The QCD radiative corrections are sizable in the Q^2 range of present experiments and their consideration is essential for the comparison and correct interpretation of data taken at different Q^2.

9.6 Spin Structure Functions in QCD

As with the unpolarized structure functions, g_1 can be expressed in terms of the flavor singlet (S) and non-singlet (NS) combinations of the polarized quark and antiquark distributions

$$\Delta\Sigma(x, t) = \sum_{i=1}^{n_f} \Delta q_i(x, t) \tag{9.64}$$

$$\Delta q_{NS}(x, t) = \frac{\sum_{i=1}^{n_f}\left(e_i^2 - \frac{1}{n_f}\sum_{k=1}^{n_f} e_k^2\right)}{\frac{1}{n_f}\sum_{k=1}^{n_f} e_k^2} \cdot \Delta q_i(x, t), \tag{9.65}$$

where $t = \ln\left(\frac{Q^2}{\Lambda^2}\right)$, Λ is the scale parameter of QCD, and n_f is the number of flavors. Now, $g_1(x, t)$ can be written in terms of the polarized quark distributions and polarized gluon distribution $\Delta G(x, t)$ as

$$g_1(x,t) = \tfrac{1}{2} \sum_{k=1}^{n_f} \frac{e_k^2}{n_f} \int_x^1 \frac{dy}{y} \left[C_q^S\left(\frac{x}{y}, \alpha_s(t)\right) \Delta\Sigma(x,t) \right.$$

$$\left. + 2n_f C_g\left(\frac{x}{y}, \alpha_s(t)\right) \Delta G(y,t) + C_q^{NS}\left(\frac{x}{y}, \alpha_s(t)\right) \Delta q_{NS}(y,t) \right].$$

$$(9.66)$$

The t dependence of both unpolarized and polarized quark and gluon distributions follows the Dokshitzer–Gribov–Lipatov–Altarelli–Parisi (DGLAP) equations. The quark singlet and gluon distributions are coupled by

$$\frac{d\Delta\Sigma(x,t)}{dt} = \frac{\alpha_s(t)}{2\pi} \int_x^1 \frac{dy}{y} \left[P_{qq}^S\left(\frac{x}{y}, \alpha_s(t)\right) \Delta\Sigma(y,t) + 2n_f P_{qg}\left(\frac{x}{y}, \alpha_s(t)\right) \Delta G(y,t) \right]$$

$$(9.67)$$

$$\frac{d\Delta G(x,t)}{dt} = \frac{\alpha_s(t)}{2\pi} \int_x^1 \frac{dy}{y} \left[P_{gq}\left(\frac{x}{y}, \alpha_s(t)\right) \Delta\Sigma(x,t) + P_{gg}\left(\frac{x}{y}, \alpha_s(t)\right) \Delta G(y,t) \right], \quad (9.68)$$

whereas the evolution of the non-singlet quark distribution is independent of both the quark singlet and gluon distributions

$$\frac{d}{dt} \Delta q_{NS}(x,t) = \frac{\alpha_s(t)}{2\pi} \int_x^1 \frac{dy}{y} P_{qq}^{NS}\left(\frac{x}{y}, \alpha_s(t)\right) \Delta q_{NS}(y,t). \quad (9.69)$$

The QCD splitting functions P_{gq} and P_{gg} are different for polarized and unpolarized parton distributions because of the presence of a soft gluon singularity at $x = 0$ in the polarized case. The ratio g_1/F_1 is therefore Q^2-dependent. However, in the valence quark region, this Q^2-dependence is expected to be small. The equations above are valid for all orders of perturbative QCD. The quark and gluon distributions, the coefficient functions, and the splitting functions depend on the mass factorization scale and on the renormalization scale. Here we adopt the simplest choice, setting both scales equal to Q^2. At leading order, the coefficient functions are $C_q^{0,S}\left(\frac{x}{y}, \alpha_s\right) = C_q^{0,NS}\left(\frac{x}{y}, \alpha_s\right) = \delta(1 - \frac{x}{y})$ and $C_g^0\left(\frac{x}{y}, \alpha_s\right) = 0$ and g_1 decouples from ΔG.

Beyond leading order, the coefficient functions and splitting functions are not uniquely defined, but depend on the renormalization and factorization scheme. Two schemes are commonly used in the analysis of polarized parton distributions: the modified minimal subtraction $(\overline{MS})$ scheme and the Adler–Bardeen (AB) scheme. In the $(\overline{MS})$ scheme, the first moment of the gluon coefficient function vanishes and the gluon density $\Delta G(x, Q^2)$ does not contribute to $\Gamma_1 = \int_0^1 g_1(x)dx$. In the AB scheme, a term proportional to $\alpha_s(Q^2)\Delta G(Q^2)$, which is attributable to the anomalous nonconservation of the singlet axial current (as discussed in Chapter 5), contributes explicitly to Γ_1. The first moments of the singlet quark distribution in the two schemes are related by

$$\Delta\Sigma_{\overline{MS}}(Q^2) = \Delta\Sigma_{AB} - n_f \frac{\alpha_s(Q^2)}{2\pi} \Delta G(Q^2), \quad (9.70)$$

$\Delta\Sigma_{AB}$ is independent of Q^2 even beyond leading order. At leading order, $\Delta G(Q^2)$ behaves as α_s^{-1}, so the scheme dependence does not vanish at large Q^2. The complete set of

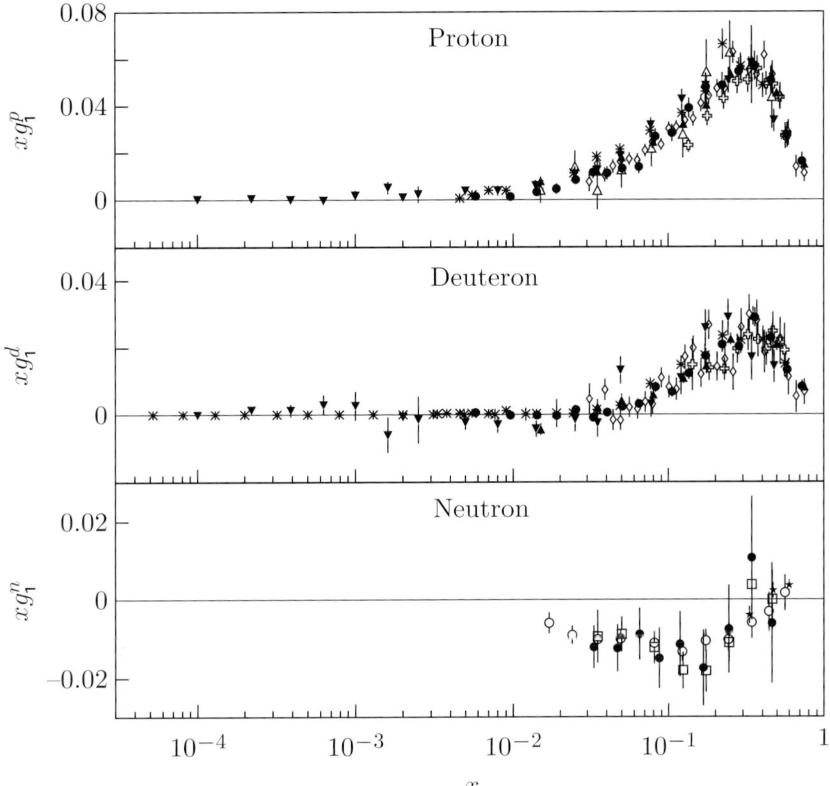

Fig. 9.4   A selection of the world data on $g_1(x)$ from [PDG14].

next-to-leading-order (NLO) coefficient and splitting functions in the $(\overline{MS})$ scheme can be used to carry out a complete NLO QCD analysis of spin-dependent structure functions.

This QCD formalism provides the means to determine the polarized quark and gluon distributions from the measured x and Q^2-dependence of the structure function g_1. Experiments over the last several decades using high-energy, polarized lepton beams incident on fixcd, polarized nucleon targets have determined g_1 over a sizable range in x and Q^2. Figure 9.4 shows the world data on $g_1^p(x)$ at two different values of Q^2. To determine parton distributions from the data, the polarized singlet, non-singlet, and gluon distributions at a fixed Q^2 are parameterized as

$$\Delta f(x, Q^2) = N(\alpha_f, \beta_f, a_f)\eta_f x^{\alpha_f}(1 - x)^{\beta_f}(1 + a_f x), \tag{9.71}$$

where $N(\alpha, \beta, a)$ is constrained by the normalization condition

$$N(\alpha, \beta, a) = \int_0^1 x^\alpha (1 - x)^\beta (1 + ax)dx = 1 \tag{9.72}$$

and Δf denotes $\Delta\Sigma$, Δq_{NS} or ΔG. With this normalization, the parameters η_g, η_{NS}, and η_S are the first moments of the gluon, non-singlet quark, and singlet quark distributions, respectively. The parton distributions are evolved to the Q^2 of the experimental data using these equations. A prediction for g_1 is computed and fitted to the experimental data using a χ^2 minimization. For the strong coupling constant, the value $\alpha_s(M_Z^2) = 0.118 \pm 0.003$ is used. The normalization of the non-singlet quark densities $\eta_{NS}^{p,n}$ is fixed using the neutron and hyperon β-decay constants and assuming $SU(3)$ flavor symmetry, $\eta_{NS}^{p,n} = \pm\frac{3}{4}\frac{g_A}{g_V} + \frac{1}{8}a_8$ with $|g_A/g_V| = F + D = 1.2601 \pm 0.0025$ and $F/D = 0.575 \pm 0.016$. Fits in the $\overline{MS}$ and AB schemes describe the data equally well; unless otherwise stated, results shown here are obtained in the AB scheme.

Global QCD analysis of the world data has been carried out [deF09] to determine the $\Delta q_f(x)$, the values of the sum rules, and the fraction of the proton spin carried by

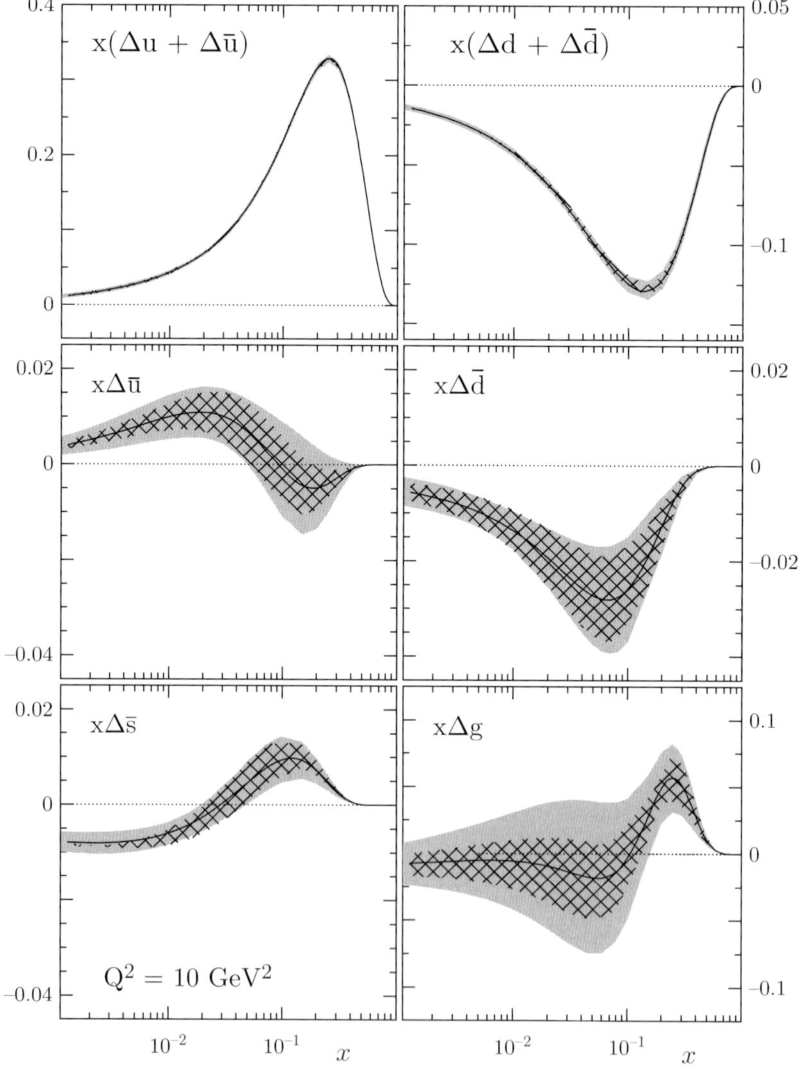

Fig. 9.5   Polarized PDFs of the proton at $Q^2 = 10$ (GeV/c)^2 in the ($\overline{MS}$) scheme along with their uncertainty bands (cross-hatched and shaded) determined in two different ways [deF09].

the quarks. Figure 9.5 shows the resulting Δq_f as well as the gluon polarization versus the momentum fraction x. In the proton, u quarks are polarized predominantly along the proton spin, while the d quarks are polarized in the opposite direction. The s quark polarization is consistent with zero. The fraction of the proton spin due to the quarks $\Delta\Sigma(Q^2 = 1) = 0.37 \pm 0.02$, which is substantially less than the 0.65 from the bag model calculation [Jaf90]. Experiments which detect final-state hadrons in coincidence with the scattered lepton, the so-called *semi-inclusive* measurements described above, can provide a flavor separation of the quark polarization. The HERMES experiment at DESY implemented the necessary particle identification to separate $\pi^{\pm}, K^{\pm}, p$, and $\bar{p}$. Assuming factorization and that the hadronization process is well described using the quark model, the polarizations of the u, d, and s quarks can be determined. Figure 9.6 shows a selection

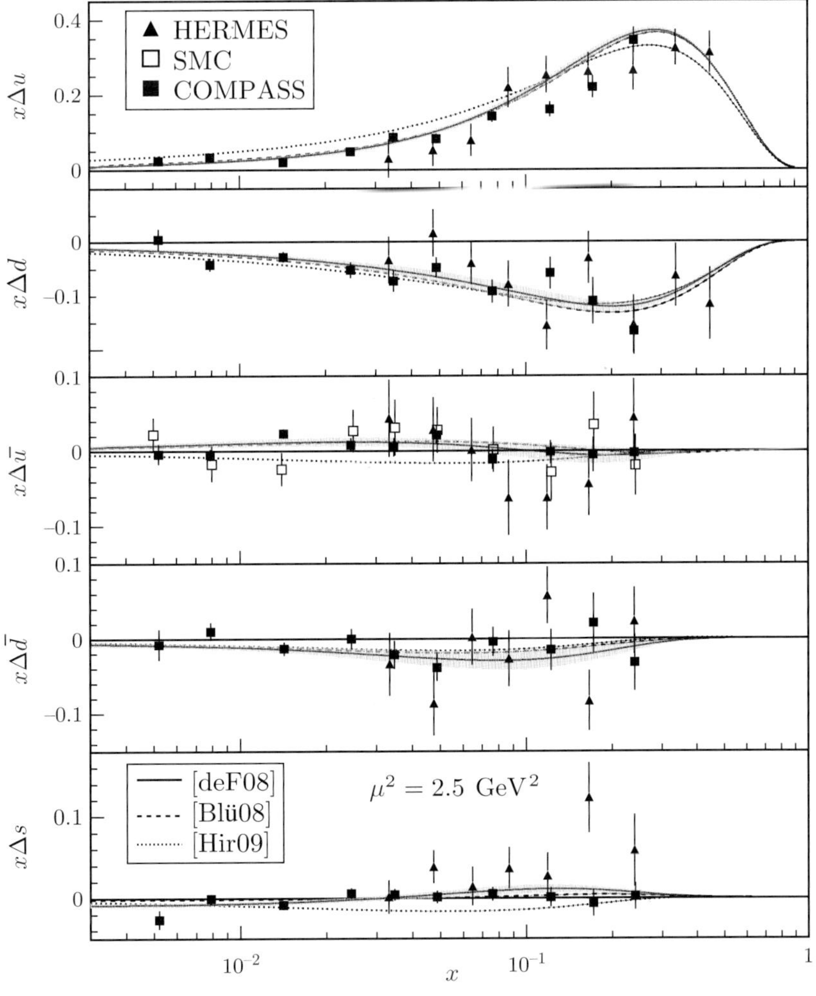

Fig. 9.6　Distributions of x times the polarized parton distributions $\Delta q(x)$ from both inclusive and semi-inclusive spin-dependent DIS from [PDG14].

of world data for the polarized valence quark distributions $x\Delta u_v(x)$ and $x\Delta d_v(x)$ as well as the sea quark distribution $x\Delta\bar{q}(x)$. The data show that the valence u quarks have a positive polarization while the valence d quarks are negatively polarized. The sea quark polarization is consistent with zero.

Traditional quark models of the nucleon are rather successful at describing the nucleon using only u and d quarks. It is important to note that, in this approximation, only the degrees of freedom associated with valence quarks are active; the effects of sea quarks are considered to be frozen as inert aspects of the effective degrees of freedom. The constituent quarks do contain internal structure associated with gluons and sea quarks (including $s\bar{s}$).

A number of experimental observations lend support to the possibility that strange quarks may play a significant role in certain aspects of nucleon structure. Neutrino (and antineutrino) scattering from the nucleon via the charged current interaction is a powerful means to probe flavor structure. Neutrinos can interact with d and s quarks by raising their charge and producing a negative lepton (e.g., $\nu_\mu + d \rightarrow \mu^- + u$ or $\nu_\mu + s \rightarrow \mu^- + c$). The c quarks produced by the s quarks then decay semi-leptonically yielding μ^+s, and so one observes $\mu^+\mu^-$ pairs from ν_μ interactions with s quarks. Similarly, antineutrinos will produce $\mu^+\mu^-$ from $\bar{s}$ quarks. 3 In this way, measurements of $s(x)$ and $\bar{s}(x)$ have been performed in deep inelastic neutrino and antineutrino experiments. The results indicate that $s(x)$ and $\bar{s}(x)$ are significant at low $x < 0.1$ and that the s and $\bar{s}$ each carry about 2% of the nucleon momentum.

9.7 Generalized Parton Distributions

In understanding the microscopic nature of matter, we rely on two main approaches. One can probe the spatial distribution of matter (charge or current) in a system through elastic scattering of electrons, photons, neutrons, etc. One measures the elastic form factors, which depend on the momentum transfer to the system, and then the Fourier transform of these form factors can provide information on the spatial distribution. Examples are the charge distribution in an atom and the atomic structure of a crystal. An alternative approach is to consider the distribution of constituents in momentum space by measuring deep inelastic knockout. Examples of this approach include the proton distribution in nuclei measured via quasielastic electron scattering (see Chapter 16) and the distribution of atoms in a quantum liquid studied using neutron scattering. With some modifications to accommodate the relativistic nature of the system, both experimental approaches have been used to explore the interior of the nucleon. The elastic nucleon form factors have been measured with ever increasing precision since the 1950s and at low momentum transfer ($<<$ nucleon mass) where the nucleon recoil effect is small, the three-dimensional Fourier transform of the form factors is interpretable as the spatial charge and current distributions of quarks. On the other hand, the parton distributions measured in DIS are the longitudinal momentum distributions of the quarks and the gluons in the infinite momentum frame. Both observables have provided great insight into the structure of the nucleon, but both have deficiencies. The form factors contain no dynamical information on the constituents, such as their velocity and angular momentum. The momentum distributions provide no information on the spatial structure.

More complete information about the microscopic structure lies in the correlation between momentum and spatial coordinates, i.e., simultaneous knowledge of a constituent's location and velocity. This knowledge is attainable for a classical system, for which one can define and study the phase-space distribution. For a quantum mechanical system, however, the notion of a phase-space distribution seems less useful because of the Heisenberg uncertainty principle: one cannot determine the position and conjugate momentum of a particle simultaneously with arbitrary precision. Nevertheless, the first quantum mechanical phase-space distribution was introduced by Wigner in 1932 and these distributions have been used in a number of areas, e.g., heavy-ion collisions, quantum molecular dynamics, signal analysis, etc.

The notion of a correlated parton position-and-momentum distribution had not been systematically explored in QCD until a few years ago. The generalized parton distribution (GPD) is a one-body matrix element that combines the kinematics of both elastic form factors and parton distributions, and is measurable in hard exclusive processes. The GPDs were introduced as objects with interesting perturbative QCD evolution, but were largely ignored because of unclear physical significance. They were rediscovered in the study of quark orbital motion and the spin structure of the nucleon in which the physics potential of the distributions began to surface.

The quest to experimentally determine the GPDs has focused on deep inelastic exclusive production of photons, mesons and lepton pairs. We consider briefly two experiments that have been extensively studied: deep virtual Compton scattering (DVCS), in which a real photon is produced, and diffractive meson production. The DVCS is cleaner, but the cross section is reduced by a factor of $1/\alpha$; meson production is easier to detect, but it is suppressed by factors of $1/Q^2$.

As nucleon matrix elements, GPDs must be calculable in the fundamental theory. At present, our only way to compute, rather than model, QCD dynamics is lattice field theory. The physical significance of GPDs was first revealed in studying the spin structure of the nucleon. One can decompose the nucleon's spin as

$$\frac{1}{2} = J_q(\mu) + J_g(\mu), \tag{9.73}$$

where the $J_{q,g}$ are the contributions from the quarks and gluons, respectively. Both contributions are gauge-invariant but renormalization scale-dependent. The $J_{q,g}$ can be expressed as matrix elements of the QCD energy–momentum tensor $T_{q,g}^{\mu\nu}$:

$$J_{q,g}(\mu) = \left\langle P1/2 \left| \int d\mathbf{x}(\mathbf{x} \times \mathbf{T}_{q,g})_z \right| P1/2 \right\rangle, \tag{9.74}$$

that can be extracted from the form factors of the quark and gluon parts of the $T_{q,g}^{\mu\nu}$. Taking the forward limit of the $\mu = 0$ component and integrating over $\mathbf{x}$, one finds that the $A_{q,g}(0)$ give the momentum fraction of the nucleon carried by the quarks and gluons, respectively, i.e., $[A_q(0) + A_g(0) = 1]$. On the other hand, one finds that [Ji97]

$$J_{q,g} = \frac{1}{2}[A_{q,g}(0) + B_{q,g}(0)]. \tag{9.75}$$

The matrix elements of the energy–momentum tensor provide the fractions of the nucleon spin carried by the quarks and gluons. Because the quark and gluon energy–momentum tensors are examples of twist-two, spin-two, helicity-independent operators, we immediately have the following sum rule for the GPDs

$$\int dxx[H_q(x,\xi,t) + E_q(x,\xi,t)] = A_q(t) + B_q(t), \tag{9.76}$$

where the ξ dependence is eliminated. If we extrapolate the sum rule to $t = 0$, the total quark contribution to the nucleon spin is obtained. The total quark contribution J_q can be decomposed gauge invariantly into the quark spin $\frac{1}{2}\Delta\Sigma$ and the orbital contribution L_q. Extracting J_q from knowledge of the GPDs and $\Delta\Sigma$, the quark orbital angular momentum can be determined. In this way, GPDs offer the potential to provide new insight into the spin structure of the nucleon.

In conclusion, it is important to place the discussion here in the context of a broader theoretical framework. Partons inside the proton can be described by Wigner distributions in six-dimensions (three position and three momentum coordinates). Wigner distributions are the quantum-mechanical constructions that are closest to a classical probability density in phase-space. However, they are not positive definite and are termed quasi-probability distributions. They can be used to compute the expectation value of physical observables. Thus, they represent the maximal knowledge of the partonic structure. They are equivalent to knowing the complete wavefunction of partons inside the nucleon. The distribution $W(x, \mathbf{b}_T, \mathbf{k}_T)$ is the master Wigner distribution. Here, $\mathbf{b}_T$ is the transverse position inside the nucleon and $\mathbf{k}_T$ is the transverse momentum of the partons inside the nucleon. By integrating over the transverse position, one is left with transverse momentum distributions (TMDs), while by integrating over transverse momentum, one obtains impact parameter distributions, whose Fourier transform yields the GPDs. The regular parton densities discussed above as well as form factors are derived by subsequent integration.

This modern, theoretical perspective has motivated an experimental program to image the three-dimensional structure of the proton using TMDs, GPDs, PDFs, and form factors. Tomography of the nucleon is a major research thrust at the upgraded Jefferson Laboratory as well as a strong motivator for a future electron–ion collider.

9.8 The Role of Partons in Nuclei

The EMC Effect

As we discuss in Chapters 13 and 14, the atomic nucleus is to a good approximation described as a system of protons and neutrons interacting in a mean field. Quantum Chromodynamics tells us that these nucleons are built from quarks and gluons. Do partons

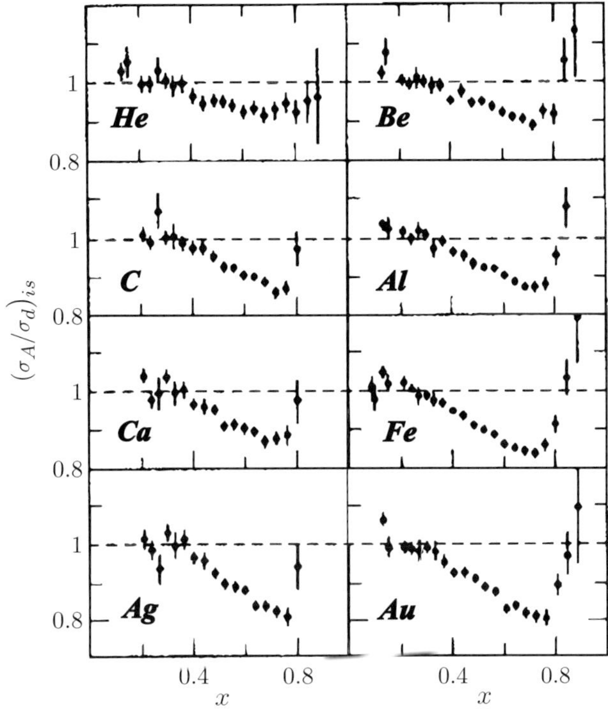

Fig. 9.7 The ratios of cross sections per average isoscalar nucleon for nuclei A compared to deuterium $(\sigma_A/\sigma_d)_{is}$ from [Gom94].

play any significant role in the structure and properties of nuclei? It has been determined that the distribution of quarks in the nucleus differs significantly from the distribution of quarks in the nucleon. This was discovered by the European Muon Collaboration (EMC) in the 1980s by measuring the ratios of structure functions of iron and deuterium in deep inelastic scattering. Figure 9.7 shows the ratio of deep inelastic inclusive cross sections for nuclei A compared to deuterium, corrected for neutron excess and averaged over Q^2. The data are plotted as a function of Bjorken x, which measures the fraction of the lightcone momentum of the nucleon carried by the struck quark in the Bjorken limit. The data have three major features. The significant reduction in the ratio from a value of 1 for $0.3 < x < 0.6$ is generally referred to as *the EMC effect*. The reduction for $x < 0.1$ is called *shadowing* and is analogous to the reduction in cross sections observed in real-photon reactions. The nucleon structure function must vanish as x approaches 1, so here and at $x > 1$ the structure function arises purely from nuclear effects, the simplest of which is the intrinsic motion of the nucleons in a nucleus at rest. The rise of the ratio above unity for $x \sim 0.1$–0.2 represents a transition region in the data and the models. In the ensuing decades, theoretical efforts to understand the EMC effect have focused on a number of different explanations without a clear consensus emerging.

In quark cluster models, it is assumed that a nucleus is partly made up of clusters containing $3N$ quarks ($N = 2, 3, 4, \ldots$). If a nucleus consists of non-interacting nucleons and multiquark clusters with some probabilities, then the valence quark distribution in a nucleus is softened in the intermediate-x region. Whether such multiquark clusters exist in the nucleus is an open question. Experimentally, clusters of mass M_c would be expected to be most significant at large $x \geq M_c/M_N$.

Dynamical rescaling models postulate that the differences between nuclear structure functions may arise from a difference in the scale of confinement of the nuclear constituents. The EMC effect can be explained by a relative shift in the scale of the measurements, i.e., $Q^2 \rightarrow \xi_A Q^2$. Quarks in heavier nuclei are confined in a region larger in size than in the free nucleon. For iron and lead, the increase in confinement radius would be 15% and 19%, respectively. However, determination of the nucleon size from quasifree electron–nucleus scattering sets an upper bound of 6% for the increase in iron.

In the conventional picture of a nucleus as a collection of bound nucleons, it is quite natural to take into account the effects of Fermi motion and nuclear binding (Chapter 16) broadens the lepton–nucleus cross section in the vicinity of $x = 1$. The binding energy per nucleon is no more than 10 MeV, is small compared to the typical energy transfer (several GeV) in lepton–nucleon deep inelastic scattering, but is significant in calculating the ratio of nuclear structure functions in the convolution model. Here the nuclear structure function $F_2^A(x)$ is given as a convolution of the free nucleon structure function $F_2^N(x)$ with the momentum distribution $f_\lambda(y)$ for the nucleon in the λth orbital:

$$F_2^A(x) = \sum_\lambda \int_x^{\frac{M_A}{M_N}} dy f_\lambda(y) F_2^N\left(\frac{x}{y}\right), \tag{9.77}$$

where y is the momentum fraction carried by the nucleon, and

$$f_\lambda(y) = \int d^4k \, \delta\left(y - \frac{Ak^+}{M_A}\right)\left(1 + \frac{k^3}{k^0}\right) S_\lambda(k), \tag{9.78}$$

where $k^+ = k^0 + k^3$. The nucleon spectral function is

$$S_\lambda(k) = |\phi_\lambda(\mathbf{k})|^2 \delta(k^0 - M_N - \epsilon_\lambda + T_R) \tag{9.79}$$

and the target nucleus is assumed to be at rest (see Chapter 16). Here, the separation (or removal) energy ϵ_λ of the struck nucleon in the orbital specified by λ in the nucleus is defined by

$$\epsilon_\lambda = M_A - M_{A-1}^\lambda - M_N \tag{9.80}$$

where M_A is the mass of the target nucleus and M_{A-1}^λ is the mass of the residual nucleus, which is usually in a highly excited hole state, say $(A-1)_\lambda$. The nucleon then has the initial energy $p_0 = M_N + \epsilon_\lambda - T_R$, where T_R is the recoil kinetic energy of the $(A-1)$ nucleus. This recoil contribution is important for light nuclear targets, but negligible for medium and heavy nuclei. If one uses a removal energy $< \epsilon_A >$ averaged over all occupied shells and an average momentum distribution of the nucleon $\phi(|\mathbf{k}|)$, then the y distribution of the nucleon becomes

$$f(y) = \left\langle \sum_{\lambda} f_{\lambda}(y) \right\rangle_{av}$$

$$= \int d^4k \left(y - \frac{Ak^+}{M_A} \right) \left(1 + \frac{k^3}{k^0} \right) |\phi(k)|^2 \delta(k^0 - m_N - <\epsilon_A> + <T_R>), \quad (9.81)$$

where $<T_R>$ is the average value of the recoil kinetic energy, $<\epsilon_{Fe}>$ is estimated at about -26 MeV. Nuclear binding and Fermi motion corrections in the convolution model account for only about 20% of the EMC effect in the mid-x region.

A nucleus consists of nucleons bound by the exchange of mesons – mostly pions at intermediate and long range. Because they are bound by mesons, the nucleons do not carry all of the momentum of the nucleus. It has been suggested that the depletion of valence quarks in the nucleus can be explained by an increase of virtual pions in the nuclear medium and thus to an enhancement of the EMC ratio at small x. Data for the ratio of structure functions consistently show a small excess above unity around $x \sim 0.15$.

In the parton model, nuclear shadowing arises when small-x gluons from different nucleons overlap in the longitudinal direction. One can therefore imagine that the shadowing begins when the gluons exceed a longitudinal size comparable to the nucleon–nucleon separation in the nucleus. The shadowing becomes stronger as x decreases and finally reaches a saturation value at $x_{sat} \simeq \frac{1}{2}R_A m_N$, where R_A is the nuclear radius. The data show that there is a definite A-dependence.

The study of hadron production in the nuclear medium is one of the methods to probe our understanding of confinement. Examples include the measurement of hadron production on nuclear targets in semi-inclusive deep inelastic lepton scattering and the jet-quenching and parton energy-loss phenomena observed in ultrarelativistic heavy-ion collisions. In each case hadron yields are observed that are different from those observed in the corresponding reactions on free nucleons.

Hadronization in Nuclei

Semi-inclusive production of hadrons in deep inelastic scattering of leptons from nuclei provides a way to investigate the spacetime development of the hadronization process. Leptoproduction of hadrons has the virtue that the energy and momentum of the struck parton are well determined, as they are tagged by the scattered lepton. By using nuclei of increasing size one can investigate the time development of hadronization. If hadronization occurs quickly, i.e., if the hadrons are produced at small distances compared to the size of atomic nuclei, the relevant interactions in the nuclear environment involve well-known hadronic cross sections such as those for pion–nucleon interactions. If, in contrast, hadronization occurs over large distances, the relevant interactions are partonic and involve the emission of gluons and quark–antiquark pairs. The two mechanisms lead to different predictions for the decrease in hadron yield, known as attenuation, on nuclei compared to the free nucleon.

The ratio R_A^h (data shown in Fig. 9.8) depends on ν, Q^2, $z = E_h/\nu$ and p_t^2, the square of the hadron momentum component transverse to the direction of the virtual photon. Thus, R_A^h can be written as

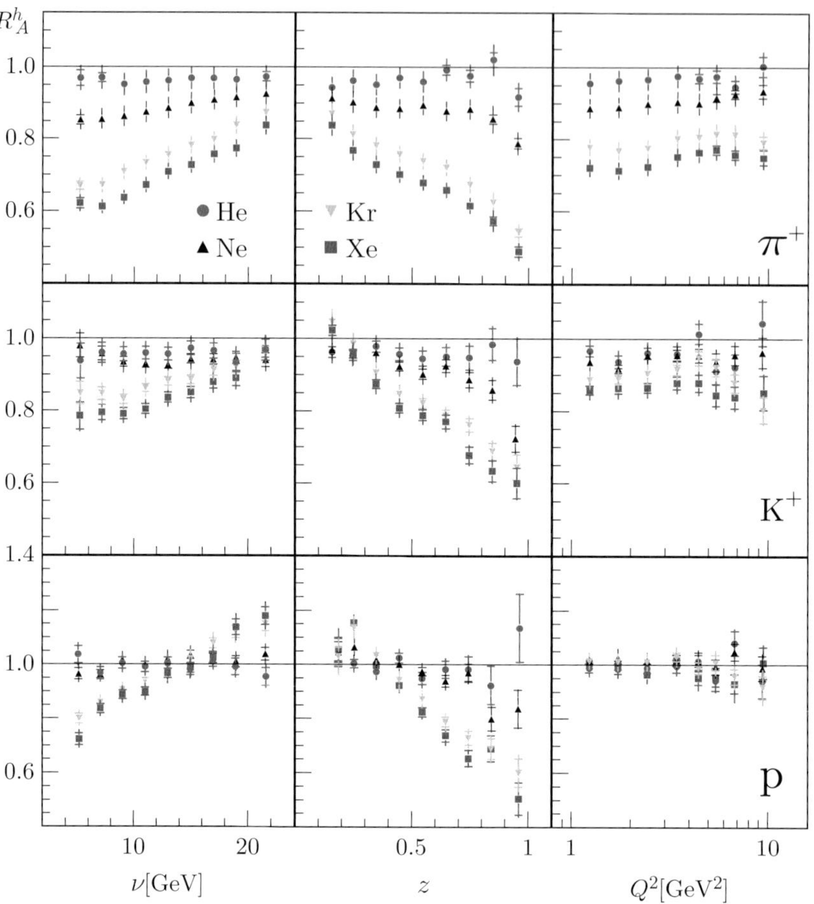

Fig. 9.8 Values of R_A^h for positively charged hadrons as a function of ν, z, and Q^2 from the HERMES experiment [Air07].

$$R_A^h(\nu, Q^2, z, p_t^2) = \frac{\left(\frac{N^h(\nu,Q^2,z,p_t^2)}{N^e(\nu,Q^2)}\right)_A}{\left(\frac{N^h(\nu,Q^2,z,p_t^2)}{N^e(\nu,Q^2)}\right)_D}, \tag{9.82}$$

with $N^h(\nu, Q^2, z, p_t^2)$ the number of semi-inclusive hadrons at given (ν, Q^2, z, p_t^2) and $N^e(\nu, Q^2)$ the number of inclusive DIS leptons at (ν, Q^2). Experiments at large values of ν give values of $R_A^h \approx 1.0$. This is interpreted as an indication that nuclear effects are negligible in that region. At lower values of ν the value of R_A^h has been found to be well below unity.

Figure 9.9 shows the p_t^2 dependence of R_A^h for different nuclei. The rise for heavier nuclei at high p_t^2, first observed by EMC, has since been seen in heavy-ion collisions and is referred to as the *Cronin effect*. The observed rise at high p_t^2 is attributed to a broadening of the p_t^2 distribution which can result from parton rescattering or from hadronic final-state interactions. The Cronin effect disappears at high z, which is consistent with a partonic

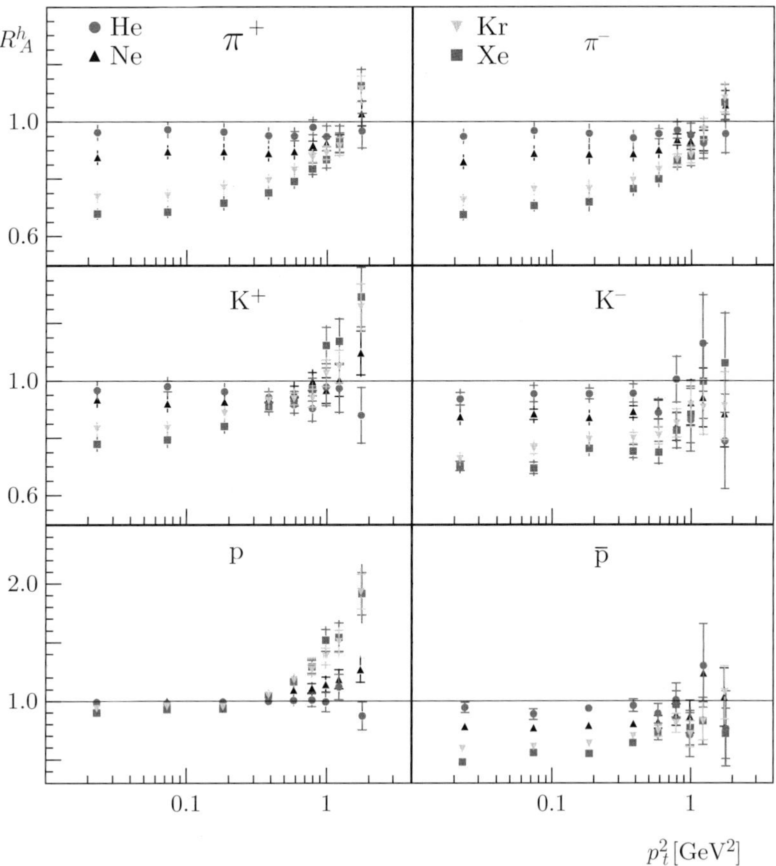

Fig. 9.9   Values of R_A^h for positively (left panel) and negatively (right panel) charged hadrons as a function of p_t^2 from the HERMES experiment [Air07].

origin for the effect. In the limit $z \to 1$ there is no room for parton rescattering because the parton is not allowed to have any energy loss.

Color Transparency

In a hard, exclusive reaction, e.g., elastic electron scattering from a proton at high energy, the scattering amplitude at large momentum transfer Q^2 is suppressed by powers of Q^2 if the proton contains more than the minimal number of constituents. This is derived from the QCD-based quark counting rules, which result from the factorization of wavefunction-like distribution amplitudes. Only the valence quarks in the proton participate in the scattering. Moreover, each quark, connected to another one by a hard gluon exchange carrying momentum of order Q, should be found within a distance of order $1/Q$. Thus, at large Q^2 one selects a very special configuration: all connected quarks are close together and form a small-size color neutral configuration. This pointlike configuration is not a stationary state and evolves until one measures combinations of normal hadrons. Such a color singlet system cannot emit or absorb soft gluons which carry energy or momentum smaller than Q.

This chapter has described how the fundamental quark and gluon structure of hadrons is studied with lepton scattering. In the next chapter, we will see how these studies can be extended using hadron beams at colliders.

Exercises

9.1   *SU(6)* **Picture of the Proton**

Consider the simple nonrelativistic $SU(6)$ wavefunction for the polarized proton

$$p \uparrow = \sqrt{\frac{2}{3}}(u \uparrow u \uparrow)d \downarrow + \sqrt{\frac{1}{3}}(u \uparrow u \downarrow)d \uparrow \ .$$

a) Show that $u \uparrow = \frac{5}{3}$, $u \downarrow = \frac{1}{3}$, $d \uparrow = \frac{1}{3}$, and $d \downarrow = \frac{2}{3}$.

b) Hence, show that in this model the total spin of the proton is carried by the quarks.

c) Show that $\Delta u - \Delta d = \frac{5}{3}$. This is equivalent to the Bjorken sum rule. Compare with data.

d) Determine the asymmetries for spin-dependent DIS from the proton and neutron in this model. Compare with data.

e) Determine the ratio of proton to neutron magnetic moment in this model. Compare with data.

9.2   **The Mandelstam Variables**

In two-body scattering $A + B \rightarrow C + D$, it is convenient to introduce the *Mandelstam variables*:

$$s \equiv (p_A + p_B)^2/c^2$$
$$t \equiv (p_A - p_C)^2/c^2$$
$$u \equiv (p_A - p_D)^2/c^2.$$

a) Show that $s + t + u = m_A^2 + m_B^2 + m_C^2 + m_D^2$.

b) Find the CM energy of A in terms of s, t, u and their masses.

c) Find the LAB energy of A, assuming B is at rest.

d) Find the total CM energy.

e) For elastic scattering of identical particles of mass m $A + A \rightarrow A + A$, $\mathbf{p}$ is the 3-momentum of the incident particle in the CM frame and θ is the scattering angle. Express s, t, u in terms of $\mathbf{p}, \theta$, and m.

9.3   **Inclusive Deep Inelastic Cross Section**

The cross section for inclusive DIS in terms of the two structure functions is written as

$$\frac{d^2\sigma}{dE'd\Omega} = \left(\frac{\alpha\hbar}{2E \sin^2(\theta/2)}\right)^2 [2W_1 \sin^2(\theta/2) + W_2 \cos^2(\theta/2)].$$

Using the Callan–Gross relation, show that the cross section can be written in terms of one structure function as follows

$$\frac{d^2\sigma}{dE'\,d\Omega} = \frac{F_1(x)}{2M}\left(\frac{\alpha\hbar}{E\sin^2(\theta/2)}\right)^2\left[1 + \frac{2EE'}{(E-E')^2}\cos^2(\theta/2)\right].$$

Using $F_1(x)$ from [PDG14], one can calculate the inclusive DIS cross section as a function of kinematics.

9.4 DIS Kinematics

Consider DIS lepton scattering.

a) On a plot of Q^2 versus ν, show the region of deep inelastic scattering from a fixed target for a beam of leptons of energy $E_0 =$ (i) 5, (ii) 10, (iii) 20, and (iv) 50 GeV.

b) With the Mandelstam variable $s = E_{\mathrm{CM}}^2$, x is the Bjorken scaling variable, and $y = \frac{p \cdot q}{p \cdot k}$, show that

$$Q^2 \approx s \cdot x \cdot y\,.$$

c) On a plot of Q^2 versus ν (log versus log), plot the accessible DIS region for (i) 11 GeV lepton beam on a proton fixed target, (ii) 27 GeV electron beam colliding head-on with a 940 GeV proton beam (HERA collider which operated from 1993 until 2007 at DESY, Hamburg, Germany), and (iii) 10 GeV electron beam colliding head-on with a 250 GeV proton beam (proposed future electron–ion collider).

9.5 Electron–Ion Collider

Neglecting electron and proton masses,

a) Show that the center-of-mass energy for a lepton beam of energy E_0 on a proton is $E_{\mathrm{CM}} = \sqrt{2ME_0}$.

b) Show that the center-of-mass energy for a lepton beam of energy E_0 colliding head-on with a proton beam of energy E_p is $E_{\mathrm{CM}} = \sqrt{4E_0 \cdot E_p}$.

c) Calculate the energy of a lepton on a proton fixed target necessary to produce the same center-of mass energy as the electron–proton collisions.

High-Energy QCD

10.1 Introduction

In Chapter 5, the non-Abelian $SU(3)_{\text{color}}$ gauge theory, QCD, was developed as the fundamental theory of the strong force in the Standard Model. Quantum chromodynamics describes hadrons as consisting of pointlike electrically charged quarks bound via color forces by the exchange of gluons. In Chapters 7, 8, and 9, lepton scattering has been used to great effect to image the nucleon at distance scales from several fm to 0.001 fm. We have interpreted these data in terms of QCD and have come to understand that the nucleon is a highly relativistic system of light quarks and massless gluons with tremendously strong forces at play. The measured momentum distributions of the quarks and gluons in the nucleon are characterized by a large rise at low x which reflects the fact that the virtual particles of QCD are playing a dominant role. By contrast, the virtual particles of QED play only a minor role in understanding the structure of the atom. Thus, understanding the structure and properties of the nucleon in terms of QCD drives experiments to high energies where the dominant, virtual particles at low x can be accessed and studied.

As the collision energy increases, regions of progressively higher gluon and sea quark density are probed. However, the density of gluons inside a nucleon, which dominate over the sea quarks, must eventually saturate to avoid untamed growth in the strength of the nucleon–nucleon cross section, which would violate the principle of unitarity. Thus far, this saturated gluon density regime has not been observed. In addition, the understanding of basic properties of the nucleon, e.g., the origin of its spin-1/2, as well as fundamental processes or hadronization that connects the experimental world of hadrons with the quarks and gluons of QCD, demand measurements at high energies. Further, it should be noted that high-energy DIS data on nuclei are sparse and characterized typically by low precision.

To frame the discussion in this chapter, a number of important questions can be formulated [EIC12]:

- *How are the sea quarks and gluons, and their spins, distributed in space and momentum inside the nucleon?* How are these quark and gluon distributions correlated with overall nucleon properties, such as spin direction?
- *Where does the saturation of gluon densities set in?* Is there a simple boundary that separates this region from that of the more dilute quark–gluon matter? If so, how do the distributions of quarks and gluons change as one crosses the boundary? Does this

saturation produce matter of universal properties in the nucleon and nuclei when viewed at high energy?

- *How does the nuclear environment affect the distribution of quarks and gluons and their interactions in nuclei?* How does the transverse spatial distribution of gluons compare to that in the nucleon? How does nuclear matter respond to a fast moving color charge passing through it? Is this response different for light or heavy quarks?

A highly desirable goal in understanding QCD is to develop a unified visualization of the subatomic world. In the case of the proton, a large body of lepton scattering data has been acquired that yields snapshots of its constituents at different spatial resolutions and shutter speeds. Together with the aid of *ab initio* lattice QCD and models, can our understanding be unified in the form of a visualization that would be comprehensible to a non-expert? In this regard, a major goal of current research is to carry out spatial imaging or tomography of the nucleon in the context of the modern theoretical picture described in Section 9.7. This will be first pursued at Jefferson Laboratory in the regime of valence quarks. It is a major motivation for a future electron–ion collider where it can be pursued in the sea quark and gluon dominated regimes.

To reach the highest energies in particle collisions, collider experiments are favored over fixed target experiments. Furthermore, high-energy circular electron (positron) beams are limited in their energy reach compared to hadron beams due to synchrotron radiation energy losses that increase dramatically with energy. Thus, only one electron (positron)–proton collider has been realized at high energy, namely HERA, which will be discussed in this chapter. The HERA collider has provided the crucial low-x precision data that underpin our current understanding of high-energy QCD. A number of high-energy hadron–hadron collider experiments have been built in Europe and the US where violent collisions of the parton constituents allow a search for new physics as well as probe our understanding of QCD.

High-energy hadron–hadron collider experiments have led to the discovery of numerous fundamental ingredients of the SM, such as the heavy electroweak W and Z bosons, the discovery of the top quark and the Higgs boson. Further, the structure of the proton, in terms of high-energy QCD, can also be investigated by these powerful accelerators, e.g., Drell–Yan experiments at FNAL to study the sea quarks and polarized proton collider experiments at RHIC to study the gluon contribution to the proton spin.

In this chapter, the detailed kinematics for different collider configurations is developed. This is then applied to three topics of current high-energy QCD research: the spin structure of the proton, the flavor asymmetry of the sea, and the low-x region. Next, the production of jets, bosons, and quarks in colliders is described. The chapter closes with a perspective on the path forward.

Hadronic Cross Sections

The leading-order Feynman diagram for the hadronic production process is shown in Fig. 10.1(a). This process is in fact the only way to produce hadrons in a purely QCD process. The production cross section is given as:

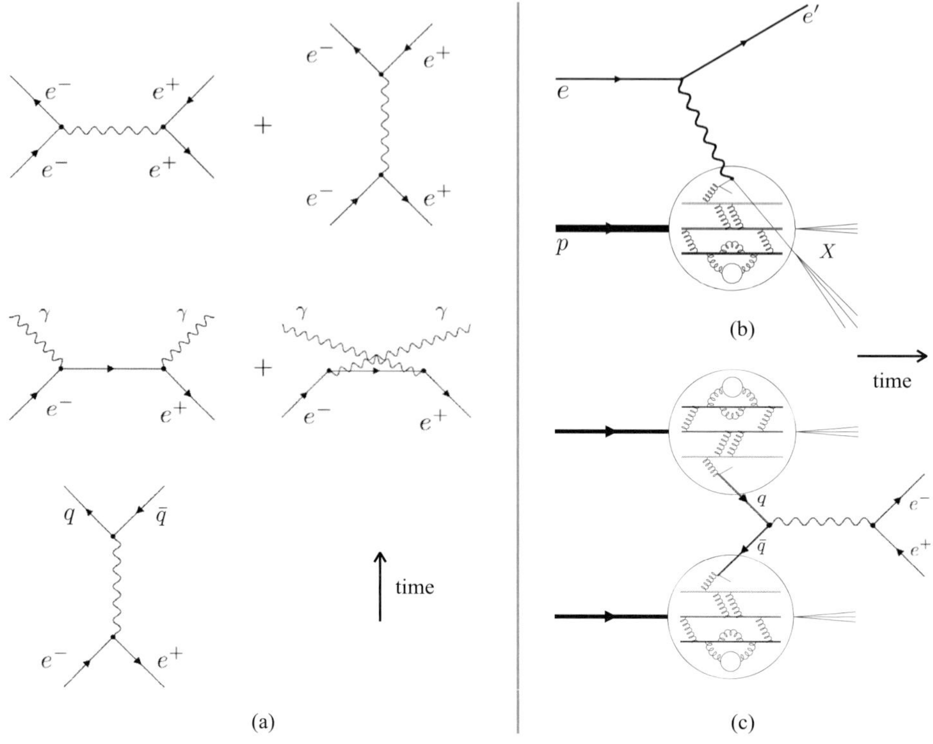

Fig. 10.1 Leading-order Feynman diagrams involving QED processes: (a) for an e^+e^- collider (Bhabha scattering, pair production, and hadron production), (b) for an ep collider (leading-order DIS) and (c) for a hadron collider (Drell–Yan production).

$$\frac{d\sigma}{d\Omega} = N_c \frac{\alpha^2}{4s} \beta \big[1 + \cos^2\theta + (1 - \beta^2)\sin^2\theta \big] Q_f^2 \tag{10.1}$$

with $\beta = v/c$, Q_f the charge of the produced fermion pair, $N_c \to 1$ for lepton pairs and $N_c \to 3$, i.e., the number of colors, for quark pairs. This relation reduces to a simple form in the limit $\beta \to 1$

$$\sigma = N_c Q_f^2 \frac{4\pi\alpha^2}{3s}. \tag{10.2}$$

The kinematic factor $4\pi\alpha^2/3s$ is identical for lepton pairs and quark pairs. It is therefore common to divide the hadronic production cross section by that for muon-pair production, which has in fact only one underlying leading-order diagram. This ratio R is therefore defined as follows:

$$R = \frac{\sigma(e^+e^- \to \text{hadrons})}{\sigma(e^+e^- \to \mu^+\mu^-)} = N_c \sum_f Q_f^2, \tag{10.3}$$

where the sum extends over all quark/antiquark pairs which contribute up to a certain center-of-mass energy $\sqrt{s}$. This ratio has been extensively measured and R is shown in Fig. 10.2 as a function of $\sqrt{s}$. At low energies, where only u, d and s quarks contribute, one expects the R value

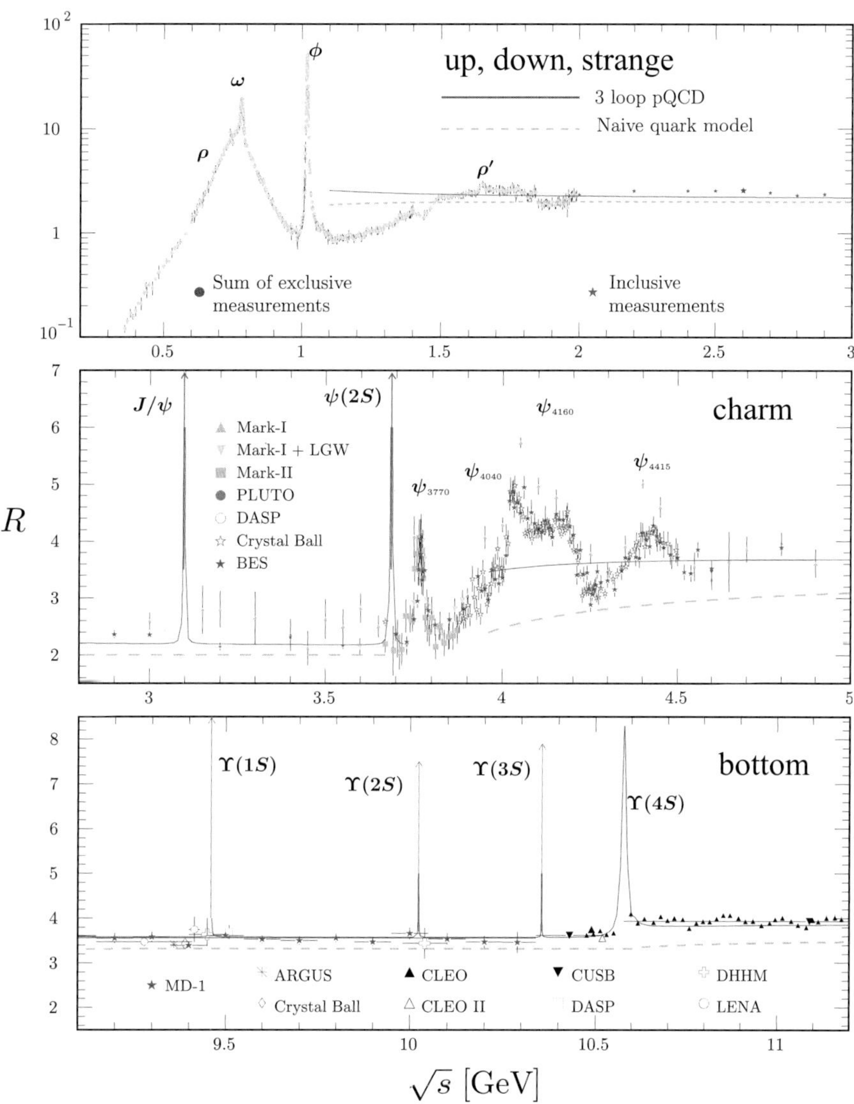

Fig. 10.2   Compilation of various R ratio measurements as a function of $\sqrt{s}$ emphasizing different quark flavor regions, i.e., light quark flavor (u, d, s) and heavy flavors (c, b); figure from [PDG14].

$$R_{u,d,s} = 3\left[\left(\frac{2}{3}\right)^2 + \left(-\frac{1}{3}\right)^2 + \left(-\frac{1}{3}\right)^2\right] = 2, \tag{10.4}$$

whereas at higher energies c and b quarks contribute, giving rise to the R values:

$$R_c = 2 + 3\left[\left(\frac{2}{3}\right)^2\right] = \frac{10}{3} \tag{10.5}$$

and

$$R_b = \frac{10}{3} + 3\left[\left(-\frac{1}{3}\right)^2\right] = \frac{11}{3}. \tag{10.6}$$

Figure 10.2 shows a compilation of various experimental results in comparison to a simple quark model calculation and higher-order QCD calculation.

Factorization

The step from an elementary process involving only initial- and final-state leptons to one involving hadrons in both the initial and final state poses a significant challenge to describing the overall measurable process in terms of underlying partonic processes. The essential idea is displayed in Fig. 10.3 for a hadron–hadron collision with initial momenta p_1 and p_2. Colliding partons for each hadron are described in terms of parton distribution functions denoted as $f_1(x_1)$ and $f_2(x_2)$. The calculable perturbative amplitude can be evaluated at higher orders, and has been worked out for numerous processes. The final step involves the conversion of partons into observable hadrons, which is referred to as fragmentation and was discussed in Section 9.4. The overall cross section is calculated as a convolution involving long-range parton distribution functions, a perturbatively

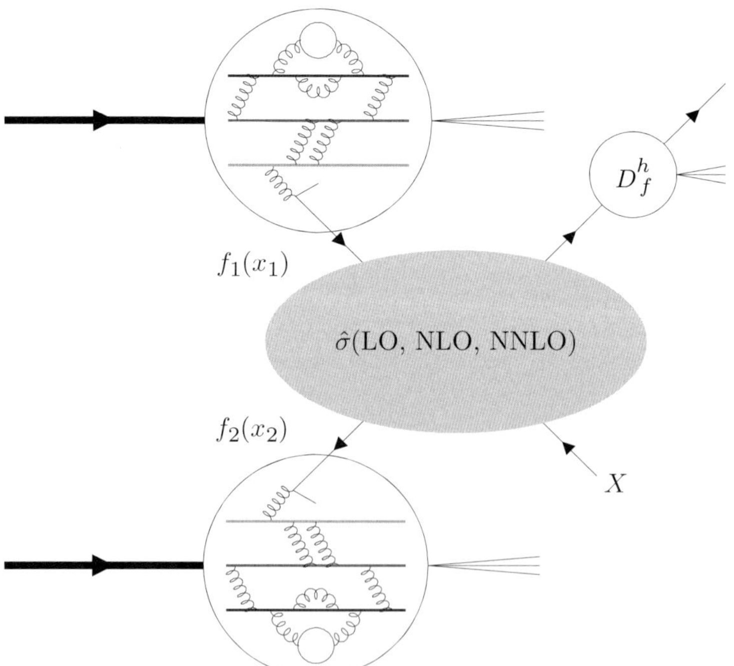

Fig. 10.3　Hadron–hadron scattering process, $p + p \rightarrow h + X$, displaying initial-state partons in terms of parton distribution functions, $f_1(x_1)$ and $f_2(x_2)$, a perturbative calculable process, $\hat{\sigma}$, and fragmentation function $D^h_f(z)$ for final-state hadron production for a parton f fragmenting into a hadron h.

calculable partonic cross section, and fragmentation functions. The universal nature of parton distribution functions allows one to predict various QCD processes.

The differential cross section in transverse momentum for hadron production, $p + p \rightarrow h + X$, is formulated below as an integral over Bjorken-x scaling variables and the momentum fraction z:

$$\left(\frac{d\sigma}{dp_T}\right)^{p+p \rightarrow h+X} = \sum_{f_1,f_2,f} \int dx_1 dx_2 dz f_1(x_1, \mu^2) f_2(x_2, \mu^2)$$

$$\times \left(\frac{d\hat{\sigma}}{dp_T}\right)(x_1 p_1, x_1 p_1, z, \mu) D_f^h(z, \mu^2) . \quad (10.7)$$

The separation into long and short-range processes involves a scale referred to as the factorization scale. For such a separation, or factorization, to hold requires the existence of universal parton distribution functions regardless of the actual process for which they are being used. The proof of this separation is known as the factorization theorem and forms an essential ingredient in the analysis of high-energy hadronic processes [Col89]. Both parton distribution and fragmentation functions are extracted from data using their predicted scale dependence involving evolution equations as discussed in Section 9.3. Generally, high-energy hadronic processes involve three fundamental ingredients: perturbative calculable cross sections and thus the behavior of asymptotic freedom in QCD, factorization, and the evolution of both parton distribution and fragmentation functions. This forms a powerful predictive framework for high-energy physics processes.

10.2 Building the Tools

Collider Kinematics

The development and verification of the Standard Model is tightly connected with the advancement of accelerator science, in particular the establishment of particle colliders, in contrast to fixed-target experimental configurations in nuclear and particle physics. A collider can provide much larger center-of-mass energies and allows one to probe various processes at much larger scales where perturbative calculations are generally well understood. Figure 10.4 shows an illustration of two different accelerator modes of a 30 GeV electron beam colliding with a stationary target (a) and with another counter-circulating proton beam of 920 GeV (b). The center-of-mass energy $\sqrt{s}$ can be evaluated as follows for a fixed-target configuration: $\sqrt{s} = \sqrt{2E_e m_p}$. For a collider configuration we

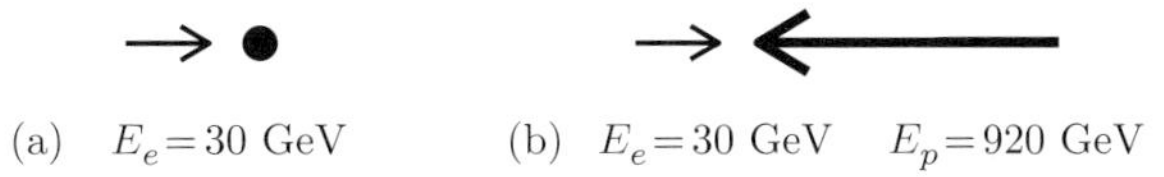

Fig. 10.4   Illustrations (a) of an ep fixed-target mode and (b) of an ep collider mode configuration. The lengths of the arrows are not to scale.

have: $\sqrt{s} = \sqrt{4E_e E_p}$. Both results can easily be obtained by evaluating the Mandelstam variable $s = (p_e + p_p)^2$ with p_e^μ and p_p^μ being the four-vectors for the initial-state electron and proton. This expression can be expressed as $s = (E_e + E_p)^2 - (\boldsymbol{p}_e + \boldsymbol{p}_p)^2$, which reduces to $s = m_e^2 + m_p^2 + 2E_e E_p(1 - \beta_e \beta_p \cos\theta)$. For a fixed-target configuration we have $\cos\theta = 0$ along with $\beta_p = 0$ whereas for a collider-mode configuration we have $\cos\theta = -1$ with $\beta_e \simeq 1$ and $\beta_p \simeq 1$.

Using the above example, one finds that the center-of-mass energy of a collider is significantly larger than its fixed target analog. A more dramatic illustration is provided by the question of what electron beam energy for a fixed-target configuration would be needed to reach the same center-of-mass energy as in a collider mode operation for the above conditions. A simple calculation reveals the astonishing result that a center-of-mass energy $\sqrt{s} = \sqrt{4E_e E_p} = 332\,\text{GeV}$ would require an electron beam energy in a fixed-target configuration of $E_e = s/2m_p = 55200\,\text{GeV}$. It is clear that such a high-energy lepton beam is far beyond the technical capabilities of current accelerator science. The above example of a 30 GeV electron beam colliding head-on with a 920 GeV proton beam was realized with the HERA accelerator at DESY.

Over the last decades, several collider programs provided a successful means of probing elementary collisions involving either lepton beams such as electron (e^-) or positron beams (e^+) or hadron beams such as proton (p) and anti-proton ($\bar{p}$) beams, which effectively provide a source of quarks and gluons taking part in a high-energy collision. These different types of initial probes provide a critical means to develop and verify the Standard Model of particle physics. An illustration in terms of leading-order Feynman diagrams involving QED processes is shown in Fig. 10.1 for a e^+e^- collider (Bhabha scattering, pair production, and hadron production), ep collider (leading-order DIS), and hadron collider (Drell–Yan production). Figure 10.5 displays three leading-order SM processes of strong interactions, while Fig. 10.6 displays three leading-order SM processes of weak interactions.

Table 10.1 provides an overview of the main accelerator parameters for past and current collider programs involving e^+e^-, $e^\pm p$, and $p\bar{p}\,/\,pp$ collisions [PDG14]. Some highlights of the physics programs for various past and current collider program are:

- e^+e^- collider PETRA (DESY): Discovery of gluon and QCD physics;
- e^+e^- collider LEP (CERN) and SLC (SLAC): Electroweak physics/QCD physics;
- $e^\pm p$ collider HERA (DESY): Proton structure involving electroweak processes probing QCD physics;
- $\bar{p}p$ collider SPS (CERN): Discovery of W/Z bosons and QCD physics;
- $\bar{p}p$ collider TEVATRON (FNAL): Top quark discovery/electroweak and QCD physics;
- pp collider LHC (CERN): Discovery of the Higgs/probing SM at highest energy and search for physics beyond the Standard Model; and
- $pp\,/\,pA\,/\,AA$ collider RHIC (BNL): Discovery of the quark-gluon plasma/probing spin phenomena of QCD/Relativistic heavy-ion program.

The relativistic invariant description of any type of particle collision is an essential element for any modern high-energy fixed-target and collider experiment, in particular involving hadron beams of any kind. In contrast to a symmetric e^+e^- collider, where the

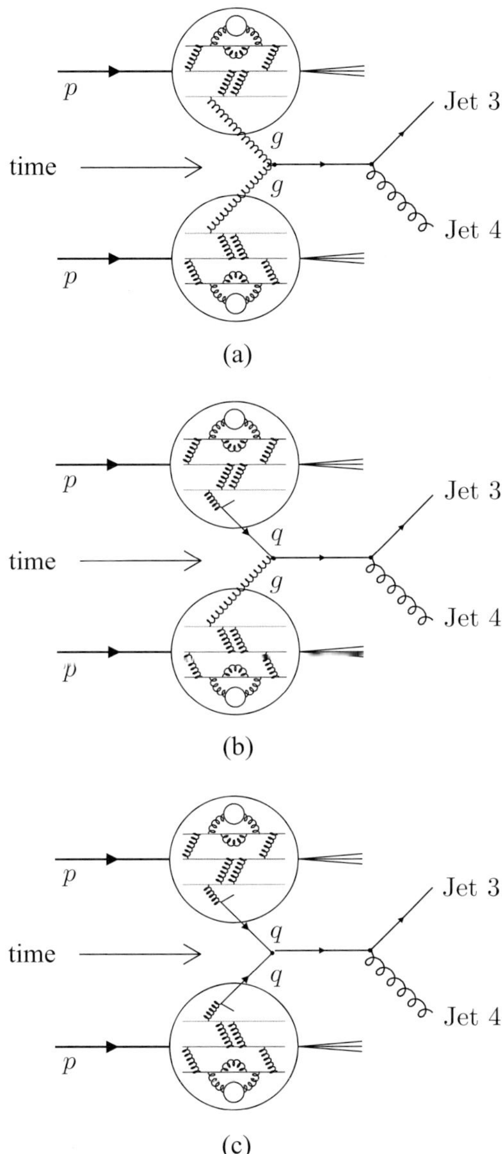

Fig. 10.5 Leading-order Feynman diagrams involving QCD processes for a hadron collider showing (a) gg, (b) qg, and (c) qq initiated processes.

longitudinal momenta of both initial-state leptons are fixed, a collision involving hadron beams provides a spectrum of longitudinal momenta of incoming partons quantified by the parton distribution functions defined and discussed in Chapter 9. Thus, a high-energy ep collider or hadron collider gives rise to a center-of-mass of the lepton–parton collision or parton–parton collision which is different for each collision. Specific relativistic invariant quantities have been introduced to describe both ep collisions and hadron–hadron collisions and their understanding is essential for all collider studies.

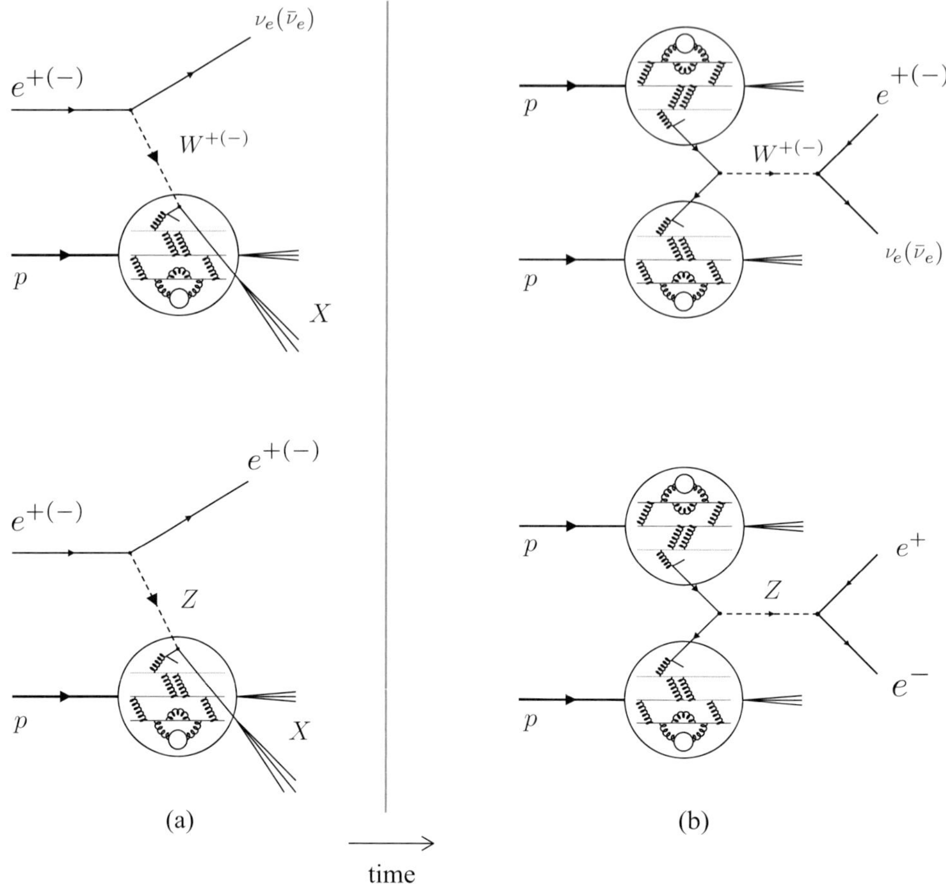

Fig. 10.6 Leading-order Feynman diagrams involving weak processes (a) for an ep collider and (b) for a hadron collider.

$e^{\pm}/p$ Kinematics

The scattering of unpolarized electrons (positrons) on unpolarized protons, as shown to first-order perturbation theory in Fig. 10.7, is described through the exchange of a Standard Model electroweak gauge boson [Sur99]:

$$e(k) + P(p) \rightarrow l(k') + X(p') \tag{10.8}$$

with the electron (positron) and proton in the initial state denoted by the four-vectors $k = (E_e, \mathbf{k}_e)$ and $p = (E_P, \mathbf{P})$, respectively. The final state consists of the scattered lepton $k' = (E'_l, \mathbf{k}'_l)$ and the hadronic final state system $p' = (E_X, \mathbf{p}_X)$.

Depending on the exchanged electroweak gauge boson, one distinguishes two classes of events:

- neutral current (NC) (electroweak gauge boson: virtual photon γ^* or Z^0 boson); and
- charged current (CC) (electroweak gauge boson: $W^{\pm}$ boson) events.

Table 10.1 Main accelerator parameters for a selection of $e^+ - e^-, e^{\pm} - p, \bar{p} - p / p - p$ past and current collider programs [PDG14]

Parameter	SLC	LEP	HERA	RHIC	TEVATRON	LHC
Colliding particles	e^+/e^-	e^+/e^-	$e^{\pm}/p$	$\vec{p}/\vec{p}$	$\bar{p}/p$	p/p
Physics start date	1989	1989	1992	2001	1987	2009
Physics end date	1998	2000	2007	ongoing	2011	ongoing
Max. beam energy (GeV)	50	104.6	30 / 920	255	980	6500
Luminosity (10^{30} cm^{-2}s^{-1})	2.5	100	75	215	431	$(1\text{–}2)\times 10^4$
Bunch crossing time (ns)	8300×10^3	22×10^3	96	107	396	24.95
Bunch length (cm)	0.1	1.0	0.83 / 8.5	60	50 / 45	9
Particles per bunch (10^{10})	4.0	45	3 / 7	18.5	26 / 9	12
Bunches per ring per species	1	4	189 / 180	111	36	2508
Average beam current (mA)	0.0008	6	40 / 90	257	70 / 24	540
Circcumferrence (km)	1.45	26.66	6.336	3.834	6.28	26.659
Interaction regions	1	4	4	6	2	4
Peak magnetic field (T)	0.597	0.135	0.274 / 5	3.5	4.4	8.3

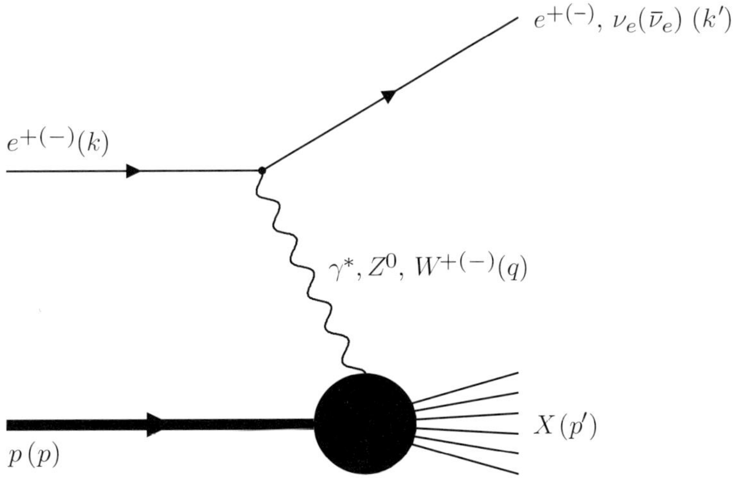

$e^{+(-)}, \nu_e(\bar{\nu}_e)\ (k')$

$e^{+(-)}(k)$

$\gamma^*, Z^0, W^{+(-)}(q)$

$X(p')$

$p(p)$

Fig. 10.7 Feynman-diagram describing unpolarized ep scattering in lowest-order perturbation theory.

Both event classes can be distinguished by the final-state lepton. In the case of NC events an electron (positron) is found in the final state ($l = e$) whereas in the case of CC events the final-state system consists of a neutrino (antineutrino) which escapes detection ($l = \nu_e$).

Collider $e^{\pm}/p$ experiments are typically designed to be able to measure the energy and direction of both the scattered lepton (in case of NC events only) and the hadronic final-state system. Two independent variables are sufficient in defining the unpolarized inelastic $e^{\pm}/p$ event kinematics at fixed beam energies, e.g., in the case of a NC event the energy E'_e and polar angle θ'_e of the scattered electron (positron).

A detailed discussion of deep inelastic scattering (DIS) is presented in Chapter 9. In the section below the emphasis is placed on the kinematic aspects. Here, in order to be self-contained, the variables necessary to provide a relativistic-invariant formulation of the unpolarized inelastic ep event kinematics are again summarized:

$$s = (k + p)^2 \simeq 4E_e E_P \tag{10.9}$$

$$t = (p - p')^2 \tag{10.10}$$

$$u = (k' - p)^2 \tag{10.11}$$

$$Q^2 = -(k - k')^2 = -(p - p')^2 = -t = -q^2 \tag{10.12}$$

$$x = \frac{Q^2}{2(p \cdot q)} \simeq -\frac{t}{u + s} \qquad 0 \leq x \leq 1 \tag{10.13}$$

$$y = \frac{p \cdot q}{p \cdot k} \simeq \frac{u + s}{s} \qquad 0 \leq y \leq 1 \tag{10.14}$$

$$W^2 = (p + q)^2 = (p')^2 = m_p^2 + \frac{Q^2}{x}(1 - x) \simeq s + t + u \tag{10.15}$$

$$\nu = \frac{p \cdot q}{m_p}. \tag{10.16}$$

The "$\simeq$" sign refers to those cases where the electron and proton masses have been neglected. As defined in Chapter 8, Q^2 is the negative square of the momentum transfer q and denotes the virtuality of the exchanged gauge boson. The momentum transfer q determines the size of the wavelength of the virtual boson and therefore the object size Δ which can be resolved in the scattering process. To resolve objects of size Δ requires the wavelength of the virtual boson λ to be smaller than Δ. The wavelength λ of the virtual boson can be written employing the Heisenberg uncertainty principle as

$$\lambda = \frac{1}{|\boldsymbol{q}|} = \frac{1}{\sqrt{\nu^2 + Q^2}} \approx \frac{2m_p x}{Q^2}. \tag{10.17}$$

Better resolution requires smaller wavelengths of the virtual boson and therefore larger momentum transfers. The maximum possible value for Q^2 is given by $Q^2_{\text{max}} = s$.

As defined in Chapter 8, W^2 is the square of the invariant mass of the hadronic final state system X. And W can be interpreted as the center-of-mass energy of the gauge boson–proton system. Small values of x correspond to large values of the invariant mass W. Recall, x is the Bjorken scaling variable and is interpreted in the quark–parton model as the fraction of the proton momentum carried by the struck parton. The limits on x follow from the fact that the square of the invariant mass W^2 has to be larger or equal to the square of the mass of the proton m_p^2, i.e., $W^2 = m_p^2 + (Q^2/x)(1 - x) \geq m_p^2$ where $x = 1$ corresponds to the elastic case for which $W = m_p$. In the proton rest frame, ν is the energy of the exchanged gauge boson

$$\nu = \frac{p \cdot q}{m_p} = \frac{m_p(E_e - E'_e)}{m_p} = (E_e - E'_e). \tag{10.18}$$

The maximum energy transfer ν_{max} is given by $\nu_{\text{max}} = s/(2m_p)$.

The quantity y is the fraction of the incoming electron energy carried by the exchanged gauge boson, also known as the inelasticity in the rest frame of the proton. The quantity y can be also written as $y = \nu/\nu_{\max}$ which yields the limits on y as given above. Finally, t denotes the momentum transfer at the hadronic vertex. The relativistic invariant variables x, y, Q^2, and s are connected through

$$Q^2 \simeq s \cdot x \cdot y, \tag{10.19}$$

where the electron and proton masses have been ignored. For fixed x and y, an ep collider allows one to reach much larger values of Q^2 as well as much lower values of x, keeping y and Q^2 fixed, due to the larger center-of-mass energy $\sqrt{s}$ compared with fixed-target experiments.

The accurate reconstruction of the Lorentz-invariant variables x, y, and Q^2 is one of the major ingredients in measuring structure functions. The final state in NC $eP \rightarrow eX$ scattering consists of the scattered electron and the hadronic final state system X. Both systems alone or any combination can be used to reconstruct the event kinematics [Ben91]. Aiming at a precision measurement of structure functions places tight constraints on the reconstruction of the kinematic variables x, y, and Q^2, on the choice of the kinematic reconstruction methods, and therefore on the measurable quantities, which in turn defines the requirements of an experimental configuration. It is therefore of vital importance to obtain a solid understanding of kinematic reconstruction methods.

Figure 10.8 shows, in lowest-order perturbation theory, the ep process with the final-state electron scattered with polar angle θ'_e having an energy E'_e. The hadronic final state consists of the current jet having angle γ and the proton remnant jet which, in the quark–parton model, originates from the fragmentation of the struck quark and the proton remnant,

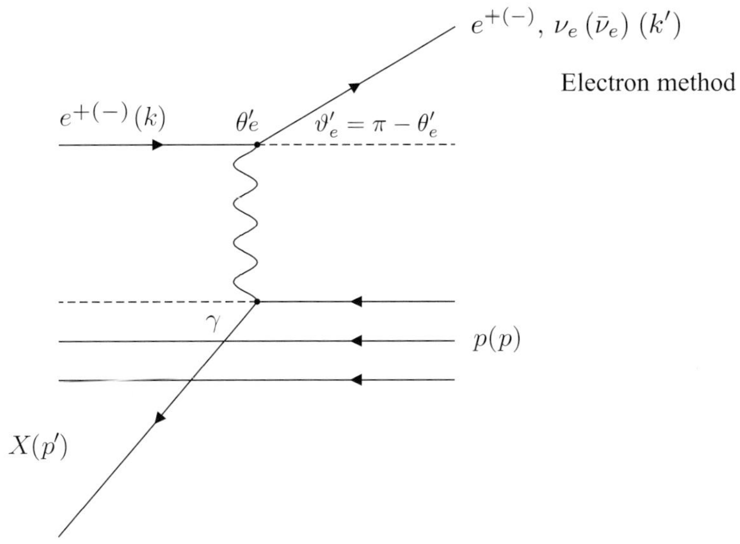

 Schematic of the neutral current ep scattering process with the final-state electron and the hadronic final state.

respectively. In a typical collider detector coordinate system, the four-vectors of the initial and final state of the process $e(k) + P(p) \to e(k') + X(p')$ are given as

$$
k = \begin{pmatrix} E_e \\ 0 \\ 0 \\ -E_e \end{pmatrix} \qquad p = \begin{pmatrix} E_P \\ 0 \\ 0 \\ E_P \end{pmatrix} \tag{10.20}
$$

$$
k' = \begin{pmatrix} E'_e \\ E'_e \sin\theta'_e \cos\phi'_e \\ E'_e \sin\theta'_e \sin\phi'_e \\ E'_e \cos\theta'_e \end{pmatrix} \qquad p' = \begin{pmatrix} \sum_h E_h \\ \sum_h p_{X,h} \\ \sum_h p_{Y,h} \\ \sum_h p_{Z,h} \end{pmatrix}, \tag{10.21}
$$

where E'_e, θ'_e, and ϕ'_e are the energy, polar angle, and azimuthal angle of the scattered electron while $\sum_h E_h$ and $(\sum_h p_{X,h}, \sum_h p_{Y,h}, \sum_h p_{Z,h})$ are the energy and momentum of the hadronic final-state system X, which requires a summation over all hadronic final-state particles h. Below we provide an overview of a reconstruction method using the scattered electron. A discussion of other reconstruction methods such as the Jaquet–Blondel method which involves the hadronic final state can be found in [Jaq79].

The *electron* method is the primary technique used in ep scattering experiments to reconstruct the event kinematics. This method relies solely on the final-state electron and is therefore applicable to NC events only. Using the above four-vectors k and k' for the initial and final-state electron, the kinematic variables x, y, and Q^2 can be written in terms of E'_e and θ'_e via

$$
x_e = \frac{Q_e^2}{s y_e} = \frac{E'_e \cos^2 \frac{\theta'_e}{2}}{E_p(1 - \frac{E'_e}{E_e} \sin^2 \frac{\theta'_e}{2})} \tag{10.22}
$$

$$
y_e = 1 - \frac{E'_e}{2E_e}(1 - \cos\theta'_e) = 1 - \frac{E'_e}{E_e} \sin^2 \frac{\theta'_e}{2} \tag{10.23}
$$

$$
Q_e^2 = 2E_e E'_e(1 + \cos\theta'_e) = 4E_e E'_e \cos^2 \frac{\theta'_e}{2} = \frac{p_{T,e}^2}{1 - y_e}. \tag{10.24}
$$

These expressions can be used to plot Q^2 as a function of x keeping either E'_e or θ'_e fixed, which is essential in order to understand what values in E'_e and θ'_e correspond to what region in the kinematic $Q^2 - x$ plane. Figure 10.9 shows lines of constant electron energy (a) and constant scattering angles (b) as well as lines of constant y values (1, 0.1, 0.01). These curves can be determined using

$$
Q_e^2(x, \theta'_e) = \frac{sx}{1 + \left(x\frac{E_p}{E_e}\right) \tan^2 \left(\frac{\theta'_e}{2}\right)} \tag{10.25}
$$

$$
Q_e^2(x, E'_e) = \frac{sx\left(1 - \frac{E'_e}{E_e}\right)}{1 - \left(x\frac{E_p}{E_e}\right)}. \tag{10.26}
$$

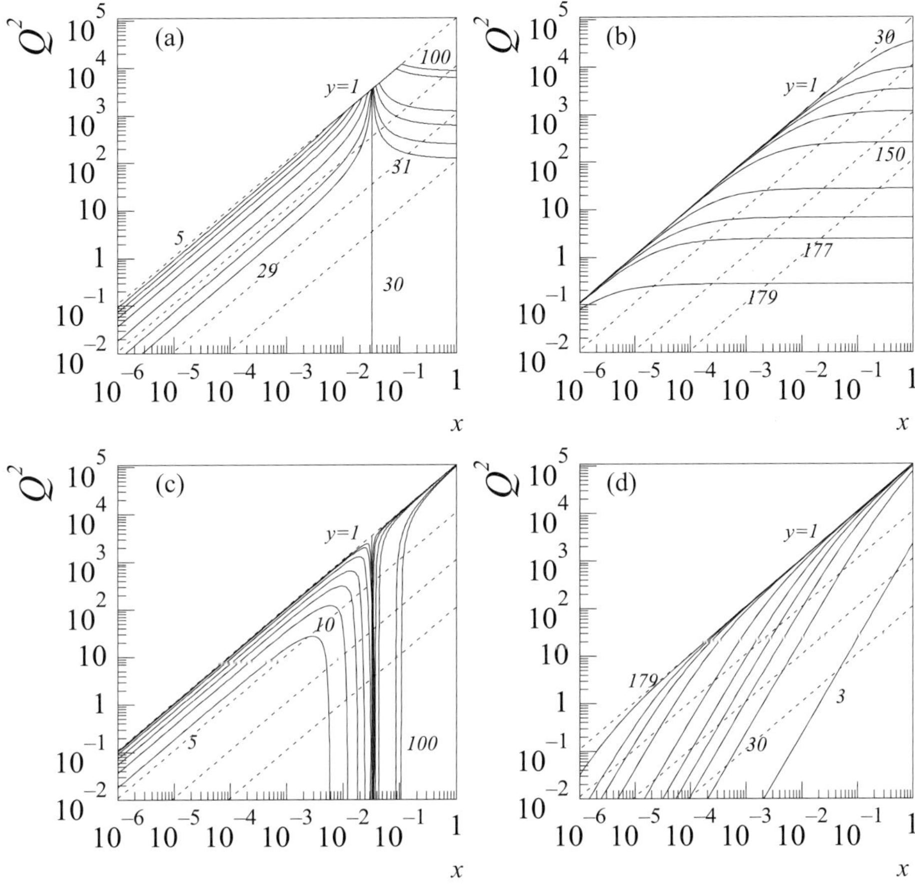

Fig. 10.9 Kinematic plane of Q^2 as a function of x for different fixed, measured variable configurations for the HERA collider configuration. The dashed lines represent lines of constant y values (1, 0.1, 0.01). The electron beam energy amounts to 27.5 GeV, whereas the proton beam energy is 920 GeV. (a) lines of constant electron energies, (b) electron scattering angles, (c) jet energies and (d) hadronic angles.

Small electron energies give rise to high y values. The kinematic limit $y = 1$ as shown in Fig. 10.9, is given by $Q^2 = sx$, i.e., the kinematic limit is determined by the center-of-mass energy. The low-Q^2 region can only be reached if one measures the final-state electron under very large angles θ'_e. For small Q^2, the energy of the scattered electron is limited to be below the electron beam energy for $x < E_e/E_P$, which is known as the kinematic-peak position $x = E_e/E_P$. In this region the lines of constant y values are essentially parallel to lines of constant energy of the scattered electron. Going from small values of the scattered electron energy towards the kinematic-peak point, the lines start to be spaced further apart from each other compared with the region of small electron energies, indicating that small changes in E'_e lead to large changes of the kinematic variables and therefore worsens the resolution.

The above expressions can be used to determine the dependencies of the kinematic variables on the measured quantities E'_e and θ'_e, viz., the relative errors of the kinematic variables are

$$\left(\frac{\delta x_e}{x_e}\right) = \left(\frac{1}{y_e}\right)\frac{\delta E'_e}{E'_e} \oplus \left[\frac{x_e}{E_e/E_P} - 1\right]\tan\left(\frac{\theta'_e}{2}\right)\delta\theta'_e \tag{10.27}$$

$$\left(\frac{\delta y_e}{y_e}\right) = \left(1 - \frac{1}{y_e}\right)\frac{\delta E'_e}{E'_e} \oplus \left(\frac{1}{y_e} - 1\right)\cot\left(\frac{\theta'_e}{2}\right)\delta\theta'_e \tag{10.28}$$

$$\left(\frac{\delta Q_e^2}{Q_e^2}\right) = \frac{\delta E'_e}{E'_e} \oplus \tan\left(\frac{\theta'_e}{2}\right)\delta\theta'_e\,. \tag{10.29}$$

The resolution in x_e and y_e diverges for $y_e \to 0$ as discussed above. The effect of $y_e \to 0$ is to enhance the relative error of the energy measurement. The electron method is therefore restricted to the region of high y values due to the singular structure of the resolution in x_e and y_e for $y_e \to 0$. The lower bound in y_e for which the resolution in x_e and y_e is still acceptable, strongly depends on the resolution with which the energy of the scattered electron can be measured.

The resolution in Q_e^2 diverges for $\theta'_e \to 180°$. The angular resolution at high values of the electron scattering angle θ'_e is the dominant contribution for the resolution in Q^2. And $\theta'_e \to 180°$ requires the capability of a particular detector to measure the energy and angle of the scattered electron at very large angles.

$p\bar{p}\,/\,pp$ Kinematics

This section will be devoted to a discussion of specific kinematic variables for a hadron–hadron collider. Both colliding hadrons with fixed initial conditions provide effectively two colliding partons, with a distribution of their longitudinal momenta quantified by their respective parton distribution functions. The partonic center-of-mass is different for every partonic collision and is boosted along the initial-state hadron beams. Relativistic kinematic variables have therefore been introduced which have well-defined and simple transformations under longitudinal boosts. Four-momenta can be formulated as

$$p^\mu = (E, p_x, p_y, p_z) = (m_T \cosh y, p_T \sin\phi, p_T \cos\phi, m_T \sinh y) \tag{10.30}$$

where the transverse mass is defined as $m_T = \sqrt{p_T^2 + m^2}$. Three key kinematic variables are introduced for a hadron collider, rapidity y, transverse momenta p_T, and azimuthal angle ϕ:

$$y = \frac{1}{2}\left(\frac{E + p_z}{E - p_z}\right)\,, \quad E_T = E\sin\theta\,. \tag{10.31}$$

The rapidity y reduces to the so-called pseudo-rapidity η in the limit of $m \to 0$ and is directly related to the polar angle θ: $\eta = -\ln\tan(\theta/2)$. The Bjorken-$x$ scaling variables are completely constrained by the pseudo-rapidity η and transverse energy E_T for a $2 \to 2$ process as shown in Fig. 10.10.

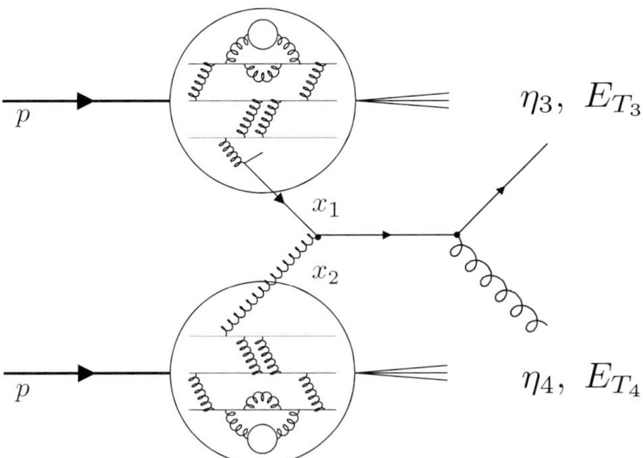

Fig. 10.10   Schematic of pp scattering proceeding through a leading-order $2 \rightarrow 2$ partonic process. The final-state is characterized by the transverse energy E_T and pseudo-rapidity η for each final-state jet.

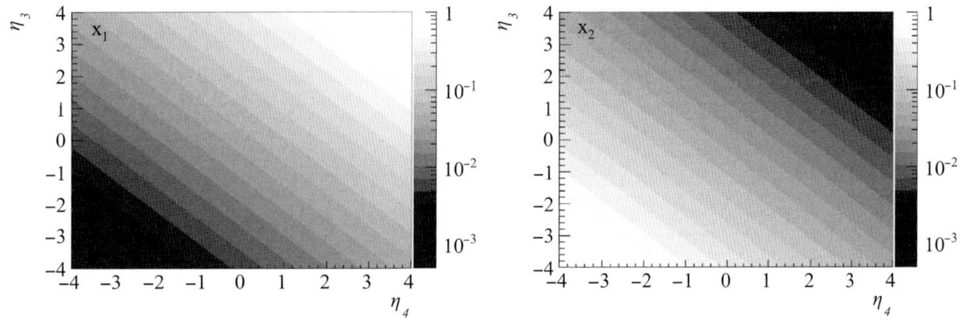

Fig. 10.11   Correlation between x_1 and x_2 as a function of η_3 and η_4 for a $2 \rightarrow 2$ partonic process of di-jet production for $\sqrt{s} = 500$ GeV and an invariant mass value of $M = 20$ GeV.

The momentum fractions x_1 and x_2 can be obtained as follows:

$$x_{1\,(2)} = \frac{1}{2} x_T \left(e^{+\,(-)\eta_3} + e^{+\,(-)\eta_4} \right), \tag{10.32}$$

where x_T is defined as $x_T = 2E_T/\sqrt{s}$. Observing two final-state particles or jets resulting from the fragmentation of partons in the central detector region, i.e. $\theta_{3,4} = 90°$ and thus $\eta_{3,4} = 0$, yields $x_1 = x_2 = x_T$, i.e., x_T quantifies the Bjorken x-values for two colliding partons, giving rise to a central production of final-state particles. Asymmetric partonic collisions, i.e., cases of $x_1 \gg x_2$ or $x_1 \ll x_2$ lead to forward/backward boosted final-state configurations such as the production of two final-state jets. The pseudo-rapidity sum provides a measure of the ratio of both Bjorken-x values via $\eta_3 + \eta_4 = \ln (x_1/x_2)$, whereas the di-jet invariant mass M provides a measure of the product of both Bjorken-x scaling variables: $M = \sqrt{s}\sqrt{x_1 x_2}$. The clear correlation between the pseudo-rapidity for two final-state jets characterized by $\eta_{3\,(4)}$ is shown in Fig. 10.11 as a grey shade for both Bjorken-x scaling variables $x_{1,\,(2)}$ for fixed values of M and $\sqrt{s}$.

The above relation for the invariant mass M can be readily obtained by using expressions for the initial-state parton four-momenta $\hat{p}_1^\mu$ and $\hat{p}_2^\mu$ in the laboratory frame:

$$\hat{p}_1^\mu = \frac{\sqrt{s}}{2}\begin{pmatrix} x_1 \\ 0 \\ 0 \\ x_1 \end{pmatrix} \quad \hat{p}_2^\mu = \frac{\sqrt{s}}{2}\begin{pmatrix} x_2 \\ 0 \\ 0 \\ -x_2 \end{pmatrix}. \tag{10.33}$$

The four-vector of the propagator term $q = \hat{p}_1 + \hat{p}_2$ for an s-channel exchange is given as

$$q^\mu = \frac{\sqrt{s}}{2}\begin{pmatrix} x_1 + x_2 \\ 0 \\ 0 \\ x_1 - x_2 \end{pmatrix}. \tag{10.34}$$

The invariant mass M is therefore obtained by evaluating the square root expression of $M = \sqrt{q^2} = \sqrt{s}\sqrt{x_1 x_2}$.

10.3　Spin Structure of the Nucleon: Polarized Proton Collider

Following the discussion of the spin structure of the nucleon as studied via lepton scattering in Chapter 9, here we describe current, successful efforts to gain further insight into the spin structure using the RHIC polarized proton collider, the world's first and only such accelerator. High energy polarized $\vec{p}\vec{p}$ collisions at a center-of-mass energy of $\sqrt{s} = 200\,\text{GeV}$ and at $\sqrt{s} = 500\,\text{GeV}$ at RHIC provide a unique way to probe the proton spin structure using very well established processes in high-energy physics, both experimentally and theoretically. The collision of polarized proton beams can be chosen to be either transverse or longitudinal, allowing for a rich physics program. The transverse polarization program allows one to study correlations between the spin of the proton and the spin and motion of partons inside the proton. The longitudinal polarization program offers a path to access the degree to which gluons and quarks/antiquarks contribute to the spin of the proton. These longitudinal studies will be discussed in more detail below.

Two important observables will be presented here which are equivalent to the cross section measurements in the unpolarized sector. A spin-dependent ratio is formed as the difference over the sum for both the longitudinal single-spin, A_L, and the longitudinal double-spin asymmetry, A_{LL}:

$$A_L = \frac{\sigma_+ - \sigma_-}{\sigma_+ + \sigma_-} \quad A_{LL} = \frac{\sigma_{++} - \sigma_{+-}}{\sigma_{++} + \sigma_{+-}}. \tag{10.35}$$

The '$+$' and '$-$' indices refer to the respective beam helicities of polarized hadron beams. Both asymmetries can be formulated using factorization among the underlying hard process involving W and Z production concerning the parity-violating asymmetry A_L and QCD processes concerning A_{LL}. Both measured asymmetries can be then used in a global analysis together with polarized DIS measurements to constrain polarized quark and gluon distribution functions.

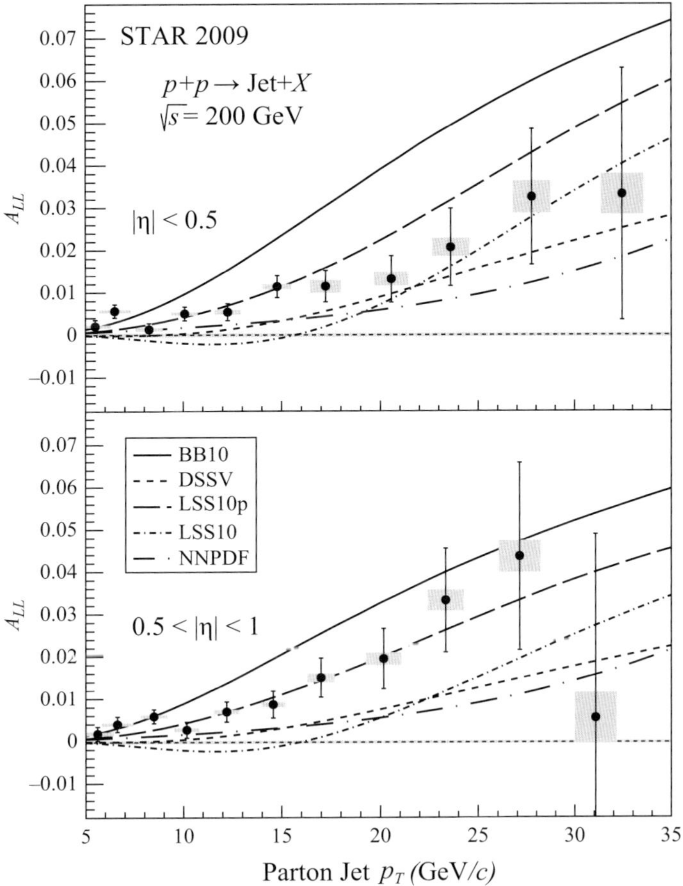

Fig. 10.12 Results of the measured longitudinal double-spin asymmetry from the STAR Collaboration adapted from [Ada15a].

Figure 10.12 displays results of longitudinal double-spin asymmetry measurements by the STAR collaboration at RHIC for inclusive jet production as a function of the respective transverse momentum. The measured asymmetries are significantly different from zero even at low transverse momentum where the underlying partonic process is dominated by gluon–gluon scattering suggesting a non–zero value of the actual gluon polarization. These data sets have been used in two independent global analyses to extract the polarized gluon distribution function.

The DSSV global analysis [deF14] provides evidence of a nonzero value of the gluon polarization $\int_{0.05}^{1} \Delta g\, dx\, (Q^2 = 10\,\text{GeV}^2) = 0.20^{+0.06}_{-0.07}$. An independent analysis by the NNPDF group [Noc14] suggests a value of $\int_{0.05}^{0.5} \Delta g\, dx\, (Q^2 = 10\,\text{GeV}^2) = 0.23^{+0.07}_{-0.07}$. The actual value in fact has a similar magnitude as the total quark polarization by itself discussed in the previous chapter.

The analysis of the 2009 data using for the first time collisions of polarized protons at $\sqrt{s} = 500$ GeV established a novel scheme to probe the quark flavor structure using W boson production. Data taken in 2012 suggest a large asymmetry of the spin contribution

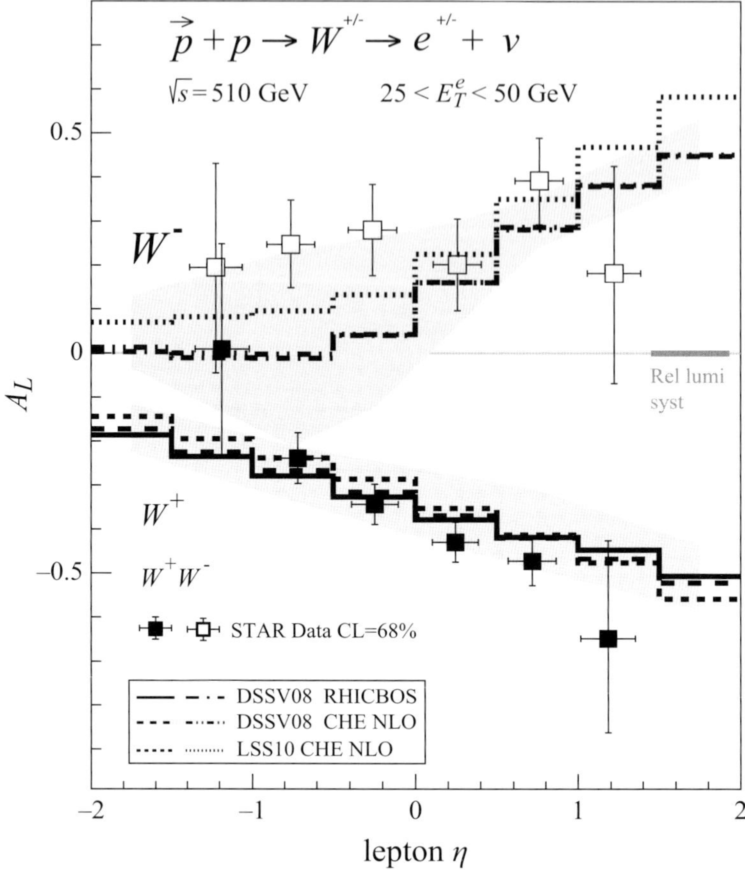

Fig. 10.13 Results of the measured longitudinal single-spin asymmetry from the STAR Collaboration adapted from [Ada14].

of $\bar{u}$ quarks and $\bar{d}$ quarks similar to the large and well-known difference between the momentum distributions of $\bar{u}$ quarks and $\bar{d}$ quarks. Figure 10.13 shows the measured longitudinal single-spin asymmetry from the STAR collaboration. The data for W^- tend to be positive, whereas the data for W^+ production are negative, which is directly related to the dominant role of negative d-quark and positive u-quark polarizations, respectively, and is consistent with the results previously obtained from lepton–nucleon semi-inclusive DIS, as discussed in Chapter 9.

10.4 Flavor Asymmetry of the Sea via the Drell–Yan Process

The Drell–Yan process [Dre70] is very similar in spirit to applying parton model ideas to deep–inelastic ep scattering and represents in fact the first successful quantitative way to account for high-energy hadron–hadron collisions. The principal partonic process is based on the s-channel of lepton-pair production $l^+ l^-$ in hadron–hadron collisions, i.e., $q + \bar{q} \rightarrow l^+ + l^-$ with an invariant mass of $M^2 = (p_{l^+} + p_{l^-})^2$.

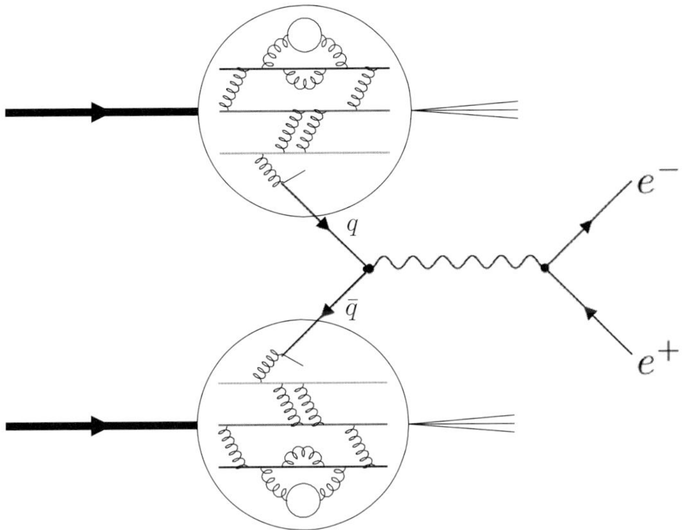

Fig. 10.14 Drell–Yan process in leading-order perturbation theory showing $q + \bar{q} \to e^+ + e^-$.

Figure 10.14 shows the basic Feynman diagram of lepton-pair production in hadron–hadron collisions. The Drell–Yan cross section can be evaluated for two colliding hadron species h_1 and h_2 as

$$\sigma_{h_1 h_2} = \sum_q \int dx_1 dx_2 f_q(x_1, M_2) f_{\bar{q}}(x_2, M^2) \hat{\sigma}\,(q + \bar{q} \to l^+ + l^-). \tag{10.36}$$

The partonic cross section at leading-order perturbation theory is given by

$$\hat{\sigma}\,(q(p_1) + \bar{q}(p_2) \to l^+ + l^-) = \frac{4\pi\alpha^2}{3\hat{s}} \frac{1}{N} Q_q^2 \tag{10.37}$$

with $\hat{s} = (p_1 + p_2)^2$ and $N = 3$.

The parton-model prediction for the double-differential cross section is readily obtained and can be written as

$$\frac{d^2\sigma}{dx_1 dx_2} = \frac{4\pi\alpha^2}{9sx_1 x_2} \left(\sum_q Q_q^2 \left[f_q(x_1) f_{\bar{q}}(x_2) + f_{\bar{q}}(x_1) f_q(x_2) \right] \right), \tag{10.38}$$

where x_1 and x_2 can be easily obtained from $x_1 = \sqrt{\tau} e^y$ and $x_2 = \sqrt{\tau} e^{-y}$ with $\tau = M^2/s$ and $y = (1/2)\ln(x_1/x_2)$. One defines the Feynman variable $x_F = x_1 - x_2$ to quantify forward ($x_F \gg 0$) and backward ($x_F \ll 0$) kinematic regions. The patron-model prediction is subject to QCD perturbative corrections similar in spirit to high-order corrections for DIS. Drell–Yan processes exhibit a similar scaling behavior to those of DIS as shown in Fig. 10.15. The basic Drell–Yan process is the starting point for W and Z boson production, which will be discussed separately.

The Gottfried sum is defined as

$$S_G \equiv \int_0^1 \frac{dx}{x} \left[F_2^p(x) - F_2^n(x) \right]. \tag{10.39}$$

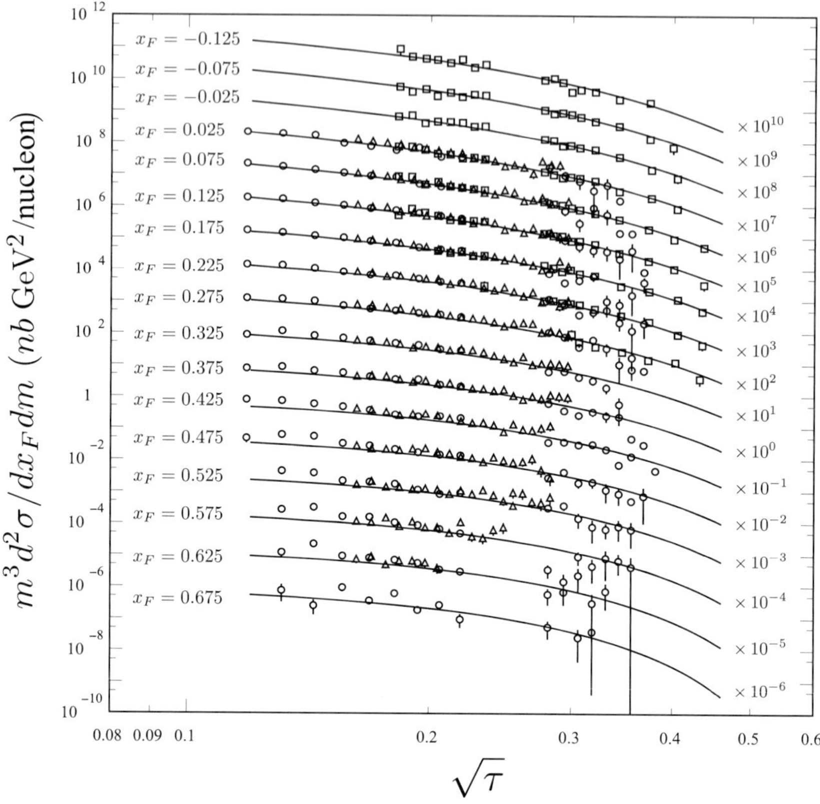

Fig. 10.15 Drell–Yan cross section $pA \rightarrow \mu^+\mu^- X$ as a function of $\sqrt{\tau}$ in comparison with NLO QCD calculations; figure adapted from [McG99].

For a flavor symmetric sea, i.e. $\bar{u}(x) = \bar{d}(x)$, we have the Gottfried sum rule $S_G = \frac{1}{3}$. However, measurements yield a significantly lower value of

$$S_G = 0.235 \pm 0.026 \tag{10.40}$$

at $Q^2 = 4$ (GeV/c)^2. Assuming that isospin symmetry is valid, a global flavor asymmetry $\int_0^1 [\bar{d}(x) - \bar{u}(x)]dx \approx 0.15$ would account for the data. Two different methods have been used to measure the x dependence: the Drell–Yan process and semi-inclusive DIS.

Experiments have been carried out on proton and deuteron targets in the kinematic regime where the Drell–Yan cross section is dominated by the annihilation of a beam quark with a target antiquark, i.e., where $x_1 > x_2$ and thus $x_F > 0$. Assuming charge symmetry and that the deuteron parton distributions can be expressed as the sum of the proton and neutron distributions, one obtains [Haw98] a simple approximate form of the Drell–Yan cross section ratio

$$\left.\frac{\sigma_{pd}^{DY}}{\sigma_{pp}^{DY}}\right|_{x_1 >> x_2} \approx \frac{1}{2}\frac{\left(1 + \frac{1}{4}\frac{d_1}{u_1}\right)}{\left(1 + \frac{1}{4}\frac{d_1}{u_1}\frac{\bar{d}_2}{\bar{u}_2}\right)}\left[1 + \frac{\bar{d}_2}{\bar{u}_2}\right]. \tag{10.41}$$

The subscripts 1 and 2 denote that the parton distributions in the proton are functions of x_1 and x_2, respectively. When $\bar{u} = \bar{d}$, the ratio is 1.

The flavor asymmetry of the light sea can also be studied using charged pion yields in SIDIS of leptons from the unpolarized proton and deuteron [Ack98a]. Consider the ratio of the differences between charged pion yields for proton and neutron targets:

$$r(x, z) = \frac{N_p^{\pi^-}(x, z) - N_n^{\pi^-}(x, z)}{N_p^{\pi^+}(x, z) - N_n^{\pi^+}(x, z)} ,$$ (10.42)

where $N^\pi(x, z)$ is the yield of pions from DIS off nucleons. We have

$$N^\pm(x, z) \propto \sum_f e_f^2 \left[q_f(x)D_f^\pm(z) + \bar{q}_f D_{\bar{f}}^\pm(z) \right] ,$$ (10.43)

from factorization. Assuming isospin symmetry as well as charge conjugation invariance, the number of light quark fragmentation functions is reduced to two: the favored and disfavored fragmentation functions. Then we can write

$$\frac{1 + r(x, z)}{1 - r(x, z)} = \frac{u(x) - d(x) + \bar{u}(x) - \bar{d}(x)}{u(x) - d(x) - \bar{u}(x) + \bar{d}(x)} J(z) ,$$ (10.44)

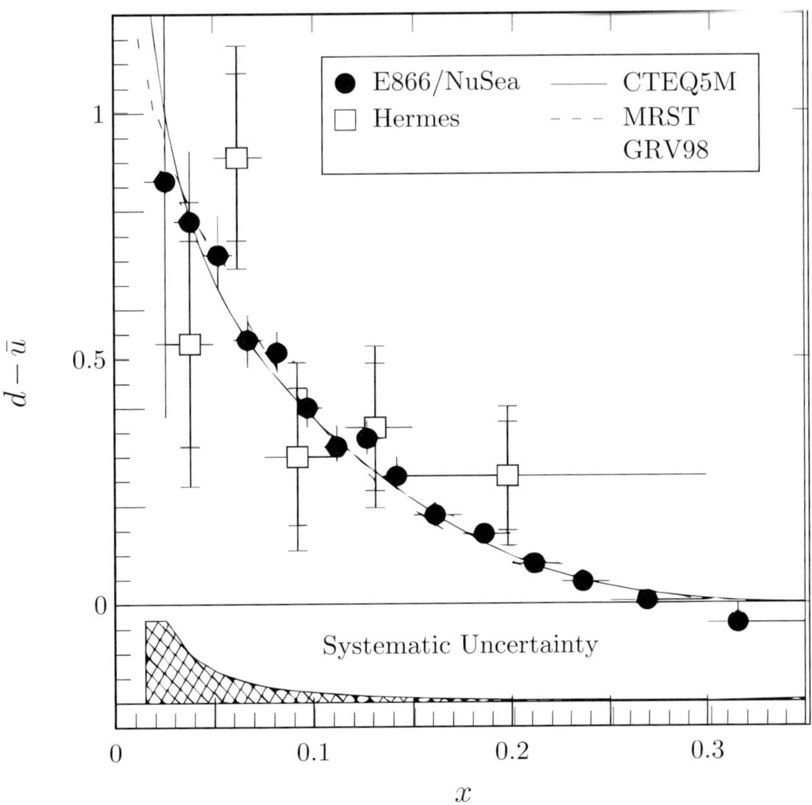

Fig. 10.16 Comparison of the E866 ($\bar{d} - \bar{u}$) results as a function of x with the parametrizations of various parton distribution functions from [Gar01]. The data are from E866 using the Drell–Yan process and from HERMES using SIDIS.

where $J(z) = \frac{3}{5}\left[\frac{\eta(z)+1}{\eta(z)-1}\right]$ and $\eta(z) \equiv \frac{D_u^{\pi^+}(z)}{D_u^{\pi^-}(z)}$. Thus, we can form the ratio

$$\frac{\bar{d}(x) - \bar{u}(x)}{u(x) - d(x)} = \frac{J(z)[1 - r(x,z)] - [1 + r(x,z)]}{J(z)[1 - r(x,z)] + [1 + r(x,z)]} . \tag{10.45}$$

Figure 10.16 shows measurements of the flavor asymmetry of the light sea $(\bar{d} - \bar{u})/(u - d)$ as a function of x using both Drell–Yan and SIDIS. Within uncertainties, the data are in reasonable agreement and show a significant flavor asymmetry in the light sea. Effects due to Pauli blocking and the pion cloud have been used to explain this asymmetry. The agreement between the two different experimental techniques validates the parton model, factorization, and the treatment of fragmentation.

10.5 Low-x Physics

In the preceding chapter, the density of sea quarks and gluons has been shown to rise dramatically at small values of x. Since $Q^2 \approx s \cdot x \cdot y$, low-$x$ is reached experimentally by going to high CM energy. For example, at HERA, experiments attained $x \sim 10^{-4}$ on the proton. The structure function F_2 and DIS cross section rise dramatically at low x values or large invariant mass W, respectively. This rise cannot continue forever, since fundamental constraints such as unitarity would be violated. The rise must eventually be tamed by an interplay between the creation and annihilation of partons. This aspect is deeply rooted in QCD, giving rise to a nonlinear behavior in contrast to electroweak processes. Electron–nucleus collisions carried out at high energies are expected to amplify the effect by a factor of $A^{1/3}$, allowing even smaller x values, $x/A^{1/3}$, to be reached.

Phenomenological models based on color-dipole interactions have been developed to describe ep scattering at low x values. Here the ep cross section is described by the interaction of a photon splitting into a $q\bar{q}$ pair and interacting with the proton through a two-gluon-exchange mechanism. While the splitting into a $q\bar{q}$ pair is a QED process, the formulation in terms of a dipole cross section relies on a phenomenological model. Several formulations have been developed that provide a successful description of the low x behavior of ep scattering.

To provide a description of a high-density partonic system, in particular that arising in the collisions of relativistic heavy ions, the idea of a color glass condensate (CGC) has been introduced [Gel10]. This condensate refers to an extreme condition in colliding relativistic heavy-ion beams which are Lorentz contracted along the direction of motion. The density of gluons in such ion beams is very large and leads to the term condensate. Gluons appear disordered in the Lorentz contracted ions: thus the term "glass." A CGC is postulated as a new universal form of high-density matter, which could be most directly observed in high-energy electron–nucleus interactions. The transition from a lower-density regime to a CGC is characterized by a scale, known as the *saturation scale.*

The DGLAP evolution equations, introduced in Chapter 9, allow one only to determine the Q^2 dependence of the parton densities, but not their x dependence. The BFKL evolution equation provides an evolution in x for fixed values of Q^2 [Bal78]. Whereas in the DLLA approximation only leading terms $\ln \frac{1}{x}$, which are accompanied by leading terms $\ln Q^2$, are summed, the BFKL (Balitsky–Fadin–Kuraev–Lipatov) evolution scheme provides a way to sum up all leading terms $\ln \frac{1}{x}$. With $xg(x, Q^2) = \int_0^{Q^2} (dk_T^2/k_T^2) f_g(x, k_T^2)$, the BFKL equation is [Ask94a]:

$$-x \frac{\partial f_g(x, k_T^2)}{\partial x} = 3 \frac{\alpha_s}{\pi} k_T^2 \int_0^\infty \frac{dk_T'^2}{k_T'^2} \left[\frac{f_g(x, k_T'^2) - f_g(x, k_T^2)}{|k_T'^2 - k_T^2|} + \frac{f_g(x, k_T^2)}{\sqrt{4k_T'^4 + k_T^4}} \right]$$

$$\equiv K \otimes f_g, \tag{10.46}$$

where K is the BFKL kernel. The BFKL equation only refers to the gluon distribution. It provides an evolution in x at fixed Q^2 with a certain starting distribution at x_0. The solution of the BFKL equation is dominated by the largest eigenvalue λ of the kernel K, which leads to a characteristic Q^2 and x dependence for the structure function $F_2(x, Q^2)$ [Ask94b]:

$$F_2(x, Q^2) \sim (Q^2)^{1/2} x^{-\lambda}, \qquad \lambda = \frac{3\alpha_s}{\pi} 4 \ln 2 . \tag{10.47}$$

The range of applicability is the region

$$\alpha_s \ln (Q^2) \ll 1, \qquad \alpha_s \ln \frac{1}{x} = \mathcal{O}(1) . \tag{10.48}$$

The BFKL power law behavior leads to a violation of unitarity in the limit $x \to 0$. The rise of F_2 through the rise of the gluon distribution at small values of x and therefore in $\sigma_{tot}^{\gamma^* p}$ is limited by the Froissart bound [Fro61]:

$$\sigma_{tot}^{\gamma^* p} \leq \frac{\pi}{m_\pi^2} \left(\ln \frac{s}{s_0} \right)^2 , \tag{10.49}$$

where m_π is the mass of the pion and s_0 is an unknown scale factor.

It is therefore expected that the rise of F_2 with decreasing x is limited. At small values in x, the density of quarks and gluons increases drastically, and quarks and gluons start to overlap. Two competing processes, the annihilation and recombination of gluons, at high parton densities will eventually limit the increase in the number of quarks and gluons. Recombination effects become important at $xg(x, Q^2) \sim r_p^2/r_g^2$, where $r_p \sim 1$ fm ~ 5 GeV^{-1} and $r_g \sim 2/Q$, yielding $xg(x, Q^2) \approx 6Q^2$, which is well outside the scope of HERA and could be only reached at much smaller values in x [Lev97].

The inclusive measurement of F_2 at HERA has shown that the evolution of structure functions through the DGLAP evolution equations are in good agreement with experimental results, although thus far, it has not been possible to discriminate between the DGLAP evolution and the BFKL equation. It has been suggested that differences in the DGLAP and the BFKL evolution equations could show up in exclusive measurements, such as in high p_t forward jets [Mue90] and in forward-energy flow [Kwi94].

Attempts have been made to achieve a unified BFKL/DGLAP description [Kwi97]. The domain of the various evolution equations is illustrated in Fig. 10.17 along with the region of a set of evolution equations (JIMWLK/BK) reaching a saturated region. Reaching such a

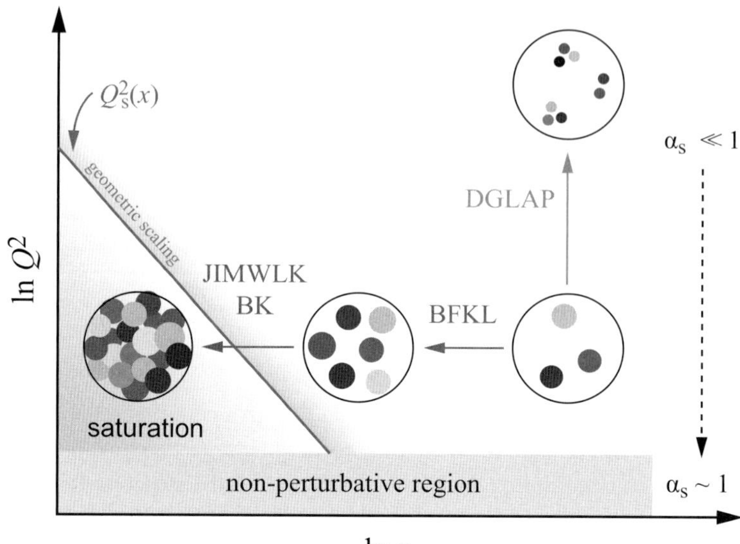

Fig. 10.17 Range of validity for various evolution equations. Increasing Q^2 leads to a better spatial resolution (DGLAP). Smaller values in x yield an increase in the parton density (BFKL). The region of high parton density is the region where saturation effects are expected to diminish the rise of F_2 with decreasing x.

saturated region is of active exploration in proton–nucleus collisions and nucleus–nucleus collisions at LHC and RHIC and forms an important component of a future EIC program.

10.6　Jets, Bosons, and Top Quarks

Jet production

The inclusive production of hadrons in e^+e^- collisions produces colored partonic objects in the final state, either two or more due to higher-order processes. The formation of collimated particles around the original direction of each final-state parton is referred to as a jet. Each particle is then detected in a detector system leaving a splash of energy in a calorimeter system or traces of tracks for charged final-state particles in a tracking system. The study of jets is essential for any modern high-energy collider experiment. Jet studies have had enormous value for the understanding of QCD. Indeed, the discovery of gluons is largely attributed to the study of three-jet events at the PETRA e^+e^- storage ring at DESY. Figure 10.18 shows a three-jet event from the TASSO experiment at PETRA, which was instrumental for the discovery of the gluon. The collimated stream of tracks can be clearly seen.

The group of objects, be it partons, particles, or segmented detector related measurements requires a well-defined algorithm to group these objects together. A discussion of

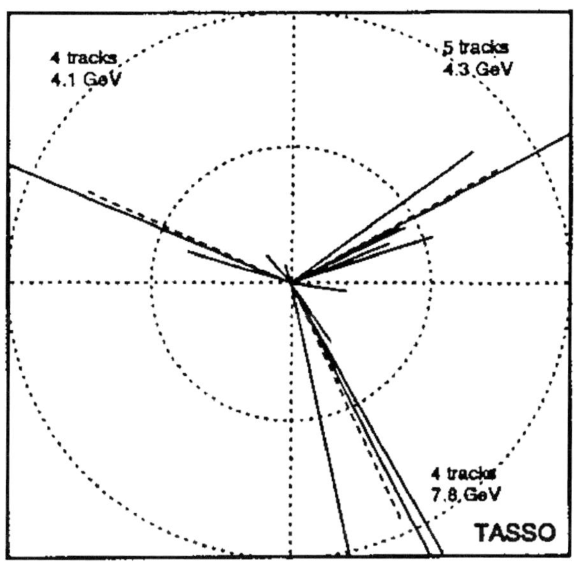

Fig. 10.18   Reproduction of the original three-jet event obtained by the TASSO collaboration [Ali10].

well-known jet reconstruction algorithms such as the cone algorithm and anti-k_T algorithm can be found in various detailed reports [Cac08, Ell93].

The following discussion will focus on $2 \to 2$ leading-order partonic processes:

$$\text{parton}_i(p_1) + \text{parton}_j(p_2) \to \text{parton}_k(p_3) + \text{parton}_l(p_4) . \tag{10.50}$$

which can be anayzed via

$$\frac{E_3 E_4 d^6 \hat{\sigma}}{d^3(p_3) d^3(p_4)} = \frac{1}{2\hat{s}} \frac{1}{16\pi^2} \overline{\sum |\mathcal{M}|^2} \delta^4(p_1 + p_2 - p_3 - p_4) . \tag{10.51}$$

It is assumed that all initial-state configurations are averaged and all final-state configurations are added. Table 10.2 shows, for all leading-order $2 \to 2$ processes, the respective matrix elements squared. All expressions are provided using Mandelstam variables, i.e., $\hat{s} = (p_1 + p_2)^2$, $\hat{t} = (p_1 - p_3)^2$ and $\hat{u} = (p_2 - p_3)^2$. Taking into account parton distribution functions for both initial-state partons i and j, the $2 \to 2$ jet cross section can be formulated as

$$\frac{d^3\sigma}{dy_3 dy_4 dp_T^2} = \frac{1}{16\pi s^2} \sum_{i,j,k,l=q,\bar{q},g} \frac{f_i(x_1, \mu^2) f_j(x_2, \mu^2)}{x_1} \frac{}{x_2} \times \overline{\sum |\mathcal{M}|^2} \frac{1}{1 + \delta_{kl}} . \tag{10.52}$$

The momentum fractions x_1 and x_2 are calculated using the basic four-vector relations:

$$x_1 = \frac{1}{2} x_T \left(e^{y_3} + e^{y_4} \right) \quad x_2 = \frac{1}{2} x_T \left(e^{-y_3} + e^{-y_4} \right) , \tag{10.53}$$

with $x_T = 2p_T \sqrt{s}$. Figure 10.19 shows a compilation of several inclusive jet cross section measurements at different center-of-mass energies in comparison with NLO QCD calculations.

Table 10.2 Leading-order $2 \to 2$ QCD processes with $g^2 = 4\pi\alpha_s$

Process	$\bar{\Sigma}|M|^2/g^4$
$qq' \to qq'$	$\dfrac{4}{9}\dfrac{\hat{s}^2 + \hat{u}^2}{\hat{t}^2}$
$q\bar{q}' \to q\bar{q}'$	$\dfrac{4}{9}\dfrac{\hat{s}^2 + \hat{u}^2}{\hat{t}^2}$
$qq \to qq$	$\dfrac{4}{9}\left(\dfrac{\hat{s}^2 + \hat{u}^2}{\hat{t}^2} + \dfrac{\hat{s}^2 + \hat{t}^2}{\hat{u}^2}\right) - \dfrac{8}{27}\dfrac{\hat{s}^2}{\hat{u}\hat{t}}$
$q\bar{q} \to q'\bar{q}'$	$\dfrac{4}{9}\dfrac{\hat{t}^2 + \hat{u}^2}{\hat{s}^2}$
$q\bar{q} \to q\bar{q}$	$\dfrac{4}{9}\left(\dfrac{\hat{s}^2 + \hat{u}^2}{\hat{t}^2} + \dfrac{\hat{t}^2 + \hat{u}^2}{\hat{s}^2}\right) - \dfrac{8}{27}\dfrac{\hat{u}^2}{\hat{s}\hat{t}}$
$q\bar{q} \to gg$	$\dfrac{32}{27}\dfrac{\hat{t}^2 + \hat{u}^2}{\hat{t}\hat{u}} - \dfrac{8}{3}\dfrac{\hat{t}^2 + \hat{u}^2}{\hat{s}^2}$
$gg \to q\bar{q}$	$\dfrac{1}{6}\dfrac{\hat{t}^2 + \hat{u}^2}{\hat{t}\hat{u}} - \dfrac{3}{8}\dfrac{\hat{t}^2 + \hat{u}^2}{\hat{s}^2}$
$gq \to gq$	$-\dfrac{4}{9}\dfrac{\hat{s}^2 + \hat{u}^2}{\hat{s}\hat{u}} + \dfrac{\hat{u}^2 + \hat{s}^2}{\hat{t}^2}$
$gg \to gg$	$\dfrac{9}{2}\left(3 - \dfrac{\hat{t}\hat{u}}{\hat{s}^2} - \dfrac{\hat{s}\hat{u}}{\hat{t}^2} - \dfrac{\hat{s}\hat{t}}{\hat{u}^2}\right)$

Table 10.3 Leading-order $2 \to 2$ prompt photon processes with $g^2 = 4\pi\alpha_s$ and $N = 3$ for $SU(3)$

Process	$\bar{\Sigma}|M|^2/g^2 e_q^2$
$q\bar{q}' \to \gamma^*, g$	$\dfrac{N^2 - 1}{N^2}\dfrac{t^2 + u^2 + 2s(s + t + u)}{tu}$
$gq \to \gamma^* q$	$-\dfrac{1}{N}\dfrac{s^2 + u^2 + 2t(s + t + u)}{su}$

Prompt photon production

Prompt or direct photon production is generally closely related to $2 \to 2$ partonic scattering, while the cross section for photon-initiated processes is substantially smaller. Prompt photon production has therefore not received as much attention as has jet production for precision QCD studies. At leading-order, prompt photon production is dominated by two subprocesses, annihilation ($q\bar{q} \to \gamma^* g$) and QCD Compton scattering ($gq \to \gamma^* q$). Table 10.3 shows the average matrix elements for leading-order Feynman prompt photon production. The QCD Compton process dominates over annihilation in proton–proton collisions, which has therefore received consideration for constraining the gluon distribution function, both in the unpolarized and polarized collider mode. Figure 10.20 shows a compilation of several prompt photon production results.

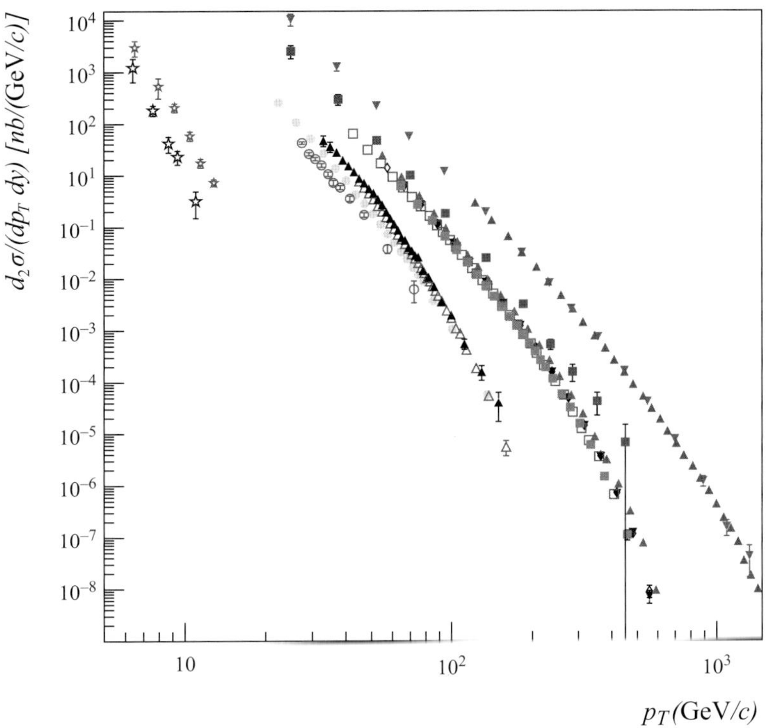

Fig. 10.19 Compilation of inclusive jet cross section measurements; figure adapted from [PDG14].

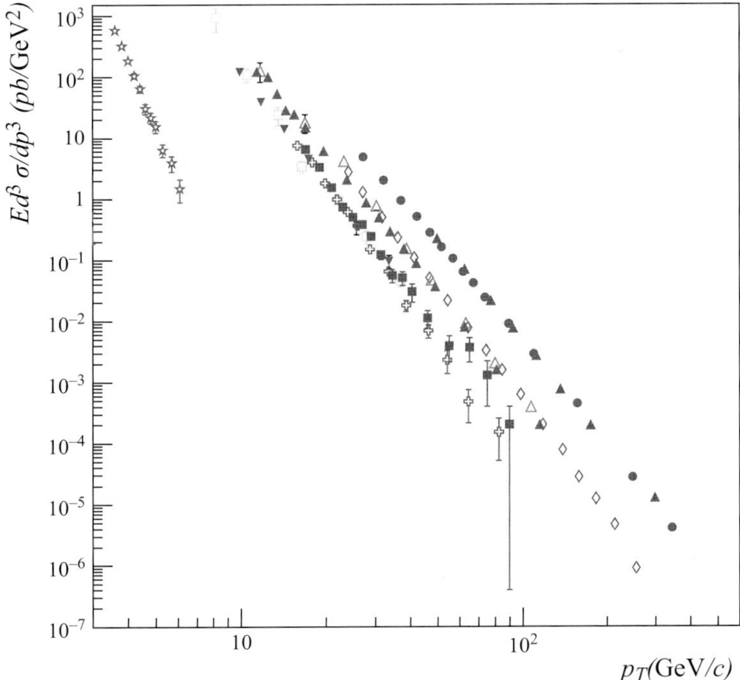

Fig. 10.20 Compilation of prompt photon cross section measurements; figure adapted from [PDG14].

Table 10.4 Event shape variables

Quantity	T	S	C
Pencil-like event	1	0	0
Spherical event	$\frac{1}{2}$	1	1

Event shape variables

Two types of topological studies have been introduced to characterize the shape of QCD processes, known as event shape variables and jet profile variables. Event shape variables enable the characterization of the shape or topology of QCD processes. These variables – thrust T, sphericity S, and the parameter C – are defined as

$$T = \max_{\mathbf{n}} \frac{\sum_i |\mathbf{p_i} \cdot \mathbf{n}|}{\sum_i |\mathbf{p_i}|} \tag{10.54}$$

$$S = \left(\frac{4}{\pi}\right)^2 \min_{\mathbf{n}} \left(\frac{\sum_i |\mathbf{p_i} \times \mathbf{n}|}{\sum_i |\mathbf{p_i}|}\right)^2 \tag{10.55}$$

$$C = \frac{3}{2} \frac{\sum_{i,j}[|\mathbf{p_i}||\mathbf{p_j}| - (\mathbf{p_i} \cdot \mathbf{p_j})^2/|\mathbf{p_i}||\mathbf{p_j}|]}{(\sum_i |\mathbf{p_i}|)^2} . \tag{10.56}$$

The vectors $\mathbf{p_i}$ are defined final-state hadron (parton) momenta and $\mathbf{n}$ is an arbitrary unit vector. The shape variables have characteristic values for pencil-like and spherical events as listed in Table 10.4. Event shape analyses thus characterize the topology of QCD type events.

The shape of jets in QCD-inspired processes, irrespective of the underlying jet finding algorithm, are characterized by the jet profile $\psi(r, R, E_T)$. This variable is defined as the ratio of the E_T sum among all final-state objects such as particles or detector objects up to a radius r divided by the E_T sum among all final-state objects such as particles or detector objects up to a radius R as a function of r.

Vector boson production and decay

The discovery of the heavy vector gauge bosons $W^\pm$ and Z^0 in $p\bar{p}$ collisions at the SPS at CERN marked the beginning of the fundamental study of the electroweak theory using high-energy colliders. These studies continued with high precision at LEP at CERN and the SLC at SLAC in e^+e^- collisions and are now being pursued as well-established probes at the LHC in unpolarized pp collisions and at RHIC in polarized pp collisions. The $W^\pm$ and Z^0 bosons have also been studied at HERA in ep collisions, opening a new way to probe the electroweak structure of the proton. Figure 10.21 shows a compilation of measured production cross sections for W and Z bosons as a function of $\sqrt{s}$ in $p + p$ and $p + \bar{p}$ collisions in comparison with higher-order SM calculations.

The leading-order Feynman diagram in hadron–hadron collisions is illustrated in Fig. 10.6 (b) for $W^\pm$ production, which clearly shows the similarity to the Drell–Yan

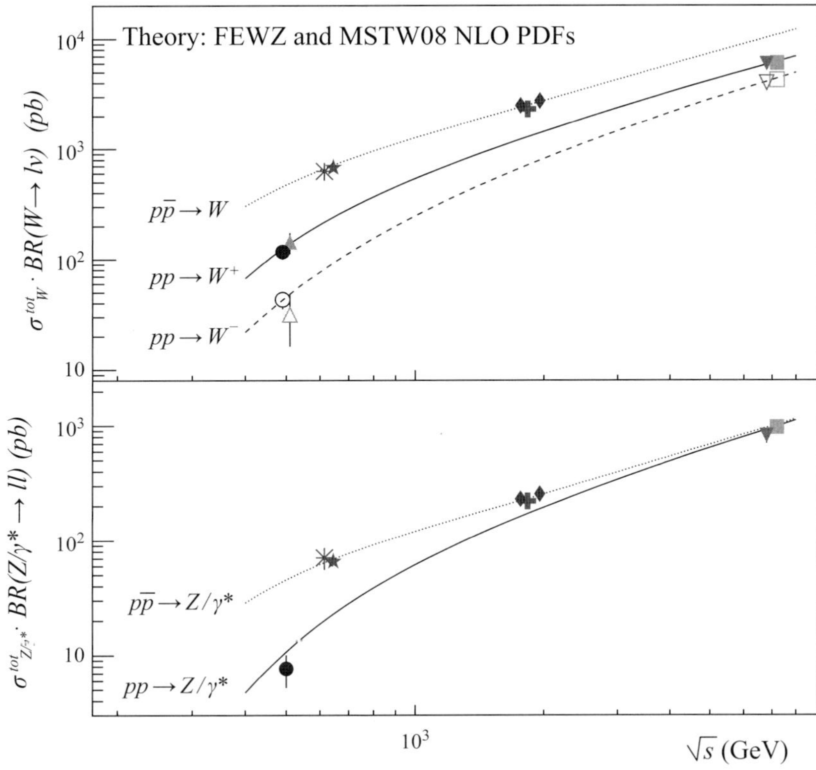

Fig. 10.21 Production cross sections for W and Z bosons as functions of $\sqrt{s}$ in $p + p$ and $p + \bar{p}$ collisions in comparison with higher-order SM calculations adapted from [Ada12].

process discussed earlier. The Z^0 production mode in hadron–hadron collisions is similar (Fig. 10.6(b)). The production of $W^{\pm}$ in e^+e^- collisions proceeds only in pair production mode, while the production of Z^0 bosons proceeds through an s-channel mode. The partial width for W and Z^0 production are given by

$$\Gamma(W \to l_i \bar{\nu}_i) = \frac{\sqrt{2} G_F m_W^3}{12\pi} \tag{10.57}$$

$$\Gamma(W \to q_i \bar{q}_j) = 3 \frac{\sqrt{2} G_F |V_{ij}|^2 m_W^3}{12\pi} \tag{10.58}$$

$$\Gamma(Z \to f\bar{f}) = N_c \frac{\sqrt{2} G_F m_Z^3}{6\pi} \times \left[(T_3 - Q_f \sin^2 \theta_w)^2 + (Q_f \sin^2 \theta_w)^2 \right]. \tag{10.59}$$

where the CKM matrix elements are denoted as V_{ij} (see Chapter 4) while N_c is 3 for the quark final state and 1 for leptonic final states. The corresponding production cross sections are

$$\frac{d\sigma}{d\Omega} = \frac{N_c^f}{N_c^i} \frac{1}{256\pi^2 s} \frac{s^2}{(s - M^2)^2 + s\Gamma^2}$$
$$\times [(L^2 + R^2)(L'^2 + R'^2)(1 + \cos^2 \theta) + (L^2 - R^2)(L'^2 - R'^2) 2 \cos \theta].$$

Table 10.5 Leading-order heavy quark production cross sections with $g^2 = 4\pi\alpha_s$

Process	$\bar{\Sigma}|M|^2/g^4$
$q\bar{q} \to Q\bar{Q}$	$\dfrac{4}{9}\left(\tau_1^2 + \tau_2^2 + \dfrac{\rho}{2}\right)$
$gg \to Q\bar{Q}$	$\left(\dfrac{1}{6\tau_1\tau_2} - \dfrac{3}{8}\right)\left(\tau_1^2 + \tau_2^2 + \rho - \dfrac{\rho^2}{4\tau_1\tau_2}\right)$

Here, M refers to the W or Z mass. The couplings are $L = (8G_F m_W^2/\sqrt{2})$ and $R = 0$ for W bosons similarly for L' and R'. For Z bosons the couplings are $L = \sqrt{8G_F m_Z^2/\sqrt{2}}\,(T_3 - \sin^2\theta_W Q)$ and $R = -\sqrt{8G_F m_Z^2/\sqrt{2}}\sin^2\theta_W Q$.

A characteristic signature for the presence of W boson production in hadron–hadron collisions is the occurrence of a peak in the transverse momentum distribution of decay leptons, which can be easily understood by considering the production of W bosons at rest. Using basic four-momentum conservation leads for the transverse momentum of the final-state lepton

$$p_T = \frac{M_W}{2}\sin\theta . \tag{10.60}$$

The transverse momentum distribution can be obtained as

$$\frac{d\sigma}{dp_T} = \frac{d\sigma}{d\cos\theta}\frac{d\cos\theta}{dp_T} = \frac{2p_T}{M_W}\frac{1}{\sqrt{(M_W/2)^2 - p_T^2}} , \tag{10.61}$$

which clearly indicates the occurrence of a peak at half of the W boson mass, known as the *Jacobian peak*. This is in striking contrast to the transverse momentum distribution for QCD initiated processes as shown in Fig. 10.19.

Heavy Quark Production and Decay

The production of heavy quarks including charm, bottom and top quarks played a critical role in the development of the Standard Model of particle physics. Various collider mode configurations have been used for these studies. The leading-order processes for the production of heavy quarks Q at mass m in a hadron–hadron collider mode configuration are a) $q(p_1) + \bar{q}(p_2) \to Q(p_3) + \bar{Q}(p_4)$ and b) $g(p_1) + g(p_2) \to Q(p_3) + \bar{Q}(p_4)$. The leading-order matrix element square expressions are provided in Table 10.5.

The scalar ratios τ_1, τ_2 and ρ, as shown in Table 10.5, are given in terms of $p_1, p_2, p_3,$ and p_4 by

$$\tau_1 = \frac{2p_1 p_3}{\hat{s}} \quad \tau_2 = \frac{2p_2 p_3}{\hat{s}} \quad \rho = \frac{4m^2}{\hat{s}} \tag{10.62}$$

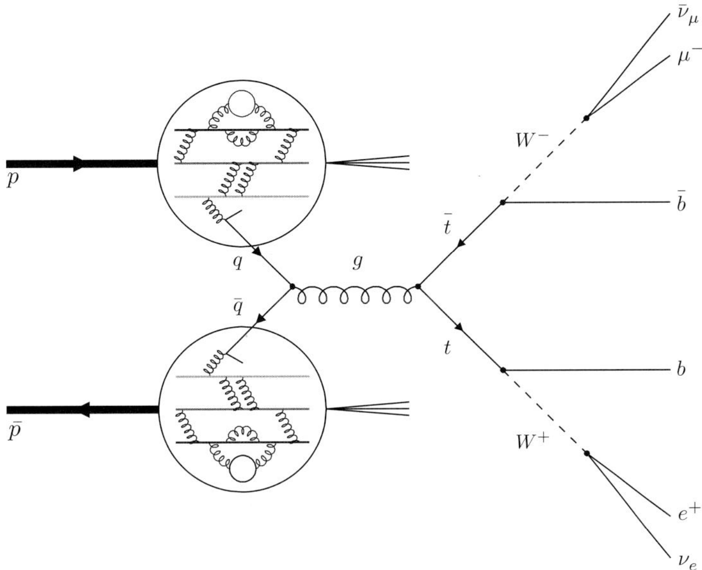

Fig. 10.22 Feynman diagram indicating top quark production.

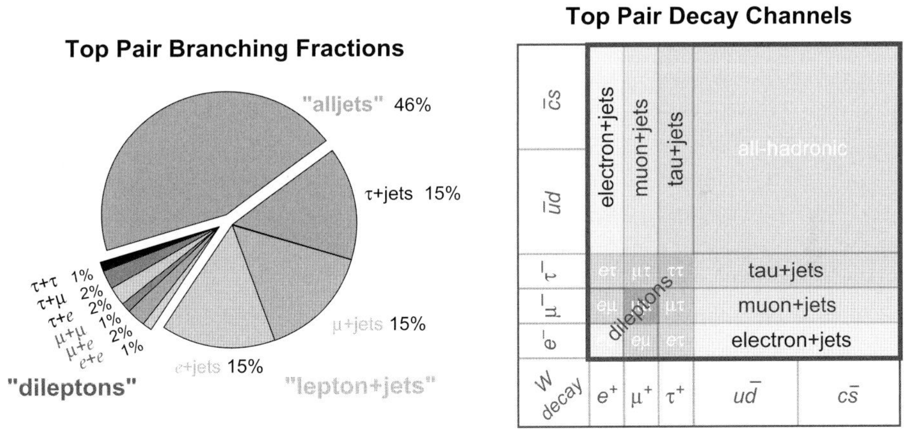

Fig. 10.23 Top quark branching ratios and decay channels.

with $\hat{s} = (p_1 + p_2)^2$ and the heavy quark mass denoted as m. The invariant cross section for heavy quark production can be written as employing factorization in $p + p$ collisions:

$$\frac{d\sigma}{dy_3 dy_4 d_T^2} = \frac{1}{16\pi^2 \hat{s}^2} \sum_{ij} x_1 f_i(x_1, \mu^2)\, x_2 f_j(x_2, \mu^2) \overline{\sum} |\mathcal{M}_{ij}|^2 . \tag{10.63}$$

The top quark was discovered at the Tevatron at Fermilab in $p\bar{p}$ collisions via the reaction displayed in Fig. 10.22. The corresponding branching fractions and decay modes are shown in Fig. 10.23.

10.7 The Path Forward

This chapter completes the first part of the book, wherein the Standard Model has been constructed and has been applied to understanding hadron structure in terms of quantum chromodynamics. This program has been very successful in that the SM has provided a rigorous theoretical framework of QCD, which has thus far passed all experimental tests. However, unique challenges remain, and our understanding of QCD is far from complete. The fundamental SM particles of QCD, the quarks and gluons, are not directly detectable. The hadronization process that connects these fundamental degrees of freedom to the experimental world of hadronic physics is not well understood. Further, important strong-interaction phenomena are emergent, i.e. they cannot be guessed from the QCD Lagrangian. An important example is confinement, discussed in Chapter 5. Reaching a fundamental and trenchant understanding of confinement has been identified as one of the top ten unsolved problems in physics.

The lack of completeness in our understanding is obvious from the discussion in this chapter. The questions posed at the beginning have been addressed with ingenious experiments, producing limited data that pose even more questions. It is accepted that a *full* understanding of hadron structure in terms of QCD will require a new electron–ion collider [Abh05], having direct access to the sea-quarks and gluons with a precision that far exceeds the current capabilities. Such a dedicated next-generation facility for the study of high-energy QCD should have the following characteristics:

- Highly polarized ($\sim$70%) electron and nucleon beams
- Ion beams from the deuteron to the heaviest nuclei
- Variable center-of-mass energies from $\sim$20 to $\sim$150 GeV
- High collision luminosity $\sim 10^{33-34}\,\mathrm{cm}^{-2}\,\mathrm{s}^{-1}$.

Such an accelerator is under consideration in China, in the US at both BNL and Jefferson Laboratory, and at CERN.

Progress in understanding QCD will also require advances in the ability to carry out *ab initio* lattice QCD calculations, requiring the development of more powerful computers and algorithms. We will see in Chapters 11 and 12 that such lattice calculations are now being successfully applied to few-body nuclei. The combination of a high luminosity electron–ion collider *and* significant advances in lattice QCD can usher in an era of precision QCD studies. The ultimate aim is to approach an understanding of the strong interaction comparable to our present precision understanding of the electroweak force, and a succesful program would enable satisfactory and complete answers to the basic questions posed at the beginning of this chapter. Finally, we would expect to realize a visualization of the subatomic world that would complement the stunning visualizations created of the large scale structure of the universe from satellite observations.

We now turn to the challenge of building a detailed understanding of the structure and properties of nuclei using hadronic degrees of freedom. The primary nuclear building blocks are the proton and neutron so we begin with a discussion of the *NN* interaction.

Exercises

10.1 Relativistic Collider Kinematics

Consider some basic quantities used to analyze high energy collider experiments.

a) Determine the electron beam energy in a fixed target ep mode configuration that would be required to reach a center-of-mass energy of $332\,\text{GeV}$?

b) How does this electron beam energy compare to the electron beam energy in a ep collider mode configuration for a proton beam energy of $920\,\text{GeV}$ to reach a center-of-mass energy of $332\,\text{GeV}$?

c) Show that for a ep collider such as HERA we have $s = (p + k)^2 \simeq 4E_e E_p$.

d) Defining x, y variables via

$$x = \frac{Q^2}{2p \cdot q}, \quad y = \frac{p \cdot q}{p \cdot k}, \quad 0 \le x, y \le 1,$$

show that we have $Q^2 \simeq sxy$, where we have defined $Q^2 = -q^2 = -(k_e - k'_e)^2$.

e) Derive the relations used in Eqs, 10.22–10.24

$$x_e = \frac{Q_e^2}{sy_e} = \frac{E'_e \cos^2 \frac{\theta'_e}{2}}{E_p \left(1 - \frac{E'_e}{E_e} \sin^2 \frac{\theta'_e}{2}\right)}$$

$$y_e = 1 - \frac{E'_e}{2E_e}(1 - \cos\theta'_e) = 1 - \frac{E'_e}{E_e} \sin^2 \frac{\theta'_e}{2}$$

$$Q_e^2 = 2E_e E'_e(1 + \cos\theta'_e) = 4E_e E'_e \cos^2 \frac{\theta'_e}{2} = \frac{p_{T,e}^2}{1 - y_e}$$

and

$$Q_e^2(x, \theta'_e) = \frac{sx}{1 + \left(x\frac{E_p}{E_e}\right) \tan^2\left(\frac{\theta'_e}{2}\right)}$$

$$Q_e^2(x, E'_e) = \frac{sx\left(1 - \frac{E'_e}{E_e}\right)}{1 - \left(x\frac{E_p}{E_e}\right)} \ .$$

f) Demonstrate that the relative errors of x_e, y_e, Q_e can be written as

$$\left(\frac{\delta x_e}{x_e}\right) = \left(\frac{1}{y_e}\right) \frac{\delta E'_e}{E'_e} \oplus \left[\frac{x_e}{E_e/E_P} - 1\right] \tan\left(\frac{\theta'_e}{2}\right) \delta\theta'_e$$

$$\left(\frac{\delta y_e}{y_e}\right) = \left(1 - \frac{1}{y_e}\right) \frac{\delta E'_e}{E'_e} \oplus \left(\frac{1}{y_e} - 1\right) \cot\left(\frac{\theta'_e}{2}\right) \delta\theta'_e$$

$$\left(\frac{\delta Q_e^2}{Q_e^2}\right) = \frac{\delta E'_e}{E'_e} \oplus \tan\left(\frac{\theta'_e}{2}\right) \delta\theta'_e.$$

10.2 Collider Kinematic Resolutions

a) Using the results of the previous problem and assuming a lepton momentum resolution of $\delta E'_e/E'_e = 1\%$ and an angular resolution $\delta\theta'_e = 5$ mrad for $E_e = 10$ GeV, $E_p = 250$ GeV determine $\delta x/x$, $\delta Q^2/Q^2$, $\delta y/y$ for lepton energies E'_e of 2, 5, and 8 GeV and scattering angles θ'_e of 30, 90, and 150 degrees.

b) For a possible future electron–ion collider where a 10 GeV electron is incident head-on with a 250 GeV proton and assuming that the direction of the proton (electron) is the forward (rear) direction, sketch on a plot of Q^2 versus x the directions of the final-state electron and current jet (along the direction of 3-momentum transfer) for (i) low x and low Q^2; (ii) high x and low Q^2; and high x and high Q^2.

10.3 Collider Physics

Now let us consider a number of physics topics studied in collider experiments.

a) Explain why the the muon cross section is used to normalize the hadron cross section in Eq. (9.28) and not the electron cross section.

b) Explain the logarithmic slope in F_2.

c) Explain the scaling phenomena in Drell–Yan production.

d) Using the definition of rapidity

$$y = \frac{1}{2}\left(\frac{E+p_z}{E-p_z}\right),$$

identify which partonic cross section is largest at mid-rapidity.

10.4 W Boson Production

The production of a W boson in a collider can be identified by its characteristic weak decay pattern.

a) Determine the Bjorken-x scaling variables for initial-state quarks/anti quarks in hadron–hadron collisions for W boson production in terms of the rapidity y, the W mass and center-of-mass energy.

b) What is the threshold energy for W boson production in a e^+e^- collider mode configuration?

c) Provide an estimate of the threshold energy for W boson production in hadron–hadron collisions.

d) Determine the ratio of the W decay-width into quarks versus leptons.

The Nucleon–Nucleon Interaction

11.1 Introduction

We have seen that the nucleon is built from the quarks and gluons of QCD. Atomic nuclei are most successfully described by a theory of hadrons, which is based on the interactions between the constituent nucleons. In Chapters 11 through 16, we develop an increasingly more sophisticated understanding of nuclei. In this chapter, we discuss the fundamental nucleon–nucleon interaction from a number of different perspectives. The nonrelativistic theory of the simplest nucleus and only bound-state of the nucleon–nucleon system, the deuteron, is developed.

11.2 Nucleon–Nucleon Scattering

We have seen in earlier chapters that the nucleon is a composite particle composed of highly energetic quarks and gluons that interact strongly via the color force of QCD. In Chapter 8, we have seen that the structure of the nucleon as probed by lepton scattering can be understood to have significant contributions from mesons. When we consider the interaction between two nucleons, it can be described, to a reasonable degree of approximation, in terms of a sum over exchanges of mesons $\pi, \sigma, \rho, \omega, \ldots$ between the nucleons. Of these exchanges, the most important at large distances comes from the exchange of the pion, since it has the lightest mass.

Since the α-particle scattering experiments of Rutherford in 1911, we know that the strong interaction has a short range of order a few femtometers (1 fm = 10^{-15} m). In analogy to the interaction between electric charges arising from the exchange of a photon, Yukawa suggested in the 1930s that the interaction between two nucleons comes about via exchange of a particle. Because nucleon interactions are observed to be short-ranged, the exchanged particle must have a significant mass. From the Heisenberg uncertainty principle, the range r and the mass m must be related by $r \sim \frac{1}{m}$ so the mass of the quanta exchanged is about 200 MeV. This particle was identified as the π meson with a mass determined to be 140 MeV.

The modern theory of interactions via particle exchange is described by quantum field theory. Thus, if we introduce a quantum field π to describe the pion field, and τ is the isospin operator, from isospin and parity invariance the pion–nucleon interaction must have the form

$$\mathcal{L}_{\text{int}} = g_{\pi NN}\bar{N}\gamma_5 \boldsymbol{\tau} \cdot \boldsymbol{\pi} N \,, \tag{11.1}$$

where $g_{\pi NN}$ is the pion–nucleon coupling constant and has the experimental value

$$\frac{g_{\pi NN}^2}{16\pi} \simeq 13.5 \,. \tag{11.2}$$

From Exercise 11.1, the one-pion exchange potential, V_{OPEP}, has the form

$$V_{\text{OPEP}}(r) = -\frac{g_{\pi NN}^2}{16\pi m_N^2}\boldsymbol{\tau}_1 \cdot \boldsymbol{\tau}_2 \left\{ \left[\left(\frac{1}{(m_\pi r)^2} + \frac{1}{m_\pi r} + \frac{1}{3} \right) \hat{S}_{12} + \frac{1}{3}\boldsymbol{\sigma}_1 \cdot \boldsymbol{\sigma}_2 \right] \frac{m_\pi^2}{r}e^{-m_\pi r} \right.$$
$$\left. -\frac{4\pi}{3}\boldsymbol{\sigma}_1 \cdot \boldsymbol{\sigma}_2 \delta^3(r) \right\} \,, \tag{11.3}$$

where

$$\hat{S}_{12} = 3\boldsymbol{\sigma}_1 \cdot \hat{\boldsymbol{r}}\boldsymbol{\sigma}_2 \cdot \hat{\boldsymbol{r}} - \boldsymbol{\sigma}_1 \cdot \boldsymbol{\sigma}_2 \tag{11.4}$$

is the tensor interaction. Thus, the OPEP potential consists of two pieces, a tensor interaction having a range $r \sim 1/m_\pi$ together with a local (zero-range) spin-spin term. $\hat{S}_{12}$, called the tensor operator, is defined as

$$\hat{S}_{12} \equiv \left[3\frac{(\boldsymbol{S} \cdot \boldsymbol{r})^2}{r^2} - \boldsymbol{S}^2 \right] \,, \tag{11.5}$$

where $\boldsymbol{S}$ is the total spin of the two nucleons. We advise the reader to work through the earlier Exercises 2.4 and 2.5 on spin coupling and central forces, where the tensor operator $\hat{S}_{12}$ is already introduced.

This potential matches the phenomenological forms deduced from experimental data at large NN distances. At smaller distances, there is also exchange of scalar so-called σ mesons (isospin 0) with mass about 500 MeV. The interaction is attractive, as we saw above. Additionally, there is also exchange of vector (spin-1) mesons, ω (isospin 0) and ρ (isospin 1), as discussed in Chapter 8, which give rise to interactions that are short-ranged and repulsive. One can build a phenomenological NN interaction based on these meson exchanges that successfully describes nucleon elastic form factors, as reviewed in [Mac01] and further discussed in Chapter 8. The short-range interactions are intrinsically model-dependent.

The OPEP interaction is physically analogous to the interaction in classical magneto-statics between two current loops. A current loop of magnetic moment $\boldsymbol{\mu}$ produces a magnetostatic potential

$$\Phi_m = -\boldsymbol{\mu} \cdot \nabla \left(\frac{1}{r} \right) \,. \tag{11.6}$$

A second loop in the external potential of the first loop will have have a potential

$$-(\boldsymbol{\mu}_1 \cdot \nabla)(\boldsymbol{\mu}_2 \cdot \nabla) \left(\frac{1}{r} \right) \,. \tag{11.7}$$

For point nucleons interacting with a pion field, this would have the form

$$-(\boldsymbol{\mu}_1 \cdot \nabla)(\boldsymbol{\mu}_2 \cdot \nabla) \left(\frac{e^{-m_\pi r}}{r} \right) \,, \tag{11.8}$$

which leads to the functional form of Eq. 11.3 without the isospin factor $\boldsymbol{\tau}_1 \cdot \boldsymbol{\tau}_2$ and the δ-function. Note that the δ-function term in V_{OPEP} is operative only for S-wave states.

Being the lightest meson, the pion is viewed as the Goldstone boson resulting from the spontaneous breaking of chiral symmetry, as discussed in Chapter 6, and is therefore a pseudoscalar. It has isospin 1 and can be represented by the vector $\boldsymbol{\pi} = (\pi_1, \pi_2, \pi_3)$ in isospin space. Combinations of $\pi_{1,2}$ generate positive and negative charged pions $\pi^{\pm}$, and π_3 corresponds to a neutral pion π^0.

In the case of two interacting nucleons, the total isospin operator is given by

$$T = t_1 + t_2 \, . \tag{11.9}$$

Ignoring the electromagnetic interaction and the mass difference of the *up* and *down* quarks, isospin is conserved by the interaction Hamiltonian and so

$$[\widehat{H}, T] = 0 \, . \tag{11.10}$$

It follows that $\widehat{H}$ can only depend on the isospin through T^2, where

$$T^2 = (t_1 + t_2)^2 \tag{11.11}$$

$$= t_1{}^2 + t_2{}^2 + 2t_1 \cdot t_2 \tag{11.12}$$

$$= \frac{3}{4} + \frac{3}{4} + \frac{1}{2}\boldsymbol{\tau}_1 \cdot \boldsymbol{\tau}_2 \, . \tag{11.13}$$

It follws that the *NN* potential naturally involves the factor $\boldsymbol{\tau}_1 \cdot \boldsymbol{\tau}_2$. States of good total T^2 are also eigenstates of $\widehat{H}$. It follows that the eigenstates with $T = 1$ (so-called *isovector* states) will be degenerate, i.e., the states

$$pp \quad nn \quad \frac{1}{\sqrt{2}}(pn + np) \tag{11.14}$$

all have identical strong interactions. The $T = 0$ eigenstate (so-called *isoscalar* state) $\frac{1}{\sqrt{2}}(pn - np)$ will be different from the three $T = 1$ states. Thus, to describe the nucleon–nucleon system, it is desirable to form states of good isospin. Note that

$$\boldsymbol{\tau}_1 \cdot \boldsymbol{\tau}_2 = 4\, \boldsymbol{T}_1 \cdot \boldsymbol{T}_2 \tag{11.15}$$

$$= 2\left[T(T+1) - \frac{3}{2} \right] \tag{11.16}$$

$$= -3, T = 0 \tag{11.17}$$

$$= +1, T = 1 \, . \tag{11.18}$$

11.3 General Form of Nucleon–Nucleon Interaction

Nucleon–nucleon scattering experiments allow us to determine the general characteristics of the interaction between two nucleons. The interaction is short-range, of order 1 fm. The interaction must be attractive over a significant region of nucleon–nucleon interspatial distance to produce bound nuclei. Experiments show clearly that the interaction depends on the spins of the nucleons. Further, as we have seen above with the tensor force, the

Table 11.1 The angular momentum states of the nucleon–nucleon system

J				
0		$^3P_0$		$^1S_0$
1	$^3S_1$	$^3D_1$	$^3P_1$	$^1P_1$
2	$^3P_2$	$^3F_2$	$^3D_2$	$^1D_2$
3	$^3D_3$	$^3G_3$	$^3F_3$	$^1F_3$
$\vdots$	$\vdots$	$\vdots$	$\vdots$	$\vdots$

interaction is non-central. In the approximation that the proton and neutron are simply different isospin states of the nucleon, the interaction is isospin-symmetric. Analysis of nucleon–nucleon scattering data shows that, at short distances, the interaction is strongly repulsive so that nuclear matter does not collapse (see Chapter 13). We shall see that the interaction includes a spin-orbit force. Finally, the nucleon–nucleon interaction conserves parity.

The short-distance potentials are parametrized, consistent with fundamental symmetries, and are fit to experimental data. For two nucleons we use $\boldsymbol{r} = \boldsymbol{r_1} - \boldsymbol{r_2}$ to represent their relative position, $\boldsymbol{p} = (\boldsymbol{p}_1 - \boldsymbol{p}_2)/2$ their relative momentum, and $\boldsymbol{s}_1$ and $\boldsymbol{s}_2$ their respective spins. The relative orbital angular momentum is $\boldsymbol{L} = \boldsymbol{r} \times \boldsymbol{p}$ and the total spin is $\boldsymbol{S} = \boldsymbol{s}_1 + \boldsymbol{s}_2$. When the spins are coupled, the total spin can either be $S = 0$ (singlet) or $S = 1$ (triplet). The total angular momentum is the sum of orbital angular momentum and total spin: $\boldsymbol{J} = \boldsymbol{L} + \boldsymbol{S}$, where the orbital angular momentum quantum number is L. In the singlet spin case, we have $J = L$ because $S = 0$. For the triplet states, $J = L - 1, L, L + 1$ if $L \neq 0$, and $J = 1$ if $L = 0$. A state with (S, L, J) is usually labeled $^{2S+1}L_J$, where $L = 0, 1, 2, 3 \ldots$ are usually called $S, P, D, F, G \ldots$ states. Table 11.1 summarizes the angular momentum states of the nucleon–nucleon system. Because of rotational symmetry (see the group $SO(3)$ in Chapter 2), no nuclear interactions can couple states with different total angular momenta so J is a good quantum number. Only the states in the first two columns can mix, since they have the same parity and angular momentum. We choose the basis states with good J as

$$|LSJM\rangle = \sum_{m_L} Y_{Lm_L} |Sm_s\rangle \langle LmSm_s| |JM\rangle , \tag{11.19}$$

where $\langle LmSm_s |LSJM \rangle$ is the Clebsch–Gordan coefficient, discussed in Chapter 2. In the coordinate representation, we use the spin-angle functions $\mathcal{Y}^M_{LSJ}$ to label the total angular momentum states. Then, the eigenfunction can be written as

$$\psi_{nLSJM} = R_{nSLJ}(r)\mathcal{Y}^M_{LSJ} , \tag{11.20}$$

where $R_{nSLJ}(r)$ is the corresponding radial wavefunction.

Consider next the Pauli exclusion principle as it applies to the two nucleon system. The total isospin T must be either 0 or 1. Consider the possible values of $T, S,$ and L to form the total wavefunction which must be antisymmetric. Assigning $+1(-1)$ to represent a symmetric (antisymmetric) wavefunction, the factors associated with spin, isospin, and orbital angular momentum resulting from exchange of the two nucleons are $(-)^{S+1}$, $(-)^{T+1}$, and $(-)^{L}$, respectively. The total wavefunction will have a factor $(-)^{L+S+T}$, which implies that $L + T + S$ must be odd to be consistent with the Pauli exclusion principle.

Let us now proceed to build the NN potential from its various possible components.

a) Consider the simplest central NN force which depends only on the relative distance between the two nucleons, $V_C(r)$. Here, different L states will give rise to different energies. For every L, the singlet and triplet spin states have the same energy. The eigenfunctions of the system can be written as

$$|nLm_L Sm_s\rangle = R_{nLS}(r) Y_{Lm_L} \chi_{Sm_S} \,.$$

b) We know from experimental data that there must be also a pure spin-dependent force. The most general form will be

$$V_S(r)\boldsymbol{\sigma}_1 \cdot \boldsymbol{\sigma}_2 = V_S(r)\left[2S^2 - 3\right] \,.$$

Hence, the matrix element of the spin operator depends on the total spin of the two particles. In the singlet state, we have $\boldsymbol{\sigma}_1 \cdot \boldsymbol{\sigma}_2 = -3$ so the potential is $V_C - 3V_S$. In the triplet state, we have $\boldsymbol{\sigma}_1 \cdot \boldsymbol{\sigma}_2 = +1$ so the potential is $V_C + V_S$. Thus, the energy now depends not only on the orbital quantum number L, but also on S.

c) We know from experiment that the NN interaction depends on the isospin of the nucleon (i.e. whether it is a proton or neutron) so we consider an isospin-dependent term in the force. The most general form is $V_I(r)\,\boldsymbol{\tau}_1 \cdot \boldsymbol{\tau}_2$. There can also be a spin-isospin-dependent term. The most general form is

$$V_{SI}(r)(\boldsymbol{\sigma}_1 \cdot \boldsymbol{\sigma}_2)(\boldsymbol{\tau}_1 \cdot \boldsymbol{\tau}_2) \,.$$

d) We know from scattering data that nucleons interact via a spin-orbit force. This is completely analogous to a similar force in atomic physics. However, in the NN interaction, it turns out that the spin-orbit force is opposite in sign to atomic physics. A spin-orbit force will be of the form

$$V_{LS}(r)\boldsymbol{L} \cdot \boldsymbol{S} \,,$$

where from angular momentum conservation

$$\boldsymbol{L} \cdot \boldsymbol{S} = \frac{1}{2}(\boldsymbol{J}^2 - \boldsymbol{L}^2 - \boldsymbol{S}^2) \,. \tag{11.21}$$

Thus, the matrix element of the operator is simple in the basis which is formed by the common eigenstates of $\boldsymbol{J}^2, J_3, \boldsymbol{L}^2, \boldsymbol{S}^2$. The potential used for solving the radial Schrödinger equation is then

$$\begin{aligned}
V_{LSJ}(r) &= V_C(r) + (\boldsymbol{\sigma_1} \cdot \boldsymbol{\sigma}_2)V_S(r) + (\boldsymbol{\tau_1} \cdot \boldsymbol{\tau}_2)V_T(r) \\
&\quad + (\boldsymbol{\sigma_1} \cdot \boldsymbol{\sigma}_2)(\boldsymbol{\tau_1} \cdot \boldsymbol{\tau}_2)V_{ST}(r) \\
&\quad + (\boldsymbol{L} \cdot \boldsymbol{S})V_{LS}(r) \,,
\end{aligned} \tag{11.22}$$

where $V_C(r)$, $V_S(r)$, $V_T(r)$, $V_{ST}(r)$ and $V_{LS}(r)$ are the radial functions for the central, spin, isospin, spin-isospin, and spin-orbit contributions, respectively.

e) Finally, we need to include the non-central tensor force developed in the previous section

$$V_{TF}(r)\left[3\frac{(\boldsymbol{\sigma}_1\cdot\boldsymbol{r})(\boldsymbol{\sigma}_2\cdot\boldsymbol{r})}{r^2}-\boldsymbol{\sigma}_1\cdot\boldsymbol{\sigma}_2\right],\tag{11.23}$$

where $V_{TF}(r)$ is the radial function which can be attractive or repulsive. Using the total spin operator we can write

$$(\boldsymbol{\sigma}_1\cdot\boldsymbol{r})(\boldsymbol{\sigma}_2\cdot\boldsymbol{r})=2(\boldsymbol{S}\cdot\boldsymbol{r})^2-r^2,\tag{11.24}$$

so the tensor structure becomes

$$\hat{S}_{12}=2\left[3\frac{(\boldsymbol{S}\cdot\boldsymbol{r})^2}{r^2}-\boldsymbol{S}^2\right].\tag{11.25}$$

In Exercise 11.4 the matrix element of the tensor operator is calculated for different states of interest. In the presence of the tensor interaction, $\boldsymbol{L}^2$ and $\boldsymbol{S}^2$ no longer commute with the Hamiltonian, and the states with the same J, but different L can mix under the interaction. However, parity is still conserved so the states can mix only when their orbital angular momenta differ by two units, e.g., $^3S_1$–$^3D_1$ (see Table 11.1). When we come to discussion of the structure of the deuteron, the tensor force plays a crucial role in giving rise to the D-state. Consequently, the deuteron ground-state properties like the magnetic and electric quadrupole moments are influenced by the non-central tensor force. These are discussed in detail in Section 11.5.

Later in the chapter, discussion of low-energy nucleon–nucleon scattering develops the concept of the phase-shift, which is measurable for each channel of Table 11.1 in elastic nucleon–nucleon scattering as a function of energy. Exercise 11.5 develops how the measured phase-shifts can inform where the different terms above in the NN interaction are significant. In particular, both the $^3S_1$ and $^1S_0$ phase-shifts change sign from positive to negative at a nucleon kinetic energy of about 250 MeV. This is clear evidence of a hard repulsive core to the NN interaction at an inter-nucleon spatial separation of about 0.3 fm. Further, the value of the $^3S_1$ phase shift at low energies is consistent with the existence of a bound state of the NN system, namely the deuteron.

11.4 The Deuteron

The deuteron is the simplest and lightest nucleus and is the only bound state of two nucleons. As we shall see in Chapter 20, the deuteron was created in the early universe via big bang nucleosynthesis and its existence is essential to the formation of all heavier elements. It has isospin $T=0$, spin-parity $J^\pi=1^+$ and binding energy $B_d=2.225$ MeV. For two spin-1/2 nucleons, only total spins of $S=0,1$ are allowed. Thus, the orbital angular momentum is restricted to the values $J-1<L<J+1$, i.e., $L=0,1,2$. Since the parity is $\pi=(-)^L=+$, only $L=0$ (S-state) and $L=2$ (D-state) are allowed, implying that $S=1$.

We now develop a nonrelativistic theory of the deuteron where in the Hamiltonian, for simplicity we consider only central and non-central tensor forces. Thus,

$$\hat{H} = -\frac{1}{M}\frac{1}{r}\frac{d^2}{dr^2}r + \frac{1}{M}\frac{L^2}{r^2} + V_C(r) + V_{TF}(r)\hat{S}_{12} \tag{11.26}$$

where M is the reduced mass of the np system. From Exercise 11.4, we have the following relations

$$\hat{S}_{12}\mathcal{Y}^M_{001} = \sqrt{8}\mathcal{Y}^M_{211}, \quad \hat{S}_{12}\mathcal{Y}^M_{211} = \sqrt{8}\mathcal{Y}^M_{011} - 2\mathcal{Y}^M_{211} \tag{11.27}$$

$$L^2\mathcal{Y}^M_{011} = 0, \quad L^2\mathcal{Y}^M_{211} = 6\mathcal{Y}^M_{211}. \tag{11.28}$$

Thus, we find the radial equations

$$\left[\frac{1}{M}\frac{d^2}{dr^2} + E - V_C(r)\right]u(r) = \sqrt{8}V_{TF}(r)w \tag{11.29}$$

$$\left[\frac{1}{M}\left(\frac{d^2}{dr^2} - \frac{6}{r^2}\right) + E + 2V_{TF}(r) - V_C(r)\right]w(r) = \sqrt{8}V_{TF}(r)u. \tag{11.30}$$

The D-state probability $P_D = \int_0^\infty dr\, w^2(r)$ is a measure of the strength of the tensor component of the NN force, even though it is a model-dependent quality with no unique, measurable value. The above equations can be solved numerically and are shown in Figs. 11.1 and 11.2 both in coordinate and momentum space for different NN potentials – see [Gil02].

Important insight into the structure of the deuteron can be obtained from consideration of both the magnetic moment μ and the quadrupole moment Q. Experimentally, these quantities have the values

$$\mu = 0.857\,\mu_N \tag{11.31}$$

$$Q = 0.286\,e \cdot \text{fm}^2, \tag{11.32}$$

where μ_N is the nucleon magneton. Since $Q \neq 0$, the deuteron ground-state cannot be pure $L = 0$. In turn, this implies that the NN interaction must have a non-central contribution. Therefore, we write the deuteron wavefunction as a linear combination of S- and D-waves

$$|\Psi_d, J = 1, M = 1> = \left[a_S \frac{u(r)}{r}\mathcal{Y}^1_{011} + a_D \frac{w(r)}{r}\mathcal{Y}^1_{211}\right]\chi^T_{00}, \tag{11.33}$$

where a_S and a_D are constants with $\sqrt{a_S^2 + a_D^2} = 1$; $u(r)$ and $w(r)$ are the radial wavefunctions for the S- and D-states, respectively; and the isospin wavefunction has the form

$$\chi^T_{00} = \frac{1}{\sqrt{2}}\left[\chi_p(1)\chi_n(2) - \chi_n(1)\chi_p(2)\right]. \tag{11.34}$$

As we shall see below, measurement of spin-dependent scattering is a sensitive probe of the D-state. Note that the D-state dominates at high nucleon momenta, in the range of 350 to 500 MeVc^{-1}.

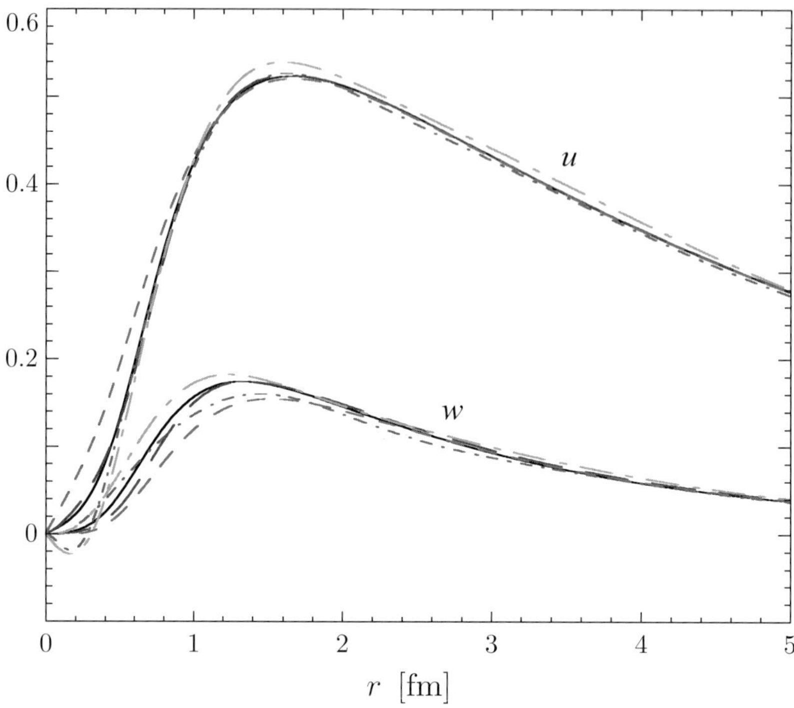

Fig. 11.1 Reduced coordinate space wavefunctions for five NN potentials: AV18 (solid), Paris (long dashed), CD Bonn (short dashed), IIB (short dot-dashed), and W16 (long dot-dashed) from [Gil02].

Deuteron Magnetic Moment

The observed magnetic moment of the deuteron provides information about the ground-state wavefunction. The total magnetic moment of the deuteron is due to the intrinsic magnetic moments of the neutron and proton, as well as the orbital angular momentum of the (charged) proton but not the (neutral) neutron. The magnetic moment operator along the 3-axis can then be written as

$$\mu_3 = \frac{1}{2}\mu_N \left(L_3 + g_p\sigma_3^p + g_n\sigma_3^n\right),\qquad(11.35)$$

where $\mu_N = e/2m_N$ is the nucleon magneton while $g_p = 5.58$ and $g_n = -3.82$ are the experimental gyromagnetic ratios for the proton and neutron respectively. (In the center-of-mass system the angular momentum of the proton is half that of the total, which is the reason for the factor of $\frac{1}{2}$ multiplying $\boldsymbol{L}$.) Taking the overall deuteron wavefunction to be

$$\left(\psi_1^1\right)^{\text{tot}}(\boldsymbol{r}) = a_S\left(\psi_1^1\right)^S(\boldsymbol{r}) + a_D\left(\psi_1^1\right)^D(\boldsymbol{r})\qquad(11.36)$$

with $|a_S|^2 + |a_D|^2 = 1$, as in Exercise 11.2, one can calculate the deuteron magnetic moment and show that

$$\left\langle\left(\psi_1^1\right)^{\text{tot}}|M_z|\left(\psi_1^1\right)^{\text{tot}}\right\rangle = |a_S|^2\left(\frac{g_p + g_n}{2}\right) + |a_D|^2\left(\frac{3 - g_p - g_n}{4}\right).\qquad(11.37)$$

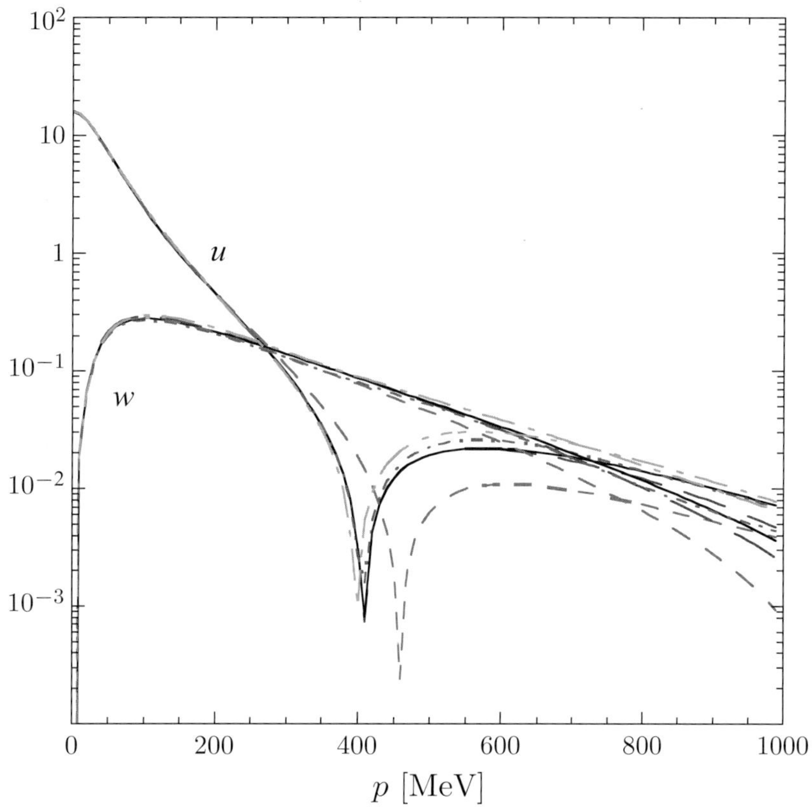

Fig. 11.2   Momentum-space wavefunctions for five models (see the caption for Fig.11.1) [Gil02].

By comparing the wavefunction calculation with the experimental value $\mu^{\mathrm{exp}} = 0.857\mu_N$, one can determine that $|a_D|^2 \simeq 0.04$, i.e., a $\sim 4\%$ D-state component in the deuteron wavefunction. This analysis of the deuteron magnetic moment must be corrected for the effects of meson-exchange currents; see Fig. 7.5. These effects are clearly observed in elastic electron–deuteron scattering and are discussed below. Including these effects of meson-exchange currents on the deuteron's magnetic moment yields values of $a_D^2 = 0.05 - 0.07$.

Deuteron Electric Quadrupole Moment

From the previous discussion, we see that an understanding of the deuteron magnetic dipole moment requires a small D-state admixture in the ground-state wavefunction. Thus, we anticipate that the deuteron is approximately, but not quite, spherical. We can examine this admixture from another perspective by looking at the deuteron electric quadrupole moment

$$e\mathcal{Q} = \frac{e}{4}\left\langle \left(\psi_1^1\right)^{\mathrm{tot}} \left| (3z^2 - r^2) \right| \left(\psi_1^1\right)^{\mathrm{tot}} \right\rangle , \tag{11.38}$$

which provides a measure of the nonsphericity of the ground-state charge distribution. (Here the factor of 4 is due to the fact that only the proton contributes to the charge distribution.)

Writing the quadrupole moment as

$$Q = \frac{1}{4} \int d\mathbf{r}\, r^2 (3\cos^2\theta - 1) \left| \left(\psi_1^1\right)^{\text{tot}}(r) \right|^2 , \qquad (11.39)$$

one can use the deuteron wavefunction to show that (see Exercise 11.3)

$$Q = \frac{1}{5\sqrt{2}} \text{Re}(a_S^* a_D) \left\langle r^2 \right\rangle_{SD} - \frac{1}{20} |a_D|^2 \left\langle r^2 \right\rangle_{DD} , \qquad (11.40)$$

where

$$\left\langle r^2 \right\rangle_{SD} = \left\langle R_S \left| r^2 \right| R_D \right\rangle \quad \text{and} \quad <r^2>_{DD} = <R_D|r^2|R_D> . \qquad (11.41)$$

Using the experimental result

$$Q_{\text{exp}} = 0.286\,\text{fm}^2 ,$$

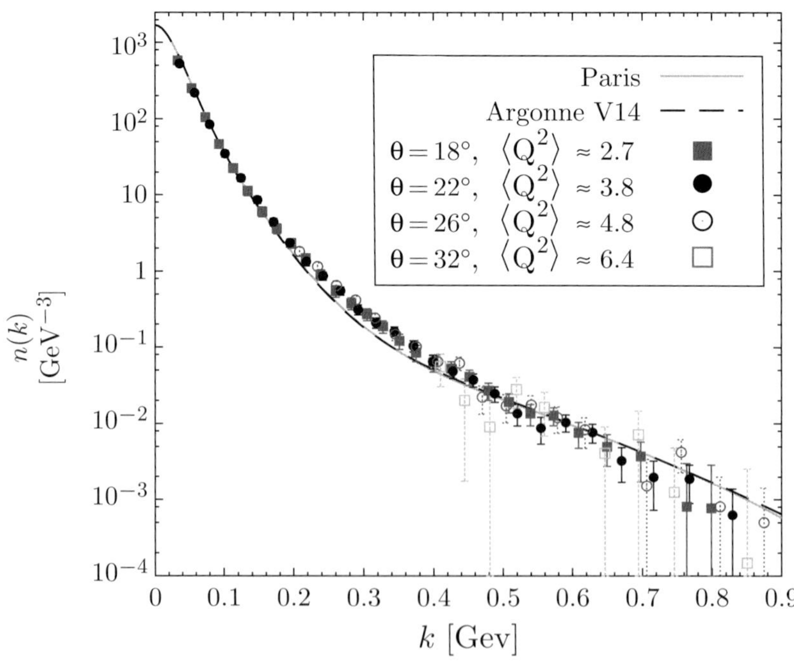

Fig. 11.3 Extracted momentum distributions from ^{2}H$(e, e'p)$ (points) and calculations (curves) using three different *NN* potentials [Fom12]. Note that Paris and AV14 are nearly indistinguishable on this scale.

and defining

$$I_1 \equiv \left(\int r^2 u(r)^2 dr \right) \tag{11.42}$$

$$I_2 \equiv \left(\int r^2 u(r) w(r) dr \right) \tag{11.43}$$

$$I_3 \equiv \left(\int r^2 w(r)^2 dr \right) , \tag{11.44}$$

we obtain

$$Q = e \left\{ \frac{\sqrt{2}}{10} \mathrm{Re}(a_S a_D^*) I_2 - \frac{|a_D|^2}{20} I_3 \right\} , \tag{11.45}$$

which yields $I_2 \simeq 10.1\,\mathrm{fm}^2$. This value is reasonable considering that the experimental value of the deuteron mean squared charge radius $< r_{\mathrm{ch}}^2 >= \frac{1}{4} I_1 \approx 4\,\mathrm{fm}^2$. Consider the quadrupole moment, as it arises from the tensor force between the two spin-1/2 nucleons. In the $M = 1$ state, the dominant $S - D$ interference term in the quadrupole moment has $M_S = 1$, implying that the spins of the two nucleons are predominantly aligned parallel to $\hat{z}$. Thus, taking $\boldsymbol{\sigma}_1 = \boldsymbol{\sigma}_2 = +\hat{z}$, we have $\boldsymbol{\sigma}_1 \cdot \boldsymbol{\sigma}_2 = +1$. Next we need to consider the relative orientation of $\hat{r}$ and we can focus on two extreme cases: $\hat{r} \parallel \hat{z}$ and $\hat{r} \perp \hat{z}$. For $\hat{r} \parallel \hat{z}$, $\boldsymbol{\sigma}_1 \cdot \hat{r} = \boldsymbol{\sigma}_2 \cdot \hat{r} = 1$, so we have $< S_{12} >= +2$ for this geometrical arrangement. This is a prolate configuration (American/rugby football shape) so we expect $Q > 0$. For $\hat{r} \perp \hat{z}$, $\boldsymbol{\sigma}_1 \cdot \hat{r} = \boldsymbol{\sigma}_2 \cdot \hat{r} = 0$, so $< S_{12} >= -1$. This is an oblate configuration (pancake shape) and we would expect that $Q < 0$. Experimentally, $Q > 0$, so that $V_T(r) < 0$, which produces an attractive force.

Given central $V_C(r)$ and tensor $V_T(r)$ potentials, the deuteron Schrödinger equation is an eigenvalue problem, with a free parameter to be determined, namely the ratio a_D/a_S. It was shown by Rarita and Schwinger [Rar41] that large classes of potentials can solve these equations with the constraints of the binding energy $B_d = 2.225\,\mathrm{MeV}$ and quadrupole moment $Q = 0.286\,\mathrm{e \cdot fm}^2$.

11.5 Low-Energy Scattering

We now turn our attention to low-energy (of order 100 to 300 MeV nucleon kinetic energy on stationary nucleon targets) NN scattering.[1] This is a beautiful example of the application of nonrelativistic scattering theory and teaches us basic aspects of the NN interaction. We begin our analysis with a brief review of conventional scattering theory as discussed in Appendix B and in [Mer98]. In the usual partial-wave expansion, we can write the scattering amplitude as

$$f(\theta) = \sum_\ell (2\ell + 1) a_\ell(k) P_\ell(\cos\theta) , \tag{11.46}$$

---

[1] Note that much of the material in this section is from [Hol99a].

where $a_\ell(k)$ has the form

$$a_\ell(k) = \frac{1}{k} e^{i\delta(k)} \sin \delta(k) = \frac{1}{k \cot \delta(k) - ik} \, . \tag{11.47}$$

A general expression for the scattering phase shift $\delta_\ell(k)$ is [Mer98]

$$\sin \delta_\ell(k) = -k \int_0^\infty dr' r' j_\ell(kr') 2m_r V(r') u_{\ell,k}(r') \, , \tag{11.48}$$

where $m_r \cong m_N/2$ is the reduced mass and

$$u_{\ell,k}(r) = r \cos \delta_\ell(k) j_\ell(kr) + kr \int_0^r dr' r' j_\ell(kr') n_\ell(kr) u_{\ell,k}(r') 2m_r V(r')$$

$$+ kr \int_r^\infty dr' r' j_\ell(kr) n_\ell(kr') u_{\ell,k}(r') 2m_r V(r') \tag{11.49}$$

is the scattering wavefunction. At low energies, one generally characterizes the analytic function $k^{2\ell+1} \cot \delta(k)$ via an effective-range expansion

$$k^{2\ell+1} \cot \delta_\ell(k) = -\frac{1}{a_\ell} + \frac{1}{2} r_\ell k^2 + \cdots \, . \tag{11.50}$$

Then from Eq. (11.48) we can write the scattering length as

$$a_\ell = \frac{1}{[(2\ell + 1)!!]^2} \int_0^\infty dr' (r')^{2\ell+2} 2m_r V(r') + \mathcal{O}(V^2). \tag{11.51}$$

At very low energy (below about $10\,\mathrm{MeV}$), the dominant nucleon–nucleon scattering happens in the S-wave. In the zero-energy limit, the nucleon scattering cross section is large, of order 20 b. We consider the physics of S-wave scattering. In the region outside of the nuclear interaction, the S-wave scattering state is described by the solution of the free Schrödinger equation,

$$\psi = A \sin(kr + \delta) \sim (e^{2i\delta} e^{ikr} - e^{-ikr}) \, , \tag{11.52}$$

where δ is the S-wave scattering phase shift. Here the first exponential represents the outgoing spherical wave and the second represents the incoming one. The total scattering cross section is

$$\sigma = \frac{4\pi}{k^2} \sin^2 \delta \, . \tag{11.53}$$

Since the cross section is finite as $k \to 0$ (the nuclear force has finite range), the phase shift vanishes as $k \to 0$. At low energy, one can make the expansion

$$k \cot \delta = -\frac{1}{a} + \frac{1}{2} r_0 k^2 + \cdots \, , \tag{11.54}$$

where a is the scattering length and r_0 is the effective range, which approximates the spatial extent of the potential. At low energy, the nuclear scattering can be effectively determined by these two numbers. At zero energy, the cross section is given by

$$\sigma = 4\pi a^2 \, , \tag{11.55}$$

which is completely defined by the scattering length. The scattering lengths for NN interactions have been extracted from scattering of both neutrons and protons from a proton target. In the former case, there are both $T = 0$ and 1 channels. From experiments on the neutron–proton system, we have

$$a_0^s = -23.715 \pm 0.015 \text{ fm}$$
$$r_0^s = 2.73 \pm 0.03 \text{ fm}$$
$$a_0^t = 5.423 \pm 0.005 \text{ fm}$$
$$r_0^t = 1.73 \pm 0.02 \text{ fm} \tag{11.56}$$

for scattering in the spin-singlet and spin-triplet channels, respectively. Thus, the experimental cross section is very large at zero energy.

The large scattering length in the spin-triplet channel indicates that there is a two-body bound state. For small but finite k, one has

$$f = \frac{1}{1/a + ik} , \tag{11.57}$$

$$\sigma = \frac{4\pi}{1/a^2 + k^2} . \tag{11.58}$$

The scattering matrix has a pole at $E = -1/2a^2 m^2$. If a is positive, this is a real bound state; if a is negative, the state is virtual. Therefore, in the $S = 1$ channel, there must be a bound state, which is the deuteron. For $T = 1$, one has a virtual state: the potential is attractive, but insufficiently strong to form a bound state. The existence of a bound state with binding energy $E_B = -\kappa^2/2m_r$ is indicated by the presence of a pole along the positive imaginary k-axis, i.e., $\kappa > 0$ under the analytic continuation $k \to i\kappa$

$$\frac{1}{a_0} + \frac{1}{2} r_0 \kappa^2 - \kappa = 0 . \tag{11.59}$$

We see then from Eq. (11.56) that there is no bound state in the np spin-singlet channel, but in the spin-triplet system there exists a solution

$$\kappa = \frac{1 - \sqrt{1 - \frac{2r_0^t}{a_0^t}}}{r_0^t} \cong 46 \text{ MeV}, \quad \text{i.e.,} \quad E_B \cong -2.2 \text{ MeV} \tag{11.60}$$

corresponding to the deuteron.

As a specific example, suppose we employ a simple square-well potential to describe the interaction

$$V(r) = \begin{cases} -V_0 & r \le R \\ 0 & r > R \end{cases} . \tag{11.61}$$

For S-wave scattering the wavefunction in the interior and exterior regions can then be written as

$$\psi(r) = \begin{cases} N j_0(Kr) & r \le R \\ N' \left[j_0(kr) \cos \delta_0 - n_0(kr) \sin \delta_0 \right] & r > R \end{cases} , \tag{11.62}$$

where j_0, n_0 are spherical Bessel functions and the interior, exterior wavenumbers are given by $k = \sqrt{2m_r E}$ and $K = \sqrt{2m_r(E + V_0)}$ respectively. The connection between the two forms can be made by matching logarithmic derivatives at the boundary, which yields

$$k \cot \delta_0 = -\frac{1}{R}\left[1 + \frac{1}{KRF(KR)}\right] \quad \text{with} \quad F(x) = \cot x - \frac{1}{x}. \tag{11.63}$$

Making the effective-range expansion, see Eq. (11.50), we find an expression for the scattering length

$$a_0 = R\left[1 - \frac{\tan(K_0 R)}{K_0 R}\right] \quad \text{where} \quad K_0 = \sqrt{2m_r V_0}. \tag{11.64}$$

Note that for weak potentials, i.e., $K_0 R \ll 1$, this form agrees with the general result Eq. (11.51)

$$a_0 = \int_0^\infty dr'\, r'^2 2m_r V(r') = -\frac{2m_r}{3} R^3 V_0 + \mathcal{O}(V_0^2). \tag{11.65}$$

When Coulomb interactions are included the analysis becomes somewhat more challenging. Suppose first that only same-charge (e.g., proton–proton) scattering is considered and that, for simplicity, we describe the interaction in terms of a potential of the form

$$V(r) = \left\{\begin{array}{ll} U(r) & r < R \\ \frac{\alpha}{r} & r > R \end{array}\right\}, \tag{11.66}$$

i.e., a strong attraction, $U(r)$, at short distances, in order to mimic the strong interaction, and the repulsive Coulomb potential α/r at large distance, where $\alpha \simeq 1/137$ is the fine structure constant. The analysis of the scattering then proceeds as above, but with the replacement of the exterior spherical Bessel functions by corresponding Coulomb wavefunctions, F_0^+, G_0^+,

$$j_0(kr) \rightarrow F_0^+(r), \qquad n_0(kr) \rightarrow G_0^+(r), \tag{11.67}$$

whose explicit form can be found in [Jac50]. For our purposes we require only the form of these functions in the limit $kr \ll 1$

$$F_0^+(r) \xrightarrow{kr\ll 1} C_\eta(\eta_+(k))\left(1 + \frac{r}{a_B} + \cdots\right)$$

$$G_0^+(r) \xrightarrow{kr\ll 1} -\frac{1}{C_\eta(\eta_+(k))}\left\{\frac{1}{kr} + 2\eta_+(k)\left[h(\eta_+(k)) + 2\gamma - 1 + \ln\frac{2r}{a_B}\right] + \cdots\right\}. \tag{11.68}$$

Here $\gamma = 0.577215\ldots$ is the Euler constant,

$$C^2(x) = \frac{2\pi x}{\exp(2\pi x) - 1} \tag{11.69}$$

is the usual Coulombic enhancement factor, $a_B = 1/m_r\alpha$ is the Bohr radius, $\eta_+(k) = 1/ka_B$, and

$$h(\eta_+(k)) = \mathrm{Re}H(i\eta_+(k)) = \eta_+^2(k)\sum_{n=1}^\infty \frac{1}{n(n^2 + \eta_+^2(k))} - \ln\eta_+(k) - \gamma, \tag{11.70}$$

where $H(x)$ is the analytic function

$$H(x) = \psi(x) + \frac{1}{2x} - \ln(x) . \tag{11.71}$$

Equating interior and exterior logarithmic derivatives we find

$$
\begin{aligned}
KF(KR) &= \frac{\cos \delta_0 F_0^{+'}(R) - \sin \delta_0 G_0^{+'}(R)}{\cos \delta_0 F_0^{+}(R) - \sin \delta_0 G_0^{+}(R)} \\[2mm]
&= \frac{k \cot \delta_0 C^2(\eta_+(k)) \frac{1}{a_B} - \frac{1}{R^2}}{k \cot \delta_0 C^2(\eta_+(k)) + \frac{1}{R} + \frac{2}{a_B}\left[h(\eta_+(k)) - \ln \frac{a_B}{2R} + 2\gamma - 1\right]} .
\end{aligned}
\tag{11.72}
$$

Since $R << a_B$, Eq. (11.72) can be written in the form

$$k \cot \delta_0 C^2(\eta_+(k)) + \frac{2}{a_B}\left[h(\eta_+(k)) - \ln \frac{a_B}{2R} + 2\gamma - 1\right] \simeq -\frac{1}{a_0} . \tag{11.73}$$

The scattering length a_C in the presence of the Coulomb interaction is conventionally defined as [Pre62]

$$k \cot \delta_0 C^2(\eta_+(k)) + \frac{2}{a_B} h(\eta_+(k)) = -\frac{1}{a_C} + \cdots \tag{11.74}$$

so that we have the relation

$$-\frac{1}{a_0} = -\frac{1}{a_C} - \frac{2}{a_B}\left(\ln \frac{a_B}{2R} + 1 - 2\gamma\right) \tag{11.75}$$

between the experimental scattering length, a_C, and that which would exist in the absence of the Coulomb interaction, a_0.

As an aside we note that, strictly speaking, a_0 is not itself an observable, since the Coulomb interaction *cannot* be turned off. However, in the case of the *pp* interaction isospin invariance requires $a_0^{pp} = a_0^{nn}$ so that one has the prediction

$$-\frac{1}{a_0^{nn}} = -\frac{1}{a_C^{pp}} - \alpha m_N \left(\ln \frac{1}{\alpha m_N R} + 1 - 2\gamma\right) . \tag{11.76}$$

While this is a model dependent result, Jackson and Blatt have shown, by treating the interior Coulomb interaction perturbatively, that a version of this result with $1 - 2\gamma \rightarrow 0.824 - 2\gamma$ is approximately valid for a wide range of strong interaction potentials [Jac50] and the correction indicated in Eq. (11.76) is essential in restoring agreement between the widely discrepant $a_0^{nn} = -18.8$ fm versus $a_C^{pp} = -7.82$ fm values obtained experimentally.

Returning to the problem at hand, the experimental scattering amplitude can then be written as

$$
\begin{aligned}
f_C^+(k) &= \frac{e^{2i\sigma_0} C^2(\eta_+(k))}{-\frac{1}{a_C} - \frac{2}{a_B} h(\eta_+(k)) - ikC^2(\eta_+(k))} \\[2mm]
&= \frac{e^{2i\sigma_0} C^2(\eta_+(k))}{-\frac{1}{a_C} - \frac{2}{a_B} H(i\eta_+(k))} ,
\end{aligned}
\tag{11.77}
$$

where $\sigma_0 = \arg\Gamma(1 - i\eta_+(k))$ is the Coulomb phase.

For completeness, though not relevant, for NN scattering, we consider the analysis of the situation involving particles of *opposite* charge is similar, except that in this case, in the absence of strong interaction effects, there will exist Coulomb bound states at momenta $k_n = i\kappa_n = i/na_B$ and energy $E_n = -\kappa_n^2/2m_r = -m_r\alpha^2/2n^2$ with $n = 1, 2, 3, \ldots$. In the presence of strong interactions between these particles, however, the energies will be shifted. One approach to the calculation of this shift is to look for the poles in the corresponding scattering process $A + B \to A + B$. Of course, in the case of oppositely charged particles the analysis of the scattering amplitude must be in terms of appropriate Coulomb wavefunctions. If, as before, we include Coulomb effects only in the exterior region, then the appropriate forms of the wavefunctions for $kr \ll 1$ are given in [Tru61] as

$$F_0^-(r) \stackrel{kr\ll 1}{\longrightarrow} C_\eta(\eta_+(k)) \left(1 - \frac{r}{a_B} + \cdots\right)$$

$$G_0^-(r) \stackrel{kr\ll 1}{\longrightarrow} -\frac{1}{C_\eta(\eta_+(k))}$$
$$\left\{\frac{1}{kr} - 2\eta_+(k)\left[H(i\eta_+(k)) - i\frac{\pi}{2}\coth(\pi\eta_+(k)) + 2\gamma - 1 + \ln i\frac{2r}{a_B}\right] + \cdots\right\}.$$

$$(11.78)$$

Equating interior and exterior logarithmic derivatives as before, we find

$$KF(KR) = \frac{\cos\delta_0 F_0^{-\prime}(R) - \sin\delta_0 G_0^{-\prime}(R)}{\cos\delta_0 F_0^-(R) - \sin\delta_0 G_0^-(R)}$$
$$= \frac{-k\cot\delta_0 C^2(-\eta_+(k))\frac{1}{a_B} - \frac{1}{R^2}}{k\cot\delta_0 C^2(-\eta_+(k)) + \frac{1}{R} - \frac{2}{a_B}(h(\eta_+(k)) - \ln\frac{a_B}{2R} + 2\gamma - 1)}.$$

$$(11.79)$$

Thus, we have

$$k\cot\delta_0 C^2(-\eta_+(k)) - \frac{2}{a_B}\left[h(\eta_+(k)) - \ln\frac{a_B}{2R} + 2\gamma - 1\right] \simeq -\frac{1}{a_0}. \qquad (11.80)$$

In this case the scattering length a_C in the presence of the Coulomb interaction is defined as

$$k\cot\delta_0 C^2(-\eta_+(k)) - \frac{2}{a_B}h(\eta_+(k)) = -\frac{1}{a_C}, \qquad (11.81)$$

so that we have the relation

$$-\frac{1}{a_0} = -\frac{1}{a_C} + \frac{2}{a_B}\left(\ln\frac{a_B}{2R} + 1 - 2\gamma\right). \qquad (11.82)$$

The corresponding scattering amplitude is then

$$f_C^-(k) = \frac{e^{-2i\sigma_0} C^2(-\eta_+(k))}{-\frac{1}{a_C} + \frac{2}{a_B}h(\eta_+(k)) - ikC^2(-\eta_+(k))}$$
$$= \frac{e^{-2i\sigma_0} C^2(-\eta_+(k))}{-\frac{1}{a_C} + \frac{2}{a_B}[H(i\eta_+(k)) - i\pi\coth\pi\eta_+(k)]}. \qquad (11.83)$$

Under the continuation $k \to i\kappa$ we have

$$H(i\eta_+(k)) - i\pi \coth(\pi \eta_+(k)) \to H(\xi) + \pi \cot \pi \xi \,, \tag{11.84}$$

where $\xi = 1/\kappa a_B$. The existence of a bound state is then indicated by

$$-\frac{1}{a_C} + \frac{2}{a_B}(H(\xi) + \pi \cot \pi \xi) = 0 \,. \tag{11.85}$$

In the limit of no strong interaction, i.e., $a_C \to 0$, we find $\xi_n = 1/\kappa_n a_B = n$ and the usual Coulomb bound state energies

$$E_n = -\frac{\kappa_n^2}{2m_r} = -\frac{m_r \alpha^2}{2n^2}, \quad n = 1, 2, 3, \dots \,, \tag{11.86}$$

while if $a_C \neq 0$ there exists a solution to Eq. (11.85)

$$\xi = \frac{1}{\kappa_n a_B} \approx n + \frac{2a_C}{a_B} \tag{11.87}$$

and a corresponding energy shift

$$\Delta E_n = -E_n \frac{4a_C}{n a_B} + \mathcal{O}\left(\frac{a_C}{a_B}\right)^2 \,, \tag{11.88}$$

which is the conventional result [Tru61, Des54].

It is important to note here that Eq. (11.88) is written in a form that relates one *experimental* quantity, the energy shift ΔE_n, to another, the scattering length a_C. Hence it is *model-independent*, even though, for clarity, we have employed a particular model in its derivation. This feature means that it is an ideal case for an effective interaction approach, as will be shown in Section 11.8.

11.6 Electromagnetic Interactions: $np \leftrightarrow d\gamma$

Another important low energy probe of the NN interaction can be found within the electromagnetic transition $np \leftrightarrow d\gamma$, which also provides evidence for the presence of MEC in the deuteron. Here the scattering np states include both spin-singlet and -triplet components. For simplicity, we represent the deuteron by purely the asymptotic form of the dominant S-wave component, which has the form

$$\psi_d(r) = \sqrt{\frac{\gamma}{2\pi}} \frac{1}{r} e^{-\gamma r}, \quad \text{or} \quad \psi_d(q) = \frac{\sqrt{8\pi \gamma}}{\gamma^2 + q^2} \,, \tag{11.89}$$

where $\gamma = \sqrt{m_N |E_B|} \simeq 45.7$ MeV is the (imaginary) momentum associated with the bound state energy E_B. Since we are considering an electromagnetic transition at very low energy we can be content to include only the lowest $E1$, $M1$, and $E2$ multipoles (see Chapter 7), which are described by the Hamiltonian

$$\hat{H} = e s_0 \hat{\boldsymbol{\epsilon}}_\gamma \cdot \left[\pm i \frac{1}{2} \boldsymbol{r} + \frac{1}{4m_N} \hat{\boldsymbol{s}}_\gamma \times \left(\mu_V(\boldsymbol{\sigma}_p - \boldsymbol{\sigma}_n) + \left(\mu_S - \frac{1}{2} \right)(\boldsymbol{\sigma}_p + \boldsymbol{\sigma}_s) \right) - \frac{i}{8} \boldsymbol{r} \boldsymbol{r} \cdot \boldsymbol{k} \right] \,. \tag{11.90}$$

Here s_γ is the photon momentum, $\mu_V, \mu_S = \mu_p \pm \mu_n$ are the isoscalar, isovector magnetic moments.[2] The $\pm$ in front of the $E1$ operator depends upon whether the $np \rightarrow d\gamma$ or $\gamma d \rightarrow np$ reaction is under consideration. The electromagnetic transition amplitude then can be written in the form

$$
\begin{aligned}
\text{Amp} = \ & \chi_f^\dagger \Big[\hat{\boldsymbol{\epsilon}}_\gamma \times \hat{\boldsymbol{s}}_\gamma \cdot \big(G_{M1V}(\boldsymbol{\sigma}_p - \boldsymbol{\sigma}_n) + G_{M1S}(\boldsymbol{\sigma}_p + \boldsymbol{\sigma}_n)\big) \\
& + G_{E1}\hat{\boldsymbol{\epsilon}}_\gamma \cdot \hat{\boldsymbol{k}} + G_{E2}\big(\boldsymbol{\sigma}_p \cdot \hat{\boldsymbol{\epsilon}}_\gamma \boldsymbol{\sigma}_n \cdot \hat{\boldsymbol{s}}_\gamma + \boldsymbol{\sigma}_n \cdot \hat{\boldsymbol{\epsilon}}_\gamma \boldsymbol{\sigma}_p \cdot \hat{\boldsymbol{s}}_\gamma\big) \Big] \chi_i \ .
\end{aligned} \tag{11.91}
$$

The leading parity-conserving transition at near-threshold energy is then the isovector $M1$ amplitude G_{M1V} which connects the $^3S_1$ deuteron to the $^1S_0$ scattering state of the np system. From Eq. (11.90) we identify

$$
G_{M1V} = \frac{es_0\mu_V}{4m_N} \int d\mathbf{r}\,\psi_{^1S_0}^{(-)*}(kr)\psi_d(r) \ . \tag{11.92}
$$

Using the asymptotic form

$$
\psi_{^1S_0}^{(-)}(kr) = \frac{e^{-i\delta_s}}{kr}(\sin kr \cos \delta_s + \cos kr \sin \delta_s) \ , \tag{11.93}
$$

the radial integral becomes

$$
\begin{aligned}
\int d\mathbf{r}\,\psi_{^1S_0}^{(-)*}(kr)\psi_d(r) &= \frac{4\pi e^{i\delta_s}}{k} \int_0^\infty d\mathbf{r}(\sin kr \cos \delta_s + \cos kr \sin \delta_s)e^{-\gamma r} \\
&= \frac{4\pi e^{i\delta_s}}{k(k^2+\gamma^2)}(k \cos \delta_s + \gamma \sin \delta_s) \ . \tag{11.94}
\end{aligned}
$$

Since by energy conservation

$$
s_0 = \frac{k^2+\gamma^2}{m_N} \ ,
$$

we can use the lowest order scattering length value for the scattering phase shift to write this result in the form

$$
G_{M1V} = \frac{e\mu_V\sqrt{8\pi\gamma}\,e^{-i\tan^{-1}ka_0^s}(1-\gamma a_0^s)}{4m_N^2\sqrt{1+k^2a_0^{s2}}} \ , \tag{11.95}
$$

where a_0^s is the spin singlet scattering length. (Note here that the phase of the amplitude is required by the Fermi–Watson theorem, which follows from unitarity.)

The $M1$ cross section is then found by squaring and mutiplying by phase space. In the case of radiative capture this is found to be

$$
\Gamma_{np\rightarrow d\gamma} = \frac{1}{|\mathbf{v}_{\text{rel}}|} \int \frac{d\mathbf{s}}{(2\pi)^3 2s_0} 2\pi\,\delta\!\left(s_0 - \frac{\gamma^2}{m_N} - \frac{k^2}{m_N}\right) \sum_{\lambda_\gamma} \frac{1}{4}\text{Tr}P_t TT^\dagger \ . \tag{11.96}
$$

Here,

$$
\begin{aligned}
\sum_{\lambda_\gamma} \frac{1}{4}\text{Tr}P_t TT^\dagger &= \frac{|G_{M1V}|^2}{4}\text{Tr}\frac{1}{4}(3+\boldsymbol{\sigma}_1\cdot\boldsymbol{\sigma}_2)\hat{\boldsymbol{\epsilon}}_\gamma^* \times \hat{\boldsymbol{s}}_\gamma \cdot (\boldsymbol{\sigma}_p - \boldsymbol{\sigma}_n)\hat{\boldsymbol{\epsilon}}_\gamma \times \hat{\boldsymbol{s}}_\gamma \cdot (\boldsymbol{\sigma}_p - \boldsymbol{\sigma}_n) \\
&= \sum_{\lambda_\gamma} |G_{M1V}|^2 \hat{\boldsymbol{\epsilon}}_\gamma^* \times \hat{\boldsymbol{s}}_\gamma \cdot \hat{\boldsymbol{\epsilon}}_\gamma \times \hat{\boldsymbol{s}}_\gamma = 2|G_{M1V}|^2 \ , \tag{11.97}
\end{aligned}
$$

---

[2] Note that the factor of two (eight) in the $E1(E2)$ component arises from the obvious identity $r_p \rightarrow \frac{1}{2}r$.

yielding

$$\sigma_{M1}(np \to d\gamma) = \frac{s_0}{2\pi |\mathbf{v}_{\text{rel}}|} 2|G_{M1V}|^2 = \frac{2\pi\alpha\mu_V^2\gamma(1 - \gamma a_0^s)^2(k^2 + \gamma^2)}{m_N^5(1 + k^2 a_0^{s2})} . \tag{11.98}$$

Putting in numbers, we find that for an incident thermal neutrons with relative velocity $|\mathbf{v}_{\text{rel}}| = 2200 \,\text{m sec}^{-1}$ the predicted cross section is about 300 mb which is about 10% smaller than the experimental value

$$\sigma_{exp} = 334 \pm 0.1 \,\text{mb} .$$

The discrepancy is due to the feature that the deuteron is not simply a system represented by a simple $|np\rangle$ Fock-space configuration. Rather a full calculation must include mesonic degrees of freedom and the 10% shortfall has been shown by Riska and Brown [Ris72] to be due to the omission of two-body effects (meson exchange currents; see also Chapter 7), specifically the diagram wherein the photon is emitted from a charged pion exchanged between proton and neutron together with the charged pion exchange diagram containing a contact $2N\pi\gamma$ (Kroll–Ruderman) term [Kro54]. In the EFT picture the "extra" piece arises from inclusion of an $M1$ four-nucleon–photon contact term connecting $^3S_1$ and $^1S_0$ states [Bea01]. A recent lattice QCD evaluation yielded a value for this contact term in good agreement with the meson exchange prediction and thereby with the experimental cross section [Bea15].

11.7 Effective Field Theory: the *NN* Interaction

In our previous discussion in Section 11.5, we employed the conventional analysis of the *NN* interaction. However, at low energies it is also useful to study the *NN* system by means of effective field theory (EFT) methods, which we describe here. If we are interested in interactions at the very lowest energies, then a scattering length description is adequate. In order to understand the basic idea of this technique, we reviewed in Sec. 11.5 the conventional way of doing quantum mechanics, which is to assume a particular form for the potential describing a particular situation and then solve the resulting Schrödinger equation. However, if one is interested in experiments in a limited energy range this is more input information than one really needs: once a potential has been chosen, the problem is completely solved at *arbitrary* energy. Using EFT methods, one selects an approximate potential in terms of an observable, the scattering length, which is valid only in the limited energy range of choice. This has a very simple form and, as we will see, is easiest to formulate in momentum space. The results are simply the consequence of quantum mechanics and must be completely consistent with the corresponding Schrödinger calculation.

First consider the situation that we have two particles A and B interacting only via a local strong interaction, so that the effective Lagrangian can be written as

$$\mathcal{L} = \sum_{i=A}^{B} \Psi_i^\dagger \left(i\frac{\partial}{\partial t} + \frac{\nabla^2}{2m_i} \right) \Psi_i - C_0 \Psi_A^\dagger \Psi_A \Psi_B^\dagger \Psi_B + \cdots . \tag{11.99}$$

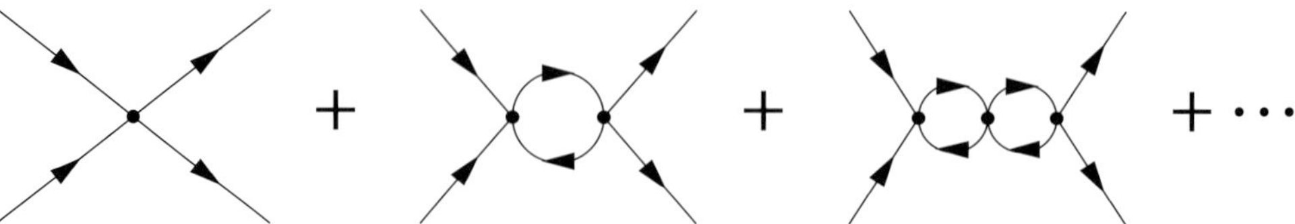

Fig. 11.4 The multiple-scattering series for NN scattering in EFT. Here the solid line designates a nucleon.

The T-matrix is then given in terms of the multiple scattering series (shown in Fig. 11.4)

$$T_{fi}(k) = -\frac{2\pi}{m_r}f(k) = C_0 + C_0^2 G_0(k) + C_0^3 G_0^2(k) + \cdots = \frac{C_0}{1 - C_0 G_0(k)} , \qquad (11.100)$$

where $G_0(k)$ is the amplitude for particles A, B to travel from zero separation to zero separation, i.e, the propagator $D_F(k; r' = 0, r = 0)$

$$G_0(k) = \lim_{r',r\to 0} \int \frac{d\mathbf{s}}{(2\pi)^3} \frac{e^{i\mathbf{s}\cdot r'} e^{-i\mathbf{s}\cdot r}}{\frac{k^2}{2m_r} - \frac{s^2}{2m_r} + i\epsilon} = \int \frac{d\mathbf{s}}{(2\pi)^3} \frac{2m_r}{k^2 - s^2 + i\epsilon} . \qquad (11.101)$$

Equivalently $T_{fi}(k)$ satisfies a Lippman–Schwinger equation

$$T_{fi}(k) = C_0 + C_0 G_0(k) T_{fi}(k) , \qquad (11.102)$$

whose solution is given in Eq. (11.100).

The complication here is that the function $G_0(k)$ is divergent and must be defined via some sort of regularization. There are a number of ways by which to do this, but perhaps the simplest is to use a cutoff regularization with $k_{\max} = \mu$, which simply eliminates the high momentum components of the wavefunction completely. Then

$$G_0(k) = -\frac{m_r}{2\pi}\left(\frac{2\mu}{\pi} + ik\right) . \qquad (11.103)$$

Other regularization schemes are similar. For example, one could subtract at an unphysical momentum point, as proposed by Gegelia [Geg98]

$$G_0(k) = \int \frac{d\mathbf{s}}{(2\pi)^3}\left(\frac{2m_r}{k^2 - s^2 + i\epsilon} + \frac{2m_r}{\mu^2 + s^2}\right) = -\frac{m_r}{2\pi}(\mu + ik) , \qquad (11.104)$$

which has been shown by Mehen and Stewart [Meh99] to be equivalent to the power divergence subtraction (PDS) scheme proposed by Kaplan, Savage, and Wise [Kap98]. In any case, the would-be linear divergence is, of course, canceled by introduction of a counterterm accounting for the omitted high-energy component of the theory, which renormalizes C_0 to $C_0(\mu)$. That $C_0(\mu)$ should be a function of the cutoff is clear because by varying the cutoff energy we are varying the amount of higher-energy physics which we are including in our effective description. The scattering amplitude then becomes

$$f(k) = -\frac{m_r}{2\pi}\left(\frac{1}{\frac{1}{C_0(\mu)} - G_0(k)}\right) = \frac{1}{-\frac{2\pi}{m_r C_0(\mu)} - \frac{2\mu}{\pi} - ik} . \qquad (11.105)$$

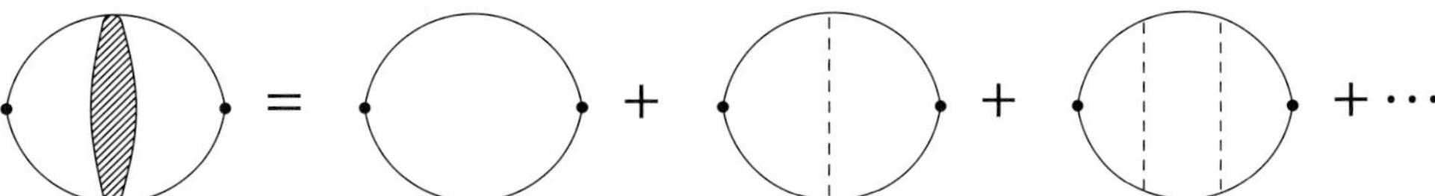

Fig. 11.5 The Coulomb corrected bubble diagram. Here the solid (dashed) line represents a proton (static photon, that is, a static Coulomb interaction).

Comparing with the usual scattering length form

$$f(k) = \frac{1}{-\frac{1}{a_0} - ik} , \tag{11.106}$$

we identify the *S*-wave scattering length a_0 as

$$-\frac{1}{a_0} = -\frac{2\pi}{m_r C_0(\mu)} - \frac{2\mu}{\pi} . \tag{11.107}$$

Of course, since a_0 is a physical observable, it is cutoff-independent, so that the μ-dependence of $1/C_0(\mu)$ is canceled by the cutoff dependence in the Green's function.

More interesting is the case where we include a Coulomb interaction between the particles. The derivatives in Eq. (11.99) then become covariant and the bubble sum is evaluated with static photon exchanges between each of the lines – each bubble is replaced by one involving a sum of zero, one, two, etc., Coulomb interactions, as shown in Fig. 11.5.

The net result in the case of same-charge scattering is the replacement of the free propagator by its Coulomb analog

$$G_0(k) \to G_C^+(k) = \lim_{r',r \to 0} \int \frac{d\mathbf{s}}{(2\pi)^3} \frac{\psi_s^+(\mathbf{r}')\psi_s^{+*}(\mathbf{r})}{\frac{k^2}{2m_r} - \frac{s^2}{2m_r} + i\epsilon}$$

$$= \int \frac{d\mathbf{s}}{(2\pi)^3} \frac{2m_r C^2(\eta_+(s))}{k^2 - s^2 + i\epsilon} , \tag{11.108}$$

where

$$\psi_s^+(\mathbf{r}) = C_\eta(\eta_+(s)) e^{i\sigma_0} e^{i\mathbf{s}\cdot\mathbf{r}} {}_1F_1(-i\eta_+(s), 1, isr - i\mathbf{s}\cdot\mathbf{r}) \tag{11.109}$$

is the outgoing Coulomb wavefunction for repulsive Coulomb scattering [Lan77]. Also, in the initial and final states the influence of static photon exchanges must be included to all orders, which produces the factor $C_\eta^2(2\pi\eta_+(k)) \exp(2i\sigma_0)$. Thus, the repulsive Coulomb scattering amplitude becomes

$$f_C^+(k) = -\frac{m_r}{2\pi} \frac{C_0 C_\eta^2(\eta_+(k)) \exp 2i\sigma_0}{1 - C_0 G_C^+(k)} . \tag{11.110}$$

The momentum integration in Eq. (11.108) can be performed as before using cutoff regularization, yielding

$$G_C^+(k) = -\frac{m_r}{2\pi} \left\{ \frac{2\mu}{\pi} + \frac{2}{a_B} \left[H(i\eta_+(k)) - \ln\frac{\mu a_B}{2\pi} - \zeta \right] \right\} , \tag{11.111}$$

where $\zeta = \ln 2\pi - \gamma$. We have then

$$
f_C^+(k) = \frac{C_\eta^2(\eta_+(k))e^{2i\sigma_0}}{-\frac{2\pi}{m_r C_0(\mu)} - \frac{2\mu}{\pi} - \frac{2}{a_B}\left[H(i\eta_+(k)) - \ln\frac{\mu a_B}{2\pi} - \zeta\right]}
$$

$$
= \frac{C_\eta^2(\eta_+(k))e^{2i\sigma_0}}{-\frac{1}{a_0} - \frac{2}{a_B}\left[h(\eta_+(k)) - \ln\frac{\mu a_B}{2\pi} - \zeta\right] - ikC_\eta^2(\eta_+(k))} . \tag{11.112}
$$

Comparing with Eq. (11.74), we identify the Coulomb scattering length as

$$
-\frac{1}{a_C} = -\frac{1}{a_0} + \frac{2}{a_B}\left(\ln\frac{\mu a_B}{2\pi} + \zeta\right) , \tag{11.113}
$$

which matches nicely with Eq. (11.75) if a reasonable cutoff $\mu \sim m_\pi \sim 1/R$ is employed. The scattering amplitude then has the simple form

$$
f_C^+(k) = \frac{C_\eta^2(\eta_+(k))e^{2i\sigma_0}}{-\frac{1}{a_C} - \frac{2}{a_B}H(i\eta_+(k))} , \tag{11.114}
$$

in agreement with Eq. (11.77).

Again for completeness we consider oppositely charged particles. In this case the analysis is parallel to that above, but there exists an important new wrinkle in that the intermediate-state sum in the Coulomb propagator must now include bound states

$$
G_0(k) \to G_C^-(k) = \lim_{r',r\to 0}\left[\sum_{n\ell m}\frac{\psi_{n\ell m}(r')\psi_{n\ell m}^*(r)}{\frac{k^2}{2m_r} + \frac{m_r\alpha^2}{2n^2}} + \int\frac{d\mathbf{s}}{(2\pi)^3}\frac{\psi_s^-(r')\psi_s^{-*}(r)}{\frac{k^2}{2m_r} - \frac{s^2}{2m_r} + i\epsilon}\right]
$$

$$
= \frac{2m_r}{\pi a_B}\sum_{n=1}^{\infty}\frac{\eta_+^2(k)}{n(n^2 + \eta_+^2(k))} + \int\frac{d\mathbf{s}}{(2\pi)^3}\frac{2m_r C_\eta^2(-\eta(s))}{k^2 - s^2 + i\epsilon} , \tag{11.115}
$$

where

$$
\psi_s^-(r) = C(-\eta_+(s))e^{-i\sigma_0}e^{i\mathbf{s}\cdot\mathbf{r}}{}_1F_1(i\eta_+(s), 1, isr - i\mathbf{s}\cdot\mathbf{r}) \tag{11.116}
$$

is the outgoing Coulomb wavefunction for attractive Coulomb scattering and

$$
\psi_{n\ell m}(r) = \left[(2a_B)^3\frac{(n - \ell - 1)!}{2n[(n + \ell)!]^3}\right]^{\frac{1}{2}}e^{-a_B r}(2a_B r)^\ell L_{n-\ell-1}^{2\ell+1}(2a_B r)Y_\ell^m(\theta,\phi) \tag{11.117}
$$

is the bound state wavefunction corresponding to quantum numbers $n\ell m$ with $L_j^i(x)$ being the associated Laguerre polynomial. Using the identity

$$
C_\eta^2(-\eta_+(k)) = -C_\eta^2(\eta_+(k)) + 2\pi\eta_+(k)\coth\pi\eta_+(k) \tag{11.118}
$$

we can write Eq. (11.115) as

$$
\begin{aligned}
G_C^-(k) = & -G_C^+(k) + 2m_r \int \frac{d\mathbf{s}}{(2\pi)^3} \frac{2\pi\,\eta_+(s)\coth\pi\eta_+(s)}{k^2 - s^2 + i\epsilon} \\
& + \frac{2m_r}{\pi a_B} \sum_{n=1}^{\infty} \frac{\eta_+^2(k)}{n(n^2 + \eta_+^2(k))} \\
= & -\frac{m_r}{2\pi}\left\{ \frac{2\mu}{\pi} - \frac{2}{a_B}\left[H(i\eta_+(k)) - i\pi\coth\pi\eta_+(k) - \ln\frac{\mu a_B}{2\pi} - \zeta \right] \right\},
\end{aligned}
$$

$$(11.119)$$

where the integration is done via contour methods and the contribution from the hyperbolic cotangent poles precisely cancels the bound state term. The resulting attractive Coulomb scattering amplitude is given by

$$
\begin{aligned}
f_C^-(k) = & \frac{C_\eta^2(-\eta_+(k))e^{-2i\sigma_0}}{-\frac{2\pi}{m_r C_0(\mu)} - \frac{2\mu}{\pi} + \frac{2}{a_B}\left[H(i\eta_+(k)) - i\pi\coth\pi\eta_+(k) - \ln\frac{\mu a_B}{2\pi} - \zeta \right]} \\
= & \frac{C_\eta^2(-\eta_+(k))e^{-2i\sigma_0}}{-\frac{1}{a_0} + \frac{2}{a_B}\left[h(\eta_+(k)) - \ln\frac{\mu a_B}{2\pi} - \zeta \right] - ikC_\eta^2(-\eta_+(k))}.
\end{aligned}
$$

$$(11.120)$$

Identifying the Coulomb scattering length via

$$
-\frac{1}{a_C} = -\frac{1}{a_0} - \frac{2}{a_B}\left(\ln\frac{\mu a_B}{2\pi} + \zeta \right),
$$

$$(11.121)$$

this reduces to the simple form

$$
f_C^-(k) = \frac{C_\eta^2(-\eta_+(k))e^{-2i\sigma_0}}{-\frac{1}{a_C} + \frac{2}{a_B}[H(i\eta_+(k)) - i\pi\coth\pi\eta_+(k)]},
$$

$$(11.122)$$

in agreement with Eq. (11.83). In order to go to the bound state limit, we can use the continuation

$$
H(i\eta_+(k)) - i\pi\coth\pi\eta_+(k) \xrightarrow{k\to i\kappa} H(\xi) + \pi\cot\pi\xi,
$$

$$(11.123)$$

in which case we find the condition

$$
0 = -\frac{1}{a_C} + \frac{2}{a_B}(H(\xi) + \pi\cot\pi\xi),
$$

$$(11.124)$$

in complete agreement with Eq. (11.85). The dependence of the energy shift on the scattering length has allowed measurement of difficult-to-measure scattering lengths in hadronic atoms such as $\pi^- p$, $\pi^- d$, and pionium ($\pi^+\pi^-$). One can improve this EFT picture by including higher-order effective range effects, but we end our discussion here.

EFT and $np \leftrightarrow d\gamma$

It is clearly interesting to compare conventional and EFT calculations of the same process and we now consider the radiative capture reaction $np \to d\gamma$ in this regard. In an EFT description of this process, we must evaluate the diagrams shown in Fig. 11.6. However,

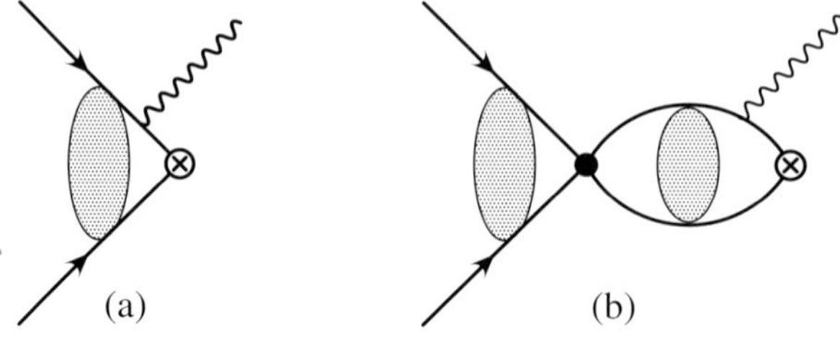

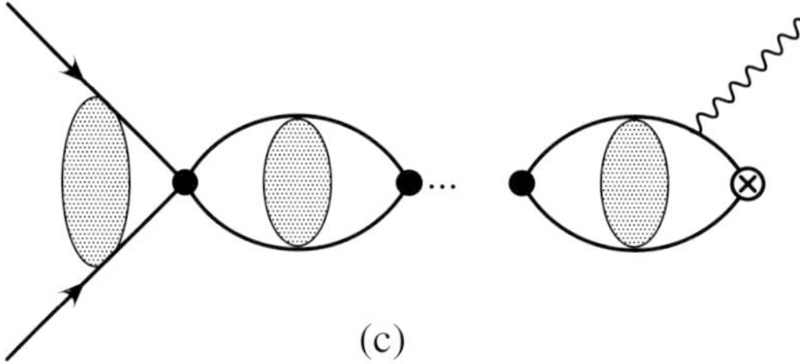

Fig. 11.6   EFT diagrams used in to calculate the radiative capture reaction $np \to d\gamma$. Here the solid (wiggly) lines represent the nucleons (photon) respectively and the cross-circle indicates the deuteron wavefunction.

there is a subtlety here which should be noted. Strictly speaking, as shown by Kaplan, Savage, and Wise [Kap99] the symbol $\otimes$ in these diagrams should be interpreted as creation or annihilation of the deuteron with wavefunction renormalization

$$\sqrt{Z} = \left(\frac{d\Sigma(E)}{dE} \right)^{-\frac{1}{2}}_{E=-B} \tag{11.125}$$

followed by propagation via

$$\frac{1}{\frac{k^2}{m_N} + \frac{\gamma^2}{m_N}} \ .$$

However, since in lowest order we have

$$\left(\frac{d\Sigma(E)}{dE} \right)^{-\frac{1}{2}}_{E=-B} = \frac{\sqrt{8\pi\gamma}}{m_N} \ , \tag{11.126}$$

we find for the product

$$\sqrt{Z} \cdot \frac{1}{\frac{k^2}{m_N} + \frac{\gamma^2}{m_N}} = \frac{\sqrt{8\pi\gamma}}{k^2 + \gamma^2} \ , \tag{11.127}$$

which is the deuteron wavefunction in momentum space. Thus, in the discussion following we shall use this substitution rather than writing the wavefunction normalization times propagator product. From Fig. 11.6 (a) then we find

$$G^a_{M1V} = \frac{e s_0 \mu_V}{4 m_N} \int \frac{d\mathbf{q}}{(2\pi)^3} \psi_k^{(0)*}(\mathbf{q}) \psi_d(\mathbf{q}) \ . \tag{11.128}$$

Since $\psi_k^{(0)*}(q) = (2\pi)^3 \delta^3(k - q)$ we have

$$G_{M1V}^{a} = \frac{e s_0 \mu_V}{4 m_N} \frac{\sqrt{8\pi\gamma}}{\gamma^2 + k^2} \,. \qquad (11.129)$$

On the other hand, from Fig. 11.6 (b) and (c) we find

$$G_{M1V}^{b+c} = \frac{e s_0 \mu_V}{4 m_N} \left(\frac{C_{0s}}{1 - C_{0s} G_0(k)} \right)^* \int \frac{d\mathbf{q}}{(2\pi)^3} G_0(r = 0, q) \psi_d(q) \,. \qquad (11.130)$$

Since

$$G_0(r = 0, q) = \frac{1}{\frac{k^2}{m_N} - \frac{q^2}{m_N} + i\epsilon} \,, \qquad (11.131)$$

this becomes

$$\begin{aligned}
G_{M1V}^{b+c} &= \frac{e s_0 \mu_V}{4 m_N} \left(\frac{C_{0s}}{1 - C_{0s} G_0(k)} \right)^* \int \frac{d\mathbf{q}}{(2\pi)^3} \frac{\sqrt{8\pi\gamma}}{(q^2 + \gamma^2)(\frac{k^2}{m_N} - \frac{q^2}{m_N} + i\epsilon)} \\
&= \frac{e s_0 \mu_V}{4 m_N} \left(\frac{C_{0s}}{1 - C_{0s} G_0(k)} \right)^* \frac{\sqrt{8\pi\gamma}}{\gamma^2 + k^2} (G_0(k) - G_0(i\gamma)) \\
&= \frac{e s_0 \mu_V}{4 m_N} \left(\frac{C_{0s}}{1 - C_{0s} G_0(k)} \right)^* \frac{\sqrt{8\pi\gamma}}{\gamma^2 + k^2} \frac{m_N}{4\pi} (-ik - \gamma) \,. \qquad (11.132)
\end{aligned}$$

Adding the two contributions we have

$$\begin{aligned}
G_{M1V}^{\text{tot}} &= \frac{e s_0 \mu_V}{4 m_N} \frac{\sqrt{8\pi\gamma}}{\gamma^2 + k^2} \left[1 - \left(\frac{-\gamma - ik}{\frac{-1}{a_0^s} - ik} \right) \right] \\
&= \frac{e s_0 \mu_V}{4 m_N} \frac{\sqrt{8\pi\gamma}}{\gamma^2 + k^2} \left(\frac{1 - \gamma a_0^s}{1 + ik a_0^s} \right) \,, \qquad (11.133)
\end{aligned}$$

which agrees completely with the result obtained earlier via conventional coordinate-space procedures. Of course, we still have a $\sim 10\%$ discrepancy with the experimental cross section, which in the EFT evaluation is handled by inclusion of a four-nucleon $M1$ counterterm (representing the meson-exchange contribution) connecting $^3S_1$ and $^1S_0$ states

$$\mathcal{L}_2^{EM} = e L_1^{M1V} (N^T \mathbf{P} \cdot \mathbf{B} N)^\dagger (N^T P_3 N) \,, \qquad (11.134)$$

where here the P_i represent relevant projection operators.

11.8 Nucleon–Nucleon Interaction from QCD

A primary goal of nuclear physicists is to understand the properties and interactions of nuclei directly from the Standard Model description of the strong force, QCD. Thus, the nucleon–nucleon interaction must ultimately be understood in terms of quarks and gluons. However, at the low energies relevant to NN interactions in nuclei, QCD is in the strong coupling regime. *Ab initio* calculations are possible only using lattice QCD.

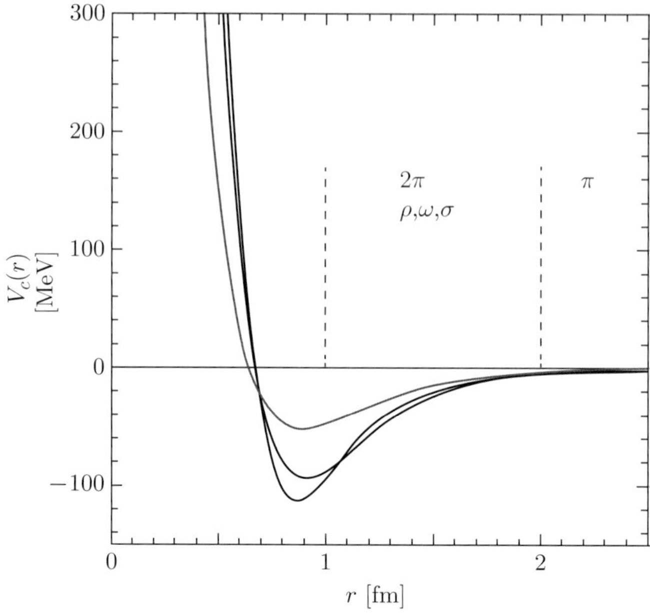

Fig. 11.7 Three examples of the modern NN potential in the $^1S_0$ (spin singlet and S-wave channel): CD-Bonn, Reid93, and AV18 from r =0.8 fm [Ish07].

Determination of the NN interaction directly from QCD using the lattice is very much a frontier area of research, still in its infancy. Calculations to date are approximate with strong assumptions invoked. Here we provide a flavor of the present status. As shown in Fig. 11.7, phenomenological potentials can be characterized by three distinct regions: the long range part ($r \geq 2$ fm) is dominated by single pion exchange; the medium range part (1 fm $\leq r \leq 2$ fm) receives significant contributions from the exchange of multiple mesons (ρ, ω, and σ); and the short range part ($r \leq 1$ fm) is known to have a strong, repulsive core. The short-range repulsive core is essential to explain the stability and saturation of nuclei (see Chapter 13) and for understanding astrophysical observations, such as neutron stars and supernova explosions, where the strong interaction plays a central role (see also Chapter 20). We expect that the repulsive core must have an explanation in terms of QCD.

Due to the steady growth of computational power and the development of innovative numerical algorithms, the calculation of static properties of light hadrons using lattice QCD has become possible, albeit with significant approximations. For example, the NN potential can be calculated on the lattice and then one can extract physical observables such as the scattering phase shift by solving the Schrödinger equation. Figure 11.8 shows results from one such calculation [Ish12] which used a lattice spacing of about 0.091 fm and a pion mass of about 700 MeV. The spatial dependence of the central potential is reasonable and the repulsive core is well reproduced. On the right panel, the calculated $^1S_0$ scattering phase shift shows qualitative behavior similar to the experimental determination. The strength is weaker, most likely due to the heavy pion mass in this calculation.

For nuclei heavier than the deuteron, it is necessary to consider the possibility of a three-nucleon force. Calculation of the binding energies of mass-3 nuclei as well as

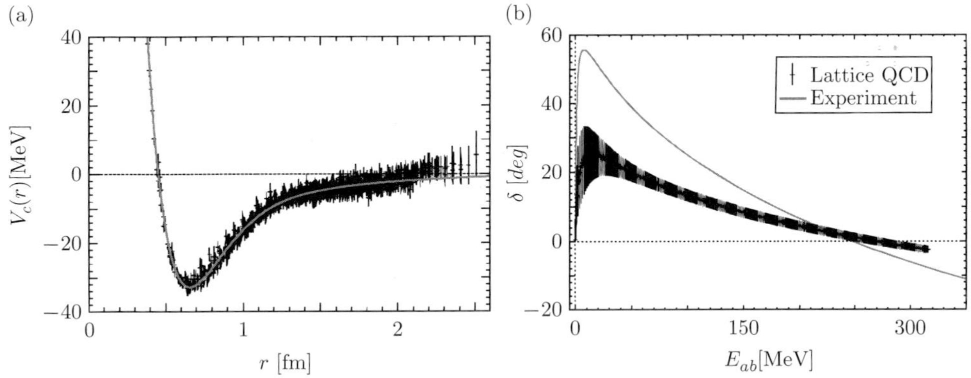

Fig. 11.8   (a) Fit of the central potential $V_C(r)$. (b) The scattering phase shift in the $^1S_0$ channel from the lattice NN potential compared with data [Ish12].

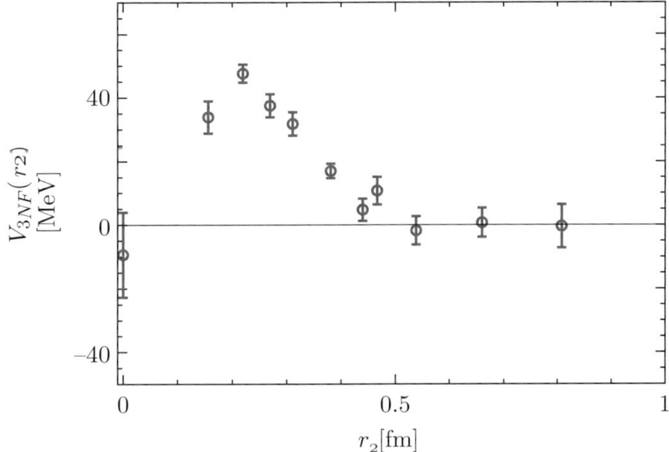

Fig. 11.9   The effective scalar-isoscalar three-nucleon force (3NF) versus r_2 (the distance between the nucleons) in the triton channel [Doi12].

precision deuteron–proton elastic scattering experiments indicate that the effects of three-nucleon forces (3NF) are non-negligible. In addition, three-nucleon forces arise naturally as higher-order terms in chiral EFT. Three-nucleon forces have been calculated in lattice QCD [Doi12] for the case of triton where the three nucleons are arranged in a simplified spatial configuration to minimize the computational cost. In this work, the lattice spacing is 2.5 fm and the pion mass is 1.13 GeV. Figure 11.9 from this work shows that the 3NF are small at large distances, which is consistent with the suppression of the two-pion exchange 3NF by the heavy pion. At short distances, a repulsive 3NF is observed. Note that a short-range, repulsive 3NF force is phenomenologically required to explain the properties of high-density matter.

Lattice calculations of heavier nuclei are also becoming possible. For example, a recent study of the famous Hoyle state in ^{12}C has been carried out [Epe12] which yields results that are qualitatively consistent with intuition based on the three-alpha cluster model of that nucleus. In Chapter 12, we focus on light nuclei, with particular emphasis on $A = 3$ and 4, which is followed by a discussion of the remainder of the periodic table in Chapters 13 through 16.

Exercises

11.1 One-Pion-Exchange Potential (OPEP)

The nucleon–nucleon interaction can be, to a reasonable degree of approximation, be described in terms of a sum over exchanges of mesons $\pi, \sigma, \rho, \omega, \ldots$ between the nucleons. Of these exchanges, the most important at large distances comes from the exchange of the pion, since it has the lightest mass. The goal of this exercise is to find the form of this long-range interaction.

From isospin and parity invariance, the pion-nucleon interaction must have the form

$$\mathcal{L}_{int} = g_{\pi NN} \bar{N} \gamma_5 \boldsymbol{\tau} \cdot \boldsymbol{\pi} N \, ,$$

where $g_{\pi NN}$ is the pion-nucleon coupling constant and has the experimental value

$$\frac{g_{\pi NN}^2}{16\pi} \simeq 13.5 \, .$$

a) Evaluate the pion-nucleon vertex in the nonrelativistic approximation $p/m_N << 1$, and show that

$$i g_{\pi NN} \bar{u}(p_f) \gamma_5 u(p_i) e^{i(p_f - p_i) \cdot x} \xrightarrow{p/m_N << 1} i g_{\pi NN} \frac{1}{2 m_N} \chi_f^\dagger \boldsymbol{\sigma} \cdot (\boldsymbol{p}_f - \boldsymbol{p}_i) \chi_i e^{i(p_f - p_i)}$$

$$= -g_{\pi NN} \frac{1}{2 m_N} \chi_f^\dagger \boldsymbol{\sigma} \cdot \boldsymbol{\nabla} \chi_i e^{i(p_f - p_i)} \, .$$

b) Calculate the NN scattering amplitude in the nonrelativistic limit and show that

$$\mathcal{M}(p_1 + p_2 \to p_1' + p_2') = g_{\pi NN}^2 \bar{u}(p_1') \gamma_5 \boldsymbol{\tau}_1 u(p_1) \cdot \frac{1}{(p_1 - p_1')^2 - m_\pi^2} \bar{u}(p_2') \gamma_5 \boldsymbol{\tau}_2 u(p_2)$$

$$\xrightarrow{p/m_N << 1} -\frac{g_{\pi NN}^2}{4 m_N^2} \chi_{1f}^\dagger \boldsymbol{\tau}_1 \boldsymbol{\sigma}_1 \cdot \boldsymbol{\nabla}_1 \chi_{1i} \frac{1}{(p_1' - p_1)^2 + m_\pi^2} \chi_{2f}^\dagger \boldsymbol{\tau}_2 \boldsymbol{\sigma}_2 \cdot \boldsymbol{\nabla}_2 \chi_{2i} \, .$$

c) Calculate the OPEP by taking the Fourier transform and show that

$$\chi_{2f}^\dagger \chi_{1f}^\dagger V_{OPEP}(r_2 - r_1) \chi_{1i} \chi_{2i} = \int \frac{d\mathbf{q}}{(2\pi)^3} e^{-i q \cdot (r_2 - r_1)} \mathrm{Amp}(\boldsymbol{q})$$

$$= -\frac{g_{\pi NN}^2}{16\pi m_N^2} \boldsymbol{\tau}_1 \cdot \boldsymbol{\tau}_2 \boldsymbol{\sigma}_1 \cdot \boldsymbol{\nabla}_1 \boldsymbol{\sigma}_2 \cdot \boldsymbol{\nabla}_2 \frac{1}{|r_2 - r_1|} \exp -m_\pi |r_2 - r_1| \, .$$

d) Perform the differentiations and show that V_{OPEP} has the form

$$V_{\text{OPEP}}(r) = -\frac{g_{\pi NN}^2}{16\pi m_N^2}\,\boldsymbol{\tau}_1\cdot\boldsymbol{\tau}_2\left\{\left[\left(\frac{1}{(m_\pi r)^2}+\frac{1}{m_\pi r}+\frac{1}{3}\right)\hat{S}_{12}+\frac{1}{3}\boldsymbol{\sigma}_1\cdot\boldsymbol{\sigma}_2\right]\frac{m_\pi^2}{r}e^{-m_\pi r}\right.$$

$$\left.-\frac{4\pi}{3}\boldsymbol{\sigma}_1\cdot\boldsymbol{\sigma}_2\delta^3(r)\right\}\;,$$

where

$$\hat{S}_{12} = 3\boldsymbol{\sigma}_1\cdot\hat{\boldsymbol{r}}\,\boldsymbol{\sigma}_2\cdot\hat{\boldsymbol{r}} - \boldsymbol{\sigma}_1\cdot\boldsymbol{\sigma}_2$$

is the tensor interaction.

The OPEP potential thus consists of two pieces, a tensor interaction having a range $r \sim 1/m_\pi$ together with a local (zero-range) spin-spin term.

11.2   The Deuteron Wavefunction

Enrico Fermi was fond of asking the question "Where's the hydrogen atom for this problem?", meaning what is the simple model that elucidates the basic physics of a given system? In the case of nuclear structure, the answer is clearly the deuteron, and it is essential to have a good understanding of this simplest of nuclear systems at both the qualitative and quantitative levels.

The deuteron is a $J^P = 1^+$ neutron–proton bound state, and we wish to construct the most general ground-state wavefunction. For definiteness, we shall construct the state $|\psi_1^1>$, but other states may be obtained by use of the angular momentum lowering operator. We begin by combining the two spins to produce a state having total spin S either 0 or 1

$$|0,0> = \sqrt{\frac{1}{2}}(|\uparrow_1>\,|\downarrow_2> - |\downarrow_1>\,|\uparrow_2>)$$

$$|1,1> = |\uparrow_1>\,|\uparrow_2>$$

$$|1,0> = \sqrt{\frac{1}{2}}(|\uparrow_1>\,|\downarrow_2> + |\downarrow_1>\,|\uparrow_2>)$$

$$|1,-1> = |\downarrow_1>\,|\downarrow_2>\;.$$

The total angular momentum is given by the sum of this total spin with the orbital angular momentum $\boldsymbol{J} = \boldsymbol{L} + \boldsymbol{S}$. There are then only two possibilities which satisfy the requirements of even parity and unit total angular momentum, $L = 0, S = 1$ and $L = 2$, $S = 1$.

a) Use Clebsch–Gordan coefficients to construct these states and show that

$$\left(\psi_1^1\right)^S(r) = R_0(r)Y_0^0(\theta,\phi)|\uparrow_1>\,|\uparrow_2>$$

$$\left(\psi_1^1\right)^D (r) = R_D(r) \left(\sqrt{\frac{3}{5}} Y_2^2(\theta,\phi)| \downarrow_1 > | \downarrow_2 > \right.$$

$$- \sqrt{\frac{3}{10}} Y_2^1(\theta,\phi) \sqrt{\frac{1}{2}}(| \uparrow_1 > | \downarrow_2 > +| \downarrow_1 > | \uparrow_2 >)$$

$$\left. + \sqrt{\frac{1}{10}} Y_2^0(\theta,\phi)| \uparrow_1 > | \uparrow_2 > \right) .$$

b) The observed magnetic moment of the deuteron provides information about the ground-state wavefunction. The total magnetic moment of the deuteron is due to the intrinsic magnetic moments of the neutron and proton, as well as the orbital angular momentum of the (charged) proton but not the (neutral) neutron. The magnetic moment operator along the 3-axis can then be written as

$$\mu_3 = \frac{1}{2} \mu_N \left(L_3 + g_p \sigma_3^p + g_n \sigma_3^n\right) ,$$

where $\mu_N = e/2m_N$ is the nucleon magneton while $g_p = 5.58$ and $g_n = -3.82$ are the experimental gyromagnetic ratios for the proton and neutron respectively. (In the center-of-mass system the angular momentum of the proton is half that of the total, which is the reason for the factor of $\frac{1}{2}$ multiplying L.) Taking the overall deuteron wavefunction to be

$$\left(\psi_1^1\right)^{\text{tot}} (r) = a_S \left(\psi_1^1\right)^S (r) + a_D \left(\psi_1^1\right)^D (r)$$

with $|a_S|^2 + |a_D|^2 = 1$, calculate the deuteron magnetic moment and show that

$$\frac{1}{\mu_N} < \left(\psi_1^1\right)^{\text{tot}} |M_z| \left(\psi_1^1\right)^{\text{tot}} > = |a_S|^2 \left(\frac{g_p + g_n}{2}\right) + |a_D|^2 \left(\frac{3 - g_p - g_n}{4}\right) .$$

c) Comparing the wavefunction calculation with the experimental value $\mu^{\text{exp}} = 0.857 \mu_N$, show that the two can be made consistent by using $|a_D|^2 \simeq 0.04$, i.e., a $\sim 4\%$ D-state component in the deuteron wavefunction.

11.3   The Deuteron Quadrupole Moment

As discussed in Exercise 11.2, understanding the deuteron magnetic dipole moment requires a small D-state admixture. Thus we anticipate that the deuteron is approximately, but not quite, spherical. We can examine this admixture from another perspective by looking at the deuteron quadrupole moment

$$eQ = \frac{e}{4} < \left(\psi_1^1\right)^{\text{tot}} |(3z^2 - r^2)| \left(\psi_1^1\right)^{\text{tot}} > ,$$

which provides a measure of the nonsphericity. (Here the factor of four is due to the fact that only the proton contributes to the charge distribution.)

a) Writing the quadrupole moment as

$$Q = \frac{1}{4} \int d\mathbf{r} r^2 (3 \cos^2 \theta - 1)| \left(\psi_1^1\right)^{\text{tot}} (r)|^2 ,$$

use the deuteron wavefunction from the previous exercise to show that

$$ \mathcal{Q} = \frac{1}{5\sqrt{2}} \mathrm{Re}(a_S^* a_D) <r^2>_{SD} - \frac{1}{20} |a_D|^2 <r^2>_{DD} \, , $$

where

$$ <r^2>_{SD} = <R_S|r^2|R_D> \quad \text{and} \quad <r^2>_{DD} = <R_D|r^2|R_D> \, . $$

b) Use the experimental result

$$ \mathcal{Q}_{\mathrm{exp}} = 0.286\,\mathrm{fm}^2 \, , $$

to extract the magnitude and sign of a_D and compare with size found in the previous exercise.

11.4   The Tensor-Force Operator $\hat{S}_{12}$

One requires matrix elements of the tensor operator

$$ \hat{S}_{12} = 3\boldsymbol{\sigma}_1 \cdot \hat{\boldsymbol{r}} \boldsymbol{\sigma}_2 \cdot \hat{\boldsymbol{r}} - \boldsymbol{\sigma}_1 \cdot \boldsymbol{\sigma}_2 $$

in order to evaluate the influence of pion exchange on the nucleon–nucleon interaction. With this in mind, demonstrate the following properties of $\hat{S}_{12}$:

a) Operating on any spin-singlet state, $\hat{S}_{12}$ yields 0.
b) From parity considerations, $\hat{S}_{12}$ applied to a spin-triplet function with $L = J \pm 1$ gives a linear combination of the triplet functions for $L = J + 1$ and $J-1$ and, applied to a triplet function with $L = J$, reproduces the same function multiplied by a constant, i.e.,

$$ \hat{S}_{12} \mathcal{Y}^M_{L1J} = a_{L,J+1} \mathcal{Y}^M_{J+1,1J} + a_{L,J-1} \mathcal{Y}^M_{J-1,1J} \quad \text{if} \quad L = J \pm 1 $$

and

$$ \hat{S}_{12} \mathcal{Y}^M_{J1J} = a_{J,J} \mathcal{Y}^M_{J1J} \, . $$

c) The operator $\hat{S}_{12}$ commutes with J_3, and hence the coefficients $a_{L,J+1}$, $a_{L,J-1}$, and $a_{J,J}$ are independent of M.
d) $a_{J+1,J-1} = a_{J-1,J+1}$.
e) By considering the particular case in which $\hat{\boldsymbol{r}}$ is along the z-axis and $M = 1$, and using the fact that $Y_{lm}(\theta = 0) = \sqrt{(2l+1)/4\pi}\,\delta_{m0}$, show that

$$ a_{J,J} = <\mathcal{Y}^M_{J1J}|\hat{S}_{12}|\mathcal{Y}^M_{J1J}> = 2 \, . $$

f) Similarly, taking the special case $\hat{\boldsymbol{r}} = \boldsymbol{k}$, $M=0$ and $M=1$, use the result obtained in b) above to find two equations between $a_{J-1,J+1}$ and $a_{J-1,J-1}$. Thus, show that

$$ <\mathcal{Y}^M_{J-1,1J}|\hat{S}_{12}|\mathcal{Y}^M_{J-1,1J}> = \frac{-2(J-1)}{2J+1} $$

$$ <\mathcal{Y}^M_{J+1,1J}|\hat{S}_{12}|\mathcal{Y}^M_{J-1,1J}> = \frac{6\sqrt{J(J+1)}}{2J+1} $$

$$ <\mathcal{Y}^M_{J+1,1J}|\hat{S}_{12}|\mathcal{Y}^M_{J+1,1J}> = \frac{-2(J+2)}{2J+1} \, . $$

11.5 Energy-Dependence of *NN* Phase Shifts

Experimentally, the phase shifts in nucleon–nucleon scattering are determined as a function of kinetic energy for the isovector ($T = 1$) and isoscalar ($T = 0$) channels. They are shown in detail in [Pia15].

a) Note that both the $^3S_1$ and $^1S_0$ phase shifts have zeros at approximately the same energy. What does this mean? Estimate the spatial separation between the nucleons at which this vanishing occurs.

b) Consider the effect of the tensor force in the P states. Assume $V(r) = V_c(r) + V_T(r)\hat{S}_{12}$. Using the results of Exercise 11.4, show that for two nucleons in the ^3P_J state

$$<^3 P_0|\hat{S}_{12}|^3P_0> = -4$$
$$<^3 P_1|\hat{S}_{12}|^3P_1> = +2$$
$$<^3 P_2|\hat{S}_{12}|^3P_2> = -\frac{2}{5}.$$

Now determine the sign of $V_T(r)$ in the P states.

c) Now consider the spin-orbit force $V_{LS}(r)\,\boldsymbol{L}\cdot\boldsymbol{S}$, which will play a significant role at higher energies. Show that the effect of this additional term is to change the slope of the $^3P_0$ phase shift if $V_{LS} < 0$.

d) Similarly, show that for two nucleons in the ^3D_J state,

$$<^3 D_1|\hat{S}_{12}|^3D_1> = -2$$
$$<^3 D_2|\hat{S}_{12}|^3D_2> = +2$$
$$<^3 D_3|\hat{S}_{12}|^3D_3> = -\frac{4}{7}.$$

Consequently, assuming central and tensor forces dominate, show that $\delta(^3D_2) > \delta(^3D_3) > \delta(^3D_1)$, which is consistent with the experimental determinations of [Pia15].

The Structure and Properties of Few-Body Nuclei

12.1 Introduction

In this chapter, few-body nuclei will be defined as those created through big bang nucleosynthesis, namely lithium, beryllium, and lighter elements. All elements heavier than beryllium and lithium were created much later, by stellar nucleosynthesis in evolving and exploding stars. This is discussed in detail in Chapter 20. In Chapter 11, the ground-state structure and properties of the deuteron have been discussed in detail. Here, the main focus will be on what can be learned from electroproduction from masses $A = 2$, 3, and 4. In addition, the topics of hypernuclei and fusion will be discussed.

Big bang nucleosynthesis (BBN) began a few minutes after the big bang, when the universe had cooled down sufficiently ($E_\gamma < 2.2$ MeV $\Rightarrow T < 10^{10}$ K) to allow deuterium nuclei to survive photodisintegration by high-energy photons. At this temperature, nucleosynthesis can take place and protons and neutrons can interact to form deuterium. Most of the deuterium then collided with other protons and neutrons to produce helium and a small amount of tritium. Lithium-7 could also form via the coalescence of one tritium and two deuterium nuclei. The BBN produced the stable nuclei ^{2}H, ^{3}He, ^{4}He, ^{6}Li, and ^{7}Li as well as the radioactive nuclei ^{3}H, ^{7}Be, and ^{8}Be. No elements heavier than beryllium were produced due to the absence of stable nuclei with five or eight nucleons. Note that this bottleneck is overcome in stars by triple collisions of ^{4}He nuclei, producing carbon. However, this process is very slow, taking tens of thousands of years to convert a significant amount of helium to carbon in stars. (See Chapter 20 for further discussion of the Big Bang and astrophysics.)

The theory of BBN predicts that roughly 25% of the mass of the universe consists of helium, with about 0.01% deuterium and smaller quantities of lithium. This prediction depends critically on the density of baryons (neutrons and protons) at the time of nucleosynthesis. The observation that helium is nowhere seen to have an abundance below 23% is strong evidence that the universe went through an early, hot phase. Further support comes from the consistency of the light element abundances for a particular value of baryon density and an independent measurement of this quantity from the anisotropies in the cosmic microwave background (CMB). There are no known post-big bang processes which can produce significant amounts of deuterium. Hence, observations about deuterium abundance suggest that the universe is finite, which is consistent with big bang theory. For ^{7}Li, there is a significant discrepancy (a factor of two to four) between the BBN prediction and observation.

Few-body nuclei play a significant role in energy generation in stars. Further, efforts have been underway for many decades to produce controlled fusion in the laboratory as a means to generate electrical power. These efforts must overcome many challenging technical obstacles. However, the promise of a power source based on a plentiful source of fuel which does not produce carbon dioxide in the Earth's atmosphere has motivated continued worldwide development of fusion.

Mature theoretical frameworks have been developed in which to calculate the properties and structure of few-body nuclei. These include: chiral effective field theory motivated by QCD, quantum Monte Carlo techniques, and Faddeev techniques. In addition, relativistic approaches have been developed. The sophisticated theoretical treatment of few-body systems motivates precision measurements which can be compared to stringent calculations.

In 1960, Faddeev formulated his mathematically rigorous scattering theory for three particles by proposing a set of coupled integral equations, which have a unique solution. In the first numerical realizations of this approach, separable interactions were introduced, reducing the three-body equations to a set of one-dimensional coupled integral equations, whose numerical solutions were feasible at the time. The standard formulation of the momentum-space Faddeev equations in the continuum presents a challenge for computation. Only in the 1980s were the Faddeev equations solved in the continuum with a realistic nucleon–nucleon force as input. Over the last several decades, calculations of nucleon–deuteron scattering experienced large improvements and refinements. Below pion production threshold, the momentum-space Faddeev equations for three-nucleon scattering can now be solved for the most modern two- and three-nucleon forces.

The nucleon mass can be calculated by treating three quarks in a Faddeev equation. The resulting current-mass evolution of the nucleon mass compares well with lattice data and deviates only by about 5% from the quark-diquark result obtained in previous studies.

12.2   Elastic Electron–Deuteron Scattering and Meson-Exchange Currents

A recurring issue in this chapter is the significant role of meson-exchange currents, These have been introduced earlier, in Chapter 7 and in Section 11.2. Detailed information about the structure of the deuteron can be obtained from electron scattering. For example, at high momentum transfers, elastic electron–deuteron scattering probes the effects of meson-exchange currents and special relativity. In the one-photon exchange approximation, elastic electron–deuteron scattering is fully described by three deuteron form factors. In its most general form, the relativistic deuteron current can be written [Gil02] as a matrix element between initial- and final-state deuteron wavefunctions

$$
- <\Psi'_d|J^\mu|\Psi_d> = \left(\left\{ G_1(Q^2)[\xi'^\star \cdot \xi] - G_3(Q^2)\frac{(\xi'^\star \cdot q)(\xi \cdot q)}{2m_d^2} \right\} (d^\mu + d'^\mu) \right.
$$

$$
\left. + G_2(Q^2)[\xi^\mu(\xi'^\star \cdot q) - \xi'^{\star\mu}(\xi \cdot q)] \right) , \tag{12.1}
$$

where the form factors $G_i(Q^2)$, $i = 1, 2, 3$, are all functions of Q^2, the four-momentum transferred by the electron with $q = d' - d$ and $Q^2 \equiv -q \cdot q$. The polarizations of the incoming (ξ) and outgoing (ξ') deuteron can be used to express the form factors in terms of helicity amplitudes. G_1, G_2, and G_3 are replaced by the charge (C), magnetic (M) and quadrupole (Q) form factors defined as

$$G_M = G_2 \tag{12.2}$$

$$G_C = G_1 + \frac{2}{3}\eta G_Q \tag{12.3}$$

$$G_Q = G_1 - G_M + (1 + \eta)G_3 , \tag{12.4}$$

where $\eta = Q^2/4m_d^2$. At $Q^2 = 0$, the form factors G_C, G_M, and G_Q give the charge, magnetic, and quadrupole moments of the deuteron

$$G_C(0) = 1 \tag{12.5}$$

$$G_Q(0) = \mathcal{Q}_d \tag{12.6}$$

$$G_M(0) = \mu_d . \tag{12.7}$$

The cross section can then be written as

$$\frac{d\sigma}{d\Omega_e} = \sigma_{\text{Mott}} \left(A(Q^2) + B(Q^2) \tan^2 \frac{\theta_e}{2} \right) , \tag{12.8}$$

where the functions $A(Q^2)$ and $B(Q^2)$ are expressed in terms of the three form factors:

$$A(Q^2) = G_C^2(Q^2) + \frac{8}{9}\eta^2 G_Q^2(Q^2) + \frac{2}{3}\eta G_M^2(Q^2) \tag{12.9}$$

$$B(Q^2) = \frac{4}{3}(1 + \eta)\eta G_M^2(Q^2) . \tag{12.10}$$

Both $A(Q^2)$ and $B(Q^2)$ can be separated by a series of measurements of the cross section as a function of electron scattering angle θ_e. The magnetic form factor can be uniquely and directly determined from $B(Q^2)$. Figures 12.1 and 12.2 confront the experimental data for $A(Q^2)$ and $B(Q^2)$ with theoretical calculations based on several phenomenological NN interactions and demonstrate the sensitivity to MEC. The Q^2 dependence of $A(Q^2)$ is reasonably described by theory. A satisfactory description of the data for $B(Q^2)$ at $Q^2 > 0.4\,(\text{GeV/c})^2$ requires the inclusion of MEC in the theoretical description. For $Q^2 > 1.5\,(\text{GeV/c})^2$ they become the dominant effect. The separation of the charge and quadrupole form factors is not possible without an additional polarization experiment. Either the spin asymmetry in electron scattering from a tensor polarized deuterium target or the final-state polarization in the electron scattering of the outgoing deuteron from an unpolarized target are required. For the case of scattering from a tensor-polarized deuterium target, the cross section is

$$\sigma = \sigma_0 \left(1 + \frac{1}{\sqrt{2}} P_{zz} A_d^T \right) , \tag{12.11}$$

where the target tensor asymmetry is

$$A_d^T = \frac{3\cos^2\theta^* - 1}{2} T_{20} - \sqrt{\frac{3}{2}} \sin 2\theta^* \cos\phi^* T_{21} + \sqrt{\frac{3}{2}} \sin^2\theta^* \cos 2\phi^* T_{22} . \tag{12.12}$$

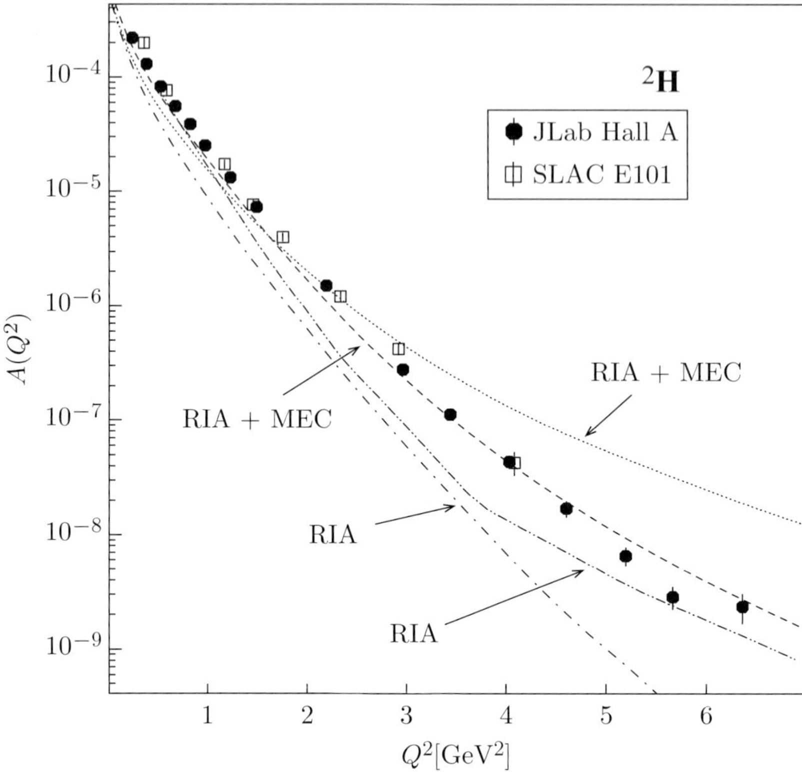

Fig. 12.1 Data for $A(Q^2)$ in elastic electron–deuteron scattering [Ale99] compared with two different theoretical calculations [Hum90, Van95] showing the sensitivity to meson-exchange currents.

Here the polarization direction is described by the polar and azimuthal angles θ^* and ϕ^*, respectively. The quantity $P_{zz} = n_+ + n_- - 2n_0$ is the target tensor polarization, where n_+, n_0, and n_- are the relative populations of the nuclear spin projections $m = +1, 0, -1$ along the direction of target polarization. Figure 12.3 shows a compilation of world data [Zha11] for T_{20} and T_{21} up to a momentum transfer of 4.5 fm^{-1}. The data are consistent with the predictions of effective field theory.

12.3 Threshold Deuteron Electrodisintegration

Deuteron electrodisintegration at threshold is another means to probe the exchange of mesons between the bound proton–neutron system. The transition from the $^3S_1 + {}^3D_1$ ground state to the barely unbound $^1S_0$ first excited state is an isovector magnetic dipole ($M1$) transition. Meson-exchange currents are enhanced relative to the nucleon currents in this transition. By performing the measurement at backward angles, where the longitudinal and elastic contributions are small, sensitivity to MEC is enhanced. Figure 12.4 shows the distribution of scattered electrons from deuterium at 155° and a beam energy of

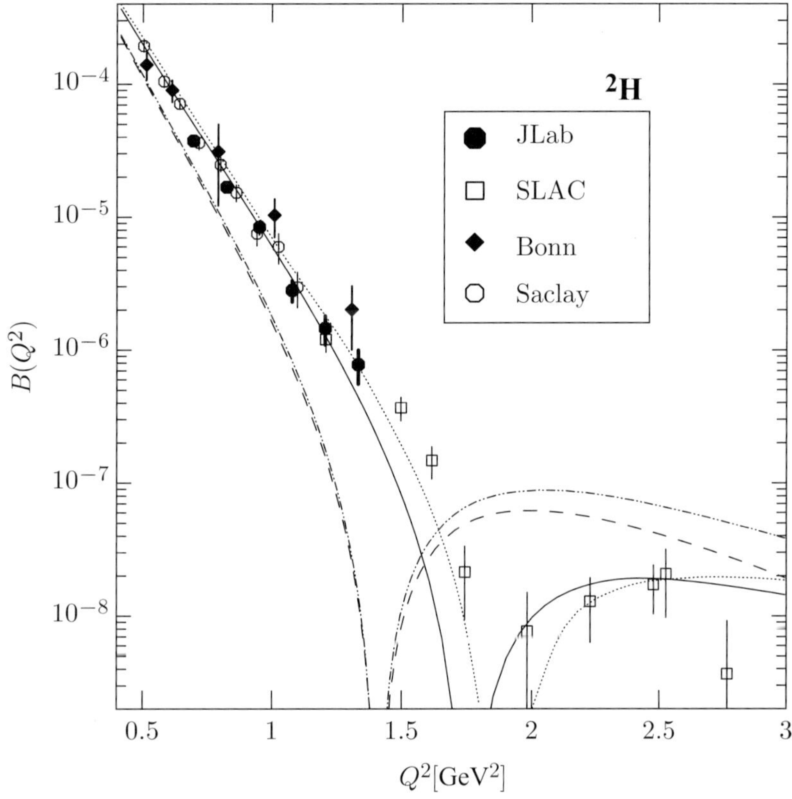

Fig. 12.2 Data for $B(Q^2)$ from elastic electron–deuteron scattering [Pet00] compared with theoretical calculations without (IA) and with (IA+MEC) meson-exchange currents.

360 MeV. The elastic peak is at zero excitation energy and the cusp due to the $^1S_0$ state is at about 3 MeV. Figure 12.5 shows threshold deuteron electrodisintegration data from [Ber81] and [Auf85] as a function of Q^2. The dotted curve is the impulse approximation, the dash-dotted curve includes π-exchange, and the dashed curve includes, in addition, ρ-exchange. The solid curve is the total result, in which the Δ-isobar is also included.

12.4 Deuteron S- and D-State Probed in Spin-dependent $(e, e'p)$ Electron Scattering

To be able to probe the D-state in scattering experiments, it is necessary to introduce polarization observables. Consider scattering of a polarized electron beam from a polarized (both vector and tensor) deuteron target. In the final state, the scattered proton is detected in coincidence with the scattered electron. The electron loses energy ω and the missing momentum p_m of the proton is defined as the difference between the measured proton momentum and the virtual photon three-vector: $\mathbf{p}_m \equiv \mathbf{p}' - \mathbf{q}$. In the case that only the

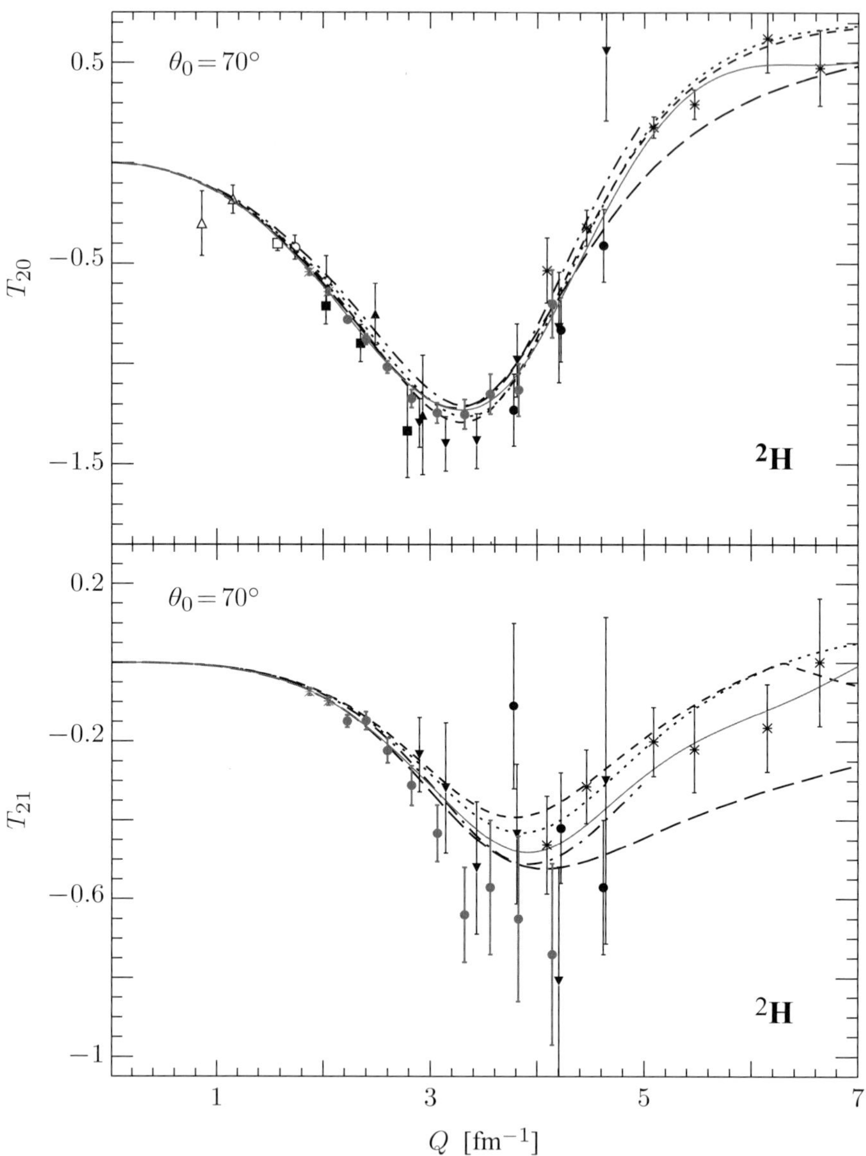

Fig. 12.3 A compilation of world data [Zha11] on T_{20} and T_{21} versus Q.

proton interacts with the photon, the missing momentum is equal to the bound relative momentum of the proton and in this way measuring the asymmetry as a function of this variable approximates probing short distances of the deuteron structure for higher p_m values. Also, to the extent that the Fermi motion of the proton can be ignored, $p_m = 0$ implies that the proton was struck "quasielastically." In this regime, one expects the plane wave impulse approximation (PWIA) to be valid, but as p_m increases, other effects such as meson-exchange currents, isobar configurations, and final-state interactions are expected to make greater contributions. (See Chapter 16 for a more complete discussion of PWIA.)

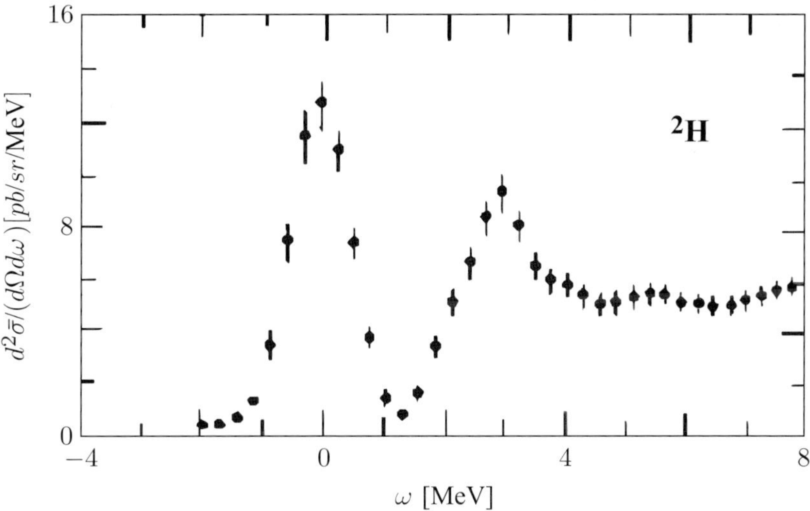

Fig. 12.4 Experimental inclusive cross section versus excitation energy from [Auf85] for electron–deuteron scattering at 360 MeV and 155°.

The total cross section can be written in terms of asymmetries diluted by various combinations of the beam helicity (h), target vector polarization (P_z), and target tensor polarization (P_{zz}) as follows:

$$\frac{d\sigma}{d\omega \, d\Omega_e \, d\Omega_{pn}^{\text{CM}}} = S_0 \left[1 + P_z A_d^V + P_{zz} A_d^T + h \left(A_e + P_z A_{ed}^V + P_{zz} A_{ed}^T\right)\right] , \qquad (12.13)$$

where $\Omega_e \equiv (\theta_e, \phi_e)$ and $\Omega_{pn}^{\text{CM}} \equiv (\theta_{pn}^{\text{CM}}, \phi_{pn})$ are the respective scattered electron and proton–neutron spherical angles. Here S_0 is the unpolarized cross section. The vector asymmetry (see Fig. 12.6) is sizable (~ 0.2 to 0.3) at zero missing momentum. (See Chapter 16 for discussion of missing energies and momenta.) This is consistent with scattering from a free proton polarized in the direction of the deuteron. At low Q^2 and high p_m, the vector asymmetry passes through zero and changes sign. The tensor asymmetry (see Fig. 12.7) vanishes at zero missing momentum and becomes sizable at about $300\,\text{MeV c}^{-1}$. This is consistent with the expectation that the D-state dominates here and in that state both protons are polarized opposite to the direction of the deuteron spin. In the S-state, which dominates at lower momenta, the two protons are anti-polarized.

12.5 The Three-Nucleon Ground State

In a three-body system there are three distinct two-body subsystems. One can sum up the two-body forces in each two-body system, and then in a second step among all three particles. Consider the Schrödinger equation for the three-body system

$$\left(H_0 + \sum_{i=1}^{3} V_i\right) \Psi = E\Psi \qquad (12.14)$$

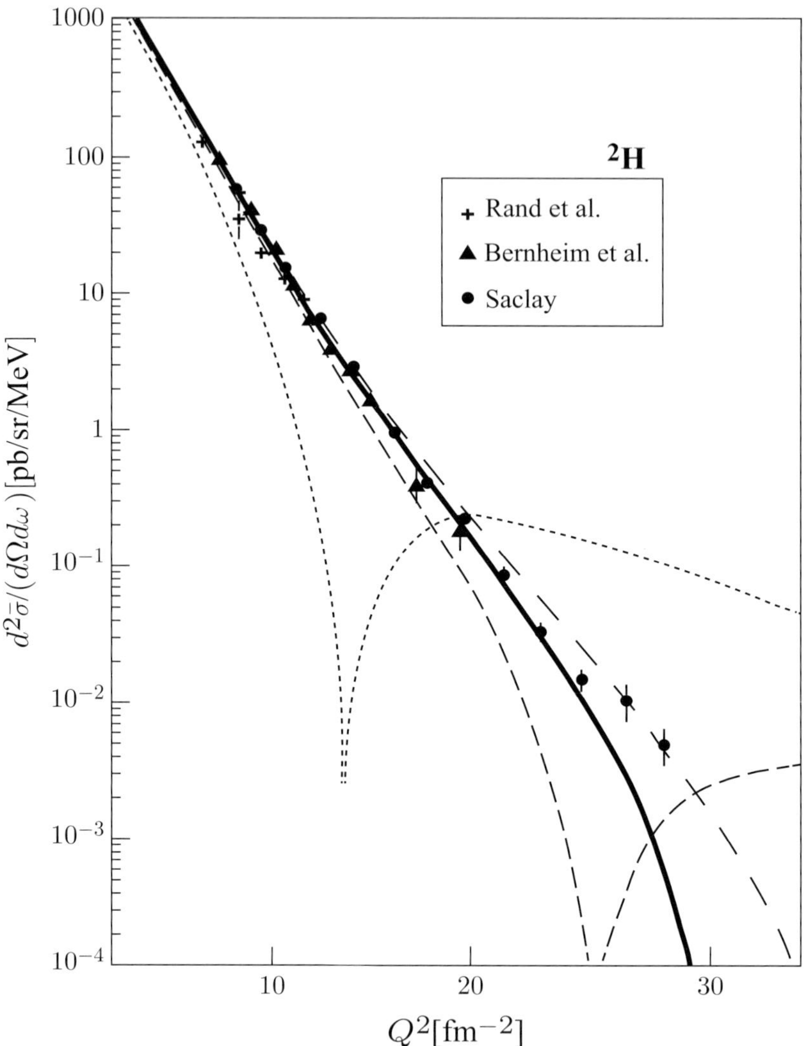

Fig. 12.5   Experimental cross sections from [Auf85] averaged over a region of energy loss extending from threshold to a relative CM energy in the np system of 3 MeV as a function of Q^2 fm^{-2}.

and define $V_1 \equiv V_{23}$, $V_2 \equiv V_{13}$, and $V_3 \equiv V_{12}$. Here, H_0 is the kinetic energy of the relative motion for the three particles. We rewrite Eq. (12.14) as an integral equation

$$\Psi = \frac{1}{E - H_0} \sum_i V_i \Psi \; . \tag{12.15}$$

One can successively iterate this equation as follows

$$\Psi = G_0 \sum_i V_i \Psi = G_0 \sum_i V_i G_0 \sum_j V_j G_0 \cdots \Psi \; . \tag{12.16}$$

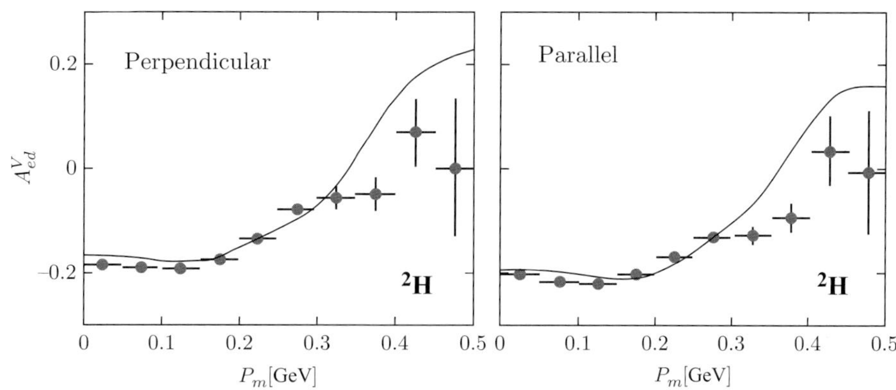

Fig. 12.6 The asymmetry A_{ed}^V in spin-dependent electron scattering from the vector polarized deuteron in the ^{2}H$(e,\ e'p)$ reaction as measured by the BLAST experiment, adapted from [DeG10]. The left (right) panel corresponds to the case where the final-state proton is detected predominantly perpendicular (parallel) to the direction of momentum transfer $\mathbf{q}$. The solid curve is the theoretical prediction. The effect of the D-state causes A_{ed}^V to turn positive at large p_m.

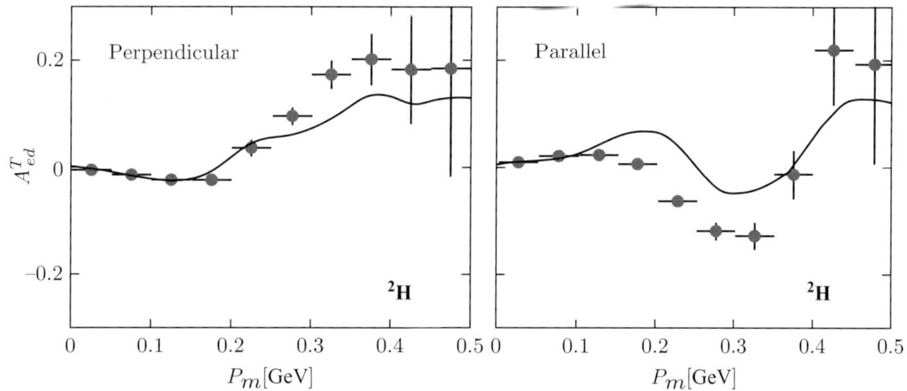

Fig. 12.7 The asymmetry A_{ed}^T in spin-dependent electron scattering from the tensor polarized deuteron in the ^{2}H$(e,\ e'p)$ reaction as measured by the BLAST experiment, adapted from [DeG10]. The left (right) panel corresponds to the case where the final-state proton is detected predominantly perpendicular (parallel) to the direction of momentum transfer $\mathbf{q}$. The solid curve is the theoretical prediction. The effect of the D-state causes A_{ed}^T to become large and positive at large p_m.

Next, we sum up all forces within each pair to infinite-order. This is achieved, according to Faddeev, by decomposing Ψ into three components, known as Faddeev components, $\Psi = \sum_i \psi_i$, with $\psi_i \equiv G_0 V_i \Psi$. Each ψ_i is that part of Ψ which has V_i as the last interaction to the left-hand side. Inserting this decomposition for Ψ on the right-hand side yields

$$\psi_i = G_0 V_i \sum_j \psi_j = G_0 V_i \psi_i + G_0 V_i \sum_{j \neq i} \psi_j \ . \tag{12.17}$$

The first term is responsible for a renewed interaction V_i, whereas in the second term, the next interaction $V_j \neq V_i$. Bringing the first term to the left side yields

$$(1 - G_0 V_i)\psi_i = G_0 V_i \sum_{j \neq i} \psi_j \,. \tag{12.18}$$

Inverting

$$\psi_i = (1 - G_0 V_i)^{-1} G_0 V_i \sum_{j \neq i} \psi_j \,. \tag{12.19}$$

The kernel can be expanded

$$\begin{aligned}
(1 - G_0 V_i)^{-1} G_0 V_i &= (1 + G_0 V_i + G_0 V_i G_0 V_i + \cdots) G_0 V_i \\
&= G_0 (V_i + V_i G_0 V_i + V_i G_0 V_i G_0 V_i + \cdots) \\
&\equiv G_0 t_i \,.
\end{aligned} \tag{12.20}$$

Where t_i is the NN T-matrix for the pair "i," which sums up V_i to infinite order and obeys the Lipmann–Schwinger equation

$$t_i = V_i + V_i G_0 t_i \tag{12.21}$$

or

$$(1 - V_i G_0) t_i = V_i \,. \tag{12.22}$$

Thus, we finally have the result

$$\psi_i = G_0 t_i \sum_{j \neq i} \psi_j \,, \tag{12.23}$$

which is a set of three coupled equations, known as the Faddeev equations. We follow here the discussion of Blankleider and Woloshyn [Bla84] where the ^{3}He wavefunction is determined using the Faddeev equations in momentum-space and the Reid soft-core nucleon–nucleon potential. The bound three-nucleon systems found in Nature are the stable ^{3}He and the radioactive ^{3}H, both of which have $J = \frac{1}{2}, T = \frac{1}{2}$, and have even parity. In the three-body system, there exist only three possible partitions (three irreducible representations of the permutation group of three objects). These are: the completely symmetric representation (S), the completely antisymmetric representation (A), and the "mixed" symmetry representation (M) – see the discussion in Chapter 3 for the qqq three-body problem. The ground state under the action of central forces only is a symmetric $^2 S_{\frac{1}{2}}$ state, i.e., $L = 0, S = \frac{1}{2}$, and a fully symmetric spatial wavefunction. We construct the three-body wavefunction in analogy to that of the deuteron. In the deuteron, we start out with six spatial coordinates for two particles which are reduced to three by working in the center-of-mass frame (the three components of $\mathbf{r}_1 - \mathbf{r}_2$). By employing a classification according to a symmetry group (rotations of space coordinates $\Rightarrow$ quantum number $\mathbf{L}$, which is not actually a good quantum number), we reduce the problem from a partial differential equation in three independent variables to a set of coupled total differential equations for the radial functions $u(r)$ and $w(r)$. This is a great simplification.

For the three-body system, we start out with nine coordinates for the positions of the three particles. Eliminating the center-of-mass from the problem reduces this to six coordinates. We can think of these six coordinates as three coordinates specifying the size and shape of the triangle formed by three particles (for example, the three sides of the triangle: $\mathbf{r}_{12}, \mathbf{r}_{23}$, and $\mathbf{r}_{31}$), and three Euler angles which are needed to specify the orientation in space of this triangle relative to some standard coordinate system. One can solve the Faddeev equation in momentum-space using a NN potential. Assuming three equal-mass nucleons with momenta $\mathbf{k}_1$, $\mathbf{k}_2$, and $\mathbf{k}_3$, one can define the Jacobi coordinates $\mathbf{P}$, $\mathbf{p}_\alpha$, and $\mathbf{q}_\alpha$, as follows

$$\mathbf{P} \equiv \mathbf{k}_\alpha + \mathbf{k}_\beta + \mathbf{k}_\gamma \tag{12.24}$$

$$\mathbf{p}_\alpha \equiv \frac{1}{2}(\mathbf{k}_\beta - \mathbf{k}_\gamma) \tag{12.25}$$

$$\mathbf{q}_\alpha \equiv \frac{1}{3}(\mathbf{k}_\beta + \mathbf{k}_\gamma - 2\mathbf{k}_\alpha) , \tag{12.26}$$

where (α, β, γ) form a cyclic ordering of the particle labels 1, 2, and 3. And $\mathbf{P}$ is, of course, the momentum of the whole system (from now on taken to vanish), $\mathbf{p}_\alpha$ is the momentum of particle β in the $\beta\gamma$ center-of-mass, and $\mathbf{q}_\alpha$ is the negative of particle α's momentum in the three-body center-of-mass. Then, the three-nucleon wavefunction

$$\psi(\mathbf{p}_\alpha, \mathbf{q}_\alpha) \equiv < \mathbf{p}_\alpha \mathbf{q}_\alpha | \psi J M_J T M_T > \tag{12.27}$$

$(J = T = \frac{1}{2}, M_T = \frac{1}{2}$ for ^{3}He, $M_T = -\frac{1}{2}$ for ^{3}H) is expressed in terms of partial waves in JJ coupling

$$\psi(\mathbf{p}_\alpha, \mathbf{q}_\alpha) = \sum_{N_\alpha, M_{N_\alpha}} < l_\alpha M_{l_\alpha} S_\alpha M_{S_\alpha} | j_\alpha M_{j_\alpha} > < L_\alpha M_{L_\alpha} \sigma_\alpha M_{\sigma_\alpha} | J_\alpha M_{J_\alpha} >$$

$$\times < j_\alpha M_{j_\alpha} J_\alpha M_{J_\alpha} | J M_J > < t_\alpha M_{t_\alpha} \tau_\alpha M_{\tau_\alpha} | T M_T >$$

$$\times Y_{l_\alpha M_{l_\alpha}}(\widehat{p}_\alpha) Y_{L_\sigma M_{L_\sigma}}(\widehat{q}_\alpha) | S_\alpha M_{S_\alpha} \sigma_\alpha M_{\sigma_\alpha} >$$

$$\times | t_\alpha M_{t_\alpha} \tau_\alpha M_{\tau_\alpha} > < p_\alpha q_\alpha; \Omega_{N_\alpha}^{TJ} | \psi > , \tag{12.28}$$

where σ_α and τ_α refer to the intrinsic spin and isospin of particle α, l_α is the relative orbital angular momentum of the pair $\beta\gamma$, L_α is the relative orbital angular momentum of the $\beta\gamma$ center-of-mass relative to α, and the remaining quantum numbers are defined by

$$\mathbf{S}_\alpha = \sigma_\beta + \sigma_\gamma \tag{12.29}$$

$$\mathbf{j}_\alpha = \mathbf{l}_\alpha + \mathbf{S}_\alpha \tag{12.30}$$

$$\mathbf{J}_\alpha = \mathbf{L}_\alpha + \sigma_\alpha \tag{12.31}$$

$$\mathbf{J} = \mathbf{j}_\alpha + \mathbf{J}_\alpha \tag{12.32}$$

$$\mathbf{t}_\alpha = \tau_\beta + \tau_\gamma \tag{12.33}$$

$$\mathbf{T} = \mathbf{t}_\alpha + \tau_\alpha . \tag{12.34}$$

The radial part of the wavefunction, $< p_\alpha q_\alpha; \Omega_{N_\alpha}^{TJ} | \psi >$, may be considered to be the overlap of the state $|\psi> \equiv |\psi J M_J T M_T >$ with $|p_\alpha q_\alpha >$ and $|\Omega_{N_\alpha}^{TJ} >$, where

$$|\Omega_{N_\alpha}^{TJ} > \equiv |(t_\alpha \tau_\alpha) T M_T > |[(l_\alpha S_\alpha) j_\alpha; (L_\alpha \sigma_\alpha) J_\alpha] J M_J > \tag{12.35}$$

and N_α represents all the labeled quantum numbers.

To understand the spin structure of the three-body system, it is illuminating to use the partial-wave decomposition of Derrick and Blatt [Der58]. In their LS coupling scheme: $\mathbf{L} = \mathbf{l}_\alpha + \mathbf{L}_\alpha$, $\mathbf{S} = \mathbf{S}_\alpha + \sigma_\alpha$. The spin-isopsin states $|(t_\alpha \tau_\alpha)TM_T > |(S_\alpha \sigma_\alpha)SM_S >$ are linearly combined to make states of definite symmetry under the interchange of any two particle labels, which is achieved by identifying the states $|(S_\alpha \sigma_\alpha)SM_S >$ [or $|(t_\alpha \tau_\alpha)TM_T >$] with basis vectors of irreducible representations of the permutation group S_3. In general, a basis vector is written $|PK >$, where P takes the values of S, A, or M depending on whether the state (and therefore the corresponding irreducible representation) is of "symmetric," "antisymmetric," or "mixed symmetry," and $K = (1$ or $2)$ labels each basis vector within a representation.

The partial-wave decomposition [Der58] involves states of definite symmetry (S, A, or M) under the simultaneous interchange of particle labels in both spin and isospin coordinates. These states $|PK >$ are constructed from the individual $|P_\tau K_\tau >$ and $|P_\sigma K_\sigma >$ according to

$$|(P_\tau P_\sigma)PK > \equiv \sum_{K_\tau K_\sigma} \begin{bmatrix} P_\tau & P_\sigma & P \\ K_\tau & K_\sigma & K \end{bmatrix} |P_\tau K_\tau > |P_\sigma K_\sigma > , \qquad (12.36)$$

involving 3-j symbols (See Chapter 2). The ^{3}He wavefunction may now be written as [Bla84]

$$\psi(\mathbf{p}_\alpha, \mathbf{q}_\alpha) = \sum_{N_\alpha} \sum_{M_L, M_S, M_{l_\alpha} M_{L_\alpha}} (-1)^R (LM_L SM_S | JM)(l_\alpha M_{l_\alpha} L_\alpha M_{L_\alpha} | LM_L)$$

$$\times Y_{l_\alpha M_{l_\alpha}}(\widehat{p}_\alpha) Y_{L_\alpha M_{L_\alpha}}(\widehat{q}_\alpha) |(M_\tau P_\sigma)PK >_{M_S}$$

$$\times \sum_{K_\tau, K_\sigma} \begin{bmatrix} M_\tau & P_\sigma & P \\ K_\tau & K_\sigma & K \end{bmatrix} \sum_{j_\alpha, J_\alpha} \begin{bmatrix} l_\alpha & L_\alpha & L \\ S_\alpha & \sigma_\alpha & S \\ j_\alpha & J_\alpha & J \end{bmatrix} < p_\alpha q_\alpha; \Omega_{N_\alpha}^{TJ} |\psi > , \qquad (12.37)$$

where $N_\alpha = \{L, S \equiv P_\sigma, l_\alpha, L_\alpha, P, K\}$, which we shall refer to as the PK channels, as opposed to the JJ channels labelled by N_α. Implicit in the above equation is a transformation from the JJ coupling scheme (in which the wavefunction is initially given) to the LS coupling scheme.

In Table 12.1 are listed the possible PK channels with $l_\alpha + L_\alpha \leq 4$, together with their percentage probabilities in the ^{3}He wavefunction. The restriction to $l_\alpha + L_\alpha \leq 4$ defines 98.6% of the wavefunction. We note that the two $L = 0$ channels with the largest probabilities (channels 1 and 4) have antisymmetric spin-isospin states (and therefore spatial states) and account for 88.6% [$\equiv P(S)$] of the wavefunction. In these channels, the contributions of the protons to the asymmetry of $^3\overrightarrow{\mathrm{He}}(\vec{e}, e')X$ is identically zero, since interchanging the two protons cannot affect the isospin part of the wavefunction, and thus the spin piece must be antisymmetric. Any contribution of the protons to the asymmetry is owing to the small components of the ^{3}He wavefunction. In this way, polarized ^{3}He can be used as an effective polarized neutron, an idea which has been used to great effect in scattering experiments to understand nucleon structure.

Table 12.1 The partial wave channels of the three-nucleon wave function within the Derrick–Blatt scheme from [Bla84]

Channel number	L	S	l_α	L_α	P	K	Probability (%)
1	0	0.5	0	0	A	1	87.44
2	0	0.5	0	0	M	2	0.74
3	0	0.5	1	1	M	1	0.74
4	0	0.5	2	2	A	1	1.20
5	0	0.5	2	2	M	2	0.06
6	1	0.5	1	1	M	1	0.01
7	1	0.5	2	2	A	1	0.01
8	1	0.5	2	2	M	2	0.01
9	1	1.5	1	1	M	1	0.01
10	1	1.5	2	2	M	2	0.01
11	2	1.5	0	2	M	2	1.08
12	2	1.5	1	1	M	1	2.63
13	2	1.5	1	3	M	1	1.05
14	2	1.5	2	0	M	2	3.06
15	2	1.5	2	2	M	2	0.18
16	2	1.5	2	1	M	1	0.37

Three-Body Forces

High-precision nucleon–nucleon potentials provide a very good description of NN scattering data up to an energy of about 350 MeV. When these potentials are used to predict binding energies of three-nucleon systems, they underestimate the experimental values of ^{3}H and ^{3}He by about 0.5 to 1 MeV. The missing binding energy can be restored by the addition of a three-body force to the nuclear Hamiltonian. Also, the study of elastic nucleon–deuteron scattering and nucleon-induced deuteron breakup revealed a number of cases where the nonrelativistic description using only pairwise forces is insufficient to explain the data. Generally, the discrepancies between a theory using only nucleon–nucleon potentials and experiment becomes larger with increasing energy of the three-nucleon system. Adding a three-nucleon force to the pairwise interactions leads in some cases to a better description of the data. The elastic nucleon–deuteron angular distribution in the region of its minimum and at backward angles is the best studied example. The clear discrepancy in these angular regions at energies up to about 100 MeV nucleon laboratory energy between a theory using only nucleon–nucleon potentials and the cross section data can be removed by adding a modern three-nucleon force to the nuclear Hamiltonian. Such a three-nucleon force must be adjusted for a given nucleon–nucleon potential to the experimental binding of ^{3}H and ^{3}He.

Three-body forces depend in an irreducible way on the simultaneous coordinates of three nucleons when only nucleon degrees of freedom are taken into account. The pion-exchange three-body forces commonly considered in nuclear physics have the schematic

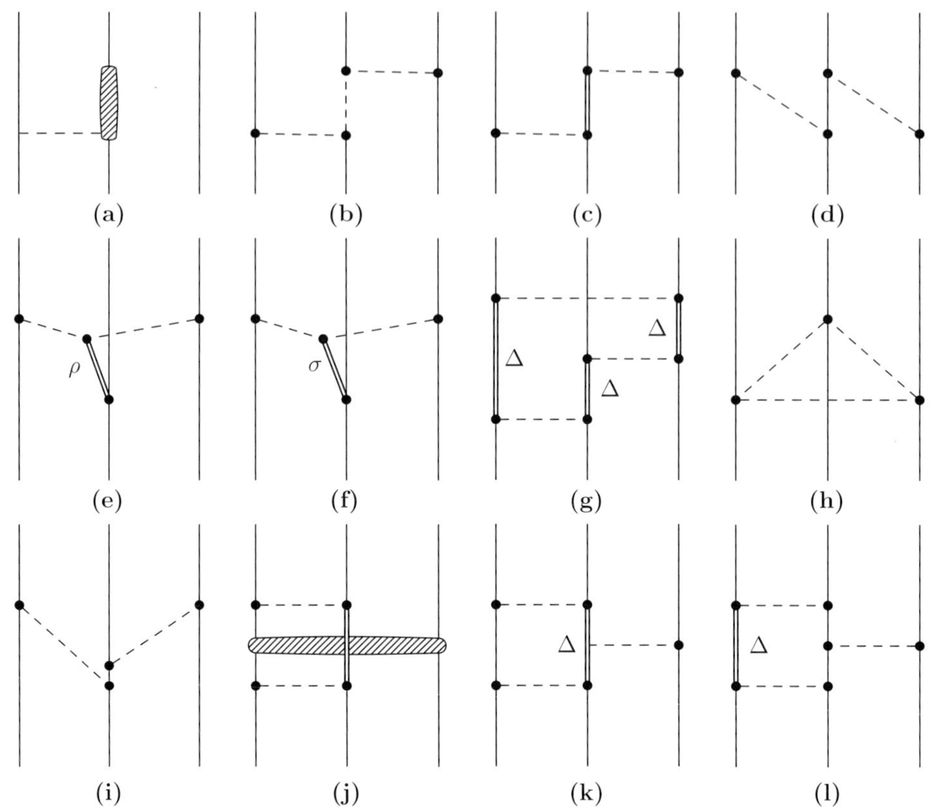

Fig. 12.8 Physical processes contributing to three-nucleon forces; figure adapted from [Fri84]. Solid, dashed, shaded, and double lines depict nucleons, pions, isobars, and heavy mesons, respectively (see text for details).

form V_π^2/m_N, where m_N is the nucleon mass and V_π is the two-body, static one-pion-exchange potential, which can take the form of a relativistic correction. On average, a nucleus is weakly bound and should be nonrelativistic, since the binding energy is of order 1% of the rest mass of the nucleus. Thus, a typical nuclear momentum is about 200 MeV c^{-1}, and $(v/c)^2$ is typically a few percent. We therefore expect that the size of any three-body force contribution to the energy should be of order 1% of the total potential energy ($\sim$50 MeV), i.e., 0.5 MeV.

Figure 12.8 from [Fri84] shows how three-nucleon forces can arise from pionic processes. These dominate because the strong short-range repulsion of the two-nucleon force produces "holes" in the nuclear wavefunction that suppresses the contributions of other operators. Diagram (a) shows the two-pion-exchange three-nucleon force. This longest-range force has a π^+ emitted by the left-most nucleon (changing from a proton to a neutron), subsequent rescattering and charge exchange of the pion on the middle nucleon, and ultimately the absorption of the π^0 on the remaining nucleon. Most of the evidence for three-nucleon forces comes from the three-nucleon system, and all of the evidence is circumstantial but nontrivial. Nonrelativistic Faddeev calculations using two-nucleon forces alone yield binding energies which are too low and produce too large

Table 12.2 Quantum Monte Carlo calculations of $A \le 6$ nuclei of energies (MeV) and radii (fm) from [Bud95]

Nucleus(J)	^{2}H(1)	^{3}H$\left(\frac{1}{2}\right)$	^{4}He(0)	^{5}He$\left(\frac{3}{2}\right)$	^{5}He$\left(\frac{1}{2}\right)$	^{6}He(0)	^{6}Li(1)	^{6}Li(3)
E(expt.)	-2.22	-8.48	-28.3	-27.2	-25.8	-29.3	-32.0	-29.8
E(calc.)	-2.22	$-8.47(2)$	$-28.3(1)$	$-26.5(2)$	$-25.7(2)$	$-28.2(8)$	$-32.4(9)$	$-28.9(6)$
R_n (calc.)	1.967	1.72	1.42(1)	3.02(3)	3.57(3)	2.62(1)	2.41(5)	2.46(7)
R_p (calc.)	1.967	1.58	1.42(1)	1.84(2)	1.99(2)	1.89(6)	2.41(5)	2.46(7)
R_p (expt.)	1.953	1.61	1.47				2.43	

a charge radius. If the tails of the wavefunctions dominate the calculation of the rms radius $< r^2 >^{\frac{1}{2}}$, one expects to find that $< r^2 >^{\frac{1}{2}} \sim E_B^{-\frac{1}{2}}$, as in the deuteron, and underbinding will produce excessively large radii. In particular, the relative diffuseness of the triton causes the long-range parts of the potential to dominate. Modern quantum Monte Carlo calculations [Bud95] using realistic models of two- and three-nucleon interactions reproduce the binding energies and radii of light nuclei up to $A = 6$, as shown in Table 12.2. The calculations are in good agreement with experiment. More recently, three-nucleon forces have been applied to heavier nuclei. Their inclusion in many-body calculations is computationally challenging and has only become feasible in recent years. Three-nucleon forces are essential to the successful calculation [Epe13] of the 7.6 MeV excited state in ^{12}C (the so-called *Hoyle state* – see discussion in Chapter 20). This state is essential for the production of oxygen from carbon in nucleosynthesis.

Elastic Electron Scattering from ^{3}He and ^{3}H

Because their ground-state wavefunctions may be calculated precisely, three-body nuclei strongly test theoretical models regarding meson-exchange currents, Δ isobars, and other non-nucleonic degrees of freedom. Also because of a destructive interference between S- and D-state components of the ground-state wavefunction, the magnetic form factor $F_M(Q^2)$ of ^{3}He is particularly sensitive to non-nucleonic effects. Calculations without non-nucleonic effects predict a diffraction minimum in $F_M(Q^2)$ near $Q^2 = 8 \, \text{fm}^{-2}$, in striking disagreement with experimental results that indicate that the minimum lies in the range $Q^2 = 17\text{–}19 \, \text{fm}^{-2}$. Figure 12.9 shows the most recent data on elastic electron scattering from ^{3}He.

Electron scattering experiments from both ^{3}He and ^{3}H nuclei offer a powerful constraint to theoretical descriptions of few-body nuclei. However, realizing a tritium target for such experiments is a formidable technical challenge. Tritium is a radioactive isotope of hydrogen, comprising one proton and two neutrons. It β-decays via the reaction ^{3}H $\to$ ^{3}He $+ \, e^- + \bar{\nu}_e$ with a half-life of 12.3 years (see Chapter 17). As we have seen, an understanding of the triton binding energy demands the inclusion of three-nucleon forces. Figure 12.10 from [Dow88] shows the determination of the longitudinal response for both ^{3}He and ^{3}H at the MIT-Bates laboratory. Quasielastic inclusive electron scattering was measured from identical cryogenic gas cells of ^{3}H and ^{3}He at two angles (54° and 134.5°) over a wide range of initial and final energies. The longitudinal (R_L, scattering from charges) and

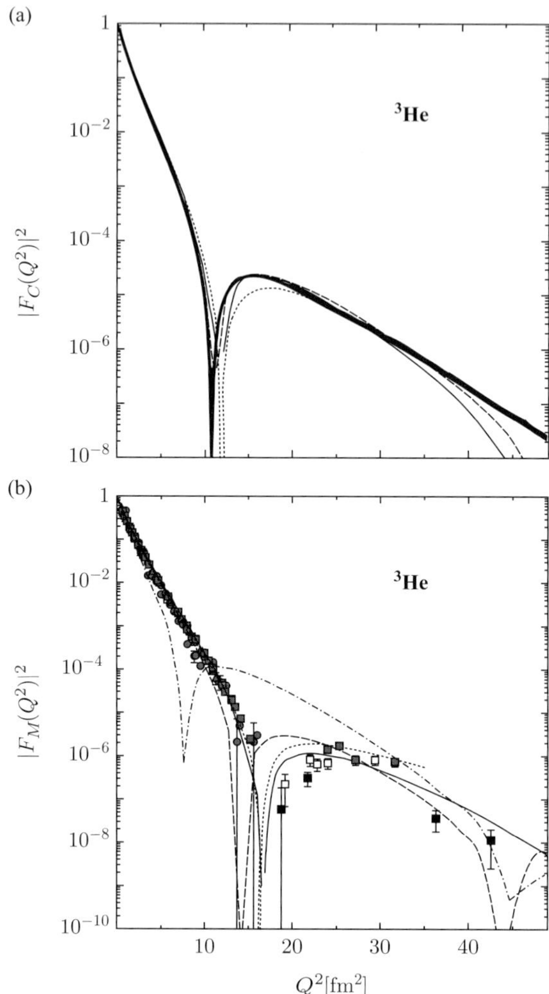

Fig. 12.9 (a) Calculated form factors for elastic charge scattering from ³He and comparison to sum-of-Gaussian fits from [Nak01a]. (b) Elastic magnetic form factor of ³He from [Nak01a].

transverse (R_T, scattering from spins and currents) contributions to the cross section can be determined by performing a Rosenbluth separation, wherein (as discussed in Chapter 7) the cross section is written as

$$\frac{d^2\sigma}{d\Omega dE_f} = \sigma_{Mott}\left[\left(\frac{Q^2}{q^2}\right)^2 R_L(q,v) + \left(\frac{1}{2}\left|\frac{Q^2}{q^2}\right| + \tan^2\frac{\theta}{2}\right) R_T(q,v)\right]. \tag{12.38}$$

An elegant means to compare data to theory, avoiding the difficulties associated with final-state wavefunctions, is to use the Coulomb sum rule (the integrated longitudinal strength at a given three-momentum transfer $q = |\mathbf{q}|$), which tests the initial-state wavefunctions, including correlation effects. These correlation effects are not expected to be important in the Coulomb sum rule when q is larger than twice the Fermi momentum, where the sum

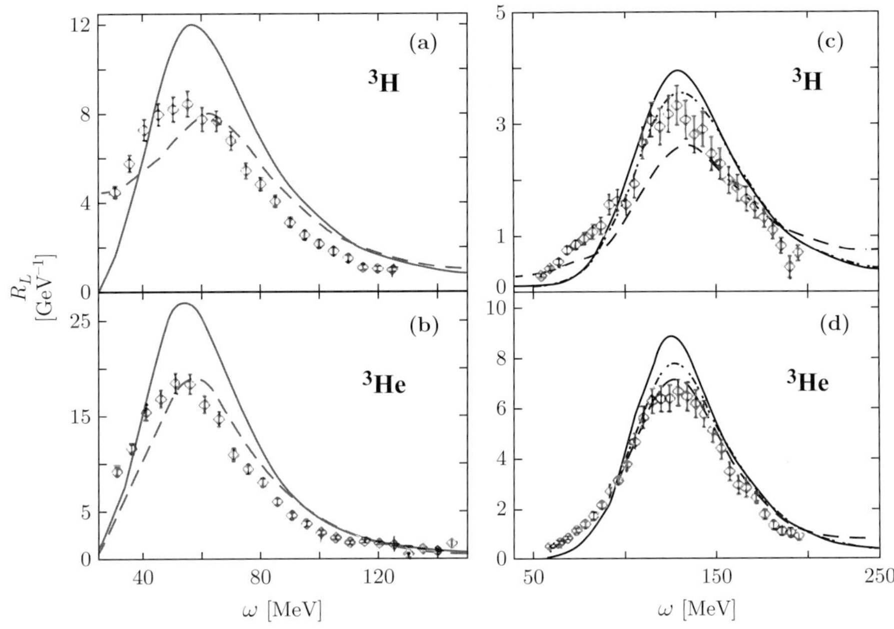

Fig. 12.10 Thc longitudinal response functions compared with theoretical calculations at $q = 300$ MeV c for (a) ^{3}H and (b) ^{3}He, and at $q = 500$ MeV c for (c) ^{3}H and (d) ^{3}He; figure adapted from [Dow88] .

rule should approach the target charge Z (plus a small contribution from the neutrons). Figure 12.11 shows the experimental Coulomb sum rules for ^{3}H and ^{3}He. For both $A = 3$ systems, the data approximately saturate at high momentum transfer (the no-correlation limit) and approach zero at low momentum transfer. The calculations are in reasonable agreement with the data.

The Alpha Particle: ^{4}He

There exists one stable mass-4 nucleus, namely the alpha particle. ^{4}He has a very large binding energy of 28.3 MeV. There is a significant probability ($\sim$12%) to find the ground state in the D-state. So-called Green's function Monte Carlo techniques have been used to great effect to determine the structure of the alpha particle [Car88]. In this approach, the ground-state wavefunction is projected out by treating the Schrödinger equation as a diffusion equation in imaginary time.

Elastic electron scattering from the spin-0 ^{4}He nucleus determines the single charge form factor $F_C(Q^2)$. Figure 12.12 shows the most recent data from Jefferson Laboratory, which rule out conclusively the applicability of the quark dimensional scaling prediction (see Chapter 8) in the measured Q^2 range. Quasielastic inclusive electron scattering has been measured on ^{4}He. The longitudinal structure function has been determined via a Rosenbluth separation at the MIT-Bates laboratory, as in the case of the three-body nuclei. The Coulomb sum rule has been formed and the results are shown in Fig. 12.13. Calculations using wavefunctions derived from two- and three-nucleon forces are in good

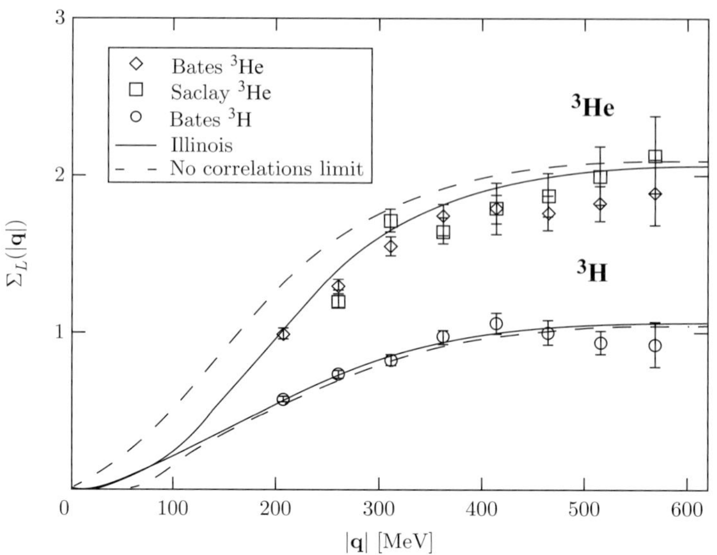

Fig. 12.11   Experimental and theoretical Coulomb sum rules for ^{3}H and ^{3}He; figure adapted from [Dow88] .

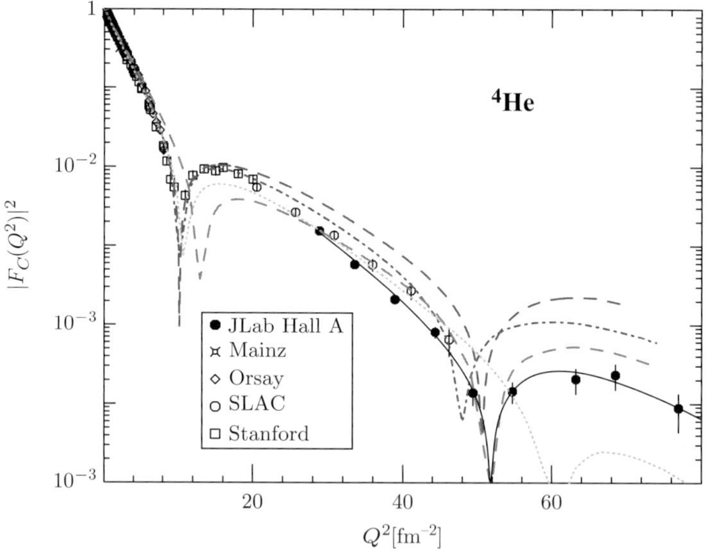

Fig. 12.12 ^{4}He charge form factor data from [Cam14] are compared to both IA and IA+MEC. Also shown are previous data from Stanford, Orsay, Mainz, and SLAC.

agreement with the data. Thus, wavefunctions assuming only nucleon degrees of freedom are able to describe the overall effect of charge correlations in these light nuclei.

In contrast to the deuteron, the helium isotopes and tritium possess no quadrupole moment, and as a consequence display directly a D-state component in their respective wavefunctions only through a nonzero ratio of D-wave to S-wave asymptotic normalization constants. For ^{4}He ($J^\pi = 0^+$), the asymptotic D-wave component is manifest only in the

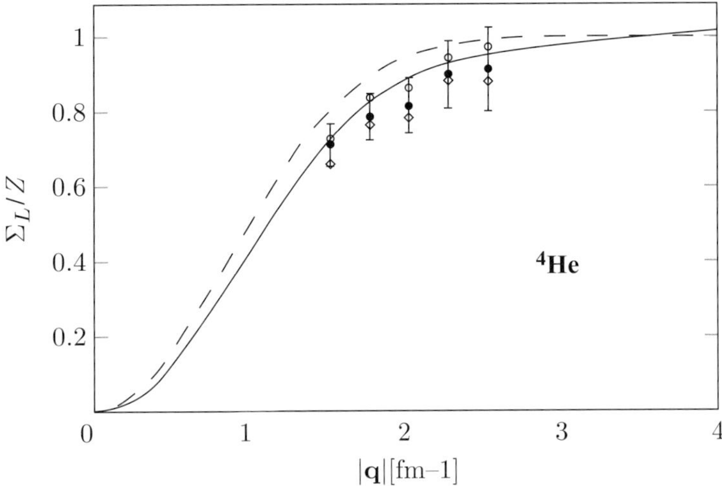

Fig. 12.13   Experimental and theoretical Coulomb sum rules for ^{4}He; figure adapted from [von90].

DD configuration where the channel spin and relative orbital angular momentum are each two and and are coupled to zero.

Green's function Monte Carlo techniques have been successfully applied to calculate the structure of the α-particle [Car88]. Using different *NN* interactions, the *D*-state probability ranges from 12% to 17.5%. The three-nucleon interaction produces a significant $\sim$3.5% increase in the *D*-state probability.

Lattice QCD Calculations of Few-Body Nuclei

Exploration of few-body nuclear structure using lattice QCD techniques has begun. For example, the magnetic moments of the lightest few nuclei have been calculated [Bea14] in a framework where the pion mass was about 800 MeV. Figure 12.14 shows the results, with impressive agreement between lattice QCD results and experiment reported. The magnetic moment of ^{3}He is very close to that of a free neutron, as expected. Analogous results are found for the triton, and the magnetic moment of the deuteron is consistent with the sum of the neutron and proton magnetic moments. The work demonstrates that QCD can be used to calculate the structure of nuclei from first principles. Calculations using these techniques at lighter quark masses and for larger nuclei are underway.

12.6   Hypernuclear Physics

The study of the strong interaction can be extended by adding *strange* quarks to the world of *up* and *down* quarks. Thus, the nucleon isospin doublet (proton and neutron) is extended to the baryon octet (p, n, $\Sigma^{\pm,0}$, Λ, $\Xi^{0,-}$) (see Chapter 3) and the interaction between them becomes much richer and more complicated. A baryon with strangeness is called a

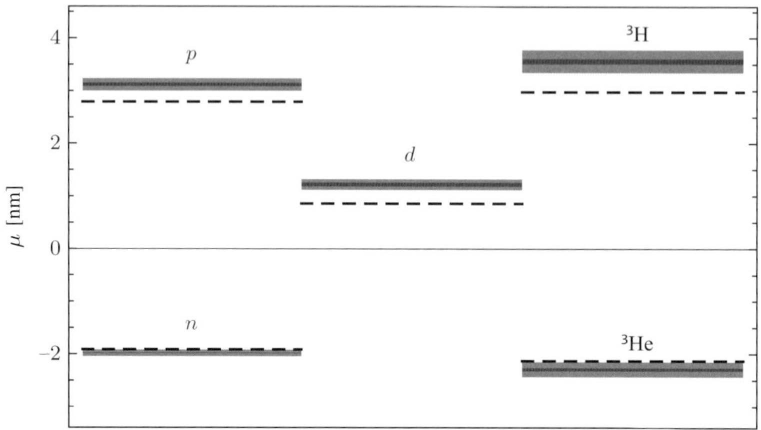

Fig. 12.14   The magnetic moments of the proton, neutron, deuteron, ^{3}He, and triton as calculated in [Bea14]. The results of the lattice QCD calculation at a pion mass of about 800 MeV, in units of lattice nucleon magnetons, are shown as the solid bands. The dashed lines show the experimentally measured values.

hyperon (Y). The hyperon–nucleon and hyperon–hyperon interactions are discussed within the context of SU(3)$_{\text{flavor}}$ symmetry, which is a natural extension of isospin symmetry. Understanding the baryon interactions in a systematic way offers the potential to bridge between the phenomenological nuclear force and fundamental QCD descriptions of the strong interaction.

The lightest baryon with a *strange* quark is the Λ particle and a nucleus containing a Λ is called a lambda hypernucleus. High resolution spectroscopy of light Λ hypernuclear structure is one of the useful ways to study the ΛN interaction. Accurate calculational techniques which have been established for normal nuclei can be applied to light Λ hypernuclei. Careful comparison of experimental results and theoretical predictions based on a YN potential can be used to refine the YN interaction model. It is interesting to consider charge symmetry breaking (CSB) effects in the ΛN interaction. Consider the binding energy difference in the hypernuclear isodoublet $^4_\Lambda$H and $^4_\Lambda$He, for which the binding energies for the ground states have been measured to be

$$B_\Lambda(^4_\Lambda\text{H}, 0^+) = 2.04 \pm 0.04 \text{ MeV} \tag{12.39}$$

$$B_\Lambda(^4_\Lambda\text{He}, 0^+) = 2.39 \pm 0.03 \text{ MeV} . \tag{12.40}$$

The excitation energies were measured by NaI γ-ray detectors and the binding energies for the first-excited states were obtained as

$$B_\Lambda(^4_\Lambda\text{H}, 1^+) = 1.00 \pm 0.06 \text{ MeV} \tag{12.41}$$

$$B_\Lambda(^4_\Lambda\text{He}, 1^+) = 1.24 \pm 0.06 \text{ MeV} . \tag{12.42}$$

The binding energy differences between the ground states and the first-excited states of the $A = 4$ hypernuclear isodoublet are

$$B_\Lambda(^4_\Lambda\text{He}, 0^+) - B_\Lambda(^4_\Lambda\text{H}, 0^+) = 0.35 \pm 0.06 \text{ MeV} \tag{12.43}$$

$$B_\Lambda(^4_\Lambda\text{He}, 1^+) - B_\Lambda(^4_\Lambda\text{H}, 1^+) = 0.24 \pm 0.08 \text{ MeV} . \tag{12.44}$$

By comparison, the energy differences for ^{3}H and ^{3}He are much smaller,

$$B(^3\text{H}) - B(^3\text{He}) - \Delta B_C = 0.764 0.693 = 0.071 \text{ MeV} , \tag{12.45}$$

where ΔB_C is a small Coulomb correction (< 0.05 MeV). Thus, the $^4_\Lambda$He-$^4_\Lambda$H CSB effect is several times larger than that of ^{3}H–^{3}He. A possible origin of this effect is the $\Lambda\Sigma$ coupling in the $NN\Lambda$ three-body force. The mass difference between Λ and Σ is only 80 MeV while the difference between the N and Δ masses is about 300 MeV. In this way, $\Lambda\Sigma$ mixing is quite important (see Chapter 3). Further, the masses of the $\Sigma^{+,0,-}$ differ significantly. For example, $M(\Sigma^-) - M(\Sigma^+) \sim 8$ MeV, which is about 10% of the $\Lambda - \Sigma$ mass difference. By comparison, the $\Delta^{++,+,0,-}$ masses are almost identical. Thus, it is reasonable to expect a significant CSB effect in the $NN\Lambda$ interaction.

Double-Λ hypernuclei ($\Lambda\Lambda$) represent a unique femto-laboratory to study the hyperon–hyperon interaction. Unfortunately, the world supply of data on double-Λ hypernuclei is very limited, even half a century after their discovery. Only a few individual events have been identified. Theoretical estimates of the binding energies of double hypernuclei have been made since the 1960s, with the aim to obtain information on the $\Lambda\Lambda$ interaction. Among possible double-Λ hypernuclei, $^6_{\Lambda\Lambda}$He has been considered to be important because it gives information not only on the $\Lambda\Lambda$ interaction, but also on the cluster structure of hypernuclei. The $^6_{\Lambda\Lambda}$He hypernucleus constitutes the lightest closed shell structure containing proton, neutron, and Λ baryons. Double-Λ hypernuclei are closely related to the potential existence of the H dibaryon, which would have a quark composition of *udsuds*. If the mass of the H dibaryon, M_H, were less than twice the mass of the Λ hyperon in a nucleus, two Λ hyperons in the nucleus would be expected to form the H. With this assumption, the lower limit of the mass of the H dibaryon can be calculated from

$$M_H > 2M_\Lambda - B_{\Lambda\Lambda} , \tag{12.46}$$

where M_Λ is the mass of a Λ hyperon in free space.

Experimentally, hypernuclei have been produced in scattering of K^- mesons from nuclear targets. The ionizing tracks of final-state hyperons and hypernuclei are detected using emulsions. For example, consider the reported observation of the $^6_{\Lambda\Lambda}$He double hypernucleus (the so-called NAGARA event) in 2001 in an experiment carried out at the KEK laboratory in Japan. Ξ^- hyperons were produced via the quasifree (K^-, K^+) process in a diamond target and brought to rest in an emulsion. The (K^-, K^+) reactions are tagged using a spectrometer system. The positions and angles of the Ξ^- hyperons at the emulsion were measured, and a mesonically decaying double-Λ hypernucleus emitted from a Ξ^- captured at rest was observed.

$$^{12}\text{C} + \Xi^- \rightarrow {}^6_{\Lambda\Lambda}\text{He} + {}^4\text{He} + {}^3\text{H} \tag{12.47}$$

$$^6_{\Lambda\Lambda}\text{He} \rightarrow {}^5_\Lambda\text{He} + \text{p} + \pi^- \tag{12.48}$$

The $\Lambda\Lambda$ interaction energy was determined to be $1.01 \pm 0.20^{+0.18}_{-0.11}$ MeV and it was deduced that the interaction was attractive.

Another interesting aspect of hypernuclei is the mechanism for their weak decay. While a free Λ decays predominantly via the weak nonleptonic modes $\Lambda \rightarrow p\pi^-$, $n\pi^0$, placing the Λ in a hypernucleus opens up a new possibility of weak nonleptonic decay via the channels $\Lambda n \rightarrow nn$ and $\Lambda p \rightarrow np$. Hypernuclear decays have been studied experimentally and it has been found that while nonleptonic pionic decay is the primary mechanism for very light systems, the $\Lambda N \rightarrow NN$ processes begins to dominate as soon as A is greater than ~ 10. There has also been a good deal of theoretical work on weak hypernuclear decay, but the ever present strong interaction, in addition to the weak interaction, makes such calculations particularly challenging.

12.7 Fusion

As will be discussed in more detail in Chapter 13, the fusion of two nuclei lighter than iron typically releases energy, while the fusion of nuclei heavier than iron absorbs energy. This is the opposite of what happens for the reverse process, nuclear fission. Thus, fusion generally occurs for few-body nuclei. The origin of the energy release in the fusion of few-body nuclei is due to an interplay of two opposing forces: namely the strong force which combines protons and neutrons, and the Coulomb force, which causes the charged protons to repel each other. For the fusion reaction to proceed, the strong force must overcome the Coulomb repulsion. Quantum tunnelling through the Coulomb barrier is essential for fusion to take place.

Fusion reactions of few-body nuclei power the stars and are an essential step in the nucleosynthesis of virtually all the elements. The fusion of light elements in stars releases energy. For example, in the fusion of two deuterium nuclei to form helium, 0.7% of the mass is carried away in the form of kinetic energy or other forms, such as electromagnetic radiation.

Research into controlled fusion, with the aim of producing fusion electrical power, has been conducted for over 60 years. At present, controlled fusion experiments have been unable to produce break-even energy production, i.e., would be self-sustaining. Typical experiments involve creating plasmas of light nuclei in intense magnetic fields and raising the temperature of the plasma to the point where nuclear fusion reactions are ignited. Fusion reactions have an energy density many times greater than nuclear fission and produce far greater energy per unit mass even though individual fission reactions are generally much more energetic than individual fusion ones. The engineering challenges of this type of experiment have not been surmounted in over six decades of sustained effort.

The cross sections for the low-energy fusion reactions are shown in Fig. 12.15. The simplest processes are the following:

$$d + {}^3\mathrm{H} \rightarrow n + {}^4\mathrm{He} + 17.58 \ \mathrm{MeV} \tag{12.49}$$

$$d + {}^3\mathrm{He} \rightarrow p + {}^4\mathrm{He} + 18.34 \ \mathrm{MeV} \ . \tag{12.50}$$

The resonant behavior of the five-nucleon reactions is clearly visible, whereas other reactions such as D+D reactions, are nonresonant. It is also evident that the cross section at lower energies is entirely dominated by the Coulomb penetrability. In order to separate the

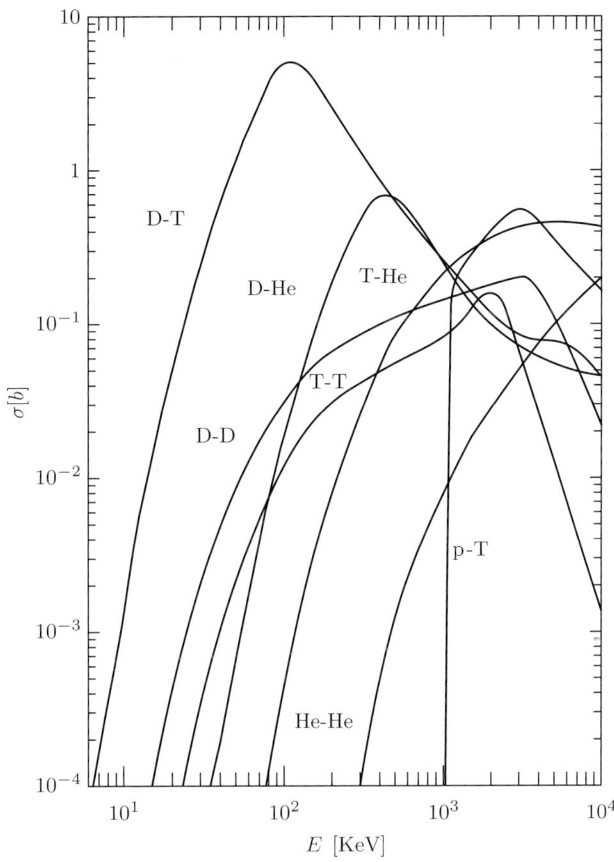

Fig. 12.15 The integrated cross sections of fusion reactions versus energy of the relative motion adapted from [Kay05].

influence of the Coulomb barrier from the nuclear reaction part it is customary to introduce the astrophysical S-factor $S(E)$ which is defined and discussed in Chapter 20.

The two mirror reactions above have some pronounced features. First, at low energies (at deuteron laboratory energies of 107 keV for $^3\mathrm{H}(d,n)^4\mathrm{He}$, and at 430 keV for $^3\mathrm{He}(d,p)^4\mathrm{He}$, respectively), both proceed via strong S-wave resonances. These resonant states are quite pure $J^\pi = \frac{3}{2}^+$ states with little admixture of a $J^\pi = \frac{1}{2}^+$ S-wave and higher wave contributions. Thus, since the transitions essentially go through one matrix element, predictions of the cross sections and polarization observables can be made quite reliably. Second, they have very high cross sections at resonance, which makes the $^3\mathrm{H}(d,n)^4\mathrm{He}$ reaction the favored choice for fusion energy production due to its greater reaction yield. The 3.5 MeV α-particles produced are useful to heat the plasma. However, the high energy (14.1 MeV) neutrons must be captured in a "blanket" wall, creating problems by material radiation damage. The neutronless mirror reaction $^3\mathrm{He}(d,p)^4\mathrm{He}$ appears ideal, except for the higher temperature needed for ignition.

It has been pointed out that spin polarizing the mass-2 and -3 nuclei in the fusion process can substantially enhance the fusion cross section. Estimates predict about a $\sim$50%

increase in the cross section at the fusion peak. Practical implementation of this attractive idea would require very large polarization rates of these nuclei and would demand that the depolarization rates in the plasma are sufficiently small. Both are formidable technical challenges.

Exercises

12.1   Pre-existing Δ-Components in the Ground State of ^{3}He

Consider an experiment to measure pre-existing Δ components in the ground state of the ^{3}He nucleus. These components can arise in interactions among three nucleons as shown. A pion exchanged between nucleons 1 and 2 excites nucleon 2 to a $\Delta(1232)$ isobar state. The Δ decays by emitting a pion which is absorbed by nucleon 3.

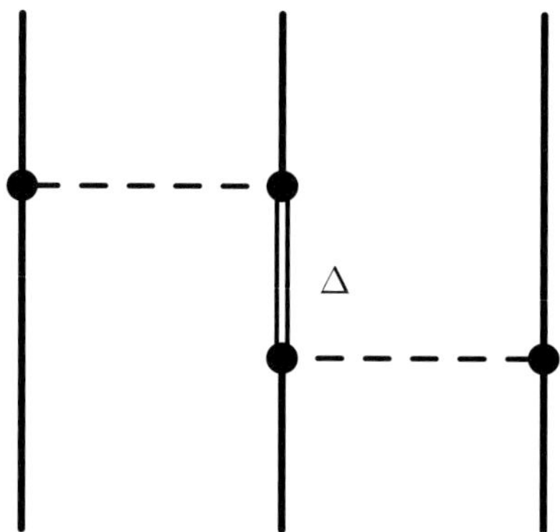

An electron beam is incident on the ^{3}He nucleus and, in the final state, the scattered electron is detected in coincidence with the pion resulting from the decay of the Δ. The reaction is

$$e + \Delta(NN) \rightarrow e' + \Delta(NN) \rightarrow e' + \pi N(NN).$$

a) Write down the different charge-state possibilities of the (spin $\frac{3}{2}$ and isospin $\frac{3}{2}$) $\Delta(1232)$.

b) Given that the $e - \Delta$ coupling is proportional to the square of the charge of the Δ, find the relative probabilities (P_Q) for knockout of the Δ in its various charge states.

c) Consider next the coupling of the pre-existing Δ with the remaining nucleons in ^{3}He (spin $= \frac{1}{2}$ and isospin $= \frac{1}{2}$). Use a table of Clebsch–Gordan coefficients to find the relative probabilities (P_Δ) for the different Δ components: $P(\Delta^{++}nn)$: $P(\Delta^+np)$: $P(\Delta^0pp)$ where n is a neutron and p is a proton.

d) Use the Table of Clebsch–Gordan coefficients to obtain the relative probabilities (P_{decay}) of the decays of the different Δ charge-states into πN states.

e) From b), c,) and d), determine the probabilities of detecting π^+, π^0, and π^- resulting from Δ knockout from ^{3}He.

12.2 Electroproduction of Δs from ^{3}He

Consider the process

$$e + N(NN) \rightarrow [\Delta(NN) \rightarrow \pi N(NN)] + e'$$

where the Δ is produced in interaction with the virtual photon.

a) Using the same analysis as in Exercise 12.1, calculate the relative probabilities of detecting π^+, π^0, and π^- resulting from Δ production on a single nucleon in ^{3}He.

b) Compare your results from 12.1 e) and 12.2 a) and suggest an experimental method for distinguishing between them. What particles in the final state should be detected for a definitive measurement?

12.3 Neutron and Proton Polarization in Polarized ^{3}He

The polarized ^{3}He nucleus is widely used as an effective polarized neutron target for scattering experiments. Here we estimate the polarization of the neutron and proton in a polarized ^{3}He nucleus following [Fri90]. We consider

$$P_{n,p}^{(\pm)} = <^3\text{He }m = +1/2|\widehat{P}_{n,p}^{(\pm)}|^3\text{He }m = +1/2> ,$$

where

$$\widehat{P}_n^{(\pm)} = \sum_i \frac{1 - \tau_3(i)}{2}\frac{1 \pm \sigma_3(i)}{2}$$

$$\widehat{P}_p^{(\pm)} = \sum_i \frac{1 + \tau_3(i)}{2}\frac{1 \pm \sigma_3(i)}{2} .$$

The isospin factor in $\widehat{P}_n(\widehat{P}_p)$ counts the number of neutrons (protons) which are aligned (+) or anti-aligned (−) with the ^{3}He spin.

a) Show that

$$P_n^{(+)} + P_n^{(-)} = 1$$
$$P_p^{(+)} + P_p^{(-)} = 2 .$$

b) We assume that the ground-state wavefunction is dominated by the S, S', and D states, i.e., we neglect the small P-state components. Assuming that

$$< m = 1/2| \sum_{i=1}^{3} \sigma_3(i)\tau_3(i)|m = 1/2 >= -\left[P(S) - \frac{1}{3}P(S') + \frac{1}{3}P(D) \right] .$$

show that

$$\langle m = 1/2|\sigma_3|m = 1/2\rangle = P(S) + P(S') - P(D)$$
$$\langle m = 1/2|\tau_3|m = 1/2\rangle = 1 = P(S) + P(S') + P(D) .$$

c) Thus, show that

$$P_n^{(+)} = 1 - \Delta$$
$$P_n^{(-)} = \Delta ,$$

where $\Delta = [P(S') + 2P(D)]/3$.

d) In addition, show that

$$P_p^{(+)} = 1 - 2\Delta'$$
$$P_p^{(-)} = 1 + 2\Delta' ,$$

where $\Delta' = [P(D) - P(S')]/6$.

e) Using the probabilities of Table 12.1, estimate the neutron and proton polarizations.

12.4 Polarized Fusion

At energies below several hundred keV, the D+^{3}He reaction is dominated by a channel which involves a single intermediate ^{5}Li state. This nearly resonant state lies 407 keV above the rest energy of D and ^{3}He and has an angular momentum and parity of $J^\pi = \frac{3}{2}^+$. The spins of D and ^{3}He are 1 and $\frac{1}{2}$, respectively, so that the total spin of the reactants is $S = \frac{3}{2}$ or $\frac{1}{2}$. At these low energies, the average orbital angular momentum is nearly zero that only the $S = \frac{3}{2}$ states contribute significantly to the $D + {}^3$He reaction. The statistical weights of the $S = \frac{3}{2}$ and $S = \frac{1}{2}$ states are four and two, respectively. Therefore, if the nuclei are unpolarized, only two-thirds of the collisions will contribute to the fusion reactions.

Consider now the effect of polarizing both the initial D and ^{3}He nuclei. We assume a magnetic field so that the fractions of D nuclei polarized parallel, transverse, and antiparallel to the magnetic field are d_+, d_0, and d_-, respectively. Correspondingly, the fractions for the polarized ^{3}He nuclei are h_+ and h_-. Then the total cross section can be written as

$$\sigma = \left[a + \frac{2}{3}b + \frac{1}{3}c \right] f\sigma_0 + \left[\frac{2}{3}b + \frac{4}{3}c \right] (1 - f)\sigma_0 , \tag{12.51}$$

where $a = d_+h_+ + d_-h_-$, $b = d_0$, $c = d_+h_- + d_-h_+$, σ_0 is the unpolarized fusion cross section, and f is the fraction of the cross section occurring due to the $\frac{3}{2}^+$ state. The magnitude of f is assumed to be close to unity.

a) Show that for unpolarized D and ^{3}He, $\sigma = \frac{2}{3}\sigma_0$.

b) Show that if the nuclei are all polarized along a magnetic field, the enhancement of the fusion cross section is $\frac{3}{2}f \cong 1.5$.

13.1 Basic Properties of Finite Nuclei

We begin the discussions of finite nuclei with some general remarks on the systematics of the nuclear ground state. For even-even nuclei (nuclei with an even number of both protons and neutrons) the nucleons making up the nucleus form pairs containing one spin-up and one spin-down nucleon to yield a net spin of zero and even parity. In Chapter 15 we consider elastic electron scattering from such nuclei as the paradigm for what follows. Odd-even and even-odd nuclei often, but not always, have the same spin as the last unpaired valence nucleon; we shall see examples when discussing elastic magnetic electron scattering and magnetic moments also in Chapter 15. Finally, odd-odd nuclei are somewhat unusual. There exist only four stable nuclei having unpaired protons *and* neutrons, namely, ^{2}H, ^{6}Li, and ^{14}N with spin-parity 1^+, and ^{10}B having 3^+. The basic characteristics of the known nuclei are the following: they occupy a region in the NZ-plane whose central valley runs roughly along the $N = Z$ line at values of A below 40 and then bends towards the region having higher values of N than of Z, as shown in Fig. 13.1. Taking cuts across the valley at either constant Z or at constant N, one climbs out of the valley, on the average moving to less bound nuclei until reaching the so-called *drip-lines* where nuclei are no longer stable to proton or neutron emission. At the bottom of the valley, where the most stable nuclei reside, one finds the binding energy per nucleon to be relatively constant for nuclei beyond $A = 40$ at a value $\sim$8.5 MeV per nucleon, as shown in Fig. 13.2 and, as the values of N and Z where stable nuclei exist become very large, this valley of stability narrows and then disappears.

One of the key questions in studies of nuclear systematics is: Do "islands of stability" exist at even higher N, Z-values, the so-called *superheavy* nuclei? An island of stability, first conjectured by Seaborg in the 1960s, is a collection of heavier isotopes of transuranic elements, expected to be more stable than those closer in atomic number to uranium with radioactive decay half-lives of minutes to days. The hypothesis has a firm basis in shell-model calculations. Recently, physicists have created element 117 independently at Dubna, Russia and Darmstadt, Germany by smashing calcium nuclei (with 20 protons) into a target of berkelium (with 97 protons). Occasionally, the nuclei fuse together to form a final-state nucleus with 117 protons. The physicists did not observe the element 117 directly, since it has a half-life of 50 msec. It decays via successive alpha decays until the isotope lawrencium 266 is formed, which had never been observed before and was found to have a very long half-life of 11 hours. This makes it one of the longest-lived superheavy isotopes known to date – it might mean that the shores of the island of stability are being approached.

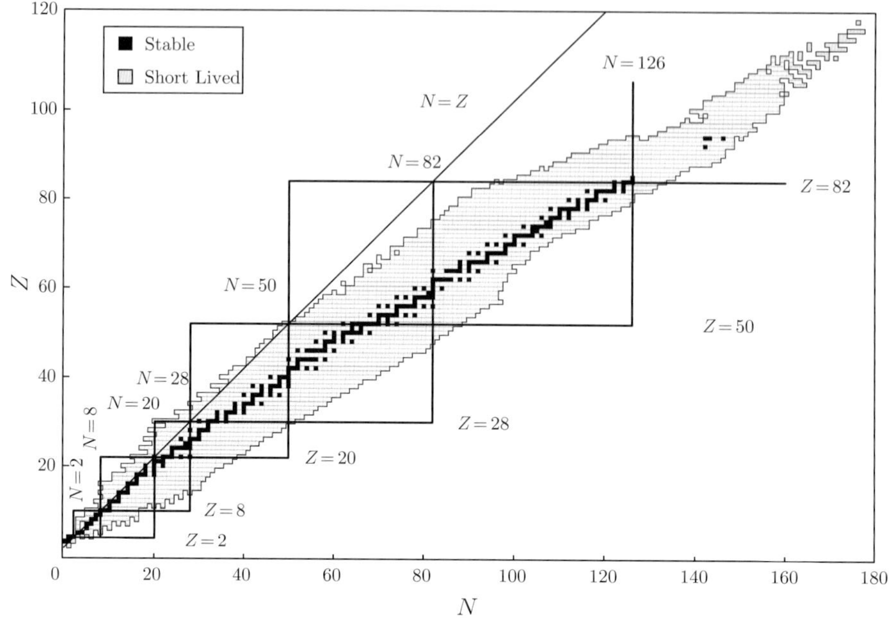

Fig. 13.1 Distribution of stable and long-lived nuclei as a function of proton and neutron numbers, Z and N, respectively, where the former are shown as filled squares. Nuclei that are unstable against β-decay, α-decay or nucleon emission are indicated by empty squares.

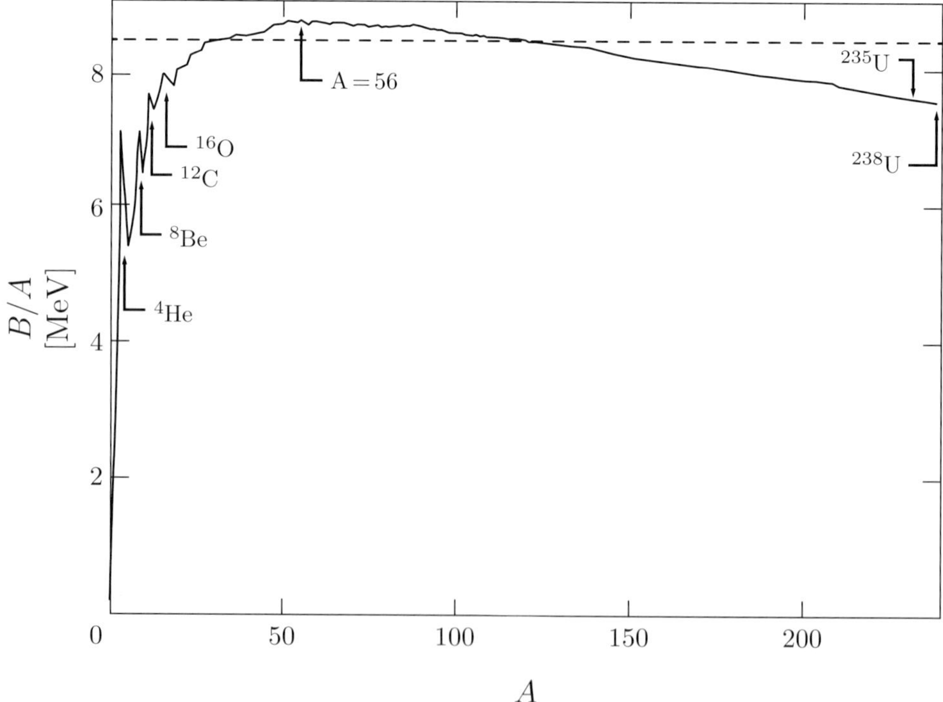

Fig. 13.2 Average binding energy per nucleon B/A versus nucleon number A for stable nuclei in the bottom of the valley of stability.

It is interesting to consider which combinations of protons and neutrons can form a bound nucleus? The answer has important consequences for nuclear structure and nuclear astrophysics. Currently, we do not know where these limits are, but we expect there are two to three times more bound nuclei than the ones we have observed and have been able to explore so far. Only 288 isotopes are stable on the timescale of the solar system. By moving away from the region of stable isotopes, by adding nucleons (either neutrons or protons), one enters the regime of short-lived radioactive nuclei. Nuclear existence ends at the drip-lines, where the last nucleon is no longer bound to the others and literally drips off. Although the proton drip-line has been reached for many elements up to $Z=83$, remarkably, the neutron drip-line is known only up to oxygen ($Z=8$). The superheavy nucleus with $Z=118$, $A=294$ marks the current upper limit of nuclear charge and mass. These borders define the currently known nuclear territory. Today, about 3000 isotopes are known to exist, less than half the number predicted by current nuclear theory. While many of the predicted nuclei are currently out of reach to be made in the laboratory, their influence cannot be ignored, as the astrophysical processes that generate many heavy elements occur relatively close to the drip-lines in nuclear *terra incognita*. The study of rare isotopes is a substantial, worldwide research endeavor. The major facilities include: Argonne Tandem Linac Accelerator System (USA), GSI Helmholtz Center for Heavy Ion Research (Germany), Grand Accélérateur National d'Ions Lourds (France), the National Superconducting Cyclotron Laboratory (USA), TRIUMF (Canada), and the Radioactive Isotope Beam Factory (RIBF) (Japan). More powerful future facilities with the aim of reaching farther away from the region of stable isotopes are FAIR at GSI, the Rare Isotope Science Project (RAON, South Korea) and the Facility for Rare Isotope Beams (FRIB, USA), presently under construction.

One of the paradigms of nuclear structure is the shell model of the nucleus, in which a common force generated by all other nucleons governs the motion of each neutron or proton. Nucleon orbits are constrained to specific energies, thereby forming shells, and nuclei having filled nucleonic shells are exceptionally well bound. The numbers of nucleons needed to fill each successive shell are called the magic numbers: The traditional ones are 2, 8, 20, 28, 50, 82, and 126, and until recently they have been assumed to be immutable. Stable magic nuclei have spherical charge distributions, almost zero intrinsic quadrupole moments, and 0^+ ground states. The first excited state due to collective motion is a 2^+ and decays are very strong to the ground state for magic nuclei. Figure 13.3 shows the strength of the 2^+ decays with the strong decays of the magic nuclei being clearly pronounced. However, a recent series of discoveries using rare isotopes with the proton-magic oxygen ($Z=8$), calcium ($Z=20$), nickel ($Z=28$), and tin ($Z=50$) isotopes has shaken the assumption that the magic numbers are fixed for nuclei away from stability. For example, spectroscopic studies of neutron-rich oxygen isotopes provide conclusive evidence for new magic numbers at $N=14$ and 16, an outcome that explains the surprising location of the drip-line for oxygen. Other examples abound. Studies of the tin isotopes have indicated that the nuclear spin-orbit force may be weakening with neutron excess [Sch04]. Studies of neutron-rich nuclei have also led to the identification of the weakly unbound nucleus ^{26}O as a possible candidate for the elusive phenomenon

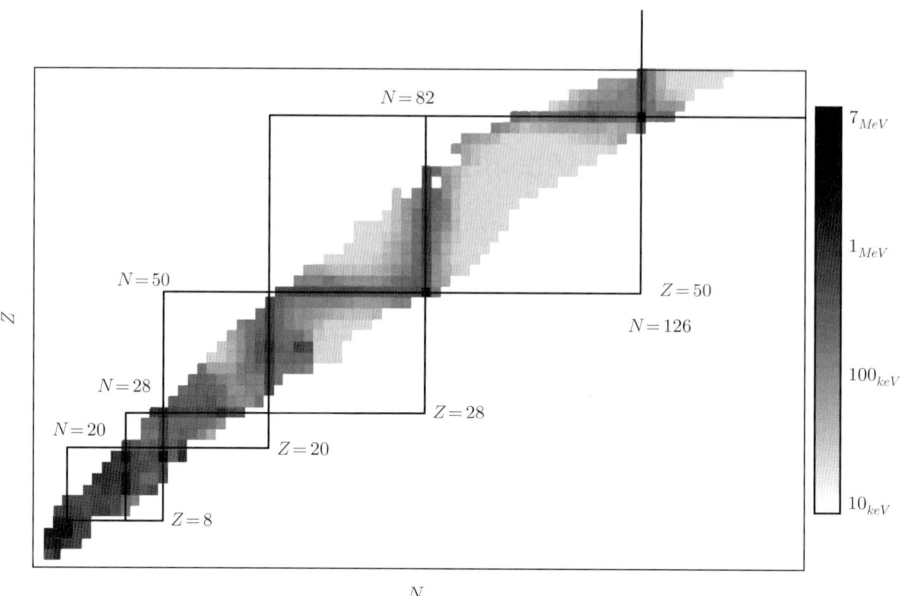

Fig. 13.3 The strength of the $E2\ 2^+ \rightarrow 0^+$ transition as a function of atomic number Z and neutron number N; figure adapted from [Sak14]. For example, the dark bands at $Z = 50$ and $N = 82$ show striking evidence for the fact that the strengths of the $E2\ \gamma$-decays correlate very strongly with the locations of the magic numbers.

of two-neutron radioactivity, which will ultimately provide a sensitive probe of neutron pairing. Evidence for a new neutron magic number 34 in the level structure of ^{54}Ca has been reported based on experimental work carried out at RIBF [Ste13]. The masses of neutron-rich calcium isotopes have also been precisely measured with on-line ion traps as far as ^{54}Ca, which contains six more neutrons than the heaviest stable calcium isotope and the first spectroscopic study has been completed for ^{60}Ti, thus providing the closest extrapolation points towards ^{60}Ca, the N=40 isotope which is key to the location of the neutron drip-line in that region of the nuclear chart. The spectroscopy of this nucleus will only be possible at FRIB.

Further progress in this area requires measurements of key isotopic chains, such as those of calcium and nickel, which encompass multiple magic numbers. With FRIB and its suite of unique instrumentation, these chains will be accessible from proton drip-line to neutron drip-line, permitting study of the N, Z dependence of the nuclear force and continuum effects. Such investigations will allow us to explore new aspects of nuclear structure in the domain where many-body correlations, rather than the average nuclear potential, dominate. Single- and even multiple-neutron emission is expected to characterize nuclei at the neutron drip-line, while β-delayed neutron decay is prevalent among neutron-excess nuclei before the drip-line is reached. Both forms of radioactivity only occur among nuclei far from stability.

The territory at, and beyond, the proton drip-line offers unique opportunities to study other exotic nuclear decays and correlations, such as ground-state one- and two-proton decay, a class of radioactivity that exists nowhere else but provides unique insight into correlation effects. The astrophysically important one-proton emitter, ^{69}Br, has already been studied and the two-proton decay of the doubly-magic ^{48}Ni nucleus has been observed for the first time. In the case of ^{69}Br, the measured (negative) proton separation energy has important consequences for the *rp*-process (see Chapter 20) occurring in type-I X-ray bursts.

The comments here concerning stability usually refer to whether or not nuclei are stable to proton or neutron emission. However, there exist other reactions where nuclei are transmuted. One such mechanism is fission wherein a heavy nucleus spontaneously breaks up into two lighter nuclei plus neutrons such as $^{238}\text{U} \rightarrow {}^{92}\text{Kr} + {}^{143}\text{Ba} + 3n$, liberating energy in the process. Another reaction is induced fission where adding a low-energy neutron to such a fissionable nucleus induces the breakup, for example, $n + {}^{235}\text{U} \rightarrow {}^{92}\text{Kr} + {}^{142}\text{Ba} + 2n$ which forms the basis for fission reactors. For light nuclei, under appropriate circumstances one can have the reverse process, namely, nuclear fusion in which two nuclei combine to form a heavier nucleus, thereby liberating energy. We shall return to discuss such processes in much more detail in Chapter 20 when we consider the *pp*-chain and the CNO cycle, which are responsible for the energy production in stars. Note that we have stated that either fissioning heavy nuclei or fusing light nuclei leads to a liberation of energy: How can this be? The answer is discussed in detail in the following section where the semi-empirical mass formula is developed. One finds that the binding energy per nucleon increases in going from light nuclei to heavier ones until reaching the Fe region ($Z \sim 26$) where the greatest binding energy per nucleon is found, after which the average binding decreases in going to still heavier species. This means that fusion is exothermic up to $Z \sim 26$ while fission is exothermic down to the same region. Indeed, a very heavy star burns its nuclear fuel via fusion and produces heavier and heavier nuclei, releasing enough energy to support the star against collapse due to gravity until the Fe region is reached. At that point the star can no longer produce the energy needed for that support and accordingly collapses, yielding a white dwarf, a neutron star, or a black hole, as discussed in Chapter 20.

Production of electricity using nuclear fission reactors that have been in operation since the 1950s offers scalable, mostly reliable, base-load power without carbon-dioxide emissions. There are over 435 commercial nuclear power reactors operating in 31 countries with over 375 GWe of total capacity worldwide. Nuclear reactors produce about 70% of the electricity in France, about 20% in the US, and about 11% worldwide. Asia, where more than 50% of the world's population lives, is leading the development of new fission reactor technologies. Advanced reactor technologies include provision of high-temperature process heat for industrial application, development of new gas-cooled, accident-tolerant reactors, and realization of small (≤ 300 MWe) modular reactors requiring lower capital investment. As fossil fuel supplies dwindle and concerns about the effects on global climate due to greenhouse gases increase, nuclear fission reactors will likely become more numerous worldwide. Safety and handling of waste are sensitive political issues, but widely accepted, sound, technical solutions exist. Harnessing fusion power to generate electricity offers significant additional advantages with more plentiful fuel and less radioactive waste produced. It has been the focus of development since the mid-twentieth century. The physics processes are reasonably well understood and megawatts of power have been produced for a few seconds. However, the engineering challenges to attaining break-even electricity production remain very formidable. The present focus is the international thermonuclear experimental reactor (ITER) at Cadarache, France which has a goal of achieving 500 MW of output power while needing 50 MW to operate. Thus, the machine aims to demonstrate the production of more energy from the fusion process than is used to initiate it. It is expected to initiate plasma experiments in the early 2020s with full deuterium–tritium fusion experiments starting by the end of that decade.

Other reactions that transmute nuclei include: α-decay in which a nucleus decays into an α-particle (a ^{4}He nucleus) plus a lighter nucleus; β-decay in which a nucleus emits a β^--particle (a negative electron) plus an electron antineutrino, $\bar{\nu}_e$, or their antiparticles, β^+ (the positive partner of the electron, the positron) and electron neutrino, ν_e, which involve the weak interaction (see Chapter 17 for a more complete discussion of β-decay); γ-decay where a nucleus in an excited state decays to a lower-energy state by emitting a photon; and reactions, such as neutron capture, (n, γ), where a neutron is captured by a nucleus and a photon emitted. The detailed abundances of the known nuclei are determined by a complicated mix of such processes which transmute nuclei into nearby species through a network of competing reactions.

A reasonable approximation for the description of nuclear ground states is to invoke the so-called *mean field* approximation wherein nucleons move roughly independently in the average potential generated by their interactions with the other nucleons. In Chapter 14 we return to make specific use of this approximation, and, using Hartree–Fock methods, show how the nucleon–nucleon interaction yields this mean field plus a weak residual interaction. For the present let us simply use this concept in discussing nuclear systematics. If some reasonable form is assumed for the central potential in which the nucleons move, say a square well or a harmonic oscillator (or perhaps something more realistic), and one solves the Schrödinger equation using this potential to obtain single-particle wavefunctions, filling all single-particle states up to the so-called Fermi level with the A nucleons in the nucleus, then one has an approximation for the nuclear ground state. However, this simple picture is not sufficient. As noted above, it is found experimentally that especially stable nuclei occur at magic numbers – N or Z equal to 2, 8, 20, 40, 50, 82, 126 – which are not predicted in this elementary approach. Rather, one must add a spin-orbit interaction to this central potential, which splits single-particle states with given orbital angular momentum ℓ into pairs with $j = \ell \pm 1/2$. The sign of this additional contribution is such that the higher-j ("stretched") configuration lies lower than the lower-j ("jack-knifed") case, and with this addition one finds good agreement with the observed magic numbers. Figure 13.4 shows a schematic diagram of the single-particle spectrum using a typical mean field of this type. Once the spin-orbit potential has been added with an appropriate strength, the magic numbers in the valley of stability emerge, as shown in the figure. Note, however, that as one progresses out of the valley and towards the drip-lines the situation changes (see the introductory discussion).

Mean-field single-particle wavefunctions may be written as a product of a radial wavefunction (the solution of the radial Schrödinger equation in the given nuclear potential), the orbital and spin angular momentum contributions, $Y_\ell^{m_\ell}(\Omega_x)$ and $\eta_{1/2}^{m_s}$, respectively, coupled

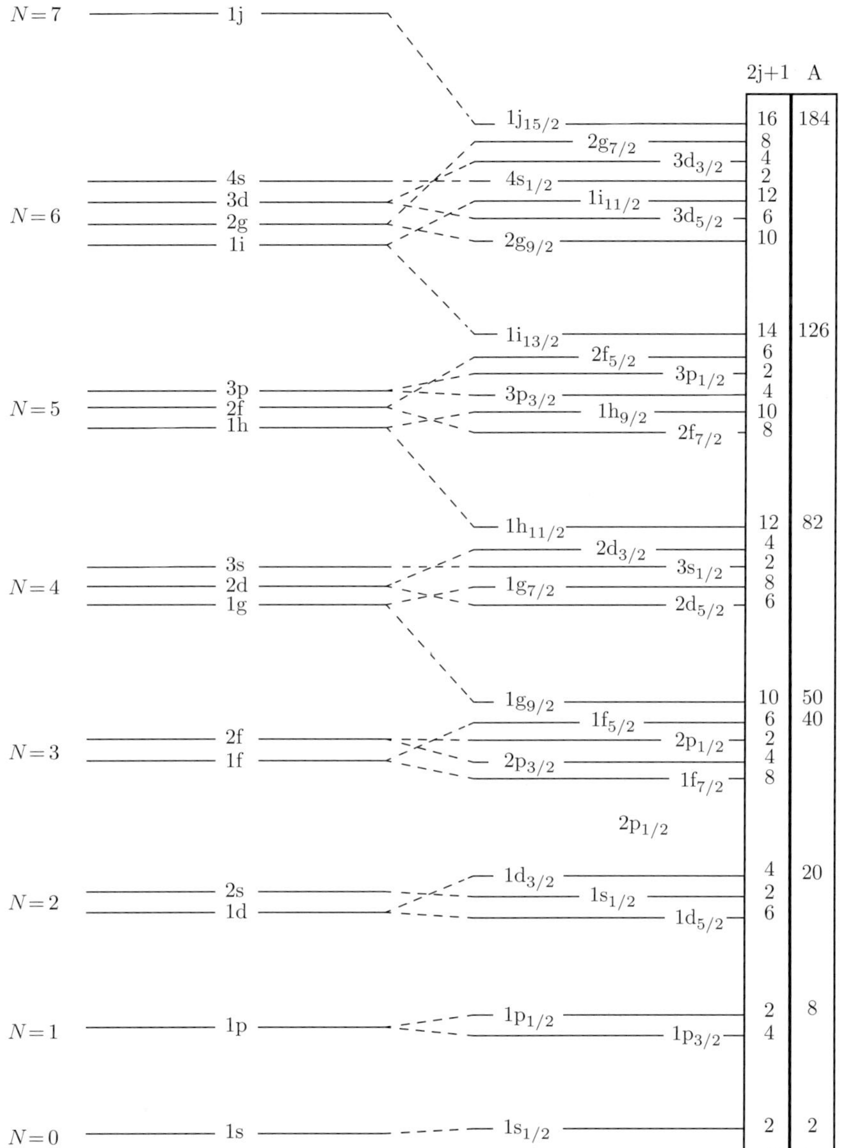

Fig. 13.4   Schematic diagram showing the single-particle energy spectrum for spherical nuclei obtained by solving the Schrödinger equation using a typical mean-field potential. The left-hand side of the figure shows typical results in the absence of a spin-orbit potential, whereas the right-hand side shows what happens when the latter is added. Clearly the groupings of the energy levels and hence the way the "magic numbers" occur depend critically on the presence of this term in the potential.

to a total angular momentum j with projection m_j, times an isospin spinor, $\xi_{1/2}^{m_t}$, for protons ($m_t = +1/2$) or neutrons ($m_t = -1/2$):

$$\varphi_{n\left(\ell\frac{1}{2}\right)j,m_j;\frac{1}{2},m_t}(\mathbf{x}) = R_{n\ell j}(x)\left[Y_\ell(\Omega_x) \otimes \eta_{1/2}\right]_j^{m_j} \xi_{1/2}^{m_t}. \tag{13.1}$$

For simplicity, we denote the complete set of single-particle quantum numbers by $\alpha = \left(n\ell j, m_j; 1/2, m_t\right)$. The many-body wavefunctions may then be written in terms of Slater determinants, which by construction are antisymmetric under exchange of particle coordinates,

$$\Phi_{\alpha_1\alpha_2\cdots\alpha_A}(1,2,\ldots,A) = \frac{1}{\sqrt{A!}} \begin{vmatrix} \varphi_{\alpha_1}(1) & \varphi_{\alpha_1}(2) & \cdots & \varphi_{\alpha_1}(A) \\ \varphi_{\alpha_2}(1) & \varphi_{\alpha_2}(2) & \cdots & \varphi_{\alpha_2}(A) \\ \vdots & \vdots & & \vdots \\ \varphi_{\alpha_A}(1) & \varphi_{\alpha_A}(2) & \cdots & \varphi_{\alpha_A}(A) \end{vmatrix}, \tag{13.2}$$

where the single-particle states are labeled with quantum numbers α_i and where the particle coordinates (spatial, spin, and isospin) are labeled $i = 1, 2, \ldots, A$. In general, the many-body wavefunctions are not simply single Slater determinants, but involve linear combinations of such quantities:

$$\Psi_k(1,2,\ldots,A) = \sum_{\{\alpha\}} C_k^{\alpha_1\alpha_2\cdots\alpha_A} \Phi_{\alpha_1\alpha_2\cdots\alpha_A}(1,2,\ldots,A), \tag{13.3}$$

where k labels the states and one should solve the many-body Schrödinger equation using a given Hamiltonian $\widehat{H}$

$$\widehat{H}\,|\Psi_k\rangle = E_k\,|\Psi_k\rangle. \tag{13.4}$$

Projecting onto the basis of Slater determinants, we have

$$\langle\Phi_\alpha|\,\widehat{H}\,|\Psi_k\rangle = E_k\,\langle\Phi_\alpha|\,\Psi_k\rangle. \tag{13.5}$$

Equivalently, one should solve the set of equations

$$\sum_{\{\beta\}} H_{\alpha\beta} C_k^\beta = E_k C_k^\alpha, \tag{13.6}$$

where $H_{\alpha\beta} \equiv \langle\Phi_\alpha|\,\widehat{H}\,|\Phi_\beta\rangle$. Naturally, it is necessary to truncate to tractable model spaces and use effective interactions tailored to a specific set of shells.

For the present purposes let us make a starting approximation and take the nuclear ground state $|0\rangle$ to be the state corresponding to the filled Fermi sea, $|F\rangle \equiv$ the state with levels completely filled for $\epsilon_\alpha \le \epsilon_F$, where ϵ_F is the energy of the Fermi surface, and levels completely empty for $\epsilon_\alpha > \epsilon_F$. In first quantization this means using a single Slater determinant as given above, and saturating the occupied levels with the lowest A single-particle orbits. In context, we note that in second quantization (occupation number representation, see [Fet71]) this approximation corresponds to taking the ground state to have no particles above the Fermi sea, that is, no particle creation operators $a_\alpha^\dagger$ with $\epsilon_\alpha > \epsilon_F$ acting on the quasi-vacuum (the filled Fermi sea) and no holes, namely no hole creation operators $b_\alpha^\dagger$ with $\epsilon_\alpha \le \epsilon_F$ acting either. With such a many-body state the ground-state matrix element of any single-particle operator (one-body operator in second quantization),

$$\mathcal{O}^{[1]} = \sum_{i=1}^{A} \mathcal{O}^{[1]}(i), \tag{13.7}$$

is then given by

$$\left\langle F \left| \mathcal{O}^{[1]} \right| F \right\rangle = \sum_{\alpha \leq F} \int d\mathbf{x}_1 \varphi_\alpha^\dagger(1) \mathcal{O}^{[1]}(1) \varphi_\alpha(1) \, , \tag{13.8}$$

where the antisymmetry of the wavefunctions has been used to permute all coordinates to the first one. Likewise, for a two-particle operator such as the potential (two-body operator in second quantization)

$$\mathcal{O}^{[2]} = \sum_{i \neq j = 1}^{A} \mathcal{O}^{[2]}(i,j) \tag{13.9}$$

one has

$$\left\langle F \left| \mathcal{O}^{[2]} \right| F \right\rangle = \tfrac{1}{2} \sum_{\alpha,\beta \leq F} \int d\mathbf{x}_1 \int d\mathbf{x}_2 \varphi_\alpha^\dagger(1) \varphi_\beta^\dagger(2) \mathcal{O}^{[2]}(1,2)$$
$$\times \left(\varphi_\alpha(1)\varphi_\beta(2) - \varphi_\beta(1)\varphi_\alpha(2) \right) \, , \tag{13.10}$$

containing direct (D) and exchange (E) contributions in the two orderings of the quantum numbers α and β.

An important example of a single-particle operator is that of the magnetic dipole $\mu^{[1]}$ whose expectation value in a nucleus having ground-state angular momentum $J_0 \geq 1/2$ yields the nuclear magnetic moment. We have already encountered both magnetic dipole moments and electric quadrupole moments (see below) in the previous two chapters. Furthermore, in Chapter 7 we discussed the electromagnetic multipole operators in general, and later, in Chapter 15, we shall return to discuss the elastic (i.e., ground-state) electron scattering form factors in more depth. For the present we note only that one has

$$\mu/\mu_N = \frac{1}{[J_0]} \sqrt{\frac{J_0}{J_0 + 1}} \langle J_0 \| \widehat{\mu}_1 \| J_0 \rangle \, , \tag{13.11}$$

where μ_N is the nucleon magneton and $[J_0] \equiv \sqrt{2J_0 + 1}$ as in Chapter 2. Let us pursue this a bit more and obtain explicit expressions for the magnetic moment in the special case of an extreme single-particle or -hole model, where one assumes that the nuclear ground state is simply a closed shell plus or minus a single particle. Then the matrix element arises entirely from the odd unpaired nucleon. The required single particle reduced matrix element has both orbital and spin contributions and involves computing

$$\langle n_0 \ell_0 j_0 \| \mu_1 \| n_0 \ell_0 j_0 \rangle = \left\langle n_0 \ell_0 j_0 \left\| \left(g_\ell L + g_s \frac{1}{2}\sigma \right) \right\| n_0 \ell_0 j_0 \right\rangle \, , \tag{13.12}$$

involving the gyromagnetic ratios (g-factors) $g_\ell = 1(0)$ and $g_s = 2\mu_p(2\mu_n)$ for protons (neutrons), respectively. The matrix elements of the orbital and spin angular momentum operators may be obtained using the so-called Landé formula (see [Won98] for a derivation). Alternatively, and instructive for readers with advanced knowledge of

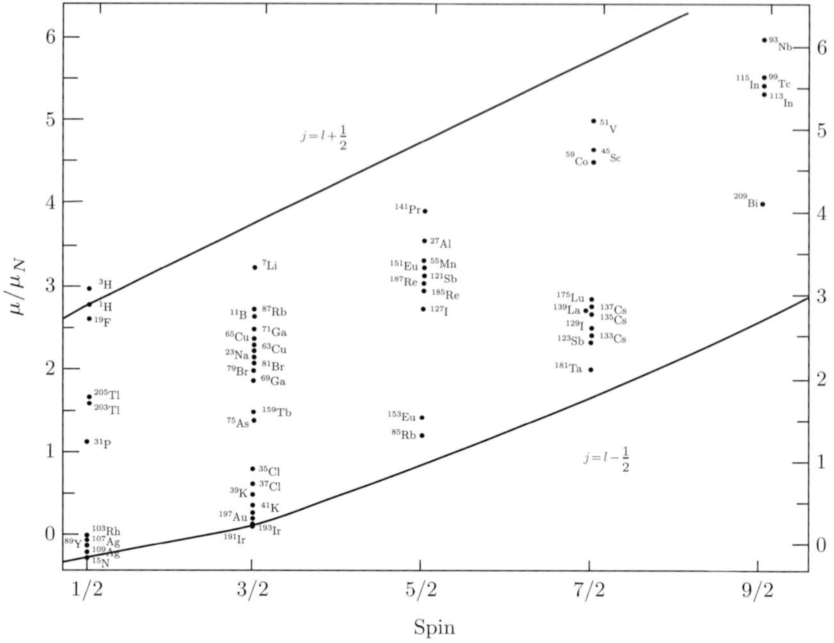

Fig. 13.5　Schmidt lines for odd-proton nuclei. The figure is adapted from [Pre62].

angular momentum recoupling, one can use expressions for matrix elements of operators in a coupled scheme operating on a subspace (see Edmonds [Edm74], eqs. 7.1.7 and 7.1.8). The results are

$$
\mu/\mu_N = \begin{cases} \ell_0 g_\ell + \frac{1}{2} g_s & j_0 = \ell_0 + 1/2 \\ \left(\frac{2\ell_0-1}{2\ell_0+1}\right)\left[(\ell_0+1)g_\ell - \frac{1}{2}g_s\right] & j_0 = \ell_0 - 1/2 \end{cases},
\tag{13.13}
$$

which are called the Schmidt lines. In Figs. 13.5 and 13.6 some typical results are shown for odd-proton and odd-neutron nuclei, respectively, and compared with the experimental values. We see that typically nuclear magnetic moments fall within the boundaries set by these Schmidt lines. In fact, when a given nucleus is one particle removed from a closed shell, (i.e., the nuclei characterized by especially tight binding) then often the magnetic dipole moments are near the Schmidt lines, providing one of the criteria that characterize the magic numbers. The type of behavior seen for the magnetic dipole moments validates the picture of the nucleus as being made from nucleons mostly forming spin-0 pairs together with one (or a few) active unpaired valence nucleons that carry the angular momentum of the nucleus and hence determine its magnetic moment. Similarly one can consider the electric quadrupole moment, defined as the projection of the ground-state charge distribution $\rho(\mathbf{x})_{00}$ onto the Legendre function $P_2(\cos\theta)$

$$
Q \equiv \int d\mathbf{x}\,\left(3z^2 - |\mathbf{x}|^2\right)\rho(\mathbf{x})_{00}\,,
\tag{13.14}
$$

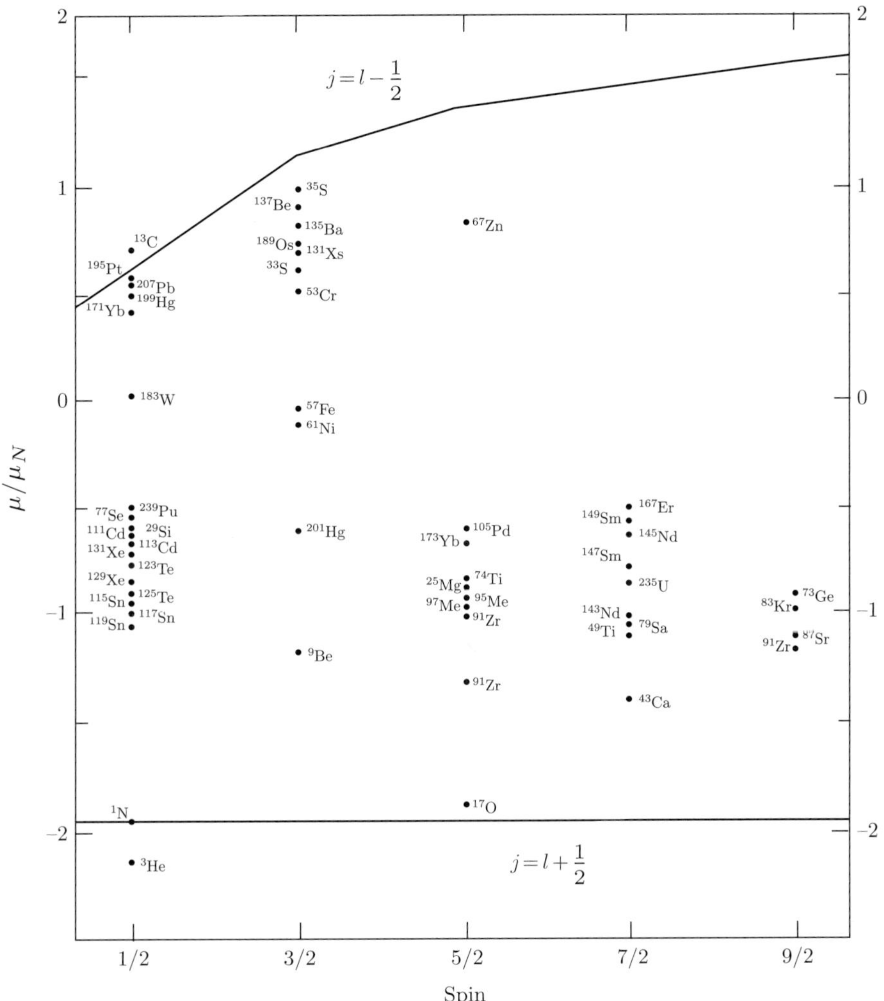

Fig. 13.6　Schmidt lines for odd-neutron nuclei. The figure is adapted from [Pre62].

and is related to the M_2^0 multipole introduced in Chapter 7 via the identity

$$x^2 Y_2^0(\Omega_x) = \sqrt{\frac{5}{16\pi}}\left(3z^2 - |\mathbf{x}|^2\right). \tag{13.15}$$

Here we follow common convention and refer to the "electric" quadrupole moment, although from the discussions earlier we realize that $\mathcal{Q}$ should actually be called the "Coulomb" quadrupole moment. Note that we require $J_0 \geq 1$ for a nonzero result. The systematics of the behavior of the quadrupole moments across the periodic table are such that near shell closures the nuclei are found to be nearly spherical, whereas mid-shell they are often deformed (see the discussions of deformed nuclei in the next chapter). Again there exist correlations with these trends and values of the magic numbers.

13.2   Nuclear and Neutron Matter

In the next chapter we return to discuss several specific models for finite nuclei and, accordingly, we leave most of those developments for later. In the remainder of the present chapter we limit our focus to the global characteristics of many-body nuclei with emphasis on nuclear and neutron matter.

Semi-Empirical Mass Formula

In Chapter 15 we shall return to consider elastic scattering of electrons from nuclei in more detail. Specifically, the so-called two-parameter Fermi charge distribution having parameters R, a and ρ_0 with one constraint, the known charge Z, is introduced there (see Eq. (15.22)). Here we draw only one observation from such experiments and also from studies of elastic hadron scattering from nuclei, namely, that the nuclear charge radius is given approximately by

$$R = a_0 A^{1/3} \, , \tag{13.16}$$

with $a_0 \simeq 1.07$ fm and that the central nuclear (matter) density $\rho(0)$ is approximately constant across the periodic table. One has

$$\rho(0) = \frac{A}{Z}\rho_0 \, , \tag{13.17}$$

where, as usual, A is the nuclear mass number and Z is the nuclear charge, with the neutron number then given by $N = A - Z$. Thus nuclei are denoted $^A_Z X_N$ as, for example, $^{13}_6 C_7$ for a carbon nucleus with $A = 13$, $Z = 6$, and $N = 7$. And so, as a starting approximation for the ground state of a nucleus let us assume that we have a uniform sphere of radius R, which implies that one has the nuclear volume

$$V = \frac{4\pi R^3}{3} \, , \tag{13.18}$$

and therefore the mean particle density

$$\rho_{\text{mean}} = \frac{A}{V} = \frac{3}{4\pi a_0^3} \simeq 1.95 \times 10^{38} \text{particles/cm}^3 \, , \tag{13.19}$$

i.e., a constant density independent of nuclear species. A second observation from elastic electron scattering (again, see Chapter 15) is that the surface thickness t (the change of radius required to go from 90% to 10% of the nuclear charge distribution at the origin ρ_0), is roughly constant for nuclei from the middle of the $2s - 1d$ shell (see below) up to ^{208}Pb. Empirically, one finds $t \simeq 2.4$ fm. We also assume that the neutron distribution has roughly the same shape as the proton (charge) distribution (see Chapter 15 where a direct lepton scattering determination of the neutron distribution has been performed; additionally, hadron scattering together with some assumptions involving the more complicated reaction mechanisms in these cases leads to roughly the same conclusions). Following Weizsäcker [Von35] we can develop an approximation for the binding energy of a general nucleus having quantum numbers A, Z, and N. One begins by taking the nucleus to be a liquid drop where the binding energy is assumed to scale with the volume of the drop ($V = 4\pi R^3/3$), so that, by Eq. (13.19) one has the so-called volume term in the binding energy

$$E_{\text{vol}} = -a_1 A \, . \tag{13.20}$$

The first correction to this bulk property comes from the fact that nucleons at the surface of the nucleus feel an asymmetric attraction from the nucleons within, leading to a surface tension contribution proportional to the nuclear surface, namely, to $4\pi R^2 \cdot \sigma$, where σ is the surface tension. Again using Eq. (13.19) this feature immediately leads to the surface energy term in the binding energy,

$$E_{\text{surface}} = 4\pi a_0^2 \sigma A^{2/3} \equiv a_2 A^{2/3} \, . \tag{13.21}$$

The next correction comes from the fact that protons in the nucleus, being charged, tend to repel one another. Taking the nucleus to have the charge uniformly distributed over a sphere of radius $R_c \equiv a_c A^{1/3}$ one can show that the total Coulomb energy involved for the $Z(Z-1)/2$ pairs of protons in the nucleus is

$$E_{\text{Coulomb}} = \frac{3}{5} \frac{Z(Z-1)\alpha}{R_c} \simeq a_3 \frac{Z(Z-1)}{A^{1/3}} \, . \tag{13.22}$$

The Coulomb energy term leads to a preference for nuclei with fewer protons than neutrons. Countering this tendency is the so-called symmetry energy contribution which can be understood by considering separate (nonrelativistic) Fermi gases for protons and neutrons and comparing the results they provide with what is obtained with a single Fermi gas for all of the nucleons in the nucleus. One has for the total particle numbers in these Fermi gases (see, for example, [Fet71])

$$Z = 2 \cdot 4\pi \left(\frac{L}{2\pi}\right)^3 \int_0^{k_F^p} k^2 dk = \left(\frac{L}{2\pi}\right)^3 \cdot \frac{8\pi}{3} \left(k_F^p\right)^3 \tag{13.23}$$

$$N = 2 \cdot 4\pi \left(\frac{L}{2\pi}\right)^3 \int_0^{k_F^n} k^2 dk = \left(\frac{L}{2\pi}\right)^3 \cdot \frac{8\pi}{3} \left(k_F^n\right)^3 \tag{13.24}$$

$$A = 4 \cdot 4\pi \left(\frac{L}{2\pi}\right)^3 \int_0^{k_F} k^2 dk = \left(\frac{L}{2\pi}\right)^3 \cdot \frac{16\pi}{3} \left(k_F\right)^3 \, , \tag{13.25}$$

where k_F^p, k_F^n, and k_F are the Fermi momenta of the three Fermi gases and the factors of 2 or 4 are the spin or spin-isospin degeneracies. The factor $L/2\pi$ arises, as usual, from converting between wavenumbers and momenta. Defining $\Delta \equiv N-Z$ and using $A = N+Z$ one therefore has

$$k_F^p = \left[1 - \frac{\Delta}{A}\right]^{1/3} k_F \tag{13.26}$$

$$k_F^p = \left[1 + \frac{\Delta}{A}\right]^{1/3} k_F \, . \tag{13.27}$$

Then, noting that the single-particle energies in the nonrelativistic Fermi gas model are given by $\epsilon(k) = k^2/2m_N$ (here as an approximation we use a common mass for protons and neutrons), the total kinetic energy of the nucleus is given by

$$\frac{E_{KE}}{(L/2\pi)^3} = 2 \cdot 4\pi \cdot \frac{1}{2m_N} \left\{ \int_0^{k_F^p} k^4 dk + \int_0^{k_F^p} k^4 dk \right\}$$

$$= \frac{4\pi}{5m_N} \left\{ (k_F^p)^5 + (k_F^n)^5 \right\}$$

$$= \frac{4\pi k_F^5}{5m_N} \left\{ \left[1 - \frac{\Delta}{A} \right]^{5/3} + \left[1 + \frac{\Delta}{A} \right]^{5/3} \right\}$$

$$= \frac{8\pi k_F^5}{5m_N} \left\{ 1 + \frac{5}{18} \left(\frac{\Delta}{A} \right)^2 + \cdots \right\}, \tag{13.28}$$

and hence

$$E_{KE} = \frac{3}{5} E_F \left\{ A + \frac{5}{18} \frac{\Delta^2}{A} + \cdots \right\}, \tag{13.29}$$

with $E_F = k_F^2/2m_N$ being the Fermi kinetic energy. In Eq. (13.28) above we have expanded, assuming Δ/A is small. In Eq. (13.29) then the first term is the standard Coulomb energy, $\frac{3E_F}{5}$, while the second, that is proportional to Δ^2/A, is the symmetry energy correction

$$E_{\text{symmetry}} = a_4 A \left(1 - \frac{2Z}{A} \right)^2. \tag{13.30}$$

Clearly, the symmetry energy term goes in the opposite direction to the Coulomb energy and therefore some compromise value of Z will occur where, for a given A, the total binding energy is maximum. Empirically we know that the valley of stability (the trajectory of greatest binding in the NZ-plane) is deepest for $N > Z$. Finally, one finds that even-even nuclei (nuclei with both Z and N even) are bound more tightly than even-odd and odd-even nuclei, and even more tightly bound than odd-odd nuclei (odd-odd nuclei are somewhat unusual, as mentioned above). See Fig. 13.7 for a sketch of the situations for odd-A and even-A nuclei. This term can be traced to the nature of the pairing interaction between the nucleons in the nucleus which favor the pairing of like types of nucleons to total angular momentum zero. Empirically this pairing term may be represented by

$$E_{\text{pairing}} = a_5 A^{-3/4} \delta, \tag{13.31}$$

where

$$\delta = \begin{cases} +1 & \text{even} - \text{even} \\ 0 & \text{even} - \text{odd} \quad \text{odd} - \text{even} \\ -1 & \text{odd} - \text{odd}. \end{cases} \tag{13.32}$$

Then finally the Weizsäcker semi-empirical mass formula for the nuclear binding energy is given by the sum of the five terms above

$$B(A, Z) = E_{\text{vol}} + E_{\text{surface}} + E_{\text{Coulomb}} + E_{\text{symmetry}} + E_{\text{pairing}}. \tag{13.33}$$

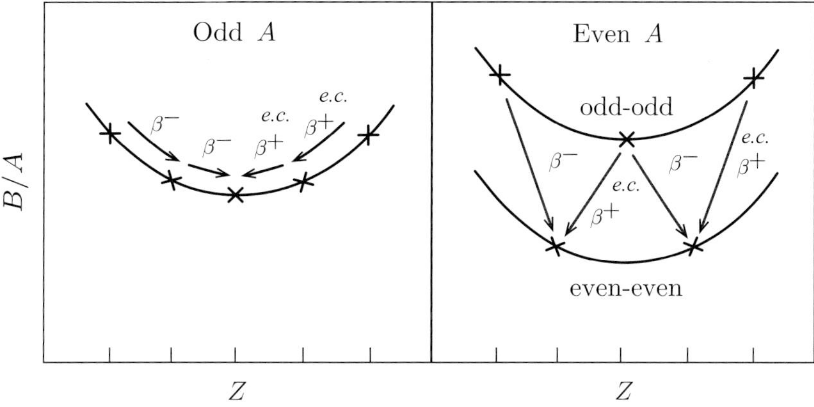

Fig. 13.7 Odd-A and even-A energy surfaces showing where allowed β-decays or electron captures (e.c.) can occur. Note that in the even-A case, situations can occur where normal β-decays are energetically forbidden and where jumps by two units are the only ones possible, the so-called double-beta-decay ($\beta\beta$) transitions discussed in Chapter 18.

The binding energy per particle is then

$$\frac{B(A,Z)}{A} = -a_1 + a_2 A^{-1/3} + a_3 Z(Z-1)A^{-4/3} + a_4 \left(1 - \frac{2Z}{A}\right)^2 + a_5 \delta A^{-7/4} \, . \quad (13.34)$$

The Weizsäcker coefficients are determined by optimizing the fit to the experimental masses of nuclei across the periodic table and are found to be [Gre53, Gre54]

$$\begin{aligned}
a_1 &= 15.75 \text{ MeV} \\
a_2 &= 17.8 \text{ MeV} \\
a_3 &= 0.71 \text{ MeV} \\
a_4 &= 23.7 \text{ MeV} \\
a_5 &= 34 \text{ MeV} \, .
\end{aligned} \quad (13.35)$$

The bottom of the valley of stability as a function of A may be found by computing the derivative of $B(A, Z)$ with respect to Z and setting that to zero to find the extremum – let us call the value of Z where the extremum is attained Z^*. Only the Coulomb energy and symmetry energy terms depend on Z and one obtains

$$Z^* = \frac{A + \frac{a_3}{4a_4}A^{2/3}}{2 + \frac{a_3}{2a_4}A^{2/3}} \, . \quad (13.36)$$

Infinite nuclear matter is defined by dropping the Coulomb energy term, calculating the binding energy per particle for symmetric matter ($N = Z$) and taking $A \to \infty$; only the volume energy term survives in this limit and one obtains

$$\left[\frac{B(A,Z)}{A}\right]_{NM} = -a_1 = -15.75 \text{ MeV} \, . \quad (13.37)$$

Infinite neutron matter with $Z = 0$ yields

$$\left[\frac{B(A,Z)}{A} \right]_{Z=0} = -a_1 + a_4 = 7.95 \text{ MeV} , \qquad (13.38)$$

and one can then see that the difference between these extremes is due to the symmetry energy term. The present interest in neutron stars, which are good approximations to pure neutron matter, depends directly on establishing the behavior in the equations-of-state for nuclear and neutron matter that relate to this contribution.

Nonrelativistic Fermi Gas Model

We now go into a more detailed discussion of nuclear matter, basing the developments on those given in [Fet71]. Being translation invariant and therefore uniform, one can employ a box of volume Ω and periodic boundary conditions so that the single-particle wavefunctions introduced above are simply plane waves accompanied by spin and isospin spinors

$$\varphi_{\mathbf{k},m_s,m_t}(\mathbf{x}) = \frac{1}{\sqrt{\Omega}} e^{i\mathbf{k}\cdot\mathbf{x}} \eta_{1/2}^{m_s} \xi_{1/2}^{m_t} . \qquad (13.39)$$

In the present section we consider only a nonrelativistic version of the many-body problem, while in the next section some discussion of relativistic mean-field theory is provided and then, later in Chapters 16 and 18 the relativistic Fermi gas model is developed for the treatment of high-energy inclusive electron scattering and neutrino reactions. Here we consider the so-called nonrelativistic degenerate Fermi gas in which all single-particle levels up to the Fermi level, the level where $k = k_F \simeq 1.42$ fm for nuclear matter, are filled with both protons and neutrons having both spin up and spin down with respect to some axis of quantization – this is the translationally invariant version of the state $|F\rangle$ discussed in Section 13.1. The same approach yields the starting approximation for discussions of condensed matter, namely, the degenerate electron gas, which is studied in an exercise at the end of this chapter. The interested reader is encouraged to look at the fine text by Fetter and Walecka [Fet71] where both problems are presented in a coherent way. Using this nuclear many-body Slater determinant with plane-waves to represent the ground state let us begin by recovering the contribution of the single-particle kinetic energy operator

$$T^{[1]} = -\sum_{i=1}^{A} \frac{\nabla_i^2}{2m_N} , \qquad (13.40)$$

which, using Eq. (13.8), yields

$$\left\langle F \left| T^{[1]} \right| F \right\rangle = K = 4 \sum_{k \leq k_F} \frac{|\mathbf{k}|^2}{2m_N} \rightarrow \frac{3}{5} E_F \cdot A , \qquad (13.41)$$

where the factor 4 comes from the spin-isospin degeneracy and the final result comes from taking the continuum limit and converting from sums to integrals. This is the result already used in Eq. (13.29), although now for protons and neutrons treated on the same footing; the ground-state matrix element of the single-particle number operator likewise yields

Eq. (13.25). To these results we want to add a two-particle potential to obtain altogether a first-order estimate of the ground-state energy. Using Eq. (13.10) one finds

$$\left\langle F \left| V^{[2]} \right| F \right\rangle \equiv v$$

$$= \sum_{k \le k_F m_s m_t} \sum_{k' \le k_F m'_s m'_t} \left[\langle k m_s m_t, k' m'_s m'_t | V | k m_s m_t, k' m'_s m'_t \rangle \right.$$

$$\left. - \langle k m_s m_t, k' m'_s m'_t | V | k' m'_s m'_t, k m_s m_t \rangle \right] / 2 \quad (13.42)$$

using obvious notation for the single-particle states. Following [Fet71] for the potential we take a combination of an ordinary central (W: Wigner) force and a space-exchange (M: Majorana) force which captures the essence of the nucleon–nucleon interaction at low energies (see the discussions in Chapter 11):

$$V^{[2]} = V(x) \left[a_W + a_M P_M \right] , \quad (13.43)$$

with P_M being the spatial exchange operator and $x = x_1 - x_2$. One then has $a_W \approx a_M$, namely nearly a so-called Serber force. We assume that the potential is non-singular so that its plane-wave matrix elements do not diverge and, in fact, can be considered to be perturbatively small. The matrix element for the first term in Eq. (13.42), the direct term, is then

$$V_D = \frac{1}{\Omega} \left[a_W \int d\mathbf{z} V(z) + a_M \int d\mathbf{z} e^{i(\mathbf{k}'-\mathbf{k})\cdot\mathbf{z}} V(z) \right] , \quad (13.44)$$

while the matrix element for the second term, the exchange term, is

$$V_E = \frac{1}{\Omega} \left[a_W \int d\mathbf{z} e^{i(\mathbf{k}'-\mathbf{k})\cdot\mathbf{z}} V(z) + a_M \int d\mathbf{z} V(z) \right] \delta_{m_s m'_s} \delta_{m_t m'_t} . \quad (13.45)$$

Upon substituting in Eq. (13.42) and again going to the continuum limit one then obtains the potential energy contribution

$$\left\langle F \left| V^{[2]} \right| F \right\rangle = \frac{\Omega}{2(2\pi)^6} \int d\mathbf{k} \int d\mathbf{k}' \theta(k_F - k) \theta(k_F - k')$$

$$\times \left\{ 16 \left[a_W \widetilde{V}(0) + a_M \widetilde{V}(\mathbf{k}' - \mathbf{k}) \right] \right.$$

$$\left. - 4 \left[a_W \widetilde{V}(0) + a_M \widetilde{V}(\mathbf{k}' - \mathbf{k}) \right] \right\} , \quad (13.46)$$

with $\widetilde{V}(\mathbf{q}) \equiv \int d\mathbf{z}\, e^{i\mathbf{q}\cdot\mathbf{z}} V(\mathbf{z})$. Using the fact that

$$\int d\mathbf{k}\, \theta(k_F - k) e^{i\mathbf{k}\cdot\mathbf{z}} = 4\pi \int_0^{k_F} dk\, k^2 j_0(kz) = 4\pi k_F^3 \frac{j_1(k_F z)}{k_F z} , \quad (13.47)$$

one then has for the kinetic and potential energy contributions per particle to the ground-state energy

$$\frac{\langle F \,|\, (T^{[1]} + V^{[2]}) \,|\, F\rangle}{A} \equiv \frac{B_1(k_F)}{A}$$

$$= \frac{3}{5} E_F + \frac{k_F^3}{12\pi^2}\left\{ (4a_W - a_M) \int dz V(z)\right.$$

$$\left. + (4a_M - a_W) \int dz V(z) \left[\frac{3j_1(k_F z)}{k_F z}\right]^2 \right\}.$$

$$(13.48)$$

This first-order result using the unperturbed ground-state wavefunction – the nonrelativistic degenerate Fermi gas – also provides a variational bound on the true ground-state energy, since application of the variational principle requires that $E \le \langle F \,|\, (T^{[1]} + V^{[2]}) \,|\, F\rangle$. If one considers $B_1(k_F)/A$ as a function of k_F or, equivalently via Eq. (13.25) as a function of density, then except at the origin, as $k_F \to \infty$ the integrand of the second term above goes to zero and the first term dominates the binding energy per nucleon. Assuming an attractive potential (which is, in fact, the case) the first term is then positive and thus $B_1(k_F)/A$ diverges as $k_F \to \infty$. In other words, for potentials of the type assumed here, the observed saturation of nuclear matter deduced from the semi-empirical mass formula discussed earlier cannot be attained, whereas saturation should occur at the value of k_F given in Eq. (13.37). If one extends the treatment to consider the full nucleon–nucleon interaction at large and intermediate distances (see Chapter 11) the same conclusion is reached. This is an important result that demonstrates the need for additional ingredients in our goal of describing nuclear matter (and also neutron matter).

Brueckner Theory

It should be clear that what is missing above is the short-range component of the nucleon–nucleon interaction. As discussed in Chapter 11, in a hadronic picture one knows that the longest-range contribution to the NN potential arises from pion exchange and that the medium-range contributions come (at least partially) from two-pion exchange, whereas heavy-meson exchange contributions are responsible for very strong short-range potentials. The latter may also be understood in terms of QCD degrees of freedom. The non-singular long- and medium-range parts of the total nucleon–nucleon potential are what we have addressed above; however, the short-range parts violate the constraints of those developments, since plane-wave matrix elements of the latter are not perturbatively small, and accordingly we need to proceed beyond perturbation theory and incorporate nonperturbative effects. This is a somewhat technical subject and goes beyond the scope of this book. We shall not proceed to a full exposition of the problem, and instead only summarize the main features of what is known as Brueckner theory. For a very complete discussion of this approach, the independent-pair approximation, the self-consistent Bethe–Goldstone equation and the (infinite) summation of the so-called ladder diagrams the reader is again referred to [Fet71]. Note that this is another example of a theme that runs through many-body nuclear theory, namely that often perturbation theory is proven to be inadequate and that one must resort to nonperturbative approaches including the summation of infinite sets of diagrams; we will encounter two other examples in the next chapter where the

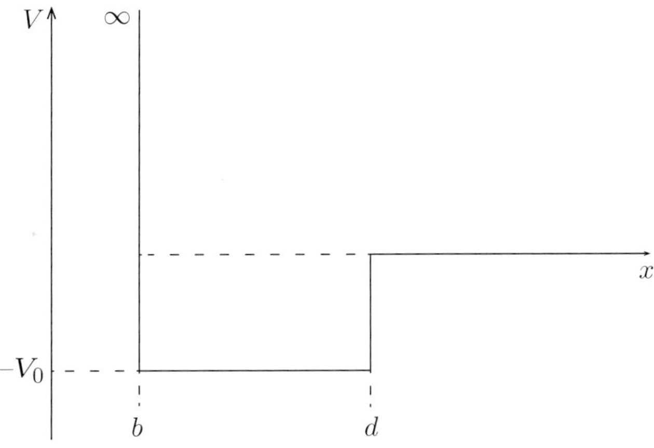

Fig. 13.8 Toy model potential having an infinite hard core of radius b outside of which, in the region $b \le x \le d$, one has an attractive potential of depth V_0.

Hartree–Fock approximation (HF) and the random-phase approximation (RPA) will both be discussed.

One begins with the Bethe–Goldstone equation for two interacting nucleons in the Fermi sea which arises when one concentrates on those two specific particles, but omits their interactions with the rest of the Fermi sea other than to incorporate the effects of the non-singular contributions discussed in the previous section, which are taken into account by allowing them to modify the single-particle energies. Specifically, one builds in the effects from the non-singular mean field in a self-energy term that contains both kinetic energy and potential energy contributions discussed above, typically by changing the nucleon mass in the nuclear medium to an effective mass. Such an approach is very commonly used in treating a wide range of quantum many-body systems, for example, in modeling electrons in condensed matter systems and leading to successful representations of their properties, including superconductivity.

The first steps in attempting a full solution to the nuclear matter problem often begins with simple "toy models" that provide some insight into how the various contributions from the NN interaction enter. For example, to capture the essence of both long/intermediate-range and short-range effects a potential such as the one shown in Fig. 13.8 can be invoked. This potential has an attractive contribution of depth V_0 acting in the region $b < x < d$ with no force for $x > d$, together with an infinite repulsive potential when $0 < x < b$. Typical numbers for the radii are $b \approx 0.4$ fm and $d \approx 2.7$ fm. The repulsive contribution then provides a "toy model" for the contributions from short-range exchanges. In fact, even this is a relatively involved problem whose solution is presented in chapter 11 of [Fet71] and so we reduce our focus to an even simpler problem, namely, discussion of the repulsive pure hard-core potential obtained by setting V_0 above to zero. One can show [Fet71] that the Bethe–Goldstone s-wave equation for this special problem is

$$\left(\frac{d^2}{dr^2} + K^2\right) u(r) = v(r)u(r) - \int_0^{\infty} dr' \chi(r, r')v(r')u(r') , \qquad (13.49)$$

where

$$\chi(r, r') = \frac{1}{\pi} \left[\frac{\sin(r - r')}{r - r'} - \frac{\sin(r + r')}{r + r'} \right],$$ (13.50)

with dimensionless variables

$$r \equiv k_F x$$
$$r' \equiv k_F x'$$
$$c \equiv k_F b \approx 0.57$$
$$u(r) \equiv k_F u(x)$$
$$v(r) \equiv \frac{V(x)}{k_F^2 / 2m_N^*},$$ (13.51)

and where m_N^* is the reduced effective mass that incorporates the mean-field effects. Note that we consider only s-waves here – we expect the contributions from the region where the short-range potential acts to be dominated by the $\ell = 0$ partial wave, since partial waves with $\ell \geq 1$ are prevented from having large contributions at short distances by their angular momentum barriers; these effects, of course, can be included in a more elaborate treatment of nuclear matter. The Bethe–Goldstone wavefunction $u(r)$ must vanish when $0 < r < c$ and has a discontinuous slope at $r = c$. Thus one can write

$$v(r)u(r) = \Gamma \delta(r - c) + w(r)\theta(c - r),$$ (13.52)

with Γ a factor to be determined. Such a form can be derived by considering a finite repulsive potential and then taking the limit as the barrier height becomes infinite. Inside the infinite repulsive core one then has

$$\frac{d^2 u}{dr^2} + K^2 u = \Gamma \left[\delta(r - c) - \chi(r, c) \right]$$
$$+ \left[w(r)\theta(c - r) - \int_0^c dr' \chi(r, r')w(r') \right].$$ (13.53)

Now one can make use of the fact that c is small; then when $r,\ r' < c$ the kernel in the equation may reasonably be approximated by

$$\chi(r, r') \approx \frac{2rr'}{3\pi},$$ (13.54)

and thus $w(r)$ is of order c^2 and its integral above in the core region is even smaller, since that goes as c^3. Making use of this smallness, the term in Eq. (13.53) involving $w(r)$ can be neglected and one arrives at a simple approximate equation for the Bethe–Goldstone wavefunction:

$$\left(\frac{d^2}{dr^2} + K^2 \right) u(r) \approx \Gamma \left[\delta(r - c) - \chi(r, c) \right]$$
$$\approx \Gamma \frac{2rc}{\pi} \int_1^\infty dq q^2 j_0(qr) j_0(qc)$$
$$\equiv F(r),$$ (13.55)

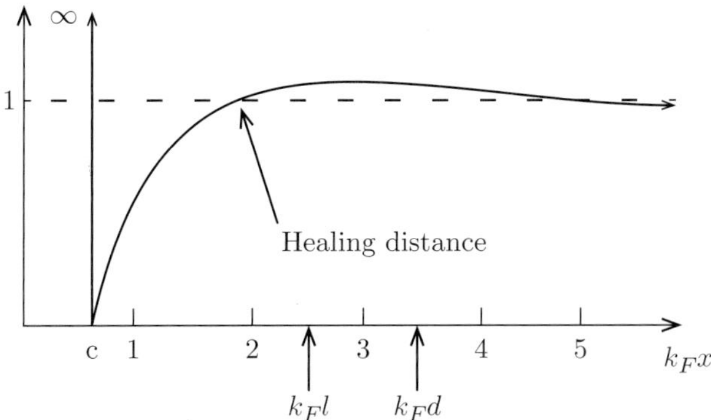

Fig. 13.9   The Bethe–Goldstone wavefunction for an s-wave pair interacting through a hard-core potential of radius b (dimensionless radius $c = k_F b$). The dimensionless average interparticle distance is $k_F \ell$. For comparison the dimensionless distance $k_F d$ (see Fig. 13.8) is also shown. The figure has been adapted from [Fet71].

subject to the constraint $u(0) = 0$. One is then left with the problem of determining Γ. The general solution to the above approximate differential equation is given by

$$
\begin{aligned}
u(r) &= \frac{1}{\Gamma} \int_0^r \sin\left[K(r - s)\right] F(s)\,ds \\
&= \frac{\sin(Kr)}{K} \int_0^r \cos(Ks)F(s)\,ds - \frac{\cos(Kr)}{K} \int_0^r \sin(Ks)F(s)\,ds\,,
\end{aligned}
\tag{13.56}
$$

with F as defined above. One can show that at large distances the second integral vanishes and, since the solution must approach a plane wave, that the first integral must approach unity, which yields the normalization condition

$$
\Gamma = \left[\cos(Kc) - \int_0^\infty \cos(Ks)F(s)\,ds\right]^{-1}\,,
\tag{13.57}
$$

hence providing the complete solution for the so-called dilute hard-sphere problem. Solving the problem numerically, one obtains the approximate Bethe–Goldstone wavefunction shown in Fig. 13.9. Clearly the wavefunction vanishes at the hard-core radius, as it should, and evolves to unity at large distances where the plane-wave solution must be reached. In between these extremes the wavefunction reaches unity at a distance $k_F x \approx 1.9$, the so-called "healing distance" where the "wound" in the interparticle wavefunction at shorter distances is "healed." This should be measured against the average interparticle distance in nuclear matter obtained using Eq. (13.25) and given by $\ell = (L^3/A)^{1/3} = (3\pi^2/2)^{1/3}k_F^{-1} \approx 1.73\,\text{fm}$ with corresponding dimensionless quantity $k_F \ell \approx 2.46$. That is, the healing distance is less than the average interparticle spacing justifying the basic assumption of the independent-pair approximation. Nuclear matter is in fact a relatively dilute system. The essential observation is the following [Fet71]: the Pauli principle suppresses the correlations introduced by the hard core and restricts its effects to short distances. In particular, the hard core cannot give rise to long-range scattering because all available energy-conserving states are already occupied. Except for the short-range correlations, a nucleon may therefore be assumed to move through nuclear matter in a plane-wave state.

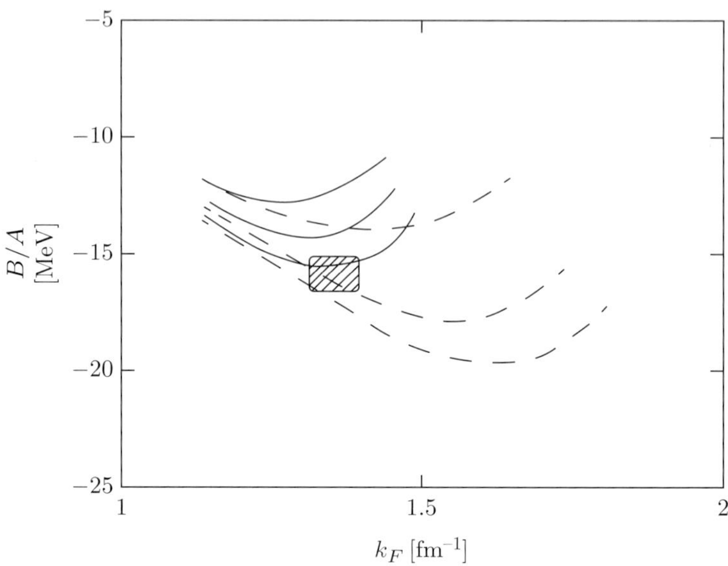

Fig. 13.10 Binding energy per nucleon B/A of symmetric nuclear matter versus the Fermi momentum k_F. The hatched region indicates what should be expected using the measured binding energies of finite nuclei and extrapolating to infinite nuclear matter. The solid lines are results obtained using several Dirac–Brueckner models, whereas the dashed lines represent results from more conventional Brueckner models. Figure adapted from [Mac86] (see also [Won98]).

After this discussion of the pure hard-core problem the next natural step is to improve the model by combining long-, intermediate-, and short-range contributions and eventually employ the full NN potential discussed in Chapter 11. Given the scope of the present book we shall not do that, but direct the reader to [Fet71] for much of the underlying formalism. As an example of how various commonly-used nucleon–nucleon interactions behave we show in Fig. 13.10 the binding energy per nucleon versus Fermi momentum. Clearly it is possible to reach the result required by the discussion earlier of the systematics of nuclear matter which is shown as a shaded region in the figure.

13.3   Relativistic Modeling of Nuclear Matter

We conclude this chapter with a brief discussion of a relativistic approach for nuclear (and neutron) matter, based on the so-called $\sigma\omega$-model. We draw upon the pioneering work of Walecka (see [Ser86]), and specifically on the simple model described in [Wal95] which has been used to motivate more sophisticated treatments of the problem. The model begins by assuming three building blocks: the baryon field ψ representing the nucleon (mass m_N), a scalar field ϕ (the "σ") with mass m_S which is coupled to the scalar density provided by the baryons, namely, to $\rho = \overline{\psi}\psi$, and a neutral vector field V_μ (the "ω") with mass m_V

which is coupled to the conserved baryon four-vector current, namely, to $B_\mu = \overline{\psi}\gamma_\mu\psi$. One assumes that the Lagrangian density for the system is

$$\mathcal{L} = \mathcal{L}^0 + \mathcal{L}^1 \, , \tag{13.58}$$

where $\mathcal{L}^0$ is the Lagrangian for the uncoupled meson fields

$$\mathcal{L}^0 = -\frac{1}{4}F_{\mu\nu}F^{\mu\nu} + \frac{1}{2}m_V^2 V_\mu V^\mu + \frac{1}{2}\left[\partial_\mu\phi\partial^\mu\phi - m_S^2\phi^2\right] \tag{13.59}$$

while $\mathcal{L}^1$ involves the baryon field and takes into account the couplings between the baryons and mesons:

$$\mathcal{L}^1 = \overline{\psi}\left[i\,\slashed{D} - (M - g_S\phi)\right]\psi \, , \tag{13.60}$$

where $D_\mu = \partial_\mu + ig_V V_\mu$ is the covariant derivative. Here the field tensor for the vector field is given as usual by

$$F_{\mu\nu} \equiv \partial_\mu V_\nu - \partial_\nu V_\mu \tag{13.61}$$

and the scalar and vector fields interact with the baryon field with coupling strengths g_S and g_V, respectively. Using the Euler–Lagrange equations for each of the field variables V^μ, ϕ, and $\overline{\psi}$, we obtain the field equations

$$\partial^\nu F_{\mu\nu} + m_V^2 V_\mu = g_V B_\mu$$
$$\left[\Box + m_S^2\right]\phi = g_S\rho$$
$$\left[i\,\slashed{D} - (m_N - g_S\phi)\right]\psi = 0 \, , \tag{13.62}$$

which are the massive vector boson analog of Maxwell's equation, the Klein–Gordon equation involving the scalar meson, and the Dirac equation for the nucleon with minimal coupling to the meson fields, respectively. One now introduces the concept of relativistic mean-field theory (RMFT). The starting point is again translational invariance and uniform matter density with baryon number B in a volume V, i.e., $\rho_B = B/V$. As the baryon density increases the source terms on the right-hand sides of Eqs. (13.62) also increase and one has a larger number of quanta present. Accordingly, one is motivated to replace the meson fields by their vacuum expectation values

$$\phi \rightarrow \langle\phi\rangle \equiv \phi_0 \tag{13.63}$$
$$V_\mu \rightarrow \langle V_\mu\rangle \equiv \delta_{\mu 0}V_0 \, , \tag{13.64}$$

where in the second equation one has used the fact that there exists no preferred direction for a uniform system at rest; hence only the $\mu = 0$ component enters. The classical fields are in fact constants, i.e., have no spacetime dependences for a uniform system at rest, allowing one to use the first of Eqs. (13.62) to write

$$V_0 = \frac{g_V}{m_V^2}\rho_B \, . \tag{13.65}$$

Substituting the classical fields into the Lagrangian then yields

$$\mathcal{L}^0_{\text{RMFT}} = \frac{1}{2}m_V^2 V_0^2 - \frac{1}{2}m_S^2\phi_0^2 \tag{13.66}$$

$$\mathcal{L}^1_{\text{RMFT}} = \overline{\psi}\left[i\,\slashed{\partial} - \gamma^0 g_V V_0 - m_N^*\right]\psi\,, \tag{13.67}$$

where the effective baryon (nucleon) mass, defined by

$$m_N^* \equiv m_N - g_S\phi_0\,, \tag{13.68}$$

also enters in the Dirac equation in Eq. (13.62). The latter is the usual equation, except for this shift in mass due to the classical scalar field and vector potential. Therefore, upon solving the Dirac equation for plane waves one finds that the energy eigenvalues are given by

$$E_\pm = g_V V_0 \pm \sqrt{k^2 + m_N^{*2}}\,, \tag{13.69}$$

with solutions $u(\mathbf{k}, \lambda)$ and $v(\mathbf{k}, \lambda)$, respectively, where λ represents the spin and isospin projections for the nucleons. Quantizing to obtain creation and destruction operators $A^\dagger_{\mathbf{k},\lambda}$, $B^\dagger_{\mathbf{k},\lambda}$ and $A_{\mathbf{k},\lambda}$, $B_{\mathbf{k},\lambda}$ for the two sets of solutions, as usual, one can then obtain the Hamiltonian for the system:

$$H = H_{\text{RMFT}} + \delta H, \tag{13.70}$$

where the RMFT contribution is given by

$$H_{\text{RMFT}} = -\frac{1}{2}m_V^2 V_0^2 + \frac{1}{2}m_S^2\phi_0^2 + g_V V_0\hat{\rho}_B$$
$$+\frac{1}{V}\sum_{\mathbf{k}\lambda}\sqrt{k^2 + m_N^{*2}}\left(A^\dagger_{\mathbf{k},\lambda}A_{\mathbf{k},\lambda} + B^\dagger_{\mathbf{k},\lambda}B_{\mathbf{k},\lambda}\right) \tag{13.71}$$

with baryon density

$$\hat{\rho}_B = \frac{1}{V}\sum_{\mathbf{k}\lambda}\left(A^\dagger_{\mathbf{k},\lambda}A_{\mathbf{k},\lambda} - B^\dagger_{\mathbf{k},\lambda}B_{\mathbf{k},\lambda}\right)\,, \tag{13.72}$$

which counts the number of baryons minus the number of antibaryons relative to the vacuum. Finally, the additional term representing the zero-point energy, the difference in energy of a filled negative-energy Dirac sea of baryons with mass m_N^* and that of a filled negative-energy sea when the baryons have mass m_N, is given by

$$\delta H = -\frac{1}{V}\sum_{\mathbf{k},\lambda}\left(\sqrt{k^2 + m_N^{*2}} - \sqrt{k^2 + m_N^2}\right)\,. \tag{13.73}$$

Everything here involves either c-numbers or is diagonal in the creation and destruction operators and thus, given the assumptions made here, the mean-field theory has been solved exactly.

As in the nonrelativistic treatment in the previous section we now assume that the ground state of symmetric nuclear matter is formed by filling all of the levels with both protons and neutrons with spin projections $+$ and $-$ (degeneracy $\gamma = 4$) up to the Fermi level having momentum k_F together with all states above this one left empty; this constitutes the degenerate ground state of the system. Immediately, upon converting to integrals, as usual, one finds the relationship

$$\rho_B = \frac{\gamma}{(2\pi)^3} \int_0^{k_F} d\mathbf{k} = \frac{\gamma}{6\pi^2} k_F^3 . \tag{13.74}$$

Then, using the above equations for the Hamiltonian and employing the first law of thermodynamics to obtain an expression for the pressure, one can show (see Exercise 13.8) that the energy density $\varepsilon \equiv E/V$ and pressure p are

$$\varepsilon(\rho_B, \phi_0) = \frac{g_V^2}{2m_V^2} \rho_B^2 + \frac{m_S^2}{2g_S^2} \left(m_N - m_N^*\right)^2 + \frac{\gamma}{(2\pi)^3} \int_0^{k_F} d\mathbf{k} \left(k^2 + m_N^{*2}\right)^{1/2}$$

$$p(\rho_B, \phi_0) = \frac{g_V^2}{2m_V^2} \rho_B^2 - \frac{m_S^2}{2g_S^2} \left(m_N - m_N^*\right)^2 + \frac{\gamma}{3(2\pi)^3} \int_0^{k_F} d\mathbf{k} k^2 \left(k^2 + m_N^{*2}\right)^{-1/2} . \tag{13.75}$$

Working at fixed volume V and baryon number B one can use the minimization condition

$$\left[\frac{\partial E}{\partial \phi_0}\right]_{V,B} = 0$$

and the energy density equation (the first of Eqs. (13.75)) to determine the classical scalar field and hence m_N^* via Eq. (13.68):

$$\phi_0 = \frac{g_S}{m_S^2} \rho_S = \frac{1}{g_S}(m_N - m_N^*) , \tag{13.76}$$

where the scalar density is given by

$$\rho_S = \frac{\gamma}{(2\pi)^3} \int_0^{k_F} d\mathbf{k} m_N^* \left(k^2 + m_N^{*2}\right)^{-1/2} . \tag{13.77}$$

Note that these two equations imply that the scalar field, or equivalently the effective mass, are determined self-consistently.

One is left with two free parameters to be adjusted to fit the saturation density and binding energy per nucleon of the system. For symmetric nuclear matter one finds that $C_S^2 \equiv (g_S M / m_S)^2 = 267.1$ and $C_V^2 \equiv (g_V M / m_V)^2 = 195.9$ [Wal95]. The energy per nucleon for nuclear matter is shown in Fig. 13.11 and the effective mass is shown in Fig. 13.12. Finally, the two figures also display results for pure neutron matter, which is simply obtained by setting γ to 2 rather than 4, whereby one can deduce some of the properties of neutron stars (see also Exercise 20.4). Following the discussion in the present chapter with its focus on infinite nuclear and neutron matter in hand, we now proceed in Chapter 14 to consider several representative models of finite nuclei.

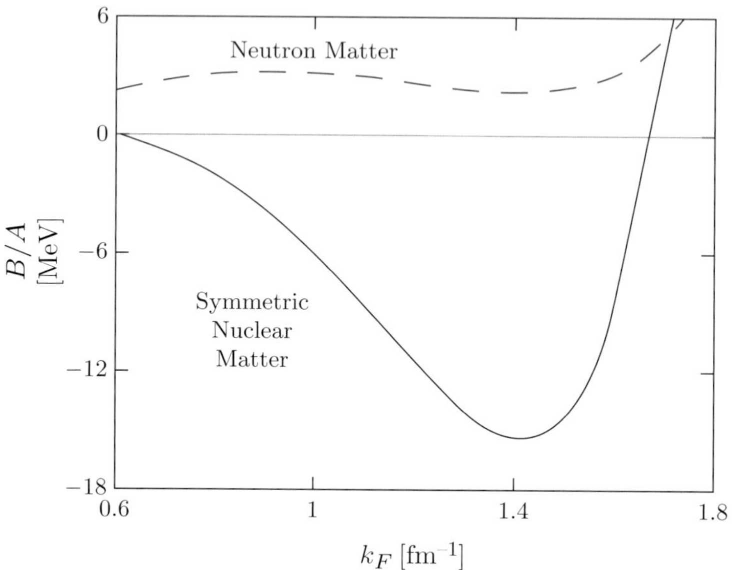

Fig. 13.11 Saturation curves for symmetric nuclear matter ($\gamma = 4$) and neutron matter ($\gamma = 2$), showing the energy per particle offset by the nucleon mass m_N versus the Fermi momentum k_F. The figure is adapted from [Ser86] and [Wal95].

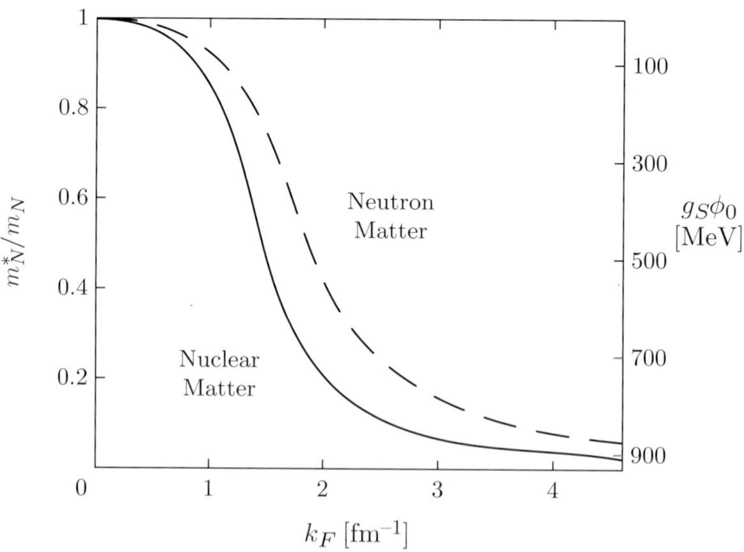

Fig. 13.12 Effective mass ratio m_N^*/m_N and $g_S\phi_0$ (see text) for symmetric nuclear matter ($\gamma = 4$) and neutron matter ($\gamma = 2$) as functions of Fermi momentum k_F. The figure is adapted from [Ser86] and [Wal95].

Exercises

13.1 Matrix Elements of One- and Two-Particle Operators

One- and two-particle operators acting in first-quantization (in contrast to one- and two-body operators acting in second-quantization, i.e., occupation number representation) have been discussed in the text. In this exercise some specific matrix elements of such operators are considered.

a) For the antisymmetrized state given as a Slater determinant

$$\Phi_{\alpha_1\alpha_2\cdots\alpha_A}(1,2,\ldots A) = \frac{1}{\sqrt{A!}} \begin{vmatrix} \varphi_{\alpha_1}(1) & \varphi_{\alpha_1}(2) & \cdots & \varphi_{\alpha_1}(A) \\ \varphi_{\alpha_2}(1) & \varphi_{\alpha_2}(2) & \cdots & \varphi_{\alpha_2}(A) \\ \vdots & \vdots & & \vdots \\ \varphi_{\alpha_A}(1) & \varphi_{\alpha_A}(2) & \cdots & \varphi_{\alpha_A}(A) \end{vmatrix},$$

following the developments in Section 13.1, show that the expressions

$$\left\langle F \left| \mathcal{O}^{[1]} \right| F \right\rangle = \sum_{\alpha \leq F} \int d\mathbf{x}_1 \varphi_\alpha^\dagger(1) \mathcal{O}^{[1]}(1) \varphi_\alpha(1) ,$$

and

$$\left\langle F \left| \mathcal{O}^{[2]} \right| F \right\rangle = \frac{1}{2} \sum_{\alpha,\beta \leq F} \int d\mathbf{x}_1 \int d\mathbf{x}_2 \varphi_\alpha^\dagger(1) \varphi_\beta^\dagger(2) \mathcal{O}^{[2]}(1,2)$$

$$\times \left(\varphi_\alpha(1)\varphi_\beta(2) - \varphi_\beta(1)\varphi_\alpha(2) \right) ,$$

emerge. Specifically, for plane waves, namely for the Fermi gas model as used in descriptions of symmetric nuclear matter, verify

$$\left\langle F \left| T^{[1]} \right| F \right\rangle = K = 4 \sum_{k \leq k_F} \frac{|\mathbf{k}|^2}{2m_N} \rightarrow \frac{3}{5} E_F \cdot A ,$$

and

$$\left\langle F \left| V^{[2]} \right| F \right\rangle \equiv v$$

$$= \sum_{k \leq k_F m_s m_t k' \leq k_F m'_s m'_t} \sum \left[\langle k m_s m_t, k' m'_s m'_t |V| k m_s m_t, k' m'_s m'_t \rangle \right.$$

$$\left. - \langle k m_s m_t, k' m'_s m'_t |V| k' m'_s m'_t, k m_s m_t \rangle \right] /2.$$

b) Making use of these results, prove the basic identity

$$\frac{\left\langle F \left| (T^{[1]} + V^{[2]}) \right| F \right\rangle}{A} \equiv \frac{B_1(k_F)}{A} = \frac{3}{5} E_F + \frac{k_F^3}{12\pi^2} \left\{ (4a_W - a_M) \int dz V(z) \right.$$

$$\left. + (4a_M - a_W) \int dz V(z) \left[\frac{3 j_1(k_F z)}{k_F z} \right]^2 \right\} .$$

for the binding energy per nucleon using the simple toy model of Section 13.2.

13.2   Schmidt Lines for Magnetic Dipole Moments of One-Particle and One-Hole Nuclei

The Schmidt lines discussed in the text are given by

$$\langle n_0 \ell_0 j_0 \| \mu_1 \| n_0 \ell_0 j_0 \rangle = \left\langle n_0 \ell_0 j_0 \left\| \left(g_\ell L + g_s \frac{1}{2} \sigma \right) \right\| n_0 \ell_0 j_0 \right\rangle ,$$

where one assumes that the nuclear ground state of an odd nucleus is to be approximated simply by a closed shell plus or minus one particle; that is, the magnetic dipole moment for the ground state of such a system will arise from the last unpaired nucleon. The single-particle state is assumed to have total angular momentum j_0 and to have orbital angular momentum ℓ_0 where the orbital and spin angular momenta are coupled ($\mathbf{J} = \mathbf{L} + \mathbf{S}$), namely, one has configuration $(\ell_0 1/2) j_0 m_{j_0}$. Since the magnetic dipole operator

$$\mu/\mu_N = \begin{cases} \ell_0 g_\ell + \frac{1}{2} g_s & j_0 = \ell_0 + 1/2 \\ \left(\frac{2\ell_0 - 1}{2\ell_0 + 1} \right) \left[(\ell_0 + 1) g_\ell - \frac{1}{2} g_s \right] & j_0 = \ell_0 - 1/2 \end{cases}$$

has both orbital and spin contributions, one must compute the matrix elements of operators acting in subspaces.

a) Perform the necessary analysis to prove the expressions for the Schmidt lines given above.

b) Calculate the ground-state magnetic moments of the following pairs of nuclei, $^{17}_{8}\text{O}$, $^{17}_{9}\text{F}$, and $^{209}_{82}\text{Pb}$, and $^{209}_{83}\text{Bi}$. Assume that they are well represented as having single particles or single holes (proton or neutrons) with respect to closed shells and compare the results with the experimental values.

13.3   Isospin Multiplets

Isobars have equal mass number, A, but different atomic number Z. Consider a set of isobars that are members of the same isospin multiplet. Their mass differences are determined by the Coulomb energies and the neutron–proton mass difference. Assume that the Coulomb energy of the nucleus of atomic number Z and radius R is

$$\frac{3}{5} \cdot \frac{Z^2 \alpha}{R} ,$$

where $R = R_0 A^{1/3}$.

a) Show that the energy difference between two adjacent nuclei of an isospin multiplet is

$$\Delta E = \frac{3\alpha}{5 R_0 A^{1/3}} (2Z - 1) - 0.78 \text{ MeV} .$$

b) We consider the $T = 1$, $J^\pi = 1^+$ isospin triplet consisting of the ground state of $^{12}_{5}\text{B}$, the 15.1 MeV excited state of $^{12}_{6}\text{C}$, and the ground state of $^{12}_{7}\text{N}$. Using the equation derived in a), and $R_0 = 1.4$ fm, show that

$$M\left(^{12}\text{C}, T = 1 \right) - M\left(^{12}\text{B}, T = 1 \right) = 2.2 \text{ MeV} .$$

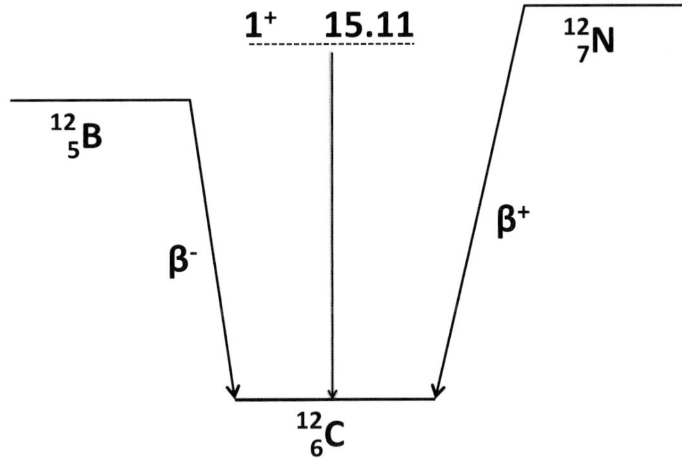

c) Knowing that $M\left(^{12}\mathrm{B}, T = 1\right) - M\left(^{12}\mathrm{C}, T = 0\right) = 12.9$ MeV, show that

$$M\left(^{12}\mathrm{C}, T = 1\right) - M\left(^{12}\mathrm{C}, T = 0\right) = 15.1 \text{ MeV} .$$

d) The same mass difference in the mass-14 isobar states is only 2.2 MeV. Can you explain why the ground state of ^{12}C has such a low energy?

13.4 Symmetry Energy

Consider the members of the $A = 135$ isobar and look up their masses using a table of mass excess. Then plot the results as a function of Z. Hence, determine the value of the symmetry energy parameter in the Weizsäcker mass formula. Perform the same calculation for members of the $A = 136$ isobar and estimate the value of the pairing parameter.

13.5 Barrier Penetration

Consider tunneling in α-decay from a heavy nucleus. The barrier penetration factor for the α-particle to pass through the Coulomb field of the residual nucleus may be estimated as follows. Consider a triangular-shaped, two-dimensional barrier of height V_0 at the radius R which decreases to the value of the kinetic energy E_α of the α-particle at the distance R_1. For a heavy nucleus with $Z = 72$, we can use $R = 9$ fm, and the height of the Coulomb barrier may be taken as the Coulomb energy at distance R. At R_1 the barrier height equals the kinetic energy of the α-particle, which we will assume to be 5 MeV. What is the probability for the α-particle to escape from the nucleus?

13.6 Brueckner Theory and the Bethe–Goldstone Equation

For the assumptions made in the discussions of nuclear matter in the text, show that the solution for the Bethe–Goldstone wavefunction inside the repulsive core is found by solving

$$\frac{d^2u}{dr^2} + K^2 u = \Gamma\left[\delta(r-c) - \chi(r,c)\right]$$

$$+ \left[w(r)\theta(c-r) - \int_0^c dr'\,\chi(r,r')w(r')\right].$$

13.7   Dirac Equation for Baryons in Uniform Nuclear and Neutron Matter

Once the mean-field approximations have been made in Section 13.3, one obtains a Lagrangian from which a Dirac equation can be constructed.

a) Perform the necessary steps to find this equation.

Since the classical scalar and vector fields enter simply as constants, it is straightforward to solve this equation.

b) Construct this solution and, as in standard treatments of quantum field theory, proceed to write the Dirac field as an expansion involving the spinor solutions, plane waves, and creation and destruction operators.

c) Using these results, prove

$$H_{\text{RMFT}} = -\frac{1}{2}m_V^2 V_0^2 + \frac{1}{2}m_S^2\phi_0^2 + g_V V_0 \widehat{\rho}_B$$

$$+ \frac{1}{V}\sum_{\mathbf{k}\lambda}\sqrt{k^2 + m_N^{*2}}\left(A_{\mathbf{k},\lambda}^\dagger A_{\mathbf{k},\lambda} + B_{\mathbf{k},\lambda}^\dagger B_{\mathbf{k},\lambda}\right)$$

with baryon density

$$\widehat{\rho}_B = \frac{1}{V}\sum_{\mathbf{k}\lambda}\left(A_{\mathbf{k},\lambda}^\dagger A_{\mathbf{k},\lambda} - B_{\mathbf{k},\lambda}^\dagger B_{\mathbf{k},\lambda}\right)$$

and

$$\delta H = -\frac{1}{V}\sum_{\mathbf{k},\lambda}\left(\sqrt{k^2 + m_N^{*2}} - \sqrt{k^2 + m_N^2}\right).$$

d) Explain how the baryon density should be interpreted in terms of the Dirac field operators.

13.8   The Equations-of-State for Symmetric Nuclear Matter and for Neutron Matter

Using the developments of relativistic nuclear and neutron matter in the text, prove the results

$$\rho_B = \frac{\gamma}{(2\pi)^3}\int_0^{k_F} d\mathbf{k} = \frac{\gamma}{6\pi^2}k_F^3$$

$$\varepsilon(\rho_B,\phi_0) = \frac{g_V^2}{2m_V^2}\rho_B^2 + \frac{m_S^2}{2g_S^2}\left(m_N - m_N^*\right)^2 + \frac{\gamma}{(2\pi)^3}\int_0^{k_F} d\mathbf{k}\left(k^2 + m_N^{*2}\right)^{1/2}$$

$$p(\rho_B,\phi_0) = \frac{g_V^2}{2m_V^2}\rho_B^2 - \frac{m_S^2}{2g_S^2}\left(m_N - m_N^*\right)^2 + \frac{\gamma}{3(2\pi)^3}\int_0^{k_F} d\mathbf{k}\,k^2\left(k^2 + m_N^{*2}\right)^{-1/2}$$

for the baryon density, energy density, and pressure, and hence for the equations of state.

13.9 Euler–Lagrange Equations for the $\sigma\omega$-Model

Write down the general form of the Euler–Lagrange equations and, using the Lagrangians

$$\mathcal{L}^0 = -\frac{1}{4}F_{\mu\nu}F^{\mu\nu} + \frac{1}{2}m_V^2 V_\mu V^\mu + \frac{1}{2}\left[\partial_\mu\phi\partial^\mu\phi - m_S^2\phi^2\right]$$

and

$$\mathcal{L}^1 = \overline{\psi}\left[i\,\slashed{D} - (M - g_S\phi)\right]\psi \, ,$$

show that the field equations

$$\partial^\nu F_{\mu\nu} + m_V^2 V_\mu = g_V B_\mu$$

$$\left[\square + m_S^2\right]\phi = g_S\rho$$

$$\left[i\,\slashed{D} - (m_N - g_S\phi)\right]\psi = 0$$

emerge.

Models of Many-Body Nuclei

We continue the discussions that started in the preceding chapter, where the general systematics of the nuclear many-body problem were introduced, and where infinite nuclear matter was the focus. In the present chapter, a selection of models that are used in descriptions of finite many-body nuclei is discussed, focusing on the nuclear ground state and on low-lying excitations in nuclei; higher-lying regions of excitation are postponed until Chapter 16. The present discussions include both microscopic approaches, where the relevant degrees of freedom are taken to be the protons and neutrons in the nucleus, and also so-called collective models where one assumes that the microscopic underlying structure can be subsumed into specific motions of a "nuclear fluid." Discussions of the latter usually begin by introducing a classical fluid, and then quantizing the harmonic oscillations of that fluid to obtain various collective quanta, "rotons," "surfons," p/n isovector oscillatory modes, etc. Both approaches have roles to play in describing the properties of the nuclear ground state and of low-lying excitations in nuclei. Ultimately in most cases, after much hard work, it has been possible to show that sufficiently sophisticated treatments of the microscopic problem in fact yield the underlying basis for the macroscopic collective description. Nevertheless, the latter continues to prove useful. The entire problem of modeling nuclear structure is quite involved and entire books have been written on it [Des74, Wal95, Won98, Row10]. In this chapter our goal is not to repeat this discussion in great depth; rather our intent is to introduce the reader to a selection of some of the issues that surround this subject of modeling many-body nuclei. This chapter is basically an introduction to some of the issues occurring in discussions of nuclear theory, while in the following chapters we then return to discuss some applications of these concepts. Specifically, we use some of these developments to illustrate what can be learned from electron scattering and to make contact with later discussions in the book. We begin with an overview of the Hartree–Fock (HF) approximation and the concept of the nuclear mean field.

14.1 Hartree–Fock Approximation and the Nuclear Mean Field

Let us now pick up the developments of the previous chapter where we briefly discussed how to represent the many-body basis in terms of Slater determinants of single-particle wavefunctions that were in turn solutions of the single-particle Schrödinger equation

$$H^{[1]}(i)\varphi_\alpha(i) = \epsilon_\alpha \varphi_\alpha(i) , \tag{14.1}$$

where $H^{(1)}(i)$ is the assumed single-particle Hamiltonian (indicated by the superscript "[1]"). The many-body Hamiltonian is taken to be

$$H = \sum_i H^{[1]}(i) + \sum_{i \neq j} v(i,j) \,, \tag{14.2}$$

namely, a single-particle term arising from the single-particle Hamiltonian plus a two-particle term containing a potential interaction $v(i,j)$.

Let us now consider what is called the Hartree–Fock approximation: this very important way of proceeding beyond perturbation theory occurs in many discussions of the quantum theory of many-particle systems, ranging from atomic, condensed matter and nuclear physics to applications in particle physics (see [Fet71] for in-depth discussions of the quantum theory of many-particle systems). The aim is to begin with the Hamiltonian written in the form

$$H = \sum_i T(i) + \sum_{i \neq j} V(i,j) \,, \tag{14.3}$$

where $T(i)$ is the single-particle kinetic energy operator and $V(i,j)$ is some approximation to the NN potential discussed in Chapter 11. One adds and subtracts some single-particle potential $U(i)$ (some average or "mean-field" potential), thereby defining the single-particle Hamiltonian above

$$H^{[1]}(i) \equiv T(i) + U(i) \,, \tag{14.4}$$

and a residual interaction

$$v(i,j) \equiv V(i,j) - U(i) \,. \tag{14.5}$$

One starts by choosing as a trial wavefunction the single Slater determinant discussed in Section 13.1 that results when the residual interaction is set to zero and only the single-particle Hamiltonian $H^{(0)}$ is taken into account. In principle, any average potential U can be assumed; however, as discussed in Chapter 13, the procedure to be developed in the following converges faster if a judicious choice is made. For instance, one can employ a central potential such as the harmonic oscillator well or one having the same shape as the two-parameter Fermi distribution plus the spin-orbit potential introduced in the previous chapter to split single-particle levels; these ideas are discussed in more detail in Chapter 15. With the residual interaction set to zero, the ground state of the nucleus will simply be the one denoted $|F\rangle = |\Phi_0\rangle = |\alpha_1 \alpha_2 \cdots \alpha_A\rangle$, where the occupied single-particle states labeled $\alpha_1 \alpha_2 \cdots \alpha_A$ are completely filled for the lowest A levels and completely empty for all higher-lying levels. The highest occupied level is the Fermi level.

Now one wants to turn on the residual interaction and improve the ground state from the degenerate approximation $|\Phi_0\rangle$ to an improved version denoted $|\Psi_0\rangle$. The variational principle tells us that the variation of the ground-state expectation value of the Hamiltonian H (i.e., the ground-state energy) must be stationary:

$$\delta \langle \Psi_0 | \widehat{H} | \Psi_0 \rangle = 0 \,, \tag{14.6}$$

or

$$\langle \delta \Psi_0 | \widehat{H} | \Psi_0 \rangle = 0 \,. \tag{14.7}$$

Here the carets are used to indicate operators in occupation number representation (second quantization), which we return to briefly in later discussions. We take the single-particle basis to be fixed, namely, we begin with the starting assumption of a complete, orthonormal set of single-particle states. Then we alter this choice by adding (hopefully) small admixtures of Slater determinants, where particles are promoted from below to above the Fermi sea, indicated by $\leq F$ and $F >$, respectively. We denote the Slater determinant obtained when the level labeled α_p above the Fermi surface is occupied (a "particle") and the level labeled α_h below the Fermi sea is empty (a "hole") by

$$|ph\rangle = |\alpha_1\alpha_2\cdots\alpha_{h-1}\alpha_{h+1}\cdots\alpha_A\alpha_p\rangle \tag{14.8}$$

$$= a_{\alpha_p}^\dagger b_{\alpha_h}^\dagger |\Phi_0\rangle \, , \tag{14.9}$$

where $a_{\alpha_p}^\dagger$ and $b_{\alpha_h}^\dagger$ are the particle and hole creation operators, respectively. These are called $1p1h$ states; in general one can also have $2p2h$, $3p3h$, etc., states and the full complexity of the nuclear many-body problem comes into play, although the restriction to only $1p1h$ states assumed here allows an important approximation to be made. The variation is then a sum over this $1p1h$ basis

$$\delta\Psi_0 = \sum_{ph} \eta_{ph} |ph\rangle \, , \tag{14.10}$$

with weights η_{ph}. Substituting Eq. (14.10) into Eq. (14.7) yields

$$\sum_{ph} \eta_{ph} \langle ph |\widehat{H}| \Phi_0\rangle = 0 \, , \tag{14.11}$$

where, since the $1p1h$ states are independent, each term in the sum must vanish, namely

$$\langle ph |\widehat{H}| \Phi_0\rangle = \left\langle ph \left|\left(\widehat{H}^{[1]} + \widehat{v}\right)\right| \Phi_0\right\rangle = 0 \, . \tag{14.12}$$

Being a one-body operator, the first term yields only single-particle matrix elements

$$H_{\alpha\beta}^{[1]} \equiv \left\langle \varphi_\alpha(1) \left|H^{[1]}(1)\right| \varphi_\beta(1)\right\rangle. \tag{14.13}$$

Note that the antisymmetry of the states and form of the one-body operator allow one to permute everything to particle number 1. Likewise, assuming the residual interaction is a two-body operator (one can also have three-body, four-body, etc., forces although here these are neglected) one has in first-quantization the two-particle matrix elements

$$v_{\alpha\beta\gamma\delta} \equiv \left\langle \varphi_\alpha(1)\varphi_\beta(2) |v(1,2)| \varphi_\gamma(1)\varphi_\delta(2)\right\rangle$$
$$- \left\langle \varphi_\alpha(1)\varphi_\beta(2) |v(1,2)| \varphi_\delta(1)\varphi_\gamma(2)\right\rangle \, , \tag{14.14}$$

using the symmetries to permute everything to particles 1 and 2. Equation (14.12) then reads

$$H_{\alpha\beta}^{[1]} + \sum_\gamma v_{\alpha\gamma\beta\gamma} = 0 \, , \tag{14.15}$$

where the sum runs over all occupied states. Of course, the original guess for U and thus $H^{[1]}$ should not be expected to satisfy this equation, the Hartree–Fock equation.

Let us proceed following [Won98] by changing notation a bit: let us use the indices α, β, γ, ... to denote the original basis, namely eigenstates of $H^{(0)}$, and indices a, b, c, ... to indicate solutions to the HF equation. The residual interaction may be expanded in the former

$$\widehat{v} = \sum_{\alpha\beta\gamma\delta} |\alpha\beta\rangle\, v_{\alpha\beta\gamma\delta}\, \langle\gamma\delta| \, , \tag{14.16}$$

yielding the HF single-particle Hamiltonian

$$\widehat{H}^{HF} = \widehat{H}^{[1]} + \sum_{b}\sum_{\alpha\beta\gamma\delta} \langle b\,|\alpha\beta\rangle\, v_{\alpha\beta\gamma\delta}\, \langle\gamma\delta\,|b\rangle \, , \tag{14.17}$$

or, equivalently, the eigenvalue equation

$$\widehat{H}^{[1]}\,|a\rangle + \sum_{b}\sum_{\alpha\beta\gamma\delta} \langle b\,|\alpha\beta\rangle\, v_{\alpha\beta\gamma\delta}\, \langle\gamma\delta\,|ba\rangle = \epsilon_a^{HF}\,|a\rangle \, , \tag{14.18}$$

where ϵ_a^{HF} is the HF single-particle energy. The above can be cast as a non-local, integro-differential equation and is usually solved by iteration. One begins with the initial choice of basis, eigenstates of $H^{[1]}$, say states labeled α_1, β_1, γ_1, ... and evaluates the matrix elements of the residual interaction as above. Equation (14.18) is then solved to obtain an improved approximation labeled a_1, b_1, c_1, ... Now one takes this improved basis to replace the initial basis; that is, one starts again with states now labeled α_2, β_2, γ_2, ... (equal to the states labeled a_1, b_1, c_1, ...) and cycles through the procedure again to obtain a second improvement, states labeled a_2, b_2, c_2, ... $\equiv \alpha_3$, β_3, γ_3, ... and so on until no further changes are seen and the final converged solution to the HF equation is obtained.

A complete discussion of the use of creation and destruction operators in treating the nuclear many-body problem is outside the scope of this book (see [Fet71] for a comprehensive treatment), nevertheless it is instructive to make contact with some of those ideas, namely, to write some of our results in this chapter in second-quantized form. One has creation operators $a_\alpha^\dagger$ where as above α labels the complete set of single-particle quantum numbers, taken here to be the final HF results. The fermion creation and destruction operators satisfy the usual anticommutation relations

$$\left\{ a_\alpha, a_\beta^\dagger \right\} = \delta_{\alpha\beta} \, , \tag{14.19}$$

and any one-body operator, for instance the usual EM current operator, may be written

$$\widehat{\mathcal{O}}^{[1]} = \sum_{\alpha\beta} \left\langle \alpha\,\left|\mathcal{O}^{[1]}\right|\,\beta \right\rangle a_\alpha^\dagger a_\beta \, , \tag{14.20}$$

where $\left\langle \alpha\,\left|\mathcal{O}^{[1]}\right|\,\beta \right\rangle$ is the single-particle matrix element similar to those discussed above. In effect, a one-body operator acting on a many-body state destroys a particle with quantum numbers β and creates one with quantum numbers α, *viz.*, changes the quantum numbers from β to α. Likewise, a two-body operator may be written

$$\widehat{\mathcal{O}}^{[2]} = \sum_{\alpha\alpha'\beta\beta'} \left\langle \alpha\alpha'\,\left|\mathcal{O}^{[2]}\right|\,\beta\beta' \right\rangle a_\alpha^\dagger a_{\alpha'}^\dagger a_{\beta'} a_\beta \, , \tag{14.21}$$

involving two-particle matrix elements and changing the quantum numbers of two particles from $\beta\beta'$ to $\alpha\alpha'$. In particular, using the above equations the nuclear Hamiltonian may be written

$$H = \sum_{\alpha\beta} H^{[1]}_{\alpha\beta} a^\dagger_\alpha a_\beta + \frac{1}{4} \sum_{\alpha\alpha'\beta\beta'} v_{\alpha\alpha'\beta\beta'} a^\dagger_\alpha a^\dagger_{\alpha'} a_{\beta'} a_\beta \,. \tag{14.22}$$

Usually in discussing the nuclear many-body problem one makes a canonical transformation from all particles above the vacuum to particles above the Fermi surface and holes below, i.e., introduces the convention

$$a_\alpha \longrightarrow \begin{cases} a_\alpha & \alpha > F \\ S_\alpha b^\dagger_{-\alpha} & \alpha \leq F, \end{cases} \tag{14.23}$$

where the notation $-\alpha$ indicates that the magnetic quantum numbers that are part of the set α have been reversed ($m_{j_\alpha} \to -m_{j_\alpha}$, $m_{t_\alpha} \to -m_{t_\alpha}$) and $S_\alpha \equiv (-)^{j_\alpha - m_{j_\alpha}} (-)^{1/2 - m_{j t_\alpha}}$ which guarantees that the hole creation operator is an irreducible tensor operator in both angular momentum and isospin (see [Fet71] for details). One sees that a_α destroys particles above the Fermi surface or creates holes ($b^\dagger_{-\alpha}$) below it. Then the Hamiltonian in Eq. (14.22) may be written in terms of zero-body (c-number), one-body and two-body contributions:

$$\begin{aligned}
\widehat{H} &= H^{[0]} + \widehat{H}^{[1]} + \widehat{H}^{[2]} \\
H^{[0]} &= \sum_{\alpha \leq F} \left(T_\alpha + \frac{1}{2} v_\alpha \right) \\
\widehat{H}^{[1]} &= \sum_{\alpha > F} \epsilon_\alpha a^\dagger_\alpha a_\alpha - \sum_{\alpha \leq F} \epsilon_\alpha b^\dagger_\alpha b_\alpha \\
\widehat{H}^{[2]} &= \frac{1}{2} \sum_{\alpha\alpha'\beta\beta'} v_{\alpha\alpha'\beta\beta'} N\left[a^\dagger_\alpha a^\dagger_{\alpha'} a_{\beta'} a_\beta \right],
\end{aligned} \tag{14.24}$$

where N indicates that the creation operators should be ordered such that all destruction operators are to the right of all creation operators, i.e., normal ordered. In the next chapter we shall return to discuss applications of HF theory to elastic electron scattering.

While the scope of this book does not extend to an in-depth treatment of the nuclear many-body problem with Green's functions, diagrammatic techniques, etc., or a full treatment of second-quantization which is only touched upon here, it may still be helpful to consider at least the basic diagrammatic ideas that underlie the HF approximation. The interested reader is directed to [Fet71] for a thorough discussion of the problem (see especially chapter 4 in that book for the HF approximation).

The basic building block of the general problem is the single-particle Green's function denoted G which allows expectation values of one-body operators to be evaluated and yields the ground-state energy E_0. The Green's function is indicated by a solid line and in lowest order, G^0, simply yields the contribution from a single particle above the Fermi surface or the absence thereof (a hole) below the Fermi surface. Saying this, we note as above that it proves useful to make a canonical transformation to a basis of particles and holes measured with respect to some chosen Fermi surface. Denoting the residual interaction v by a dashed line and first considering only diagrams where a single interaction

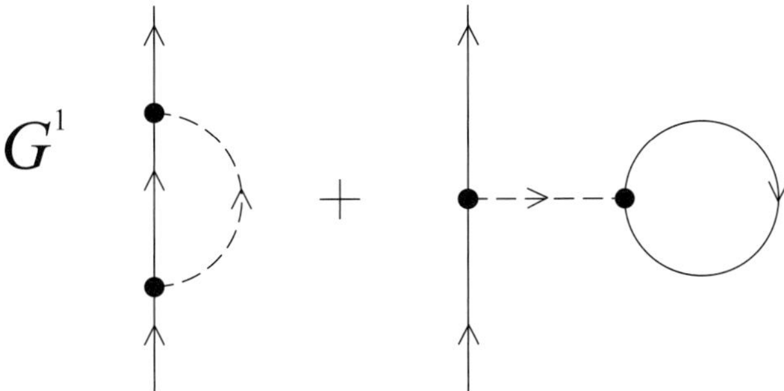

Fig. 14.1 First-order contribution to the single-particle Green's function.

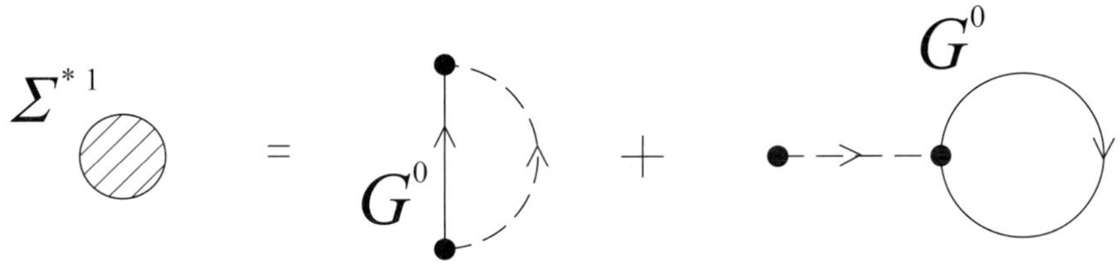

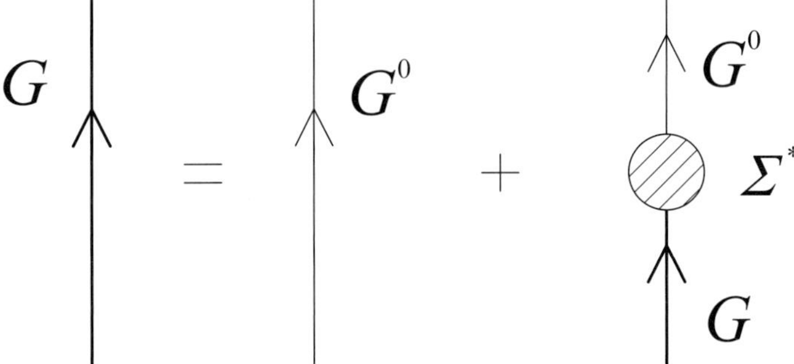

Fig. 14.2 First-order self-energy insertion with direct (right-hand diagram) and exchange (left-hand diagram) contributions.

Fig. 14.3 Dyson's equation for the full single-particle Green's function G involving the full proper self-energy insertion Σ^*.

occurs, one then has the first-order Green's function represented diagrammatically in Fig. 14.1. where, isolating the piece where the interaction occurs, one has the so-called first-order self-energy insertion containing two contributions, a direct term and an exchange term, as shown in Fig. 14.2. This is first-order perturbation theory.

Next we illustrate how to renormalize the Green's function to obtain a nonperturbative approximation for the dynamics. The general technique is shown in Fig. 14.3, where the new Green's function is obtained by solving the effective integral equation indicated diagrammatically in the figure. The renormalized Green's function G is obtained as shown on the right-hand side of the equation by adding to the first-order answer the result of having that same renormalized result interact via a generalized proper self-energy insertion Σ^* and connect to the outcome in first-order. Clearly this is a non-linear problem: the

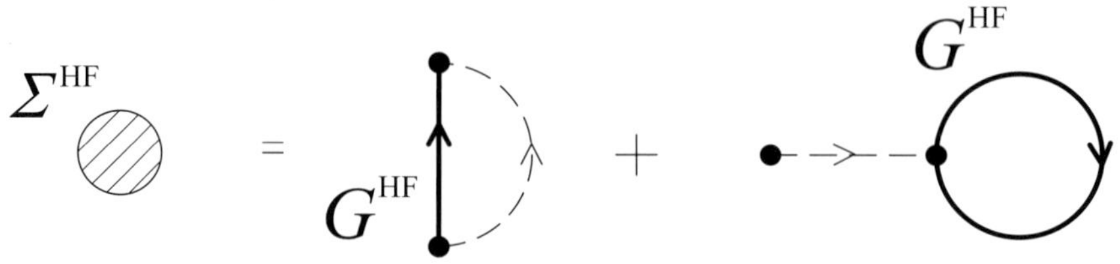

Fig. 14.4 Self-energy insertion which has HF Green's function lines in direct and exchange terms as indicated.

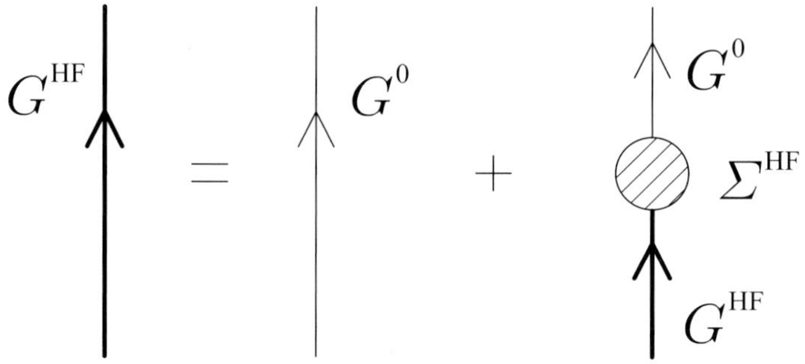

Fig. 14.5 Dyson's equation in the HF approximation.

equation that corresponds to the diagrams shown in the figure is called Dyson's equation. If one takes the contributions term by term it is clear that a very complex structure results. One can verify that taking the left-hand side of the equation and inserting it for G on the right-hand side, doing it again, etc., generates the infinite sum of all diagrams with the pieces of the insertion inserted in all ways (one can try this with the first-order insertion to generate the sums of all terms with the two contributions, namely, the direct so-called tadpole and exchange contributions – see Exercise 14.8). At this stage one has gone beyond perturbation theory and made a nonperturbative approximation.

This is a general statement and its solution would correspond to the full single-particle Green's function. Now we make some assumptions that can be shown to correspond to the HF approximation and to yield for G what we will call G^{HF}. This involves what we will call the HF self-energy insertion Σ^{HF}, shown diagrammatically in Fig. 14.4, which contains G^{HF}. In turn, the latter is obtained as indicated diagrammatically in Fig. 14.5 and involves both G^{HF} and Σ^{HF} on the right-hand side of Dyson's equation. In effect, this self-consistent approach is not only nonperturbative, but nonlinear. As in the developments presented above the problem is usually attacked by iteratively improving from some starting point to a stable numerical solution.

Note that the HF approximation is one of the very important nonperturbative approaches employed in atomic, condensed matter and nuclear physics. Other nonperturbative approaches are discussed, for instance, in Chapter 6. Also note that extensions to HF are denoted by other names, although they are based on the same types of ideas; for instance the BCS theory of superconductivity, used in condensed matter, nuclear, and even particle physics can be formulated as above.

So far in this chapter we have summarized the basic approach to mean-field theory using as active degrees of freedom the nucleons themselves. We have developed the basic model widely used as a starting point for microscopic descriptions of nuclei, and specifically, for the nuclear ground state. In the following chapter, we show results for elastic electron scattering where data are compared with ground-state charge and neutron distributions of spherical nuclei obtained in the HF approximation using typical nuclear interactions. In the present chapter, we next turn to a brief discussion of what happens when, as is usually the case, the nuclear ground state is not spherical.

14.2 Rotational Model of Deformed Nuclei

Near closed shells typically nuclei are found to be spherical; however, away from these regions this is not the case and nuclei are usually deformed. The simplest type of deformation is quadrupolar, meaning that the nuclear surface radius is given by

$$R(x, \theta, \phi) = R_0 \left\{ 1 + \sum_{M=-2}^{+2} a_{2,M} Y_2^M(\theta, \phi) \right\} \tag{14.25}$$

with five shape coefficients $a_{2,M}$. Following [Won98] let us see what constraints can be placed on these coefficients. The orientation in space of a deformed nucleus can be characterized by the Euler angles (α, β, γ) (see [Edm74] for more detailed discussions) and, since the absolute orientation in space is immaterial, one can impose constraints on the five shape coefficients by recognizing that the relationship between the laboratory frame and the nuclear "body-fixed" frame is accomplished via the rotation matrices (again, see [Edm74]):

$$a_{L,M} = \sum_{M=-2}^{+2} \mathcal{D}_{MM}^L(\alpha, \beta, \gamma) a_{L,M} , \tag{14.26}$$

yielding $a_{2,-1} = a_{2,+1} = 0$ and $a_{2,-2} = a_{2,+2}$, namely three constraints. Thus, using more common notation the remaining two shape coefficients may be written $a_{2,0} \equiv \beta \cos \gamma$ and $\sqrt{2} a_{2,+2} \equiv \beta \sin \gamma$, i.e., in terms of the Hill–Wheeler variables β and γ. Using explicit expressions for the spherical harmonics, this leads from Eq. (14.25) to

$$R(x, \theta, \phi) = R_0 \left\{ 1 + \beta \sqrt{\frac{5}{16\pi}} \left[\cos \gamma (3 \cos^2 \theta - 1) + \sqrt{3} \sin \gamma \sin^2 \theta \cos 2\phi \right] \right\} . \tag{14.27}$$

The parameter β characterizes the amount of deformation, ranging from oblate (flattened at the poles) when β is negative to prolate (pointed at the poles like an American football) when β is positive. The other parameter γ characterizes the departure from axial symmetry. See, also, the discussion for the deuteron in Chapter 11.

Classically, for rotations one has $\mathbf{J} = I\omega$, where I is the moment of inertia and where the energy is given by

$$E_J = \frac{1}{2}I\omega^2 = \frac{\mathbf{J}^2}{2I} . \tag{14.28}$$

The quantum analog is

$$H_{\text{rot}} = \sum_{i=1}^{3} \frac{1}{2I_i}\mathbf{J}_i^2 . \tag{14.29}$$

Often one finds that nuclei are axially symmetric, i.e., that one has (say) $I_1 = I_2 \equiv I$, but $I_3 \neq I$, so that

$$H_{\text{rot}} = \frac{1}{2I}(\mathbf{J}^2 - J_3^2) + \frac{1}{2I_3}J_3^2 . \tag{14.30}$$

As seen above, classically a rotating body requires the three Euler angles to specify its orientation in space, whereas quantum mechanically one must characterize the states involved using a complete set of quantum numbers. In this present situation these can be J, where the eigenvalue of $\mathbf{J}^2$ is $J(J+1)$, M, the projection of $\mathbf{J}$ along the axis of quantization in the laboratory system labeled z, and K, the projection of $\mathbf{J}$ along the body-fixed axis of quantization labeled 3 [Won98, Row10]. The rotational wavefunction may be written as a product of an intrinsic part ψ_{int} and a rotational part which is purely geometric, involving the rotation $\mathcal{D}$-functions [Edm74] which relate spherical harmonics in one system to those in another:

$$Y_J^K(\theta',\phi') = \sum_M \mathcal{D}_{MK}^J(\alpha,\beta,\gamma)Y_J^M(\theta,\phi) , \tag{14.31}$$

where the direction specified by the angles (θ',ϕ') is related to the direction specified by (θ,ϕ) via the Euler angles (α,β,γ). Under a parity transformation one may show that $\mathcal{D}_{MK}^J(\alpha,\beta,\gamma)$ goes to $(-)^{J+K}\mathcal{D}_{M,-K}^J(\alpha,\beta,\gamma)$ and, since one has both a phase factor and a change of sign of K, the $\mathcal{D}$-function in general does not have definite parity. To obtain a state of definite parity one takes a linear combination of states with $\pm K$ and uses only $|K|$ to label the states:

$$|JMK>_{\text{rot}}= \sqrt{\frac{2J+1}{16\pi^2(1+\delta_{K0})}}\left[\mathcal{D}_{MK}^J(\alpha,\beta,\gamma) \pm (-)^{J+K}\mathcal{D}_{M,-K}^J(\alpha,\beta,\gamma)\right] , \tag{14.32}$$

with $\pm$ for positive (negative) parity, respectively. The factors inside the square root are chosen to yield the proper normalization (see [Edm74] for properties of the $\mathcal{D}$-functions). In the more general triaxial case where $I_1 \neq I_2 \neq I_3$ a linear combination of states with fixed J and M, but different K is required and K is no longer a good quantum number.

A nucleus may have a given intrinsic state but different amounts of angular momentum in the laboratory systemi, i.e., different values of J. A set of states of this type forms a rotational band: the states in a given band are all related to one another in energy, static moments and electromagnetic transition rates. Using the wavefunction in Eq. (14.32) one can show that for positive-parity $K = 0$ states, only even integer values of J can occur and the reverse, for negative-parity $K = 0$ states, only odd integer values of J can occur; for $K > 0$ one has all states with $J \geq K$. That is, for $K > 0$ one has states with

$$J = K,\ K+1,\ K+2\ldots, \tag{14.33}$$

with energies

$$E_J = \frac{1}{2I}J(J+1) + E_K, \tag{14.34}$$

where E_K is the contribution from the intrinsic part of the wavefunction. This pattern is beautifully realized in many cases with the rotational band structure extending up to quite high values of J. Without proof we note here the so-called intrinsic mass quadrupole moment of the nucleus can be related to β. Note that the quadrupole moment discussed in Chapter 11 for deuterium and considered in the following chapter for elastic electron scattering is, in contrast, the charge quadrupole moment arising from electromagnetic interactions with the nuclear charge, rather than with all of the nucleons in the nucleus. Additionally, the strength of gamma-decays between neighboring states in a deformed nucleus can also be expressed in terms of β; recall that via Fig. 13.3 this was identified as a way to find where the magic numbers are to be found.

In closing this discussion, we note that even more can be done on the symmetry characterizations of deformed nuclei (see, for instance, the excellent discussion in [Iac03]). Additionally, one can proceed from spherical HF modeling to deformed HF treatments, although this lies outside the scope of the book.

14.3 Vibrational Model

Another type of excitation at low energy that nuclear matter may possess is the so-called vibrational mode in which the shape of a nucleus changes while the density remains constant (incompressible) due to the relative "stiffness" of nuclear matter. We follow the developments given in [Def66]. Using the analogy to vibrations of a liquid drop, where the velocity field is assumed to be irrotational, the departures from a spherical shape may be described by expanding the classical drop radius in spherical harmonics:

$$R(\theta,\phi) = R_0 \left\{ 1 + \sum_{L,M} \alpha_{L,M} Y_L^M(\theta,\phi) - \frac{1}{4\pi} \sum_{L,M} |\alpha_{L,M}|^2 \right\}, \tag{14.35}$$

which involves a set of expansion coefficients $\{\alpha_{L,M}\}$, where R_0 is the radius of the equivalent sphere and $R(\theta,\phi)$ is the radius of the nuclear surface given as a function of spherical polar angles θ and ϕ. For each mode of order L one has $2L+1$ parameters, $M = -L,\ldots,+L$. We assume that $\alpha_{L,M}^* = (-)^M \alpha_{L,-M}$ which guarantees that the radius is real. We shall consider small oscillations and work only to second-order in α. For such small oscillations the volume of the drop is given by $V = 4\pi R_0^3/3 + \mathcal{O}(\alpha^3)$ and the change in the surface area of the liquid drop is found to be

$$\Delta S = S - 4\pi R_0^2 = \frac{1}{2} R_0^2 \sum_{L,M} (L-1)(L+2) |\alpha_{L,M}|^2, \tag{14.36}$$

which vanishes for $L = 1$ (dipole) oscillations corresponding simply to translations of the nucleus as a whole and not to excitations. Accordingly, we need only to consider oscillations with $L \geq 2$. For $L = 2$ one has quadrupole vibrations of the nucleus, with prolate when the deformation is positive (outwards) and oblate when negative (inwards). Thus the quadrupole vibration mode corresponds to the motion spherical $\rightarrow$ prolate $\rightarrow$ spherical $\rightarrow$ oblate $\rightarrow$ spherical, etc. Even octupole ($L = 3$) vibrations can be important for low-lying states in nuclei.

Writing the velocity field as $\mathbf{v} = \nabla \psi$ the physical solution to Laplace's equation is of the form

$$\psi = \sum_{L,M} \beta_{L,M} r^L Y_L^M(\theta, \phi) \,, \tag{14.37}$$

which yields the kinetic energy to second-order in the expansion coefficients

$$T = \int dV \frac{1}{2} \rho \mathbf{v}^2 = \frac{1}{2} \rho R_0 \sum_{L,M} L R_0^{2L} \left| \beta_{L,M} \right|^2 + \mathcal{O}(\beta^3) \,, \tag{14.38}$$

where ρ is the mass density of the drop. Equating the velocity field at the surface to the velocity of the surface, namely $\partial R / \partial t = \partial \psi / \partial r$ yields $\partial \alpha_{L,M} / \partial t \equiv \dot{\alpha}_{L,M} = L R_0^{2L-2} \beta_{L,M}$, one has

$$T = \frac{1}{2} \rho R_0^5 \sum_{L,M} \frac{1}{L} \left| \dot{\alpha}_{L,M} \right|^2 \,. \tag{14.39}$$

Taking σ as the surface tension of the drop and using Eq. (14.36) one then obtains the drop Lagrangian:

$$L = T - V = \frac{1}{2} \rho R_0^5 \sum_{L,M} \frac{1}{L} \left| \dot{\alpha}_{L,M} \right|^2 - \frac{1}{2} \sigma R_0^2 \sum_{L,M} (L - 1)(L + 2) \left| \alpha_{L,M} \right|^2 \,. \tag{14.40}$$

In fact, as discussed in [Def66], the Coulomb interaction can also be included, although for simplicity we do not do so here. With $p_{L,M}$ taken to be the momenta conjugate to the displacement coefficients $\alpha_{L,M}$, this yields the Hamiltonian

$$H = \sum_{\substack{L,M \\ L \geq 2}} \left[\frac{1}{2 B_L} \left| p_{L,M} \right|^2 + \frac{1}{2} C_L \left| \alpha_{L,M} \right|^2 \right] \,, \tag{14.41}$$

where one has $B_L \equiv \rho R_0^5 / L$, $C_L \equiv \sigma R_0^2 (L - 1)(L + 2)$, $\omega_L^2 \equiv C_L / B_L$, and $p_{L,M} = B_L (-)^L \dot{\alpha}_{L,-M}$. The constants can be obtained from the semi-empirical mass formula in Chapter 13. We thus have the Hamiltonian in the form of an infinite set of uncoupled harmonic oscillators. This is a very important approach used throughout treatments of the many-body problem. Often there is good reason to suspect that a particular type of motion in classical descriptions of the system (for instance, here the nuclear fluid) is important for specific regions of excitation energy. Assuming small-amplitude oscillations, rotations, dilations, etc., of this fluid as in the present example, one can proceed in the familiar way to quantize the system. Other similar types of motion can be studied using exactly the same approach; here we focus on these so-called surfon excitations to illustrate the procedure.

Writing the expansion coefficients in the usual form for problems of this type

$$\alpha_{L,M} = \frac{1}{\sqrt{2\sqrt{B_L C_L}}} \left[\alpha_{L,M} e^{-i\omega_t t} + (-)^M \alpha^*_{L,M} e^{i\omega_t t} \right] , \tag{14.42}$$

one may then treat the coefficients as creation and destruction operators, creating and destroying surface oscillations or surfons, and write the Hamiltonian in standard form

$$H = \sum_{\substack{L,M \\ L \geq 2}} \frac{1}{2}\omega_L \left(\alpha^\dagger_{L,M} \alpha_{L,M} + \alpha_{L,M} \alpha^\dagger_{L,M} \right) \tag{14.43}$$

with commutation relations

$$\left[\alpha_{L,M}, \alpha^\dagger_{L',M'} \right] = \delta_{LL'} \delta_{MM'} \tag{14.44}$$

$$\left[\alpha_{L,M}, \alpha_{L',M'} \right] = \left[\alpha^\dagger_{L,M}, \alpha^\dagger_{L',M'} \right] = 0 . \tag{14.45}$$

The matrix elements between states differing by one surfon follow immediately:

$$\left\langle n_{L,M} + 1 \left| \alpha^\dagger_{L,M} \right| n_{L,M} \right\rangle = \sqrt{n_{L,M} + 1} \tag{14.46}$$

$$\left\langle n_{L,M} - 1 \left| \alpha_{L,M} \right| n_{L,M} \right\rangle = \sqrt{n_{L,M}} \tag{14.47}$$

and the spectrum is given by

$$\omega_L^2 \equiv \frac{1}{8} L(L-1)(L+2)\omega_2^2 , \tag{14.48}$$

as illustrated in Fig. 14.6; one may show that the parity of the surfon states is $(-)^L$. It is now simple to employ this collective model to describe electroexcitation of these states – but this we postpone until Chapter 15.

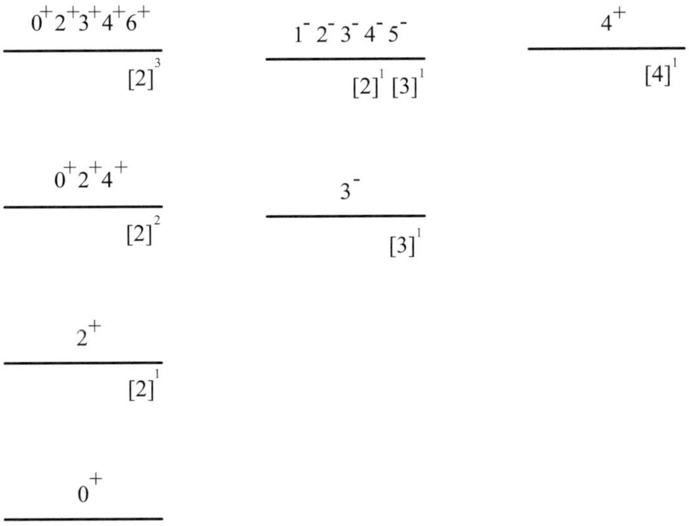

Fig. 14.6   Low-lying surfon spectrum for single, double, and triple quadrupole surfon states, a state with one octupole surfon, a state having one quadrupole with one octupole, and a single hexadecapole surfon state.

As noted above, these surfon excitations are examples of a wide class of motions of the nuclear fluid that can be studied using similar techniques to proceed from classical, to semi-classical and then to quantized descriptions of the problem. For instance, one can have compressional modes where the nucleus in a sense "breathes." Furthermore, much very elegant work has, for example, gone into developing the so-called interacting boson approximation (IBA) with interesting aspects of group theory used to characterize the nature of the bosons – these discussions go beyond the scope of the book; the interested reader is directed to [Iac87, Iac91] for detailed presentation of the underlying ideas. We next turn to higher-lying excitations addressed both from a collective point of view and using more microscopic degrees of freedom.

14.4 Single-Particle Transitions and Giant Resonances

Finally, in this chapter we consider an important type of excitation found at somewhat higher energies than those discussed above. In particular, one typically finds relatively large cross sections in processes such as nucleon scattering, (p, n) reactions, pion scattering and charge-exchange reactions, electron scattering and gamma-decay (plus others) when probing the excitation region lying at tens of MeV in nuclei, specifically, at about 20–25 MeV in light nuclei, falling to about 15 MeV in heavy nuclei such as ^{208}Pb. The total strength found is much larger than is predicted on the basis of a transition from the ground state to a $1p1h$ state (a state where one particle is promoted to a higher-lying single-particle state in the mean field, leaving behind a single hole, to be discussed in the following). This is the region of the so-called "giant resonances" including the important isovector $E1$ (electric dipole) excitation – the giant dipole resonance or GDR – which is found at an energy of about $E^*(\mathrm{GDR}) \sim 78A^{-1/3}$ MeV, that is, lies higher in energy than the typical shell spacing of $41\,A^{-1/3}$ MeV discussed in Chapter 13. Similar considerations occur for other multipolarities and one sees clear evidence for both giant quadrupole ($E2$) and giant octupole ($E3$) resonances. The large strengths and locations of these excitations point to the fact that giant resonances are collective excitations, involving the cooperative participation of several nucleons, in contrast to $1p1h$ excitations. The residual interaction acting between the nucleons in the nucleus (i.e., what is left over from the NN interaction when the mean-field approximation is invoked; see above) shifts the strength upwards (for isovector excitations) and spreads it out, leading to a width of roughly 6 MeV. For all types of giant resonances one may develop semi-classical models, as well as more microscopic descriptions based on the shell model. In the following we take both points of view.

Here we begin with a specific collective semi-classical description and follow this presentation by a more microscopic treatment. The former is the Goldhaber–Teller model [Gol48, Def66] where, in its simplest form, one assumes that the resonance is formed when the neutrons move opposite to the protons in a collective way providing a large electric dipole moment. Again, as we did earlier, take the Hamiltonian for this excitation to be a simple harmonic oscillator whose restoring force is due to the decreased overlap of the proton/neutron fluids, and is related to the symmetry energy discussed in Chapter 13. Using the system's reduced mass $\mu = NZm_N/A$ with m_N the nucleon mass, then one has

$$H = \frac{1}{2\mu}\mathbf{p}^2 + \frac{1}{2}\mu\omega_0^2\mathbf{r}^2 \,, \tag{14.49}$$

where $\mathbf{r}$ is the coordinate for the relative separation of the fluids, $\mathbf{p}$ is the relative momentum with the system's center-of-mass at rest, and ω_0 is the oscillation frequency discussed above. We are then immediately in a position to quantize $\mathbf{r}$ and $\mathbf{p}$, yielding the (spherical tensor) operators $\widehat{r}_m$ and $\widehat{p}_m$ with $m = 0, \pm 1$ which may be written in terms of creation and destruction operators, as usual. The excited states are then formed by acting on the ground state to build one-quantum, two-quantum, etc., states with $J^\pi = 1^-$ ($1\omega_0$), $0^+, 2^+$ ($2\omega_0$), etc., the single-quantum excitation being the isovector GDR with $\omega_0 = E^*(\text{GDR})$. In Chapter 15, we continue with this model to provide a simple treatment of excitation of the GDR via electron scattering or photoexcitation. Here we continue by turning to more microscopic descriptions.

Let us begin our discussion by considering the ground state to be a single Slater determinant with all single-particle states filled to the Fermi level and all higher-lying states being empty; specifically, this could be the HF ground state discussed above. Then we consider excited states where a single nucleon is promoted from below the Fermi surface to above it, leaving behind a single hole in the Fermi sea, i.e., we consider pure one-particle, one-hole ($1p1h$) states. We label the excited state by the single-particle and single-hole quantum numbers, $n_p\ell_p j_p$ and $n_h\ell_h j_h$, respectively, $\left|(n_p\ell_p j_p)(n_h\ell_h j_h)^{-1} J^\pi M_J; T M_T\right\rangle$. Here orbital angular momenta ℓ_p and ℓ_h are vector-coupled with the nucleon's spin of 1/2 to form j_p and j_h, respectively, and these are in turn vector-coupled to form the state's angular momentum JM_J. The nucleon isospin 1/2 is likewise vector-coupled to form states with total isospin $T = 0, 1$ and projection M_T. For example, if we consider ^{16}O to be a closed-shell 0^+ isospin-0 nucleus with filled $1s_{1/2}$, $1p_{3/2}$, and $1p_{1/2}$ shells, and if particles are promoted to the $1d_{5/2} - 2s_{1/2} - 1d_{3/2}$ shell, then $1p1h$ states with the following quantum numbers can be formed. The allowed values of J^π are given in Table 14.1 and T may be either 0 or 1; the table is arranged so that one expects the lowest-lying states to be the upper left-hand entries while the highest-lying are to the lower right. Since only the active particle (the one promoted from below to above the Fermi level) can be involved for a one-body operator, the many-body reduced matrix element is given simply by the single-particle reduced matrix element with single-particle states having the quantum numbers of the $1p1h$ state:

$$\left\langle (n_p\ell_p j_p)(n_h\ell_h j_h)^{-1} J^\pi \left\|\widehat{\mathcal{O}}_J\right\| 0^+ \right\rangle = \left\langle n_p\ell_p j_p \left\|\mathcal{O}_J\right\| n_h\ell_h j_h \right\rangle \,, \tag{14.50}$$

Table 14.1 Pure $1p1h$ states in the $2s$-$1d$ (particle) and $1s$-$1p$ (hole) shells

	$1d_{5/2}$	$2s_{1/2}$	$1d_{3/2}$
$1p_{1/2}^{-1}$	$2^- 3^-$	$0^- 1^-$	$1^- 2^-$
$1p_{3/2}^{-1}$	$1^- 2^- 3^- 4^-$	$1^- 2^-$	$0^- 1^- 2^- 3^-$
$1s_{1/2}^{-1}$	$2^+ 3^+$	$0^+ 1^+$	$1^+ 2^+$

suppressing the isospin quantum numbers for clarity. In Chapter 15, we present a simple model for the electron scattering form factors involved in photo- and electro-excitation of these pure $1p1h$ states. In fact, some, such as the high-spin states discussed above, will be seen to be reasonably well represented as $1p1h$ states; however, others, such as the GDR, are not. The latter states are found to be collective and thus require a more sophisticated treatment. This brings us to the last selected topic in this chapter, namely, the so-called random-phase approximation (RPA).

Random Phase Approximation

Again we start by considering excited states obtained by admixing particle–hole states. Following [Def66, Fet71], we begin with the particle–hole creation operator with good total angular momentum and isospin

$$\widehat{\xi}^\dagger(abJM_J, T_{M_T}) \equiv \sum_{m_{j_\alpha} m_{j_\beta}} \langle j_a m_{j_\alpha} j_b m_{j_\beta} | j_a j_b JM_J \rangle$$

$$\times \sum_{m_{t_\alpha} m_{t_\beta}} \left\langle \frac{1}{2} m_{t_\alpha} \frac{1}{2} m_{t_\beta} \Big| \frac{1}{2} \frac{1}{2} TM_T \right\rangle a_\alpha^\dagger b_\beta^\dagger \, , \tag{14.51}$$

where we use the short-hand notation $\alpha \leftrightarrow (a, m_{j_\alpha} m_{t_\alpha})$ so that the latin indices refer only to total angular momenta and isospin, but not to the magnetic quantum numbers, $a \leftrightarrow \left(n_a \left(\ell_a \frac{1}{2}\right) j_a, t_a = \frac{1}{2}\right)$. Forming the commutator with the Hamiltonian and computing the matrix elements between the ground state $|\Psi_0\rangle$ (here assumed to have zero angular momentum and isospin for simplicity) and an exact eigenstate of the many-body nuclear system with good angular momentum and isospin quantum numbers $\left|\Psi_{JM_J TM_T}^K\right\rangle$, where K labels the specific excited state with the selected many-body quantum numbers, we have

$$\left\langle \Psi_{JM_J TM_T}^K \left| \left[\widehat{H}, \widehat{\xi}^\dagger(abJM_J, T_{M_T})\right] \right| \Psi_0 \right\rangle = (E_{JT} - E_0) \, \psi_{JT}^K(ab) \, , \tag{14.52}$$

where $E_{JT}^* \equiv E_{JT} - E_0$ is the excitation energy and

$$\psi_{JT}^K(ab) \equiv \left\langle \Psi_{JM_J TM_T}^K \left| \widehat{\xi}^\dagger(abJM_J, T_{M_T}) \right| \Psi_0 \right\rangle \, , \tag{14.53}$$

together with the corresponding matrix element of the ph destruction operator

$$\phi_{JT}^K(ab) \equiv (-)^{J-M_J}(-)^{T-M_T} \left\langle \Psi_{JM_J TM_T}^K \left| \widehat{\xi}(abJ - M_J, T - M_T) \right| \Psi_0 \right\rangle \, . \tag{14.54}$$

It may be shown that the normalization condition for these ph matrix elements is

$$\sum_{ab} \left(\psi_{JT}^K(ab) \psi_{J'T'}^{K'}(ab) - \phi_{JT}^K(ab) \phi_{J'T'}^{K'}(ab) \right) = \delta_{KK'} \delta_{JJ'} \delta_{TT'} \, . \tag{14.55}$$

On the other hand, if we evaluate the commutator directly using the second-quantized form of the Hamiltonian given above, we obviously obtain a very complicated expression. As a simplification, we can make the approximation of retaining only terms that contain either a particle–hole creation or particle–hole destruction operator, namely, we linearize the equations of motion to invoke the so-called random-phase approximation (RPA).

The name is somewhat obscure in the present context and the interested reader is encouraged to explore the various ways of arriving at this approximation following different paths. Indeed, at the end of this chapter we indicate diagrammatically what the approximation means, thereby taking a still different point of view. The result of linearizing the equation of motion and taking matrix elements between the ground and excited states yields [Def66]

$$(E_{JT} - E_0)\, \psi_{JT}^K(ab) = (\epsilon_a - \epsilon_b)\, \psi_{JT}^K(ab)$$
$$+ \sum_{a'b'} \left\{ X_{ab;a'b'}^{JT}\, \psi_{JT}^K(a'b') + Y_{ab;a'b'}^{JT}\, \phi_{JT}^K(a'b') \right\}$$

$$(14.56)$$

$$-(E_{JT} - E_0)\, \phi_{JT}^K(ab) = (\epsilon_a - \epsilon_b)\, \phi_{JT}^K(ab)$$
$$+ \sum_{a'b'} \left\{ X_{ab;a'b'}^{JT}\, \phi_{JT}^K(a'b') + Y_{ab;a'b'}^{JT}\, \psi_{JT}^K(a'b') \right\} ,$$

$$(14.57)$$

the RPA equations, where $X_{ab;a'b'}^{JT}$ involves matrix elements of the residual interaction and $Y_{ab;a'b'}^{JT} = (-)^{j_a + j_{a'} - J} (-)^{1-T} X_{ab;a'b'}^{JT}$ (we omit the details here; the interested reader will find more detail in [Fet71]). The usual approach taken is to assume some model space (some set of single-particle states), to assume a form for the residual interaction, and to solve the above linearized RPA matrix equations using standard techniques to obtain $\psi_{JT}^K(ab)$ and $\phi_{JT}^K(ab)$ for ab in that model space. The doubly-reduced (reduced in both angular momentum and isospin and indicated by $\langle\ |||\ \ |||\ \rangle$) matrix element of any one-body multipole operator is then immediately given in terms of doubly-reduced single-particle matrix elements of the sort discussed above and in Chapter 13:

$$\left\langle \Psi_{JM_J TM_T}^K \left|\left|\left| \widehat{\mathcal{O}}_{JT}^{[1]} \right|\right|\right| \Psi_0 \right\rangle = \sum_{a'b'} \left\{ X_{ab;a'b'}^{JT} \left\langle a \left|\left|\left| \mathcal{O}^{[1]} \right|\right|\right| b \right\rangle \right.$$
$$\left. + (-)^{j_a + j_{a'} - J} (-)^{1-T} Y_{ab;a'b'}^{JT} \left\langle b \left|\left|\left| \mathcal{O}^{[1]} \right|\right|\right| a \right\rangle \right\} ,$$

$$(14.58)$$

which can be simplified for specific operators using the symmetries in the single-particle matrix elements when $a \leftrightarrow b$. When all is said and done, one finds that the states obtained using the RPA may in some cases still be close to pure $1p1h$ configurations – the high-spin, so-called stretched states are examples of these – while others become nontrivial admixtures of several $1p1h$ configurations, thereby becoming collective and providing the microscopic basis for semi-classical models such as the Goldhaber–Teller model. Again we postpone the discussions of the photo- and electroexcitation of all of these states to Chapter 15 and end here with a treatment of how to view the problem diagrammatically.

As in the first section of this chapter, where diagrammatic techniques were discussed for the HF approximation, we briefly mention the nature of diagrammatic techniques for the RPA. The interested reader is directed to the full treatment of the problem in [Fet71]. If we indicate the residual interaction (i.e., after shifting to a mean-field description such

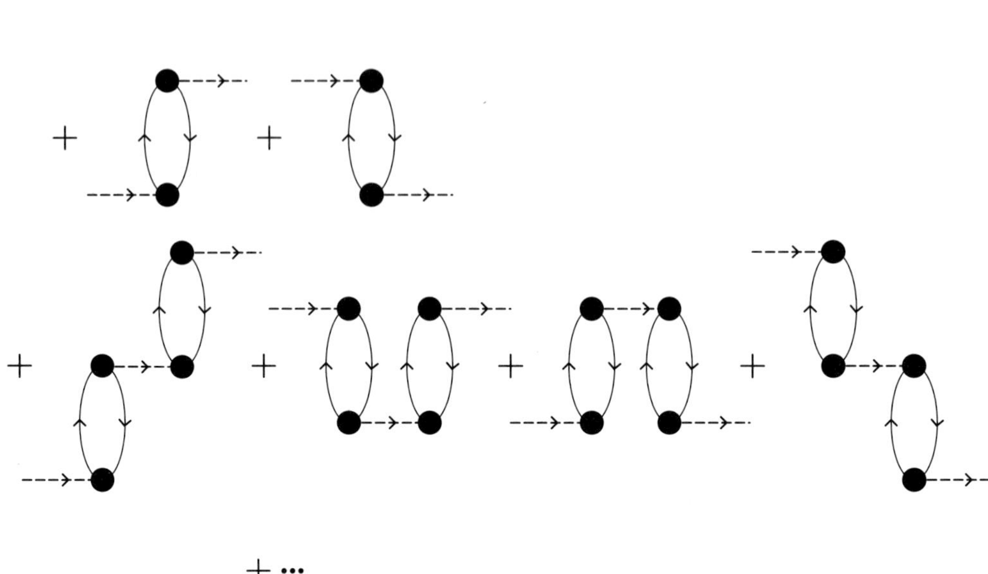

The random-phase approximation or sum of the ring diagrams.

as HF where the average behavior of the two-body interaction has already been taken into account) by a dashed line, and indicate the creation and destruction of particle–hole states by *ph* bubbles or rings, namely lines going up for particles and down for holes, then the RPA corresponds to the sum of all of the diagrams shown in Fig. 14.7. On the right-hand side of the equation the first term is just the bare residual interaction; the second term denotes the creation of a single *ph* pair (and involves RPA amplitude X); the third term designates the destruction of a single *ph* pair (with amplitude Y); and so on. Sometimes this is called the summation of the ring diagrams, for obvious reasons. The RPA refers more to the algebraic developments introduced earlier where, when cast in a matrix form, the approximation is to treat diagonal matrix elements of the interaction one way, but to assume that the off-diagonal matrix elements occur with random phases and so can be treated differently.

We depart this chapter having made a brief tour of several typical models used in nuclear physics. Of course, this summary merely touches the surface and, as noted earlier, there exist entire books written on the subject. Our intent has been to limit the presentation to a few basic concepts, such as how to proceed from classical descriptions of the nuclear fluid to semi-classical models that can be quantized as harmonic oscillators and, importantly, how one attempts to make reasonable approximations to what is largely a nonperturbative problem. In Chapter 15, we proceed to use these formal developments in a specific context, that of elastic charge and magnetic electron scattering and to electroexcitation of low-lying states in nuclei, after which, in Chapter 16, the somewhat different problem of electron scattering in the so-called quasielastic and resonance regions will provide the focus.

Exercises

14.1   Quantum Numbers of the Nucleus

From the level scheme given by the independent-particle shell model, if $^{25}_{12}\text{Mg}$ is a spherical nucleus, what are the most likely spin, parity, and isospin quantum numbers of its ground state? If, instead, it is a deformed nucleus with prolate deformation, what are the most likely spin and parity?

14.2   Quadrupole Moment and Shape

Consider the quadrupole moment of a non-spherical nucleus in a rotational model. This is discussed in detail in [Won98]. Assuming axial symmetry, the quadrupole moment is characterized by the parameter δ, the difference between R_3 and $R_\perp$, defined by

$$\delta = \frac{3}{2} \cdot \frac{R_3^2 - R_\perp^2}{R_3^2 + 2R_\perp^2} \, ,$$

where R_3 is the radius of the nucleus along the body-fixed symmetric 3-axis and $R_\perp$ is the radius in the direction perpendicular to it. Consider now the ground state of $^{152}_{63}\text{Eu}$ which is known to be 3^- with an electric (Coulomb) quadrupole moment of $+3.16 \times 10^2$ e·fm^2. Determine δ and find the intrinsic quadrupole moment of the nucleus in its the ground state. What is the shape of this nucleus?

14.3 α-Particle Model of the Nucleus

Consider the α-particle model of the nuclei ^{12}C (three α-particles) and ^{16}O (four α-particles).

a) Evaluate the difference between the total binding energy of ^{12}C and the total binding energy of the three α-particles.

b) Evaluate the difference between the total binding energy of ^{16}O and the total binding energy of the four α-particles.

c) By counting the number of $\alpha\alpha$ bonds in each of the two nuclei, estimate the bond energy between two α-particles in this model.

14.4   Rotational Model of the Nucleus

For an axially-symmetric nucleus of mass M with density given by

$$\rho(r) = \begin{cases} \rho_0, & \text{for } r \leq R_0(1 + \beta Y_{2,0}(\theta, \phi)) \\ 0 & \text{otherwise} \, , \end{cases}$$

show

a) that the intrinsic (charge) quadrupole moment up to second-order in deformation parameter β is given by

$$Q_0 = \frac{1}{\sqrt{5\pi}} Z R_0^2 \beta (1 + 0.36\beta) \, ,$$

and

b) the moment of inertia about the laboratory z-axis is given by

$$I = \frac{2}{5} M R_0^2 (1 + 0.31\beta) \, .$$

14.5 Hartree and Hartree–Fock Equations with the Coulomb Interaction

While our main focus in this book is placed on issues of nuclear and particle physics where the strong interaction plays a dominant role, it is still informative to see how other systems, such as those relevant in atomic and condensed matter physics, also employ the same many-body methods discussed in this chapter. For example, in the case of atomic physics, start with the Hamiltonian

$$H = \sum_s \frac{1}{2m} \int d\mathbf{r} \nabla \psi_s^\dagger(\mathbf{r}) \cdot \nabla \psi_s(\mathbf{r}) + \sum_s \int d\mathbf{r} \psi_s^\dagger(\mathbf{r}) V_c(\mathbf{r}) \psi_s(\mathbf{r})$$

$$+ \frac{1}{2} \sum_{ss'} \int d\mathbf{r} \int d\mathbf{r}' \psi_s^\dagger(\mathbf{r}) \psi_{s'}^\dagger(\mathbf{r}') V(\mathbf{r} - \mathbf{r}') \psi_{s'}(\mathbf{r}') \psi_s(\mathbf{r}) ,$$

where $V_c(r) = -Ze^2/r$ and $V(\mathbf{r}' - \mathbf{r}) = e^2/|\mathbf{r}' - \mathbf{r}|$ and use this to calculate $\langle \psi | H | \psi \rangle$.

a) If $|\psi\rangle$ is taken to be the state

$$\langle \mathbf{r}_1 \mathbf{r}_2 ... \mathbf{r}_N | \psi \rangle = \phi_1(\mathbf{r}_1) \phi_2(\mathbf{r}_2) .... \phi_N(\mathbf{r}_N) ,$$

minimize $\langle \psi | H | \psi \rangle$ and show that what results are the Hartree equations

$$\left(-\frac{\nabla^2}{2m} - \frac{Ze^2}{r} \right) \phi_i(\mathbf{r}) + \int d\mathbf{r}' \frac{e^2}{|\mathbf{r} - \mathbf{r}'|} \sum_{j \neq i} |\phi_j(\mathbf{r}')|^2 \phi_i(\mathbf{r}) = \epsilon_i \phi_i(\mathbf{r}) .$$

b) If $|\psi\rangle$ is taken to be the fully antisymmetrized state

$$|\psi\rangle = \frac{1}{\sqrt{N!}} \begin{vmatrix} \phi_1(\mathbf{r}_1) & \cdots & \phi_1(\mathbf{r}_N) \\ \cdot & \cdots & \cdot \\ \cdot & \cdots & \cdot \\ \cdot & \cdots & \cdot \\ \phi_N(\mathbf{r}_1) & \cdots & \phi_N(\mathbf{r}_N) \end{vmatrix}$$

minimize $\langle \psi | H | \psi \rangle$ and show that what results are the HF equations,

$$\left(-\frac{\nabla^2}{2m} - \frac{Ze^2}{r} \right) \phi_i(\mathbf{r}) + \int d\mathbf{r}' \frac{e^2}{|\mathbf{r} - \mathbf{r}'|} \sum_{j \neq i} \phi_j^*(\mathbf{r}') [\phi_j(\mathbf{r}') \phi_i(\mathbf{r}) - \delta_{s_i, s_j} \phi_j(\mathbf{r}) \phi_i(\mathbf{r}')]$$

$$= \epsilon_i \phi_i(\mathbf{r}) .$$

Note that in both parts a) and b) the states $\phi_i(\mathbf{r})$ are assumed to be orthogonal normalized states:

$$\int d\mathbf{r} \phi_i(\mathbf{r}) \phi_j(\mathbf{r}) = \delta_{ij} .$$

14.6 Example of an Infinite System: The Electron Gas

Continuing the theme of the previous exercise, we again use the Coulomb interaction. Now, however, we consider an infinite system, a "gas" of electrons, as is appropriate in many studies of condensed matter physics [Fet71]. Indeed, the electron gas is the beginning paradigm for electrons in condensed matter systems. Note that the developments here are very closely related to those in Chapter 13 where nuclear matter was the focus. As noted previously, the treatment of nuclear (and neutron) matter presented there can be extended to a full HF analysis.

The Hamiltonian for an electron gas may be written as

$$H = H_0 + H_1 + H_2 \, ,$$

where

$$H_0 = -\frac{1}{2m} \sum_{i=1}^{N} \nabla_i^2$$

$$H_1 = \sum_{i=1}^{N} U(r_i)$$

$$H_2 = \frac{1}{2} \sum_{i \neq j} \frac{e^2}{|r_i - r_j|} \exp\left(-\kappa |r_i - r_j|\right) \, ,$$

where $U(r_i)$ is a uniform potential of magnitude $-U_0$ over a box of volume Ω. And $U(r_i)$ may be thought of as the average effect of the positive ions if the electron gas is in a solid. The parameter κ is the inverse of a scattering radius, which may be permitted to become infinite later ($\kappa \to 0$).

Begin by choosing plane waves with periodic boudary conditions as a basic orthonormal set of single-particle states (inside the box) conditions:

$$\phi_k(r) = \frac{1}{\sqrt{\Omega}} e^{ik \cdot r} \, .$$

Ignore the spin degrees of freedom.

a) Show that in terms of creation and destruction operators

$$H_0 = \sum_k \frac{k^2}{2m} a_k^\dagger a_k$$

$$H_1 = -U_0 \sum_k a_k^\dagger a_k$$

$$H_2 = \frac{2\pi e^2}{\Omega} \sum_{kk'\ell\ell'} a_{\ell'}^\dagger a_{k'}^\dagger a_k a_\ell \frac{1}{Q^2 + \kappa^2} \delta_{k'+\ell',k+\ell} \, ,$$

where $Q = k - k' = \ell' - \ell$ and δ is a Kronecker delta.

b) Neglecting the interaction H_2, the ground state $|g.s.\rangle$ is clearly given by a degenerate Fermi gas; i.e., the occupation numbers are

$$n_k = \begin{cases} 1 & |k| \leq k_F \\ 0 & |k| > k_F, \end{cases}$$

where k_F is the Fermi momentum. Show that in this state

$$\langle H_2 \rangle = \langle g.s.|H_2|g.s.\rangle = E_{\text{dir}} + E_{\text{exc}} \, ,$$

where E_{dir} is the direct Coulomb interaction energy of all the electrons

$$\frac{E_{\text{dir}}}{\Omega} = \frac{2\pi e^2}{\kappa^2} \frac{N}{\Omega}$$

and E_{exc} is the exchange energy

$$\frac{E_{\text{exc}}}{\Omega} = -\frac{e^2 \Omega}{(2\pi)^5 N} \int \int \frac{dk d\ell}{|k - \ell|^2} \, ,$$

where N, the total number of electrons in the gas, is very large and the two integrals are both over the Fermi sphere. The term E_{dir} can be canceled by a suitable choice of U_0. Note that in obtaining the last formula we let $\kappa \to 0$ and made the replacement

$$\sum_k \to \Omega \int \frac{dk}{(2\pi)^3} \, .$$

c) Show that

$$\frac{E_{\text{exc}}}{N} = -\frac{3e^2}{8\pi} k_F \, ,$$

and estimate E_{exc}/N for a typical metal.

14.7 Zero Temperature Interacting Bose Gas

Now consider another condensed matter system, a Bose gas. Consider the binding energy per particle of a zero temperature dilute Bose gas with a repulsive interaction. Dilute means that the scattering length a is much less than the average particle spacing $n^{-\frac{1}{3}}$, where $n = N/\Omega$ is the number density. In this limit S-wave scattering dominates and the relevant low momentum matrix elements of the potential are independent of the momentum, so that the interaction can be represented by a simple contact term,

$$H = \sum_p \frac{p^2}{2m} a_p^\dagger a_p + \frac{U_0}{2\Omega} \sum_{p_1 p_2 p_3 p_4} a_{p_4}^\dagger a_{p_3}^\dagger a_{p_2} a_{p_1} \, .$$

If there were no interaction the system at zero temperature would be represented by a simple wavefunction with all N particles in the ground state,

$$\left\langle g.s. \left| \hat{N} \right| g.s. \right\rangle = N \, ,$$

where $\hat{N} = \sum_p a_p^\dagger a_p$ is the number operator.

However, with the interaction turned on this will change. Still, provided that U_0 is not too large we expect that the occupation number of any particular excited state will be $\mathcal{O}(1/N)$ compared to the ground state and that interactions between pairs of excited states will be $\mathcal{O}(1/N^2)$ relative to the ground state and can be ignored. We expect then that

$$a_0^\dagger a_0 \, |g.s.\rangle = N_0 \, |g.s.\rangle \sim \mathcal{O}(N) \, |g.s.\rangle$$

$$a_0 a_0^\dagger \, |g.s.\rangle = (a_0^\dagger a_0 + 1) \, |g.s.\rangle = (N_0 + 1) \, |g.s.\rangle \, .$$

But, since $N_0 \gg 1$, the 1 can be neglected and we can pretend that both a_0 and $a_0^\dagger$ are c-numbers

$$a_0 \simeq a_0^\dagger = N_0^{\frac{1}{2}} \, .$$

Now rewrite the interaction by separating out excited and ground state terms.

a) Show that:

$$N_0 = N - \sum_{p \neq 0} a_p^\dagger a_p$$

$$a_0^\dagger a_0^\dagger a_0 a_0 = N^2 - 2N \sum_{p \neq 0} a_p^\dagger a_p \, .$$

b) Show in addition that

$$\sum_{p \neq 0} (a_p^\dagger a_{-p}^\dagger a_0 a_0 + a_0^\dagger a_0^\dagger a_p a_{-p} + a_0^\dagger a_p^\dagger a_p a_0 + a_p^\dagger a_0^\dagger a_0 a_p + a_p^\dagger a_0^\dagger a_p a_0)$$

$$\approx N \sum_{p \neq 0} (a_p^\dagger a_{-p}^\dagger + a_p a_{-p} + 4 a_p^\dagger a_p)$$

and that this leads to the result

$$H \simeq \frac{N^2 U_0}{2\Omega} + \frac{N U_0}{2\Omega} \sum_{p \neq 0} (a_p^\dagger a_{-p}^\dagger + a_p a_{-p} + 2 a_p^\dagger a_p) + \sum \frac{p^2}{2m} a_p^\dagger a_p \, .$$

c) Now perform a Bogliubov transformation, by defining

$$b_p = \frac{a_p + \alpha_p a_{-p}^\dagger}{\sqrt{1 - \alpha_p^2}} \qquad b_p^\dagger = \frac{a_p^\dagger + \alpha_p a_{-p}}{\sqrt{1 - \alpha_p^2}}$$

and show that

$$[b_p, b_{p'}^\dagger] = \delta_{pp'}$$

$$b_p^\dagger b_p = \frac{1}{1 - \alpha_p^2} (a_p^\dagger a_p + \alpha_p a_p^\dagger a_{-p}^\dagger + \alpha_p a_p a_{-p} + \alpha_p^2 (1 + a_{-p}^\dagger a_{-p})) \, .$$

d) Pick α_p so that the Hamiltonian can be written as

$$H = E_0 + \sum_{p \neq 0} \epsilon(p) b_p^\dagger b_p$$

and find both $\epsilon(p)$ and E_0. Show that

$$E_0 = \frac{N^2 U_0}{2\Omega} + \sum_{p \neq 0} \left(\epsilon(p) - \frac{p^2}{2m} - \frac{N U_0}{2\Omega} \right)$$

$$\epsilon(p) = \sqrt{\left(\frac{p^2}{2m} \right)^2 + \frac{N U_0 p^2}{\Omega m}} \, .$$

Note here then that E_0 is the new ground state energy while $\epsilon(p)$ is the energy of excitations (phonons) in the system.

e) Show that E_0 can be expressed by merely adding to the lowest order energy the vacuum energy of the Bogliubov oscillators (with appropriate terms included to subtract the divergences).

f) Now convert from the potential U_0 (which is *not* an observable) to the S-wave scattering length (a). Use second-order perturbation theory to show that

$$\frac{4\pi}{m}a = U_0 - \frac{U_0^2}{\Omega}\sum_{p\neq 0}\frac{1}{\frac{p^2}{m}}$$

and rewrite your expressions for $\epsilon(p)$ and E_0 in terms of a.

g) Show that as $p \to 0$ the phonon excitation energy is of the form

$$\epsilon(p) = pv\,,$$

where v is the phonon velocity and thus also the velocity of sound in the system.

h) By integrating over the excitation spectrum using $\sum_p = \Omega \int \frac{d^3p}{(2\pi)^3}$ show that the ground state energy is of the form

$$\frac{E_{g.s.}}{\Omega} = \frac{\langle g.s.\,|H|\,g.s.\rangle}{\Omega} = \frac{2\pi a n^2}{m}\left[1 + \frac{128}{15\sqrt{\pi}}(na^3)^{\frac{1}{2}} + \cdots\right].$$

i) Find the number of particles in the lowest energy state, $N_0 = \langle g.s.\,|a^\dagger a_0|\,g.s.\rangle$, and show that

$$\frac{N_0}{N} = 1 - \frac{8}{3\sqrt{\pi}}(na^3)^{\frac{1}{2}} + \cdots.$$

j) Find the speed of sound via the usual thermodynamic identity

$$v_s = \sqrt{\frac{\Omega^2}{mN}\frac{\partial^2 E}{\partial\Omega^2}}$$

and show that this yields the same result as found using the phonon velocity.

k) Show that an explicit expression for the ground state is

$$|g.s.\rangle = \prod_p \sqrt{1 - \alpha_p^2}\,\exp\left[-\alpha_p a_p^\dagger a_{-p}^\dagger\right]|0\rangle\,,$$

i.e., verify that the state is normalized to unity and that $b_p\,|g.s.\rangle = 0\,.$

14.8   Diagrammatic Expansion of the HF Green's Function
In the text the basics of expressing the HF approximation in terms of single-particle Green's functions were summarized. Taking the form of the Dyson's equation involved for G^{HF} and the diagrammatic representation of the HF self-energy insertion (Figs. 14.5 and 14.4, respectively), start making expansions counting diagrams with specific numbers of interaction lines. Do this up to three interaction lines and detail the number of topologically distinct diagrams found in each order.

14.9   The Vibrational Model of Low-Lying Nuclear Excitations
Following the steps outlined in the text, fill in the following details:

a) Obtain the Lagrangian

$$L = T - V = \frac{1}{2}\rho R_0^5 \sum_{L,M}\frac{1}{L}\left|\dot{\alpha}_{L,M}\right|^2 - \frac{1}{2}\sigma R_0^2 \sum_{L,M}(L-1)(L+2)\left|\alpha_{L,M}\right|^2\,.$$

and Hamiltonian

$$H = \sum_{\substack{L,M \\ L \geq 2}} \left[\frac{1}{2B_L} \left| p_{L,M} \right|^2 + \frac{1}{2} C_L \left| \alpha_{L,M} \right|^2 \right],$$

respectively, where

$$B_L \equiv \rho R_0^5/L, \ C_L \equiv \sigma R_0^2 (L-1)(L+2), \omega_L^2 \equiv C_L/B_L \text{ and } p_{L,M} = B_L(-)^L \dot{\alpha}_{L,-M}.$$

b) Show that the quantized Hamiltonian

$$H = \sum_{\substack{L,M \\ L \geq 2}} \frac{1}{2} \omega_L \left(\alpha^\dagger_{L,M} \alpha_{L,M} + \alpha_{L,M} \alpha^\dagger_{L,M} \right)$$

emerges.

c) Prove that the matrix elements of the creation and destruction operators involved are

$$\left\langle n_{L,M} + 1 \left| \alpha^\dagger_{L,M} \right| n_{L,M} \right\rangle = \sqrt{n_{L,M} + 1}$$

$$\left\langle n_{L,M} - 1 \left| \alpha_{L,M} \right| n_{L,M} \right\rangle = \sqrt{n_{L,M}},$$

and obtain the oscillation frequency

$$\omega_L^2 \equiv \frac{1}{8} L(L-1)(L+2)\omega_2^2.$$

d) Confirm that the parity of these surfon states is $(-)^L$.

In this chapter, we discuss the ground and low-lying excited states of nuclei, focusing on electron scattering studies, beginning with elastic scattering from nuclei. The simplicity of light nuclei is ideal for a detailed study of the nucleon–nucleon interaction in the nucleus, as discussed in Chapters 11 and 12. However, *few-body* systems do not exhibit the special characteristics of nuclear matter, such as saturation of the nuclear force and the constant binding energy per nucleon (see Chapter 13). In addition, collective behavior is totally absent; such features can only be studied in heavy nuclei. One is faced with the problem of isolating the properties of nuclear matter from the effects due to the nuclear surface. Nuclear physicists are faced with the challenge of having to derive bulk properties from the behavior of nuclei, as was discussed in some detail in Chapter 13. The superb spatial resolution and penetrability of the electron scattering probe facilitates the study of the interior of heavy nuclei, thus enabling the isolation of bulk properties from surface effects.

We start our discussions of electron scattering from the ground and low-lying excited states of nuclei, employing the modeling discussed in the previous chapter, by considering elastic scattering, wherein the energy transfer (allowing for recoil) to a nucleus of mass M_i is related to the three-momentum transfer by

$$\omega = \epsilon - \epsilon' = \sqrt{q^2 + M_i^2} - M_i = |Q^2|/2M_i \,, \tag{15.1}$$

since $Q^2 = q^2 - \omega^2 > 0$ as usual. We then have

$$\left|\frac{Q^2}{q^2}\right| = \left\{1 + \frac{|Q^2|}{4M_i^2}\right\}^{-1} \,, \tag{15.2}$$

which is close to unity if the momentum transfer is small compared with twice the target mass; under these circumstances the electron scattering kinematic factors introduced earlier become approximately $v_L \cong 1$ and $v_T \cong 1/2 + \tan^2 \theta_e/2$. The cross section for electron scattering from discrete states in nuclei was obtained in Chapter 7:

$$\frac{d\sigma}{d\Omega_e} = 4\pi \sigma_M f_{\text{rec}}^{-1} F^2(q) \tag{15.3}$$

with nuclear recoil correction

$$f_{\text{rec}} = 1 + \frac{\epsilon}{2M_i} \sin^2 \theta_e/2 \,, \tag{15.4}$$

where the total form factor (for unpolarized scattering) can be decomposed into longitudinal (charge) and transverse contributions

$$F^2(q) = v_L F_L^2(q) + v_T F_T^2(q) \,. \tag{15.5}$$

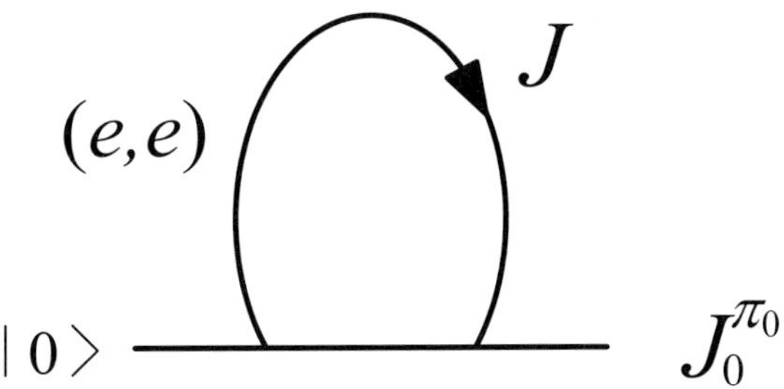

Fig. 15.1   Elastic electron scattering; see text for notation.

With $|f\rangle = |i\rangle$ we have $J_f = J_i \equiv J_0$ and $\pi_f = \pi_i \equiv \pi_0$ (see Fig. 15.1), and recalling the discussions of parity and time-reversal symmetry in Chapter 7, we have

$$\text{parity: } J = \text{even } \quad CJ/EJ \tag{15.6}$$

$$J = \text{odd} \quad MJ \,, \tag{15.7}$$

and

$$\text{time-reversal: } J = \text{even} \quad CJ \tag{15.8}$$

$$J = \text{odd} \quad EJ/MJ \,. \tag{15.9}$$

Hence only even-J Coulomb multipoles ($C0, C2, C4, \ldots$) and odd-J magnetic multipoles ($M1, M3, M5, \ldots$) can occur for elastic scattering if these basic symmetries are respected. The form factors are the ground-state matrix elements of the appropriate multipole operators:

$$F_{CJ}(q) \equiv \frac{1}{[J_0]} \langle J_0 \| \widehat{M}_J \| J_0 \rangle \tag{15.10}$$

$$F_{MJ}(q) \equiv \frac{1}{[J_0]} \langle J_0 \| i\widehat{T}_J^{\text{mag}}(q) \| J_0 \rangle \,. \tag{15.11}$$

The squares of the total form factors are then

$$F_L^2(q) = \sum_{J \geq 0,\text{even}} F_{CJ}^2(q) \tag{15.12}$$

$$F_T^2(q) = \sum_{J \geq 1,\text{odd}} F_{MJ}^2(q) \,, \tag{15.13}$$

where the sums truncate at the highest multipole allowed by angular momentum conservation. Some examples are given in Table 15.1.

15.1 Parity-Conserving Elastic Electron Scattering from Spin-0 Nuclei

Let us now focus on elastic charge scattering and begin with the simplest situation where $J_0 = 0$, in which case only the Coulomb monopole $C0$ is permitted. Introducing a special normalization for this particular form factor, we define $F_0(q)$ to be

Table 15.1 Examples of allowed multipoles in elastic electron scattering

Ground-state spin J_0	Allowed multipoles	Examples
0	$C0$	^{16}O, ^{208}Pb
1/2	$C0$, $M1$	p, ^3He, ^3H
1	$C0$, $M1$, $C2$	^2H, ^{14}N
3/2	$C0$, $M1$, $C2$, $M3$	^7Li, ^{11}B
$\vdots$	$\vdots$	$\vdots$
9/2	$C0$, $C2$, $C4$, $C6$, $C8$, $M1$, $M3$, $M5$, $M7$, $M9$	^{209}Bi

$$F_0(q) \equiv \frac{\sqrt{4\pi}}{Z} \langle 0 \,\|\widehat{M}_0(q)\|\, 0\rangle \tag{15.14}$$

$$= \frac{\sqrt{4\pi}}{Z} \langle 00 \,|\widehat{M}_{00}(q)|\, 00\rangle \,, \tag{15.15}$$

so that

$$F_L^2(q) = \frac{Z^2}{4\pi} F_0^2(q) \,. \tag{15.16}$$

The Coulomb monopole operator is given by (see Eq. (7.75))

$$\widehat{M}_{00}(q) = \int d\mathbf{x}\, M_0^0(q\mathbf{x})\widehat{\rho}(\mathbf{x}) \tag{15.17}$$

with

$$M_0^0(q\mathbf{x}) = j_0(qx)Y_0^0(\Omega_x) \,. \tag{15.18}$$

Of course, $j_0(qx) = \sin qx/qx$ and $Y_0^0(\Omega_x) = 1/\sqrt{4\pi}$ so the form factor becomes

$$F_0(q) = \frac{1}{Z} \int d\mathbf{x}\, \frac{\sin qx}{qx}\rho(\mathbf{x})_{00} \,, \tag{15.19}$$

where $\rho(\mathbf{x})_{00} \equiv \langle 0\,|\hat{\rho}(\mathbf{x})|\,0\rangle$ is the ground-state charge density. The normalization above has been chosen such that $\int d\mathbf{x}\, \rho(\mathbf{x})_{00} = Z$. One can extend these ideas to define higher moments of the radius (x^2, $x^4,\ldots$ in addition to x^0 as in Eq. (15.19)); in particular, the mean-square radius of the nuclear charge distribution

$$\frac{1}{Z} \int d\mathbf{x}\, x^2 \rho(\mathbf{x})_{00} = \left\langle x^2\right\rangle_{00} \tag{15.20}$$

proves to be useful. As $q \to 0$ we have $\sin qx/qx \to 1 - (1/3!)(qx)^2 + (1/5!)(qx)^4 - \cdots$ and accordingly may write

$$F_0(q) \xrightarrow[q\to 0]{} 1 - \frac{1}{6}(qR_{\mathrm{ch}})^2 + \cdots \,, \tag{15.21}$$

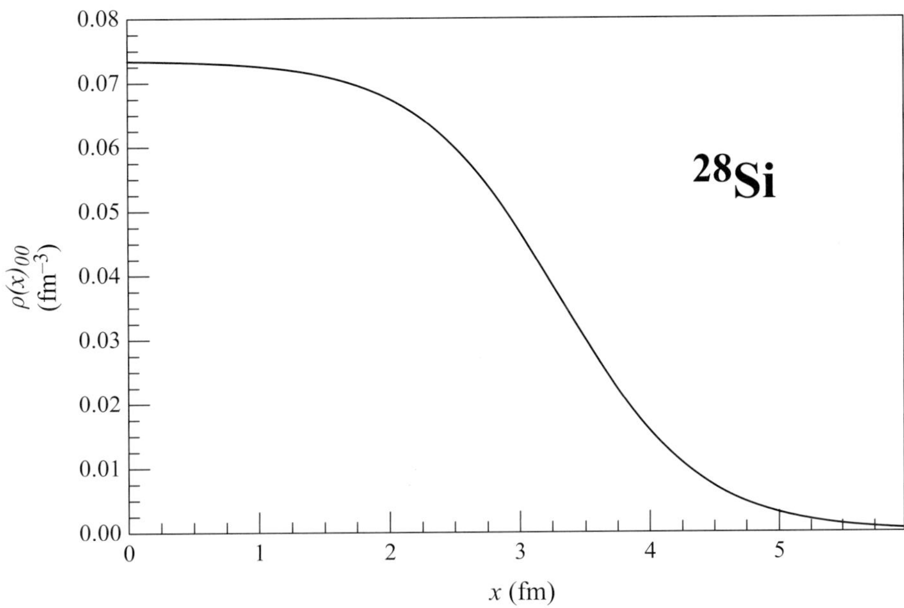

Fig. 15.2   Two-parameter Fermi (2PF) distribution for $^{28}_{14}$Si with $R = 3.30$ fm and $a = 0.54$ fm.

where the nuclear root-mean-square (RMS) charge radius is given by $\sqrt{\langle x^2\rangle_{00}} \equiv R_{\mathrm{ch}} = \cong a_0 A^{1/3}$, with $a_0 \cong 1.07$ fm. Equivalently in momentum space it is possible to characterize the typical scale of momentum transfer in elastic scattering by

$$q_{\mathrm{nuc}} \equiv \pi/R_{ch} \cong 3A^{-1/3} \ \mathrm{fm}^{-1} \cong 580 A^{-1/3} \ \mathrm{MeV}/\mathrm{c} \ .$$

This quantity may be used to specify what is considered to be "low-q" $\leftrightarrow q \ll q_{\mathrm{nuc}}$ as contrasted with the opposite extreme where $q \gg q_{\mathrm{nuc}}$.

A rough phenomenological fit to the charge distribution (in the spherical case) may be made using the so-called two-parameter Fermi (2PF) distribution shown in Fig. 15.2

$$\rho(x)_{00} = \frac{\rho_0}{1 + \exp\left(\frac{x-R}{a}\right)} , \tag{15.22}$$

with ρ_0 fixed by the normalization condition above. As already noted in Chapter 13, from results of fitting the form factor calculated using this charge distribution to data, one finds that $\frac{A}{Z}\rho_0 \cong$ central nuclear density, which is roughly constant for a wide range of nuclei. Also, the 90% to 10% surface thickness t, which is related to the nuclear diffuseness by $t = (4\ln 3)a$ is found to be about 2.4 fm for a wide range of nuclei, implying that $a \cong 0.54$ fm. For instance, considering $^{28}_{14}$Si and taking $a = 0.54$ fm, it is straightforward to compute the Fourier transform of the ground-state density in Eq. (15.22) as a function of R. From the above RMS charge radius relationship with $a_0 = 1.07$ fm, one has $R_{\mathrm{ch}} = 3.25$ fm and finds that this is attained when the radial parameter R defined via (Eq. 15.22) is 3.30 fm. The resulting Fourier transform is shown in Fig. 15.3 and provides a basic model for elastic electron scattering from silicon.

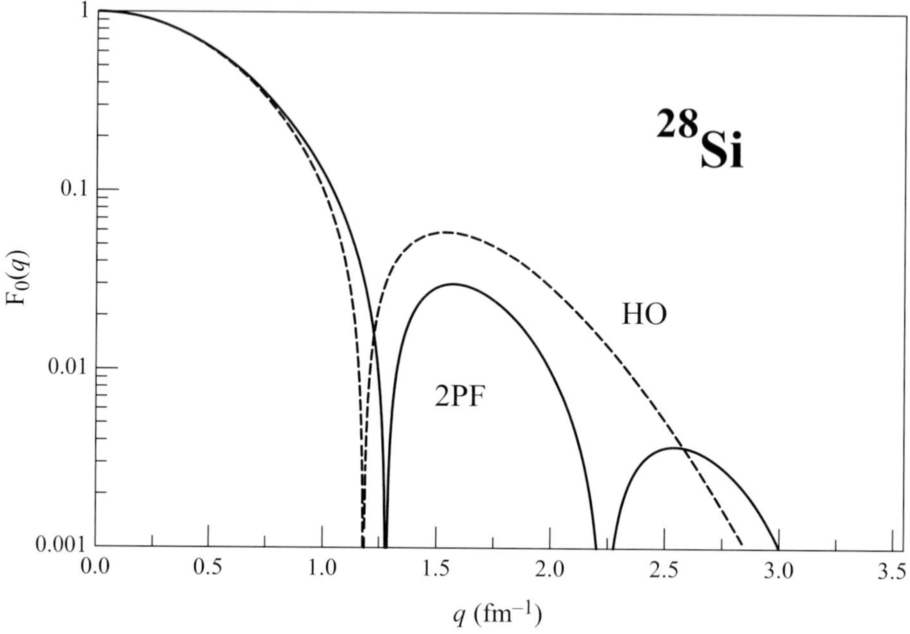

Fig. 15.3 Fourier transform of the two-parameter Fermi (2PF) distribution of Fig. 15.2 together with the harmonic oscillator (HO) result discussed in the text.

In passing, at this point we note that parity-violating electron scattering from spin-0 nuclei (or whenever the coherent elastic $C0$ multipole dominates) is essentially a probe of the ground-state neutron distribution [Don89]; we return later to discuss these extensions. Furthermore, while details of the hadronic interaction make the nature of the probe somewhat different, valuable information has been accumulated using elastic scattering of hadrons from nuclei. In particular, proton and neutron scattering have been intensively studied in the past, scattering of alpha particles and heavy ions is also useful and continues to be pursued, and scattering of pions has provided unique insights into the isospin structure of the ground-state density.

Let us now return to more general aspects of elastic charge electron scattering. First, the charge density operator in first quantization may be written in the form

$$\rho(\mathbf{x}) = \sum_{i=1}^{A} \delta(\mathbf{x} - \mathbf{x}_i) \frac{1}{2}(1 + \tau_{3i}) \; ; \tag{15.23}$$

here only pointlike protons are considered – this will be improved upon later. Then the ground-state charge density $\rho(\mathbf{x})_{00}$ is given by

$$\langle F | \widehat{\rho}(\mathbf{x}) | F \rangle = \sum_{\alpha \leq F} \varphi_\alpha^\dagger(\mathbf{x}) \frac{1}{2}(1 + \tau_3) \varphi_\alpha(\mathbf{x}) \tag{15.24}$$

$$= \sum_{\substack{n\ell jm_j \\ \leq F}} \left| \varphi_{n\left(\ell\frac{1}{2}\right)j,m_j;\frac{1}{2},+\frac{1}{2}}(\mathbf{x}) \right|^2 \; . \tag{15.25}$$

Inserting the single-proton wavefunctions into the matrix element, we obtain the elastic $C0$ form factor

$$F_0(q) = \frac{1}{Z} \int_0^\infty x^2 dx \frac{\sin qx}{qx} \sum_{n\ell j \leq F} (2j + 1) \left| R_{n\ell j}(x) \right|^2 . \tag{15.26}$$

An improvement is to assume that the total charge density is a convolution of the distribution of point nucleons with the intrinsic nucleon charge distribution. Since the Fourier transform of a convolution is the product of the Fourier transforms of the two contributions, this implies that the pointlike form factor in Eq. (15.26) above should simply be multiplied by the nucleon charge form factor. In fact, since the neutron does have a form factor, even though its total charge is zero (see Chapter 8), it is straightforward to sum over all nucleons by including as weighting factors $G_E^{p,n}$. Furthermore, since an independent-particle model with an average potential fixed in coordinate space is not the full story and the center-of-mass of the nuclear many-body ground state must undergo zero-point fluctuations with respect to the origin of the potential used to obtain the wavefunctions employed above, a so-called CM correction must be included (see later).

One is now in a position to use some specific mean field such as the HF approximation discussed in the Chapter 14 to obtain the radial wavefunctions and hence the single-particle matrix elements required above. Any reasonable form can easily be handled numerically; however, it is useful to have analytic expressions to work with. The classic procedure is to use harmonic oscillator wavefunctions [Def66, Don79]. The interested reader is encouraged to explore the literature where the entire problem is presented in depth, although we follow a different route and exploit results of performing the necessary algebra, namely, we summarize the discussions and tables to be found in [Don79a]. There it is shown that for harmonic oscillator (HO) wavefunctions all of the basic electroweak single-particle matrix elements involve seven operators M, Δ, Δ', Σ, Σ', Σ'', and Ω (see [Def66, Don75, Don79] for their definitions). The single-particle reduced matrix elements may be written

$$\langle n'\ell' j' \| T_J(q) \| n\ell j \rangle = \frac{1}{\sqrt{4\pi}} y^{(J-K)/2} e^{-y} p(y) , \tag{15.27}$$

where $y \equiv (bq/2)^2$, with b the oscillator parameter. We can write $b = 1/\sqrt{m_N \omega_0}$, where ω_0 is the oscillator energy, and one roughly finds $\omega_0 \approx 41/A^{1/3}$ MeV. Here $K = 2$ for the natural parity operators M_J, Δ'_J, and Σ_J and $K = 1$ for the rest, the non-natural parity operators. The tables yield polynomials in the form

$$p(y) = \sqrt{J_1} \frac{J_2}{J_3} \left[\frac{I_1}{K_1} + \frac{I_2}{K_2} y + \frac{I_3}{K_3} y^2 + \cdots \right] , \tag{15.28}$$

truncating as expansions in powers of y as explained in [Don79a], where the quantities in Eq. (15.28) are also defined.

Let us illustrate the use of the tables by considering $C0$ elastic scattering from nuclei having $Z_{1s} \leq 2$ protons in the $1s$-shell, $Z_{1p} \leq 6$ in the $1p$-shell, $Z_{1d} \leq 10$ in the $1d$-shell and $Z_{2s} \leq 2$ in the $2s$-shell, with $Z = Z_{1s} + Z_{1p} + Z_{1d} + Z_{2s}$, i.e., nuclei up to ^{40}Ca (the procedures can easily be extended to heavier nuclei). Tables I, V and XI of [Don79a] immediately yield the following,

$$\langle 1\ell \,|j_0(qx)|\, 1\ell \rangle = e^{-y} \begin{cases} 1 & \ell = 0 \\ 1 - \frac{2}{3}y & \ell = 1 \\ 1 - \frac{4}{3}y + \frac{4}{35}y^2 & \ell = 2 \end{cases} \tag{15.29}$$

$$\langle 20 \,|j_0(qx)|\, 20 \rangle = e^{-y}\left\{ 1 - \frac{4}{3}y + \frac{2}{3}y^2 \right\} \tag{15.30}$$

from which the $C0$ form factor is easily obtained

$$
\begin{aligned}
F_0(q) &= \frac{1}{Z}\left[Z_{1s} + Z_{1p}\left(1 - \frac{2}{3}y\right) + Z_{1d}\left(1 - \frac{4}{3}y + \frac{4}{35}y^2\right) \right. \\
&\qquad \left. + Z_{2s}\left(1 - \frac{4}{3}y + \frac{2}{3}y^2\right) \right] e^{-y} f_{\mathrm{CM}}(y) G_E^p \tag{15.31} \\
&= \left[1 - \frac{1}{Z}\left\{ (Z_{1p} + 2Z_{1d} + 2Z_{2s})\,y - \left(\frac{4}{35}Z_{1d} + \frac{2}{3}Z_{2s}\right)y^2 \right\} \right] \\
&\qquad \times e^{-y} f_{\mathrm{CM}}(y) G_E^p , \tag{15.32}
\end{aligned}
$$

where in the HO model the CM correction is simply given by $f_{\mathrm{CM}}(y) = \exp(y/A)$ and where we have inserted the proton form factor, as discussed above. Clearly the limit as q goes to 0 is unity, as it should be, and it is straightforward to use Eq. (15.21) to obtain the RMS charge radius in this model. An example is given for $^{28}_{14}$Si in Fig. 15.3 using $b = 1.90$ fm for the oscillator parameter to have the same RMS charge radius as the 2PF model used above. Later we again employ the tables.

We conclude these discussions of parity-conserving elastic electron scattering from spin-0 nuclei by showing some typical results using the more sophisticated models introduced in Chapter 14. For example, the elastic form factor of ^{208}Pb is shown in Fig. 15.4 together with the corresponding ground-state charge density whose uncertainty is shown as a band. The latter is compared with density-dependent Hartree–Fock (DDHF) calculations indicating the reasonable, but not perfect, understanding of the nuclear ground-state charge density in terms of mean-field theory. Note that the measured form factor has been determined over an astonishing 12 orders-of-magnitude! In Fig. 15.5 the ground-state charge distributions for doubly closed shell nuclei deduced from elastic electron scattering are shown together with mean-field predictions, indicating the extent of the agreement across the periodic table for such nuclei; again the bands indicate the uncertainties in extracting the charge densities.

15.2　Parity-Violating Elastic Electron Scattering from Spin-0 Nuclei

As noted above, coherent elastic parity-violating (PV) electron scattering plays a special role when combined with coherent elastic parity-conserving electron scattering in providing information not only on the nuclear ground-state proton distribution, but also on the

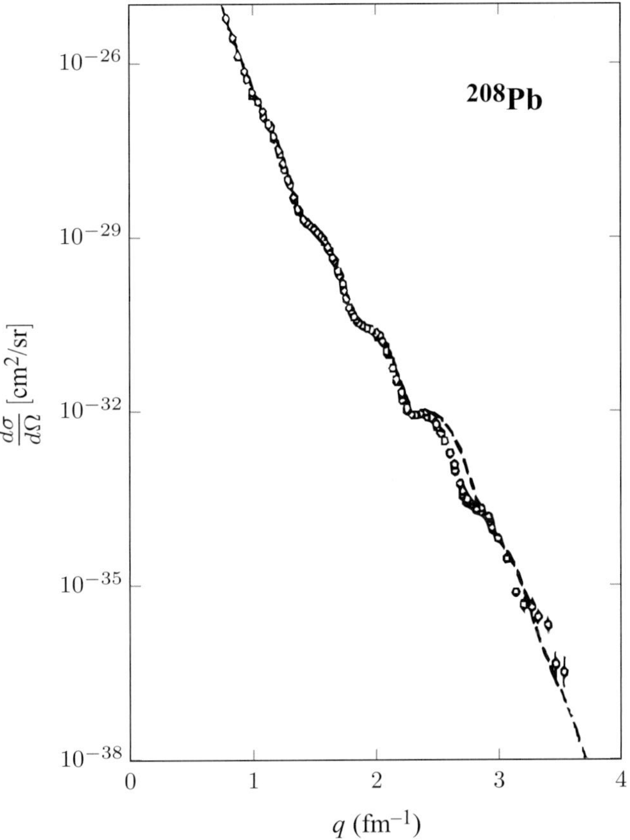

Fig. 15.4 The elastic electron scattering cross section from ^{208}Pb as a function of momentum transfer [Fro87], together with a mean-field prediction [Dec80] (dashed curve).

nuclear ground-state neutron distribution, as well as interesting information on hadronic structure. Let us see how this comes about. As discussed in Chapter 7, when one considers the inclusive scattering of longitudinally polarized electrons from unpolarized nucleons or nuclei the PV helicity asymmetry

$$\mathcal{A} = \frac{d\sigma^+ - d\sigma^-}{d\sigma^+ + d\sigma^-} \tag{15.33}$$

is parity-violating and involves the interference between a diagram where a γ is exchanged and one where a Z^0 is exchanged (see Fig. 7.6). In the plane-wave Born approximation (PWBA) the asymmetry may be written

$$\mathcal{A} = \frac{G_F Q^2}{2\pi\alpha\sqrt{2}} \cdot \frac{W^{\text{PV}}}{W^{\text{PC}}} , \tag{15.34}$$

where G_F is the Fermi constant, α is the fine-structure constant, and the ratio of the parity-violating (PV) to parity-conserving (PC) hadronic responses enters. As usual, the inclusive PC response has two contributions, one longitudinal (L) and one transverse (T):

$$W^{\text{PC}} = v_L W_L + v_T W_T , \tag{15.35}$$

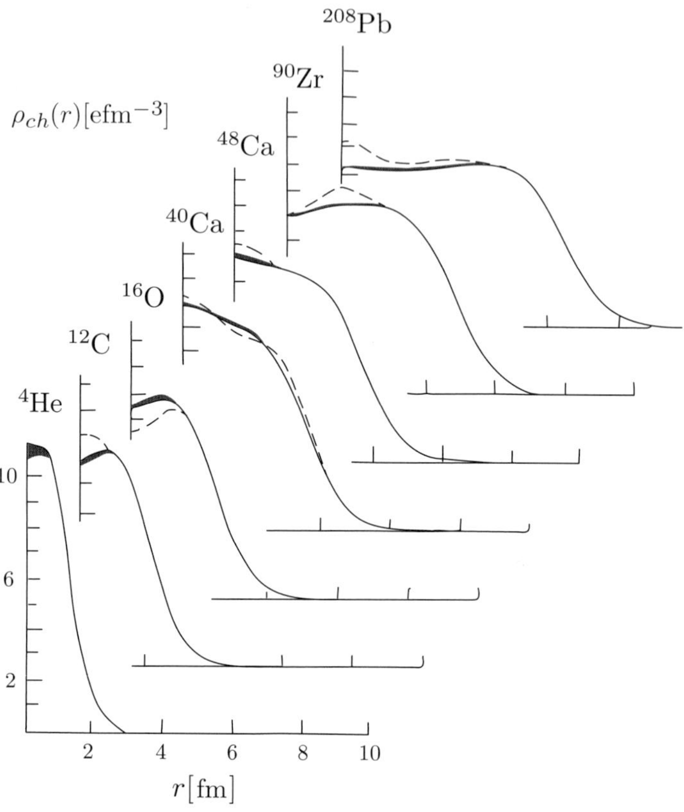

Fig. 15.5 Experimentally determined charge density distributions for doubly closed shell nuclei ranging from ^{4}He to ^{208}Pb compared with mean-field predictions (see Fig. 15.4 for references). Experimental data are shown as dark bands, while the mean-field charge densities are indicated with dashed lines.

weighted by the usual Rosenbluth factors v_L and v_T. The inclusive PV response has three contributions, as seen in Chapter 7:

$$W^{\mathrm{PV}} = a_A\left[v_L\widetilde{W}_L + v_T\widetilde{W}_T\right] + a_V\left[v_{T'}\widetilde{W}^{T'}\right], \tag{15.36}$$

where the tildes over these responses indicate that they arise from interferences between the electromagnetic (EM) and weak neutral current (WNC) matrix elements. The weak neutral current leptonic couplings a_V (polar-vector) and a_A (axial-vector) at tree-level in the Standard Model are

$$a_V = 4\sin^2\theta_W - 1 \simeq -0.08 \tag{15.37}$$

$$a_A = -1, \tag{15.38}$$

using $\sin^2\theta_W \simeq 0.23$.

In Chapter 8 we considered the case of PV electron scattering from the nucleon; here we specialize the discussion to the case of elastic scattering from spin-0 nuclei [Don89, Mor09, Mor14, Mor15]. In this case there are no transverse responses and only monopole matrix elements of the electroweak current can occur

$$\frac{W^{\mathrm{PV}}}{W^{\mathrm{PC}}} = a_A \cdot \frac{\widetilde{W}_L(\mathrm{monopole})}{W_L(\mathrm{monopole})} = a_A \cdot \frac{\widetilde{F}_{C0}}{F_{C0}} \,, \tag{15.39}$$

where F_{C0} is the usual EM elastic form factor discussed above and where $\widetilde{F}_{C0}$ is the analogous quantity for the WNC. If in addition we now restrict our attention to $N = Z$ nuclei, assume that these are eigenstates of isospin ($T = 0$), and assume that strangeness can be ignored (all of these restrictions are released below), then the EM and WNC matrix elements are proportional [Don89]

$$\widetilde{F}_{C0} = \beta_V^{(0)} F_{C0} \,, \tag{15.40}$$

where at tree-level in the Standard Model one has $\beta_V^{(0)} = -2\sin^2\theta_W$ for the isoscalar hadronic WNC coupling. Under these conditions a very simple result occurs:

$$\frac{W^{\mathrm{PV}}}{W^{\mathrm{PC}}} = a_A \beta_V^{(0)} \tag{15.41}$$

and the PV asymmetry is completely devoid of any hadronic structure issues:

$$\mathcal{A} = \frac{G_F Q^2}{2\pi\alpha\sqrt{2}} \cdot a_A \beta_V^{(0)} \simeq 3.22 \times 10^{-6} Q^2 \tag{15.42}$$

with Q^2 in fm^{-2}, a result obtained four decades ago [Fei75, Wal77]. If the restrictions assumed before are valid, then measuring this elastic scattering PV asymmetry would constitute a test of the Standard Model, specifically the WNC contributions. It should be realized that elastic scattering has an important advantage, namely, the figure-of-merit (the "doability" of such measurements; see [Mus94]) goes as the product of the asymmetry squared times the EM cross section and for elastic scattering, as we saw above, the latter is coherent and therefore goes as Z^2, making the figure-of-merit atypically large.

Of course, the restrictions assumed above are not valid at some level; the issue is to know what that level is in practice. First, one can go beyond the PWBA and consider the Coulomb distortion of the incoming and scattered electron, which is done in all modern work. Second, the ground states of $N = Z$ nuclei are not exact eigenstates of isospin; their projection of isospin is zero, since the charge and neutron numbers are presumed to be good quantum numbers, but the total isospin is not strictly $T = 0$ because (at least) the Coulomb interaction between the protons in the nucleus breaks the symmetry. Such effects have been modeled in the past [Don89, Mor09, Mor14], although we will not pursue them here, other than to say that when included, the asymmetry is multiplied by a factor $1 + \Gamma^{(I)}$, where $\Gamma^{(I)}$ characterizes the isospin mixing (see the following point). Third, strangeness in the nucleons in the nucleus (and also in other non-nucleonic degrees of freedom in the nucleus [Mus94]; these are not considered here) while small from our current understanding, as discussed in Chapter 8, may not be zero. Assuming that the former is dominant at low momentum transfers, and therefore neglecting strangeness effects in meson-exchange currents, one has a modification of the hadronic ratio in Eq. (15.41), namely it becomes multiplied by a factor $1 + \Gamma^{(S)}(q)$

$$\mathcal{A} = \frac{G_F Q^2}{2\pi\alpha\sqrt{2}} \cdot a_A \beta_V^{(0)} \left[1 + \Gamma^{(S)}(q)\right], \tag{15.43}$$

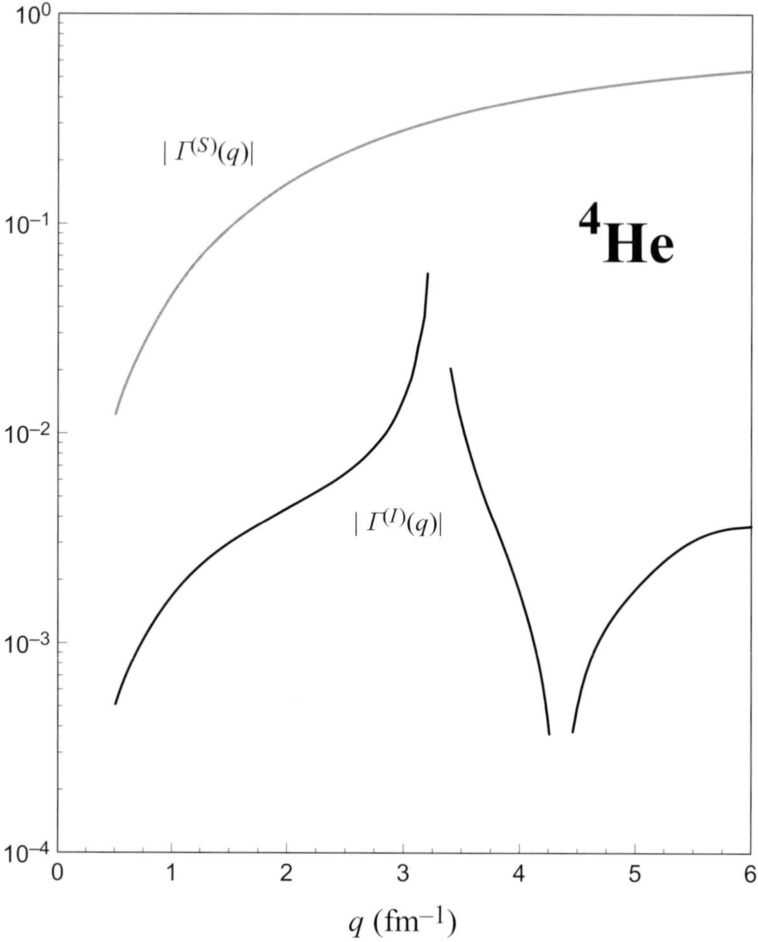

Fig. 15.6　The predicted deviation functions for PV elastic electron scattering from ^{4}He: $\Gamma^{(S)}$ for strangeness content and $\Gamma^{(I)}$ for isospin mixing (see text for the definitions of these quantities).

where

$$\Gamma^{(S)}(q) = \frac{G_E^{(0)}(\tau)}{2\sin^2\theta_W G_E^{(s)}(\tau)} \simeq 2.2\rho_s\tau$$

(see Chapter 8). The PV asymmetry has been measured at Jefferson Laboratory in the so-called HAPPEX-He experiment [Ani06, Ach07] on ^{4}He to place limits on the electric strangeness parameter ρ_s. Figure 15.6 shows the predicted deviations of the PV asymmetry due to strangeness content in the nucleon and to isospin mixing. From these one expects that studies of ^{4}He should be more sensitive to the former. In the future one might hope to tighten these limits and to address the other issues introduced above (isospin breaking, beyond tree-level contributions) by comparing results for PV elastic scattering from ^{4}He with those from ^{12}C using the so-called MESA facility, which is at the planning stage [Aul11].

To conclude this part of the discussion, let us turn to heavy nuclei where $N > Z$. In these cases it proves useful to shift from using isospin quantum numbers $T = 0, 1$ to a description in terms of protons and neutrons. The basic formalism was introduced in [Don89] and here we summarize the essentials. First we convert the isoscalar and isovector hadronic WNC couplings $\beta_V^{(0)} = -2\sin^2\theta_W$ and $\beta_V^{(1)} = 1 - 2\sin^2\theta_W$ into the equivalent proton/neutron couplings

$$\beta_V^p = \frac{1}{2}\left(\beta_V^{(0)} + \beta_V^{(1)}\right) = \frac{1}{2}\left(1 - 4\sin^2\theta_W\right) \simeq 0.04 \tag{15.44}$$

$$\beta_V^n = \frac{1}{2}\left(\beta_V^{(0)} - \beta_V^{(1)}\right) = -\frac{1}{2}, \tag{15.45}$$

where the numbers are the tree-level Standard Model values. Note that the neutron's WNC coupling is roughly 12 times that of the proton. In effect, the WNC (vector) hadronic coupling is mostly due to neutrons and strongly suppressed for protons, in contrast to the EM charge contributions which are mostly the reverse, namely, due to protons with much smaller effects from neutrons via the minimal contributions from G_E^n (see Chapter 8). To exploit these facts, one can write the PV hadronic ratio above in the form [Don89]

$$\frac{W^{\mathrm{PV}}}{W^{\mathrm{PC}}} \equiv a_A \left(\beta_V^p + \frac{N}{Z}\beta_V^n\right)\left[1 + \Gamma_{np}(q)\right], \tag{15.46}$$

and then one has

$$\Gamma_{np}(q) = -\beta_V''\left[1 - \frac{\int d\mathbf{x}\, j_0(qx)\rho_n(x)/N}{\int d\mathbf{x}\, j_0(qx)\rho_p(x)/Z}\right], \tag{15.47}$$

where

$$\beta_V'' \equiv \frac{N\beta_V^n}{Z\beta_V^p + N\beta_V^n}. \tag{15.48}$$

In deriving these results we have assumed that the coherent monopole form factors for the proton and neutron distributions are the only relevant ones. Of course, one can study the other elastic multipoles ($C2, C4, \ldots, M1, M3, M5, \ldots$, as will be discussed later) if the particular nucleus of interest does not have spin-0; however, for heavy nuclei the coherent monopole matrix elements are clearly the dominant ones. Note that, if the proton and neutron distributions are mathematically similar (i.e., $\rho_n(x)/N = \rho_p(x)/Z$), then $\Gamma_{np}(q) = 0$ and the hadronic ratio in Eq. (15.46) becomes independent of q; thus any momentum transfer dependence in this quantity is a signature for how the neutron and proton distributions differ. As discussed in [Don89], at low momentum transfer, PV elastic scattering from a heavy nucleus such as ^{208}Pb is sensitive to the difference between the RMS radii of the proton and neutron distributions (see Fig. 15.7). This is being realized in the so-called PREX and CREX experiments at Jefferson Laboratory. While small, the deviations are large enough to lead to a determination of the proton–neutron nuclear radius difference. We note in passing that in detailed modeling, especially for high-Z nuclei, the Coulomb distortion of the initial- and final-state electrons in the scattering is always taken into account.

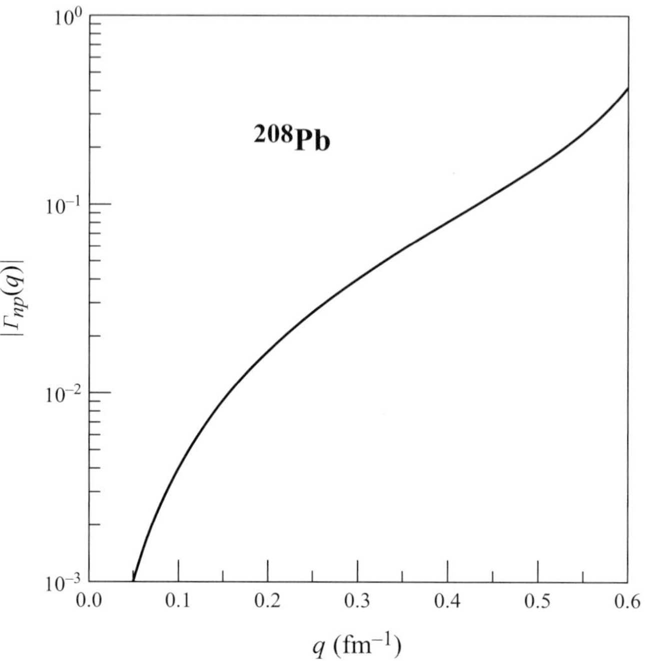

Fig. 15.7   The deviation function Γ_{np} for PV elastic electron scattering from ^{208}Pb (see text for the definition of this quantity).

15.3 Elastic Scattering from Non-Spin-0 Nuclei: Elastic Magnetic Scattering

The discussions earlier, of elastic charge scattering from spin-0 nuclei, may straightforwardly be extended to include cases with $J_0 \neq 0$. Then, of course, the complete set of allowed multipoles enters, $C0, C2, C4\ldots, M1, M3, M5\ldots$, as permitted by angular momentum conservation.

Elastic Charge Scattering from Non-Spin-0 Nuclei

The complete charge form factor is the following (we discuss elastic magnetic electron scattering later):

$$F_{\text{ch}}^2(q) \equiv \frac{4\pi}{Z^2} F_L^2(q) = \frac{4\pi}{Z^2} \left[\frac{1}{2J_0 + 1} \sum_{J \geq 0,\text{even}} \left| \langle J_0 \| \widehat{M}_J(q) \| J_0 \rangle \right|^2 \right]$$

$$(15.49)$$

$$\equiv F_0^2(q) + F_2^2(q) + \cdots + \begin{cases} F_{2J_0}^2(q) & J_0 = \text{integer} \\ F_{2J_0-1}^2(q) & J_0 = \text{half-integer} \,, \end{cases} \quad (15.50)$$

where "ch" denotes charge scattering. Here $F_0^2(q)$ is as above, while the $C2, \ldots$ cases are given by

$$F_2^2(q) = \frac{4\pi}{Z^2} \frac{1}{2J_0 + 1} \left| \langle J_0 \| \widehat{M}_2(q) \| J_0 \rangle \right|^2 \ldots . \tag{15.51}$$

In the Coulomb quadrupole case we have

$$\widehat{M}_{2M}(q) = \int d\mathbf{x} \, M_2^M(q\mathbf{x}) \widehat{\rho}(\mathbf{x}) , \tag{15.52}$$

with

$$M_2^M(q\mathbf{x}) = j_2(qx) Y_2^M(\Omega_x) . \tag{15.53}$$

Again taking the low-q or long wavelength limit (LWL), using the expansion of the spherical Bessel function

$$j_2(qx) \xrightarrow[LWL]{} \frac{(qx)^2}{5!!} + \cdots , \tag{15.54}$$

we find that

$$\langle J_0, J_0 | \widehat{M}_{20}(q) | J_0, J_0 \rangle = \begin{pmatrix} J_0 & 2 & J_0 \\ -J_0 & 0 & J_0 \end{pmatrix} \langle J_0 \| \widehat{M}_2(q) \| J_0 \rangle$$

$$\xrightarrow[LWL]{} \frac{q^2}{5!!} \sqrt{\frac{5}{16\pi}} \mathcal{Q} , \tag{15.55}$$

where the usual definition of the electric quadrupole moment,

$$\mathcal{Q} \equiv \int d\mathbf{x} \left(3z^2 - |\mathbf{x}|^2 \right) \rho(\mathbf{x})_{00} , \tag{15.56}$$

has been introduced using the identity

$$x^2 Y_2^0(\Omega_x) = \sqrt{\frac{5}{16\pi}} \left(3z^2 - |\mathbf{x}|^2 \right) . \tag{15.57}$$

Here we follow common convention and refer to the "electric" quadrupole moment, although from the discussions above we realize that it should actually be called the "Coulomb" quadrupole moment (see also the discussions in Chapter 11). Note that we must have $J_0 \geq 1$ for a nonzero result, in which case we find that

$$F_2^2(q) \xrightarrow[LWL]{} \frac{1}{180(2J_0 + 1) \begin{pmatrix} J_0 & 2 & J_0 \\ -J_0 & 0 & J_0 \end{pmatrix}^2} \left[\frac{\mathcal{Q}}{Z} \right]^2 q^4 . \tag{15.58}$$

We must remember that the $C0$ form factor has terms that also go as q^4,

$$F_0^2(q) = \left\{ 1 - \frac{1}{3!} \langle x^2 \rangle_{00} q^2 + \frac{1}{5!} \langle x^4 \rangle_{00} q^4 + \cdots \right\}^2 \tag{15.59}$$

and thus the q^4 term in $F_{\text{ch}}^2(q)$ at low-q contains inseparable contributions from $F_0^2(q)$ and $F_2^2(q)$. The analysis proceeds similarly for the $J = 4, 6, \ldots$ cases.

Elastic Magnetic Electron Scattering

Continuing the above discussion, here we consider elastic magnetic electron scattering. We have already seen that the transverse contributions involve only odd-J magnetic multipoles. This is for unpolarized electrons scattering from unpolarized targets; the general situation with polarizations is beyond the scope of this book, but may be found in [Don86] and [Don84].

$$
F_T^2(q) = \frac{1}{2J_0 + 1} \sum_{J \geq 1, \text{odd}} \left| \langle J_0 \| i\widehat{T}_J^{\text{mag}}(q) \| J_0 \rangle \right|^2
$$

$$
\equiv F_1^2(q) + F_3^2(q) + \cdots + \begin{cases} F_{2J_0}^2(q), & J_0 = \text{half-integer} \\ F_{2J_0-1}^2(q), & J_0 = \text{integer}, \end{cases}
\tag{15.60}
$$

where

$$
\widehat{T}_{JM}^{\text{mag}}(q) = \int d\mathbf{x} \, \mathbf{M}_{JJ}^M(q\mathbf{x}) \cdot \widehat{\mathbf{J}}(\mathbf{x})
\tag{15.61}
$$

with

$$
\mathbf{M}_{JL}^M(q\mathbf{x}) \equiv j_L(qx) \mathbf{Y}_{JL1}^M(\Omega_x)
\tag{15.62}
$$

and

$$
\widehat{\mathbf{J}}(\mathbf{x}) = \widehat{\mathbf{j}}(\mathbf{x}) + \nabla \times \widehat{\mu}(\mathbf{x}) .
\tag{15.63}
$$

In first quantization with point nucleons we have a convection current

$$
\mathbf{j}(\mathbf{x}) = \sum_{i=1}^{A} e(i) \left\{ \frac{\mathbf{p}(i)}{m_N} \delta\left(\mathbf{x} - \mathbf{x}_i\right) \right\}_{\text{sym}} ,
\tag{15.64}
$$

with $\mathbf{p}(i) = -i\nabla(i)$, where $e(i) \equiv \frac{1}{2}(1 + \tau_3(i))$ and m_N is the nucleon mass, and a magnetization

$$
\mu(\mathbf{x}) = \sum_{i=1}^{A} \mu(i)\sigma(i)\delta\left(\mathbf{x} - \mathbf{x}_i\right) ,
\tag{15.65}
$$

where $\mu(i) = \frac{1}{2}(1 + \tau_3(i))\mu_p + \frac{1}{2}(1 - \tau_3(i))\mu_n$. Here, as usual, $\mu_{p,n}$ are the magnetic moments of the proton and neutron, respectively. Thus, for the ith nucleon in the sums used above in constructing the single-particle magnetic multipole operators, we have

$$
T_{JM}^{\text{mag}}(i) = T_{JM}^{\text{mag}}(i)_o + T_{JM}^{\text{mag}}(i)_m
\tag{15.66}
$$

with

$$
iT_{JM}^{\text{mag}}(i)_o = \frac{q}{m_N} e(i) \Delta_J^M(q\mathbf{x}_i)
\tag{15.67}
$$

$$
iT_{JM}^{\text{mag}}(i)_m = -\frac{q}{2m_N} \mu(i) \Sigma_J'^M(q\mathbf{x}_i) ,
\tag{15.68}
$$

where, following [Wal75, Don75, Don79, Don79a], we define two multipole operators

$$\Delta_J^M(q\mathbf{x}_i) \equiv \mathbf{M}_{JJ}^M(q\mathbf{x}_i) \cdot \frac{1}{q}\nabla(i)$$

$$= -\frac{1}{\sqrt{J(J+1)}}\left(\frac{1}{q}\nabla M_J^M(q\mathbf{x}_i)\right) \cdot \mathbf{L}(i) \tag{15.69}$$

$$\Sigma_J'^M(q\mathbf{x}_i) \equiv \left(-\frac{i}{q}\nabla \times \mathbf{M}_{JJ}^M(q\mathbf{x}_i)\right) \cdot \sigma(i) . \tag{15.70}$$

Here the result in Eq. (15.69) is obtained using vector identities [Edm74] together with the orbital angular momentum operator given by $\mathbf{L} = -i(\mathbf{x} \times \nabla)$. The case where $J = 1$ deserves a bit more attention: one can show that in the long wavelength limit (LWL) where $q \to 0$ we have

$$\Delta_1^M(q\mathbf{x}_i) \xrightarrow[LWL]{} -\frac{1}{\sqrt{6}}\mathbf{M}_{10}^M(q\mathbf{x}_i) \cdot \mathbf{L}(i) \tag{15.71}$$

$$\Sigma_1'^M(q\mathbf{x}_i) \xrightarrow[LWL]{} \frac{2}{\sqrt{6}}\mathbf{M}_{10}^M(q\mathbf{x}_i) \cdot \sigma(i) , \tag{15.72}$$

where $\mathbf{M}_{10}^M(q\mathbf{x}_i) = j_0(qx_i)Y_0^0\mathbf{e}_M$. In the LWL, this becomes simply $\frac{1}{\sqrt{4\pi}}\mathbf{e}_M$ and therefore one has

$$-iT_{1M}^{\mathrm{mag}}(i) \xrightarrow[\mathrm{LWL}]{} \frac{1}{\sqrt{6\pi}}\left(\frac{q}{2m_N}\right)\mu_{1M}(i) , \tag{15.73}$$

where, with the conventional choice of sign, $\widehat{\mu}_{1M}$ is the magnetic dipole moment operator. Since at low-q the $M3, M5, \dots$ contributions all fall with higher powers of momentum transfer than the result in Eq. (15.73), we find accordingly that

$$F_T^2(q) \xrightarrow[LWL]{} \frac{1}{6\pi}\left(\frac{J_0 + 1}{J_0}\right)\left(\frac{q}{2m_N}\right)^2 \mu^2 , \tag{15.74}$$

where μ is the nuclear magnetic dipole moment; we return to this below. We point the reader to Chapters 11 and 12, where the magnetic dipole moment has already been discussed for few-body nuclei, and to Chapter 13 where the Schmidt lines have been introduced.

In an extreme single-particle model one has one unpaired particle above a closed core, or a single hole in a closed shell, namely $|J_0, M_0\rangle = a_\alpha^\dagger |F\rangle$ with $\epsilon_\alpha > \epsilon_F$ and $|J_0, M_0\rangle = b_\alpha^\dagger |F\rangle$ with $\epsilon_\alpha \leq \epsilon_F$, respectively, where $\alpha = n_0\ell_0 j_0, m_{j_0}$ with $J_0, M_0 = j_0, m_{j_0}$. Using the above formalism for matrix elements of single-particle operators, one has only the contribution from that unpaired particle or hole:

$$\langle J_0 \| \hat{T}_J^{\mathrm{mag}} \| J_0\rangle = \langle n_0\ell_0 j_0 \| T_J^{\mathrm{mag}} \| n_0\ell_0 j_0\rangle . \tag{15.75}$$

Some examples are the following:

$$\text{Single-particle } |^{17}\text{F}; 5/2^+\rangle \simeq a_{1d5/2,p}^\dagger |^{16}\text{O}; 0^+\rangle$$
$$|^{17}\text{O}; 5/2^+\rangle \simeq a_{1d5/2,n}^\dagger |^{16}\text{O}; 0^+\rangle$$
$$\text{Single-hole } |^{15}\text{N}; 1/2^-\rangle \simeq b_{1p1/2,p}^\dagger |^{16}\text{O}; 0^+\rangle$$
$$|^{15}\text{O}; 1/2^-\rangle \simeq b_{1p1/2,n}^\dagger |^{16}\text{O}; 0^+\rangle .$$

Table 15.2 Transverse polynomials from the Donnelly–Haxton tables for a pure $1d5/2$ configuration

Multipolarity	Convection $p_{\Delta_J}(y)$	Magnetization $p_{\Sigma'_J}(y)$
$J = 1$	$-\dfrac{2\sqrt{7}}{\sqrt{5}}\left(1 - \tfrac{2}{5}y\right)$	$\dfrac{2\sqrt{7}}{\sqrt{5}}\left(1 - \tfrac{8}{5}y + \tfrac{12}{35}y^2\right)$
$J = 3$	$\dfrac{4\sqrt{3}}{5\sqrt{5}}$	$-\dfrac{16\sqrt{3}}{5\sqrt{5}}\left(1 - \tfrac{1}{3}y\right)$
$J = 5$	0	$\dfrac{8\sqrt{2}}{\sqrt{105}}$

Let us conclude these discussions by considering a specific example, namely, one where the unpaired nucleon has $J_0 = j_0 = \frac{5}{2}$, $\ell_0 = 2$, $n_0 = 1$ where $M1, M3$, and $M5$ magnetic multipoles occur. Several examples occur in the $1d_{5/2}$ shell; one of those will be discussed in more detail next. Using the tables in [Don79a] (in fact this case is done in that reference to illustrate the procedures for finding the single-particle matrix elements of the general set of electroweak operators) we find that

$$\langle 1d_{5/2} \| \Delta_J(q\mathbf{x}) \| 1d_{5/2} \rangle_{\mathrm{HO}} = \frac{1}{\sqrt{4\pi}} y^{(J-1)/2} e^{-y} p_{\Delta_J}(y) \tag{15.76}$$

$$\langle 1d_{5/2} \| \Sigma'_J(q\mathbf{x}) \| 1d_{5/2} \rangle_{\mathrm{HO}} = \frac{1}{\sqrt{4\pi}} y^{(J-1)/2} e^{-y} p_{\Sigma'_J}(y) \, , \tag{15.77}$$

where the polynomials are those given in Table 15.2.

As before, including the CM correction and the nucleon form factors to incorporate the finite distributions of charge and spin, we have

$$\left\langle \frac{5}{2} \left\| i\hat{T}_J^{\mathrm{mag}}(q) \right\| \frac{5}{2} \right\rangle = \frac{q}{m_N} \left\{ G_E \langle 1d_{5/2} \| \Delta_J(q\mathbf{x}) \| 1d_{5/2} \rangle \right.$$
$$\left. - \frac{1}{2} G_M \langle 1d_{5/2} \| \Sigma'_J(q\mathbf{x}) \| 1d_{5/2} \rangle \right\} f_{\mathrm{CM}}(y) \, , \tag{15.78}$$

where $G_{E,M}$ are the form factors of protons or neutrons, depending on the flavor of the odd particle. A specific example is shown in Fig. 15.8 for the case of ^{27}Al viewed in the extreme single-particle model as a proton hole in a filled $1d_{5/2}$ shell, namely, the simple model used above for ^{28}Si. Clearly, one has a rather good understanding of the elastic magnetic $M1$, $M3$, and $M5$ form factors of ^{27}Al. Indeed, as discussed at length in a review article on the subject [Don84], good agreement occurs for many cases across the periodic table. One should note that as soon as qR is of order unity, where R is the nuclear radius, the LWL no longer pertains, one cannot expand the spherical Bessel functions involved in obtaining the form factors as illustrated above, and hence all allowed multipoles can occur with similar strengths. The example presented here involves up to $M3$ since the ground-state spin of ^{27}Al is 5/2; however, more extreme cases such as ^{209}Bi, which has a ground-state spin of 9/2, occur, in that case requiring $M1, M3, M5, M7$, and $M9$ magnetic multipoles, as well as $C0, C2, C4, C6$, and $C8$ Coulomb multipoles (see [Don84]).

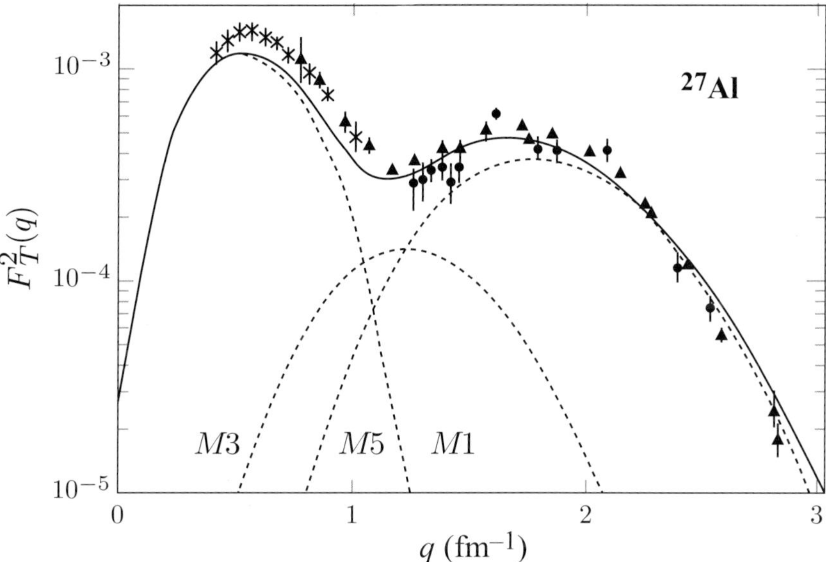

Fig. 15.8   Elastic magnetic electron scattering from ^{27}Al. For more discussion of elastic magnetic electron scattering and references to the data see [Don84].

We conclude this discussion by commenting that the same ideas can be applied to any one-body probe of the nuclear ground state. All nonrelativistic multipole operators can be reduced to the set discussed in this chapter and, accordingly, the same nuclear structure content underlies elastic or inelastic scattering of electrons, neutrinos, axions, dark matter particles, etc. – see, for example, the coherent ν scattering discussions of Chapter 18.

15.4 Electroexcitation of Low-Lying Excited States

Next we turn to electroexcitation of low-lying excited states in nuclei and the closely related processes of photoexcitation and gamma-decay. In Chapter 14, several models were introduced to deal with such excitations; these were both microscopic models and models based on semi-classical approximations for various oscillatory modes of the "nuclear fluid." We begin with the semi-classical surfon model discussed in Section 14.3.

Semi-classical models

For discussion of electron scattering, we need the charge and current density operators. The former is given by

$$\rho(\mathbf{r}) = \frac{3Z}{4\pi R_0^3}\theta\left[R(\theta,\phi) - r\right]\ , \tag{15.79}$$

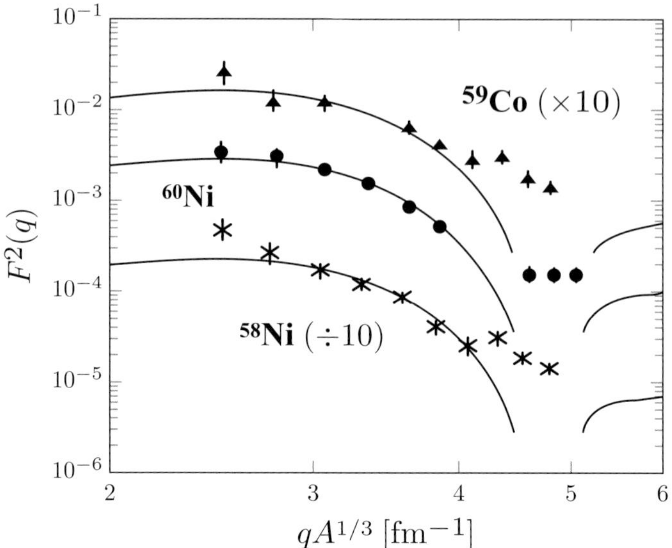

Fig. 15.9　The square of the inelastic form factors for excitation of the lowest 2^+ one-surfon states in ^{58}Ni, ^{59}Co, and ^{60}Ni compared with the model discussed in the text (see Section 14.3). The data are from [Cra61].

using Eq. (14.35) for $R(\theta, \phi)$, while the latter is given, to linear order, by

$$\mathbf{j}(\mathbf{r}) = \frac{3Z}{4\pi R_0} \sum_{\substack{L,M \\ L \geq 2}} \frac{1}{L} \frac{d}{dt} \alpha_{L,M} \left[\nabla \left(\frac{r}{r_0} \right)^L Y_L^M(\theta, \phi) \right] \theta \left[R(\theta, \phi) - r \right] . \tag{15.80}$$

These expressions allow us to compute the multipole matrix elements using the developments discussed in Chapter 7. For our present purposes we restrict our attention to Coulomb excitations of single-surfon states having angular momentum L where one finds that

$$\left| \left\langle (L)^1, L^\pi \, \| M_L(q) \| \, 0 \right\rangle \right|^2 = \frac{9Z^2}{32\pi^2} (2L+1) \frac{1}{\sqrt{B_L C_L}} \, |j_L(qR_0)|^2 . \tag{15.81}$$

Here one sees that the square of the form factor is proportional to $|j_L(qR_0)|^2$, in quite good agreement with experiment, one example of which is shown in Fig. 15.9.

Similarly, for the Goldhaber–Teller model of the giant electric dipole resonance, we have for the Coulomb operator (see Eq. (7.75))

$$\widehat{M}_{1M}(q) = \int d\mathbf{x} M_1^M(q\mathbf{x}) \widehat{\rho}(\mathbf{x})$$

$$\cong -\frac{N}{A} \widehat{\mathbf{r}} \cdot \int d\Omega_x Y_1^M(\Omega_x) \int x dx \mathbf{x} j_1(qx) \frac{\partial \rho_0}{\partial x}$$

$$= \sqrt{\frac{4\pi}{3}} \frac{N}{A} q \widehat{r}_m \int_0^\infty x^2 dx \rho_0(x) j_0(qx) , \tag{15.82}$$

where the last line arises from partially integrating and using properties of the spherical Bessel functions. Comparing with what we found for elastic scattering (see Eq. (15.17))

$$M_{00}(q) = \sqrt{4\pi} \int_0^\infty x^2 dx \rho_0(x) j_0(qx) , \tag{15.83}$$

one arrives at the result that the square of the GDR Coulomb matrix element is proportional to the square of the elastic scattering Coulomb matrix element

$$4\pi \left| \langle 1^- \| \widehat{M}_1(q) \| 0 \rangle \right|^2 \cong \left(\frac{N}{A} \right)^2 \left(\frac{q^2}{2\mu} \right) \frac{1}{\omega_0} \left| \langle 0 \| \widehat{M}_0(q) \| 0 \rangle \right|^2 , \tag{15.84}$$

i.e., the squares of the form factors are seen to be proportional to one another with the factor $(N/A)^2 q^2 / 2\mu\omega_0$. Here $\mu = (NZ/A)m_N$ is the reduced mass.

In a simple model in which only convection of the proton fluid is taken into account, the current density operator can be written

$$\widehat{\mathbf{j}}(\mathbf{x}) = \rho_0(x) \frac{N}{A} \frac{d}{dt} \widehat{\mathbf{r}} , \tag{15.85}$$

and the electric dipole operator is obtained

$$\widehat{T}_{1M}^{\mathrm{el}}(q) = \int d\mathbf{x} \left[\frac{1}{q} \nabla \times \mathbf{M}_{11}^M(q\mathbf{x}) \right] \cdot \widehat{\mathbf{j}}(\mathbf{x})$$

$$= i \sqrt{\frac{8\pi}{3}} \frac{d}{dt} \widehat{r}_M \int_0^\infty x^2 dx \rho_0(x) j_0(qx) . \tag{15.86}$$

Using identities on vector spherical harmonics [Edm74], one obtains the following:

$$\left| \langle 1^- \| \widehat{T}_1^{\mathrm{el}}(q) \| 0 \rangle \right|^2 \cong 2 \left(\frac{\omega_0}{q} \right)^2 \left| \langle 1^- \| \widehat{M}_1(q) \| 0 \rangle \right|^2 , \tag{15.87}$$

where the square of the Coulomb matrix element is given in Eq. (15.84), which immediately yields the GDR form factors and thus the excitation cross section. In the dashed curves in Figs. 15.10 and 15.11 the squares of the GT $C1$ and $E1$ form factors are shown for a typical case, namely, for ^{16}O using $\omega_0 = 18.4$ MeV (see the following section).

Pure 1p1h States and the RPA

Let us consider a more microscopic model. We focus on the pure $1p1h$ states in ^{16}O discussed in Section 14.4 and listed in Table 14.1 for two specific cases. First, consider two $1p1h$ negative-parity dipole states (1) $\left| 1d_{5/2} 1p_{3/2}^{-1} 1^- \right\rangle$ and (2) $\left| 2s_{1/2} 1p_{1/2}^{-1} 1^- \right\rangle$, which may be either isoscalar ($T = 0$) or isovector ($T = 1$). Transitions from the ground state to these states then have $C1$ and $E1$ multipole matrix elements. Using harmonic oscillator single-particle wavefunctions with oscillator parameter b chosen to be 1.5 fm to agree with the choice for ω_0 made previously, using the fact that the oscillator energy is given by $\omega_0 = 1/m_N b^2$, and employing the tables discussed earlier, one finds that

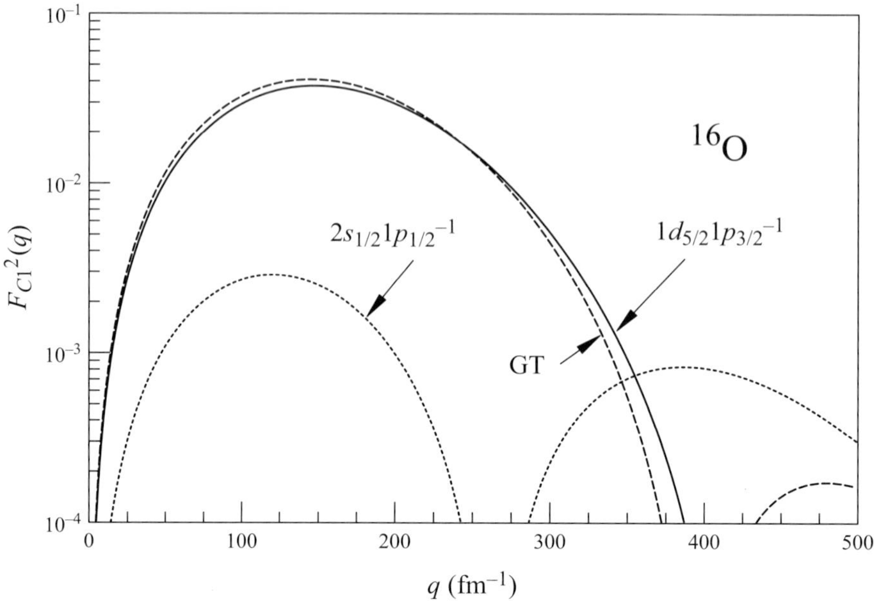

Fig. 15.10   C1 form factors for the Goldhaber–Teller (GT) model (dashed) and for the two pure $T = 1$, $1p1h$ configurations discussed in the text.

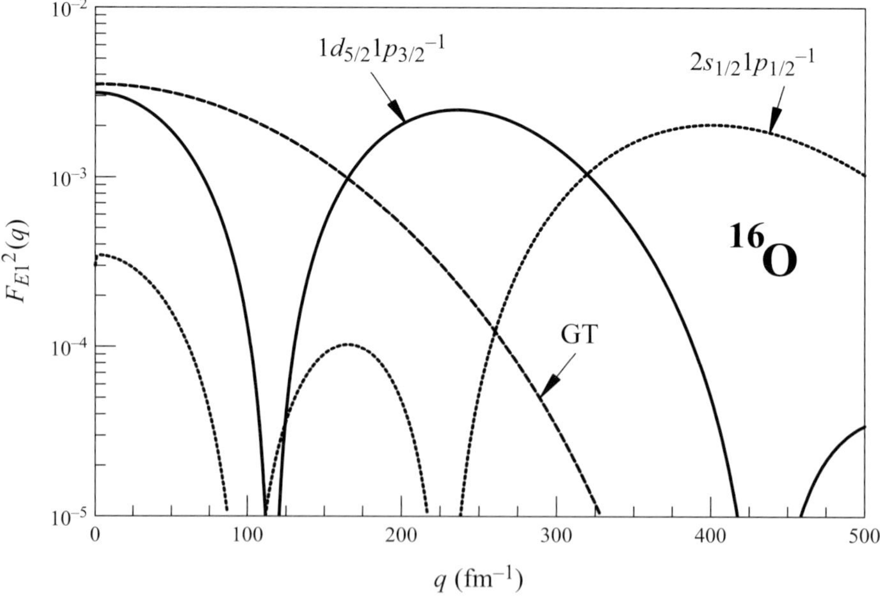

Fig. 15.11   As for Fig. 15.10, but now showing the corresponding E1 form factors.

$$\langle 1d_{5/2} \| M_1(q) \| 1p_{3/2}\rangle = \frac{1}{4\pi} 2\sqrt{2} G_E^{(T)} \left(1 - \frac{2}{5}y\right) y^{1/2} e^{-y} f_{CM}(y), \tag{15.88}$$

$$\langle 1d_{5/2} \| T_1^{el}(q) \| 1p_{3/2}\rangle = -\frac{1}{4\pi}\left(\frac{q}{m_N}\right)\left[G_E^{(T)}\left(1 - \frac{4}{5}y\right) - 2G_M^{(T)} y\left(1 - \frac{2}{5}y\right)\right]$$
$$\times y^{-1/2} e^{-y} f_{CM}(y), \tag{15.89}$$

$$\langle 2s_{1/2} \| M_1(q) \| 1p_{1/2}\rangle = \frac{1}{4\pi}\frac{2\sqrt{2}}{3} G_E^{(T)} (1 - y) y^{1/2} e^{-y} f_{CM}(y), \tag{15.90}$$

$$\langle 2s_{1/2} \| T_1^{el}(q) \| 1p_{1/2}\rangle = -\frac{1}{4\pi}\frac{1}{3}\left(\frac{q}{m_N}\right)\left[G_E^{(T)}(1 + y) - 2G_M^{(T)} y(1 - y)\right]$$
$$\times y^{-1/2} e^{-y} f_{CM}(y), \tag{15.91}$$

yielding the $0^+ \, T = 0 \to 1^- \, T = 1$ form factors displayed in Figs. 15.10 and 15.11 (solid curves in the figures). Here

$$G_{E,M}^{(T)} = \begin{cases} G_{E,M}^p + G_{E,M}^n & T = 0 \\ G_{E,M}^p - G_{E,M}^n & T = 1 \end{cases}, \tag{15.92}$$

and, in contrast to the tables in [Don79a], we employ Sachs form factors. Note that in the LWL one has

$$\langle 1d_{5/2} \| T_1^{el}(q) \| 1p_{3/2}\rangle \xrightarrow[\text{LWL}]{} -\sqrt{2}\left(\frac{\omega_0}{q}\right)\langle 1d_{5/2} \| M_1(q) \| 1p_{3/2}\rangle \tag{15.93}$$

$$\langle 2s_{1/2} \| T_1^{el}(q) \| 1p_{1/2}\rangle \xrightarrow[\text{LWL}]{} -\sqrt{2}\left(\frac{\omega_0}{q}\right)\langle 2s_{1/2} \| M_1(q) \| 1p_{1/2}\rangle. \tag{15.94}$$

For the $C1$ form factors one sees that the agreement between the GT results and those from using pure $1p1h$ state number (2) is quite good, whereas state (1) provides a very different result. On the other hand, the $E1$ form factors are all different except for the low-q behavior of the GT and state (2) form factors. This last fact is likely because the GT model is too simple, having only a convection-current contribution, whereas the more microscopic ph model has both convection and magnetization contributions; the latter dominate at high momentum transfers.

As two other examples, this time for abnormal-parity transitions of higher multipolarity, we consider excitation to the $T = 1$ $1p1h$ negative-parity 2^- and 4^- states $\left|1d_{5/2}1p_{3/2}^{-1}2^-\right\rangle$ and $\left|1d_{5/2}1p_{3/2}^{-1}4^-\right\rangle$, respectively, with their corresponding $M2$ (magnetic quadrupole) and $M4$ (magnetic hexadecapole) multipole form factors

$$\langle 1d_{5/2} \| iT_2^{mag}(q) \| 1p_{3/2}\rangle = -\frac{1}{4\pi}\frac{2\sqrt{7}}{3\sqrt{5}}\left(\frac{q}{m_N}\right)\left[G_E^{(T)} + \frac{3}{2}G_M^{(T)}\left(1 - \frac{10}{21}y\right)\right]$$
$$\times y^{1/2} e^{-y} f_{CM}(y), \tag{15.95}$$

and

$$\langle 1d_{5/2} \| iT_4^{mag}(q) \| 1p_{3/2}\rangle = \frac{1}{4\pi}\frac{2}{\sqrt{7}}\left(\frac{q}{m_N}\right) G_M^{(T)} y^{3/2} e^{-y} f_{CM}(y), \tag{15.96}$$

shown in Fig. 15.12. Note that the high-spin, so-called "stretched configurations" such as the 4^- state considered here can be important at moderate-to-high momentum transfers.

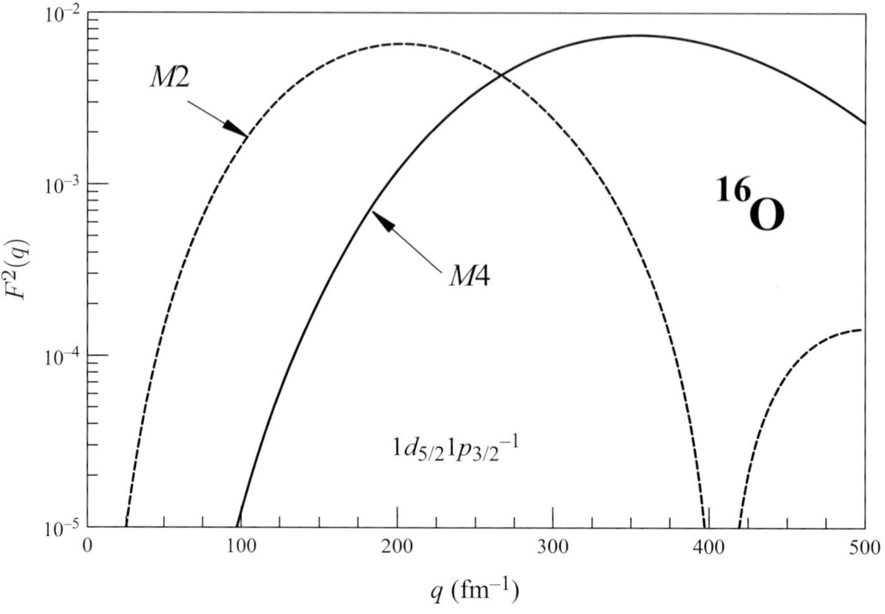

Fig. 15.12   M2 (dashed curve) and M4 (solid curve) form factors for the pure $1d_{5/2}1p_{3/2}^{-1}\,T=1\;1p1h$ configuration discussed in the text.

The concept of allowed and forbidden multipoles that pertains for real-photon reactions (where $q=\omega$ is small) and also for low-q processes such as beta decay (see Chapter 17) does not do so once qR is of order unity, since then the spherical Bessel functions used above in the definitions of the multipole operators cannot be expanded as they can in the LWL. Evidence for such high-spin stretched configurations was first found in the late 1960s [Don68] and subsequent experimental studies found many other examples ranging up to $0^+ \rightarrow 14^-$ $M14$ transitions; note that an $M14$ multipole is a $2^{14}=16384$ pole multipole!

An observation here that has some relevance for studies of $\beta\beta$-decay is the following: for $2\nu\beta\beta$-decay the momentum transfer involved is very small and only the lowest multipolarity transitions are involved. On the other hand, for $0\nu\beta\beta$-decay typical modeling indicates that rather large momentum transfers occur, i.e., similar to those discussed here for electron scattering. This means that $2\nu\beta\beta$ and $0\nu\beta\beta$-decays are different, and that the latter involve transitions through high-spin states in the intermediate nucleus involved ($\beta\beta$-decay is discussed in more detail in Chapters 17 and 18).

In general, a microscopic model with residual interactions such as the RPA discussed in Chapter 14 will require a linear combination of purely $1p1h$ states of the type considered here. Accordingly the full microscopic form factor will arise from a sum of individual $1p1h$ form factors with weighting coefficients determined by solving the RPA equations, the so-called one-body density matrix elements (see [Row72] for a treatment of the open-shell RPA, OSRPA). In fact, the RPA results are generally seen to reflect the q-dependence of the data for momentum transfers ranging up to about 400 MeV c^{-1} and to show the general characteristics of the excitation spectrum, as shown in Fig. 15.13. In the figure the results of using an OSRPA model for excitations in ^{12}C [Don72a] are compared with a sketch

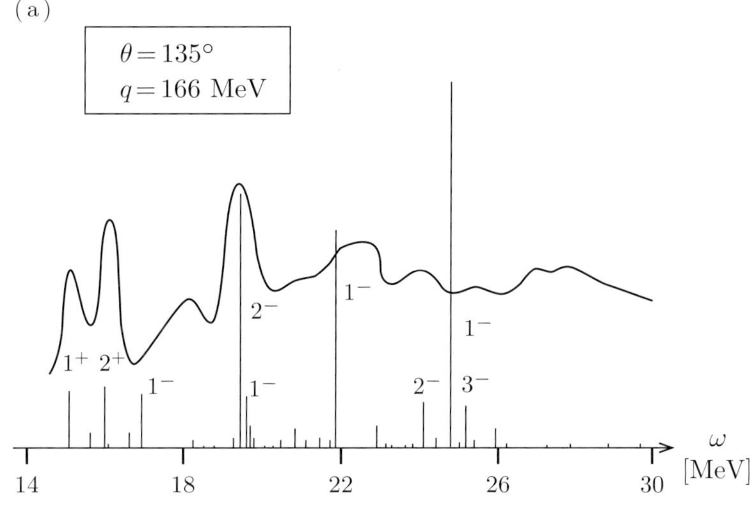

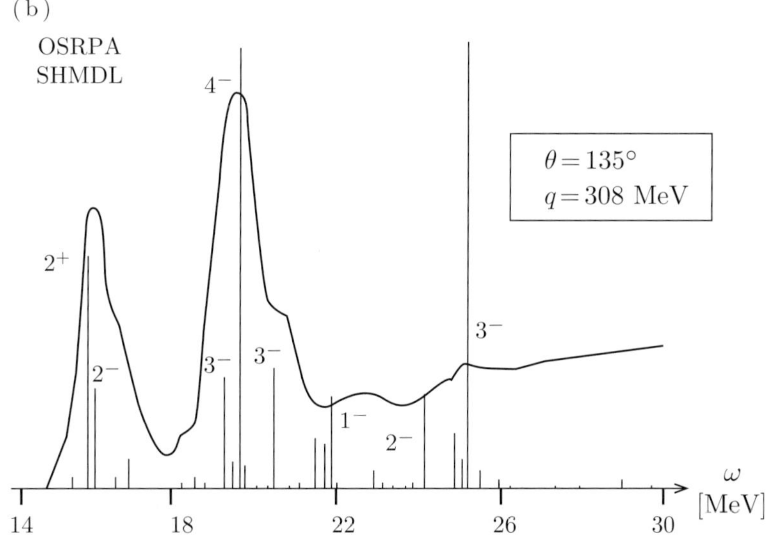

Fig. 15.13 Open-shell RPA (OSRPA) plus shell-model (SHMDL) results for excitations in ^{12}C compared with a sketch of electron scattering data for two choices of kinematics: (a) shows results for modest values of momentum transfer q, (b) while the lower panel is for high values of q. See the text for discussion and citations.

of electron scattering data [Yam71] for two choices of kinematics: Fig. 15.13(a) shows results for modest values of momentum transfer q, while Fig. 15.13(b) is for high values of q. Clearly the former situation is dominated by low-multipolarity transitions, whereas the latter shows the importance of the high-spin states and the relative unimportance of the low-multipolarity transitions. Note that the apparent lack of agreement beyond about 24 MeV is due to the fact that the continuum becomes important there, whereas the modeling has been done using only bound-state wavefunctions; when the continuum is taken into account and spreading to multi-particle/multi-hole states can occur, the localized strength seen in the figure becomes distributed in energy.

Clearly this discussion of low-lying nuclear excitations merely scratches the surface; indeed, entire books are devoted to treatments of the wide range of models that have been applied to electron scattering, photoexcitation, γ-decay, and to decay or excitation mediated by other types of interactions, for instance, β-decay (see Chapter 17) or neutrino excitation (see Chapter 18). In the present context we leave this subject and move on to higher excitation energies, typically above $\sim$30 MeV, where a different class of models proves to be appropriate and where new issues arise. As we shall see in Chapter 16, at very high energy and momentum transfers one must confront the problem that nonrelativistic modeling is inadequate and attempts to incorporate relativistic effects are needed.

Exercises

15.1   Elastic Scattering: Parity and Time Reversal

Symmetries have been discussed in Chapter 2 and, in the present chapter, we have seen that parity and time-reversal invariance, together with conservation of angular momentum, limit the multipoles allowed for elastic scattering; this exercise is focused on proving this statement.

a) Using the properties of the elastic multipoles discussed in Chapter 7, show that when spatial inversion is assumed to be a good symmetry, one has only even-J Coulomb and electric multipoles together with only odd-J magnetic multipoles.

b) If, additionally, time reversal symmetry is invoked, show that the even-J electric multipoles also cannot occur and that one is left only with even-J Coulomb and odd-J magnetic multipoles.

c) Supposing the ground state of a given nucleus to have angular momentum $J_0 = 7/2$ (for example, the ground state of ^{41}Ca) what multipoles occur for elastic electron scattering? Which are coherent and which are not?

15.2   Elastic Charge Form Factor of a Spin-0 Nucleus

Using the formalism presented in the present chapter, this exercise involves a specific example of how to implement a model for the ground-state charge distribution and to find the related elastic $C0$ form factor.

a) Using the phenomenological model for the ground-state charge distribution of ^{28}Si discussed in the text

$$\rho(x)_{00} = \frac{\rho_0}{1 + \exp\left(\frac{x-R}{a}\right)},$$

compute the elastic charge form factor $C0$ using

$$F_0(q) = \frac{1}{Z} \int d\mathbf{x} \, \frac{\sin qx}{qx} \rho(\mathbf{x})_{00}.$$

b) Compare this result with that obtained using harmonic oscillator wavefunctions in a closed-shell $1s_{1/2}^4 1p_{3/2}^8 1p_{1/2}^4 1d_{5/2}^{12}$ model (the tables of Donnelly and Haxton [Don79a] may be used for the latter; as an approximation, consider only protons and neglect the contributions from neutrons).

c) Plot the results for momentum transfers up to 4 fm^{-1}. What value of the oscillator parameter yields the best agreement between the two models? What are the RMS charge radii obtained?

15.3 Elastic Form Factors of a Non-Spin-0 Nucleus

This exercise proceeds from Exercise 15.2 to consideration of a non-spin-0 nucleus. Assume a $1s_{1/2}^4 1p_{3/2}^8 1p_{1/2}^4 1d_{5/2}^{12} 2s_{1/2}^4 1d_{3/2}^7$ model for the ground state of ^{39}K.

a) Using harmonic oscillator wavefunctions with oscillator parameter $b = 2$ fm and the Donnelly–Haxton tables [Don79a], obtain expressions for all of the elastic form factors that enter, assuming angular momentum and parity are good quantum numbers and that one has time reversal invariance.

b) What are the RMS charge radius, magnetic dipole moment and quadrupole moment that result?

15.4 Parity-Violating Elastic Electron Scattering I

Building on the results of the preceding two exercises, here PV elastic electron scattering is considered. Assume that ^{40}Ca is a closed $1s_{1/2}^4 1p_{3/2}^8 1p_{1/2}^4 1d_{5/2}^{12} 2s_{1/2}^4 1d_{3/2}^8$-shell nucleus and that ^{48}Ca is built from that closed shell plus eight neutrons in the $1f_{7/2}$-shell.

a) Obtain the parity-conserving elastic scattering cross section when the single-particle wavefunctions are taken to be harmonic oscillators with oscillator parameter $b = 2.2$ fm (again the Donnelly–Haxton tables [Don79a] may be used).

b) Take into account both protons and neutrons in computing the elastic form factor. How important are the neutrons?

c) Obtain the parity-violating elastic electron scattering asymmetry (assume that strangeness can be neglected).

d) Suppose that one has a facility with an 85% polarized electron beam of luminosity $\mathcal{L} = 10^{39}$ cm^{-2} sec^{-1} and a detector of 100 msr, how long will it take to measure the PV asymmetry to $\delta A/A \sim 5\%$?

15.5 Parity-Violating Elastic Electron Scattering II

In this exercise, a different approach to PV elastic electron scattering is considered.

a) Prove

$$\frac{W^{\mathrm{PV}}}{W^{\mathrm{PC}}} \equiv a_A \left(\beta_V^p + \frac{N}{Z} \beta_V^n \right) \left[1 + \Gamma_{np}(q) \right] ,$$

where

$$\Gamma_{np}(q) = -\beta_V'' \left[1 - \frac{\int d\mathbf{x}\, j_0(qx)\rho_n(x)/N}{\int d\mathbf{x}\, j_0(qx)\rho_p(x)/Z} \right] ,$$

and where

$$\beta_V'' \equiv \frac{N\beta_V^n}{Z\beta_V^p + N\beta_V^n} .$$

b) Consider the 2PF function

$$\rho(x)_{00} = \frac{\rho_0}{1 + \exp\left(\frac{x-R}{a}\right)}$$

discussed in the text as an approximation for the proton and neutron distributions of ^{208}Pb. Assume that these have slightly different radius parameters. Compute the transforms

$$\Gamma_{np}(q) = -\beta_V'' \left[1 - \frac{\int d\mathbf{x} j_0(qx)\rho_n(x)/N}{\int d\mathbf{x} j_0(qx)\rho_p(x)/Z} \right],$$

where

$$\beta_V'' \equiv \frac{N\beta_V^n}{Z\beta_V^p + N\beta_V^n}$$

and hence the deviation function $\Gamma_{np}(q)$ as a function of the difference in radius parameter. (Assume that the average radius parameter is 3 fm and the surface diffuseness parameter is 0.5 fm; of course the distributions must be normalized to the correct numbers of protons and neutrons.)

15.6   Inelastic Electron Scattering in ^{39}K

In Exercise 15.3 a model for the ground state of ^{39}K was introduced. Suppose that an excited state of the same nucleus is obtained by promoting a proton from the $1d_{3/2}$ shell in the ground state to the $1f_{7/2}$ shell.

a) What multipoles are allowed?
b) Obtain the form factor for the lowest multipolarity allowed, remembering that a proton transition is assumed.
c) Obtain the form factors for the highest multipolarity allowed.

The tables cited above may be used; take $b = 2$ fm.

15.7   Inelastic Electron Scattering in ^{16}O

In the text the case of pure $1p1h$ excitations in ^{16}O was considered, with focus placed on two specific configurations, namely, the $T = 1$ states of the type $1d_{5/2}1p_{3/2}^{-1}$ and $2s_{1/2}1p_{1/2}^{-1}$.

a) Compute form factors for the analogous $T = 0$ states and compare the results with those already found.
b) Extend the analysis to all $T = 1$ $1p1h$ states involving particles in the $2s - 1d$ shell and holes in the $1p$ shell.

The tables cited above may be used; take $b = 1.5$ fm.

15.8   Parity-Violating Inelastic Electron Scattering

Consider an inelastic transition $J_i^{\pi_i} T_i = 0^+0 \rightarrow J_f^{\pi_f} T_f = 1^+0$ in a nucleus such as ^{12}C where a single nucleon is excited from a $1p_{3/2}$ level to a $1p_{1/2}$ level.

a) Compute the inelastic PC electron scattering cross section for this transition and extend this calculation to compute the PV asymmetry (the tables of [Don79a] may be used).

b) Assuming that the same "typical" experimental conditions as in Exercise 15.4 are available, and that PV elastic scattering from ^{12}C yields a significant determination of the elastic asymmetry in 1000 hours, by computing both the elastic and inelastic figures-of-merit provide an estimate of how much beam time would be required to determine the inelastic asymmetry to the same precision as the elastic asymmetry.

Electroexcitation of High-Lying Excitations of the Nucleus

16.1 Introduction

Let us begin the discussion of electron scattering for kinematics where the dominant reaction mechanism has a nucleus undergoing a transition from its ground state to the region where a nucleon is ejected "quasi-freely" from the nucleus. For *semi-inclusive* reactions one has both the scattered electron and the ejected particle (in the present case, a nucleon) detected in coincidence, while for *inclusive* scattering only the scattered electron is detected. Both of these processes will be discussed in more detail below.

The kinematic variables involved have already been introduced in Chapter 7 (see Fig. 7.2); here, for convenience we summarize the conventions introduced there. The incident electron has four-momentum $k^\mu = (\epsilon, \mathbf{k})$, the scattered electron $k'^\mu = (\epsilon', \mathbf{k}')$ and thus the virtual photon carries the four-momentum transfer $q^\mu = (\omega, \mathbf{q}) = k^\mu - k'^\mu = (\epsilon - \epsilon', \mathbf{k} - \mathbf{k}')$, conserving four-momentum transfer at the $ee'\gamma$ vertex. As usual one has $Q^2 = q^2 - \omega^2 > 0$ (spacelike). On the nuclear side the initial state labeled A has four-momentum $p_A^\mu = (M_A, 0)$ in the laboratory system, while the final state has $p_f^\mu = (E_f, \mathbf{p}_f) = (M_A + \omega, \mathbf{q})$ by four-momentum conservation. So far, all of this is completely general. Referring to the lower panel of Fig. 7.1 we see that several typical, but not completely distinct, excitation regions can be identified, namely, the low-energy region discussed in Chapter 15, the quasielastic (QE) and Δ-dominated regions and at high excitation energies the regime of deep inelastic scattering (DIS) discussed in Chapter 9. In the present chapter, QE electron scattering and Δ excitation in nuclei form the focus. The former is assumed to be dominated by ejection of single protons and neutrons, with small corrections from ejection of clusters of nucleons (deuterons, alphas, etc.). In fact, if the nucleus were merely a collection of protons and neutrons at rest and one scattered electrons elastically from these constituents, then the cross section would be very simple, just a δ-function at the QE condition $\omega_{QE} \equiv Q^2/2m_N$, where m_N is the nucleon mass (for simplicity we take $m_p \simeq m_n \equiv m_N$); in other language this corresponds to $x = 1$ (Bjorken x). Of course, the nucleons in the nucleus are moving and interacting with one another in both their initial and final states and so the δ-function representation is too crude. In fact a broad peak occurs in the experimental spectrum and the width of the peak is related to the mean momentum in the distribution of nucleons in the nucleus, in other words, to the Fermi momentum, k_F, introduced earlier. Moreover, the interactions between the nucleons in both initial nuclear ground-state and final nuclear (continuum) wavefunctions cause small shifts away from the QE peak condition to occur. Both of these facts will be employed here in setting the basic scales in the problem when addressing the physics of the QE region. For excitations above the QE peak (that is, at ω significantly above ω_{QE}), one expects other modes of excitation to play a role, namely, production of mesons, excitation of baryon resonances and ejection of (at least) two nucleons via two-body meson-exchange currents (MEC). Ultimately, at very high momentum transfers and very high excitation energies, the regime of DIS should be approached (see the discussions in Chapter 9).

16.2 Quasielastic Electron Scattering and the Fermi Gas Model

Beginning with the basics, let us presume the inclusive (e, e') cross section (at least in the QE region) to be made up of the coincidence $(e, e'p)$ and $(e, e'n)$ cross sections (see the next section), each integrated over the kinematic variables that characterize the outgoing protons and neutrons, and summed with weighting factors Z and N, the proton and neutron numbers, respectively. What it means to integrate over the final-state nucleon kinematics requires some discussion, since, of course, all of this is to be done for fixed electron scattering kinematics, which constrains the final state to some degree. As in the previous discussions of parity-conserving inclusive electron scattering the double-differential unpolarized cross section may be written in the standard form

$$\frac{d^2\sigma}{d\Omega_e d\epsilon'} = \sigma_{Mott} \left(v_L R_L(q, \omega) + v_T R_T(q, \omega) \right) , \tag{16.1}$$

where the Mott cross section is given as before by

$$\sigma_{Mott} = \left[\frac{\alpha \cos \theta_e/2}{2\epsilon \sin^2 \theta_e/2} \right]^2 , \tag{16.2}$$

and the kinematic Rosenbluth factors are

$$v_L = \left| \frac{Q^2}{q^2} \right|^2 \tag{16.3}$$

$$v_T = \frac{1}{2} \left| \frac{Q^2}{q^2} \right| + \tan^2 \theta_e/2 , \tag{16.4}$$

(see also Chapter 15). The longitudinal (L) and transverse (T) response functions $R_{L,T}(q, \omega)$ are functions of the three-momentum transfer q and energy transfer ω. Following standard procedures [Alb88], these are in turn given by specific Lorentz components of the hadronic (nuclear) tensor $W^{\mu\upsilon}$ discussed below

$$R_L(q, \omega) = W^{00}(q, \omega) \tag{16.5}$$

$$R_T(q, \omega) = W^{11}(q, \omega) + W^{22}(q, \omega) , \tag{16.6}$$

where the z-axis has been chosen to lie in the $\mathbf{q}$ direction.

Next, let us develop a very simple model for the hadronic tensor. In general this requires a treatment of the complexities of the nuclear many-body problem for both the initial and final states involved; additionally, typically the interest in QE scattering is focused on

GeV energies and momenta, and so, relativistic aspects of the problem cannot be ignored. However, despite the limitations of the model, the easiest approach is the one we shall adopt here, namely, the relativistic Fermi gas (RFG) model [Van78, Alb88]. In due course we shall show that this picture is reasonably successful (typically at $\sim$25–30% level), but that if better predictive power is required, more sophisticated models must be invoked and these fall beyond the scope of this book. The nonrelativistic Fermi gas model was used in Chapter 13 when discussing the semi-empirical mass formula. Here we need to model excitations from the (relativistic) Fermi gas ground state. The latter is assumed to have on-shell nucleon spinors for all single-particle states labeled by momenta $\mathbf{p}$ (on-shell implying that the nucleon energies are given by $E(\mathbf{p}) = (\mathbf{p}^2 + m_N^2)^{1/2}$) completely filled up to the Fermi surface characterized by Fermi momentum k_F and completely empty for all higher momenta. That is, we assume a θ-function (step-function) momentum distribution. The excitations are then assumed to be pure $1p1h$ (one-particle one-hole) in Nature, meaning that a nucleon is promoted from below the Fermi surface to a single-particle state above the Fermi surface, as in the previous chapter. The hole has energy $E(\mathbf{p})$ and the particle has energy $E(\mathbf{p} + \mathbf{q})$, since the three-momentum must be conserved, and the energy must also be conserved, implying that $\omega = E(\mathbf{p} + \mathbf{q}) - E(\mathbf{p})$. The full nuclear result must then be built from an integral over the entire Fermi sea. For each $1p1h$ excitation we must include the electron-nucleon tensor (specifically the 00 and 11+22 components discussed earlier) that were introduced in Chapter 8:

$$W^{\mu\nu}_{\text{nucleon}}(p+q,p) = -W_1(\tau)\left[g^{\mu\nu} + \frac{q^\mu q^\nu}{Q^2}\right] + W_2(\tau)V^\mu V^\nu , \qquad (16.7)$$

where

$$V^\mu \equiv \frac{1}{m_N}\left[p^\mu + \left(\frac{p \cdot q}{Q^2}\right)q^\mu\right] . \qquad (16.8)$$

Note that this result is for a moving nucleon, in contrast to the discussions in Chapter 8 where the rest frame was selected for the initial-state nucleon. The invariant responses are given as usual in terms of the nucleon form factors by

$$W_1(\tau) = \tau G_M^2(\tau) \qquad (16.9)$$

$$W_2(\tau) = \frac{1}{1+\tau}\left[G_E^2(\tau) + \tau G_M^2(\tau)\right] , \qquad (16.10)$$

where as before $\tau \equiv Q^2/4m_N^2$ is the dimensionless four-momentum transfer. When summing over nucleons we will have Z cases of Eqs. (16.9, 16.10) with proton form factors and N cases with neutron form factors. Accordingly we define

$$\widetilde{W}^{\mu\nu}_{\text{nucleon}}(p+q,p) \equiv ZW_p^{\mu\nu}(p+q,p) + NW_n^{\mu\nu}(p+q,p) , \qquad (16.11)$$

and corresponding quantities $\widetilde{G}_{E,M}^2$ and $\widetilde{W}_{1,2}$ summed over protons and neutrons, and weighted by Z and N as in Eq. (16.11). Putting all of this together, the hadronic tensor for QE electron scattering in the RFG model is given by

$$W^{\mu\nu}_{\mathrm{RFG}}(q,\omega) = \frac{3m_N^2}{4\pi k_F^3} \int \frac{d\mathbf{p}}{E(\mathbf{p})E(\mathbf{p}+\mathbf{q})} d\mathbf{p}\, \delta[\omega - (E(\mathbf{p}+\mathbf{q}) - E(\mathbf{p}))]$$

$$\times \theta(k_F - |\mathbf{p}|)\theta(|\mathbf{p}+\mathbf{q}| - k_F)\widetilde{W}^{\mu\nu}_{\mathrm{nucleon}}(p+q,p)\,, \qquad (16.12)$$

where the energies in the denominator arise from the relativistic phase space and the factors before the integral come from properly normalizing the momentum distributions (the θ-function forms), such that they integrate to Z protons and N neutrons (see [Van78, Alb88]); also, for simplicity, we take the Fermi momenta for protons and neutrons to be the same. Note that the nucleon with momentum $\mathbf{p}$ (the hole) must come from below the Fermi surface, while the nucleon with momentum $\mathbf{p}+\mathbf{q}$ (the particle) must be above the Fermi surface, i.e., this is where Pauli blocking arises in this simple model. All of the integrals can be done analytically, yielding

$$R_{L,T}(q,\omega) = \frac{3}{4m_N\kappa\eta_F^3}(\epsilon_F - \Gamma)\,\theta\,(\epsilon_F - \Gamma)$$

$$\times \begin{cases} \frac{\kappa^2}{\tau}\left[\widetilde{G}_E^2(\tau) + \widetilde{W}_2(\tau)\Delta\right] & \text{for } L \\[2mm] 2\tau\widetilde{G}_M^2(\tau) + \widetilde{W}_2(\tau)\Delta & \text{for } T\,. \end{cases} \qquad (16.13)$$

Here, for use in the following, we use dimensionless variables $\kappa \equiv q/2m_N$, $\lambda \equiv \omega/2m_N$ so that $\tau = \kappa^2 - \lambda^2 > 0$, and $\eta_F \equiv k_F/m_N$ together with $\epsilon_F = (1 + \eta_F)^{1/2}$ and $\xi_F \equiv \epsilon_F - 1$. Typically k_F is 200–250 MeV/c (see below) and so η_F is roughly 1/4, allowing one to make reasonable expansions in η_F (but not, of course, in κ, λ, or τ, since these quantities may all be large). And Γ and Δ are given by

$$\Gamma(\kappa,\lambda) \equiv \max\left[(\epsilon_F - 2\lambda),\left(\kappa\sqrt{1 + 1/\tau} - \lambda\right)\right] \qquad (16.14)$$

$$\Delta(\kappa,\lambda) \equiv \frac{\tau}{\kappa^2}\left[\frac{1}{3}\left(\epsilon_F^2 + \epsilon_F\Gamma + \Gamma^2\right) + \lambda(\epsilon_F + \Gamma) + \lambda^2\right] - (1+\tau)\,. \qquad (16.15)$$

These equations constitute the complete analytic result for the RFG in both Pauli-blocked (first situation in Eq. (16.14)) and non-Pauli-blocked (second situation in Eq. (16.14)) regimes.

Having derived the RFG we now turn to a brief discussion of the concept of scaling. We will find that the RFG provides a reasonable model for treatments of quasielastic electron scattering, although it is too simple to capture some of the details seen in the data, and more sophisticated models or phenomenological scaling approximations are required when a high level of agreement between theory and experiment are required. Importantly, we shall return in Chapter 18 to discuss the simple extensions of the RFG model to include treatments of charge-changing QE neutrino and antineutrino reactions on nuclei; there we will find that more sophisticated modeling than that provided by the RFG is required for a quantitative understanding of the quasielastic region in charge-changing neutrino reactions. Note also that more insights into the RFG model can be attained by doing the exercises in this chapter and in Chapter 18.

16.3 Inclusive Electron Scattering and Scaling

Let us now turn to another example, inclusive electron scattering from nuclei. The focus will be placed on relatively high excitation energies, namely, away from elastic scattering and excitation of low-lying discrete states or collective modes such as giant resonances, as discussed in Chapter 15. Furthermore, let us consider the situation in which $q \geq 2k_F$ ($\kappa \geq \eta_F$) where Pauli blocking does not occur in the RFG model. Then, using Eq. (16.14), the factor $\epsilon_F - \Gamma$ in Eq. (16.13) becomes $\xi_F(1 - \psi^2)$, where $\lambda = \lambda_0 = \frac{1}{2}\left[\sqrt{1 + 4\kappa^2} - 1\right]$ $\left(\text{which is equivalent to } \omega = \omega_0 = \sqrt{q^2 + m_N^2} - m_N\right)$ is where the QE peak occurs, and we have defined the following dimensionless quantity:

$$\psi \equiv \sqrt{\frac{1}{\xi_F}\left[\kappa\sqrt{1 + 1/\tau} - (1 + \lambda)\right]} \times \begin{cases} +1 & \lambda \geq \lambda_0 \\ -1 & \lambda < \lambda_0 \end{cases} \tag{16.16}$$

$$= \frac{1}{\sqrt{\xi_F}}\frac{\lambda - \tau}{\sqrt{(1 + \lambda)\tau + \kappa\sqrt{\tau(1 + \tau)}}} . \tag{16.17}$$

The QE peak condition $\lambda = \lambda_0$ also corresponds to $\lambda = \tau$, and as is clear from Eq. (16.17), the variable ψ is zero at the QE peak. Indeed, one has Bjorken $x = 1$ (<1, >1) when $\psi = 0$ (> 0, < 0); while the former is appropriate for electron scattering from the proton (see Chapter 8), the latter is usually better suited to discussions of electron scattering from nuclei. One may use the pair of variables (κ, ψ) instead of (κ, λ) (or equivalently (q, ω)) to characterize inclusive electron scattering, by writing λ and thereby τ as functions of (κ, ψ).

If one gathers all of the single-nucleon content in Eq. (16.13) into

$$D_{L,T}(\kappa, \psi) \equiv \frac{\xi_F}{m_N \kappa \eta_F^3} \times \begin{cases} \frac{\kappa^2}{\tau}\left[\widetilde{G}_E^2(\tau) + \widetilde{W}_2(\tau)\Delta\right] & \text{for } L \\ 2\tau\widetilde{G}_M^2(\tau) + \widetilde{W}_2(\tau)\Delta & \text{for } T, \end{cases} \tag{16.18}$$

then the RFG implies that in the non-Pauli-blocked region one can define dimensionless reduced responses by

$$f_{L,T}(\psi) \equiv \frac{R_{L,T}(\kappa, \psi)}{D_{L,T}(\kappa, \psi)} . \tag{16.19}$$

For the RFG model a very simple result is obtained:

$$f_L^{\text{RFG}}(\psi) = f_T^{\text{RFG}}(\psi) \equiv f^{\text{RFG}}(\psi) = \frac{3}{4}\left(1 - \psi^2\right)\theta\left(1 - \psi^2\right) . \tag{16.20}$$

Several observations emerge. The first equality implies that, upon dividing out the single-nucleon content as in Eq. (16.19), a single reduced response results: this universality is called scaling of the zeroth kind. The fact that this universal reduced response is a function only of ψ and not of κ (i.e., is independent of momentum transfer q) is called scaling of the first kind; ψ is called the scaling variable and $f(\psi)$ the scaling function. By including the dependence on the particular nucleus being selected, namely the k_F dependence contained in the factors ξ_F, η_F, and Δ in Eq. (16.18) one finds that the scaling function depends on a universal scaling variable (i.e., is independent of nuclear

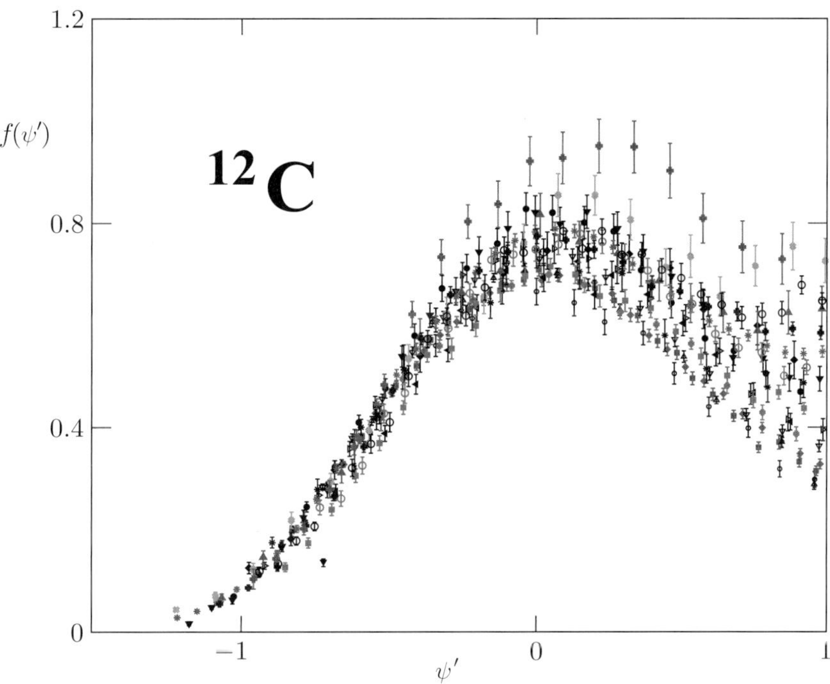

Fig. 16.1 The scaling function introduced in Eq. (16.19) versus the scaling variable in Eqs. (16.16–16.17) for inclusive electron scattering from ^{12}C (see [Day15] for the data). The various data points come from different electron scattering energies and angles that correspond to specific values of ψ', but different values of q. The good coalescence of the results in the scaling region ($\psi' < 0$) is known as scaling of the first kind. On the other hand, when $\psi' > 0$ pion production and excitation of baryon resonances begin to play a significant role and the scaling is not observed. The latter are discussed in the next section.

species); which is called scaling of the second kind. When scaling of both the first and second kinds occurs, as it does in the RFG model, one says that superscaling occurs. Finally, one has assumed that protons and neutrons scale the same way and this is called scaling of the third kind.

These ideas of scaling have proven to be very useful in bringing inclusive electron scattering data into forms that are more universal. That is, in the QE region the cross section *factorizes* into 1) single-nucleon content and 2) the specific properties of the nuclear response of pointlike nucleons. In effect, one has achieved a separation of the hadronic content in the individual form factors from the specifics of the nuclear many-body problem. While motivated by simple models as discussed above, the observations are that scaling behavior is a useful concept.

Scaling of the first kind is illustrated in Fig. 16.1. Data for the full response

$$R^{\text{expt}}(q, \omega) \equiv v_L R_L^{\text{expt}}(q, \omega) + v_T R_T^{\text{expt}}(q, \omega) \tag{16.21}$$

(see Eq. (16.1)) are divided by

$$D(\kappa, \psi') \equiv v_L D_L(\kappa, \psi') + v_T D_T(\kappa, \psi') \tag{16.22}$$

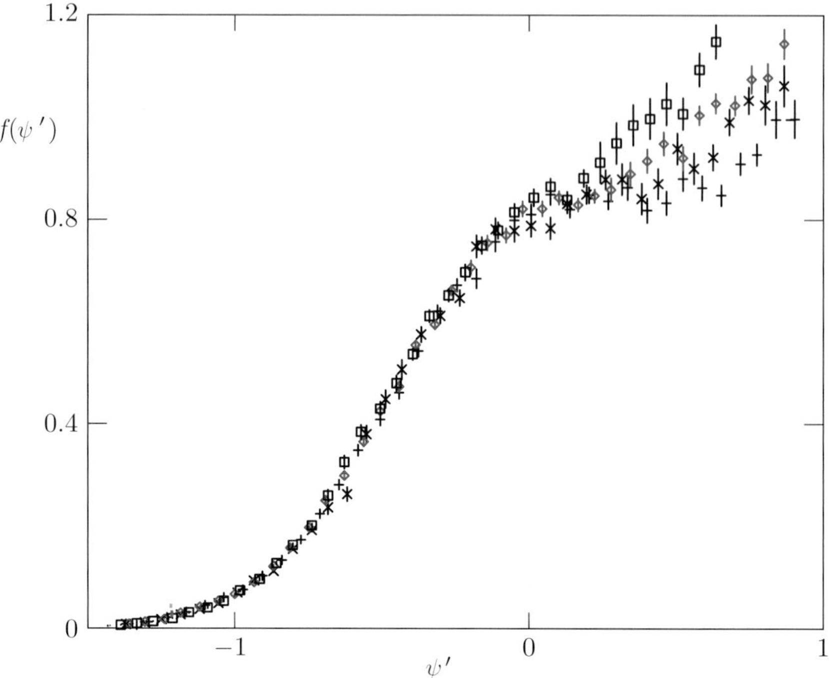

Fig. 16.2   The scaling function $f(\psi')$ versus ψ' for electron scattering with $\epsilon = 3.6$ GeV and $\theta_e = 15^o$, for four different nuclei, ^{12}C, ^{27}Al, ^{56}Fe, and ^{197}Au, adapted from [Don99, Don99a]. The coalescence of the scaled results for the different targets when $\psi' < 0$ provides a demonstration of scaling of the second kind.

to form

$$f^{\text{expt}}(q, \psi') \equiv \frac{R^{\text{expt}}(q, \omega)}{D(\kappa, \psi')} \; . \tag{16.23}$$

Here one follows the procedure usually employed and makes a small, purely phenomenological shift of the energy transfer by an amount E_{shift} defining $\omega' \equiv \omega - E_{\text{shift}}$ and hence λ', τ', and, by a simple substitution of the primed variables into Eqs. (16.16) and (16.17), ψ'.

Furthermore, in Fig. 16.2 some results are displayed for scaling of the second kind. By choosing k_F and E_{shift} for the four nuclei in the figure, the data are seen to have a universal behavior in the region below the QE peak ($\psi' < 0$). Perhaps this is not so surprising, since one has empirical parameters to adjust, although it is not at all obvious that the data would line up over the entire scaling region. Initial expectations were that the region where the mean-field dynamics are dominant, namely, within the "Fermi cone," meaning $|\psi'| \ll 1$, would have this behavior, but that outside this region where, for instance, short-range correlation effects are believed to play a significant role, this second-kind scaling might begin to break down. For the region below the QE peak in Fig. 16.3 the cross section outside the Fermi cone is so small that it is hard to learn anything. However, if the results are displayed on a semilog scale as in Fig. 16.4 a very clear picture emerges: scaling of the second kind at relatively large momentum transfers is very good in the region below the quasielastic peak. Indeed, it is significantly better than scaling of the first kind. Just

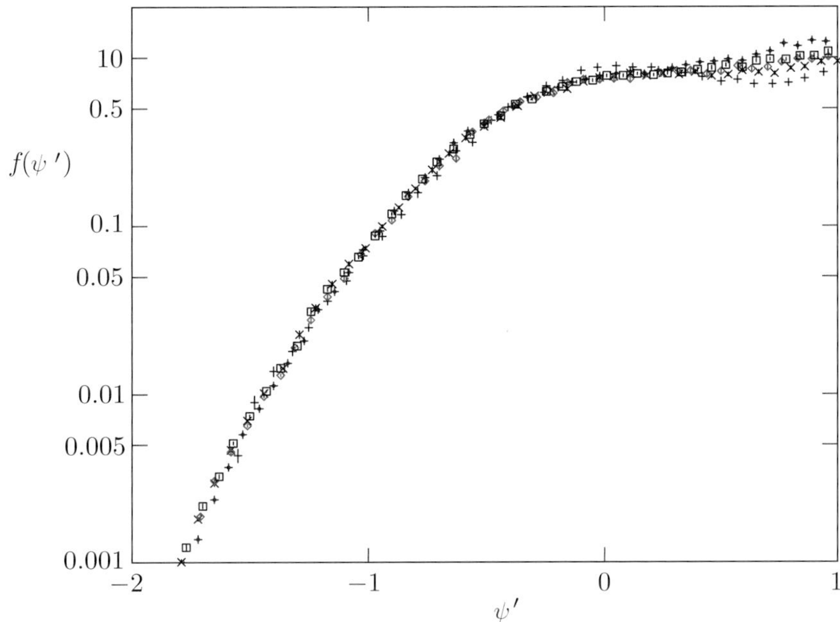

Fig. 16.3 As for Fig. 16.2, but now on a semilog scale showing the remarkably good second-kind scaling behavior in the region where $\psi' < 0$. Note that five nuclei are included here, ranging from ^{4}He to ^{197}Au.

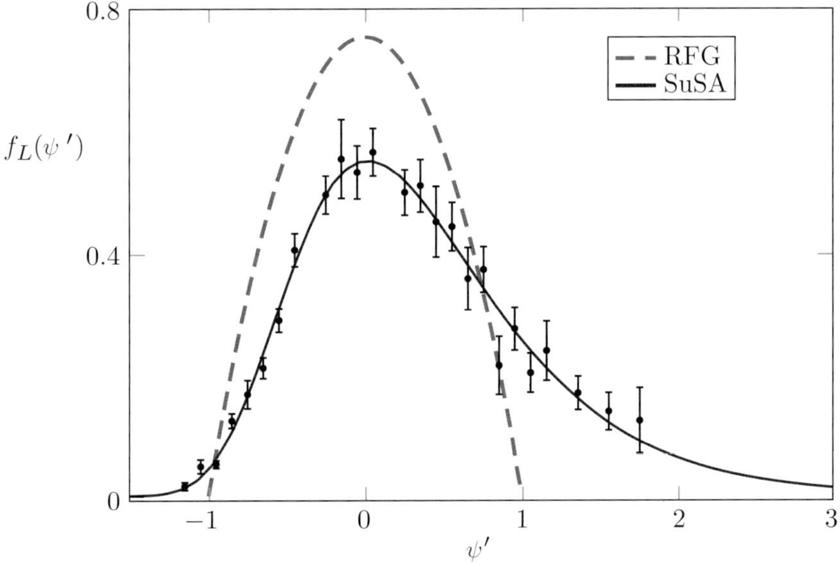

Fig. 16.4 Longitudinal scaling function $f_L(\psi')$ versus ψ' for three nuclei (^{12}C, ^{40}Ca, and ^{56}Fe) and for three different momentum transfers ($q = 300, 380,$ and 570 MeV c^{-1}), demonstrating superscaling. For a discussion of the solid curve shown in the figure and a reference to the data see [Don99a].

Table 16.1 Fermi momenta and energy shifts from scaling analyses

Nucleus	k_F (MeV/c)	E_{shift} (MeV)
^{2}H	55	2
^{3}He	150	5
^{4}He	200	20
^{7}Li	165	15
^{12}C	228	20
Mg	230	25
^{27}Al	236	18
^{40}Ca	241	28
^{56}Fe	241	23
Ni	245	30
Sn	245	28
Pb	248	31

how successful this concept is requires a closer examination of the empirical parameters and the sensitivity of the pattern to their choices.Scaling of both first and second kinds, *i.e,* superscaling, occurs for the longitudinal part of the response, as illustrated in Fig. 16.4.

The adjusted parameters are given in Table 16.1, which is adapted from [Mai02]. Note the anomalous value for lithium, curious because the k_F used for ^{4}He is $200\,\text{MeV}\,\text{c}^{-1}$. This anomaly is easily explained if one assumes that ^{6}Li is essentially a deuteron (with $k_F = 55\,\text{MeV}\,\text{c}^{-1}$) plus an alpha-particle. Taking the weighted mean $[\{4 \times (200)^2 + 2 \times (55)^2\}/6]^{1/2}$ gives $166\,\text{MeV}\,\text{c}^{-1}$, which is very close to the fit value of $165\,\text{MeV}\,\text{c}^{-1}$. A word of caution should be said here: the values of the Fermi momenta given in the table are found by fitting QE electron scattering data at relatively high energies. In contrast, the Fermi momenta used in Chapter 13 are appropriate for nonrelativistic descriptions of nuclear ground states. Since QE scattering involves (at least) the high-energy excitation of particles from the ground state to $1p1h$ states, where their final-state properties play a role, the Fermi momenta used in this chapter are effective quantities. Or, said another way, each nucleus apparently has a characteristic momentum scale which is captured here in the variable k_F. Furthermore, the phenomenological shift given in the table is not necessarily simply connected to some property of the nucleus, such as the separation energy or mean binding energy, although surely both of those aspects presumably play a role in determining E_{shift}.

Having seen that scaling is, in fact, observed to a greater or lesser extent in studies of inclusive electron scattering, let us return to mention a few reasons why this is a useful concept. First, the fact that it works reasonably well is a rather direct proof that the idea that nucleons constitute the basic effective degrees of freedom in the nucleus is correct. Scaling in the QE region depends on having the electroweak interaction take place via elastic eN scattering (hence the use of the word "quasielastic"), factorized from the nuclear physics issues of how those nucleons are distributed, the latter being captured by the scaling function. Second, the scaled responses can be explored using a variety of nuclear models that are more sophisticated than the RFG model used here. Many of these models are seen to provide some scaling behavior; however, most do not work quantitatively at high momentum transfers. For instance, one sees clear evidence of the need for relativistic effects, and these are often not easily incorporated. A third motivation for studying scaling is provided by the observation that much of what is seen in inclusive electron scattering also form a basis for the more general electroweak response. Specifically, inclusive charge-changing neutrino reactions, to be discussed in Chapter 18, contain some of the same scaling functions as those discussed here. Accordingly, to some extent one can avoid the limitations of direct nuclear modeling by using the phenomenology of scaling.

16.4 Δ-Excitation in Nuclei

The RFG for quasielastic electron scattering discussed in Section 16.2 may be extended to a relativistic Fermi gas model for use in the region where pion production, production of N^*s and Δs, etc., can occur. Although it is relatively straightforward to consider the general situation within the context of a factorized approximation, as invoked above for the RFG, here we shall restrict our attention to the region where electroexcitation of the Δ is the most important contribution. The procedures followed in Section 16.2 are simple to extend to this case. First, in Eq. (16.12) the energy of the outgoing particle is now $E_\Delta(\mathbf{p} + \mathbf{q}) = (m_\Delta^2 + (\mathbf{p} + \mathbf{q})^2)^{1/2}$ (here we neglect the fact that the delta is unstable and hence has a width), and secondly, the single-nucleon (elastic) tensor $\widetilde{W}^{\mu\nu}_{\text{nucleon}}(p + q, p)$ must be replaced by its analog $\widetilde{W}^{\mu\nu}_{N \to \Delta}(p + q, p)$ for excitation of a nucleon to a Δ. We have the same dimensionless variables as before, but now, since the final-state kinematics are different, it is convenient to introduce in addition

$$\mu_\Delta \equiv \frac{m_\Delta}{m_N} \tag{16.24}$$

$$\rho_\Delta \equiv 1 + \frac{\mu_\Delta^2 - 1}{4\tau} \; ; \tag{16.25}$$

note that if $m_\Delta \to m_N$ then $\mu_\Delta = \rho_\Delta = 1$. Just as for the RFG model of QE scattering, for this case one can obtain an *inelastic* scaling variable

$$\psi_\Delta(\kappa, \lambda) = \sqrt{\frac{1}{\xi_F}\left[\kappa\sqrt{\rho_\Delta^2 + \frac{1}{\tau}} - (1 + \rho_\Delta \lambda)\right]} \times \begin{cases} +1 & \lambda \geq \lambda_0^\Delta \\ -1 & \lambda < \lambda_0^\Delta \end{cases} , \tag{16.26}$$

where the peak of the response is located at

$$\lambda_0^\Delta = \frac{1}{2}\left[\sqrt{\mu_\Delta^2 + 4\kappa^2} - 1\right] , \tag{16.27}$$

which is the same as the condition $\omega_0^\Delta = \sqrt{m_\Delta^2 + q^2} - m_N$. The analog to Eq. (16.7) is

$$W^{\mu\nu}_{N \to \Delta}(p + q, p) = -W_1^\Delta(\tau)\left[g^{\mu\nu} + \frac{q^\mu q^\nu}{Q^2}\right] + W_2^\Delta(\tau)V^\mu V^\nu , \tag{16.28}$$

with new invariant functions $W_{1,2}^{\Delta}(\tau)$ (see below). Also, as in Eq. (16.11), it is convenient to assemble the total tensor from its proton and neutron components

$$\widetilde{W}_{N\to\Delta}^{\mu\nu}(p+q,p) \equiv ZW_{p\to\Delta^+}^{\mu\nu}(p+q,p) + NW_{n\to\Delta^0}^{\mu\nu}(p+q,p) \tag{16.29}$$

and likewise for $\widetilde{W}_{1,2}^{\Delta}(\tau)$. Equation (16.15) must also be extended to take into account the changed final-state kinematics:

$$\Delta(\kappa,\lambda) \to \mathcal{D}(\kappa,\lambda)$$
$$\equiv \frac{\tau}{\kappa^2}\left[\frac{1}{3}\left(1+\psi_\Delta^2+\psi_\Delta^4\right)\xi_F^2 + (1+\rho_\Delta\lambda)\left(1+\psi_\Delta^2\right)\xi_F + (1+\rho_\Delta\lambda)^2\right]$$
$$-(1+\rho_\Delta^2\tau)\,. \tag{16.30}$$

The inclusive $N\to\Delta$ response functions may then be shown to be

$$R_{L,T}(q,\omega) = \frac{3\xi_F}{4m_N\kappa\eta_F^3}\left(1-\psi_\Delta^2\right)\theta\left(1-\psi_\Delta^2\right) \tag{16.31}$$
$$\times\begin{cases} \frac{\kappa^2}{\tau}\left[(1+\rho_\Delta^2\tau)\widetilde{W}_2^\Delta(\tau) - \widetilde{W}_1^\Delta(\tau) + \widetilde{W}_2^\Delta(\tau)\mathcal{D}(\kappa,\lambda)\right] & \text{for } L \\ 2\widetilde{W}_1^\Delta(\tau) + \widetilde{W}_2^\Delta(\tau)\mathcal{D}(\kappa,\lambda) & \text{for } T\,. \end{cases}$$

All that remains is to provide a model for the two invariant functions $\widetilde{W}_{1,2}^\Delta(\tau)$. One approach is to invoke some parametrization for them in analogy with the EM form factors of the nucleon. Another is to consider specific models, such as the constituent quark model discussed in Chapter 3, together with some assumed dependence on Q^2. Alternatively, one might use lattice QCD to simulate these quantities, as discussed in Chapter 5. Here we simply assume the phenomenological approach taken in [Ama99] and show some typical results from that study; see Fig. 16.5.

16.5 Nuclear Spectral Function and the Nucleon Momentum Distribution

Let us now consider semi-inclusive electron scattering where, in addition to detecting the scattered electron, one also has some other particle detected in coincidence. Such processes are denoted $A(e,e'x)$ meaning that the scattered electron e' and particle x are detected (typically) in two spectrometers. Particle x may be anything that can be knocked out of the nucleus or produced at the kinematics under consideration. For instance, with increasing excitation energy one may produce pions, knock out several nucleons, etc. In the following we specialize to the situation where a single nucleon is knocked out of the nucleus via the $(e,e'p)$ or $(e,e'n)$ reactions.

Therefore, let us begin this discussion by looking at the kinematics of the $(e,e'N)$ reaction, with $N = p$ or n. Referring to Fig. 16.6, we have as usual the four-momentum transfer $q^\mu = (\omega,\mathbf{q})$, with three-momentum transfer $q = |\mathbf{q}|$, energy transfer ω and electron scattering angle θ_e. In the laboratory frame the target nucleus has four-momentum

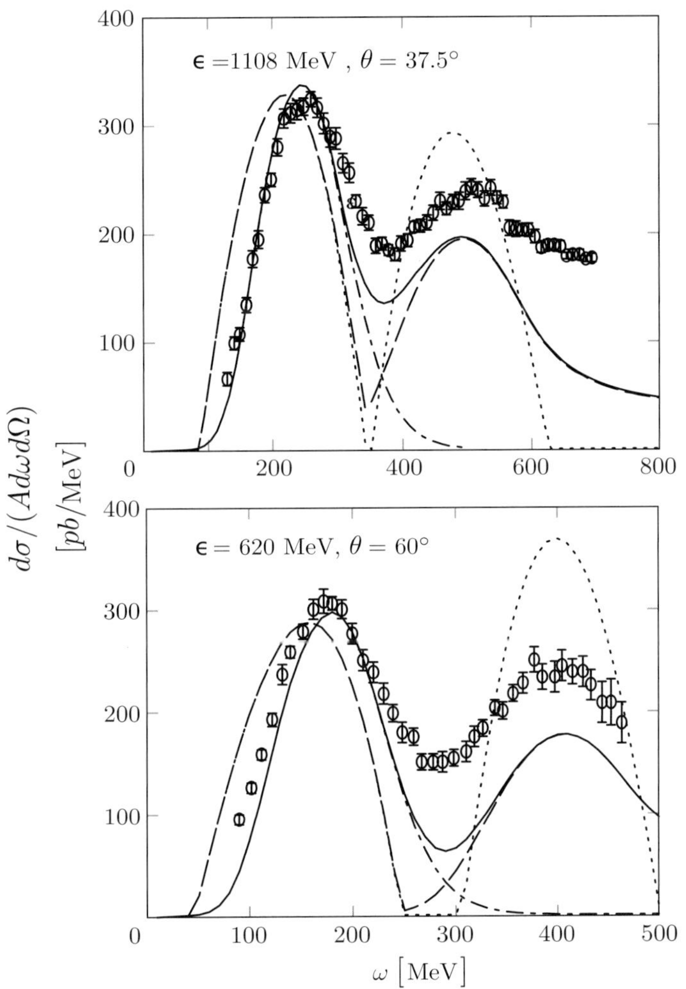

Fig. 16.5 Comparisons of models with data [Day15] for electron scattering from ^{12}C: left-hand dashed curves (RFG QE), dotted curves (RFG Δ), dash-dotted curves (SuSA QE), right-hand dashed curves (SuSA Δ), and solid curves (SuSA total), as discussed in the text.

$p_A^\mu = (M_A^0, 0)$. The nucleon detected in coincidence with the electron has four-momentum $p_N^\mu = (E_N, \mathbf{p}_N)$, with $E_N = \sqrt{m_N^2 + p_N^2}$. The three-vector momentum $\mathbf{p}_N$ involves the azimuthal angle ϕ_N between the plane in which the electrons lie and that containing the momentum transfer and outgoing nucleon momentum, plus two variables specifying the remaining kinematics of the outgoing nucleon: the three-momentum of the nucleon $p_N = |\mathbf{p}_N|$ or its energy $E_N = \sqrt{m_N^2 + p_N^2}$ and its polar angle θ_N, the angle between $\mathbf{q}$ and $\mathbf{p}_N$. Then four-momentum conservation fixes

$$p_{A-1}^\mu = (E_{A-1}, \mathbf{p}_{A-1}) = q^\mu + p_A^\mu - p_N^\mu \,, \tag{16.32}$$

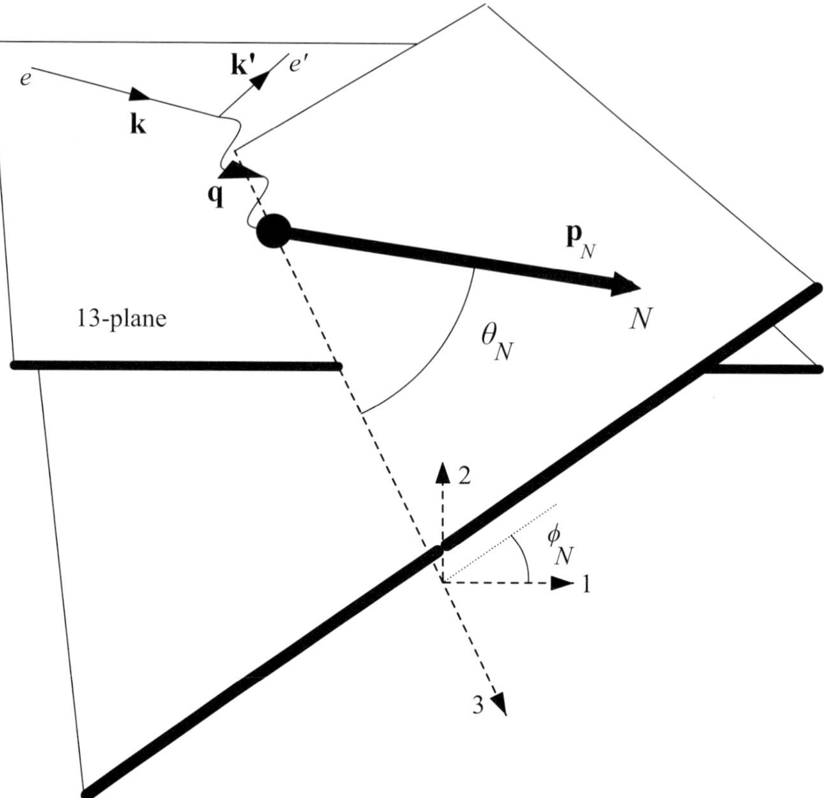

Fig. 16.6   Semi-inclusive $(e, e'N)$ electron scattering. The electron scattering plane defines the coordinate system with the z-axis along the three-momentum transfer $\mathbf{q}$, the y-axis normal to the lepton plane and the x-axis forming the orthogonal coordinate system, as shown. The nucleon is assumed to be detected in coincidence with the scattered electron. The three-momentum for this, $\mathbf{p}_N$, lies in the plane defined by polar angle θ_N and azimuthal angle ϕ_N, also as shown.

and so one can use q, ω (but see later) and θ_e together with ϕ_N; but instead of p_N (or E_N) and θ_N, alternatively one may use the missing-momentum $\mathbf{p}_m = \mathbf{p}_{A-1} \equiv -\mathbf{p}$, specifically its magnitude $p = |\mathbf{p}| = |\mathbf{p}_N - \mathbf{q}|$, and a variable to characterize the degree of excitation of the residual system; for the latter we use

$$\mathcal{E}(p) \equiv \sqrt{(M_{A-1})^2 + p^2} - \sqrt{(M^0_{A-1})^2 + p^2} \,, \qquad (16.33)$$

where M_{A-1} is the (in general) excited recoiling system mass, while M^0_{A-1} is that system's mass when in its ground state. The target mass is denoted M^0_A and the separation energy relates three of the masses via $E_S \equiv M^0_{A-1} + m_N - M^0_A$. The variable $\mathcal{E}$ is essentially the familiar missing-energy minus the separation energy, $\mathcal{E} \simeq E_m - E_S$. One has that $\mathcal{E} \geq 0$ and takes on the value zero when the daughter nucleus is left in its ground state.

Energy conservation requires that

$$
\begin{aligned}
M_A^0 + \omega &= E_N + E_{A-1} \\
&= \sqrt{m_N^2 + p_N^2} + E_{A-1}^0 + \mathcal{E} \\
&= \sqrt{m_N^2 + (\mathbf{q} + \mathbf{p})^2} + \sqrt{(M_{A-1}^0)^2 + p^2} + \mathcal{E} \,,
\end{aligned}
\tag{16.34}
$$

which is equivalent to

$$
\begin{aligned}
\mathcal{E} &= M_A^0 + \omega - \sqrt{(M_{A-1}^0)^2 + p^2} - \sqrt{m_N^2 + q^2 + p^2 + 2pq\cos\theta} \\
&= \mathcal{E}(q,\omega;p,\cos\theta) \,,
\end{aligned}
\tag{16.35}
$$

where θ is the angle between $\mathbf{p}$ and $\mathbf{q}$. Alternatively, we can write $\cos\theta$ as a function of $(q,\omega;p,\mathcal{E})$ using Eq. (16.35).

In summary so far, we have q, ω, θ_e, ϕ_N, p, and $\mathcal{E}$ as independent variables with which to describe coincidence nucleon knockout. This development is completely general; no approximations such as PWIA (see below) are made here. Of course, there are constraints on the ranges over which these quantities are physical. Specifically, one has $\mathcal{E} \geq 0$; for $\mathcal{E} = 0$ the lowest value of the missing momentum that can be reached is $p_{\min} = |y|$, where

$$
\begin{aligned}
y = \frac{1}{2W^2} \Bigg\{ &-q\left[W^2 + \left(M_{A-1}^0\right)^2 - m_N^2 \right] \\
&+ \left(M_A^0 + \omega\right)\sqrt{W^2 - \left(M_{A-1}^0 + m_N\right)^2}\sqrt{W^2 - \left(M_{A-1}^0 - m_N\right)^2} \Bigg\}
\end{aligned}
\tag{16.36}
$$

with as usual

$$
W = \sqrt{\left(M_A^0 + \omega\right)^2 - q^2} \,.
\tag{16.37}
$$

The maximum attainable value of $p = p_{\max} \equiv Y$ is given by the above with $-q \to q$, which leads to the kinematic restrictions shown in Figs. 16.7 and 16.8. If ω lies below the QE peak, then one finds the situation in Fig. 16.7 with $y < 0$ and where the physical region lies at $0 \leq \mathcal{E}(p) \leq \mathcal{E}_M(p)$ and hence at $-y \leq p \leq Y$. The variable $\mathcal{E}_M(p)$ is obtained from Eq. (16.35) by setting $\theta = 180°$, which yields

$$
\mathcal{E}_M = M_A^0 + \omega - \sqrt{(M_{A-1}^0)^2 + p^2} - \sqrt{m_N^2 + (q-p)^2} \,.
\tag{16.38}
$$

On the other hand, if ω is above the QE peak, then Fig. 16.8 pertains, $y > 0$ and the physical region lies at $\mathcal{E}_M(-p) \leq \mathcal{E}(p) \leq \mathcal{E}_M(p)$ for $0 \leq p \leq +y$ and $0 \leq \mathcal{E}(p) \leq \mathcal{E}_M(p)$ for $+y \leq p \leq Y$.

Note that if one neglects the knockout of clusters from the nucleus (deuterons, α-particles, fission fragments, etc.), which are expected to be small contributions at high energies and momenta, then the total cross section may be expected to be dominated by knockout of protons and neutrons. What complicates matters is the fact that one believes that knockout of more than one nucleon plays a not-insignificant role. That is, when one considers say $(e, e'p)$ reactions, the final nuclear state is not just a residual daughter nucleus, but can (*must* in some circumstances) involve the emission of other nucleons beyond the one presumed to be detected. Nevertheless, if nucleon knockout is dominant, then one can

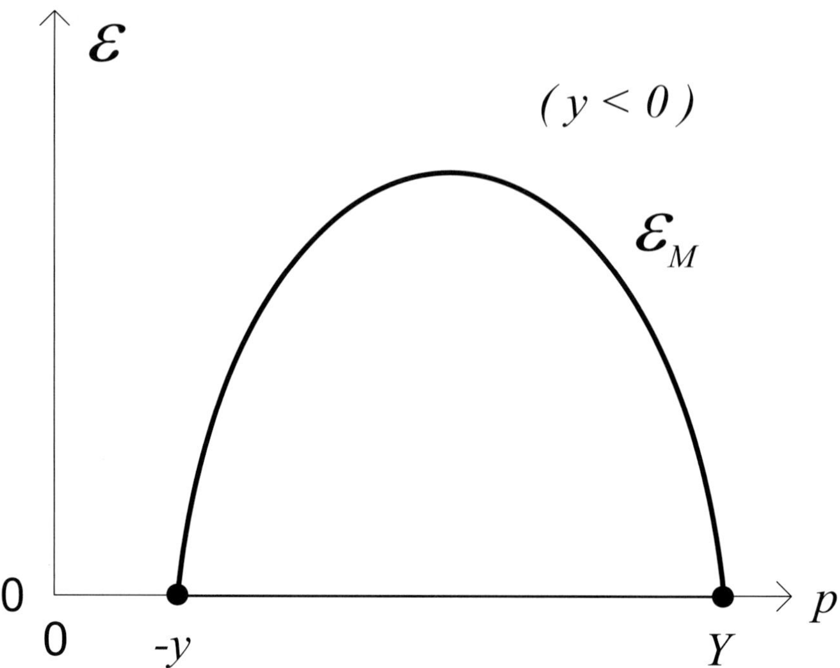

Fig. 16.7 The allowed kinematical region in the $(\mathcal{E},p)$-plane in situations where $y < 0$, i.e., for excitation energies below the QE peak. The region is bounded by the heavy lines, namely, where $0 \leq \mathcal{E} \leq \mathcal{E}_M(p)$.

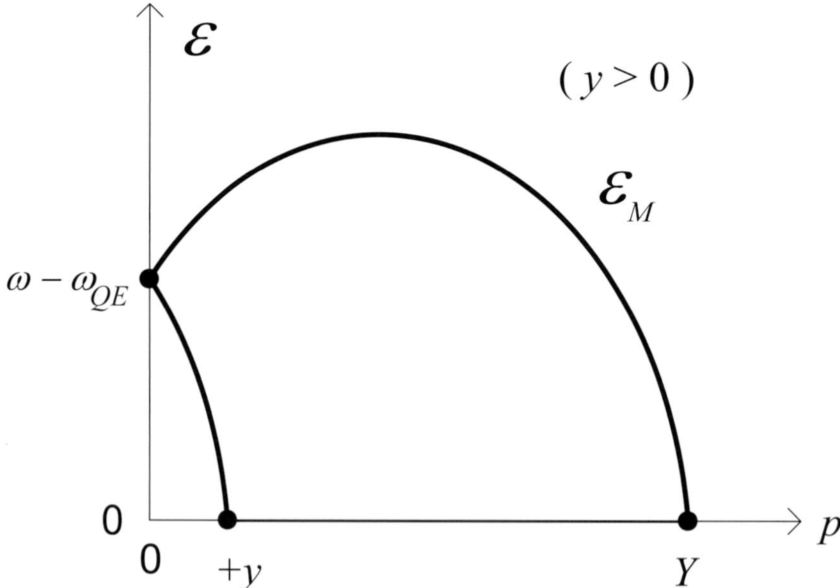

Fig. 16.8 As for Fig. 16.7, but now for $y > 0$, i.e., above the QE peak. The region is bounded by the heavy lines, namely, where $\mathcal{E}_M(-p) \leq \mathcal{E} \leq \mathcal{E}_M(+p)$ for $0 \leq p \leq y$ and where $0 \leq \mathcal{E} \leq \mathcal{E}_M(+p)$ for $y \leq p$.

say that the inclusive cross section arises primarily from the sum of the integral over the $(e, e'p)$ cross section plus the integral over the $(e, e'n)$ cross section minus the integral over the $(e, e'pn)$ cross section to eliminate double-counting. The integrals range over the physically allowed regions shown in Figs. 16.7 and 16.8.

The general formalism for semi-inclusive electron scattering from nuclei including electron and target polarization, and with determinations of the polarizations of some outgoing particle x has been presented in [Ras89], namely, for reactions such as $A(\vec{e}, e'x)B$, $\vec{A}(e, e'x)B$, $A(e, e'\vec{x})B$, $\vec{A}(\vec{e}, e'x)B$, $A(\vec{e}, e'\vec{x})B$, and so on. For studies with polarized electrons, but without other polarizations, the semi-inclusive $A(e, e'x)B$ and $A(\vec{e}, e'x)B$ reactions have been cast in their general form using the basic symmetry properties of the leptonic and hadronic tensors involved, as discussed in [Don85]. In fact, in this cited work, the formalism for even more complicated situations, such as $A(e, e'xy)B$ and $A(\vec{e}, e'xy)B$ reactions, is also developed. Such discussion goes beyond the scope of this book. In the present context we simply draw on the cited work for the most basic case of unpolarized semi-inclusive scattering. Note also that in some of the discussions in Chapters 8, 9, 11, and 12 we have already employed some of these ideas where polarizations are involved.

The $A(e, e'x)$ reaction can be written

$$\frac{d\Sigma}{d\epsilon' d\Omega_e dp_x d\Omega_x} = \Sigma_0 \cdot \mathcal{R} \,, \tag{16.39}$$

where Σ_0 is the elementary cross section including the appropriate kinematic factors (see [Ras89]). The general response function $\mathcal{R}$ may be written in the form

$$\mathcal{R} = v_L R_L + v_T R_T + v_{TL} R_{TL} \cos \phi_x + v_{TT} R_{TT} \cos 2\phi_x \,, \tag{16.40}$$

where the kinematic factors ("Rosenbluth-like" factors) v_K, $K = L, T, TL, TT$ were developed in Chapter 7. In Eq. (16.40) all of the dependence on the electron scattering angle θ_e and the out-of-plane angle ϕ_x (see Fig. 16.6) is explicit, and accordingly, the response functions R_K, $K = L, T, TL, TT$ depend only on the remaining dynamical variables, namely, on $(q, \omega, p_x, \theta_x)$. Clearly the general problem is much more complicated than what we found for inclusive electron scattering. In general one must have a model for the ground state of the nucleus, for the (in general off-shell) electromagnetic operators including meson-exchange current contributions, and for the final-state system. This complete analysis has yet to be accomplished as well as one might like and typically rather radical simplifications are invoked; we continue below with the most basic of these approaches to help the reader get some feeling for where the issues lie.

To help clarify the issues involved in semi-inclusive nucleon knockout reactions, it is useful to consider a simplified model which, while not completely quantitative, does capture the basic ideas that are central in studies of this type. Specifically, one often makes the approximation that the outgoing nucleon that is detected in coincidence with the scattered electron is a plane wave. If one invokes the PWIA for the reaction (see Fig. 16.9); then a nucleon of energy

$$E(p, \mathcal{E}) = M_A^0 - \sqrt{(M_{A-1}^0)^2 + p^2} - \mathcal{E} \tag{16.41}$$

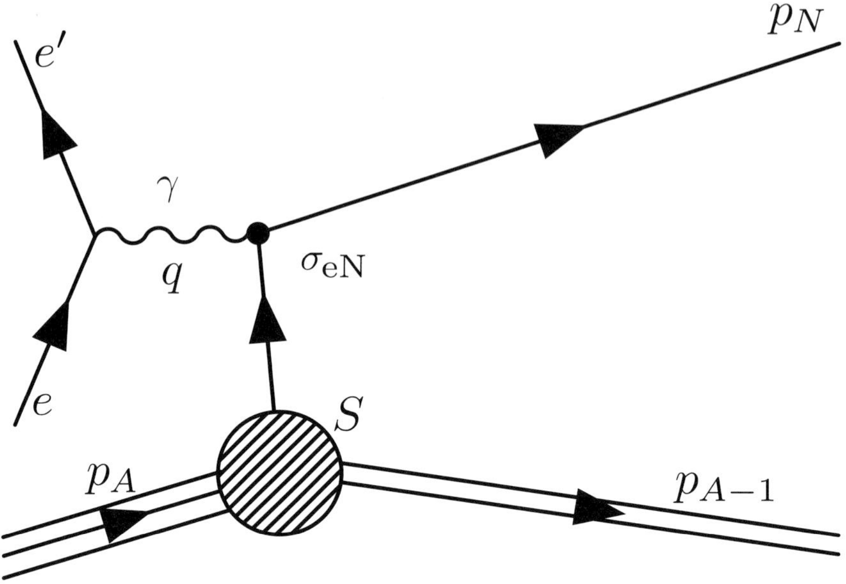

Fig. 16.9 The plane-wave impulse approximation (see text for details).

and momentum p is struck by the virtual photon and is ejected from the nucleus as a plane-wave (on-shell) with energy E_N and momentum p_N (see Fig. 16.7). The kinematics of the reaction require the struck nucleon to be off-shell; that is, $E \neq \bar{E}$, where $\bar{E} \equiv (m_N^2 + p^2)^{1/2}$.

This assumption dramatically simplifies the description and leads to a factorized approximation for the (unpolarized) semi-inclusive cross section involving the product of an off-shell single-nucleon cross section $\sigma_{eN}(q, \omega; p, \mathcal{E})$ times the so-called spectral function $S(p_m, E_m)$:

$$\frac{d^6\sigma}{dk'_e d\Omega'_e dp_N d\Omega_N} = \frac{p_N m_N M^0_{A-1}}{\sqrt{(M^0_{A-1})^2 + p^2}} \sigma_{eN}(q, \omega; p, \mathcal{E}) S(p_m, E_m) \,, \tag{16.42}$$

where $N = p$ or n. The terminology and prescriptions for the off-shell single-nucleon cross section may be found in [Def69]. One should remember that the missing energy and momentum variables are simply related to p and $\mathcal{E}$ and, accordingly, both are used in the expressions given here. The spectral function, $S(p_m, E_m)$ is the joint probability of finding a nucleon of momentum p_m in the nucleus and of reaching final states in the daughter nucleus specified by missing energy E_m. In general, this is a complicated function of the two variables in the (E_m, p_m)-plane. However, for the extreme independent-particle shell model one has a tractable form:

$$S(p_m, E_m) = \sum_\alpha N_\alpha |\phi_\alpha(p_m)|^2 \delta(M_A + \epsilon_\alpha - E_m) \,, \tag{16.43}$$

where $\phi_\alpha(p_m)$ is the single-particle wavefunction in momentum space for the state with single-particle quantum numbers α and associated energy eigenvalue ϵ_α, and N_α is the so-called spectroscopic factor which includes the occupancy of the single-particle level in the nuclear many-body wavefunction for the initial nucleus. When integrating only over a small region of missing energy around the energy eigenvalue ϵ_α, one has

$$n_\alpha(p_m) = \int_{M_A+\epsilon_\alpha-\epsilon}^{M_A+\epsilon_\alpha+\epsilon} S(p_m, E_m)dE_m = N_\alpha|\phi_\alpha(p_m)|^2 , \tag{16.44}$$

namely, a result proportional to the momentum distribution for the specific single-particle state labeled α; we return later to discuss this in more detail. Before continuing, we note that, while the PWIA provides a good starting point for discussions of the spectral function and momentum distributions, it is only an approximation and must be extended in practical applications. First, there are no final-state interactions (FSI) included in the PWIA and these are typically approximated by extending the analysis to the so-called distorted-wave impulse approximation (DWIA) where some aspects of FSI are taken into account by using distorted waves for the outgoing nucleon. The distorted waves are obtained by invoking an optical potential that makes contact with nucleon scattering from nuclei at similar kinematics to those required for the $(e, e'N)$ reaction. This is, of course, an approximation and does not fully take into account, for instance, coupled-channel effects; the latter are very difficult to incorporate, especially if relativistic modeling is demanded. Moreover, the very basic assumption of factorization in Eq. (16.42) is also an approximation which is known to be violated to some degree in certain kinematical regions. Finally, the EM interaction of the virtual photon with off-shell nucleons in the nucleus – they *must* be off-shell – is treated only in a phenomenological way. Despite these qualifications, the approach being presented here has proven to be very useful in probing the energy–momentum distributions of nucleons in the nucleus.

In Figs. 16.10 and 16.11 some typical results are shown for the spectral function for the reaction $^{208}\mathrm{Pb}(e, e'p)^{207}\mathrm{Tl}$.

Let us conclude with a brief discussion of how *inclusive* electron scattering can provide useful information on the total momentum distribution $n(k)$. We have seen above that, using *semi-inclusive* quasielastic scattering $A(e, e'p)$, the spectral function $S(E_m, p_m)$ can be determined, at least for some range of missing energy and missing momentum. If one could determine $S(E_m, p_m)$ for all values of the missing energy, then the momentum distribution could be found:

$$n(k) = \int_{E_{\min}}^{\infty} S(k, E_m)dE_m . \tag{16.45}$$

However, in fact, semi-inclusive electron scattering is not actually capable of providing the spectral function for missing energies above some finite value, namely, the maximum, $\mathcal{E}_M(p)$, given in Eq. (16.38); recall that $E_m \simeq \mathcal{E} + E_S$. Indeed, even when $q \to \infty$ this function approaches a finite limit [Don99a]. Thus, the application of Eq. (16.45) depends on the spectral function having little strength beyond this finite maximum value of E_m. Assuming convergence of the above integral, Fig. 16.12 shows the extracted momentum distributions for $^2\mathrm{H}$ (where the above is not a restriction), $^3\mathrm{He}$, $^4\mathrm{He}$, $^{12}\mathrm{C}$, $^{56}\mathrm{Fe}$, obtained from both inclusive and exclusive quasielastic electron scattering as well as an extrapolation to nuclear matter.

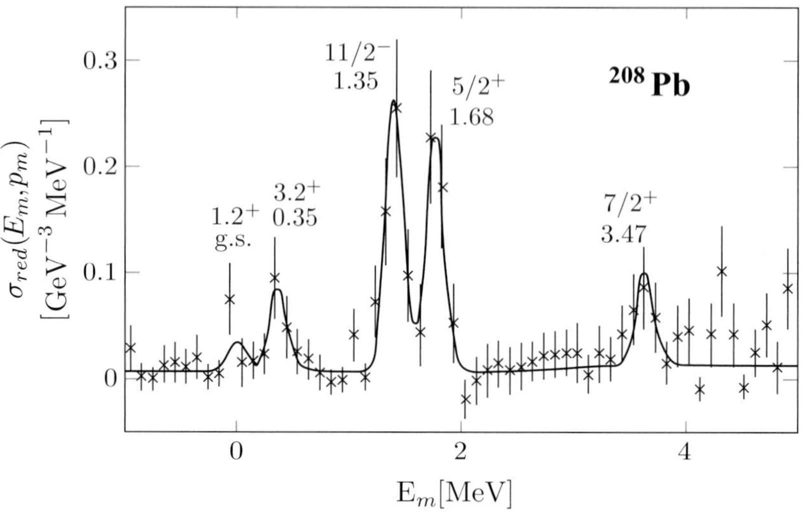

Fig. 16.10 The reduced cross section of the ^{208}Pb$(e, e'p)$ reaction, defined as the sixfold differential cross section, divided by the off-shell electron–proton cross section, and by the appropriate kinematic factor, at an average missing momentum of 340 MeV c^{-1}. One observes the knockout of valence protons to discrete states in ^{207}Tl, labeled by their spin, parity, and excitation energy. The solid curve is the result of a fit to the spectrum; figure adapted from [Bob94].

We note in passing that, upon integrating the semi-inclusive cross sections for $(e, e'p)$ and $(e, e'n)$ reactions, adding the integrals together and eliminating double-counting via subtraction of the $(e, e'pn)$ cross section, one can obtain an alternative approximation for the inclusive quasielastic cross section. The integration regions are those shown in Figs. 16.7 and 16.8. In fact, the variable y discussed above and indicated in the figures presents a different choice for the scaling variable. One has $\psi' \simeq y/k_A$, where k_A is a typical nuclear scale, roughly the Fermi momentum k_F given in Table 16.1. We have now arrived at an alternative form of scaling based on the y-variable defined earlier [Day90, Wes75, Kaw75]. For the present discussion we note only that the y-scaling function in the PWIA for $y < 0$ can be written

$$f(q, y) = k_F \times 2\pi \int_{-y}^{Y} k\, dk \int_{0}^{\mathcal{E}_M} d\mathcal{E}\, S(k, \mathcal{E}) . \tag{16.46}$$

For scaling to be realized fully, one must go to large momentum transfers and, if q is large, then Y can reasonably be taken to be infinite, whereas $\mathcal{E}_M$ goes to its asymptotic value which still depends on y (see Exercise 16.5). But if extending that finite limit to infinity is a good approximation (which it may, or may not be), then the expression on the right-hand side of Eq. (16.46) only depends on y via the lower limit in the momentum integral. Clearly, under these circumstances, taking the derivative $[\partial f(q, y)/\partial y]_q$ will yield the momentum distribution, and hence useful information on $n(k)$ can also be obtained using *inclusive* quasielastic electron scattering [Mon71]. Note that this approach to obtaining the momentum distribution includes both neutrons and protons.

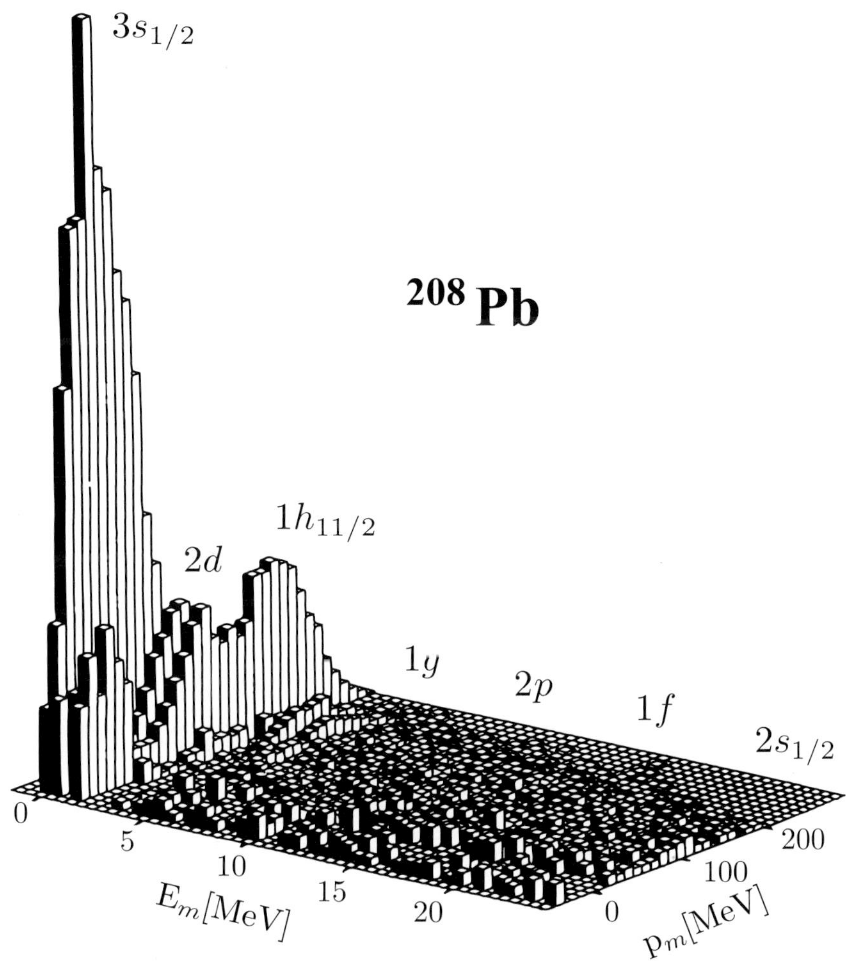

Fig. 16.11 The ^{208}Pb spectral function from [Qui88], but now showing a representation in the missing energy–missing momentum plane (see Fig. 13.4 for the shell-model configurations involved).

The advent of high duty-factor electron accelerators in the 1990s produced high-quality quasielastic $(e, e'p)$ data for targets from deuterium to lead. In this way, the momentum distributions of many single-particle states in nuclei have been studied. Theoretical treatments of initial- and final-state interactions of both electrons and protons have reached a high level of sophistication. Higher beam energies reduce the effects of FSI so that systematic uncertainties of order 5% have been attained. In Fig. 16.13 are shown distorted momentum distributions for ^{16}O, ^{40}Ca, ^{90}Zr, and ^{206}Pb. The k-dependence gives a precise measurement of the momentum distribution of the initially-bound nucleon, while the overall amplitude of the momentum distribution measures accurately the normalization of the quasi-hole state, often called the *spectroscopic factor*. The $(e, e'p)$ reaction is sensitive to the nuclear interior, so these measurements do approach a genuine determination of this quantity. Figure 16.14 shows the summed spectroscopic strength for valence orbitals. The

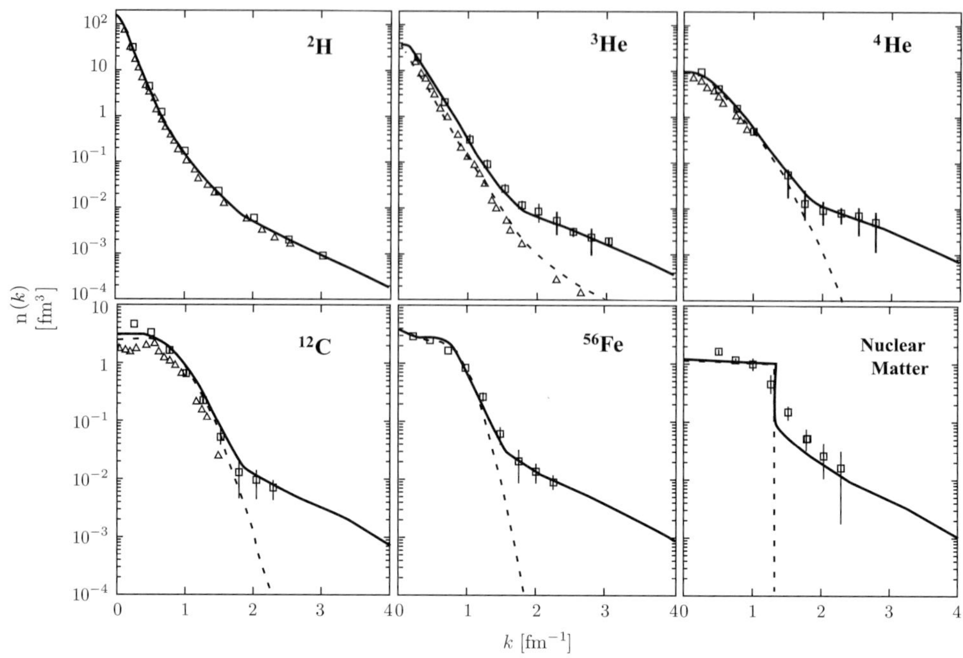

Fig. 16.12 The momentum distributions $n(k)$ versus k of several nuclei and nuclear matter extracted from analysis of inclusive, $A(e, e')X$ (open squares), and semi-inclusive $A(e, e'p)$ (full and open triangles) cross sections; figure adapted from [Att91].

strength is integrated over a small energy level (typically up to 20 MeV). The average value for the closed shell nuclei studied is about 65%, which is appreciably lower than the 90% occupation for valence orbitals predicted by Hartree–Fock and RPA. The missing strength is attributed to the effect of short-range correlations (SRC), which lead to a population of states above k_F and a depopulation of states below. As a consequence, the nucleon momentum distribution $n(k)$ attains a high-momentum tail that extends far above k_F (see Fig. 16.12). In the spectral function, the correlated strength at high k only appears at high missing energy E_m. The character of the SRC can be understood in terms of the basic characteristics of the nucleon–nucleon interaction described in Chapter 11, namely, the short-range repulsion and contributions of intermediate- to long-range tensor character. These induce strong spatial-spin-isospin NN correlations, as we have seen previously. For example, the two-nucleon density distributions in states with pairs having spin $S = 1$ and isospin $T = 0$ are very small at short distances and exhibit strong anisotropies depending on the spin projection. These correlations impact strongly the momentum distribution $n(k)$ of NN pairs and lead to large differences in the np versus pp momentum distributions [Sch07]. Measurements at Jefferson Laboratory [Sub08] indicate a clear dominance of $^{12}C(e, e'pn)$ events over $^{12}C(e, e'pp)$ events, as expected from the tensor nature of the nucleon–nucleon interaction. The present understanding of the structure of ^{12}C is that it is composed of about 80% mean-field nucleons and 20% SRC pairs, where the latter is composed of about 90% np-SRC pairs and 5% pp and nn SRC pairs each [Hen13]. In closing, we return to mention

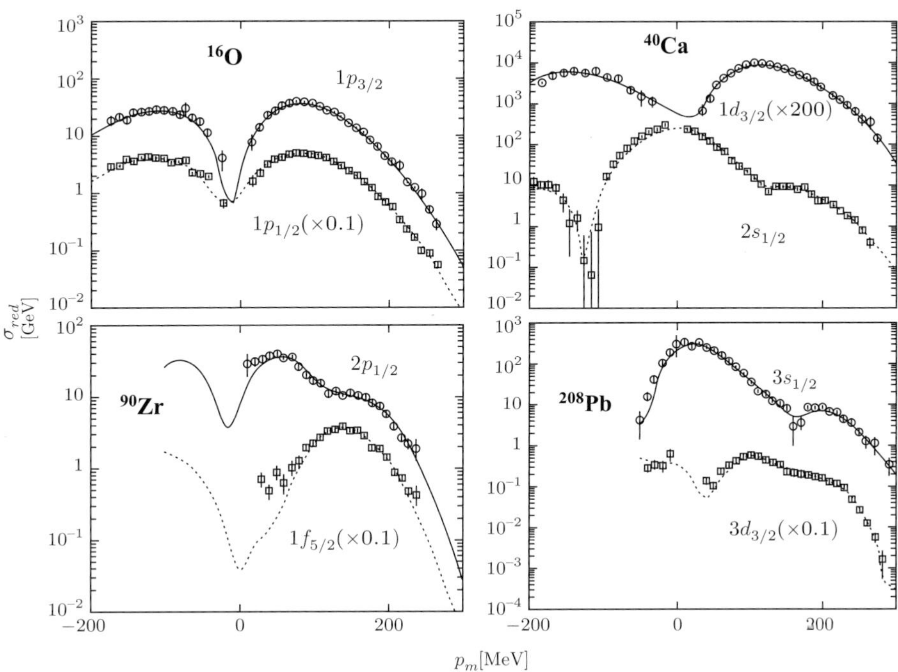

Fig. 16.13　The experimental momentum distributions for transitions in the reaction $(e, e'p)$ on ^{16}O, ^{40}Ca, ^{90}Zr, and ^{208}Pb involving knockout from the valence shell (upper data) and the next deeper (sub)shell (lower data); figure adapted from [Lap93].

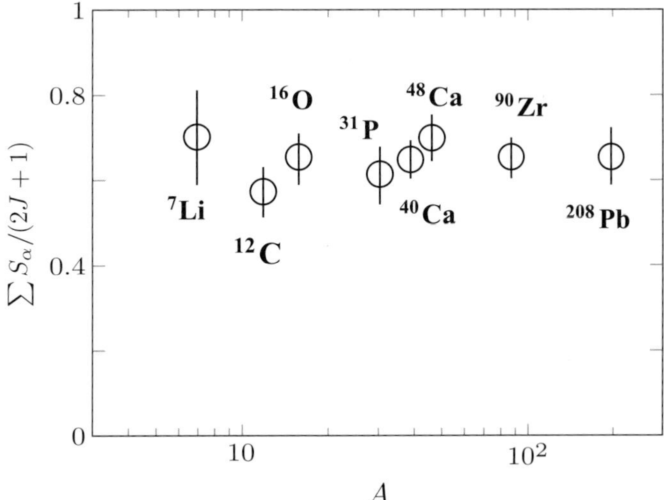

Fig. 16.14　The quasi-particle strength $\sum S_\alpha/(2J+1)$ for proton valence orbitals determined from $(e, e'p)$ experiments versus A from [Lap93].

the EMC effect, described in Sect. 9.8. The convolution model involves an integral over the momentum distribution of the nucleons in the nucleus. Thus, $n(k)$ enters directly and SRC are important. Indeed, analysis of world data shows that the magnitude of the EMC effect in nucleus A is linearly related to the probability that a nucleon in the nucleus is part of a two-nucleon SRC-pair [Wei11]. At this time, there is no universally accepted explanation of this interesting relationship.

We now turn from these chapters whose focus has been placed on issues of nuclear physics, including electron scattering studies over a range of excitation energies spanning topics from elastic scattering to electroexcitation of nuclei at high energies, to the weak interaction in Chapters 17 and 18.

Exercises

16.1 The Relativistic Fermi Gas

The basic definition of the RFG model is given by Eq. (16.12),

$$W_{\text{RFG}}^{\mu\nu}(q,\omega) = \frac{3m_N^2}{4\pi k_F^3} \int \frac{d\mathbf{p}}{E(\mathbf{p})E(\mathbf{p}+\mathbf{q})} \delta[\omega - (E(\mathbf{p}+\mathbf{q}) - E(\mathbf{p}))]$$
$$\times \theta\,(k_F - |\mathbf{p}|)\,\theta(|\mathbf{p}+\mathbf{q}| - k_F)\widetilde{W}_{\text{nucleon}}^{\mu\nu}(p+q,p) \,,$$

with the on-shell single-nucleon EM responses given by

$$W_{\text{nucleon}}^{\mu\nu}(p+q,p) = -W_1(\tau)\left[g^{\mu\nu} + \frac{q^\mu q^\nu}{Q^2}\right] + W_2(\tau)V^\mu V^\nu \,,$$

where

$$V^\mu \equiv \frac{1}{m_N}\left[p^\mu + \left(\frac{p\cdot q}{Q^2}\right)q^\mu\right] \,.$$

a) Substituting the latter into the former, show that

$$R_{L,T}(q,\omega) = \frac{3}{4m_N\kappa\eta_F^3}\,(\epsilon_F - \Gamma)\,\theta\,(\epsilon_F - \Gamma)$$
$$\times \begin{cases} \frac{\kappa^2}{\tau}\left[\widetilde{G}_E^2(\tau) + \widetilde{W}_2(\tau)\Delta\right] & \text{for } L \\[2mm] 2\tau\widetilde{G}_M^2(\tau) + \widetilde{W}_2(\tau)\Delta & \text{for } T \end{cases}$$

results both when one has Pauli blocking and when Pauli blocking is absent.

b) Perform the steps outlined in Section 16.3 to prove that the universal scaling functions

$$f_L^{\text{RFG}}(\psi) = f_T^{\text{RFG}}(\psi) \equiv f^{\text{RFG}}(\psi) = \frac{3}{4}(1 - \psi^2)\theta(1 - \psi^2)$$

emerge when Pauli blocking is absent.

16.2   **The RFG Model for Excitations Involving Baryon Resonances**

In Section 16.4 electroexcitation of the Δ in the RFG model of nuclei was summarized. Generalize this discussion for any baryon resonance X of given mass M_X.

a) Show that the inelastic scaling variable, the extension of

$$\psi_\Delta(\kappa,\lambda) = \sqrt{\frac{1}{\xi_F}\left[\kappa\sqrt{\rho_\Delta^2 + \frac{1}{\tau}} - (1 + \rho_\Delta\lambda)\right]} \times \begin{cases} +1 & \lambda \geq \lambda_0^\Delta \\ -1 & \lambda < \lambda_0^\Delta \end{cases} ,$$

naturally emerges from considerations of the kinematics.

b) Prove that, in the non-Pauli-blocked region.

$$R_{L,T}(q,\omega) = \frac{3\xi_F}{4m_N\kappa\eta_F^3}\left(1 - \psi_\Delta^2\right)\theta\left(1 - \psi_\Delta^2\right)$$

$$\times \begin{cases} \frac{\kappa^2}{\tau}\left[(1 + \rho_\Delta^2\tau)\widetilde{W}_2^\Delta(\tau) - \widetilde{W}_1^\Delta(\tau) + \widetilde{W}_2^\Delta(\tau)\mathcal{D}(\kappa,\lambda)\right] & \text{for } L \\ 2\widetilde{W}_1^\Delta(\tau) + \widetilde{W}_2^\Delta(\tau)\mathcal{D}(\kappa,\lambda) & \text{for } T \end{cases} ,$$

and generalize $(\Delta \rightarrow X)$ to yield the RFG model of quasielastic baryon production.

16.3   **Quasielastic Scattering and the Width of the QE Peak**

Consider inclusive quasielastic electron scattering from a nucleus in the Fermi gas model. The Fermi momentum of the nucleons, each of mass m_N, is $\mathbf{k}_F$ and the energy and three-momentum of the virtual photon are ν and $\mathbf{q}$, respectively. Show that the width of the quasielastic peak is given by

$$\Delta E' \approx \frac{2|\mathbf{q}| \cdot |\mathbf{k}_F|}{m_N} ,$$

assuming the proton is nonrelativistic, i.e., that the nonrelativistic Fermi gas model (NRFG) is being employed. Compare with what happens when the RFG model discussed in the text is used. Note that experimental data agree with the RFG at large values of q, but not with the NRFG. Typically at what momentum transfer do the two models differ significantly, say by about 20%?

16.4   **Semi-Inclusive Electron Scattering**

Consider the kinematic boundaries for $(e, e'N)$ reactions discussed in Section 16.5.

a) Verify the boundaries for energy transfers below the QE peak.

b) Verify the boundaries for energies above the peak.

c) Determine where the peak occurs and obtain a relationship between this result and the peak position found in the RFG model.

16.5   **Inclusive Quasielastic Scattering: the Scaling Function and the Spectral Function**

In Section 16.5 the scaling function was written in terms of the spectral function assuming the validity of the PWIA, namely, as

$$f(q,y) = k_F \times 2\pi \int_{-y}^{Y} k\,dk \int_0^{\mathcal{E}_M} d\mathcal{E}\, S(k, \mathcal{E}) .$$

Consider the large-q limit on that expression, the regime where scaling is supposed to be applicable.

a) By applying the Leibnitz theorem take the derivative with respect to y for fixed q.
b) Assume that q becomes very large compared with y and provide expressions for Y and $\mathcal{E}_M$. Obtain the $q \to \infty$ limit for the latter and show that this tends to a finite limit. What does this imply for the derivative obtained in part a)?

16.6 **General Characteristics of $(e, e'x)$ Reactions**

In Chapter 7, the general form for inclusive electron scattering was obtained by building the hadronic tensor in terms of the available four-momenta. Extend that analysis to semi-inclusive $(e, e'x)$ reactions, where now another four-momentum comes into play, namely, p_x^μ.

a) Build the most general symmetric and antisymmetric hadron tensors.
b) Invoke the properties of the polar-vector EM current to eliminate some of the contributions found in Part a).
c) Impose the continuity equation to further limit the possibilities. How many terms survive?
d) Contract the result above with the general (polarized as well as unpolarized) leptonic tensor obtained in Chapter 7. In particular, show that

$$\mathcal{R}_{fi} = v_L R_{fi}^L + v_T R_{fi}^T + v_{TL} R_{fi}^{TL} \cos \phi_x + v_{TT} R_{fi}^{TT} \cos 2\phi_x$$

results when the electrons are unpolarized. What happens when the electron beam is linearly polarized?

16.7 **The Real-Photon Limit of Semi-Inclusive Electron Scattering: $(\vec{\gamma}, N)$ Reactions**

In Chapter 7 the basics for studies of unpolarized inclusive electron scattering from unpolarized targets were introduced, while in this chapter some extensions to unpolarized, but semi-inclusive electron scattering were considered. In this exercise, the real-photon limit for (γ, x) reactions, those where some outgoing particle x is presumed to be detected, is taken as the limit of semi-inclusive $(e, e'x)$ reactions where the electron scattering angle goes to zero and hence where $Q^2 \to 0$. While this could be done directly using a Feynman diagram with one external (on-shell) photon line, it can also be obtained as the limit of electron scattering. As seen in Eq. (16.40) the general form for (unpolarized) semi-inclusive electron scattering has a response that goes as

$$\mathcal{R}_{fi} = v_L R_{fi}^L + v_T R_{fi}^T + v_{TL} R_{fi}^{TL} \cos \phi_x + v_{TT} R_{fi}^{TT} \cos 2\phi_x \ .$$

For unpolarized real-γ reactions we have already seen that only the transverse response R^T survives (see Eqs. (7.102) and (7.103)); i.e., real-γ reactions have purely transverse photons, while in electron scattering, where virtual photons enter, all four terms (L, T, TL, and TT) occur. For *polarized* real-γ reactions the L and TL terms can be shown to be absent, although now both transverse contributions, T and TT, play a role. The point of this exercise is to explore what happens in this more general situation.

We begin by focusing on the two combinations of the matrix elements of the transverse projections of the current that can occur (see Eqs. (7.65) and (7.67)),

$$W_{fi}^{T} = \left|J_{fi}(+1)\right|^2 + \left|J_{fi}(-1)\right|^2 \equiv R_{fi}^{T}$$
$$W_{fi}^{TT} = 2\mathrm{Re}\left(J_{fi}(+1)^* J_{fi}(-1)\right) \equiv R_{fi}^{TT}\cos\phi_x \,,$$

where in the TT case the explicit dependence on ϕ_x has been extracted (the T case has no ϕ_x dependence), as in the text. Noting that from Eqs. (7.69) and (7.71) one has

$$v_T = \frac{1}{2}\rho + \tan^2\theta_e/2$$
$$v_{TT} = -\frac{1}{2}\rho \,,$$

where $\rho \equiv Q^2/q^2$, one then has for the transverse parts of the response

$$\mathcal{R}_{fi}^{\text{trans}} = v_T R_{fi}^{T} + v_{TT} R_{fi}^{TT}\cos\phi_x$$
$$\equiv v_T R^{\text{real photons}} \,,$$

and thus for the effective response for real-γ kinematics,

$$R_{fi}^{\text{real photons}} = R_{fi}^{T}(q=\omega) + \frac{v_{TT}}{v_T}R_{fi}^{TT}(q=\omega)\cos\phi_x \,,$$

showing that for real photons in situations where both T and TT terms contribute one must generalize R^T to $R^{\text{real photons}}$.

a) Assuming the ERL and taking the $Q^2 \to 0$ limit show that

$$\frac{v_{TT}}{v_T} \to -\left[\frac{2\epsilon\epsilon'}{\epsilon^2 + \epsilon'^2}\right] .$$

For $\epsilon \gg \omega$, as required when keeping ω fixed while taking the limit as the scattering angle goes to zero (see Eq. (7.5) and Exercise 7.7), this becomes -1, and hence one has

$$R_{fi}^{\text{real photons}} = R_{fi}^{T} - R_{fi}^{TT}\cos\phi_x \,.$$

b) Recalling the definitions of the helicity projections of the transverse current matrix elements (see the developments in Section 7.3), obtain expressions for the Cartesian projections (x and y, corresponding to 1 and 2, respectively, using the conventional coordinate system), and hence rewrite R^T and R^{TT} using the latter basis.

c) In Eq. (7.92) the general multipole expansion involving transverse projections of the current were given. Assuming that only $E1$ and $M1$ multipoles are important, derive expressions for J_{fi}^{x} and J_{fi}^{y}.

d) Finally, consider transitions from a 0^+ ground state to excited states with spin-parity 1^- ($E1$) or 1^+ ($M1$). Write the matrix elements that occur above in terms of reduced matrix elements of the $E1$ and $M1$ multipole operators. For *linearly polarized* real photons use the expressions derived above to determine which multipoles enter and what ϕ_x-dependence occurs.

We note the relevance of these developments: for real-γ facilities such as the HIγS at Duke University where an intense source of polarized real photons exists many very interesting measurements can be undertaken; here the ability to separate contributions of $E1$ from $M1$ contributions in photoexcitation provides the prime motivation.

Beta Decay

17.1 Introduction

Before getting into the primary material of this chapter, nuclear beta decay, and its use as a possible probe for beyond the standard model physics, we review a few basic tenets of the weak interaction.

We have seen that the electromagnetic interaction can be written in terms of the coupling of a gauge boson, the photon, with the electromagnetic current, written in terms of light quarks, i.e., u, d:

$$J_\mu^{em} = \frac{2}{3}\bar{u}\gamma_\mu u - \frac{1}{3}\bar{d}\gamma_\mu d = T_3 + \frac{Y}{2} .$$ (17.1)

As in Chapter 2, we define the isospinor,

$$\psi = \begin{pmatrix} u \\ d \end{pmatrix} ,$$

so that $Y = \frac{1}{3}\bar{\psi}\gamma_\mu\psi$ is the hypercharge operator and identifies the isoscalar term of the current, while $T_3 = \frac{1}{2}\bar{\psi}\gamma_\mu\tau_3\psi$ is the third component of the isospin operator and labels the isovector term above. Likewise the weak interaction can be succinctly written in terms of currents interacting with gauge bosons. In this case the charged weak current, in the light quark sector, has the form

$$J_\mu^\pm = \bar{\psi}_L\gamma_\mu\tau_\pm\psi_L = \bar{\psi}\gamma_\mu(1-\gamma_5)\frac{1}{2}\tau_\pm\psi ,$$ (17.2)

while the neutral weak current can be written as

$$J_\mu^Z = \frac{1}{4}\bar{\psi}\gamma_\mu(1-\gamma_5)\tau_3\psi - \sin^2\theta_w J_\mu^{em} .$$ (17.3)

Clearly, the vector component of the charge raising/lowering weak current forms an isotriplet together with the isovector piece of the electromagnetic current and is (neglecting the small isospin violation) conserved. This proposition, due to Feynman and Gell-Mann [Fey58], is called the conserved vector current hypothesis (CVC) and has important consequences for hadronic weak interactions. Also due to Feynman and Gell-Mann is the proposal that the charged weak current is purely left-handed, $J_\mu^\pm = V_\mu^\pm - A_\mu^\pm$. This feature can also be carefully tested experimentally, as we shall discuss.

Of course, an important difference between the electromagnetic and weak interaction is the fact that the electromagnetic gauge boson, the photon, is massless, while the $W^\pm$, Z^0 bosons mediating the weak interaction are (quite) massive

$$M_W^\pm = 80.385 \pm 0.015 \text{ GeV}, \qquad M_Z = 91.188 \pm 0.002 \text{ GeV} ,$$

meaning that the photon propagator

$$\frac{-ig_{\mu\nu}}{q^2}$$

is replaced by

$$\frac{-i(g_{\mu\nu} - \frac{1}{M^2}q_\mu q_\nu)}{q^2 - M^2} .$$

We can see the consequence of this replacement by noting that in the nonrelativisitic limit the Born approximation gives the transition amplitude in terms of the Fourier transform of the potential,

$$T(\boldsymbol{p}_f - \boldsymbol{p}_i) = <f|T|i> \simeq <f|V|i> = \int d\mathbf{r}\, e^{i(\boldsymbol{p}_f - \boldsymbol{p}_i)\cdot \boldsymbol{r}} V(\boldsymbol{r}) . \tag{17.4}$$

We can determine the potential then by taking the inverse Fourier transform of the transition amplitude

$$V(\boldsymbol{r}) = \int \frac{d\mathbf{q}}{(2\pi)^3} e^{-i\boldsymbol{q}\cdot\boldsymbol{r}} T(\boldsymbol{q}) . \tag{17.5}$$

In the nonrelativistic limit, wherein the idea of a potential makes sense, the energy transfer is much smaller than the corresponding momentum transfer,

$$q_0 = E_f - E_i \simeq \frac{p_f^2}{2M} - \frac{p_i^2}{2M} = \frac{(\boldsymbol{p}_f + \boldsymbol{p}_i)}{2M} \cdot (\boldsymbol{p}_f - \boldsymbol{p}_i) \simeq |\boldsymbol{q}| \times \mathcal{O}\left(\frac{v}{c}\right) . \tag{17.6}$$

Hence, in the case of the electromagnetic interaction the potential has its familiar Coulomb form

$$V_{em}(\boldsymbol{r}) = -\int \frac{d\mathbf{q}}{(2\pi)^3} e^{-i\boldsymbol{q}\cdot\boldsymbol{r}} \frac{e_1 e_2}{q_0^2 - \boldsymbol{q}^2} \simeq \int \frac{d\mathbf{q}}{(2\pi)^3} e^{-i\boldsymbol{q}\cdot\boldsymbol{r}} \frac{e_1 e_2}{\boldsymbol{q}^2} = \frac{e_1 e_2}{4\pi |\boldsymbol{r}|} . \tag{17.7}$$

Here the long range of the potential is due to the fact that the photon is massless. On the other hand, in the case of the weak interaction, the gauge bosons are *very* massive. The corresponding potential has a Yukawa form

$$V_{wk}(\boldsymbol{r}) \sim -\int \frac{d\mathbf{q}}{(2\pi)^3} e^{-i\boldsymbol{q}\cdot\boldsymbol{r}} \frac{g_1 g_2}{q_0^2 - \boldsymbol{q}^2 - M^2} \simeq \int \frac{d\mathbf{q}}{(2\pi)^3} e^{-i\boldsymbol{q}\cdot\boldsymbol{r}} \frac{g_1 g_2}{\boldsymbol{q}^2 + M^2} = \frac{g_1 g_2 e^{-M|\boldsymbol{r}|}}{4\pi |\boldsymbol{r}|}$$
$$\tag{17.8}$$

and is extremely short-ranged. Because of the exponential, the weak potential effectively vanishes beyond a distance $r \sim 1/M \sim 2 \times 10^{-3}$ fm.

Another consequence of this coupling to very massive gauge bosons is that, for low-energy weak interactions such as nuclear beta decay, wherein $\boldsymbol{q}^2 << M^2$, the propagator effectively becomes $\sim ig_{\mu\nu}/M^2$. Then, writing the fundamental weak hadronic current as

$$\frac{g_w}{2\sqrt{2}} V_{ud}\bar{u}\gamma_\mu(1 - \gamma_5)d ,$$

the basic charged-current semileptonic weak interaction assumes its traditional low-energy form

$$\mathcal{H}_w \simeq \frac{g_w^2}{8M_W^2} V_{ud}\bar{u}\gamma_\mu(1 - \gamma_5)d\, \bar{e}\gamma^\mu(1 - \gamma_5)\nu_e . \tag{17.9}$$

The combination $g_w^2/8M_W^2$ is usually designated by the symbol $G_F/\sqrt{2}$, where G_F is the Fermi coupling constant and has the dimensions of inverse energy squared: $G_F \sim 10^{-5}\,\mathrm{GeV}^{-2}$.

Muon Decay

In order to determine G_F, we require an experiment involving weak decay. The simplest such reaction is one involving only leptons, so that the complications introduced by the presence of strong interactions can be avoided. Consider then muon decay, $\mu^- \to e^- \bar{\nu}_e \nu_\mu$, for which the decay amplitude becomes

$$\mathcal{M} = \frac{G_F}{\sqrt{2}} \bar{u}_e(p_4)\gamma_\alpha(1-\gamma_5)v_{\nu_e}(p_2)\bar{u}_{\nu_\mu}(p_3)\gamma^\alpha(1-\gamma_5)u_\mu(p_1) . \tag{17.10}$$

Since the muon is so much heavier than the electron, $m_\mu \sim 200\, m_e$, we can neglect the electron mass. The sum over spins then becomes

$$\frac{1}{2}\sum_{s_i}|\mathcal{M}|^2 = 64G_F^2 p_1 \cdot p_2 p_3 \cdot p_4 \tag{17.11}$$

and the decay rate can be found from Fermi's golden rule

$$\begin{aligned}
d\Gamma_\mu = \;& \frac{1}{2m_\mu}\int \frac{d\mathbf{p}_2}{(2\pi)^3 2E_2}\frac{d\mathbf{p}_3}{(2\pi)^3 2E_3}\frac{d\mathbf{p}_4}{(2\pi)^3 2E_4} \\
& \times (2\pi)^4\delta^4(p_1 - p_2 - p_3 - p_4)64G_F^2 p_1 \cdot p_2 p_3 \cdot p_4 .
\end{aligned} \tag{17.12}$$

Performing the integration over the neutrino variables, we determine

$$\frac{d\Gamma_\mu}{dE_e} = \frac{64G_F^2 m_\mu^2 E_e^2}{(4\pi)^3}\left(1 - \frac{4E_e}{3m_\mu}\right), \tag{17.13}$$

which provides an excellent fit to the experimental muon decay spectrum.

The total decay rate is then

$$\Gamma_\mu = \frac{64G_F^2 m_\mu^2}{(4\pi)^3}\int_0^{\frac{1}{2}m_\mu} dE_e E_e^2\left(1 - \frac{4E_e}{3m_\mu}\right) = \frac{G_F^2 m_\mu^5}{192\pi^3}. \tag{17.14}$$

When augmented by a small radiative correction term, this result can be compared with the experimental muon lifetime in order to extract the value of the Fermi coupling

$$G_F = (1.1663787 \pm 0.0000006)\times 10^{-5}\,\mathrm{GeV}^{-2}. \tag{17.15}$$

Neutron Decay

The same technique can be used in order to analyze the decay of the neutron

$$n \to p + e^- + \bar{\nu}_e .$$

However, there is an important difference. Since the energy release in this decay, $m_n - m_p \sim 1.3\,\mathrm{MeV}$, is comparable to the electron rest mass, it is not appropriate to neglect m_e. If we take

$$< p|J_\mu^+|n >= \bar{u}_p(p_3)\gamma_\mu(1 - \gamma_5)u_n(p_1) \,, \tag{17.16}$$

we have

$$\frac{1}{2}\sum_{s_i}|\mathcal{M}|^2 = 64G_F^2 p_1 \cdot p_2 p_3 \cdot p_4 \tag{17.17}$$

as before and

$$d\Gamma_n = \frac{1}{2m_n}\int \frac{d\mathbf{p}_2}{(2\pi)^3 2E_2}\frac{d\mathbf{p}_3}{(2\pi)^3 2E_3}\frac{d\mathbf{p}_4}{(2\pi)^3 2E_4}(2\pi)^4\delta^4(p_1 - p_2 - p_3 - p_4)$$

$$\times 64G_F^2 p_1 \cdot p_2 p_3 \cdot p_4 \equiv \frac{32G_F^2}{(4\pi)^3}J(E_e)dE_e \,, \tag{17.18}$$

where

$$J(E_e) = \frac{1}{2}(m_n^2 - m_p^2 - m_e^2)(E_+^2 - E_-^2) - \frac{2}{3}m_n(E_+^3 - E_-^3) \tag{17.19}$$

with

$$E_\pm = \frac{\frac{1}{2}(m_n^2 - m_p^2 - m_e^2) - m_n E_e}{m_n - E_e \mp p_e}. \tag{17.20}$$

Expanding to lowest-order in the small quantity $(m_n - m_p)/m_n$, we find

$$J(E_e) \simeq 4E_e p_e[(m_n - m_p) - E_e]^2 \,. \tag{17.21}$$

Thus the electron spectrum becomes

$$\frac{d\Gamma_n}{dE_e} = \frac{G_F^2}{\pi}E_e\sqrt{E_e^2 - m_e^2}[(m_n - m_p) - E_e]^2 \tag{17.22}$$

and, defining $r = (m_n - m_p)/m_e$, the total decay rate is

$$\Gamma_n = \frac{G_F^2 m_e^5}{4\pi^3}\left[\frac{1}{15}(2r^4 - 9r^2 - 8)\sqrt{r^2 - 1} + r\log(r + \sqrt{r^2 - 1})\right] \,. \tag{17.23}$$

Putting in numbers, we find $\tau_n = 1/\Gamma_n = 1316\,\text{sec}$, while experimentally $\tau_n^{\text{exp}} = 880.3 \pm 1.1\,\text{sec}$.

What went wrong here? Aside from small effects, such as our neglect of electromagnetic effects, there are two primary causes. The first is that, due to quark mixing, the normalization of the leptonic weak interaction, which is responsible for muon decay and the semileptonic weak interaction, which determines neutron decay, differ by the CKM matrix element $V_{ud} \simeq 0.97$. However, more important is that, since the neutron and proton are strongly- interacting particles, the replacement Eq. (17.16) is naive. Rather, we should write

$$< p|J_\mu^+|n >= \bar{p}(p_3)\gamma_\mu(g_V - g_A\gamma_5)n(p_1) \,. \tag{17.24}$$

Now for the vector current, the close relationship to its conserved electromagnetic sibling means that the normalization in going from quark to hadron is unchanged, i.e., $g_V(q^2 = 0) = 1$. However, this is *not* true for the axial-vector current, since $\partial_\mu A^\mu \neq 0$, and we must allow for a different normalization, $g_A \neq 1$, and it is up to experiment to determine the size of the renormalized value. Repeating the above analysis, but now with this modification, we find

$$\Gamma_n = \frac{G_F^2 V_{ud}^2 m_e^5}{16\pi^3}\left[\frac{1}{15}\left(2r^4 - 9r^2 - 8\right)\sqrt{r^2-1} + r\log(r + \sqrt{r^2-1})\right]\left(g_V^2 + 3g_A^2\right),$$

(17.25)

which yields

$$g_V^2 + 3g_A^2 \simeq 6.10, \quad \text{i.e.,} \quad g_A \approx 1.3 .$$

(17.26)

Careful inclusion of Coulomb and other effects leads to the presently accepted value [PDG14]

$$g_A = -1.2723 \pm 0.0023 .$$

(17.27)

Pion and Kaon Decay

A third important application of this formalism involves the simplest $(P_{\ell2})$ semileptonic decay of the charged pion and kaon,

$$\pi^+ \to \mu^+ \nu_\mu,\ e^+ \nu_e .$$

$$K^+ \to \mu^+ \nu_\mu,\ e^+ \nu_e .$$

In this case the quarks involved in the weak interaction are both bound in the decaying particle. The form of the weak decay amplitude is then

$$\mathcal{M} = \frac{G_F}{\sqrt{2}} V_{ud} \bar{u}_{\nu_\ell}(p_2)\gamma^\mu(1 - \gamma_5)v_\ell(p_3) < 0|J_\mu^-|P_{p_1}^+ > .$$

(17.28)

From spin-parity considerations, only the axial-vector current can contribute and, since the only four-vector available is the four-momentum p_1, we must have

$$< 0|A_\mu^-|P_{p_1}^+ > \equiv \sqrt{2}F_P p_{1\mu} ,$$

(17.29)

where F_P is the pseudoscalar-decay constant. Calculating the decay rate, we find [Hol90a]

$$\Gamma_P = \frac{G_F^2 F_P^2}{256\pi m_P^3} m_\ell^2 (m_P^2 - m_\ell^2)^2$$

(17.30)

and, comparing with the experimental numbers (and including small electromagnetic effects), we find

$$F_\pi = (92.2 \pm 0.3)\ \text{MeV} \simeq 0.83 F_K,$$

(17.31)

which is an important number in the chiral symmetry analysis performed in Chapter 6.

Focusing first on the pion, an interesting feature is the $e^+ \nu_e$ decay, which has a far larger phase space than its muonic analog $\mu^+ \nu_\mu$, and therefore might naively be expected to represent the dominant mode. However, calculation reveals a very strong suppression

$$\left(\frac{\Gamma(\pi^+ \to e^+ \nu_e)}{\Gamma(\pi^- \to \mu^+ \nu_\mu)}\right)^{\text{theo}} = \frac{m_e^2(m_\pi^2 - m_e^2)^2}{m_\mu^2(m_\pi^2 - m_\mu^2)^2} = 1.28 \times 10^{-4} .$$

(17.32)

Electromagnetic corrections modify this prediction to $(1.2352 \pm 0.0001) \times 10^{-4}$ [Cir12] which are in excellent agreement with the experimental number [PDG14]

$$\left(\frac{\Gamma(\pi^+ \to e^+ \nu_e)}{\Gamma(\pi^= \to \mu^+ \nu_\mu)} \right)^{\text{exp}} = (1.230 \pm 0.004) \times 10^{-4} . \tag{17.33}$$

The origin of this suppression of the electron mode has to do with the purely left-handed structure of the weak current. If the positron were truly massless then it would have to be purely right-handed, while the corresponding neutrino emitted in the decay would have to be purely left-handed. Since the two leptons fly off in opposite directions, this configuration carries unit spin along the decay direction and cannot result from the disintegration of a spinless pion. The excellent agreement between experimental and theoretical ratios for the electron and muon decay modes of the pion strongly confirms the left-handed structure of the weak interaction proposed by Feynman and Gell-Mann [Fey58].

A similar situation obtains for the kaon. In this case the suppression factor

$$\left(\frac{\Gamma(Ki^+ \to e^+ \nu_e)}{\Gamma(K^- \to \mu^+ \nu_\mu)} \right)^{\text{theo}} = \frac{m_e^2 (m_K^2 - m_e^2)^2}{m_\mu^2 (m_K^2 - m_\mu^2)^2} = 2.60 \times 10^{-5} . \tag{17.34}$$

Again including electromagnetic and other corrections, we find a prediction $(2.477 \pm 0.001) \times 10^{-5}$ [Cir12], which agres well with the experimental value

$$\left(\frac{\Gamma(K^+ \to e^+ \nu_e)}{\Gamma(K^+ \to \mu^+ \nu_\mu)} \right)^{\text{exp}} = (2.49 \pm 0.01) \times 10^{-5} . \tag{17.35}$$

Of course, what is actually determined in these $\ell \nu_\ell$ decay mode experiments is the product of the psedoscalar coupling constant F_P and the relevant CKM element V_{ij}. By taking a ratio of the pion and kaon $\ell \nu_\ell$ decay rates one determines

$$\frac{V_{us} F_K}{V_{ud} F_\pi} = 0.2758 \pm 0.0005 \tag{17.36}$$

then, using a lattice gauge theory evaluation

$$\frac{F_K}{F_\pi}^{\text{lattice}} = 1.193 \pm 0.006 , \tag{17.37}$$

we obtain the ratio of CKM elements

$$\frac{V_{us}}{V_{ud}} = 0.2312 \pm 0.0013. \tag{17.38}$$

When combined with the value $V_{ud} = 0.97425 \pm 0.00022$ obtained from the analysis of Fermi decays. as described in Section 17.4 we determine

$$V_{us} = 0.2253 \pm 0.0009 . \tag{17.39}$$

An alternative approach to this determinationof V_{us} comes from analysis of the $(K_{\ell 3})$ decay modes $K \to \pi \ell \nu_\ell$, wherein there no longer exists helicity suppression so that the electron decay mode now dominates over its muonic analog. One now requires the polar vector matrix element

$$< \pi_{p_2} |V_\mu| K_{p_1} > = f_+(q^2)(p_1 + p_2)_\mu + f_-(q^2) q_\mu \tag{17.40}$$

where $q = p_1 - p_2$ is the four-momentum transfeer. In the case of the dominant electron decay mode the form factor $f_-(q^2)$ can be neglected, since the identity

$$q^\mu \bar{v}_e(p_e)\gamma_\mu(1 - \gamma_5)u_{v_e}(p_v) >= m_e \bar{v}_e(p_e)(1 - \gamma_5)u_{v_e}(p_v) \tag{17.41}$$

means that any f_- contribution must be accompanied by the tiny electron mass. In the case of the form factor $f_+(q^2)$ $SU(3)$ symmetry requires

$$f_+^{K^+\pi^0}(q^2 = 0) = \sqrt{2}f_+^{K^0\pi^-}(q^2 = 0) = 1 \tag{17.42}$$

and it can be shown that any corrections to Eq. (17.42) must be of second order in $SU(3)$ breaking and are therefore small. Lattice gauge theory estimates predict

$$f_+(0) = 0.959 \pm 0.005 \tag{17.43}$$

which can be used, together with the measured branching ratio for $K_{\ell 3}$ decays to produce a value

$$V_{us} = 0.2254 \pm 0.0013 \,, \tag{17.44}$$

which is seen to be in good agreement with the number obtained in Eq. (17.39).

Similar analyses, augmented by "heavy quark" techniques, can be used to study D and B meson channels and to glean information about CKM elements involving c or b quarks. However, we shall end our discussion here, and with this material in hand, we now transition to our main focus, which is nuclear beta decay and its use as a probe for physics beyond the Standard Model.

17.2 Nuclear Beta Decay

Nuclear beta decay is one of the primary decay mechanisms involving nuclei and has three forms

  i) electron emission: $(Z, N) \to (Z + 1, N - 1) + e^- + \bar{v}_e$.
 ii) positron emission: $(Z, N) \to (Z - 1, N + 1) + e^+ + v_e$.
iii) electron capture: $(Z, N) + e^- \to (Z - 1, N + 1) + v_e$.

Electron capture is the process by which an atomic electron is captured by the nucleus. If the capture occurs from the K-shell then it is referred to as K-capture, etc. If $M(Z, N)$ designates the atomic mass, then we see that the energy released in electron emission or capture is

$$E_0^{\beta^-} = M(Z, N) - M(Z + 1, N - 1) \,,$$
$$E_0^{ec} = M(Z, N) - M(Z - 1, N + 1)$$

while that for positron decay is

$$E_0^{\beta^+} = M(Z, N) - M(Z - 1, N + 1) + 2m_e \,.$$

Comparing E_0^{ec} and $E_0^{\beta^+}$, we note that there exists a range of atomic mass differences for which electron capture can occur, but positron emission cannot.

The energy releases in a nuclear β-decay process range from tens of keV, as in the case of tritium decay, to as much as 18 MeV, as in the decay of ^{8}B. In either case the wavelength of the beta particle is much larger than the size of the nucleus and the effects of finite nuclear size are minimal, i.e., the variation of the wavefunction over the nuclear volume is tiny. Also small and usually negligible is the recoil energy

$$E_R = \frac{(\boldsymbol{p}_e + \boldsymbol{p}_v)^2}{2m_N A} \leq \frac{E_0^2}{2M_N A}$$

of the daughter nucleus. Indeed for a typical energy release of ~ 1 MeV the recoil energy for a nucleus with $A = 100$ would be less than 15 eV. However, it is important to retain such effects in some high-precision tests of β-decay, such as when studying electron–neutrino correlations, as described later. Coulomb effects are also important and must be included. This is done by replacing the electron wavefunction at the origin, which is unity for a plane wave, by its value in the presence of the Coulomb field due to the nucleus. In the nonrelativistic limit, this has the familiar Sommerfeld form

$$F(Z, E_e) = \left| \frac{2\pi \eta}{1 - \exp(-2\pi \eta)} \right| \quad \text{with} \quad \eta = \pm \frac{Z\alpha E_e}{p_e} , \tag{17.45}$$

where $F(Z, E_e)$ is the Fermi function. The form actually used by Fermi is somewhat different. Since the relativistic density for a point nucleus diverges at $r = 0$, he used the value of the density at the nuclear surface

$$F(Z, E_e) = 2(2p_e R)^{2(s-1)} \frac{1+s}{s^2 + \eta^2} \left| \frac{\exp(\pi \eta)}{2) \frac{\Gamma(s+1+i\eta)}{\Gamma(2s+1)}} \right| \quad \text{with} \quad s^2 = 1 - Z^2\alpha^2. \tag{17.46}$$

The decay probability is obtained using Fermi's golden rule

$$d\Gamma_\beta \simeq \int \frac{d\boldsymbol{p}_e}{(2\pi)^3} \frac{d\boldsymbol{p}_v}{(2\pi)^3} |\mathcal{H}_{fi}|^2 2\pi \delta(E_0 - E_e - E_v) . \tag{17.47}$$

Writing

$$|\mathcal{H}_{fi}|^2 \simeq F(Z, E_e) \frac{G_F^2}{2} |V_{ud}|^2 |M_{fi}|^2, \tag{17.48}$$

where $|M_{fi}|^2$ is dimensionless and $\mathcal{O}(1)$, we find

$$d\Gamma_\beta = F(Z, E_e) \frac{G_F^2 |V_{ud}|^2}{2\pi^3} |M_{fi}|^2 p_e E_e (E_0 - E_e)^2 dE_e , \tag{17.49}$$

The classic method by which to study β-decay spectra is via the Kurie plot

$$K(E_e) = \left[\frac{\frac{d\Gamma_\beta}{dE_e}}{F(Z, E_e) E_e p_e} \right]^{\frac{1}{2}} . \tag{17.50}$$

Where $d\Gamma_\beta / dE_e$ is the number of beta decays as a function of the electron energy, i.e., the differential decay rate. Figure 17.1 shows the measured beta spectrum from ^{187}Re decay. If the weak matrix element $\mathcal{M}_{fi}$ is energy-independent, then $K(E_e) \propto E_0 - E_e$, i.e., one should find a linear plot versus E_e intersecting the electron energy axis at the end point energy E_0.

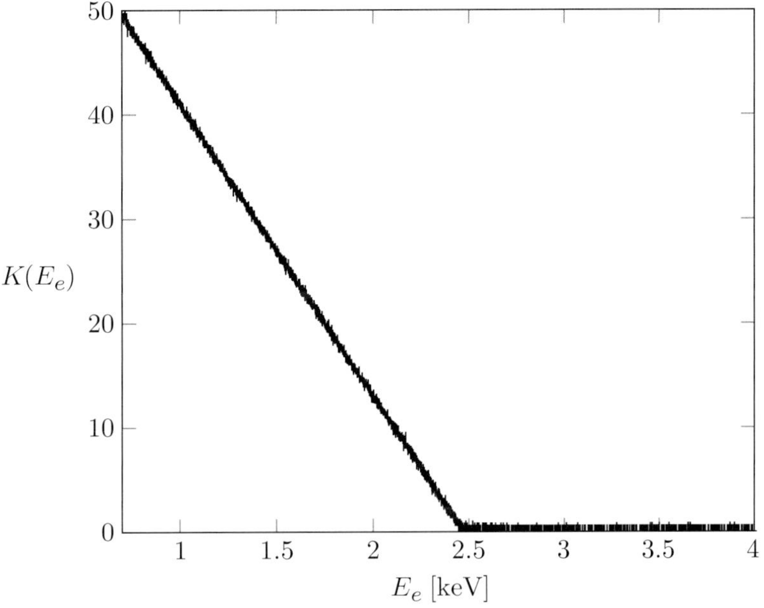

Fig. 17.1 Kurie plot showing the measured beta spectrum from ^{187}Re decay (from the MARE collaboration) [Sis04]. The spectrum is normalized to the Fermi distribution of Eq. (17.50).

Deviations from this simple straight line shape would indicate energy dependence of the weak matrix element, or nonzero neutrino mass, etc.. The total decay rate is given by

$$\Gamma_\beta = \frac{G_F^2 |V_{ud}|^2}{2\pi^3} |M_{fi}|^2 f(Z, R, E_0) , \tag{17.51}$$

with

$$f(Z, R, E_0) = \int_{m_e}^{E_0} dE_e \, \rho(Z, R, E_e)(E_0 - E_e)^2 E_e p_e . \tag{17.52}$$

In the absence of the Coulomb field, the integration can be performed exactly

$$f(0, R, E_0) = \sqrt{E_0^2 - m_e^2} \left[\frac{E_0^4}{30} - \frac{3E_0^2 m_e^2}{20} - \frac{2m_e^4}{15} \right]$$

$$+ \frac{E_0 m_e^4}{4} \log \left(\frac{E_0}{m_e} + \sqrt{\frac{E_0^2}{m_e^2} - 1} \right) \xrightarrow{E_0 \gg m_e} \frac{E_0^5}{30} . \tag{17.53}$$

Thus, the decay rate rises very rapidly with the energy release. In this way it is easy to understand the difference between the very short (2.2×10^{-6} sec) lifetime of the muon versus the relatively long (880 sec) lifetime of the neutron as being primarily due to the difference in the energy release

$$E_0^\mu / E_0^n \sim 75 \quad \text{so} \quad (\tau_n/\tau_\mu)^{\text{kin}} \sim 2 \times 10^9 \quad \text{versus} \quad (\tau_n/\tau_\mu)^{\text{exp}} \sim 4 \times 10^9 .$$

This large kinematic factor makes it difficult to conveniently compare the dynamical features of such decays, so it is common to define the so-called $ft_{\frac{1}{2}}$-value

$$ft_{\frac{1}{2}} = \frac{2\pi^3 \log 2}{G_F^2 |V_{ud}|^2 |M_{fi}|^2} \,, \tag{17.54}$$

where here $t_{\frac{1}{2}}$ is the half-life and is related to the decay rate via

$$t_{\frac{1}{2}} = \frac{\log 2}{\Gamma_\beta} \,.$$

Thus, since f is a calculable function and the half-life is measurable, it becomes possible to determine the dynamical quantities $|M_{fi}|^2$ for each nuclear transition. The results of this analysis can be characterized via

$$\log ft_{\frac{1}{2}} \sim \begin{cases} < 3.5 & \text{superallowed} \\ 5.7 \pm 1.1 & \text{allowed} \\ > 6.8 & \text{forbidden} \,. \end{cases}$$

where here the terms *allowed, superallowed, forbidden* often appear in the literature and so it is important to understand their meaning. An allowed transition is one wherein the outgoing lepton pair carries off zero orbital angular momentum, so that the parity of the initial and final nuclear states must be the same. The term superallowed is used when the initial and final nuclei are members of the same isomultiplet. The term forbidden designates cases in which the lepton pair carries off nonzero angular momentum. When this angular momentum is $\ell = 1, 3, 5, \ldots$ the parity of initial and final states must change, while when this angular momentum is $\ell = 2, 4, 6, \ldots$ there is no parity change but the initial and final spins can differ by values larger than unity. We shall discuss only the case of allowed decays here. Nevertheless there exists a vast literature on forbidden decays and there has been recent work that suggests that inclusion of forbidden transitions is essential to the understanding of reactor neutrino spectra [Gar14]. Forbidden transitions are characterized by weak decays that vanish ($|M_{fi}|^2 = 0$) in lowest-order and will not be further considered here.

The remainder of our discussion will then be focused on allowed and superallowed decays. In order to understand the meaning of these terms, we discuss a bit of historical notation. Assuming that the operators responsible for β-decay are single-particle operators, then in lowest-order (i.e., neglecting momentum factors) they must be constructed from the operators

$$1, \quad \sigma_i, \quad \tau_i \,.$$

Since the operators must transform a neutron into a proton or *vice versa*, the dependence on τ_i is fixed, namely

$$\tau_i^+ : \quad \beta^- \text{ emission} \qquad \tau_i^- : \quad \beta^+ \text{ emission} \,.$$

Because $|M_{fi}|^2$ must be rotationally invariant, the weak matrix element in lowest-order involves two contributions

$$|M_{fi}|^2 = \frac{1}{2J_i + 1} \sum_{m_f,m_i} \left(|g_V|^2 | <f| \sum_k \tau_k^{\pm} |i> |^2 + |g_A|^2 | <f| \sum_k \tau_k^{\pm} \sigma_k |i> |^2 \right)$$

$$\equiv |g_V|^2 |M_F|^2 + |g_A|^2 |M_{GT}|^2 \,, \tag{17.55}$$

where g_V, g_A are constants signifying polar-vector, axial-vector hadronic couplings, respectively and the sum is over all final, initial spin projections m_f, m_i. The weak transition proportional to g_V is called a "Fermi" transition, while that proportional to g_A is called a Gamow–Teller transition. By looking at the associated operators, we see that there exist certain selection rules for both. Those for Fermi transitions are the strictest. Since

$$\sum_k \tau_k^{\pm} = 2T^{\pm},$$

where T is the isospin operator, we see that the weak Fermi matrix element vanishes unless the initial and final states are isotopic analogs, such as the pairs $n - p, {}^3\text{H} - {}^3\text{He}$, etc., in which case

$$| < T, T_3 \pm 1 | T_+ | T, T_3 > |^2 = T(T + 1) - T_3(T_3 + 1) \,. \tag{17.56}$$

The selection rules can then be characterized by $\Delta J = \Delta T = 0$ and $\Delta T_3 = \pm 1$ with no change in parity.

On the other hand, in the case of a Gamow–Teller transition, since the operator

$$\sum_k \tau_k^{\pm} \sigma_k$$

transforms as a vector in both spin *and* isospin space, we see that a nonvanishing matrix element requires $\Delta J = 0, \pm 1 \ (0 \rightarrow 0 \text{ forbidden}), \Delta T = 0, \pm 1 \ (0 \rightarrow 0 \text{ forbidden})$, $\Delta T_3 = \pm 1$.

We then see that in a Fermi transition the electron and neutrino carry off zero total angular momentum, while in the case of a Gamow–Teller transition they carry off unit total angular momentum. When the Gamow–Teller matrix element is nonvanishing, but the Fermi matrix element is zero, the transition is called "allowed," while if the Fermi matrix element is nonvanishing the decay is termed "superallowed." Of course, there are many superallowed decays wherein the Gamow–Teller matrix element is also nonzero. An example is neutron decay, for which

$$|M_{GT}|^2 = \frac{1}{2} \sum_{m_i,m_f=-\frac{1}{2}}^{\frac{1}{2}} \sum_{k=1}^{3} | < m_f |\sigma_k| m_i > |^2 = 3 \,. \tag{17.57}$$

Thus, we find

$$|M_{fi}|^2 \propto |g_V|^2 + 3|g_A|^2 \,, \tag{17.58}$$

as found earlier, cf. Eq. (17.25).

While we could continue discussing the basic phenomenology of such β-decays, it is more useful for our purposes to switch at this point to analyze the ways in which such decays can be used to probe the fundamental structure of the weak interaction. In this way the nucleus becomes a laboratory wherein fine points of weak interaction structure can be precisely examined.

17.3 The Nucleus as a Laboratory

The basic structure of the electroweak interaction has been developed earlier. Here we wish to examine in some detail how nuclear beta decay can be used both to confirm this structure as well as to seek possible Beyond-Standard Model (BSM) effects.[1] Additional ways to probe for possible BSM effects will be sketched in Chapter 21.

When one writes down the effective low-energy charged-current interaction relevant for nuclear beta decay, muon capture, and neutrino scattering

$$\mathcal{H}_w = \frac{G_F}{\sqrt{2}} V_{ud} \bar{u} \gamma_\mu (1 - \gamma_5) d \bar{e} \gamma^\mu (1 - \gamma_5) \nu_e + \text{h.c.} ; \tag{17.59}$$

the form is elegant and concise. However, hidden in this structure are numerous assumptions, each of which is subject to experimental verification. Among these are

i) Pure $V_\mu - A_\mu$ structure, i.e., no scalar, pseudoscalar, or tensor interactions.

ii) Time reversal invariance.

iii) G-parity: Defining $G = C \exp(-i\pi T_2)$ (see also Chapter 2), the weak currents satisfy

$$G(V_\mu - A_\mu)G^{-1} = V_\mu + A_\mu .$$

This requirement is generally called "no second-class currents."

iv) CVC: The weak vector current is related to the electromagnetic current via a simple isospin rotation

$$V_\mu^\pm = \mp[T_\pm, V_\mu^{em}] ,$$

where $T_\pm = T_1 \pm iT_2$ are isospin raising/lowering operators. This condition is termed the "conserved vector current" or CVC hypothesis.

v) PCAC: The axial-vector current would also be conserved were this symmetry not broken spontaneously. However, due to this breaking and because the axial-vector divergence is a pseudoscalar, it can be used as an interpolating field for the pion

$$\partial^\mu A_\mu = F_\pi m_\pi^2 \phi_\pi + \mathcal{O}(\phi_\pi^3) . \tag{17.60}$$

This requirement is called the "partially conserved axial-vector current" or PCAC hypothesis and is closely connected to chiral symmetry (see Chapter 6).

vi) Unitarity: The weak coupling constant $G_F^\beta V_{ud}$ responsible for nuclear beta decay is identical to that in nuclear muon capture and is related to that responsible for muon decay, G_F^μ, via

$$G_F^\beta = G_F^\mu, \tag{17.61}$$

---

[1] Much of the material in this section is taken from [Hol14], where use of the beta decay precision frontier in order to seek possible deviations from the Standard Model is detailed, together with other similar articles.

where V_{ud} is related by unitarity to other mixing angles via

$$|V_{ud}|^2 + |V_{us}|^2 + |V_{ub}|^2 = 1 \, . \tag{17.62}$$

In order to see how such symmetries can be tested, it is useful to restrict our attention to the allowed decays. An example is neutron beta decay, which takes place between two $J^\pi = \frac{1}{2}^+$ states. As is well known, Lorentz invariance and parity considerations prescribe the most general form of the corresponding vector and axial-vector matrix elements to be

$$< p_{p'}|V_\mu|n_p > \ = \ \bar{u}_p(p') \left(\gamma_\mu f_1(q^2) - i\sigma_{\mu\nu} q^\nu \frac{f_2(q^2)}{2m_N} + q_\mu \frac{f_3(q^2)}{2m_N} \right) u_n(p)$$

$$< p_{p'}|A_\mu|n_p > \ = \ \bar{u}_p(p') \left(\gamma_\mu g_1(q^2) - i\sigma_{\mu\nu} q^\nu \frac{g_2(q^2)}{2m_N} + q_\mu \frac{g_3(q^2)}{2m_N} \right) \gamma_5 u_n(p) \, ,$$

$$\tag{17.63}$$

where $q_\mu = p_\mu - p'_\mu$ is the momentum transfer and $m_N = (m_p + m_n)/2$ is the average nucleon mass. Here, f_1, g_1 are the usual vector, axial-vector form factors, whose structure mirrors that of the underlying quark current. The remaining forms are often called "induced," since they arise from the presence of strong interaction effects. The induced structures f_2, g_2 represent weak magnetism, weak electricity couplings, while f_3, g_3 represent induced scalar, pseudoscalar terms. If the strong interactions obey isospin and charge conjugation symmetry, then the forms f_3 and g_2 must vanish, as will be shown in Section 17.5.

In a typical nuclear β-decay, we have $q_0 \leq 10$ MeV. Thus, recoil effects are $\mathcal{O}(q/m_N) \sim 1\%$, and one can employ a form which is valid to $\mathcal{O}(1/m_N)$ for arbitrary allowed $(J^\pm \to J^\pm, (J \pm 1)^\pm)$ transitions [Hol74]:

$$\ell^\mu < \beta|V_\mu|\alpha > = \ \left(a(q^2) \frac{P \cdot \ell}{2M} + e(q^2) \frac{q \cdot \ell}{2M} \right) \delta_{JJ'} \delta_{MM'}$$

$$+ i\frac{b(q^2)}{2M} C^{M'k;M}_{J'1;J} (\boldsymbol{q} \times \boldsymbol{\ell})_k$$

$$+ C^{M'k;M}_{J'2;J} \left[\frac{f(q^2)}{2M} C^{nn';k}_{11;2} \ell_n q_{n'} \right.$$

$$\left. + \frac{g(q^2)}{(2M)^3} P \cdot \ell \sqrt{\frac{4\pi}{5}} Y^k_2(\hat{\boldsymbol{q}}) q^2 + \cdots \right]$$

$$\ell^\mu < \beta|A_\mu|\alpha > = \ C^{M'k;M}_{J'1;J} \epsilon_{ijk} \epsilon_{ij\lambda\eta} \frac{1}{4M} \left[c(q^2) \ell^\lambda P^\eta - d(q^2) \ell^\lambda q^\eta \right.$$

$$\left. + \frac{1}{(2M)^2} h(q^2) q^\lambda P^\eta q \cdot \ell \right]$$

$$+ C^{M'k;M}_{J'2;J} C^{nn';k}_{12;2} \ell_n \sqrt{\frac{4\pi}{5}} Y^{n'}_2(\hat{\boldsymbol{q}}) \frac{q^2}{(2M)^2} j_2(q^2)$$

$$+ C^{M'k;M}_{J'3;J} C^{nn';k}_{12;3} \ell_n \sqrt{\frac{4\pi}{5}} Y^{n'}_2(\hat{\boldsymbol{q}}) \frac{q^2}{(2M)^2} j_3(q^2) + \cdots \tag{17.64}$$

where $P = p_1 + p_2$ and $q = p_1 - p_2$. Here each term corresponds to one in the analogous neutron transition via

$$
\begin{aligned}
a &\rightarrow f_1, & c &\rightarrow g_1 \\
b &\rightarrow f_2, & d &\rightarrow g_2 \\
e &\rightarrow f_3, & h &\rightarrow g_3 .
\end{aligned}
\tag{17.65}
$$

In addition, there exist terms f, g, j_2, j_3, which have no $J = \frac{1}{2} \rightarrow J' = \frac{1}{2}$ analog, since they involve $\Delta J = 2, 3$.

Each of these symmetries has been carefully studied in many experiments and good agreement with the Standard Model predictions is found in each case. For this reason present interest in such processes is often to see if they can be used as a probe for possible BSM physics. The idea here is a simple one. The Standard Model charged-current weak interaction is purely $V - A$. However, BSM interactions are not required to have this form and are presumably mediated by the exchange of as yet unseen heavy bosons. The existence of new high-intensity sources together with the availability of magnetic-optical-trapping (MOT) techniques means that high-precision measurements of beta-decay reactions are not only possible but are at hand. In order to analyze such low-energy processes, we use a general weak contact interaction of the form originally proposed by Lee and Yang [Lee56]

$$
\begin{aligned}
\mathcal{H}_{\text{eff}} = \ & \bar{p}n\bar{e}(C_S + C'_S\gamma_5)\nu_e \\
& + \bar{p}\gamma^\mu n\bar{e}\gamma_\mu(C_V + C'_V\gamma_5)\nu_e \\
& + \frac{1}{2}\bar{p}\sigma^{\mu\nu}n\bar{e}\sigma_{\mu\nu}(C_T + C'_T\gamma_5)\nu_e \\
& + \bar{p}\gamma^\mu\gamma_5 n\bar{e}\gamma_\mu(C_A + C'_A\gamma_5)\nu_e \\
& + \bar{p}\gamma_5 n\bar{e}(C_P + C'_P\gamma_5)\nu_e + h.c.
\end{aligned}
\tag{17.66}
$$

Here the Standard Model values are

$$
C_S = C'_S = C_T = C'_T = C_P = C'_P = 0
\tag{17.67}
$$

and

$$
\frac{G_F^{(0)}}{\sqrt{2}}V_{ud} = C_V = C'_V = \frac{1}{\lambda}C_A = \frac{1}{\lambda}C'_A ,
\tag{17.68}
$$

where $G_F^{(0)} = g^2/8M_W^2$ is the Fermi constant, with g being the $SU(2)_L$ gauge coupling, V_{ud} is the CKM matrix element which connects u, d quarks, and $\lambda = g_A/g_V \simeq 1.27$ is the ratio of axial-vector to vector couplings in neutron beta decay cf. Eq. (17.27).

Recently, a model-independent way has been developed to analyze such processes within the context of effective field theory. In this approach, one defines an effective interaction in terms of the conventional V–A weak interaction accompanied by a complete set of (small) phenomenological four-fermion dimension-six contact terms representing the possible exchange of heavy bosons [Cir13]:

$$
\begin{aligned}
\mathcal{H}_{\text{eff}} = \frac{G_F^{(0)}V_{ud}}{\sqrt{2}} \Big[& (1 + \delta_\beta)\bar{u}\gamma^\mu(1 - \gamma_5)d\bar{e}\gamma_\mu(1 - \gamma_5)\nu_e \\
& + \epsilon_L\bar{u}\gamma^\mu(1 - \gamma_5)d\bar{e}\gamma_\mu(1 - \gamma_5)\nu_e + \tilde{\epsilon}_L\bar{u}\gamma^\mu(1 - \gamma_5)d\bar{e}\gamma_\mu(1 + \gamma_5)\nu_e \\
& + \epsilon_R\bar{u}\gamma^\mu(1 + \gamma_5)d\bar{e}\gamma_\mu(1 - \gamma_5)\nu_e + \tilde{\epsilon}_R\bar{u}\gamma^\mu(1 + \gamma_5)d\bar{e}\gamma_\mu(1 + \gamma_5)\nu_e
\end{aligned}
$$

$$
\begin{aligned}
&+ \ \epsilon_S \bar{u} d \bar{e}(1 - \gamma_5)\nu_e + \tilde{\epsilon}_S \bar{u} d \bar{e}(1 + \gamma_5)\nu_e \\
&- \ \epsilon_P \bar{u}\gamma_5 d \bar{e}(1 - \gamma_5)\nu_e - \tilde{\epsilon}_P \bar{u}\gamma_5 d \bar{e}(1 + \gamma_5)\nu_e \\
&+ \ \epsilon_T \bar{u}\sigma^{\mu\nu}(1 - \gamma_5) d \bar{e}\sigma_{\mu\nu}(1 - \gamma_5)\nu_e \\
&+ \ \tilde{\epsilon}_T \bar{u}\sigma^{\mu\nu}(1 + \gamma_5) d \bar{e}\sigma_{\mu\nu}(1 + \gamma_5)\nu_e \Big] + h.c.
\end{aligned}
\tag{17.69}
$$

Here δ_β is an electromagnetic correction, and $\epsilon_i, \tilde{\epsilon}_i$ represent BSM couplings, which have canonical size

$$
\epsilon_i, \tilde{\epsilon}_i \sim v^2 / \Lambda_{\mathrm{BSM}}^2 \,,
\tag{17.70}
$$

where $v = (2\sqrt{2}G_F^{(0)})^{-\frac{1}{2}} \sim 170\,\mathrm{GeV}$ and Λ_{BSM} is the scale of the new physics. If $\Lambda_{\mathrm{BSM}} \sim 1\,\mathrm{TeV}$, then the natural size might be expected to be $\epsilon_i \sim 10^{-3}$, but this is a matter which must be settled experimentally. The relation between the two effective Hamiltonian bases is

$$
\begin{aligned}
C_i &= \frac{G_F^{(0)} V_{ud}}{\sqrt{2}} \bar{C}_i \\
\bar{C}_V &= g_V(1 + \delta_\beta + \epsilon_L + \epsilon_R + \tilde{\epsilon}_L + \tilde{\epsilon}_R) \\
\bar{C}_V' &= g_V(1 + \delta_\beta + \epsilon_L + \epsilon_R - \tilde{\epsilon}_L - \tilde{\epsilon}_R) \\
\bar{C}_A &= g_A(1 + \delta_\beta + \epsilon_L - \epsilon_R - \tilde{\epsilon}_L + \tilde{\epsilon}_R) \\
\bar{C}_A' &= g_A(1 + \delta_\beta + \epsilon_L - \epsilon_R + \tilde{\epsilon}_L - \tilde{\epsilon}_R) \\
\bar{C}_S &= g_S(\epsilon_S + \tilde{\epsilon}_S) \\
\bar{C}_S' &= g_S(\epsilon_S - \tilde{\epsilon}_S) \\
\bar{C}_P &= g_P(\epsilon_P - \tilde{\epsilon}_P) \\
\bar{C}_P' &= g_P(\epsilon_P + \tilde{\epsilon}_P) \\
\bar{C}_T &= 4g_T(\epsilon_T + \tilde{\epsilon}_T) \\
\bar{C}_T' &= 4g_T(\epsilon_T - \tilde{\epsilon}_T) \,,
\end{aligned}
\tag{17.71}
$$

where the constants g_i represent the renormalizations which occur when transforming between the quark and nucleon representations. That is, the general neutron–proton matrix elements can be written as

$$
\begin{aligned}
<p(p_2)|\bar{u}\gamma_\mu d|n(p_1)> &= \bar{u}_p(p_2)\left[g_V(q^2)\gamma_\mu - g_M(q^2)\frac{i}{2M}\sigma_{\mu\nu}q^\nu + \tilde{g}_S(q^2)\frac{1}{2M}q_\mu\right]u_n(p_1) \\
<p(p_2)|\bar{u}\gamma_\mu\gamma_5 d|n(p_1)> &= \bar{u}_p(p_2)\left[g_A(q^2)\gamma_\mu - g_E(q^2)\frac{i}{2M}\sigma_{\mu\nu}q^\nu + \tilde{g}_P(q^2)\frac{1}{2M}q_\mu\right]\gamma_5 u_n(p_1) \\
<p(p_2)|\bar{u} d|n(p_1)> &= g_S(q^2)\bar{u}_p(p_2)u_n(p_1) \\
<p(p_2)|\bar{u}\gamma_5 d|n(p_1)> &= g_P(q^2)\bar{u}_p(p_2)\gamma_5 u_n(p_1) \\
<p(p_2)|\bar{u}\sigma_{\mu\nu} d|n(p_1)> &= \bar{u}_p(p_2)\Big[g_T(q^2)\sigma_{\mu\nu} + g_T^1(q^2)\frac{1}{2M}\left(q_\mu\gamma_\nu - q_\nu\gamma_\mu\right) \\
&\qquad + g_T^2(q^2)\frac{1}{4M^2}\left(q_\mu P_\nu - q_\nu P_\mu\right) \\
&\qquad + g_T^3(q^2)\frac{1}{2M}\left(\gamma_\mu \slashed{q}\gamma_\nu - \gamma_\nu \slashed{q}\gamma_\mu\right)\Big]u_n(p_1)
\end{aligned}
\tag{17.72}
$$

and can be evaluated experimentally in the case of g_V, g_A, namely $g_V(0) = 1, g_A(0) \simeq 1.27$, and from the lattice in the case of $g_S(0), g_T(0)$, which has determined $g_S = 0.8 \pm 0.4$ and $g_T = 1.05 \pm 0.35$ [Gre12].

If we neglect recoil effects, which is in general a very good approximation since they are typically $\mathcal{O}(q/m_N) \sim 1\%$, there exist only the leading Fermi and Gamow–Teller forms $a = g_V M_F$ and $c = g_A M_{GT}$, and many experiments are analyzed using only these two quantities. A general expression describing the differential decay rate for allowed beta decay can be written as

$$
\begin{aligned}
\frac{d\Gamma}{dE_e d\Omega_e d\Omega_\nu} = {} & \frac{1}{(2\pi)^5} p_e E_e (E_0 - E_e)^2 \xi \left\{ 1 + a_{e\nu}(E_e) \frac{\boldsymbol{p}_e \cdot \boldsymbol{p}_\nu}{E_e E_\nu} + b_F \frac{m_e}{E_e} \right. \\
& + F_{e\nu}(E_e) \left(\frac{\boldsymbol{p}_e \cdot \hat{\boldsymbol{n}} \boldsymbol{p}_\nu \cdot \hat{\boldsymbol{n}}}{E_e E_\nu} - \frac{1}{3} \frac{\boldsymbol{p} \cdot \boldsymbol{p}_\nu}{E_e E_\nu} \right) \left(\frac{3 <(\boldsymbol{J} \cdot \hat{\boldsymbol{n}})^2> - J(J+1)}{J(2J-1)} \right) \\
& + \frac{<\boldsymbol{J} \cdot \hat{\boldsymbol{n}}>}{J} \hat{\boldsymbol{n}} \cdot \left(A_e(E_e) \frac{\boldsymbol{p}_e}{E_e} + B_\nu(E_e) \frac{\boldsymbol{p}_\nu}{E_\nu} + D_{e\nu}(E_e) \frac{\boldsymbol{p}_e \times \boldsymbol{p}_\nu}{E_e E_\nu} \right) \\
& + \boldsymbol{\sigma} \cdot \left[G_e(E_e) \frac{\boldsymbol{p}_e}{E_e} + H_\nu(E_e) \frac{\boldsymbol{p}_\nu}{E_\nu} + K_{e\nu}(E_e) \frac{\boldsymbol{p}_e}{E_e + m_e} \left(\frac{\boldsymbol{p}_e \cdot \boldsymbol{p}_\nu}{E_e E_\nu} \right) \right. \\
& + L_e(E_e) \frac{\boldsymbol{p}_e \times \boldsymbol{p}_\nu}{E_e E_\nu} + N_e(E_e) \hat{\boldsymbol{n}} \frac{<\boldsymbol{J} \cdot \hat{\boldsymbol{n}}>}{J} \\
& + Q_e(E_e) \frac{\boldsymbol{p}_e}{E_e + m_e} \left(\hat{\boldsymbol{n}} \cdot \frac{\boldsymbol{p}_e}{E_e} \frac{<\hat{\boldsymbol{n}} \cdot \boldsymbol{J}>}{J} \right) \\
& \left. \left. + R_e(E_e) \frac{<\hat{\boldsymbol{n}} \cdot \boldsymbol{J}>}{J} \hat{\boldsymbol{n}} \times \frac{\boldsymbol{p}_e}{E_e} \right] \right\}
\end{aligned}
\tag{17.73}
$$

and leading-order results for the various correlation functions have been given by Jackson, Treiman, and Wyld in terms of the Lee–Lang parameters [Jac57].

The dependence on the Beyond-Standard Model (BSM) couplings C_S, C'_S, C_T, C'_T is quadratic unless accompanied by the chirality-changing factor m_e/E_e or the electron spin, in which case the dependence is linear. Because of this feature, although some analyses assert restrictions which are quadratic in BSM quantities, as a practical matter experiments which quote values for a given correlation such as the beta-spin correlation parameter A_e measure the difference between events in which the beta is emitted parallel and antiparallel to the polarization direction, which means that what is actually measured is

$$
A_e^{ex}(E_e) = \frac{A_e(E_e)}{1 + b_F < \frac{m_e}{E_e} >},
\tag{17.74}
$$

where b_F, defined by

$$
\zeta b_F = \pm 2\mathrm{Re} \left[|M_F|^2 (C_S C_V^* + C'_S C_V'^*) + |M_{GT}|^2 (C_T C_A^* + C'_T C_A'^*) \right]
\tag{17.75}
$$

with

$$
\begin{aligned}
\zeta = {} & \sqrt{1 - Z^2 \alpha^2} \left[|M_F|^2 (|C_V|^2 + |C'_V|^2 + |C_S|^2 + |C'_S|^2) \right. \\
& \left. + |M_{GT}|^2 (|C_A|^2 + |C'_A|^2 + |C_T|^2 + |C'_T|^2) \right]
\end{aligned}
\tag{17.76}
$$

is called the Fierz interference parameter.

As discussed in Section 17.4, direct measurement of the Fierz term can be made in superallowed Fermi ($0^+ - 0^+$ analog) decays. Since, in the Standard Model, the vector weak charge is also the isospin-lowering operator (all such transitions are positron emitters), the Fermi matrix element M_F must vanish unless taken between parent-daughter states which are isotopic analogs. Since both parent and daughter are analog states having $J^\pi = J'^{\pi'} = 0^+$, the Gamow–Teller matrix element must vanish, $M_{GT} = 0$, and the Fermi matrix element between states having unit isospin is just

$$M_F = I_- = \sqrt{(T - T_3)(T + T_3 + 1)} = \sqrt{2} \qquad (17.77)$$

and has the same value for the case that the parent nucleus has $T_3 = -1$ such as ^{10}C, ^{14}O, ^{22}Mg or $T_3 = 0$ such as ^{26m}Al, ^{38m}K, ^{74}Rb. Equation (17.77) is an exact prediction up to small isospin-breaking effects. Since the Fermi coupling G_F is known from muon decay, Eq. (17.77) allows the extraction of the CKM matrix element V_{ud} from the experimental study of such transitions.

For allowed decays that are *not* superallowed Fermi decay ($0^+ \to 0^+$), no such symmetry allows a prediction for the value of the Gamow–Teller matrix element, and therefore M_{GT} must be determined either i) empirically, from lifetime/correlation measurements, or ii) calculated using nuclear wavefunctions, though the latter can only be done reliably if meson-exchange effects are included. In the case of BSM matrix elements involving possible scalar or tensor exchange, wavefunction calculations involving impulse approximation estimates from the corresponding nucleon quantities are required.

The development of atomic trapping techniques has permitted the detection of daughter nucleus recoil from an unpolarized parent, which allows the measurement of the electron–neutrino correlation parameter, a_{ev}, which is defined via

$$\frac{d\Gamma}{dE_e d\Omega_v} \sim 1 + a_{ev}(E_e)\frac{\boldsymbol{p}_e \cdot \boldsymbol{p}_v}{E_e E_v} + \cdots . \qquad (17.78)$$

Neglecting recoil terms, we have the simple prediction

$$\zeta a_{ev}(0) = |M_F|^2(|C_V|^2 + |C_V'|^2 - |C_S|^2 - |C_S'|^2)$$
$$- \frac{1}{3}|M_{GT}|^2(|C_A|^2 + |C_A'|^2 - |C_T|^2 - |C_T'|^2), \qquad (17.79)$$

where ζ was defined in Eq. (17.75).

In the Standard Model we then have for Gamow–Teller decays

$$^{SM}a_{ev}^{GT}(0) = -\frac{1}{3}, \qquad (17.80)$$

while for a pure Fermi transition we have

$$^{SM}a_{ev}^{F}(0) = 1 . \qquad (17.81)$$

For mixed Fermi/Gamow–Teller transitions there also exists a precise Standard Model prediction,

$$^{SM}a_{ev}^{\text{mixed}}(0) = \frac{1 - \frac{1}{3}\rho^2}{1 + \rho^2} , \qquad (17.82)$$

once the lifetime has yielded the ratio of Fermi to Gamow–Teller strengths, i.e., $\rho = c/a = g_A M_{GT}/g_V M_F$. Strong, $\mathcal{O}(1\%)$, limits on BSM scalar and tensor couplings from such experiments on nuclei such as ^{6}He [Joh63] and ^{21}Na [Vet08] have been established and future improvements are expected. These will be discussed next.

A second type of experiment which is important for symmetry tests involves correlations between final-state electrons or positrons and the polarization or alignment of the parent state, for which the differential decay rate has the form

$$\frac{d\Gamma}{dE_e d\Omega_e d\Omega_\nu} \sim 1 + A_e(E_e) P_{\hat{n}} \frac{\hat{n} \cdot \boldsymbol{p}_e}{E_e} + V_e(E_e) \Lambda_{\hat{n}} \left(\frac{1}{E_e^2} \hat{n} \cdot \boldsymbol{p}_e \hat{n} \cdot \boldsymbol{p}_e - \frac{p_e^2}{3E_e^2} \right) , \quad (17.83)$$

where $P_{\hat{n}} = <\boldsymbol{J} \cdot \hat{n}> /J$ is the polarization and $\Lambda_{\hat{n}} = 1 - 3(<\boldsymbol{J} \cdot \hat{n}^2> /J(J+1)$ is the alignment. The beta-spin correlation, $A_e(E_e)$, receives both leading-order and recoil corrections, while the alignment correlation $V(E_e)$ is purely a recoil effect. Both quantities have been measured as functions of E_e and provide measures of the recoil form factors. In this regard, it is important to point out that the axial-tensor form factor $d(q^2)$, which vanishes in the Standard Model from isospin invariance in the case of transitions between isotopic analog states such as in neutron beta decay or tritium decay, is in general *nonvanishing* and is comparable in size to the weak magnetism term $b(q^2)$.[2] In the case of mirror transitions the size of the axial-tensor form factor is identical for electron and positron decays, and this can be tested experimentally. A theoretical prediction for the weak magnetism term $b(0)$ in terms of the difference between parent and daughter magnetic moments exists for transitions between isotopic analogs, while in the case of mirror transitions the size of the weak magnetism is given in terms of the electromagnetic $M1$ width of the transition from the excited isotopic analog state of the daughter nucleus. Since the corrections due the recoil effects for all observables has been calculated, together with electromagnetic corrections, these Standard Model predictions are all testable [Hol74].

An alternative route for testing the Standard Model recoil predictions is to employ transitions wherein the daughter state is unstable and itself decays:

i) electromagnetically, such as in the mirror transitions in the $A = 20$ system to the 2^+ 1.63 MeV excited state of ^{20}Ne, which in turn decays via photon emission to the 0^+ ground state [Min11]; and

ii) strongly, such as in the mirror transitions in the $A = 8$ system to the 2^+ 2.90 MeV excited state of ^{8}Be, which in turn decays via the emission of two alpha particles [Sum11].

In the former case there exists a beta-gamma correlation

$$\frac{d\Gamma}{dE_e d\Omega_e d\Omega_\gamma} \sim 1 + \frac{1}{2} Y_e(E_e) \left(\left(\frac{\boldsymbol{p}_e \cdot \hat{\boldsymbol{p}}_\gamma}{E_e} \right)^2 - \frac{p_e^2}{3E_e^2} \right), \quad (17.84)$$

---

[2] Note that the vanishing of $d(q^2)$ for transitions between isotopic analog states is a Standard Model prediction and is violated by so-called second-class currents, which can arise if quarks have an additional quantum number [Hol76].

while in the latter there is a beta-alpha correlation

$$\frac{d\Gamma}{dE_e d\Omega_e d\Omega_\alpha} \sim 1 + Y_e(E_e)\left(\left(\frac{\boldsymbol{p}_e \cdot \hat{\boldsymbol{p}}_\alpha}{E_e}\right)^2 - \frac{p_e^2}{3E_e^2}\right) - 2\frac{\boldsymbol{p}_e \cdot \hat{\boldsymbol{p}}_\alpha}{Mv^*}, \qquad (17.85)$$

where v^* is the velocity of the alpha particle in the daughter rest frame. Here the form of the decay correlation coefficient, $Y_e(E_e)$, is purely recoil-order and is sensitive to the form factors $b(q^2)$, $d(q^2)$ [Hol74].

As will be discussed in the next section, by exploiting a series of correlations such as discussed here, it is possible to probe the Standard Model for possible BSM effects at the limits of experimental accuracy, which are being pushed to the $\mathcal{O}(0.1\%)$ level.

17.4 Experimental Constraints

Moving on to actual experiments, we begin with muon decay, $\mu^- \to e^- \bar{\nu}_e \nu_\mu$, which is described by the effective Hamiltonian

$$\mathcal{H}_{\text{eff}} = \frac{G_F^{(0)}}{\sqrt{2}}(1 + \delta_\mu + \epsilon_\mu)\bar{e}\gamma_\alpha(1 - \gamma_5)\nu_e\nu_\mu\gamma^\alpha(1 - \gamma_5)\mu + h.c. \qquad (17.86)$$

Again δ_μ is a radiative correction, while ϵ_μ represents possible BSM effects. On the experimental side $G_\mu = G_F^{(0)}(1 + \delta_\mu + \epsilon_\mu)$ has been measured to be $G_\mu = 1.1663787(6) \times 10^{-5}\,\text{GeV}^{-2}$ [PDG14].

In the case of semileptonic decay, one requires the size of the CKM matrix element V_{ud}. This quantity has been measured in a number of ways. The most precise determination is via the systematic study of Fermi decays. In the Standard Model such decays involve only the polar-vector hadronic current and connect $J^\pi = 0^+$ levels, which are members of a common isospin triplet, leading to the prediction $a(0) = \sqrt{2}$ for all such transitions. The measurements can be characterized by the so-called $\mathcal{F}t_{\frac{1}{2}}$ value, which is defined via

$$\mathcal{F}t \equiv ft_{\frac{1}{2}}(1 + \delta_R)(1 + \delta_{NS} - \delta_C) = \frac{K}{G_\mu^2 V_{ud}^2}, \qquad (17.87)$$

where, as before, f is the phase space factor, $t_{\frac{1}{2}}$ is the half-life, and the factors δ_i represent various correction factors. The terms δ_R and δ_{NS} are small transition-dependent corrections, the former arising from electromagnetic corrections and the latter from nuclear structure [Mar06]. The term δ_C is due to the breaking of isospin invariance due to the feature that the "last" proton in the parent nucleus is bound less strongly than the "last" neutron in the daughter. The constant K is given by

$$K = \frac{2\pi^3 \log 2}{m_e^5} = 8120.2787(11) \times 10^{-10}\,\text{GeV}^{-4}\,\text{sec} \qquad (17.88)$$

and experiments on a series of such nuclei have yielded a remarkably consistent set of $\mathcal{F}t$ values as seen in Table 17.1. Fitting to these and additional such Fermi decays yields

Table 17.1 Partial list of $0^+ - 0^+$ Fermi transitions, their Q-values, and their measured $\mathcal{F}t$ values. These numbers are from review articles by Hardy and Towner [Tow10, Tow15] where data from additional transitions can also be found

Nucleus	E_0(KeV)	$\mathcal{F}t$(sec)
^{10}C	885.87(11)	3076.7(4.6)
^{14}O	1809.24(23)	3071.5(3.3)
^{26m}Al	3210.66(06)	3072.4(1.4)
^{34}Cl	4469.64(23)	3070.2(2.1)
^{38m}K	5022.40(11)	3072.5(2.4)
^{42}Sc	5404.28(30)	3072.4(2.7)
^{46}V	6030.49(16)	3073.3(2.7)
^{50}Mn	6612.45(07)	3070.9(2.8)
^{54}Co	7222.37(28)	3069.9(3.2)

$< \mathcal{F}t >= 3071.81(83)$ sec with $\chi^2/v = 0.28$, suggesting that uncertainties are well under control [Tow10, Tow09]. The corresponding value of the CKM matrix element is

$$V_{ud} = 0.97425(22) \,, \tag{17.89}$$

making this the most precisely determined mixing element, where the uncertainties are dominated by the ability to estimate the isospin-breaking terms δ_C.

The consistency of these values can also be used to provide a limit on possible BSM effects. Specifically, the value of the CKM matrix element measured in Fermi decays, $\bar{V}_{ud}^{0^+ - 0^+}$, is related to its elementary value, V_{ud}, via

$$|\bar{V}_{ud}^{0^+ - 0^+}|^2 = |V_{ud}|^2 \left[1 + 2\mathrm{Re}(\epsilon_L + \epsilon_R - \epsilon_\mu) + c_{0^+}^S (Z) g_S \mathrm{Re}\epsilon_S \right] \,, \tag{17.90}$$

where

$$c_{0^+}^S (Z) = -2\zeta < \frac{m_e}{E_e} > \tag{17.91}$$

with

$$< \frac{m_e}{E_e} >= \frac{I_1(Q_{EC}/m_e)}{I_0(Q_{EC}/m_e)} \,. \tag{17.92}$$

Here we have defined

$$I_n(x_0) = \int_1^{x_0} dx x^{1-n}(x - x_0)^2 \sqrt{x^2 - 1} \,, \tag{17.93}$$

where $Q_{EC} = M_\alpha - M_\beta$ is the Q-value. The last term in Eq. (17.90) arises from the so-called Fierz interference term b_F. The agreement of the measured $\mathcal{F}t$ values over a large range of nuclei can be used to provide a strong limit on the absence of scalar interactions via the absence of $< m_e/E_e >$-dependence [Tow14]

$$- 1.0 \times 10^{-3} < g_S \mathrm{Re}\epsilon_S < 3.2 \times 10^{-3} \,. \tag{17.94}$$

One can also place a very strong limit on the factor $\epsilon_L + \epsilon_R - \epsilon_\mu$ by employing the value of V_{us} obtained from analysis of $\Delta S = 1$ weak processes such as $K_{\ell 2}, K_{\ell 3}$ decay [Cir11, Ant10],

$$V_{us} = 0.2256(9) \,, \tag{17.95}$$

which leads to the three-generation unitarity check

$$|V_{ud}|^2 + |V_{us}|^2 + |V_{ub}|^2 = 1.00008(56) \,, \tag{17.96}$$

and thereby

$$\mathrm{Re}(\epsilon_L + \epsilon_R - \epsilon_\mu) < 5 \times 10^{-4} \,. \tag{17.97}$$

There exists an alternative way to measure V_{ud}, by the use of the $0^- - 0^-$ transition, $\pi^+ \rightarrow \pi^0 e^+ \nu_e$, which has the advantage of being insensitive to the electromagnetic and isospin-breaking effects which affect the Fermi decay analysis. This decay has been studied in the PIBETA experiment at PSI and has yielded a CKM element [Poc04]

$$V_{ud} = 0.9728(30) \,, \tag{17.98}$$

assuming the absence of BSM physics, which is quite consistent with that obtained in Eq. (17.89), but has considerably larger uncertainty, due to the very small branching ratio of the pion beta-decay reaction,

$$\frac{\Gamma(\pi^+ \rightarrow \pi^0 e^+ \nu_e)}{\Gamma_{tot}(\pi^+)} = 1.036(6) \times 10^{-8} \,, \tag{17.99}$$

and the corresponding inability to generate large statistics.

The PIBETA experiment was also used to study the radiative decay, $\pi^+ \rightarrow e^+ \nu_e \gamma$, which provides a strong limit on the absence of possible tensor interactions. Defining

$$< \gamma(\epsilon,k)|\bar{u}\sigma_{\mu\nu}\gamma_5 d|\pi^+ > = \frac{1}{2}\epsilon_{\mu\nu\alpha\beta} < \gamma(\epsilon,k)|\sigma^{\alpha\beta}|\pi^+ > = -\frac{e}{2}f_T(k_\mu\epsilon_\nu - k_\nu\epsilon_\mu) \,, \tag{17.100}$$

the PIBETA analysis provided the strong limit [Poc04]

$$-2.0 \times 10^{-4} < f_T \mathrm{Re}\epsilon_T < 2.6 \times 10^{-4}. \tag{17.101}$$

A large-N_c theoretical analysis by Portoles and Mateu estimated $f_T(\mu = 1 \text{ GeV}) = 0.24(4)$ [Mat07]. Evolving f_T to $\mu = 2$ GeV, we find the limit

$$-1.1 \times 10^{-3} < \mathrm{Re}\epsilon_T < 1.36 \times 10^{-3}. \tag{17.102}$$

Ordinary beta decay can also be used to provide tensor limits. We begin with the neutron. Specifically, defining $\lambda = g_A(0)/g_V(0)$, the neutron lifetime yields

$$|\bar{V}^n_{ud}| \simeq |V_{ud}|^2 \left[1 + 2\mathrm{Re}(\epsilon_L + \epsilon_R - \epsilon_\mu) \right.$$

$$\left. + \frac{2}{1 + 3\lambda^2}(g_S \mathrm{Re}\epsilon_S + 12\lambda g_T \mathrm{Re}\epsilon_T)\left(\frac{6\lambda^2}{1 + 3\lambda^2} - <\frac{m_e}{E_e}>\right) \right] \tag{17.103}$$

At the present time there exists some uncertainty in the neutron lifetime, outside the quoted error bars, but the PDG recommends the value [PDG14]

$$\tau_n = 880.3(1.1) \text{ sec} \,. \tag{17.104}$$

Table 17.2 Recent measurements of neutron decay parameters. Shown are the recommended Particle Data Group averages together with the experimental data used in generating these values

Quantity	Value
$< \tau_n >_{\text{PDG}}$	885.7(0.8) sec [PDG14]
τ_n	888.4(3.2) sec [Nes92]
τ_n	886.3(3.4) sec [Nic05]
τ_n	878.5(0.8) sec [Ser05]
τ_n	889.2(4.8) sec [Byr02]
τ_n	882.6(2.7) sec [MaM93]
τ_n	887.6(3.0) sec [Mam89]
τ_n	891(9) sec [Spi88]
$< A_e^n >_{\text{PDG}}$	$-0.1173(13)$ [PDG14]
A_e^n	$-0.1189(7)$ [Abe02]
A_e^n	$-0.1160(15)$ [Lia97]
A_e^n	$-0.1135(14)$ [Ero97]
A_e^n	$-0.1146(19)$ [Bop86]

If one assumes the absence of BSM contributions, then V_{ud} can be determined from

$$f_n \tau_n \log 2 (1 + \delta_R)(1 + 3\lambda^2) = \frac{K}{G_F^2 V_{ud}^2} \,, \tag{17.105}$$

provided the the neutron beta-spin correlation parameter $A_e^n(0)$ is used to provide a value of λ via

$$A_e^n(0) = 2 \frac{\lambda - \lambda^2}{1 + 3\lambda^2} \,. \tag{17.106}$$

Note that this correlation is independent of V_{ud} and involves cancelation between the VA and AA terms. Unfortunately, here too, as seen from Table 17.2, there currently exists uncertainly outside the quoted errors. However, if we use the PDG recommended value, $A_e^n(0) = -0.1176(11)$, we find

$$\lambda = 1.2701(25) \tag{17.107}$$

and

$$V_{ud} = 0.9746(19) \tag{17.108}$$

assuming the validity of the Standard Model, in good agreement with the value in Eq. (17.89) found in Fermi decays.

On the other hand, we can set limits on the tensor current if the value of V_{ud} from Fermi decay is assumed. In this way, assuming the absence of scalar interactions, one determines

$$g_T \text{Re} \epsilon_T \leq 4 \times 10^{-2} \,. \tag{17.109}$$

Finally, the value of V_{ud} has also been determined by analysis of various allowed decays of light nuclei which are members of a common isospin multiplet. In this case the Standard

Model was assumed and the value of the Gamow–Teller matrix element was determined by use of correlation measurements in the same system. The nuclei used were ^{19}Ne, ^{21}Na, ^{29}P, ^{35}Ar, and ^{37}K. The electron-neutrino correlation parameter a_{ev} was used in the case of ^{21}Na($\rho = -0.7136(72)$), the polarization-beta correlation parameter A_β was used for ^{19}Ne ($\rho = 1.5995(45)$), ^{29}P($\rho = -0.593(104)$), and ^{35}Ar($\rho = -0.279(16)$), while the polarization-neutrino parameter B_ν was employed in the case of ^{37}K($\rho = 0.561(27)$). The value of $\rho = c/a = g_A M_{GT}/g_V M_F$ is extracted via

$$a_{ev}(0) = \left(1 - \frac{1}{3}\rho^2\right) \Big/ \left(1 + \rho^2\right)$$

$$A_e(0) = \frac{\rho^2 - 2\rho\sqrt{J(J+1)}}{(1+\rho^2)(J+1)}$$

$$B_\nu(0) = -\frac{\rho^2 + 2\rho\sqrt{J(J+1)}}{(1+\rho^2)(J+1)}. \tag{17.110}$$

Combining these values with precise measurements of the various ft values yielded the result [Nav09]

$$V_{ud} = 0.9719(17) , \tag{17.111}$$

again in agreement with the value from Fermi decays. The use of the Standard Model is required here, and hence any dependence on possible BSM effects cannot be extracted.

Correlations

An alternative technique to obtain limits on possible BSM physics comes from correlation measurements in nuclear beta decay.[3] There are various possibilities in this regard which have been employed. In order to avoid having to measure the absolute polarization, one can employ the electron/neutrino correlation parameter a_{ev}, which has the Standard Model value

$$a_{ev} = \frac{a^2 - \frac{1}{3}c^2}{a^2 + c^2} \tag{17.112}$$

for a general β-decay, if recoil effects are neglected. The form of such recoil corrections is known, and these are generally included in precision experimental analyses. In the case of Fermi decays, a TRIUMF collaboration has reported the result [Gor05]

$$a_{ev}^F(^{38m}\text{K}) = 0.9981(45) \tag{17.113}$$

for the decay of ^{38m}K, in good agreement with the Standard Model value of unity, while an ISOLDE measurement of the electron-neutrino correlation for the decay of ^{32}Ar has given a similar value, with slightly less precision [Ade99]

$$a_{ev}^F(^{32}\text{Ar}) = 0.9989(65) . \tag{17.114}$$

---

[3] For simplicity, in this section we shall assume Standard Model values for the vector, axial-vector currents, time reversal invariance, and will use the notation of Lee and Yang for BSM effects. The connection with the model-independent notation in Eq. (17.69) can be made by use of Eq. (17.68).

Using the expression,

$$^{SM}a_{ev}^F - {}^{exp}a_{ev}^F \simeq \frac{C_S^2}{C_V^2} + \frac{C_S'^2}{C_V^2} - \zeta \frac{C_S + C_S'}{C_V} < \frac{m_e}{E_e} >,$$ (17.115)

derived from assuming Standard Model values for C_V, C_V', we can rewrite Eq. (17.115) as

$$\left(\frac{C_S}{C_V} - \frac{1}{2}\zeta < \frac{m_e}{E_e} > \right)^2 + \left(\frac{C_S'}{C_V} - \frac{1}{2}\zeta < \frac{m_e}{E_e} > \right)^2$$
$$= \frac{1}{2}\zeta^2 < \frac{m_e}{E_e} >^2 + {}^{SM}a_{ev}^F - {}^{exp}a_{ev}^F.$$ (17.116)

We then see that the experimental limits are described in terms of circles offset from the origin in the C_S, C_S' plane. There are two such precise experiments that we will consider: ^{38m}K with $< m_e/E_e >= 0.133$ and ^{32}Ar with $< m_e/E_e >= 0.191$. The corresponding limits are shown in Fig. 17.2. Note that there are very strong $\sim 10^{-3}$ constraints on the combination $\frac{C_S}{C_V} + \frac{C_S'}{C_V}$, but much looser $\sim 10^{-2}$ restrictions on $\frac{C_S}{C_V} - \frac{C_S'}{C_V}$. This is because

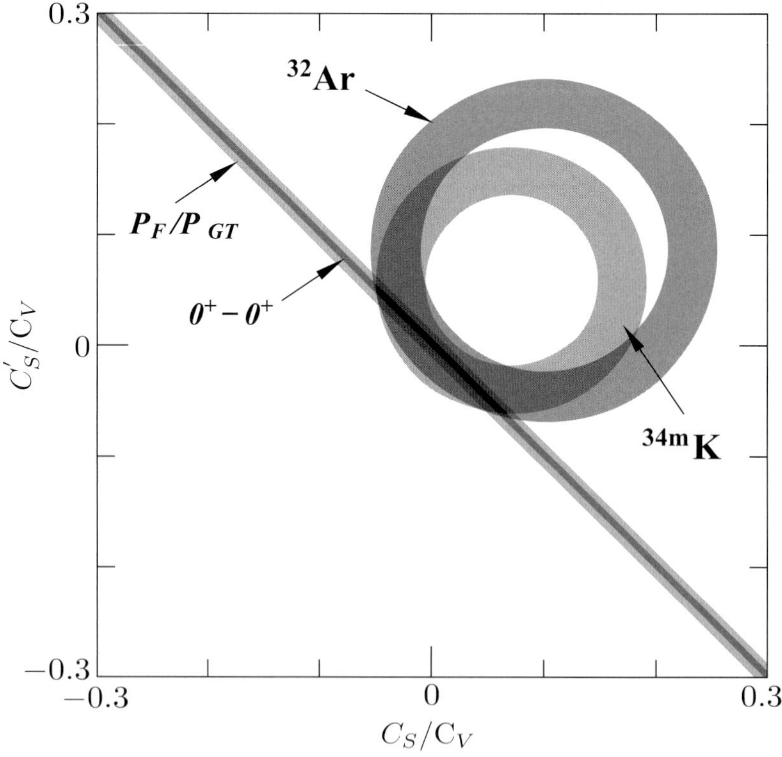

Fig. 17.2   Shown are the 1σ limits on possible weak scalar couplings C_S, C_S' from various experiments [Hol14]. Shown in circles are the limits from the electron-neutrino correlation parameter a_{ev} in the Fermi decays ^{38m}K and ^{32}Ar respectively, while the straight lines indicate the limits from superallowed $0^+ - 0^+$ decays and P_F/P_{GT} measurements in ^{10}C, ^{14}O. The common overlap region is indicated by the black band.

$C_S = -C'_S$ denotes a right-handed leptonic current and therefore $\frac{C_S}{C_V} - \frac{C'_S}{C_V}$ appears quadratically in the decay spectrum, while $C_S = C'_S$ indicates a left-handed leptonic current, which can interfere linearly with the leading $V - A$ weak interaction.

In the case of Gamow–Teller decay, a measurement of ^{6}He decay, with $< m_e/E_e > = 0.286$ has yielded [Joh63, Glu98]

$$a_{ev}^{GT}(^6\text{He}) = -0.3308(30) \,, \tag{17.117}$$

again in good agreement with the Standard Model value of $-1/3$. Comparing with

$$-3(^{SM}a_{ev}^{GT} - ^{exp}a_{ev}^{GT}) = \frac{C_T^2}{C_A^2} + \frac{C_T'^2}{C_A^2} + \zeta \frac{C_T + C_T'}{C_A} < \frac{m_e}{E_E} > \tag{17.118}$$

obtained by assuming Standard Model values for C_A, C'_A, we find the constraint

$$\left(\frac{C_T}{C_A} + \frac{1}{2}\zeta < \frac{m_e}{E_e} > \right)^2 + \left(\frac{C'_T}{C_A} + \frac{1}{2}\zeta < \frac{m_e}{E_e} > \right)^2$$
$$= \frac{1}{2}\zeta^2 < \frac{m_e}{E_e} >^2 -3(^{SM}a_{ev}^{GT} - ^{exp}a_{ev}^{GT}) \,, 2 \tag{17.119}$$

so that again the limits are given by offset circles in the C_T, C'_T plane, as shown in Fig. 17.3.

Finally, an experiment on ^{21}Na, which involves a mixture of Fermi and Gamow–Teller transitions, yielded

$$a_{ev}^{\text{mixed}}(^{21}\text{Na}) = 0.5502(38)(46)(54) \,, \tag{17.120}$$

in good agreement with the Standard Model value

$$^{SM}a_{ev}(^{21}\text{Na}) = 0.553(2) \,. \tag{17.121}$$

However, the limits provided on possible scalar/tensor interactions are more difficult to analyze, since this is a mixed transition. Also, they are somewhat weaker than given by the above experiments, so we do not plot them.

A second method which is sensitive to possible scalar/tensor terms is the comparison of longitudinal polarizations of nearby Fermi and Gamow–Teller decays. In the case of ^{14}O, ^{10}C, the experiments have given [Car91]

$$P_F/P_{GT} = 0.9996(37) \,, \tag{17.122}$$

while, in the case of ^{26m}Al, ^{30}P, the experimental result is [Wic87]

$$P_F/P_{GT} = 1.003(18) \,. \tag{17.123}$$

The corresponding theoretical expressions for longitudinal polarization are

$$P_F = \frac{p_e}{E_e} \frac{2\text{Re}(C_S^* C'_S - C_V^* C'_V)}{\left[|C_V|^2 + |C'_V|^2 + |C_S|^2 + |C'_S|^2 \right] (1 + b_F < \frac{m_e}{E_e} >)} \tag{17.124}$$

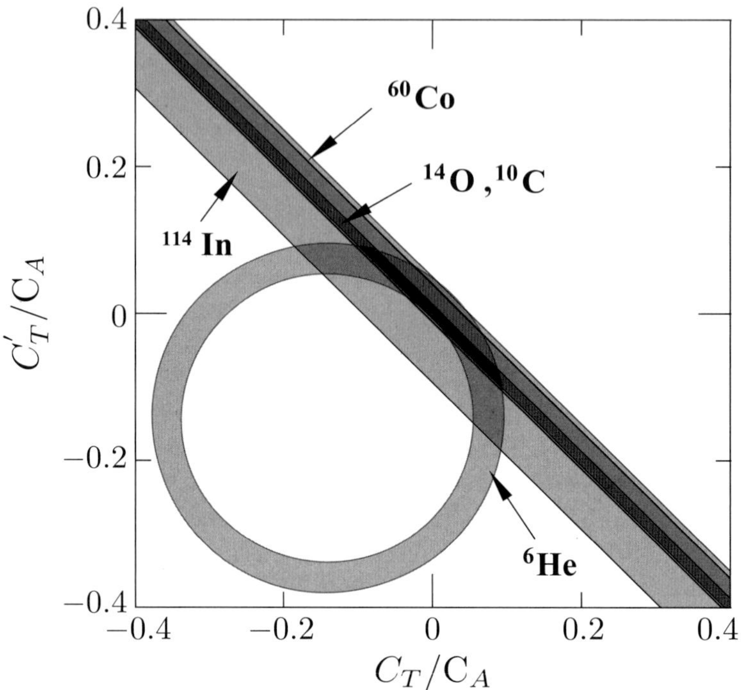

Fig. 17.3   Shown are the 1σ limits on possible weak tensor couplings C_T, C'_T from various experiments [Hol14]. Shown by the circles are the limits from the electron–neutrino correlation parameter a_{ev} in the Gamow–Teller decay ^{6}He. The straight lines represent limits from P_F/P_{GT} in the decays ^{10}C, ^{14}O, respectively, and from the beta-spin correlation parameter A_e in the Gamow–Teller decays ^{60}Co and ^{114}In. The common overlap region is indicated by the black band.

for Fermi decay and

$$P_{GT} = \frac{p_e}{E_e} \frac{2\mathrm{Re}(C_T^* C'_T - C_A^* C'_A)}{\left[|C_A|^2 + |C'_A|^2 + |C_T|^2 + |C'_T|^2\right](1 + b_F < \frac{m_e}{E_e} >)} \tag{17.125}$$

for Gamow–Teller transitions. Thus, assuming Standard Model values for the vector, axial-vector couplings and keeping only the linear terms, we find

$$^{exp}(P_F/P_{GT}) -^{SM}(P_F/P_{GT}) \simeq \zeta < \frac{m_e}{E_e} > \left(\frac{C_S + C'_S}{C_V} - \frac{C_T + C'_T}{C_A} \right), \tag{17.126}$$

yielding

$$-1.1 \times 10^{-2} < \frac{C_S + C'_S}{C_V} - \frac{C_T + C'_T}{C_A} < 1.4 \times 10^{-2} \tag{17.127}$$

for ^{14}O, ^{10}C with $< m_e/E_e >= 0.292$ and

$$-6.9 \times 10^{-2} < \frac{C_T + C'_T}{C_A} - \frac{C_S + C'_S}{C_V} < 9.7 \times 10^{-2} \tag{17.128}$$

for ^{26m}Al, ^{30}P with $< m_e/E_e >= 0.216$. These limits are shown in Figs. 17.2 and 17.3 assuming only scalar or only tensor currents are applicable.

A third tack involves measurement of the polarization-lepton momentum correlation parameter A_e, for which, again omitting recoil terms, we have the theoretical value

$$\tau A_e = \pm \lambda_{JJ'} |M_{GT}|^2 2\mathrm{Re}(C_T^* C_T' - C_A^* C_A')$$
$$+ \delta_{JJ'} \sqrt{\frac{J}{J+1}} M_F M_{GT} 2\mathrm{Re}(C_S^* C_T' + C_T^* C_S' - C_V^* C_A' - C_A^* C_V') \,,$$

where τ is given in Eq. (17.79).

Recent measurements have employed the Gamow–Teller decays ^{60}Co and ^{114}In, for which the results are [Wau10]

$$A_e(^{60}\mathrm{Co}) = -1.014(12)(16) \tag{17.129}$$

and [Wau09]

$$A_e(^{114}\mathrm{In}) = -0.994(10)(10) \,. \tag{17.130}$$

Since both of these transitions are electron Gamow–Teller decays with $J' = J - 1$, then keeping only the linear terms, the theoretical expression can be written as

$$^{SM}A_e^{GT} - ^{exp}A_e^{GT} = \zeta \frac{C_T + C_T'}{C_A} \left\langle \frac{m_e}{E_e} \right\rangle \,, \tag{17.131}$$

assuming that the axial-vector couplings C_A, C_A' have their Standard Model values. We then find the constraint

$$- 0.9 \times 10^{-2} < \frac{C_T + C_T'}{C_A} < 4.8 \times 10^{-2} \tag{17.132}$$

for ^{60}Co with $< m_e/E_e >= 0.704$ and

$$- 9.6 \times 10^{-2} < \frac{C_T + C_T'}{C_A} < 3.8 \times 10^{-2} \tag{17.133}$$

for ^{114}In with $\zeta < m_e/E_e >= 0.209$, and these limits are shown in Fig. 17.3.

We then see that there is no evidence for the existence of a nonzero tensor interaction. However, as in the scalar case, there exist very strong $\sim 10^{-3}$ constraints on the combination, $\frac{C_T}{C_A} + \frac{C_T'}{C_A}$ but much looser $\sim 10^{-2}$ restrictions on $\frac{C_T}{C_A} - \frac{C_T'}{C_A}$. This is because $C_T = -C_T'$ denotes a right-handed leptonic current and therefore $\frac{C_T}{C_A} - \frac{C_T'}{C_A}$ appears quadratically in the decay spectrum, while $C_T = C_T'$ indicates a left-handed leptonic current, which can interfere linearly with the leading $V - A$ weak interaction.

Right-handed Currents

Above we have looked at precision beta decay experiments as a probe of possible scalar, pseudoscalar, and tensor currents. A different and particularly interesting type of BSM effect is that of a right-handed hadronic/leptonic current, as would arise in a theory which seeks to explain the left-handed nature of weak interactions. The idea is that at high energy the theory becomes parity-conserving with equal mixtures of left- and right-handed currents. When the left-right symmetry is broken, if the right-handed W boson is much heavier than the left-handed W, then right-handed effects are suppressed by the ratio $M_{W_R}^2/M_{W_L}^2$ and this is a possible explanation for the dominance of left-handed effects. Of course, in general, there will also be a mixing angle involved, so that the W_L couples to a mixture of left- and right-handed currents. Thus, in a general parameterization we can write

$$W_L = W_1 \cos\zeta + W_2 \sin\zeta, \qquad W_R = -W_1 \sin\zeta + W_2 \cos\zeta , \tag{17.134}$$

where W_1, W_2 are the mass eigenstates.[4] Then the previously discussed experiments can also be analyzed seeking possible right-handed effects.

Writing the basic interaction as

$$\mathcal{L} = \frac{g}{2\sqrt{2}} \left[W_L^{-\mu}(V_\mu^+ - A_\mu^+) + W_R^{-\mu}(V_\mu^+ + A_\mu^+) + h.c. \right] , \tag{17.136}$$

the effective low-energy weak Hamiltonian then becomes [Beg77]

$$\mathcal{H}_{\text{eff}} = \frac{G_{\text{eff}}}{\sqrt{2}} \left[V_\mu^+ V^{-\mu} + \eta_{AA} A_\mu^+ A^{-\mu} - \eta_{AV}(V_\mu^+ A^{-\mu} + A_\mu^+ V^{-\mu}) \right] , \tag{17.137}$$

where

$$\frac{G_{\text{eff}}}{\sqrt{2}} = \left[\frac{g^2}{8m_1^2}(\cos\zeta - \sin\zeta)^2 + \frac{g^2}{8m_2^2}(\cos\zeta + \sin\zeta)^2) \right]$$

$$\eta_{AA} = \frac{\lambda^2 m_2^2 + m_1^2}{\lambda^2 m_1^2 + m_2^2}$$

$$\eta_{VA} = \frac{\lambda(m_1^2 - m_2^2)}{\lambda^2 m_1^2 + m_2^2} \tag{17.138}$$

and

$$\lambda = \frac{1 + \tan\zeta}{1 - \tan\zeta} . \tag{17.139}$$

It is useful to define the related right-handed quantities x, y via [Hol77]

$$x = \frac{1 + \eta_{VA}}{1 - \eta_{VA}} \qquad y = \frac{\eta_{AA} + \eta_{VA}}{\eta_{AA} - \eta_{VA}} . \tag{17.140}$$

Then, in terms of this notation, limits can be placed on each of these quantities. From the longitudinal polarization of outgoing betas we find,

$$P_F = \frac{p_e}{E_e}\left(\frac{1 - x^2}{1 + x^2}\right) \simeq \frac{p_e}{E_e}(1 - 2x^2) , \tag{17.141}$$

---

[4] In terms of the model-independent formalism in Eq. (17.69), this possibility corresponds to

$$\epsilon_\mu = \epsilon_L = 0, \qquad \epsilon_R = \tilde{\epsilon}_L = -\zeta, \qquad \tilde{\epsilon}_R = \frac{M_{W_1}^2}{M_{W_2}^2} . \tag{17.135}$$

for Fermi decay and

$$P_{GT} = \frac{p_e}{E_e}\left(\frac{1-y^2}{1+y^2}\right) \simeq \frac{p_e}{E_e}(1 - 2y^2)\,, \tag{17.142}$$

for Gamow–Teller transitions. For the ratio of Fermi and Gamow–Teller polarizations we then find

$$^{exp}(P_F/P_{GT}) - {}^{SM}(P_F/P_{GT}) \simeq 2(y^2 - x^2)\,, \tag{17.143}$$

which implies the bounds

$$-2.0 \times 10^{-3} < y^2 - x^2 < 1.7 \times 10^{-3} \tag{17.144}$$

from the results from ^{14}O,^{10}C and

$$-0.8 \times 10^{-2} < y^2 - x^2 < 1.0 \times 10^{-2} \tag{17.145}$$

in the case of the ^{26m}Al,^{30}P results.

From the Michel ρ-parameter in muon decay we determine

$$^{exp}\rho - {}^{SM}\rho \simeq -\frac{3}{8}(x - y)^2\,. \tag{17.146}$$

Use of the PDG value [PDG14]

$$^{exp}\rho = 0.7503(4)\,, \tag{17.147}$$

which leads to the bound

$$(x - y)^2 < 3 \times 10^{-4}\,. \tag{17.148}$$

In the case of the Michel parameter ξ we have

$$^{exp}\xi - {}^{SM}\xi \simeq -(x^2 + y^2) \tag{17.149}$$

and the PDG value [PDG14]

$$^{exp}\xi = 1.0007(35) \tag{17.150}$$

yields the bound

$$x^2 + y^2 < 2.8 \times 10^{-3}\,. \tag{17.151}$$

Another limit comes from comparison of the beta-spin correlation parameter A_e with the ft value in a situation where significant cancelation is involved, such as in ^{19}Ne decay, where we have

$$A_e^{^{19}\mathrm{Ne}} = \frac{2}{3}\frac{c^2 + \sqrt{3}ac - y^2c^2 - \sqrt{3}xyac}{a^2 + c^2 + x^2a^2 + y^2c^2}$$
$$\frac{(ft)^{0^+ - 0^+}}{ft^{^{19}\mathrm{Ne}}} = \frac{a^2 + c^2 + x^2a^2 + y^2c^2}{2(1 + x^2)a^2}\,. \tag{17.152}$$

Using the CVC value $a = 1$ and the Gamow–Teller term $c = -1.5995(45)$ obtained from the ^{19}Ne lifetime, the Standard Model prediction for the beta-spin correlation is found to be

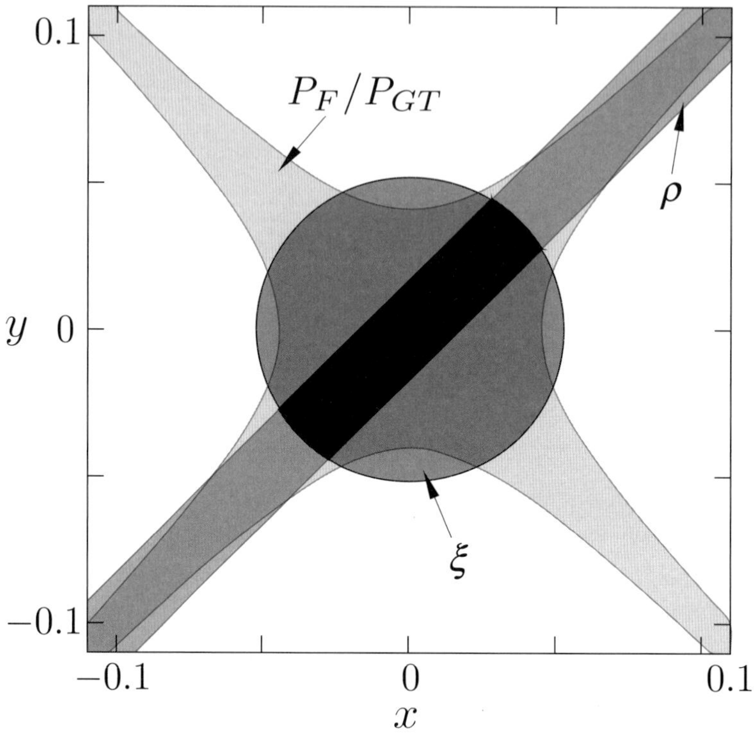

Fig. 17.4 Shown are the limits on right-handed couplings x, y from various experiments [Hol14]. Shown by the straight lines and circle are the 1σ limits from the muon decay Michel parameters ρ and ξ respectively, while the limits from P_F/P_{GT} in ^{10}C and ^{14}O are indicated by the hyperbolas. The common overlap region is displayed by the black band.

$$^{SM}A_e^{^{19}\text{Ne}} = -(3.97 \pm 0.14) \times 10^{-2}.$$ (17.153)

Comparing with the experimental number

$$^{exp}A_e^{^{19}\text{Ne}} = -(3.91 \pm 0.14) \times 10^{-2},$$ (17.154)

we find the limits

$$-0.45y^2 + 0.52xy + 0.01x^2 = (-0.6 \pm 2.0) \times 10^{-3}.$$ (17.155)

The constaints in the x, y plane are shown in Fig. 17.4.[5] We observe that the limits on x and y are generally in the vicinity of 5%, although the experiments themselves are of 0.1% precision. This is a phenomenon that we have seen before and arises because of the feature that right-handed currents do not interfere with their left-handed counterparts, meaning the experimental dependence on the parameters x and y must be quadratic.

---

[5] The ^{19}Ne constraints are not shown, as they are not competitive with the alternatives.

17.5 Second-Class Currents

A different type of BSM interaction is represented by the concept of second-class currents, which were suggested by Weinberg [Wei58]. The idea is that the ordinary $\Delta S = 0$ polar-vector and axial-vector currents have definite properties under the G-parity transformation via

$$GV_\mu G^{-1} = V_\mu \qquad GA_\mu G^{-1} = -A_\mu. \tag{17.156}$$

These are said to be currents of the "first class" and Weinberg designated currents having the opposite behavior under G-parity

$$GV_\mu G^{-1} = -V_\mu \qquad GA_\mu G^{-1} = A_\mu \tag{17.157}$$

as being "second-class currents." The G-parity behavior of these currents has consequences for their nucleon matrix elements [Hol71]. Specifically, in the general representation of nucleon polar-vector and axial-vector matrix elements given in Eq. (17.63):

i) the induced scalar and axial tensor (weak electricity) form factors, $f_3(q^2)$ and $g_2(q^2)$, arise from second-class currents; and
ii) the leading vector and axial-vector form factors, $f_1(q^2)$ and $g_1(q^2)$, as well as the induced pseudoscalar and weak magnetism form factors, $g_3(q^2)$ and $f_2(q^2)$, arise from first-class currents.

Thus, the absence of second-class currents requires the vanishing of $f_3(q^2)$ and $g_2(q^2)$.

Since the usual quark-model forms $\bar{u}\gamma_\mu d$ and $\bar{u}\gamma_\mu\gamma_5 d$ are of the first-class, the concept of second-class currents does not arise in the general BSM expansion given in Eq. (17.69). In order to produce a second-class current in the quark model, one procedure involves the introduction of derivative couplings such as $i\partial_\mu \bar{u} d$ and $i\partial^\nu \bar{u}\sigma_{\mu\nu}\gamma_5 d$. Since these are anathema to renormalization, derivative forms are generally rejected. If derivative currents are not allowed, it is still possible to generate second-class currents, provided one introduces a new quantum number in addition to color and flavor (Feynman called this quantum number "smell"). If we label the new quantum number by an integer $j = 1, 2$ then currents

$$^IV_\mu^+ = \delta_{ij}\bar{u}_i\gamma_\mu d_j \quad \text{and} \quad ^IA_\mu^+ = \delta_{ij}\bar{u}_i\gamma_\mu\gamma_5 d_j$$

are first-class while

$$^{II}V_\mu^+ = \epsilon_{ij}\bar{u}_i\gamma_\mu d_j \quad \text{and} \quad ^{II}A_\mu^+ = \epsilon_{ij}\bar{u}_i\gamma_\mu\gamma_5 d_j$$

are second-class currents [Hol71]. Neither one of these forms is expected to be present in a realistic description of the weak interaction. Nevertheless, the possible presence of such second-class effects can be sought in beta decays and this program has been carried out for the past forty years by increasingly precise experiments. There are two ways to accomplish this search.

a) One is to employ the nuclear generalization of the nucleon form factor G-parity behavior which, in the absence of second-class currents, requires the vanishing of the nuclear-scalar and axial-tensor form factors $e(q^2)$ and $d(q^2)$ between nuclear states which are members of the same isotopic multiplet. In the case of the induced-scalar contribution, the absence of second-class effects is strongly constrained by the outstanding agreement between $\mathcal{F}t$-values of Fermi decays over a wide range of nuclei, as discussed above, so most attention has been focused on the weak electric form factor $d(q^2)$. A precision measurement is required here, since the presence of a weak electric term is a recoil term and is accompanied by a factor E_e/M, where E_e is the beta energy and M is the mean nuclear mass. Since such a linear dependence on E_e also arises from the weak magnetism term $b(q^2)$, what is measured is a linear combination of d and b, and thus it is necessary to separate them. This can be achieved due to the fact that for isotopic analog states the weak magnetism form factor is predicted in terms of the difference of magnetic moments of the parent and daughter states via

$$b = \pm A M_F (\mu_\beta - \mu_\alpha) \sqrt{\frac{J+1}{J}} . \tag{17.158}$$

This program has been carried out for the $\frac{1}{2}^+ - \frac{1}{2}^+$ transition $^{19}\mathrm{Ne} \to {}^{19}\mathrm{F} + e^+ + \nu_e$, where the positron energy dependence of the beta-spin correlation parameter $A_e(E_e)$

$$A_e(E) = f_4(E_e)/f_1(E_e), \tag{17.159}$$

was measured, where [Hol74]

$$f_4(E_e) = \frac{2}{\sqrt{3}} \mathrm{Re} a^* \left[c - \frac{E_0}{3M}(c - b - d_{II}) + \frac{E_e}{3M}(7c - b - d_{II}) \right]$$
$$+ \frac{2}{3}\mathrm{Re} c^* \left[c - \frac{2E_0}{3M}(c - b - d_{II}) + \frac{E_e}{3M}(11c - 5b + d_{II}) \right] \tag{17.160}$$

and

$$f_1(E_e) = |a|^2 + |c|^2 - \frac{2}{3}\frac{E_0}{3M}\mathrm{Re} c^*(c - b - d_{II})$$
$$+ \frac{2}{3}\frac{E_e}{3M}(3|a|^2 + 5|c|^2 - 2\mathrm{Re} c^* b) - \frac{1}{3}\frac{m_e^2}{ME_e}(\mathrm{Re})c^*(2c - 2b - d_{II}) .$$
$$\tag{17.161}$$

The experiment yielded a nonzero value for the second-class tensor form factor d_{II} [Cal75]

$$\frac{b}{Ac} = 4.90 \pm 0.01, \qquad \frac{d_{II}}{Ac} = -8.2 \pm 3.3 , \tag{17.162}$$

although the experimental systematic uncertainties were large.

b) The second procedure involves comparison of mirror electron and positron decays for the case of a common daughter state and parents which are members of the same isotopic multiplet. Examples of such decays exist for the $2^+ - 2^+$ transitions $^8\mathrm{Li} \to {}^8\mathrm{Be}^*(2.90\ \mathrm{MeV}) + e^- + \bar{\nu}_e$, $^8\mathrm{B} \to {}^8\mathrm{Be}^*(2.90\ \mathrm{MeV}) + e^+ + \nu_e$, for the $1^+ - 0^+$ transitions $^{12}\mathrm{B} \to {}^{12}\mathrm{C} + e^- + \bar{\nu}_e$, $^{12}\mathrm{N} \to {}^{12}\mathrm{C} + e^+ + \nu_e$, and the $2^+ - 2^+$ transitions

^{20}F $\rightarrow^{20}$ Ne*(1.63 MeV) $+ e^- + \bar{\nu}_e$, ^{20}Na $\rightarrow^{20}$ Ne*(1.63 MeV) $+ e^+ + \nu_e$, and all three mirror transition beta decays have been carefully studied, especially by the Osaka group. The idea in the case of mass-12 is to align the parent nuclei and then to measure the beta-alignment correlation $V(E_e)$ for both electron and positron branches, cf. Eq. (17.80). From the expression

$$V(E_e) = \frac{E_e}{3m_N}\left(\pm\frac{b}{Ac} \mp \frac{d_{II}}{Ac} - \frac{d_I}{Ac}\right), \tag{17.163}$$

where we have decomposed the axial-tensor into its first- and second-class components, d_I and d_{II}, we see that this correlation is purely a recoil effect and involves the difference/sum of b/Ac and d_I/Ac for the electron/positron branches in the absence of second-class currents. The weak magnetism form factor b can be determined from the $M1$ width of the photon decay from the 1^+ analog 15.11 MeV excited state of ^{12}C to the 0^+ ground state. The absence of second-class currents does not require the vanishing of the axial-tensor form factor d, but equality of the axial-tensor terms for the electron and positron decay. The result from the Osaka experiments is that [Min01]

$$\frac{b}{Ac} = 4.04\pm0.03, \quad \frac{d_I}{Ac} = 4.96\pm0.09\pm0.05, \quad \frac{d_{II}}{Ac} = -0.15\pm0.12\pm0.05 . \tag{17.164}$$

In the case of the $A = 8$ and $A = 20$ systems the daughter states are unstable. The 2.90 MeV excited state of ^{8}Be decays to a pair of α particles and what is measured is the beta-alpha correlation parameter $Y(E_e)$ (cf. Eq. (17.81)), while the 1.63 MeV excited state of ^{20}Ne decays via photon emission to the ground state and what is measured is the same parameter $Y(E_e)$, but now for measurement of the beta-gamma correlation (cf. Eq. (17.82)). A complication in these cases, which is not present in the $A = 12$ decay, is the presence of higher-order form factors, which must be taken into account. The result for $Y(E_e)$ is

$$\begin{aligned}
Y(E_e) = \quad -\frac{E_e}{3m_N}\Bigg[&\frac{1}{A} \pm \frac{b}{Ac} - \frac{d_I}{Ac} - \frac{d_{II}}{Ac} \\
&+ \frac{(-)^s}{\sqrt{14}}\left(\pm\frac{f}{Ac}\frac{E_0 + 2E_e}{E_0} + \frac{3}{2}\frac{j_2}{A^2c}\frac{E_0 - 2E_e}{m_N}\right) \\
&- \frac{3}{\sqrt{35}}\frac{j_3}{A^2c}\frac{E_e}{m_N}\Bigg].
\end{aligned} \tag{17.165}$$

The result for the $A = 20$ experiments is [Min11]

$$\frac{b}{Ac} = 8.58 \pm 0.28, \quad \frac{d_I}{Ac} = 8.00 \pm 0.64 \pm 0.35, \quad \frac{d_{II}}{Ac} = 0.18 \pm 0.42 \pm 0.24 \tag{17.166}$$

with higher-order form factors

$$\frac{j_2}{A^2c} = -21 \pm 83 \pm 100, \quad \frac{j_3}{A^2c} = -1273 \pm 195 \pm 82 , \tag{17.167}$$

while that for the $A = 8$ system is [Sum11]

$$\frac{b}{Ac} = 7.5 \pm 0.2, \quad \frac{d_I}{Ac} = 5.5 \pm 1.7, \quad \frac{d_{II}}{Ac} = -0.28 \pm 0.46 \pm 0.19 \tag{17.168}$$

with higher-order form factors

$$\frac{j_2}{A^2 c} = -490 \pm 70, \quad \frac{j_3}{A^2 c} = -980 \pm 280, \quad \frac{f}{Ac} = 1.0 \pm 0.3 . \tag{17.169}$$

The overall conclusion is that there exists no evidence for the existence of second-class currents at the level of about 5–10% of weak magnetism.

17.6 Time Reversal Tests

In the previous pages we have generally assumed for simplicity that time reversal invariance is respected, so that all form factors are relatively real. In this case we see that there are no contributions to the T-violating triple correlations such as D_{ev}, L_{ev}, R_e, unless there is a relative phase between some of the decay amplitudes, except for a small T-invariant contribution to these quantities, which arises from final-state beta-nucleus electromagnetic interactions. The D_{ev} parameter in particular, has been used to provide very precise limits on the absence of T-violation in beta decay. Although in the Standard Model there exists no leading T-violation in this case from final-state electromagnetic interactions, there does exist a small contribution from electromagnetic interactions at recoil order, as first calculated by Callen and Treiman [Cal67]. The general form for the electromagnetic scattering contributions to the D_{ev} parameter is [Hol72]

$$D_{ev}^{em} = -3\frac{Z\alpha E_e^2}{8 M p_e}\theta_{JJ'}\frac{\left[|c|^2 \mp \mathrm{Re} c^*(b - d)\right]}{|a|^2 + |c|^2} , \tag{17.170}$$

where

$$\theta_{JJ'} = \begin{cases} -(J + 1)/(2J - 1) & J = J' + 1 \\ 1 & J = J' \\ -J/(2J + 3) & J = J' - 1 . \end{cases} \tag{17.171}$$

Precise experimental measurements of the D_{ev} parameter exist both for the case of neutron and $^{19}\mathrm{Ne}$ decay, in which the weak electric form factor d vanishes, so that the Coulomb scattering effect arises from interference between the Gamow–Teller and weak magnetism terms. The Coulomb contributions are

$$^n D_{ev}^{em} = 2 \times 10^{-5} \quad \text{and} \quad ^{19}\mathrm{Ne} D_{ev}^{em} = 2 \times 10^{-4} , \tag{17.172}$$

respectively, in these two cases, which is smaller than the current experimental results

$$^n D_{ev} = (-0.94 \pm 1.89 \pm 0.97) \times 10^{-4} \tag{17.173}$$

for neutron decay [Chu12] and

$$^{19}\mathrm{Ne} D_{ev} = (-0.5 \pm 0.001) \times 10^{-3} \tag{17.174}$$

in the case of $^{19}\mathrm{Ne}$ [Bal77].

The R_e parameter has also been measured both in $^8\mathrm{Li}$ and neutron decay. In the latter case a result [Koz09]

$$^n R_e = 0.004 \pm 0.012 \pm 0.005 \tag{17.175}$$

was obtained, together with

$$^{n}N_e = 0.067 \pm 0.011 \pm 0.005 \,. \tag{17.176}$$

In the case of ^{8}Li it was found that [Hub03]

$$^{8\text{Li}}R_e = 0.0009 \pm 0.0022 \,, \tag{17.177}$$

so that no evidence to T-violation in semileptonic weak decay exists, even in the case of non-standard (BSM) interactions.

We close this discussion here, having shown above how precision beta decay experiments can and have been used as a probe for the existence of possible BSM effects. In Chapter 21 we will discuss some of the ways by which such BSM physics could be manifested.

Exercises

17.1   **Beta-Decay Classification**

Classify the following β-decays as allowed, forbidden, Fermi, or Gamow–Teller transitions:

a) $^3\text{H} \to {}^3\text{He} + e^- + \bar{\nu}_e,$
b) $^8\text{Li}(2^+) \to {}^8\text{Be}^*(2^+) + e^- + \bar{\nu}_e,$
c) $^{12}\text{N}(1^+) \to {}^{12}\text{C}(0^+) \to e^+ + \nu_e,$
d) $^{16}\text{N}^*(0^-) \to {}^{16}\text{O}(0^+) + e^+ + \nu_e,$
e) $^{21}\text{Na}^*(\frac{3}{2}^+) \to {}^{21}\text{Ne}(\frac{3}{2}^+) + e^+ + \nu_e.$

17.2   **Second-Class Currents**

In this problem we explore the second-class currents discussed in Section 17.5. Using the general definition of weak form factors $a, b, c, \ldots j_3$ given in Eq. 17.64,

a) Suppose that the parent and daughter states of a nuclear beta decay are members of a common isotopic multiplet, i.e., $|i> = |T, T_3>$ and $|f> = |T, T_3 \pm 1>$. Examples are the decay of ^{19}Ne to ^{19}F or of ^{3}H to ^{3}He. Suppose that the weak currents are purely first-class. That is, suppose that the G-parity of the polar (axial) vector current is positive (negative), respectively. Show that the form factors $d(q^2)$, $e(q^2)$, $f(q^2)$, $j_2(q^2)$ must vanish, while there is no constraint on the remaining form factors.

b) Consider an $SU(3)$ generalization of this result. That is, suppose we consider a strangeness-changing beta decay such as $\Xi^- \to \Lambda + e^- + \bar{\nu}_e$. Show that in the $SU(3)$ symmetry limit the induced-tensor and induced-scalar form factors $g_2(q^2)$ and $f_3(q^2)$ form factors are required to vanish.

c) Of course, G-parity and its $SU(3)$ generalization are only approximate symmetries. Estimate the size of the induced tensor for a realistic beta decay, both $\Delta S = 0$ and $\Delta S = 1$.

d) Consider the case of mirror beta decays, that is, electron and positron decays which occur from initial states which are members of the same isotopic multiplet to final states which are members of a common isotopic multiplet. Examples are the decay of ^{12}B and ^{12}N to the ^{12}C ground state or of ^{8}B and ^{8}Li to the $J^P = 2^+$ first-excited state of ^{8}Be. Suppose that the weak currents are purely first-class. Demonstrate that each of the electron-decay form factors is required to have the same size as its positron-decay counterpart.

17.3   Testing CVC

Even before the development of the Standard Model, it was theorized that the strangeness-conserving component of the weak polar-vector current was related to the electromagnetic current by a simple isospin rotation

$$[T_\pm, V_\mu^{em}] = V_\mu^\pm \,,$$

which is called the conserved vector current (CVC) hypothesis. Various tests of this suggestion were proposed and verified. One of the first was a proposal by Gell-Mann to measure the electron energy dependence of the unpolarized

$$^{12}\text{B} \rightarrow\, ^{12}\text{C} + e^- + \bar\nu_e \quad \text{and} \quad ^{12}\text{N} \rightarrow\, ^{12}\text{C} + e^+ + \nu_e$$

beta-decay spectra.

a) Since both ^{12}B and ^{12}N have $J^\pi = 1^+$, while ^{12}C is an even-even nucleus and has $J^\pi = 0^+$, show that, neglecting the electron mass compared to the $\sim 12\,\text{MeV}$ decay energy, the only weak form factors which contribute to these decays are the Gamow–Teller axial-vector term c, the weak magnetism term b, and the weak electricity term d, and that the size of these terms should be identical for β^+ and β^- transitions.

b) Evaluate the decay spectrum and show that it has the expected form

$$\frac{d\Gamma^\pm}{dE_e} \sim |c|^2 \pm \frac{4}{3}\text{Re}(c^*b)\frac{E}{M} \,.$$

c) Calculate the Gamow–Teller amplitudes c^+ and c^- from the measured ^{12}B and ^{12}N lifetimes and show that they have the same magnitude.

d) Show that the size of the weak magnetism term b is predicted by CVC in terms of the M1 electromagnetic width of the decay of the 15.11 MeV $J^\pi = 1^+$ excited state of ^{12}C to the ^{12}C ground state via

$$\Gamma_{M1} = \frac{1}{6M^2}\alpha_{em}|b|^2\Delta^3,$$

where $\Delta = 15.11\,\text{MeV}$ is the excitation energy.

This prediction was tested in the experiment of Lee *et al.*, and was one of the first experimental verifications of the CVC hypothesis [Lee63].

17.4   Muon Capture

Besides beta decay and neutrino scattering, another manifestation of the semileptonic weak interaction is muon capture, in which a negatively-charged muon is captured by an atom. Typically, the muon is captured in a high angular momentum state, but then cascades down to the $1S$ state, which has a significant overlap with the nucleus. Because of this overlap, at some point the muon interacts with the nucleus of charge

Z and is captured, changing the nucleus to one with charge $Z - 1$ and releasing a muon neutrino in the process.

For simplicity, suppose we are dealing with capture by a proton, which after capture becomes a neutron. Using the hadronic current matrix element

$$< n(p_2)|V_\mu + A_\mu|p(p_1) > = \bar{u}(p_2)\left[g_V(q^2)\gamma_\mu - g_M(q^2)\frac{i}{2m_N}\sigma_{\mu\nu}q^\nu + g_A(q^2)\gamma_\mu\gamma_5 \right.$$

$$\left. - g_P(q^2)q_\mu\gamma_5 + g_{II}(q^2)\frac{i}{2m_N}\sigma_{\mu\nu}q^\nu\gamma_5 \right]u(p_1)\,,$$

where $q = p_1 - p_2$ is the momentum transfer and $m_N = \frac{1}{2}(m_p + m_n)$ is the mean mass,

a) Show that the muon capture on the proton takes place at

$$q^2 = -m_\mu^2\left[1 - 2\frac{m_n - m_p}{m_\mu^2} + \frac{m_n - m_p - m_\mu}{M_1 + m_\mu}\left(1 - \frac{m_n - m_p}{m_\mu} \right) \right] = -0.88m_\mu^2\,.$$

b) In the approximation that the muon is at rest, show that the T-matrix can be written in the form

$$T = \frac{G}{\sqrt{2}}V_{ud}\chi_2^\dagger\chi_\nu^\dagger(1 - \boldsymbol{\sigma}\cdot\hat{\boldsymbol{k}})$$

$$\times\left[G_V 1\cdot 1' - G_A\boldsymbol{\sigma}\cdot\boldsymbol{\sigma}' - G_P\boldsymbol{\sigma}\cdot\hat{\boldsymbol{k}}\boldsymbol{\sigma}'\cdot\hat{\boldsymbol{k}} \right]\chi_\mu\chi_1\,,$$

where the unprimed operators act on the lepton spinors and the primed operators act in the hadron spinor space and

$$G_V = g_V(q^2)\left(1 + \frac{k_0}{E_2 + M_2} \right) - g_M(q^2)\frac{m_\mu k_0}{2M(E_2 + M_2)}$$

$$G_A = g_A(q^2) + g_V(q^2)\frac{k_0}{E_2 + M_2} + g_M(q^2)\frac{k_0}{2M}\left(1 + \frac{k_0 - m_\mu}{E_2 + M_2} \right)$$

$$-g_{II}(q^2)\left(\frac{m_\mu - k_0}{2M} - \frac{k_0^2}{2M(E_2 + M_2)} \right)$$

$$G_P = \frac{k_0}{E_2 + M_2}\left[g_A(q^2) - g_V(q^2) - m_\mu g_P(q^2) \right]$$

$$-\frac{k_0}{2M}\left[g_{II}(q^2)\left(1 + \frac{m_\mu}{E_2 + M_2} \right) + g_M(q^2)\left(1 + \frac{k_0}{E_2 + M_2} \right) \right],$$

with

$$k_0 = \frac{m_\mu M_1}{M_1 + m_\mu}\left(1 + \frac{M_1^2 - M_2^2 + m_\mu^2}{2m_\mu M_1} \right)$$

being the neutrino energy.

c) In gaseous hydrogen most of the capture takes place in the hyperfine singlet state. In this case, by taking the muon to be in a simple Coulomb field, show that

$$\Gamma_{H-\text{singlet}} = \frac{G^2 V_{ud}^2 m_\mu^3 \alpha^3}{2\pi^2} \frac{E_2 + M_2}{2E_2} \frac{k_0^2}{1 + \frac{k_0}{E_2}} \left(\frac{M_1}{M_1 + m_\mu}\right)^3 |G_V + 3G_A + G_P|^2 .$$

d) Calculate the triplet capture rate and show that

$$\Gamma_{H-\text{triplet}} = \frac{G^2 V_{ud}^2 m_\mu^3 \alpha^3}{2\pi^2} \frac{E_2 + M_2}{2E_2} \frac{k_0^2}{1 + \frac{k_0}{E_2}} \left(\frac{M_1}{M_1 + m_\mu}\right)^3$$
$$\times \left[\text{Re}(G_V - G_A)^*(3G_V - 3G_A - 2G_P) + 3|G_P|^2\right] .$$

e) Assuming the axial-tensor induced-pseudoscalar form factor to be dominated by the pion pole show that

$$g_P(q^2) = \frac{2Mg_A(q^2)}{m_\pi^2 - q^2} .$$

f) Show that CVC requires that the weak magnetic form factor have the value

$$g_M(0) = \kappa_p - \kappa_n = 3.70 ,$$

where $\kappa_p = 1.79$ and $\kappa_n = -1.91$ are the anomalous magnetic moments of the proton and neutron respectively in units of the nucleon magneton.

g) Show that G-parity requires the vanishing of $g_{II}(q^2)$.

h) Calculate the expected singlet and triplet capture rates and show that

$$\Gamma_{H-\text{singlet}} \simeq 710\,\text{s}^{-1}$$

$$\Gamma_{H-\text{triplet}} \simeq 15\,\text{s}^{-1} .$$

Neutrino Physics

18.1 Introduction

In this chapter we study properties of the neutrino as well as its interactions. The existence of the neutrino was first postulated in 1930 by Pauli in order to retain the conservation of energy, momentum, and angular momentum in nuclear beta decay [Pau33], but it was not until 1956 that the neutrino was actually observed by Reines and Cowan [Cow56].

In 1934 Fermi used the neutrino to construct his current-current model for the beta decay process, which has led to the very successful present picture of the weak interactions [Fer34]. In this picture we can understand why the neutrino was so difficult to detect, since its interactions are so tiny. Consider the process

$$\bar{\nu}_e + p \rightarrow e^+ + n \,,$$

which is the reaction wherein the neutrino was first observed. The associated cross section is then

$$\sigma \sim \left| \frac{G_F}{\sqrt{2}} \right|^2 \int \frac{d\mathbf{p}_e}{(2\pi)^3} 2\pi \, \delta(m_p + E_\nu - m_n - E_e) \, |M_{\rm w}|^2 \sim \frac{G_F^2}{2\pi} p_e E_e \, |M_{\rm w}|^2 \,, \qquad (18.1)$$

where $G_F \simeq 10^{-5} m_N^2$ is the Fermi constant defined in Chapter 4. For incident neutrino energies large compared with the electron rest mass, Eq. (18.1) becomes $\sigma \sim G_F^2 E_\nu^2 |M_{\rm w}|^2 / 2\pi \sim 10^{-44} \, {\rm cm}^2$ for $E_\nu \sim 1$ MeV. Taking the matrix element $M_{\rm w} \sim 1$, if we consider a single neutrino passing through a slab of material having a target density ρ ($\sim 10^{23}$ atoms/cm^3 for typical materials), it would travel a distance

$$\Delta x \sim 1/(\rho \sigma) \sim 10^{21} \, {\rm cm} \qquad (18.2)$$

before interacting, a distance comparable to 100 billion Earth radii. Neutrino reactions are indeed *weak*! The way to circumvent this dilemma is to have lots of low-energy neutrinos, say from a nuclear reactor or an accelerator, and/or huge targets.

We now know that there exist *three* distinct neutrino flavors, an electron type ν_e, a muon type ν_μ, and a tau type ν_τ, with the weak current coupling electrons only to ν_e, muons only to ν_μ, and taus only to ν_τ,

$$J_\mu^{\rm lep} = \psi_e^\dagger \mathcal{O}_\mu \psi_{\nu_e} + \psi_\mu^\dagger \mathcal{O}_\mu \psi_{\nu_\mu} + \psi_\tau^\dagger \mathcal{O}_\mu \psi_{\nu_\tau}, \qquad (18.3)$$

where $\mathcal{O}_\mu = \gamma_\mu(1 - \gamma_5)$, namely, V–A. Measurements of the decay width of the neutral Z-boson [Don14] (see also Exercise 4.1) and astrophysical arguments based on the helium abundance in the universe [Ste78] (see Chapter 20) suggest that this list may exhaust the set of lepton–neutrino pairs: there appear to be no more light neutrinos beyond ν_τ.

The feature that the lepton current involves the chirality operator $1 - \gamma_5$ has an important consequence with respect to neutrinos. From the form of the plane-wave solutions to the Dirac equation

$$\psi(x) = \sqrt{\frac{E+m}{2E}} \left(\begin{array}{c} \chi \\ \frac{\sigma \cdot p}{E+m} \chi \end{array} \right) \exp(-ip \cdot x) \,, \tag{18.4}$$

where χ is the usual two-component Pauli spinor, we find that, in the limit as $m \to 0$ and $E \to |p|$, Eq. (18.4) becomes

$$\psi(x) \xrightarrow{m \to 0} \psi_0(x) = \frac{1}{\sqrt{2}} \left(\begin{array}{c} \chi \\ \sigma \cdot \hat{p} \chi \end{array} \right) \exp(-ip \cdot x) \,, \tag{18.5}$$

so that

$$(1 - \gamma_5)\psi_0(x) = \left\{ \begin{array}{ll} \psi_0(x) & \text{if } \sigma \cdot \hat{p} \chi = -\chi \\ 0 & \text{if } \sigma \cdot \hat{p} \chi = \chi. \end{array} \right. \tag{18.6}$$

This result is important, as will be discussed in the next section, because the neutrino is *extremely* light and therefore, since the neutrino interacts only via interactions which involve the chirality operator, essentially *all* neutrinos must be left-handed. Similarly, it is easy to see that essentially all antineutrinos must be right-handed. With this introductory background out of the way, we now move to consider aspects of the neutrino that have recently been studied, namely masses and mixings.

18.2 Neutrino Mass

The issue of whether the neutrino has a nonzero mass, or not, has been long of interest. In the simplest Standard Model, as postulated in Chapter 4, the neutrino is purely left-handed and therefore must be massless. However, the existence of neutrino mass is not a theoretical axiom. Rather, this question is a subject for *experimental* study. That any such mass must be small is known from the feature that the maximum electron energy measured in a β-decay spectrum agrees to high precision with the mass difference of initial and final nuclear states. However, Fermi, in his seminal paper on beta decay [Fer34], noted that this question could be answered more definitively by carefully examining the spectral shape near the electron endpoint, i.e., it is possible to plot the spectrum in such a way that a nonzero mass would be revealed as a distortion at the endpoint tangent to (perpendicular to) the energy axis (see Fig. 18.1). The spectral dependence on the neutrino mass can be inferred directly by studying the phase space responsible for the β-decay electron energy distribution. Recall the beta-decay spectrum introduced in Eq. (17.51):

$$f(Z, R, E_0) = \int_{m_e}^{E_0} dE_e \rho(Z, R, E_e)(E_0 - E_e)^2 E_e \sqrt{E_e^2 - m_e^2} \,. \tag{18.7}$$

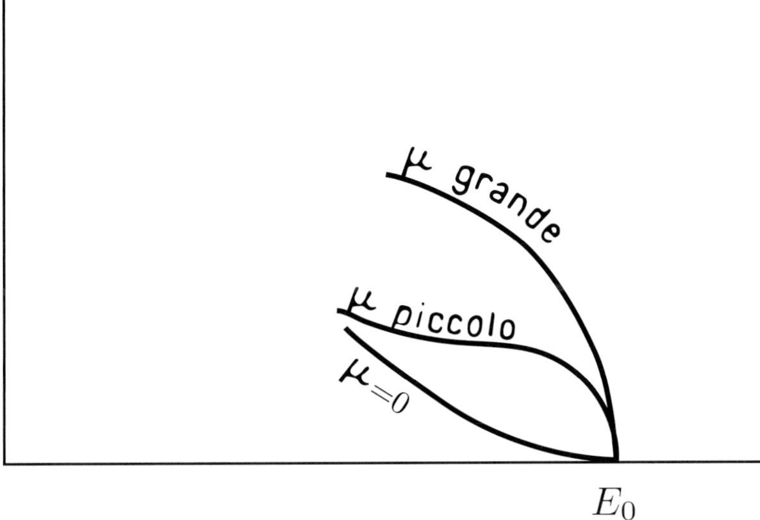

Fig. 18.1 Excerpt from Fermi's seminal paper, where the impact of neutrino mass on the beta-decay spectrum is already sketched out [Fer34].

In the limit that the neutrino is fully relativistic, the phase space is proportional to $f(Z, R, E_0) \sim p_e E_e E_\nu^2$. If we relax this condition, we find the phase space is really proportional to $p_e E_e p_\nu E_\nu$, thus allowing the following modification to Eq. (17.51):

$$f(Z, R, E_0) = \int_{m_e}^{E_0} dE_e \rho(Z, R, E_e) E_e \sqrt{E_e^2 - m_e^2} (E_0 - E_e) \sqrt{(E_0 - E_e)^2 - m_\nu^2} . \quad (18.8)$$

Providing a slight modification due to the presence of different mass eigenstates, the differential energy spectrum from β-decay can be written as:

$$\frac{df(Z, R, E_0)}{dE_e} = \sum_i |U_{ei}|^2 \rho(Z, R, E_e) E_e \sqrt{E_e^2 - m_e^2} (E_0 - E_e) \sqrt{(E_0 - E_e)^2 - m_{\nu_i}^2} , \quad (18.9)$$

where U_{ei} correspond to the mixing elements relevant to each of the neutrino mass eigenstates (see the discussion in Chapter 4 of the PMNS mixing matrix).

In the years since Fermi's paper there have taken place a series of such measurements, with steadily increasing precision. It is clear that use of a β-decay parent nucleus with a relatively low Q-value is helpful, since a larger fraction of the total spectrum then resides within a given interval from the endpoint. (Of course, a corresponding challenge is presented by the fact that the event rate is going to zero at the endpoint.) Most experimenters have selected tritium, which has an 18.6 keV endpoint. The present limit from such measurements is

$$m_{\bar{\nu}_e} < \begin{cases} 2.3 \, \text{eV} \; \text{Mainz [Kra05]} \\ 2.1 \, \text{eV} \; \text{Troitsk [Ase11]} . \end{cases} \quad (18.10)$$

Both the Mainz and Troitsk experiments used electromagnetic spectrometers (known as magnetic adiabatic collimation with electrostatic filtering, or MAC-E Filters) to set the above limits. The ultimate incarnation of this technique is being pursued by the KATRIN experiment [Ang05], located in Karlsruhe, Germany. KATRIN aims to improve this limit by an order of magnitude. Other techniques, including calorimetric techniques (the ECHo [Bla13] and HOLMES [Alp14] experiments) and frequency-based techniques (Project 8 [Mon09, Asn14]), are also starting to be pursued to provide further sensitivity on experimental determinations of the neutrino mass scale.

Because of the difficulty of such measurements, it is interesting that a remarkable event occurred that was able to establish an *independent* bound on the $\bar{\nu}_e$ mass, the detection of Supernova 1987A. This event resulted from the explosion of a star in the Large Magellanic Cloud about $170,000$ years ago, the light (and neutrinos) from which finally reached Earth in 1987 and was the first such optical supernova in our vicinity of the galaxy in nearly 400 years, the previous occurrence having been noted by Kepler in 1604! Some understanding of stellar evolution is necessary in order to make the connection between SN1987A and neutrinos. A more complete discussion of supernovae and of stellar evolution is provided in Chapter 20.

Let us consider how these supernova neutrino events limit the neutrino mass. If $m_\nu \neq 0$, neutrinos of different energy will travel with different velocities according to

$$v = \frac{p_\nu}{\sqrt{p_\nu^2 + m_\nu^2}} \simeq 1 - \frac{m_\nu^2}{2E_\nu^2} \tag{18.11}$$

rather than with the fixed velocity c. Consequently, the time of arrival of higher-energy neutrinos will be earlier than their lower-energy counterparts. The difference in arrival times is

$$\frac{\delta t}{t} \sim \frac{\delta v}{v} \sim \frac{m_\nu^2}{E_\nu^2} \frac{\delta E_\nu}{E_\nu}, \tag{18.12}$$

where t is the total time in transit from the source. Eleven events were seen in the Kamiokande detector in Japan and eight others were seen in a mine in Ohio. These were detected with an energy spread of $\delta E_\nu \sim 10\,\mathrm{MeV}$ and found to have an average energy 2–3 times larger than that. The events arrived over a roughly 10 second time span. Thus we find in a simple analysis

$$m_{\bar{\nu}_e} \leq E_\nu \left(\frac{\delta t}{t} \frac{E_\nu}{\delta E_\nu} \right)^{\frac{1}{2}} \sim 10\,\mathrm{MeV} \left(\frac{10\,\mathrm{sec}}{10^{13}\,\mathrm{sec}} \right)^{\frac{1}{2}} \sim 10\,\mathrm{eV}. \tag{18.13}$$

A more careful analysis, which takes into account the time and energy distribution, yields a similar limit, $m_{\bar{\nu}_e} \leq 5.7\,\mathrm{eV}$. It is remarkable that a handful of neutrino events from a star that lived and died long before the advent of civilization can place a bound on the neutrino mass comparable to that gained from many years of terrestrial high-precision weak interaction studies.

Indirect information on neutrino mass is obtained via measurement of the Cosmic microwave background (CMB) anisotropy. Precision data (as described in Chapter 21) are used to determine fundamental cosmological parameters, including an upper bound for the mass due to neutrinos. A global fit to world data yields an upper bound of $0.2\,\mathrm{eV}$ for the sum of light neutrino masses.

We have focused thus far on *electron* neutrino mass measurements because such experiments have achieved the greatest precision. Bounds have also been established on the masses of the ν_μ and ν_τ,

$$m_{\nu_\mu} \leq 190\,\text{keV}, \qquad m_{\nu_\tau} \leq 18\,\text{MeV}, \tag{18.14}$$

from careful kinematic analyses of decays, such as $\pi^+ \to \mu^+ + \nu_\mu$ and $\tau \to 5\pi + \nu_\tau$. The large energy releases in decays of the π and τ and limited energy resolution of the detectors employed results in less precise neutrino mass bounds in these cases. It is worth noting that some theories of neutrino mass predict that the ν_τ will be the heaviest neutrino; e.g., in some models, neutrino masses scale as the squares of the masses of the corresponding charged leptons. From this perspective the limits directly above are not necessarily less significant as tests of the underlying particle physics than the tighter bounds established on the $\bar{\nu}_e$.

This discussion can be succinctly summarized: there exists at present *no* evidence for nonzero neutrino mass from *direct* measurements. Naturally, when increasingly precise experiments continue to yield values consistent with zero, it is tempting to assume that the number really *is* zero. Indeed, as mentioned above, in the simplest version of the Standard Model of elementary particles, the neutrino mass is taken to vanish. This result does not follow from any fundamental principle, however, and can be modified accordingly by adding right-handed fields $\psi_{\nu_i R}$ if the evidence were to change. This is precisely what has evolved over the past decade, as we now describe.

18.3 Neutrino Oscillations

Significant improvements in the "direct" neutrino mass measurements described above require considerable time and effort. For this reason, there is great interest in measurements of a different type, exploiting neutrino mixing, which might be able to probe far smaller masses, albeit in a less direct fashion. The essential idea, as outlined in Chapter 4, can be traced to the seminal paper of Pontecorvo [Pon68], who first pointed out that neutrino oscillations would occur if the neutrino states of definite mass do not coincide with the weak interaction eigenstates. Subsequently, large underground experiments were constructed which demonstrated that neutrinos do indeed oscillate. These experiments were recognized by the award of the 2015 Nobel Prize in Physics.

To understand such oscillations it is helpful to first consider the more familiar phenomenon of Faraday rotation, the rotation around the beam direction of the polarization vector of linearly-polarized light as it propagates through a magnetized material. Faraday rotation occurs because the index of refraction (and thereby the potential acting on the light) depends on the state of circular polarization – i.e., the two states of definite circular polarization $\left(|\chi_\pm> = \sqrt{\tfrac{1}{2}}(|+x> \pm i|+y>) \right)$ propagate in time with distinct phases. Thus, if $|\chi(t=0)> = |+x> = \sqrt{\tfrac{1}{2}}(|\chi_+> + |\chi_->)$, we find

$$|\chi(t)> = \sqrt{\frac{1}{2}}\left(|\chi_+> \exp\left(-i\frac{\omega_0}{n_+}t\right) + |\chi_-> \exp\left(-i\frac{\omega_0}{n_-}t\right)\right)$$

$$= \exp\left(-i\frac{\omega_0}{2}\frac{n_+ + n_-}{n_+ n_-}t\right)\left(|+x> \cos\left(\frac{\omega_0}{2}\frac{n_+ - n_-}{n_+ n_-}t\right)\right.$$

$$\left. - |+y> \sin\left(\frac{\omega_0}{2}\frac{n_+ - n_-}{n_+ n_-}t\right)\right), \tag{18.15}$$

so that the polarization vector rotates with frequency

$$\omega_{\text{pol}} = \frac{\omega_0}{2}\frac{n_+ - n_-}{n_+ n_-}. \tag{18.16}$$

That is, the linear polarization states $|+x>$ and $|+y>$ "mix" as the light passes through the magnetic medium.

Now we generalize this idea to the case of neutrinos and we review the oscillation phenomenon introduced in Chapter 4. Since there exist three neutrino generations, a full mixing discussion for neutrinos involves the 3×3 PMNS matrix

$$U = \begin{pmatrix} c_{12}c_{13} & s_{12}c_{13} & s_{13}e^{-i\delta} \\ -s_{12}c_{23} - c_{12}s_{23}s_{13}e^{i\delta} & c_{12}c_{23} - s_{12}s_{23}s_{13}e^{i\delta} & s_{23}c_{13} \\ s_{12}s_{23} - c_{12}c_{23}s_{13}e^{i\delta} & -c_{12}s_{23} - s_{12}c_{23}s_{13}e^{i\delta} & c_{23}c_{13} \end{pmatrix}, \tag{18.17}$$

where $c_{ij}, s_{ij} = \cos\theta_{ij}, \sin\theta_{ij}$. However, although a full discussion involves three neutrino families, the essential physics of oscillations is illustrated most clearly by considering the interactions of only two flavors, which we choose to be ν_e and ν_μ. In this limit the situation is similar to the two-state Faraday rotation problem discussed above. The relevant component of the weak lepton current involves the combination $\bar{e}\nu_e + \bar{\mu}\nu_\mu$ and is given by Eq. (18.3). This interaction effectively *defines* the ν_e and ν_μ as weak interaction eigenstates, the neutrinos accompanying the electron and muon, respectively, when these charged leptons are weakly produced. Yet there exists a second Hamiltonian, the free Hamiltonian describing the propagation of an isolated neutrino. The eigenstates of this second Hamiltonian are the mass eigenstates. If these two mass eigenstates are distinct (and thus at least one is nonzero), then in general the eigenstates diagonalizing the mass Hamiltonian will *not* diagonalize the weak interaction. If we label the mass eigenstates as $|\nu_1 >$ and $|\nu_2 >$, then

$$|\nu_1 > = \cos\theta|\nu_e > + \sin\theta|\nu_\mu > \quad \text{with mass } m_1$$

$$|\nu_2 > = -\sin\theta|\nu_e > + \cos\theta|\nu_\mu > \quad \text{with mass } m_2, \tag{18.18}$$

where θ is a mixing angle which connects the mass and weak eigenstates. Now suppose that at time $t = 0$ an electron neutrino is produced with given momentum $\boldsymbol{p}$

$$|\psi(t = 0)> = |\nu_e> = \cos\theta|\nu_1 > - \sin\theta|\nu_2 > . \tag{18.19}$$

The mass eigenstates propagate with simple phases, since they are the eigenstates of the free Hamiltonian, so that at a distance $\sim ct$ from the source the neutrino state has become

$$|\psi(t)> = \cos\theta|\nu_1 > e^{-iE_1 t} - \sin\theta|\nu_2 > e^{-iE_2 t}, \tag{18.20}$$

where $E_i = \sqrt{p^2 + m_i^2}$. Projecting back upon weak eigenstates, we have

$$< v_e|\psi(t) > \; = \cos^2\theta e^{-iE_1 t} + \sin^2\theta e^{-iE_2 t}$$
$$< v_\mu|\psi(t) > \; = \cos\theta \sin\theta (e^{-iE_1 t} - e^{-iE_2 t}) \,. \tag{18.21}$$

Noting that

$$E_1 \sim p + \frac{m_1^2 + m_2^2}{4p} - \frac{\delta m^2}{4p}$$

$$E_2 \sim p + \frac{m_1^2 + m_2^2}{4p} + \frac{\delta m^2}{4p} \,, \tag{18.22}$$

where $\delta m^2 = m_2^2 - m_1^2$, we find at $t > 0$ a probability

$$p(t) = |< v_\mu|\psi(t) >|^2 = \sin^2 2\theta \sin^2 \frac{\delta m^2 t}{4p} \tag{18.23}$$

that the v_e will have transformed into a v_μ. This change of neutrino identity is called a "neutrino oscillation," due to the time- or distance-dependent behavior of $p(t)$ and is the neutrino version of Faraday rotation discussed above. This phenomenon is a sensitive probe for neutrino masses, given non-degenerate neutrinos and a non-vanishing mixing angle θ. Since the simplest standard electroweak model developed in Chapter 4 predicts massless neutrinos, any observation of neutrino oscillation constitutes definitive evidence for physics beyond the minimal Standard Model. Experimentally, the phenomenon can be probed either by looking for a deficit of neutrinos of a specific neutrino flavor from a predicted flux (*disappearance* measurements), or observing an excess of a particular flavor over prediction (*appearance* measurements).

Neutrino oscillations in free space were discussed many years ago as a solution to an observed deficit of neutrinos coming from the Sun [Bar81a]. However, in the meantime, it has been realized that one of the properties of our Sun is that it can greatly enhance oscillations, even if mixing angles are small, an effect known as the Mihkeyev–Smirnov–Wolfenstein (MSW) mechanism. The starting point is a generalization of the vacuum neutrino oscillation discussion given above. Previously, we discussed the case where the initial neutrino had a definite flavor. But we could have considered the somewhat more general initial state

$$|v(t = 0)\rangle = a_e(t = 0)|v_e\rangle + a_\mu(t = 0)|v_\mu\rangle \,. \tag{18.24}$$

and, as before, could expand this wavefunction in terms of the mass eigenstates, which propagate simply, as

$$i\frac{d}{dx}\begin{pmatrix} a_e \\ a_\mu \end{pmatrix} = \frac{1}{4E}\begin{pmatrix} -\delta m^2 \cos 2\theta_v & \delta m^2 \sin 2\theta_v \\ \delta m^2 \sin 2\theta_v & \delta m^2 \cos 2\theta_v \end{pmatrix}\begin{pmatrix} a_e \\ a_\mu \end{pmatrix} \,. \tag{18.25}$$

(Note that a common phase has been ignored, as it can be absorbed into the overall phase of the coefficients a_e and a_μ, and thus has no consequence.) We have also labeled the mixing angle as θ_v, to emphasize that it is the *vacuum* value, and equated $x = t$, that is, set $c = 1$.

The view of neutrino oscillations changed when Mikheyev and Smirnov [Mik85] showed in 1985 that the density dependence of the neutrino effective mass, a phenomenon first discussed by Wolfenstein [Wol78] in 1978, could greatly enhance oscillation probabilities: a ν_e can be adiabatically transformed into a ν_μ as it traverses a critical density within the Sun. It became clear that the Sun was not only an excellent neutrino source, but also a natural regenerator for enhancing the effects of flavor mixing.

The effects of matter alter the neutrino evolution equation in an apparently simple way [Hax86]

$$i\frac{d}{dx}\begin{pmatrix} a_e \\ a_\mu \end{pmatrix} = \begin{pmatrix} \frac{G_F\rho(x)}{\sqrt{2}} - \frac{\delta m^2 \cos 2\theta_v}{4E} & \frac{\delta m^2 \sin 2\theta_v}{4E} \\ \frac{\delta m^2 \sin 2\theta_v}{4E} & -\frac{G_F\rho(x)}{\sqrt{2}} + \frac{\delta m^2 \cos 2\theta_v}{4E} \end{pmatrix}\begin{pmatrix} a_e \\ a_\mu \end{pmatrix} , \tag{18.26}$$

where $\rho(x)$ is the solar electron density. The term ocurring in the diagonal elements, $G_F\rho(x)/\sqrt{2}$, represents the effective contribution to m_ν^2 that arises from neutrino–electron scattering, as derived in Exercise 18.1. The indices of refraction of electron and muon neutrinos differ because the former scatter by charged *and* neutral currents, while the latter have only neutral-current interactions, since the Sun contains electrons but no muons. The difference in the forward scattering amplitudes determines the density-dependent splitting of the diagonal elements of the new time-evolution equation.

It is helpful to rewrite Eq. (18.26) in a basis consisting of the light and heavy local mass eigenstates (i.e., the states that diagonalize the right-hand side of the equation),

$$|\nu_L(x)\rangle = \cos\theta(x)|\nu_e\rangle - \sin\theta(x)|\nu_\mu\rangle$$
$$|\nu_H(x)\rangle = \sin\theta(x)|\nu_e\rangle + \cos\theta(x)|\nu_\mu\rangle . \tag{18.27}$$

The local mixing angle is defined by

$$\sin 2\theta(x) = \frac{\sin 2\theta_v}{\sqrt{X^2(x) + \sin^2 2\theta_v}}$$

$$\cos 2\theta(x) = \frac{-X(x)}{\sqrt{X^2(x) + \sin^2 2\theta_v}} , \tag{18.28}$$

where $X(x) = 2\sqrt{2}G_F\rho(x)E/\delta m^2 - \cos 2\theta_v$. Thus $\theta(x)$ ranges from θ_v to $\pi/2$ as the density $\rho(x)$ goes from 0 to ∞.

If we define

$$|\nu(x)\rangle = a_H(x)|\nu_H(x)\rangle + a_L(x)|\nu_L(x)\rangle , \tag{18.29}$$

the neutrino propagation can be rewritten in terms of the local mass eigenstates

$$i\frac{d}{dx}\begin{pmatrix} a_H \\ a_L \end{pmatrix} = \begin{pmatrix} \lambda(x) & i\alpha(x) \\ -i\alpha(x) & -\lambda(x) \end{pmatrix}\begin{pmatrix} a_H \\ a_L \end{pmatrix} \tag{18.30}$$

with the splitting determined by

$$2\lambda(x) = \frac{\delta m^2}{2E}\sqrt{X^2(x) + \sin^2 2\theta_v} \tag{18.31}$$

and mixing of the eigenstates governed by the density gradient

$$\alpha(x) = \left(\frac{E}{\delta m^2}\right) \frac{\sqrt{2}\,G_F \frac{d}{dx}\rho(x)\sin 2\theta_v}{X^2(x) + \sin^2 2\theta_v}. \tag{18.32}$$

The results above are intriguing: the local mass eigenstates diagonalize the matrix if the density is constant, that is, if $\alpha = 0$. In this constant-density limit, the problem is no more complicated than the original vacuum oscillation case, although the mixing angle is modified because of the presence of the matter. But if the density is not constant, the mass eigenstates in fact evolve as the density changes and this is the crux of the MSW effect. Note that the splitting achieves its minimum value, $\frac{\delta m^2}{2E}\sin 2\theta_v$, at a critical density $\rho_c = \rho(x_c)$

$$2\sqrt{2}EG_F\rho_c = \delta m^2\cos 2\theta_v \tag{18.33}$$

that defines the point where the diagonal elements of the matrix in Eq. (18.26) cross. Also note that the MSW effect is not blind to the sign of δm^2. Indeed, matter effects have the potential to disentangle the hierarchy of the neutrino masses (which vacuum oscillations clearly cannot). The MSW effect has already been used effectively for this purpose to disentangle the hierarchy between m_1 and m_2 in studying neutrino oscillation stemming from the Sun.

This local-mass-eigenstate form of the propagation equation can be trivially integrated if the splitting of the diagonal elements is large compared to the off-diagonal elements, so that the effects of $\alpha(x)$ can be ignored. Then

$$\gamma(x) = \left|\frac{\lambda(x)}{\alpha(x)}\right| = \frac{\sin^2 2\theta_v}{\cos 2\theta_v}\frac{\delta m^2}{2E}\frac{1}{\left|\frac{1}{\rho_c}\frac{d\rho(x)}{dx}\right|}\frac{[X(x)^2 + \sin^2 2\theta_v]^{3/2}}{\sin^3 2\theta_v} \gg 1\,, \tag{18.34}$$

a condition that becomes particularly stringent near the crossing point, where $X(x)$ vanishes,

$$\gamma_c = \gamma(x_c) = \frac{\sin^2 2\theta_v}{\cos 2\theta_v}\frac{\delta m^2}{2E}\frac{1}{\left|\frac{1}{\rho_c}\frac{d\rho(x)}{dx}\big|_{x=x_c}\right|} \gg 1\,. \tag{18.35}$$

The resulting adiabatic electron–neutrino survival probability [Bet86], valid when $\gamma_c \gg 1$, is

$$p_{v_e}^{\mathrm{adiab}} = \frac{1}{2} + \frac{1}{2}\cos 2\theta_v\cos 2\theta_i\,, \tag{18.36}$$

where $\theta_i = \theta(x_i)$ is the local mixing angle at the density where the neutrino was produced. Thus, if $\theta_v \ll 1$ and if the starting solar core density is sufficiently high, so that $\theta_i \sim \pi/2$, we find $p_{v_e}^{\mathrm{adiab}} \sim 0$.

The physical picture behind this derivation is that, if one makes the usual assumption that in vacuum the v_e is almost identical to the light mass eigenstate, $v_L(0)$, i.e., $m_1 < m_2$ and $\cos\theta_v \sim 1$. But as the density increases, the matter effects make the v_e heavier than the v_μ, with $v_e \to v_H(x)$ as $\rho(x)$ becomes large. The special property of the Sun is that it produces v_es at high density that then propagate to the vacuum where they are measured. The adiabatic approximation tells us that if initially $v_e \sim v_H(x)$, the neutrino will remain in the heavy-mass eigenstate provided the density changes slowly. That is, if the solar density gradient is sufficiently gentle, the neutrino will emerge from the Sun in the heavy vacuum eigenstate, $\sim \nu_\mu$. This guarantees nearly complete conversion of ν_es into ν_μs, producing a flux that cannot be detected by conventional solar neutrino detectors (for a footprint of the effect, see Fig. 18.4).

Neutrino Sources and Their Role in Neutrino Oscillation Measurements

As Fig. 18.2 illustrates, the universe is filled with copious sources of neutrinos stemming from the early big bang, astrophysical sources, fission reactors to modern accelerators. We focus on how the different mass splittings have been uncovered by experimentalists exploiting these common sources of neutrinos.

Solar Neutrinos and θ_{12}

The earliest indication of neutrino oscillations came from observation of high-energy electron neutrinos which were produced in the Sun. The origin of the energy produced

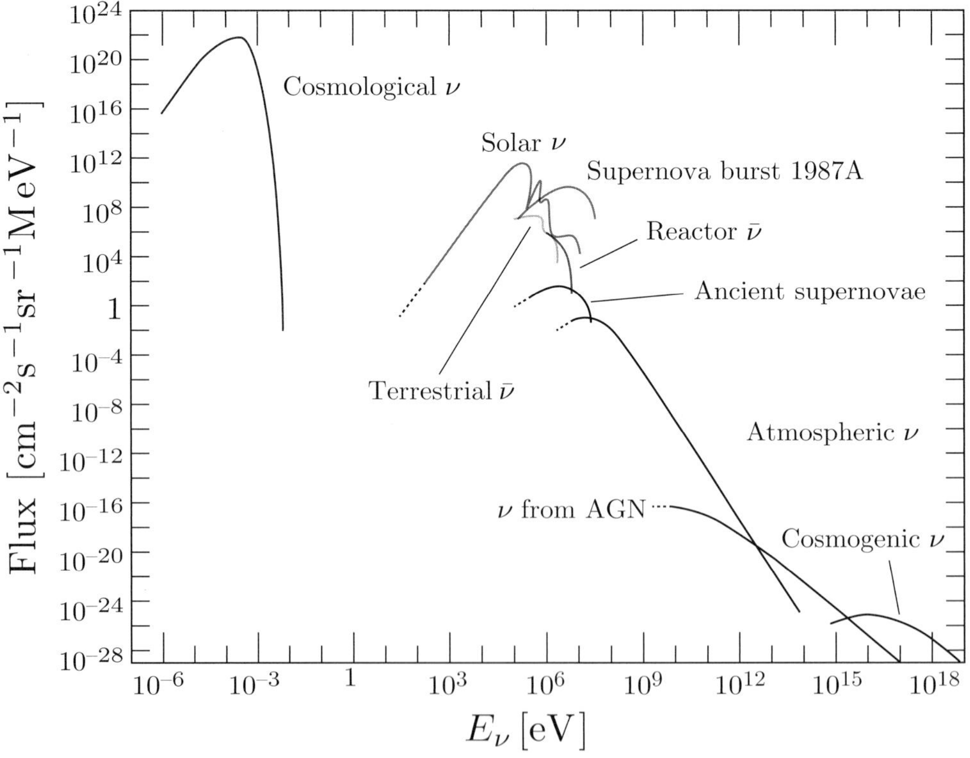

Fig. 18.2　Flux of neutrinos by source versus neutrino energy [Kat12]. Note that the terrestrial antineutrinos arising from β-decay deep within the Earth were detected for the first time by the KamLAND experiment [Gan13] during the long-term shutdown of Japanese nuclear reactors following the March 2011 Fukushima nuclear accident.

by the Sun (or any star) is thermonuclear reactions in the solar core and the theoretical model which describes this process is called the standard solar model (SSM) and is described in detail in Chapter 20. The SSM picture makes use of four basic assumptions:

i) The Sun evolves in hydrostatic equilibrium, involving a balance between the gravitational force and the pressure gradient. Here one must specify the equation-of-state as a function of temperature, density, and solar composition.

ii) Energy is transported through the solar medium by radiation and convection. While the solar envelope is convective, radiative transport dominates in the core region where the thermonuclear reactions take place. Here the opacity depends sensitively on the solar composition, especially on the abundances of heavier elements.

iii) The thermonuclear energy chains involve processes by which four protons are converted into ^{4}He

$$4p \rightarrow {}^4\text{He} + 2e^+ + 2\nu_e.$$

The SSM predicts that the lion's share (98%) of this reaction occurs via the pp chain, as shown in Fig. 18.3, with the CNO cycle accounting for the rest. In this picture the core temperature is $T_c \sim 1.5 \times 10^7$ K and the electron density is $N_e \sim 6 \times 10^{25}\text{cm}^{-3}$.

iv) The model is constrained to produce the current radius, mass, and luminosity. Here the initial ^{4}He/H ratio is adjusted to reproduce the current luminosity at the Sun's 4.6 billion year age.

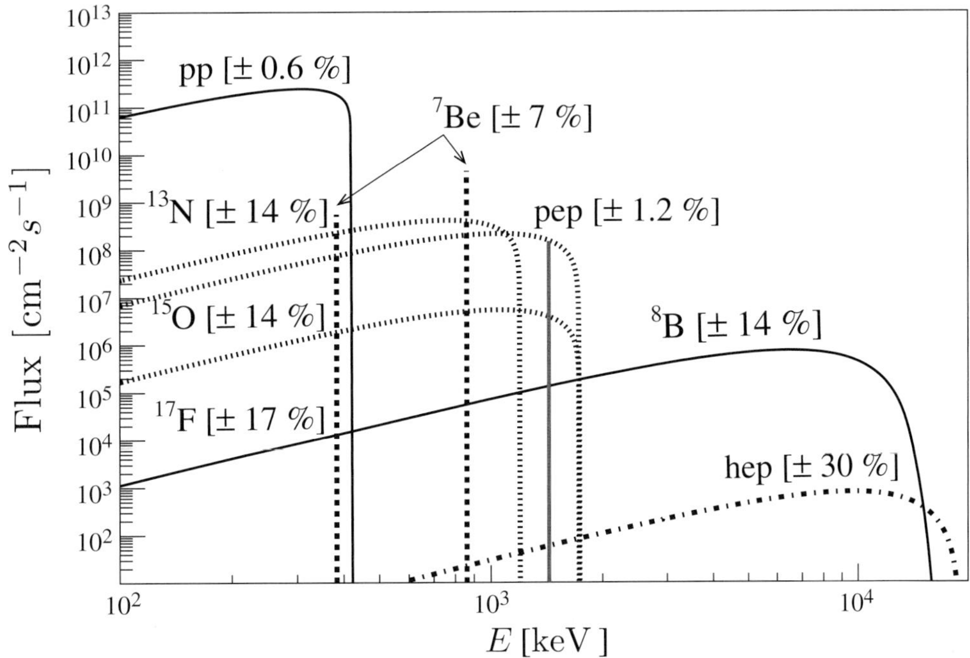

Fig. 18.3   The calculated solar neutrino energy spectrum from a variety of decay chain progenies for a particular solar model [Bel14].

Table 18.1 Neutrino fluxes predicted in the standard solar model

Reaction	$E_{\nu_e}^{\max}$ (MeV)	Flux(10^{10} cm^{-2} s^{-1})
$p + p \rightarrow d + e^+ + \nu_e$	0.42	5.9
$^7\mathrm{Be} + e^- \rightarrow {}^7\mathrm{Li} + \nu_e$	0.86(90%)	4.8×10^{-1}
	0.36(10%)	
$^8\mathrm{B} \rightarrow {}^8\mathrm{Be} * + e^+ + \nu_e$	14.06	5.2×10^{-4}

The SSM is very successful at predicting the flux of neutrinos produced in the reactions which take place in the core once matter-mediated oscillations are taken into account, yielding the results shown in Fig. 18.3. There exist three categories of neutrinos which are produced during this process, with:

  i) high energy: $E_{\nu_e} > 2$ MeV;
 ii) medium energy: $E_{\nu_e} \sim 1$ MeV; and
iii) low energy: $E_{\nu_e} < 0.42$ MeV,

which are identified by the location in the *pp* chain at which they are produced. The SSM-predicted flux associated with these neutrinos is given in Table 18.1.

The first detection of solar neutrinos was by the Davis experiment in the Homestake gold mine in South Dakota, which involved a detector containing 615 tons of the cleaning fluid C_2Cl_4. The experiment identified neutrinos via the charged current reaction $^{37}\mathrm{Cl}(\nu_e, e^-)^{37}\mathrm{Ar}$ and is primarily (80%) sensitive to the high-energy $^8\mathrm{B}$ neutrinos [Cle98]. The measured capture rate of $2.56 \pm 0.16 \pm 0.16$ SNU (1 SNU$=10^{-36}$ captures/atoms/sec) is about a factor of three smaller than the SSM-predicted rate of 7.6 ± 1.2 SNU. The reason for this discrepancy is that the electron neutrinos produced in the solar core have oscillated into a mixture of muon- and tau-neutrinos which do not possess the energy required to produce charged-current events in the detector. Taking the energy of the detected neutrinos as the spectral maximum, which occurs at about half the endpoint energy, ~ 7 MeV, we find

$$2\sqrt{2}E_{\nu_e} G_F N_{ec} \simeq 1.1 \times 10^{-4} \text{ eV}^2 \gg \Delta m_{12}^2 \cos 2\theta_{12} \simeq 2.8 \times 10^{-5} \text{ eV}^2 , \tag{18.37}$$

so that the condition for adiabatic MSW propagation is satisfied. Thus, the local mixing angle at the production point of the neutrino is near $\pi/2$ and the resultant probability to generate a ν_e event in the Homestake detector is

$$P_{\nu_e} \simeq \frac{1}{2}(1 - \cos 2\theta_{12}) = \sin^2 \theta_{12} \simeq \frac{1}{3} , \tag{18.38}$$

as found experimentally (see Fig. 18.4).

A second class of solar neutrino experiment involved the use of gallium, wherein neutrinos can be detected via the $^{71}\mathrm{Ga}(\nu_e, e^-)^{71}\mathrm{Ge}$ reaction. Such a detector is primarily sensitive to the low-energy *pp* neutrinos and medium-energy $^7\mathrm{Be}$ neutrinos. There were two such experiments, SAGE and GALLEX, and these experiments found that the measured signal, when a small $^8\mathrm{B}$ contribution is subtracted, is only about 55% of the rate predicted by the SSM [Bah97]. Using a neutrino energy $E_{\nu_e} \sim 0.4$ MeV, we now have

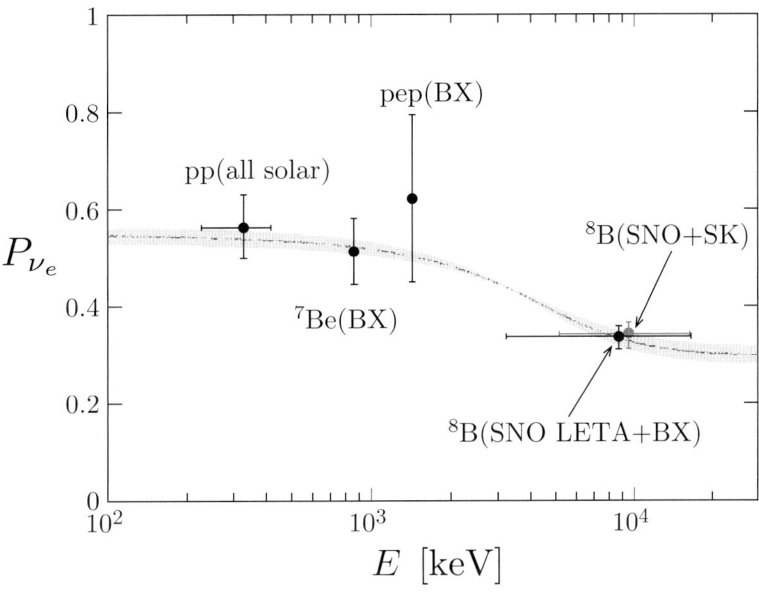

Fig. 18.4 The electron neutrino survival probability as a function of neutrino energy assuming matter-enhanced oscillations. The data points indicate measurements as made by the SNO and Super-Kamiokande experiments (for ^{8}B), Borexino, and radio-chemical experiments [Bel14].

$$2\sqrt{2}E_{\nu_e}G_F N_{ec} \simeq 6.3 \times 10^{-6}\ \text{eV}^2 \ll \Delta m_{12}^2 \cos 2\theta_{12} \simeq 2.8 \times 10^{-5}\ \text{eV}^2, \tag{18.39}$$

so that the MSW effect can be neglected and only vacuum mixing needs to be considered. The survival rate in the two-channel mixing case is then

$$P_{\nu_e} = 1 - \sin^2 2\theta_{12} \sin^2 \frac{\Delta m_{12}^2 L}{4E_{\nu_e}} \tag{18.40}$$

and since the oscillation distance $L \sim 2\pi E_{\nu_e}/m_{12}^2 \sim 6$ km is much smaller than the radius of the solar core, the distance L can be averaged, yielding $< P_{\nu_e} > = 1 - \frac{1}{2}\sin^2 2\theta_{12} \simeq 0.57$, in good agreement with experiment.

A confirmation both of the standard solar model and of neutrino mixing was provided by the Sudbury neutrino observatory (SNO) experiment, which involved the use of a heavy water (D_2O) target [Aha05]. The electron neutrino was detected via the charged-current reaction

$$\nu_e + d \rightarrow p + p + e^-\,,$$

while the neutral current weak interaction could be detected via the reaction

$$\nu_x + d \rightarrow p + n + \nu_x\,,$$

which is permitted for all neutrino flavors. The SNO sensitivity was primarily to the high-energy ^{8}B neutrinos, due to its energy threshold. The result of their measurement was that

$$\frac{\phi(\nu_e)}{\phi(\nu_e) + \phi(\nu_\mu) + \phi(\nu_\tau)} = 0.34 \pm 0.023 \pm 0.030\,, \tag{18.41}$$

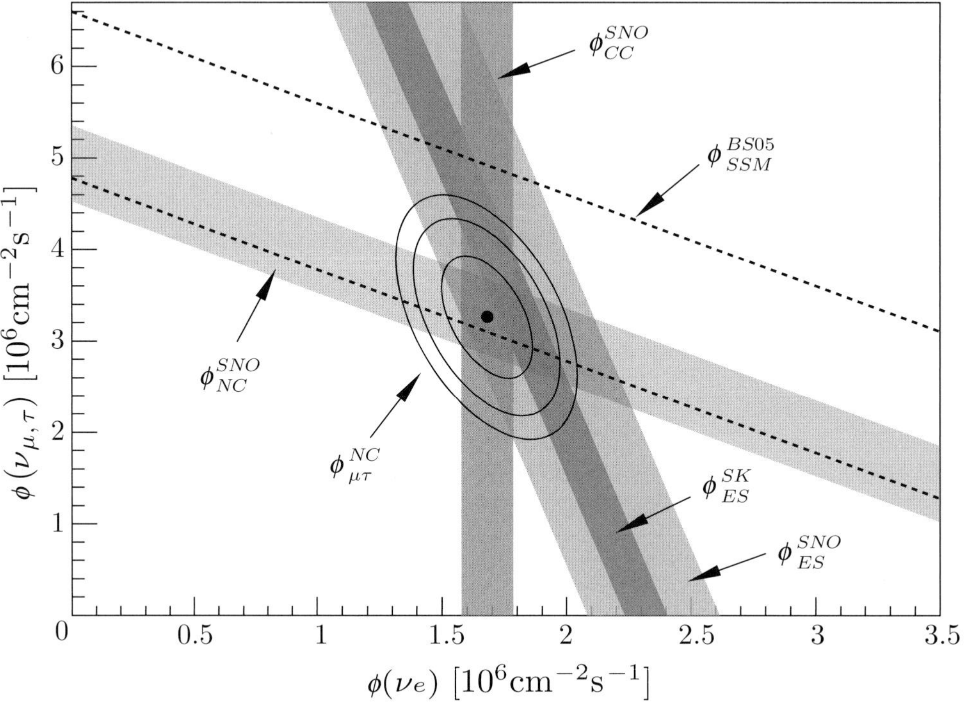

Fig. 18.5 Fluxes of ^{8}B solar neutrinos, $\phi(\nu_e)$, and $\phi(\nu_{\mu,\tau})$, deduced from the SNO's charged-current (CC), ν_e elastic scattering (ES), and neutral-current (NC) results. The Super-Kamiokande ES flux is from [Fuk02]. The BS05(OP) standard solar model prediction [Bah05] is also shown. The bands represent the 1σ uncertainty, while the contours show the 68%, 95%, and 99% joint probability for $\phi(\nu_e)$ and $\phi(\nu_{\mu,\tau})$. Note that the prediction for no neutrino oscillations is off-scale in this figure. This figure is taken from [Aha05].

which clearly shows that the electron neutrino flux is only a third of the total flux, in agreement with the Davis experiment. In addition, the SNO experiment measured the total neutrino flux as

$$\phi_{\text{tot}}(\nu) = (4.94 \pm 0.21 \pm 0.36) \times 10^6 \text{ cm}^{-2} \text{ sec}^{-1} , \tag{18.42}$$

which is in good agreement with the prediction of the SSM, shown in Table 18.1. Figure 18.5 shows the SNO and Super-Kamiokande results which consistently support the neutrino oscillation resolution of the solar neutrino puzzle.

Even stronger confirmation of the oscillation phenomena observed in the solar sector came through the measurements of the KamLAND experiment in Japan. KamLAND was an experiment which employed a 1 kiloton liquid scintillator detector located at the Kamioka mine mentioned previously. There exist 55 power reactors in the vicinity and the operating records of the reactors were used by the investigators to predict the electron antineutrino flux in the detector. Due to its energy and baseline, KamLAND provided a striking orthogonal measurement of the oscillation parameters $(\Delta m_{12}^2, \theta_{12})$. Since a nearby detector was not used, the data (see Fig. 18.6) were plotted as a function of $L/E_{\bar{\nu}_e}$, where $L = 180$ km is the flux-weighted average distance to the reactors. A very good fit was

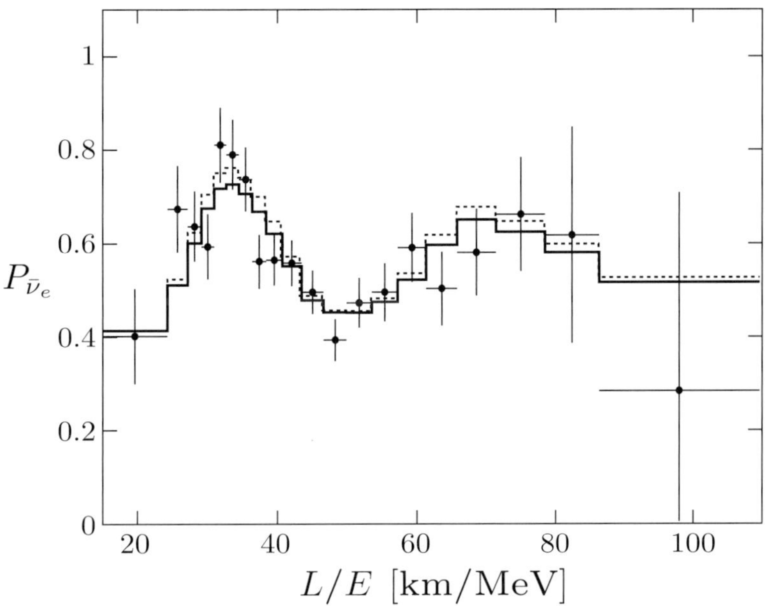

Fig. 18.6 The antineutrino survival probability versus L/E from the KamLAND experiment, figure adapted from [Gan11].

obtained using the mass splitting $\Delta m_{12}^2 = (7.6 \pm 0.2) \times 10^{-5}$ eV^2 and mixing angle $\tan^2 \theta_{12} = 0.47 \pm 0.06$ [Abe08].

Figure 18.6 shows the ratio of the background and geo-neutrino subtracted antineutrino spectrum from the KamLAND experiment [Abe08] to the expectation for no oscillation as a function of the quantity L/E. The quantity L is an effective baseline taken as a flux-weighted average ($L = 180$ km). The histogram and curve show the expectation accounting for the distances to the individual reactors, time-dependent flux variations, and efficiencies. The figure shows the behavior expected from neutrino oscillation, where the electron antineutrino survival probability is

$$P_{ee} = 1 - \sin^2 2\theta \, \sin^2(\Delta m^2 L/E) \,. \tag{18.43}$$

Atmospheric Neutrinos, Accelerator Neutrinos and Measuring the Δm_{23}^2 Splitting

When high-energy cosmic rays strike the Earth's atmosphere, a multitude of secondary particles is produced, most of which travel at nearly the speed of light in the same direction as the incoming cosmic ray (see Chapter 20 for a detailed discussion of cosmic rays). Many of the secondaries are pions and kaons, which decay into electrons and muons, and their associated neutrinos and antineutrinos. The fluxes are large but, because the neutrinos interact weakly, only a very large detector can provide evidence for their existence. An example is Super-Kamiokande, which is a $50,000$ ton water detector located in the Kamioka mine [Wen10]. The energies of these neutrinos (typically ~ 1 GeV) are high enough to produce either electrons or muons, depending on the neutrino flavor and, using known cross sections and decay rates, one anticipates about twice as many muon neutrinos as electron neutrinos. This is because, for example, a pion decays predominantly via

$$\pi^+ \to \mu^+ + \nu_\mu \to e^+ + \nu_e + \nu_\mu + \bar{\nu}_\mu \, ,$$

so that two muon-type neutrinos (antineutrinos) are produced for each electron neutrino.

When the neutrinos pass through the detector, the resulting high-energy electrons and muons produce Cerenkov radiation, and the relatively diffuse electron Cerenkov ring compared with its relatively clean muon analog allows one to distinguish electrons from muons with about 98% accuracy. Also, since the charged lepton tends to travel in the same direction as the incident neutrino which produced it, one can identify the original neutrino direction. What is found is that the electron neutrino rate is about what one would expect. However, there is a considerable zenith angle dependence for the muon neutrinos, with a very strong suppression of the rate associated with muon neutrinos which enter the detector from below, having been produced by cosmic rays which have entered the atmosphere on the opposite side of the globe and which have traveled through the Earth before interaction in the Super-Kamiokande detector. The deficit is due to mixing of the muon neutrino with the tau neutrino, since the $\sim 1\,\text{GeV}$ energies are insufficient to permit production of the τ lepton. The strong suppression of the muon neutrino flux is associated with maximal $\theta_{23} \sim \pi/4$ and the zenith angle dependence indicates that the oscillation length is of order the Earth's diameter, $\Delta m_{23}^2 \sim 2 \times 10^{-3}\,\text{eV}^2$, and yields a good fit to the Super-Kamiokande data.

Complementary information on the same mass splitting can be extracted using particle accelerators, rather than cosmic rays, as the source of neutrinos. A particle accelerator which can generate a high-energy charged pion or kaon beam is also a source of neutrinos, which are produced in the direction of the pion or kaon flux. Such neutrinos can travel to a distant detector and one can verify whether the falloff is consistent with the predicted distance attenuation or if neutrino mixing is in play. This type of experiment involves a nearby detector to determine the neutrino flux together with a larger detector at the distant location. Because of the large distances involved, the beam passes through the Earth's mantle so that density-modified mixing must be included, although, since the density is constant, no adiabaticity assumption is necessary. This type of program has been initiated at three locations around the world:

i) MINOS with the Fermilab main injector as the neutrino source, a 580 ton detector located several hundred meters from the proton target, together with a 5400 ton detector located in the Soudan mine, which is 735 km away in northern Minnesota [Ada13]. The neutrino beam consists primarily of muon neutrinos with a small background from electron neutrinos. By comparing the muon neutrino flux at the two locations, the experimenters determined that there was a deficit, which is consistent with a mass difference $\Delta m_{23}^2 = (2.3 \pm 0.1) \times 10^{-3}\,\text{eV}^2$ and a mixing angle $\sin^2 2\theta_{23} > 0.9$.

ii) T2K with the J-PARC accelerator at Tokai, Japan sending muon neutrinos to the 50,000 ton Super-Kamiokande detector, which is 295 km distant. The nearby detector consists of a set of segmented neutrino detectors and is located 280 m from the graphite target [Abe14]. The muon neutrino flux comparison again determined a deficit which can be fit with a mass difference $\Delta m_{23}^2 = (2.6 \pm 0.4) \times 10^{-3}\,\mathrm{eV}^2$ and a mixing angle $\sin^2 2\theta_{23} > 0.84$.

iii) OPERA uses the SPS muon neutrino beam at CERN together with 1300 tons of photographic emulsion "bricks" located in the Gran Sasso tunnel, which is located 730 km away in central Italy [Aga14]. In this case, the experimenters look for the conversion of muon neutrinos to tau neutrinos, which are signified by the appearance of tau mesons. So far, several tau neutrinos have been detected, and data-taking continues.

Reactor Neutrinos and the Measurement of θ_{13}

Since the splittings of Δm_{12} and Δm_{23} strongly constrain Δm_{13} (assuming that number of neutrinos is three), experiments can be specifically optimized to measure this last mixing angle. Reactor neutrinos with baselines of order a few kilometers provides the ideal L/E_ν probe to measure θ_{13}.

A nuclear reactor acts as a strong source of electron antineutrinos with a spectrum of energies in the $\sim$MeV range. Several such reactor complexes around the world have been used in order to study neutrino oscillations:

i) Chooz and Double Chooz, located off of the Chooz nuclear power plant in France, provided the first strong hint of a non-vanishing value of θ_{13}. Double Chooz consists of two detectors situated 400 meters and 1050 meters from the main reactors. Double Chooz is a successor to the CHOOZ experiment; one of its detectors occupies the same site as its predecessor. The result of the Double Chooz experiment was a measurement of the mixing angle θ_{13}: $\sin^2 2\theta_{13} = 0.090 \pm 0.034$.

ii) Daya Bay, about 30 miles north of Hong Kong in southern China is the site of six 3 gigawatt power reactors. The Daya Bay experiment uses a set of eight 20 ton liquid scintillator detectors located in three underground halls in the vicinity of the reactors. The nearest halls are only 0.5 km from the reactors, while the "distant" hall is 1.8 km distant [An12]. The signal which is sought is the disappearance of the antineutrino signal, which was observed at the level of about 6%. The interpretation is that the electron antineutrino has oscillated into a tau antineutrino, which is unobservable. Of course, mixing with the muon neutrino also occurs, as can be seen in the expression

$$P_{ee}(L) \approx 1 - \sin^2 2\theta_{13} \sin^2 \frac{\Delta m_{13}^2}{4E} L - \cos^4 \theta_{13} \sin^2 2\theta_{12} \sin^2 \frac{\Delta m_{12}^2}{4E} L \,. \tag{18.44}$$

However, since $\Delta m_{13}^2 \simeq \Delta m_{23}^2 \gg \Delta m_{12}^2$ the corresponding distance at which the minimum associated with $\bar{\nu}_e - \bar{\nu}_\mu$ mixing occurs ($\sim$60 km) is much larger than that associated with $\bar{\nu}_e - \bar{\nu}_\tau$ mixing. The result of the Daya Bay experiment was a measurement of the mixing angle θ_{13}: $\sin^2 2\theta_{13} = 0.090 \pm 0.008$.

iii) RENO is a Korean experiment which is also sensitive to $\bar{\nu}_e - \bar{\nu}_\tau$ mixing. In this case six reactors located at the Yonggwang power plant on the west coast of the Korean peninsula were used along with two gadolinium-loaded liquid scintillator detectors, one at 290 m and the other at 1.4 km from the center of the reactor array [Ahn12]. The idea here is similar to that used to analyze the Daya Bay experiment and the results are also comparable, $\sin^2 2\theta_{13} = 0.113 \pm 0.013 \pm 0.019$, in good agreement with the Daya Bay result.

The above list is certainly not inclusive, but demonstrates the range of techniques being employed to probe neutrino mixing. We observe that all the above methods have provided meaningful input to neutrino mixing phenomenology, and all different classes of experiments are ongoing, so that significantly improved limits can be expected in the not too distant future.

Summary: The Mass Hierarchy Problem

Neutrino oscillations in vacuum are proportional to $\sin^2(1.267\,\Delta m^2 L/E)$ (where Δm^2 is measured in eV^2, L in km, and E in GeV), and are therefore insensitive to the *sign* of Δm^2. Consequently, for θ_{23}, i.e., mixing between muon and tau neutrinos, we do not know whether mass state 2 is lighter or heavier than mass state 3. This is known as the mass hierarchy problem and is shown schematically in Fig. 18.7. In the normal hierarchy, the mass m_3 is heavier than the masses m_1 and m_2. In the inverted hierarchy, this is reversed.

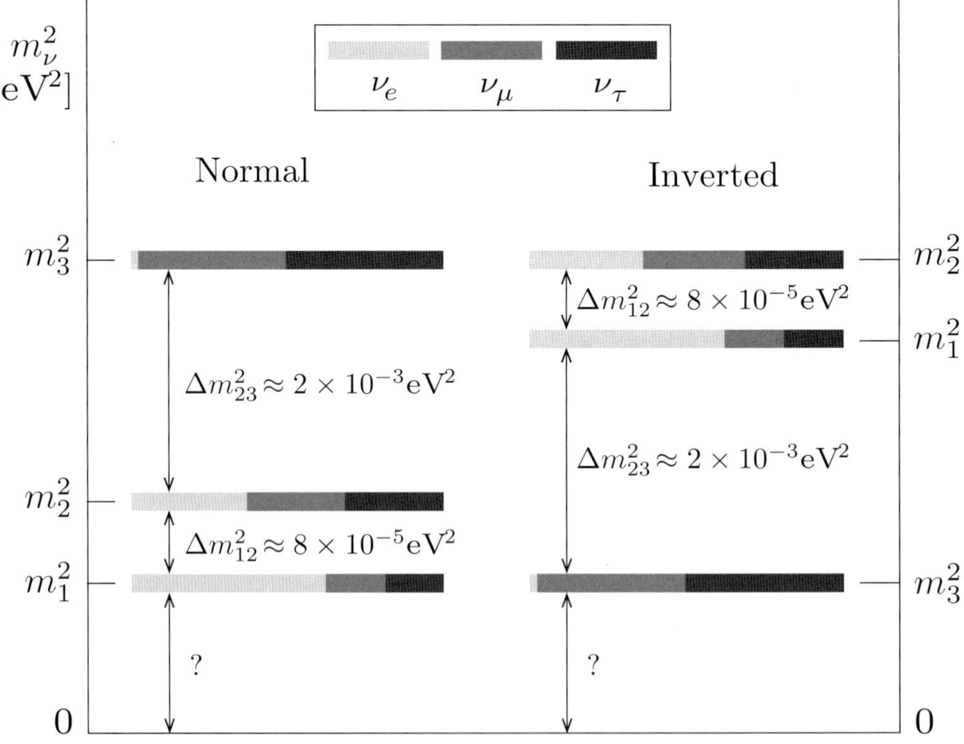

Fig. 18.7 The mass hierarchy problem, whereby the ordering and scale of neutrino masses remains unknown. The Δm_{ij}^2 values have been determined from experiments involving solar, reactor, and atmospheric neutrinos. The absolute vertical scale is constrained by the experimental limits on neutrino mass which are about 2 eV from tritium endpoint data and about 0.2 eV from interpretations of temperature fluctuations in the cosmic microwave background; see Section 21.3.

By contrast, matter-enhanced mixing (the MSW effect) is sensitive to the sign of the mass difference, because normal matter has a high *electron* density (but not mu or tau density). Neutrino experiments with baselines of order 1000 km or more can in principle disentangle the mass hierarchy: electron neutrino appearance is enhanced for normal hierarchy ($m_1 < m_3$), but suppressed for the inverted hierarchy ($m_1 > m_3$). This effect is very small for T2K, with a baseline of about 300 km, but is significant for NOνA (baseline 810 km), and even more so for the proposed experiment between Fermilab and the Sanford underground research facility (SURF:1300 km). Note that this works only for the θ_{13}, not for the larger θ_{23} mixing.

Despite the greatly improved knowledge of neutrinos and their mixing obtained during the past few years, a number of important questions remain and are being studied. One is the absolute values and patterns of neutrino mass. As we have seen, two of the three mixing angles, θ_{12}, θ_{23}, are near to being maximal, in that the angles are close to $\sin^2 2\theta = 1$, which produces the largest neutrino oscillations. The third angle, θ_{13} is considerably smaller, with $\sin^2 \theta_{13} \sim 0.1$. This pattern is then quite different from that of the quarks, wherein the mixing angles are all small (see Chapter 4). This neutrino mixing pattern is referred to as *bi-maximal*.

As yet, we do not have an absolute measure of the neutrino masses, only two mass *differences*. If the neutrino mass pattern behaves similarly to that of quark masses, then we would expect that $m_1 < m_2 < m_3$, referred to as *normal hierarchy*. However, we do not yet know if this is the pattern which is realized by Nature. Another possibility is the *inverted hierarchy* with $m_1 \sim m_2 \ll m_3$. The *degenerate* option refers to the situation that the neutrino masses are similar in magnitude and have small mass splittings between them, so that $m_1 \sim m_2 \sim m_3 \gg \Delta m_{23}$.

The fact that the neutrino mass pattern is qualitatively different from that of the quarks remains a puzzle which is not addressed by the Standard Model.

$\beta\beta$-Decay and Majorana Neutrinos

As a neutral spin-$\frac{1}{2}$ particle, the neutrino is unique in having the possibility to have a Majorana component to its mass in addition to (or instead of) its Dirac from. This important issue is being studied in neutrinoless $\beta\beta$-decay experiments [Giu12]. There exist many situations in Nature wherein one has a nucleus for which ordinary beta decay,

$$(Z,N) \rightarrow (Z+1, N-1) + e^- + \bar{\nu}_e \,,$$

is kinematically forbidden, but which is allowed to decay via emission of two lepton pairs,

$$(Z,N) \rightarrow (Z+2, N-2) + e^- + e^- + \bar{\nu}_e + \bar{\nu}_e \,.$$

Because this $\beta\beta$-decay involves five-body phase space as well as two factors of the weak coupling constant G_F this process is very rare but it *has* been observed in many nuclei, with lifetimes of order 10^{20} years, as shown in Table 18.2.

The rate for such processes is

$$\Gamma_{2\nu} = m_e^{11} F_2 \left(\frac{Q}{m_e} \right) \left| g_A^2 M_{GT} - g_V^2 M_F \right|^2 \left(\frac{\mathcal{F}(Z)}{E_i - <E_n> - \frac{1}{2}E_0} \right), \tag{18.45}$$

Table 18.2 Half-lives of two-neutrino double beta emitters, $T^{2\nu}_{\frac{1}{2}}$ and upper bounds on half-lives of neutrinoless double beta emitters and corresponding limits on $< m_\nu >$

Nucleus	Q-value (MeV)	$T^{2\nu}_{\frac{1}{2}}$ (10^{19} yr)	$T^{0\nu}_{\frac{1}{2}}$ (yr)	$< m_\nu >$ (eV)
^{48}Ca	4.276	$3.9 \pm 0.7 \pm 0.6$	$> 1.14 \times 10^{22}$	< 7.2
^{76}Ge	2.039	170 ± 20	$> 1.6 \times 10^{25}$	< 0.33
^{82}Se	2.992	$9.6 \pm 0.3 \pm 1.0$	$> 1.9 \times 10^{23}$	< 1.3
^{100}Mo	3.034	$0.711 \pm 0.002 \pm 0.054$	$> 5.8 \times 10^{23}$	< 0.8
^{116}Cd	2.804	$2.8 \pm 0.1 \pm 0.3$	$> 1.7 \times 10^{23}$	< 1.7
^{128}Te	0.876	20000 ± 1000	$> 7.7 \times 10^{24}$	< 1.1
^{130}Te	2.529	$76 \pm 15 \pm 8$	$> 3 \times 10^{23}$	< 0.46
^{136}Xe	2.467		$> 4.4 \times 10^{23}$	< 1.8
^{150}Nd	3.368	$0.92 \pm 0.03 \pm 0.07$	$> 1.2 \times 10^{21}$	< 7

where $\mathcal{F}(Z)$ is a Fermi function,

$$F_2(x) = x^7 \left(1 + \frac{x}{2} + \frac{x^2}{9} + \frac{x^3}{90} + \frac{x^4}{1980} \right) \tag{18.46}$$

is a kinematic factor and

$$M_F = \left\langle f \left| \frac{1}{2} \sum_{ij} \tau_i^+ \tau_j^- \right| i \right\rangle$$

$$M_{GT} = \left\langle f \left| \frac{1}{2} \sum_{ij} \tau_i^+ \tau_j^- \, \boldsymbol{\sigma}_i \cdot \boldsymbol{\sigma}_j \right| i \right\rangle \tag{18.47}$$

are nuclear matrix elements (see Chapter 17). Here the closure approximation has been made to represent a sum over intermediate states via an average excitation energy $< E_n >$. The experimental 2ν decay rates can then be used to determine these matrix elements, which are difficult to calculate reliably.

However, if the neutrino has a Majorana mass component, it is possible to have a *neutrinoless $\beta\beta$-decay*

$$(Z, N) \rightarrow (Z + 2, N - 2) + e^- + e^-$$

and the experimental observation of such a process would unambiguously answer the question of whether the neutrino has a Majorana mass component or not. The observation of such a decay would also provide a measure of the Majorana mass itself, or more precisely. to the weighted quantity

$$< m_{\beta\beta} >= \sum_{i=1}^{3} U_{ie}^2 m_i \, . \tag{18.48}$$

Note that it is the square of the matrix element U_{ie}, not the usual combination $UU^\dagger$, that enters this reaction, because both weak currents lead to e^- emission in the final state.

The decay rate for such a neutrinoless decay is given by

$$\Gamma_{0\nu} \sim m_e^7 F_0 \left(\frac{Q}{m_e}\right) \left| G_A^2 \widetilde{M}_{GT} - g_V^2 \widetilde{M}_F \right|^2 \frac{<m_{\beta\beta}>^2}{m_e^2}, \tag{18.49}$$

where

$$F_0(x) = x \left(1 + 2x + \frac{4x^2}{3} + \frac{x^3}{3} + \frac{x^4}{30}\right) \tag{18.50}$$

is a kinematic factor and

$$\widetilde{M}_F = \left\langle f \left| \frac{1}{2} \sum_{ij} \tau_i^+ \tau_j^- \frac{1}{r_{ij}} \right| i \right\rangle$$

$$\widetilde{M}_{GT} = \left\langle f \left| \frac{1}{2} \sum_{ij} \tau_i^+ \tau_j^- \boldsymbol{\sigma}_i \cdot \boldsymbol{\sigma}_j \frac{1}{r_{ij}} \right| i \right\rangle \tag{18.51}$$

are nuclear matrix elements. The factor of $1/r_{ij}$ comes from the spatial dependence associated with neutrino propagation in the limit in which the neutrino mass is neglected.

As of yet no such 0ν mode has been observed and the current limits on such decays are listed in Table 18.2. As can be seen from this table, present limits on $<m_\nu>$ are currently at the electron volt level, but experiments are planned that aim to lower this bound by as much as a factor of about a hundred. The $0\nu\beta\beta$-decay half-life can be related to an effective Majorana mass according to

$$\frac{1}{T_{1/2}^{0\nu}} = G_{0\nu} g_A^4 |M^{(0\nu)}|^2 <m_{\beta\beta}>^2, \tag{18.52}$$

where $G_{0\nu}$ is a phase space factor, m_i is the mass of the neutrino mass eigenstate ν_i, and $M_{0\nu}$ is the transition nuclear matrix element. The matrix element has significant nuclear theoretical uncertainties. In the standard three-massive-neutrino paradigm,

$$<m_{\beta\beta}>^2 = |\cos^2\theta_{12}\cos^2\theta_{13}m_1 + e^{2i\lambda_2}\sin^2\theta_{12}\cos^2\theta_{13}m_2 + e^{2i(\lambda_3-\delta_{CP})}\sin^2\theta_{13}m_3|^2. \tag{18.53}$$

If none of the neutrino masses vanishes, $<m_{\beta\beta}>$ is a function of not only the mixing angles θ_{ij}, and the neutrino masses m_i, but also the CP-violating phase δ_{CP}, as well as two Majorana phases $\lambda_{2,3}$. We have only partial knowledge of the parameters controlling $<m_{\beta\beta}>$. The mixing angles and mass splittings (modulo the mass hierarchy problem) are known; however, we do not know the ordering and the absolute mass scale of the spectrum. Further, the phases δ_{CP} and $\lambda_{2,3}$ are unknown. Figure 18.8 shows $<m_{\beta\beta}>$ in meV plotted versus the mass of the lightest neutrino, also in meV.

As described above, the focus of current experiments is to increase the sensitivity by increasing the size of the experiment, which dictates increasing the amount of the selected isotope. Within about a decade, a ton-scale experiment is anticipated and, as shown in Fig. 18.8, discovery of $0\nu\beta\beta$-decay should be feasible if the inverted hierarchy holds or the mass of the lightest neutrino is larger than about 50 meV. On that timescale, we also expect direct information on the mass ordering (from accelerator experiments studying oscillations), on the absolute mass scale (from KATRIN), and from cosmology.

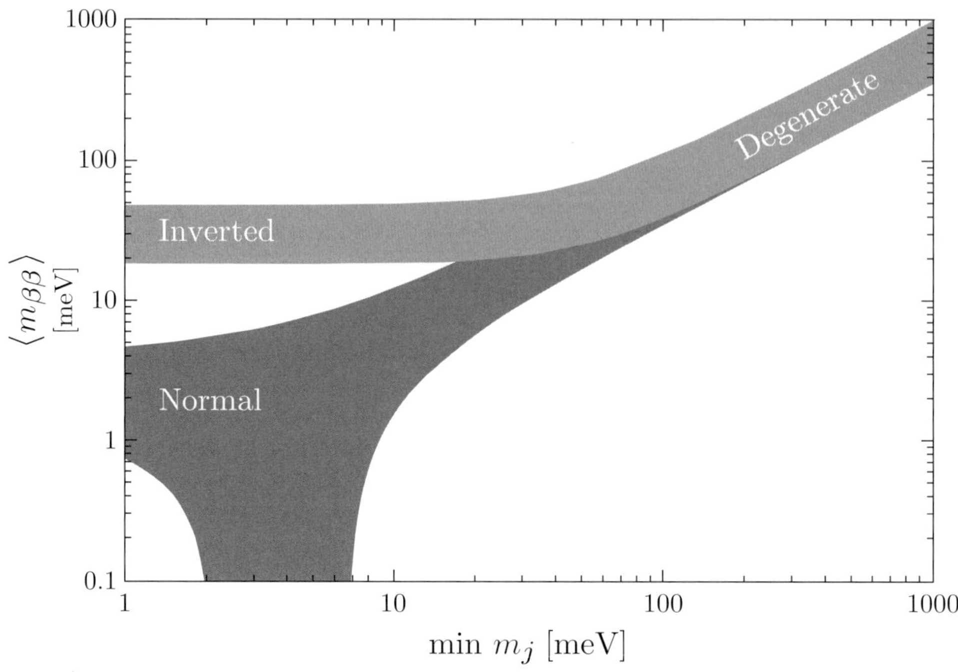

Fig. 18.8   The effective Majorana mass $|\langle m_{\beta\beta}\rangle|$ as a function of $\min(m_j)$ (from [PDG14]). Note the split in the mass observable due to the ordering (normal, inverted) and the mass scale in comparison to the mass splitting (degenerate). Current measurements on neutrinoless $\beta\beta$-decay lifetime place a (model-dependent) upper limit on the $m_{\beta\beta}$ parameter of 350 meV.

It is expected that this additional information will contribute to the interpretation of both positive or null $0\nu\beta\beta$ results.

Thus far, we have discussed experiments which have probed neutrino properties. However, the neutrino can also be used as a probe to study other systems and that is the subject of the remainder of this chapter.

18.4 Neutrino Reactions

Researchers further investigate the nature of the neutrino mass or explore whether neutrinos can help explain the matter–antimatter asymmetry of the Universe. At the heart of many of these experiments is the need to understand how neutrinos interact with other Standard Model particles. An understanding of these basic interactions (cross sections) is often an understated but truly essential element of any experimental neutrino program.

The Standard Model provides guidance for how to model such cross sections properly. However, the exact implementation of such varies widely depending on the energy, type of target or level of precision one wishes to achieve. Often the nature of the interaction and the underlying assumptions one can make are determined by the total momentum exchange of the interaction (for which the neutrino energy is a poor but often used proxy). We begin the discussions here with a treatment of neutrino–lepton scattering, a purely leptonic reaction for which the familiar rules of quantum field theory can be used to obtain the cross section. This is then followed by discussions of charge-changing or neutral current neutrino reactions (CC and NC, respectively) with nucleons and nuclei. Indeed, the latter are essential, since the targets/detectors employed in experiments worldwide are very large and accordingly typically must be constructed from materials like water or mineral oil, i.e., from nuclei.

A crude subdivision of the types of reactions involving nucleons or nuclei is the following:

i) *Neutrino–nucleon scattering, especially inverse beta decay.*

ii) *Elastic scattering*: This is the neutral current neutrino scattering analog of elastic electron scattering, as presented in Chapter 15. As discussed there and below, for sufficiently "soft" scatters (i.e., at very small momentum transfer), the neutrino probes the nucleus as a single coherent object.

iii) *CCν and NCν reactions involving low-lying states in nuclei*: The latter are the counterparts to inelastic electron scattering (see Chapter 15), whereas the former are the analogous charge-changing reactions which initiate transitions between neighboring nuclei, similar to beta decay (see Chapter 17).

iv) *Quasielastic scattering (CCQE and NCQE)*: Neutrinos can interact with nuclei in a similar way to electrons, except via the weak interaction rather than the EM interaction. In Chapter 16 the topic of quasielastic electron scattering was discussed in some detail and below these ideas are extended to include the CCQE reaction.

v) *Resonance production*: Neutrinos can excite the target nucleon to a resonance state. The resulting baryonic resonance (Δ, N^*) then decays to a variety of possible mesonic final states producing combinations of nucleons and mesons. This can happen on a nucleon or on a nucleus (see Chapter 16 for the analogous excitation of the Δ in nuclei).

vi) *Deep inelastic scattering*: Given enough energy, the neutrino can resolve the individual quark constituents of the nucleon, in which case this is called deep inelastic scattering (DIS) (cf., Chapter 9) and is manifested in the creation of a hadronic shower.

As a result of these competing processes, the products of neutrino interactions include a variety of final states ranging from the emission of nucleons to more complex final states including pions, kaons, and collections of mesons. This energy regime is often referred to as the "transition region" because it corresponds to the boundary between quasielastic scattering on the one end and deep inelastic scattering on the other. Historically, adequate theoretical descriptions of QE, resonance-mediated, and DIS have been formulated, but there exists no uniform description which globally describes the transition between these processes or how they should be combined.

Neutrino–Lepton Scattering

We will begin with perhaps one of the simplest interactions to describe, neutrino–lepton scattering. As a purely leptonic interaction, neutrino–lepton scattering allows us to

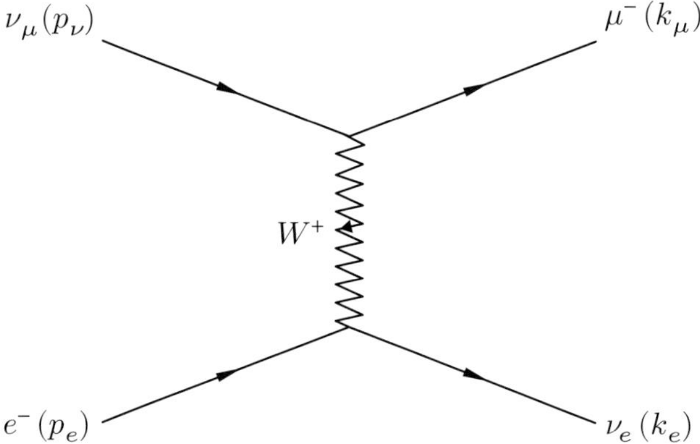

Fig. 18.9 Diagram of two-body scattering between an incoming muon neutrino with four-momentum p_ν and an electron at rest with four-momentum p_e. See text for details.

establish the formalism and terminology without introducing some of the complexity that can accompany neutrino scattering from hadronic systems. The general form of the two-body scattering process is governed by the dynamics of the process encoded in the matrix elements and the phase space available in the interaction. Figure 18.9 shows the tree-level diagram of a neutrino–lepton charged current interaction, known as inverse muon decay, wherein a muon neutrino with four-momentum p_ν (aligned along the z-direction) scatters from an electron with four-momentum p_e at rest in the laboratory frame, producing an outgoing muon with four-momentum k_μ and a scattered electron neutrino with four-momentum k_e. In the laboratory frame, the components of these quantities can be written as:

$$
\begin{aligned}
p_\nu &= (E_\nu, \boldsymbol{p}_\nu) \\
k_\mu &= (E_\mu, \boldsymbol{k}_\mu) \\
p_e &= (m_e, \boldsymbol{0}) \\
k_e &= (E_e, \boldsymbol{k}_e) \,.
\end{aligned}
$$

Here the first term of the four-vector represents the zeroth component corresponding to the energy portion of the energy–momentum vector, with the usual energy–momentum relation $E_i^2 = |\boldsymbol{k}|_i^2 + m_i^2$. From these four-vector quantities, we can define the conventional Lorentz-invariant quantities Q^2, y, and s, as we have done for Chapters 7 and 8.

In the case of two-body collisions between an incoming neutrino and a (stationary) target lepton, the cross section is given in general by the formula [Ber74]:

$$
\frac{d\sigma}{dq^2} = \frac{1}{16\pi} \frac{|\mathcal{M}^2|}{(s - (m_e + m_\nu)^2)(s - (m_e - m_\nu)^2)}, \tag{18.54}
$$

which, in the context of very small neutrino masses, simplifies to

$$
\frac{d\sigma}{dq^2} = \frac{1}{16\pi} \frac{|\mathcal{M}^2|}{(s - m_e^2)^2} \,. \tag{18.55}
$$

Here, $\mathcal{M}$ is the matrix element associated with our particular interaction (Fig. 18.5). In the laboratory frame, it is always possible to express the cross section in alternative ways by making use of the appropriate Jacobian. For example, to determine the cross section as a function of the muon scattering angle, θ_μ, the Jacobian is

$$\frac{dq^2}{d\cos\theta_\mu} = 2|\boldsymbol{p}_\nu||\boldsymbol{k}_\mu|\,, \tag{18.56}$$

while the Jacobian written in terms of the fraction of the neutrino energy imparted to the outgoing lepton energy (y) is given by

$$\frac{dq^2}{dy} = 2m_e E_\nu\,. \tag{18.57}$$

Depending on what physics one is interested in studying, the differential cross sections can be recast to highlight a particular dependence or behavior.

The full description of the interaction is encoded within the matrix element, and the Standard Model readily provides a prescription to describe neutrino interactions via the leptonic charged current and neutral current in the weak interaction Lagrangian. Within the framework of the Standard Model (see Chapter 4), a variety of neutrino interactions are readily described [Wei67]. Such interactions all fall within the context of the general gauge theory of $SU(2)_L \times U(1)_Y$ and can be divided into three broad categories. The first is mediated by the exchange of a charged W boson, otherwise known as a charged-current (CC) exchange. The leptonic charged weak current, $J_\mu^\pm$, is given by

$$J_\mu^\pm = \bar{\psi}_L \gamma_\mu \tau_\pm \psi_L = \bar{\psi}\gamma_\mu(1-\gamma_5)\frac{1}{2}\tau_\pm \psi\,. \tag{18.58}$$

The second type of interaction, known as the neutral current (NC) exchange, is similar in character to the charged-current case. The leptonic neutral current term, J_μ^Z, describes the exchange of the neutral Z^0 boson, with

$$J_\mu^Z = \frac{1}{4}\bar{\psi}\gamma_\mu(1-\gamma_5)\tau_3\psi - \sin^2\theta_w J_\mu^{em}\,. \tag{18.59}$$

Although the charged leptonic fields are of a definite mass eigenstate, this is not necessarily so for the neutrino fields, giving rise to the phenomena of neutrino oscillations discussed above.

Historically, the neutrino–lepton CC and NC interactions have been used to study the structure of the weak force in great detail. Returning to the calculation of the charged and neutral current reactions, the previously defined currents enter directly into the Lagrangian via their coupling to the heavy gauge bosons, $W^\pm$ and Z^0

$$\mathcal{L}_{CC} = -\frac{g}{\sqrt{2}}(J_\mu^+ W^\mu + J_\mu^- W^\mu) \tag{18.60}$$

$$\mathcal{L}_{NC} = -\frac{g}{2\cos\theta_W}J_Z^\mu Z_\mu\,. \tag{18.61}$$

Here, W_μ and Z_μ represent the heavy gauge boson field, g is the $SU(2)_L$ coupling constant while θ_W is the weak mixing angle. It is customary to represent these exchanges by the use of Feynman diagrams, as shown in Fig. 18.10. Using this formalism, it is possible to

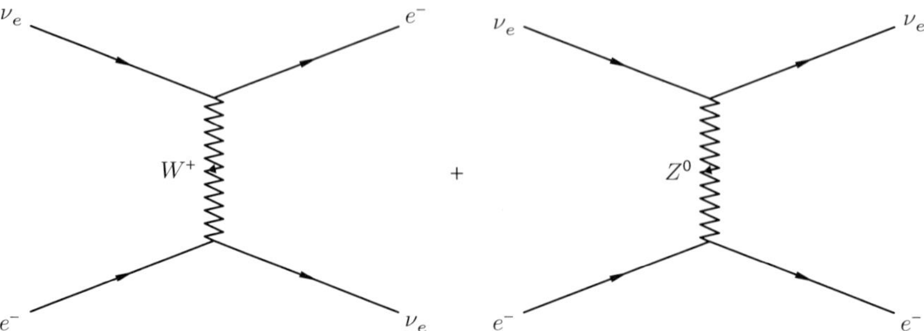

Fig. 18.10 Feynman tree-level diagram for charged and neutral current components of $\nu_e + e^- \to \nu_e + e^-$ scattering.

articulate all neutrino interactions [Hoo71] within this simple framework. We begin by examining one of the simplest manifestations of the above formalism, where the reaction involves a purely charged-current interaction:

$$\nu_l + e^- \to l^- + \nu_e \quad (l = \mu \text{ or } \tau) . \tag{18.62}$$

The corresponding tree-level amplitude can be calculated from the above expressions, and in the case of $\nu_l + e$ (sometimes known as inverse muon or inverse tau decays), one finds that

$$\mathcal{M}_{CC} = -\frac{G_F}{\sqrt{2}} \{ [\bar{l}\gamma^\mu (1 - \gamma_5)\nu_l][\bar{\nu}_e \gamma_\mu (1 - \gamma_5)e] \} . \tag{18.63}$$

Here, and in all future cases unless otherwise specified, we assume that the four-momentum of the intermediate boson is much smaller than its mass (i.e., $|q^2| \ll M_{W,Z}^2$), so that propagator effects can be ignored. In this approximation, the coupling strength is then dictated primarily by the Fermi constant, G_F:

$$G_F = \frac{g^2}{4\sqrt{2}M_W^2} = 1.1663787(6) \times 10^{-5} \text{ GeV}^{-2}. \tag{18.64}$$

By summing over all polarizations, and integrating over all unobserved momenta, one obtains the differential cross section with respect to the fractional energy imparted to the outgoing lepton,

$$\frac{d\sigma(\nu_l e \to \nu_e l)}{dy} = \frac{2m_e G_F^2 E_\nu}{\pi} \left(1 - \frac{(m_l^2 - m_e^2)}{2m_e E_\nu} \right), \tag{18.65}$$

where E_ν is the energy of the incident neutrino and m_e and m_l are the masses of the electron and outgoing lepton, respectively. The dimensionless inelasticity parameter, y, reflects the kinetic energy of the outgoing lepton, which in this particular example is

$$y = \frac{E_l - \frac{(m_l^2 + m_e^2)}{2m_e}}{E_\nu} .$$

The limits of y are such that

$$0 \le y \le y_{\max} = 1 - \frac{m_l^2}{2m_e E_\nu + m_e^2} . \tag{18.66}$$

(Note that, in this derivation, we have neglected the contribution from neutrino mass, which in this context is too small to be observed kinematically.) The above cross section has a threshold energy imposed by the kinematics of the system,

$$E_\nu \geq \frac{(m_l^2 - m_e^2)}{2m_e} .$$

In the case where $E_\nu \gg E_{\text{thresh}}$, integration of the above expression yields a simple result for the total neutrino cross section as a function of neutrino energy,

$$\sigma \simeq \frac{2m_e G_F^2 E_\nu}{\pi} = \frac{G_F^2 s}{\pi} , \tag{18.67}$$

where s is the square of the collision center-of-mass energy. Note that the cross section grows linearly with the energy of the neutrino.

Because of the different available spin states, the equivalent expression for the inverse lepton decay of antineutrinos,

$$\bar{\nu}_e + e^- \rightarrow \bar{\nu}_l + l^- \quad (l^- = \mu^- \text{ or } \tau^-) , \tag{18.68}$$

has a different dependence on y than its neutrino counterpart, although the matrix elements are equivalent.

$$\frac{d\sigma(\bar{\nu}_e e^- \rightarrow \bar{\nu}_l l^-)}{dy} = \frac{2m_e G_F^2 E_\nu}{\pi} \left((1-y)^2 - \frac{(m_l^2 - m_e^2)(1-y)}{2m_e E_\nu} \right) . \tag{18.69}$$

Upon integration, the total cross section is approximately a factor of three lower than the neutrino cross section, the suppression arising entirely from helicity considerations.

Having considered a charged-current example, we turn our attention to a purely neutral-current exchange, such as

$$\bar{\nu}_l + e \rightarrow \bar{\nu}_l + e \quad (1 = \mu \text{ or } \tau) . \tag{18.70}$$

In such neutral-current interactions, we are must consider both left-handed *and* right-handed leptonic couplings. As a result, one obtains a more complex expression for the relevant matrix element (for a useful review, see [Ada09]):

$$\mathcal{M}_{NC} = -\sqrt{2} G_F \{ [\bar{\nu}_l \gamma^\mu (g_V^\nu - g_A^\nu \gamma_5) \nu_l] [\bar{e} \gamma_\mu (g_V^f - g_A^f \gamma_5) e] \} . \tag{18.71}$$

Here we have expressed the coupling strengths in terms of polar-vector and axial-vector coupling constants (g_V and g_A, respectively). An equivalent representation can be given using purely left- and right- handed couplings:

$$\begin{aligned}
\mathcal{M}_{NC} = -\sqrt{2} G_F \{ & [g_L^\nu \bar{\nu}_l \gamma^\mu (1 - \gamma_5) \nu_l + g_R^\nu \bar{\nu}_l \gamma^\mu (1 + \gamma_5) \nu_l] \\
& \times \{ [g_L^f \bar{e} \gamma^\mu (1 - \gamma_5) e + g_R^f \bar{e} \gamma^\mu (1 + \gamma_5) e] \} .
\end{aligned} \tag{18.72}$$

The values of these coupling constants are dictated by the Standard Model

$$g_L^\nu = \sqrt{\rho} \left(\frac{1}{2} \right)$$

$$g_R^\nu = 0$$

$$g_L^f = \sqrt{\rho}(T_3^f - Q^f \sin^2 \theta_W)$$

$$g_R^f = \sqrt{\rho}(-Q^f \sin^2 \theta_W) \,,$$

or, equivalently,

$$g_V^\nu = g_L^\nu + g_R^\nu = \sqrt{\rho}\left(\frac{1}{2}\right)$$

$$g_A^\nu = g_L^\nu - g_R^\nu = \sqrt{\rho}\left(\frac{1}{2}\right)$$

$$g_V^f = g_L^f + g_R^f = \sqrt{\rho}(T_3^f - 2Q^f \sin^2 \theta_W)$$

$$g_A^f = g_L^f - g_R^f = \sqrt{\rho}(I_3^f) \,.$$

Here, T_3^f and Q^f are the weak isospin and electromagnetic charge of the target lepton, ρ is the relative coupling strength between charged- and neutral-current interactions (at tree level, $\rho \equiv 1$), while θ_W is the weak mixing angle. Recall that the Standard Model defines the relation between the electroweak couplings and gauge boson masses M_W and M_Z to be

$$\sin^2 \theta_W \equiv 1 - \frac{M_W^2}{M_Z^2} \,. \tag{18.73}$$

We observe that the cross sections for the neutral current reactions highlighted above are directly sensitive to the values of these neutral current couplings. In the literature, the cross section is often expressed in terms of the vector and axial-vector quantities

$$g_V \equiv (2g_L^\nu g_V^f)$$

$$g_A \equiv (2g_L^\nu g_A^f) \,,$$

so that

$$\frac{d\sigma\,(\nu_l e \to \nu_l e)}{dy} = \frac{m_e G_F^2 E_\nu}{2\pi}\left((g_V + g_A)^2 + (g_V - g_A)^2(1 - y)^2 - (g_V^2 - g_A^2)\frac{m_e y}{E_\nu}\right),$$

$$\frac{d\sigma\,(\bar{\nu}_l e \to \bar{\nu}_l e)}{dy} = \frac{m_e G_F^2 E_\nu}{2\pi}\left((g_V - g_A)^2 + (g_V + g_A)^2(1 - y)^2 - (g_V^2 - g_A^2)\frac{m_e y}{E_\nu}\right).$$

Although we have limited ourselves above to discussing neutrino–lepton scattering, the rules governing the coupling strengths are completely determined by the Standard Model and can be used to describe neutrino–quark interactions as well. A full list of the different coupling strengths for the known fermion fields is given in Table 18.3. A more in-depth discussion of these topics can be found in a variety of introductory textbooks. In particular, [Giu07] is an excellent in-depth resource.

Neutrino–electron scattering has been a powerful probe of the weak interaction, both in terms of the total cross section as well as the energy dependence [Mar03]. Before leaving neutrino-lepton interactions completely, we turn our attention to the last possible reaction archetype, wherein the charged-current and neutral-current amplitudes interfere with one

Table 18.3 Values for the g_V (vector), g_A (axial), g_L (left), and g_R (right) coupling constants for the known fermion fields

Fermion	g_L^f	g_R^f	g_V^f	g_A^f
ν_e, ν_μ, ν_τ	$+\dfrac{1}{2}$	0	$+\dfrac{1}{2}$	$+\dfrac{1}{2}$
e, μ, τ	$-\dfrac{1}{2}$	$+\sin^2\theta_W$	$-\dfrac{1}{2}+2\sin^2\theta_W$	$-\dfrac{1}{2}$
u, c, t	$+\dfrac{1}{2}-\dfrac{2}{3}\sin^2\theta_W$	$-\dfrac{2}{3}\sin^2\theta_W$	$+\dfrac{1}{2}-\dfrac{4}{3}\sin^2\theta_W$	$+\dfrac{1}{2}$
d, s, b	$-\dfrac{1}{2}+\dfrac{1}{3}\sin^2\theta_W$	$+\dfrac{1}{3}\sin^2\theta_W$	$-\dfrac{1}{2}+\dfrac{2}{3}\sin^2\theta_W$	$-\dfrac{1}{2}$

another. Such a combined exchange is realized in $\nu_e + e^- \to \nu_e + e^-$ scattering. A feature of electron–neutrino scattering is that it is highly directional in nature. The outgoing electron is emitted at very small angles with respect to the incoming neutrino direction. A simple kinematic argument shows that

$$E_e \theta_e^2 \leq 2m_e \,. \tag{18.74}$$

This feature has been exploited extensively in various neutrino experiments, particularly for solar neutrino detection. The Kamiokande neutrino experiment was the first to use this reaction to reconstruct B^8 neutrino events from the Sun and point back to the source. The Super-Kamiokande experiment later expanded the technique, creating a photograph of the Sun using neutrinos [Fuk98],[1] and the technique was later used by other solar experiments, such as SNO [Ahm01, Ahm02a, Ahm02b] and BOREXINO [Ali02, Arp08].

Next let us turn to discussions of neutrino reactions with nucleons and nuclei, beginning with a brief treatment of so-called inverse beta decay.

Inverse Beta Decay

The simplest nuclear interaction that we can study is antineutrino–proton scattering, otherwise known as inverse β-decay

$$\bar{\nu}_e + p \to e^+ + n \,, \tag{18.75}$$

and represents one of the earliest reactions to be studied, both theoretically [Bet34] and experimentally [Rei76]. This reaction is measured using neutrinos produced from fission in nuclear reactors with neutrino energies ranging from threshold ($E_\nu \geq 1.806\,\text{MeV}$) to about $10\,\text{MeV}$.[2] As the inverse beta-decay reaction plays an important role in understanding supernova explosion mechanisms, its relevance at somewhat higher energies (10–$20\,\text{MeV}$) is also of importance. Here we follow the formalism of [Bea99], who expanded the cross

---

[1] This elegant image, in fact, serves as the cover of this book.
[2] The neutrino energy threshold E_ν^{thresh} in the laboratory frame is defined by $[(m_n + m_e)^2 - m_p^2]/2m_p$.

section on the proton to first-order in nucleon mass to study the angular dependence of the cross section. In this approximation, all relevant form factors approach their zero-momentum values and the relevant matrix element is given by

$$\mathcal{M} = \frac{G_F V_{ud}}{\sqrt{2}} \left[\langle \bar{n} | \left(\gamma_\mu g_V(0) - \gamma_\mu \gamma_5 g_A(0) - \frac{i g_M(0)}{2M_n} \sigma_{\mu\nu} q^\nu \right) | p \rangle \langle \bar{v}_e | \gamma^\mu (1 - \gamma_5) | e \rangle \right],$$

$$(18.76)$$

where g_V, g_A, and g_M are nuclear vector, axial-vector, and Pauli (weak magnetism) form factors evaluated at zero momentum transfer (for greater detail on the form factor behavior, see Section 17.3). To first-order, the differential cross section can therefore be written as

$$\frac{d\sigma(\bar{v}_e p \rightarrow e^+ n)}{d\cos\theta} = \frac{G_F^2 |V_{ud}|^2 E_e p_e}{2\pi} \left[g_V^2(0)(1 + \beta_e \cos\theta) + 3g_A^2(0) \left(1 - \frac{\beta_e}{3} \cos\theta \right) \right],$$

$$(18.77)$$

where E_e, p_e, β_e, and $\cos\theta$ refer to the electron energy, momentum, velocity, and scattering angle, respectively. Note that the formalism is quite similar to what was discussed for neutron beta decay in Chapter 17.

A few properties in the above formula immediately attract our attention. First and foremost is that the cross section neatly divides into two distinct pieces: a vector-like component, often called the Fermi transition, and an axial-vector-like component, often referred to as the Gamow–Teller term. This classification was previously discussed in greater detail back in Chapter 17. A second striking feature is its angular dependence. The vector portion has a clear $(1 + \beta_e \cos\theta)$ dependence, while the axial-vector component has a $(1 - \frac{\beta_e}{3} \cos\theta)$ behavior, at least to first-order in the nucleon mass. For antineutrino–proton interactions the angular distribution is backward peaked, indicating that the vector and axial-vector terms both contribute with comparable amplitudes. This weak dependence is less pronounced for cases where the interaction is dominantly Gamow–Teller in nature, such as for vd reactions. In such cases, the angular distribution is significantly more prominent in the backward direction. Such angular distributions have been posited as an experimental tag for supernova detection [Bea02].

Next let us turn to brief treatments of neutrino reactions with nuclei. As discussed in Chapter 17, the weak interaction has long been studied using nuclei via beta decay, and via the inverse beta decay process discussed above, namely, the prototypical (anti)neutrino–hadron reaction. Reactions with nuclei gained in interest in the 1970s with the advent of high-intensity, medium-energy accelerators (LAMPF at Los Alamos, TRIUMF in Vancouver and PSI (aka SIN) in Switzerland). In particular, at LAMPF the possibilities of both a stopped-pion decay facility and a decay-in-flight facility were exploited. The former is attained by having the primary proton beam hit the beam stop, produce pions (π^+) which come to rest and decay to muons (μ^+) and mono-energetic muon neutrinos (v_μ), followed by which the muons come to rest and decay to positrons, electron neutrinos and muon antineutrinos ($e^+ \, v_e \, \bar{v}_\mu$), with energies given by the Michel spectrum.

These experimental opportunities stimulated theoretical studies in the early 1970s. For instance, many of the issues in studies of CC neutrino reactions with nuclei were identified in [OCo72, Don72, Don73, Don73a, Wal75] and in studies of NC neutrino scattering in [Don74, Don75]. For the latter, the reader is encouraged to examine a review article [Don79] at the end of the decade specifically focused on neutral current neutrino reactions (together with a brief section on PV electron scattering; see also Chapter 15) – there one can gain some insight into the thinking at a time when the Standard Model had not yet been identified as the potentially correct model and other gauge theory alternatives were still in play.

We begin a brief summary of the various energy regimes listed above with a discussion of elastic neutrino scattering. For CCν reactions one may say that they involve "nothing in, but something out," meaning the final-state charged lepton can be detected. However, as far as weak neutral current (WNC) scatterings are concerned, of course, it is impractical to detect the scattered neutrino and so one has "nothing in and nothing out," although something can happen such as the excitation and subsequent decay of a nucleus. Indeed this idea motivated much of the early work on WNC in nuclei [Don79]. The ultimate is what we discuss in more detail below, namely, elastic scattering where there is "nothing in, nothing out and nothing happens," except that the nucleus in its ground state recoils.

Coherent Scattering

A treatment of elastic neutrino scattering closely parallels the developments presented in Chapter 15 for elastic electron scattering. As before, one has the familiar allowed (polar-vector) Coulomb and magnetic multipoles, $C0$, $C2$, $C4$, etc., and $M1$, $M3$, $M5$, etc., with isoscalar and isovector components, now, however, weighted by the WNC coupling constants (see [Don79] for details). In addition, there is now the axial-vector WNC to deal with. The development of the corresponding multipoles proceeds completely in parallel with the procedures presented in Chapter 7, although this takes us beyond the scope of the book and so is not given here; the reader is directed to [Wal75, Don75, Don79, Don79a] for details.

In general all allowed multipoles play a role. However, just as in elastic electron scattering, the Coulomb monopole ($C0$) has a special role to play. This is the only type of multipole that is coherent, meaning that all nucleons can contribute, whereas for the other allowed multipoles typically only one (or a few) of the nucleons in the nucleus plays a role. Examples already discussed in Chapter 15 are the electron scattering elastic charge form factor which goes as Z in the long wavelength limit and the elastic magnetic dipole form factor which goes as the magnetic dipole moment in the LWL, the latter coming typically from a single unpaired nucleon (see Chapter 13). The same situation occurs for elastic neutrino scattering: whereas all allowed vector and axial-vector multipole form factors contribute, only the Coulomb monopole is coherent. For elastic neutrino scattering this form factor is similar to its electron scattering analog, except, because of the detailed nature of the WNC couplings, in this case it is nearly proportional to the neutron number N, rather than Z. The situation is a bit more complicated for elastic neutrino scattering in that beyond this coherent Coulomb monopole contribution there is also an (incoherent) axial-vector dipole contribution that does not vanish in the LWL. Thus, if one has relatively low-energy neutrinos, such as the stopped-pion situation mentioned above, and therefore low momentum transfer q (where "low" means that $qR \ll 1$, where R is roughly the nuclear radius), the LWL pertains and only these two types of form factors play any significant role. Furthermore, while for very light nuclei the incoherent axial-vector dipole contributions can compete with the coherent vector monopole contribution, for heavy nuclei the coherence is likely to make the latter dominant, roughly by the factor $N^2 : 1$. Let us assume this to be the case, and briefly discuss coherent neutrino scattering from nuclei [Fre74, Don75, Don76, Don82a, Don85].

Given a recoil kinetic energy T and an incoming neutrino energy E_ν, the differential cross section for such a neutrino–nucleus interaction can be written compactly as

$$\frac{d\sigma}{dT} = \frac{G_F^2}{4\pi} Q_W^2 M_A \left(1 - \frac{M_A T}{2E_\nu^2} \right) F(q)^2 , \tag{18.78}$$

where M_A is the target mass ($M_A \simeq A m_N$), $F(q)$ is an appropriate elastic $C0$ form factor, and Q_W is the weak charge

$$Q_W = N - Z(1 - 4\sin^2 \theta_W) . \tag{18.79}$$

The cross section essentially scales quadratically with neutron (N) and proton (Z) number, although the latter is highly suppressed due to the factor $1 - 4\sin^2 \theta_W$, which is small, and thus the cross section goes roughly as N^2, as stated above. The form factor $F(q)$ encodes the coherence across the nucleus and drops quickly as qR becomes large.

Despite the strong coherent enhancement enjoyed by this process, this particular cross section has yet to be detected experimentally. Part of the obstacle stems from the extremely small energies of the emitted recoil. The maximum recoil energy from such an interaction is limited by the kinematics of the elastic collision

$$T_{\max} = \frac{E_\nu}{1 + \frac{M_A}{2E_\nu}} , \tag{18.80}$$

similar to that of any elastic scattering where the mass of the incoming particle is negligible. Several experiments have been proposed to detect this coherent enhancement, often taking advantage of advances in recoil detection typically utilized by dark matter experiments [Sch06, For12]. The interaction has also been proposed as a possible mechanism relevant for detection of cosmic relic neutrinos, due to its nonzero cross section at zero momentum transfer. However, the G_F^2 suppression makes detection beyond the reach of any realistic experiment.

Finally, we note an interesting relationship with elastic coherent parity-conserving and -violating electron scattering [Mor15a]: at tree-level in the extreme relativistic limit for electrons, neglecting Coulomb distortion effects on the electron (these can easily be taken into account) and assuming single-Z^0 exchange (the SM) one has

$$\left[\frac{d\sigma}{d\Omega} \right]_{(\nu,\nu)}^{\text{coherent elastic}} = \left[\frac{d\sigma}{d\Omega} \right]_{(e,e)}^{\text{coherent elastic, ERL}} \mathcal{A}_{PV}^2 , \tag{18.81}$$

where the left-hand side of the equation has the differential cross section for coherent elastic neutrino scattering from which Eq. (18.78) can easily be obtained through a simple change of variables, where the differential cross section on the right-hand side is the familiar parity-conserving coherent elastic electron scattering cross section (which is typically very well measured, namely, at roughly the 1% level) and where $\mathcal{A}_{PV}$ is the PV asymmetry discussed in Chapter 15. The last has been measured for the cases of ^{4}He (HAPPEX-He) and ^{208}Pb (PREX) and one can project for the future that other cases can be measured at the level of a few parts per thousand. Thus, any deviation from this simple relationship would signal something that violates the underlying assumptions made here. The basic fact behind this simple relationship is that, however a Z^0 interacts coherently with a nucleus (e.g., with strangeness content, with isospin mixing, etc.), it does so the same way when initiated by electron or neutrino scattering – any differences must come from the lepton couplings being unusual or from higher-order contributions, such as $\gamma - Z^0$ box diagrams in the PV electron scattering case.

Neutrino Reactions and Low-Lying Excited States in Nuclei

In passing, we mention an example of the opportunities for exciting low-lying states in nuclei in NCν scatterings or the analogous CCν counterparts. As discussed earlier, these reactions provided some of the earliest points of focus for both theoretical and experimental studies of neutrino–nucleus physics.

A special transition of this type is provided by the $A - 12$ system involving the 0^+ $T = 0$ ground state of ^{12}C together with the $T = 1$ triplet of states including the 1^+ ground states of ^{12}B ($M_T = -1$) and ^{12}N ($M_T = +1$) with the 1^+ 15.11 MeV excited state of ^{12}C ($M_T = -1$). The CC neutrino reactions in particular are especially noteworthy, since the ground state of ^{12}N which is reached, for instance, in the ^{12}C(ν_e, e^-)^{12}N reaction, is the only state of ^{12}N that is stable to proton emission. Accordingly, the beta decay of that state back to the ground state of ^{12}C provides a clean signal that the CC neutrino reaction involving these two states, and no other, has taken place. This set of transitions was modeled using a limited basis of one-body density matrix elements together with the known electron scattering $0^+ \rightarrow 1^+$ cross section, the two beta-decay rates for the decays of the ground states of ^{12}B and ^{12}N back to the ground state of ^{12}C and the muon capture rate for the transition between the ground states of ^{12}C and ^{12}B (see [Don74, Don75, Don79] for details). With this "calibration," the CC and NC cross sections could then be predicted. In fact, subsequently both were measured and found to agree very well with theory, verifying what was known about the weak interaction at the time. See the cited literature for other examples of this type of analysis.

Next we consider higher-energy neutrino reactions with nuclei where the ideas presented in Chapter 16 are relevant.

Charge-Changing Quasielastic Neutrino Scattering

For neutrino-nuclear reactions at energies of order a few GeV, rather than involving the nuclear ground state or low-lying excited states as above, one now finds that the cross sections arise from nucleon knockout and from the production of mesons and excited baryons. This is the regime discussed in Chapter 16 for electron scattering, where quasielastic scattering and excitations for kinematics so that the Δ plays a role, are important. One should note at the outset a different usage of the word "quasielastic" in the two fields: the electron scattering community typically means the knockout of single nucleons, while the neutrino community really takes the word to mean the "no pion" cross section, namely, those events where no signature for an emitted pion is found. Unfortunately, this can lead to some confusion, since the former community then classifies processes where, say, two nucleons are emitted via interactions involving two-body meson-exchange currents as being beyond quasielastic. Also, those neutrino events where no pion is observed can actually arise from pion production followed by absorption of the pion before it is detected, and accordingly requiring a model-dependent correction. One might ask why one does not do as in electron scattering and take the complete inclusive cross section. The reason is that one wants to isolate those "quasielastic" events as they are well-suited to neutrino oscillation studies at the energies of the main facilities in the field, namely, FNAL and KEK.

With these caveats let us proceed to discuss the inclusive charge-changing quasielastic (CCQE) reaction (the neutral current quasielastic (NCQE) reaction can be developed in a similar way). In a CCQE reaction with incoming neutrinos, a neutron in the target nucleus is converted to a proton, which is then ejected. In the case of an antineutrino scattering, the reverse happens where a proton is converted into a neutron:

$$\nu_\mu\, n \to \mu^-\, p\,, \qquad \bar{\nu}_\mu\, p \to \mu^+\, n\,. \tag{18.82}$$

Such simple reactions were extensively studied in the 1970–1990s, primarily using deuterium-filled bubble chambers. The main interest at the time was in testing the V–A nature of the weak interaction and in measuring the axial-vector form factor of the nucleon, topics that were considered particularly important in providing an anchor for the study of NC interactions.

In predicting the CCQE cross section, early experiments relied heavily on the formalism of Llewellyn-Smith in 1972 [Lle72]. In the case of elastic scattering off free nucleons, the differential cross section can be expressed as:

$$\frac{d\sigma}{dQ^2} = \frac{G_F^2 m_N^2}{8\pi E_\nu^2}\left[A \pm \frac{(s-u)}{m_N^2}B + \frac{(s-u)^2}{m_N^4}C \right], \tag{18.83}$$

where $(-)+$ refers to (anti)neutrino scattering and $(s-u) = 4m_N E_\nu - Q^2 - m_l^2$. The factors A, B, and C are functions of the familiar vector (F_1 and F_2), axial-vector (F_A) and pseudoscalar (F_M) form factors of the nucleon introduced in earlier chapters (their Q^2 dependence is implied):

$$A = \frac{(m_l^2 + Q^2)}{m_N^2}\Bigg[(1+\tau)F_A^2 - (1-\tau)F_1^2 + \tau(1-\tau)F_2^2 + 4\tau F_1 F_2$$

$$-\frac{m_l^2}{4m_N^2}\left((F_1+F_2)^2 + (F_A+2F_M)^2 - \left(\frac{Q^2}{m_N^2}+4\right)F_M^2 \right)\Bigg] \tag{18.84}$$

$$B = \frac{Q^2}{m_N^2}F_A(F_1+F_2) \tag{18.85}$$

$$C = \frac{1}{4}\left(F_A^2 + F_1^2 + \eta F_2^2 \right), \tag{18.86}$$

where $\tau = Q^2/4m_N^2$.

As stated earlier, present neutrino experiments use nuclei as targets and, as a result, since the nucleons in the nucleus are interacting, nuclear effects become much more important and produce sizable modifications to the CCQE differential cross section from Eq. (18.83). One simple approach that can be taken to get at least a rough idea of the importance of these effects is based on what has been discussed in Chapter 16, namely, the relativistic Fermi gas model (RFG). Early work of this nature was undertaken by [Smi72]; here we employ the formalism used in Section 16.3 together with [Ama05].

The CCQE response functions are simply the extensions of $R_{L,T}$ derived in Chapter 16:

$$R_X^Y = R_0 \widetilde{f}_{\mathrm{RFG}} U_X^Y \,, \tag{18.87}$$

where the superscripts Y represent the vector and axial-vector nature of the responses (VV, AA, VA) and the subscripts X represent their spacetime projections ($CC, CL, \ldots$). The overall factor here, R_0, is as in Eq. (16.13), and now, instead of having both proton and neutron single-nucleon responses weighted by Z and N, respectively (see Eq. (16.11)), one has only the $n \to p$ case weighted by N and the $p \to n$ case weighted by Z for neutrinos and antineutrinos, respectively. The purely isovector factors U are given by

$$
\begin{aligned}
U_L^{VV} &= \frac{\kappa^2}{\tau}\left((G_E^{(1)})^2 + W_2^{(1)}\Delta\right) \\
U_T^{VV} &= 2\tau(G_M^{(1)})^2 + W_2^{(1)}\Delta \\
U_{CC}^{AA} &= \frac{\kappa^2}{\tau}\left(\frac{\lambda^2}{\kappa^2}(G_A'^{(1)})^2 + (G_A^{(1)})^2\Delta\right) \\
U_{LL}^{AA} &= \frac{\kappa^2}{\tau}\left((G_A'^{(1)})^2 + \frac{\lambda^2}{\kappa^2}(G_A^{(1)})^2\Delta\right) \\
U_{CL}^{AA} &= -\frac{\kappa\lambda}{\tau}\left((G_A'^{(1)})^2 + (G_A^{(1)})^2\Delta\right) \\
U_T^{AA} &= (G_A^{(1)})^2\left[2(1+\tau) + \Delta\right] \\
U_{T'}^{VA} &= 2\sqrt{\tau(1+\tau)}H_{VA}^{(1)}\left[1 + \Delta'\right],
\end{aligned}
\tag{18.88}
$$

where, in addition to Δ defined in Chapter 16, the following definition

$$\Delta' = \frac{1}{\kappa}\sqrt{\frac{\tau}{1+\tau}}\left(\lambda + \frac{1}{2}(\epsilon_F + \Gamma)\right) - 1 \tag{18.89}$$

has been used. The isovector nucleon form factors $G_E^{(1)} = G_{E_p} - G_{E_n}$, $G_M^{(1)} = G_{M_p} - G_{M_n}$, $G_A^{(1)}$ and $G_P^{(1)}$ enter here through the combinations

$$
\begin{aligned}
W_2^{(1)} &= \frac{1}{1+\tau}\left((G_E^{(1)})^2 + \tau(G_M^{(1)})^2\right) \\
G_A'^{(1)} &= G_A^{(1)} - \tau G_P^{(1)} \\
H_{VA}^{(1)} &= G_M^{(1)}G_A^{(1)}.
\end{aligned}
\tag{18.90}
$$

One also defines the following components of the matrix element squared

$$X_L = V_L R_L^{VV}$$
$$X_{C/L} = V_{CC} R_{CC}^{AA} + 2V_{CL} R_{CL}^{AA} + V_{LL} R_{LL}^{AA}$$
$$X_T = V_T (R_T^{VV} + R_T^{AA})$$
$$X_{T'} = 2V_{T'} R_{T'}^{VA} , \tag{18.91}$$

which contain leptonic tensor components V

$$V_{CC} = 1 - \tan^2 \tilde{\theta}/2 \cdot \delta^2$$
$$V_{CL} = v + \tan^2 \tilde{\theta}/2 \cdot \frac{\delta^2}{\rho'}$$
$$V_{LL} = v^2 + \tan^2 \tilde{\theta}/2 \cdot \left(1 + \frac{2v}{\rho'} + \rho\delta^2\right) \cdot \delta^2$$
$$V_L = V_{CC} - 2v V_{CL} + v^2 V_{LL}$$
$$V_T = \frac{1}{2}\rho + \tan^2 \tilde{\theta}/2 - \frac{1}{\rho'} \tan^2 \tilde{\theta}/2 \cdot \left(v + \frac{1}{2}\rho\rho'\delta^2\right) \cdot \delta^2$$
$$V_{T'} = \frac{1}{\rho'} \tan^2 \tilde{\theta}/2 \cdot (1 - v\rho'\delta^2) \tag{18.92}$$

and also involve the additional dimensionless variables:

$$v \equiv \omega/q$$
$$\rho \equiv 1 - v^2$$
$$\rho' \equiv q/(\epsilon + \epsilon')$$
$$\delta = m'/\sqrt{|Q^2|}$$
$$v_0 \equiv (\epsilon + \epsilon')^2 - q^2$$
$$\tan^2 \tilde{\theta}/2 = |Q^2|/v_0 . \tag{18.93}$$

Note that, in this case, one in general should not invoke the extreme relativistic limit (ERL) for an outgoing muon, but should retain its mass, leading to the somewhat more complicated lepton kinematic factors above. The ERL is recovered simply by setting δ to zero and $\tilde{\theta}$ to θ. The CCQE cross section in the RFG model is then

$$\frac{d\sigma}{d\Omega dk'} = \frac{G_F^2 \cos^2 \theta_C}{2\pi^2} \frac{k'^2 v_0}{4\epsilon\epsilon'} \left[X_L + X_{C/L} + X_T + \chi X_{T'}\right], \tag{18.94}$$

where θ_C is the Cabbibo angle (see also Chapter 4 where the quark-mixing parameters are discussed) and where $\chi = 1$ for neutrino scattering and $\chi = -1$ for antineutrino scattering.

All of the developments here can easily be extended to include RFG modeling for neutrino excitation of the Δ using the results discussed in Section 16.4 (see also the exercises).

This simple model is still being used in Monte Carlo event generators where modern experiments are being simulated. However, as we saw in Chapter 16 the RFG model is not

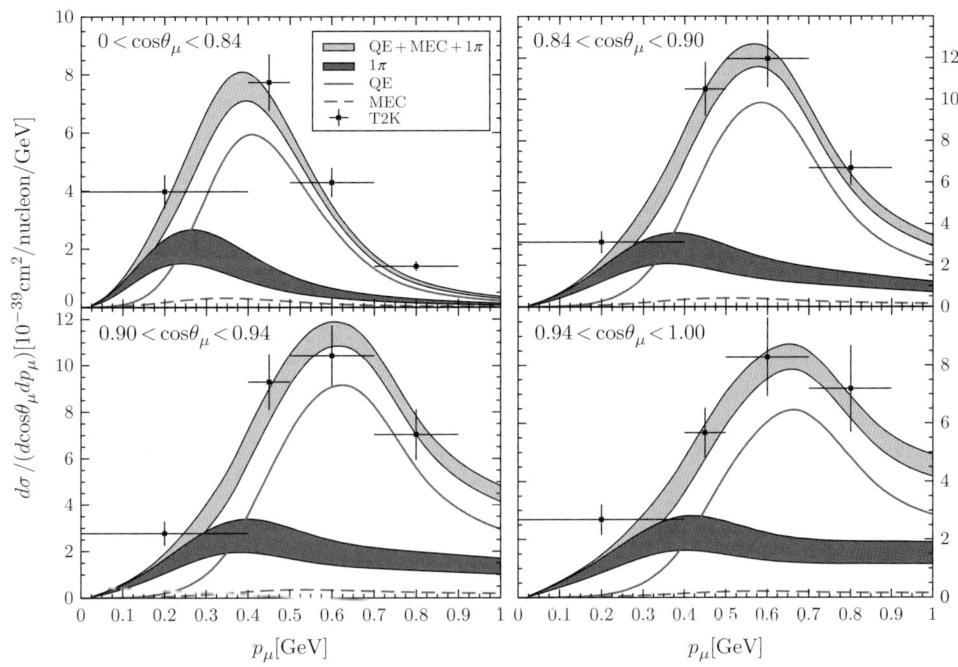

Fig. 18.11　The charge-changing T2K experiment flux-folded $\nu_\mu - {}^{12}\mathrm{C}$ double-differential cross section per nucleon evaluated in the SuSAv2 + MEC + 1π model of [Iva15]. The data are from [Abe13].

really capable of reproducing electron scattering cross sections in the QE region and more sophisticated approaches should be followed (scaling analyses, relativistic modeling such as via relativistic mean field theory, approaches using spectral functions, nonrelativistic *ab initio* approaches when they can be used, etc.; see the discussions in Chapter 16). A full exposition of this subject goes beyond the scope of the book and thus we end this part by presenting some typical results of recent experiments and modern theoretical modeling – other models yield rather similar results to those shown here.

In Fig. 18.11 we show recent results from the T2K experiment in Japan for inclusive muon neutrino reactions with ${}^{12}\mathrm{C}$, together with modern theory [Iva15]. The latter involves the so-called SuperScaling Analysis (SuSA) version 2 for the quasielastic contribution, together with $2p2h$ MEC (vector) contributions and pion production modeled as proceeding through the Δ resonance (see also [Ama14a]). Given that the range of energies involved here is relatively low, one sees excellent agreement between theory and experiment. Typically attempts have been made to isolate the "QE" contribution (as discussed above, strictly speaking the QE and MEC contributions) from pion production; some high-energy results for MINERνA and NOMAD at FNAL are shown in Figs. 18.12 and 18.13 and again the agreements at high energies are quite good, indicating that for such kinematics a reasonably high level of understanding exists. On the other hand, for intermediate situations such as the MiniBooNE results in Figs. 18.13 and 18.14, there is a shortfall for some kinematics, possibly due to the fact that axial-vector MEC contributions have not yet been included in the modeling.

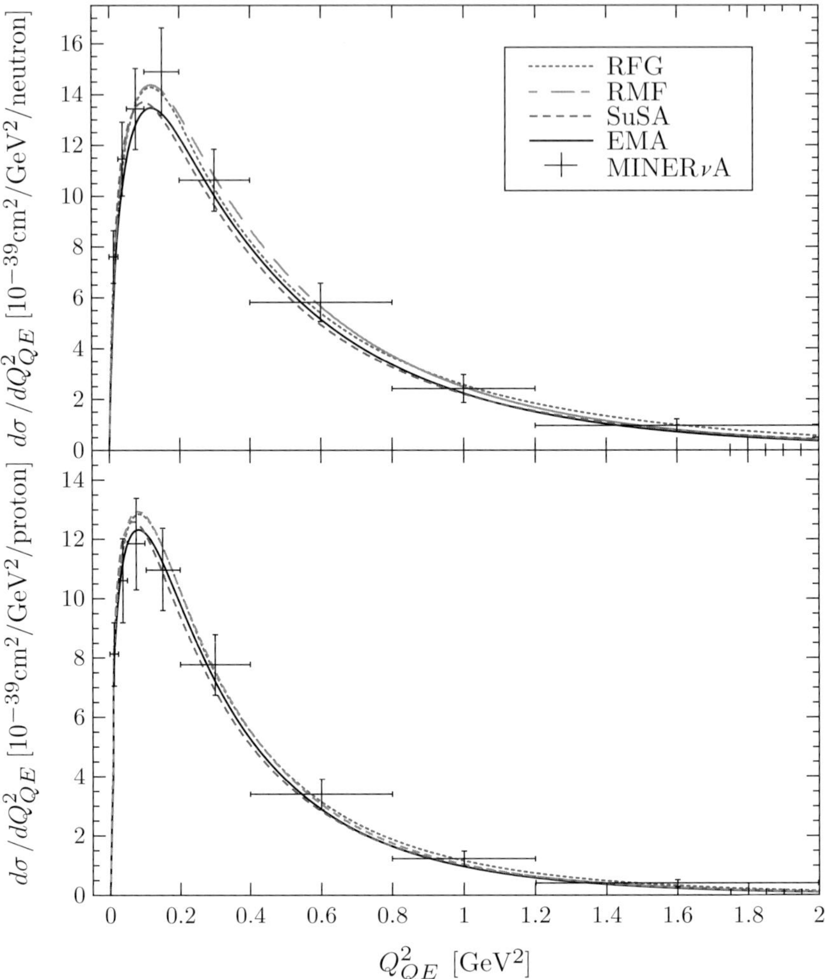

Fig. 18.12 Flux-folded $\nu_\mu - {}^{12}C$ (upper panel) and $\bar{\nu}_\mu - {}^{12}C$ plus $\bar{\nu}_\mu - {}^{1}H$ (lower panel) charge-changing neutrino cross sections per nucleon using various models [Meg14]. Data from the MINERνA experiment are from [Fio13, Fie15].

These developments of inclusive neutrino reactions can be extended to include semi-inclusive processes where some particle is detected in coincidence with the final-state charged lepton, just as inclusive electron scattering was generalized to semi-inclusive scattering in Chapter 16. For instance, one can consider reactions like ${}^{A}X(\nu_\mu, \mu^- p)^{A-1}X$. Recently the formalism for studies of this type has been presented [Mor14a] and applied to the special case of neutrinodisintegration of deuterium [Mor15b].

Neutrino Deep Inelastic Scattering

Neutrino deep inelastic scattering (DIS) has long been used to validate the Standard Model and probe nucleon structure. Over the years, experiments have measured cross

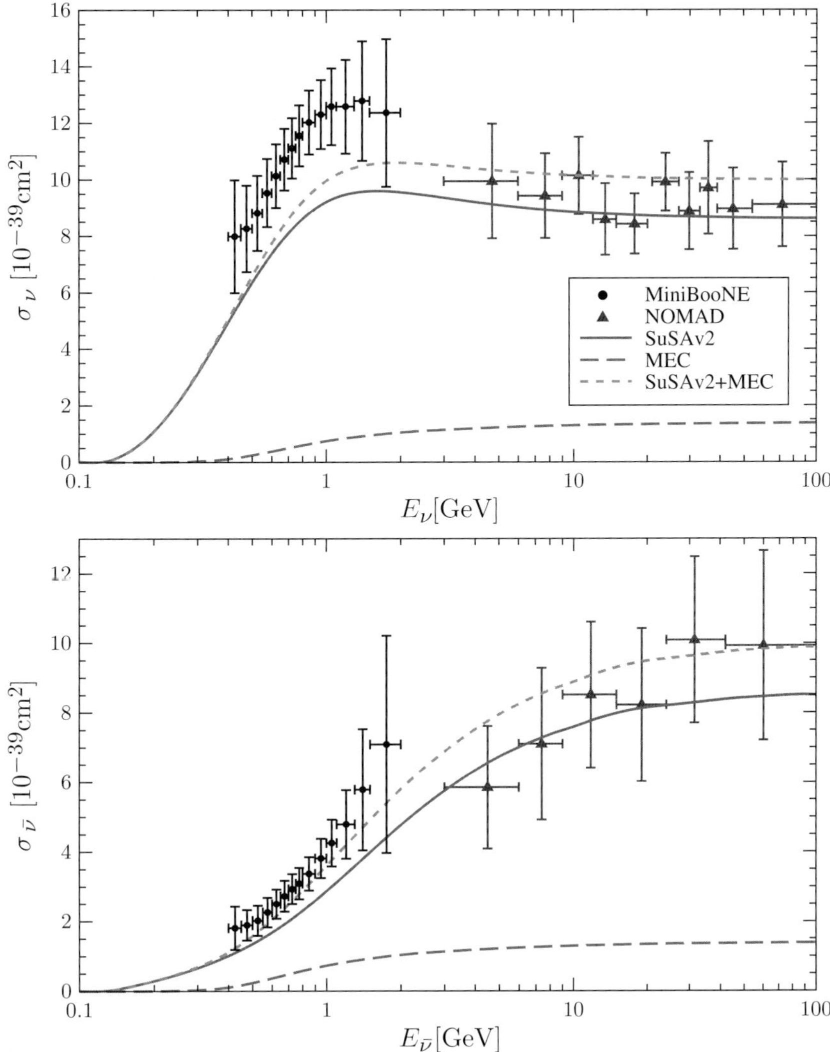

Fig. 18.13 Charge-changing quasielastic (CCQE) neutrino and antineutrino cross sections in the SuSAv2 + MEC model of [Meg15]. Data from the MiniBooNE and NOMAD experiments are also shown [Agu10, Agu13].

sections, electroweak parameters, coupling constants, nucleon structure functions and scaling variables using such processes. In DIS, the neutrino scatters off a quark in the nucleon via the exchange of a virtual W or Z boson producing a lepton and a hadronic system in the final state. Both CC and NC processes are possible:

$$\begin{aligned} \nu_\mu N &\rightarrow \mu^- X & \bar{\nu}_\mu N &\rightarrow \mu^+ X \\ \nu_\mu N &\rightarrow \nu_\mu X & \bar{\nu}_\mu N &\rightarrow \bar{\nu}_\mu X \,. \end{aligned} \tag{18.95}$$

Here, we restrict ourselves to the case of ν_μ scattering, as an example, although ν_e and ν_τ DIS reactions are also possible.

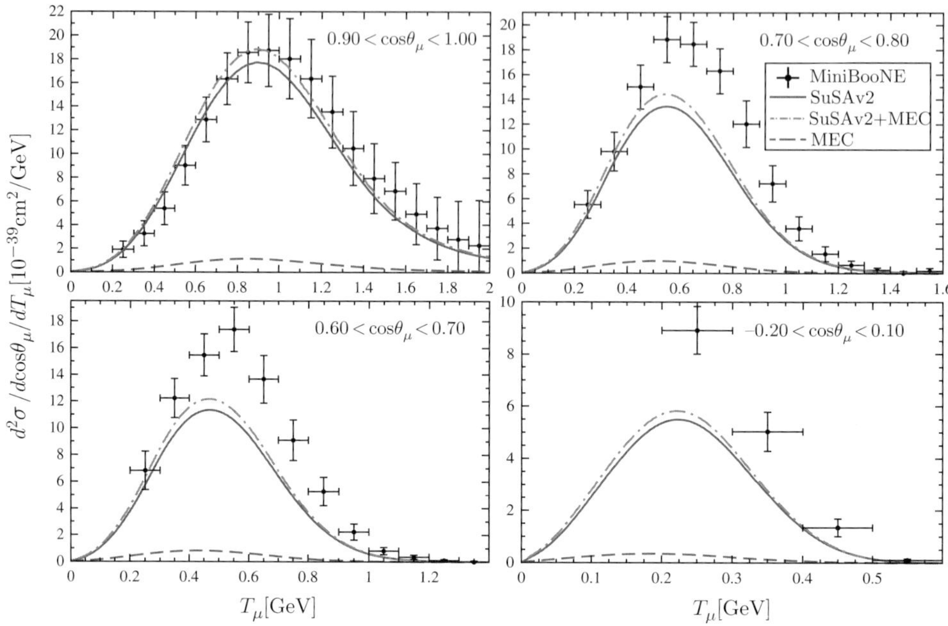

Fig. 18.14 Flux-integrated double-differential charge-changing quasielastic (CCQE) $\nu_\mu - {}^{12}\mathrm{C}$ results from the MiniBooNE experiment compared with the SuSAv2 + MEC model of [Meg15]. The data are from [Agu10].

Following the formalism introduced above for muon neutrino–lepton scattering, DIS processes can be completely described in terms of three dimensionless kinematic invariants Q^2, x, and y. The Bjorken scaling variable (x) plays a particularly prominent role in deep inelastic scattering, and was discussed extensively in Chapter 9 for the case of electron scattering where the electron mass was neglected. For CC DIS scattering of muon neutrinos from the nucleon, we have

$$Q^2 = -m_\mu^2 + 2E_\nu (E_\mu - p_\mu \cos\theta_\mu) , \qquad (18.96)$$

where E_ν is the incident neutrino energy, and m_μ, E_μ, p_μ, and $\cos\theta_\mu$ are the mass, energy, momentum, and scattering angle of the outgoing muon in the laboratory frame. In the case of NC scattering, the outgoing neutrino is not reconstructed. Thus, experimentally, all of the event information must be inferred from the hadronic shower in that case.

Using these variables, the inclusive cross section for DIS scattering of neutrinos and antineutrinos can then be written as:

$$\frac{d^2\sigma^{\nu,\bar{\nu}}}{dx\,dy} = \frac{G_F^2 M E_\nu}{\pi\,(1 + Q^2/M_{W,Z}^2)^2} \left[\frac{y^2}{2} 2x F_1(x, Q^2) + \left(1 - y - \frac{Mxy}{2E}\right) F_2(x, Q^2) \pm y\left(1 - \frac{y}{2}\right) x F_3(x, Q^2) \right], \qquad (18.97)$$

where the $+(-)$ sign in the last term refers to neutrino(antineutrino) interactions. In the above expression, $F_i(x, Q^2)$ are the dimensionless nucleon structure functions that encode the underlying structure of the target. As discussed in Chapter 9 for DIS electron scattering, there exist two such structure functions, while for neutrino scattering there is additionally a third structure function, $xF_3(x, Q^2)$, which represents the VA interference term.

Assuming the quark-parton model (see also Chapter 9), in which the nucleon consists of partons (quarks and gluons), $F_i(x, Q^2)$ can be expressed in terms of the quark composition of the target. They depend on the target and the type of reaction, and are functions of x and Q^2. In the simplest case, the nucleon structure functions can then be expressed as the sum of the probabilities,

$$F_2(x, Q^2) = 2 \sum_{i=u,d,\dots} (xq(x, Q^2) + x\bar{q}(x, Q^2))$$

$$xF_3(x, Q^2) = 2 \sum_{i=u,d,\dots} (xq(x, Q^2) - x\bar{q}(x, Q^2)) , \tag{18.98}$$

where the sum is over all quark flavors. As discussed in Chapter 9, $F_2(x, Q^2)$ measures the sum of the quark and antiquark PDFs in the nucleon, while $xF_3(x, Q^2)$ measures their difference and is therefore sensitive to the valence quark PDFs. The third structure function, $2xF_1(x, Q^2)$, is commonly related to $F_2(x, Q^2)$, as also discussed in Chapter 9.

Some characteristic results are shown in Fig. 18.15. Most notable is the constancy of the cross section with energy, indicating the success of the parton description at these high energies. Reactions of charged-current and neutral-current ratios at these high energies have been successfully exploited to extract information on the coupling parameters of the weak force. Neutrino and antineutrino scattering measurements (particularly on iron targets) have also allowed a unique extraction of the structure function $xF_3(x, Q^2)$ in beautiful detail (see Fig. 18.16).

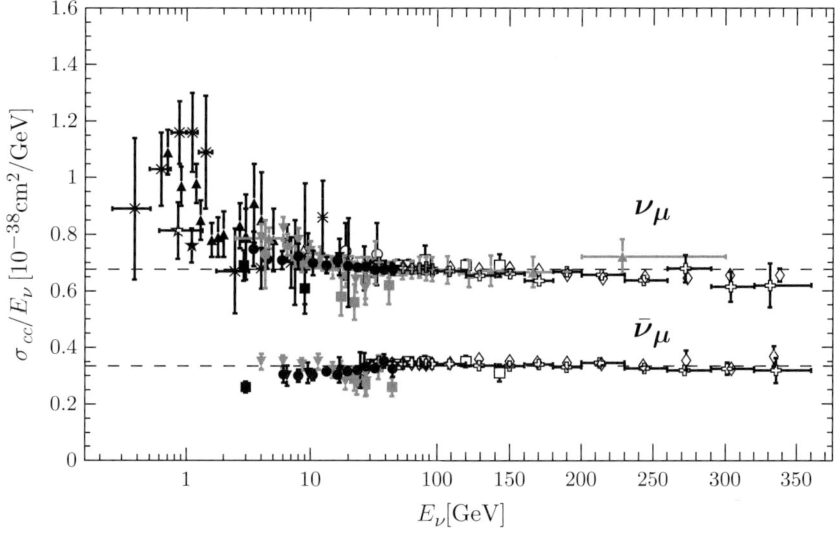

Measurements of the inclusive neutrino and antineutrino charge-changing (CC) cross sections ($\nu_\mu N \to \mu^- X$ and $\bar{\nu}_\mu N \to \mu^+ X$) divided by neutrino energy plotted as a function of neutrino energy. Here, N refers to an isoscalar nucleon within the target. The dotted lines indicate the world-averaged cross sections, [PDG14].

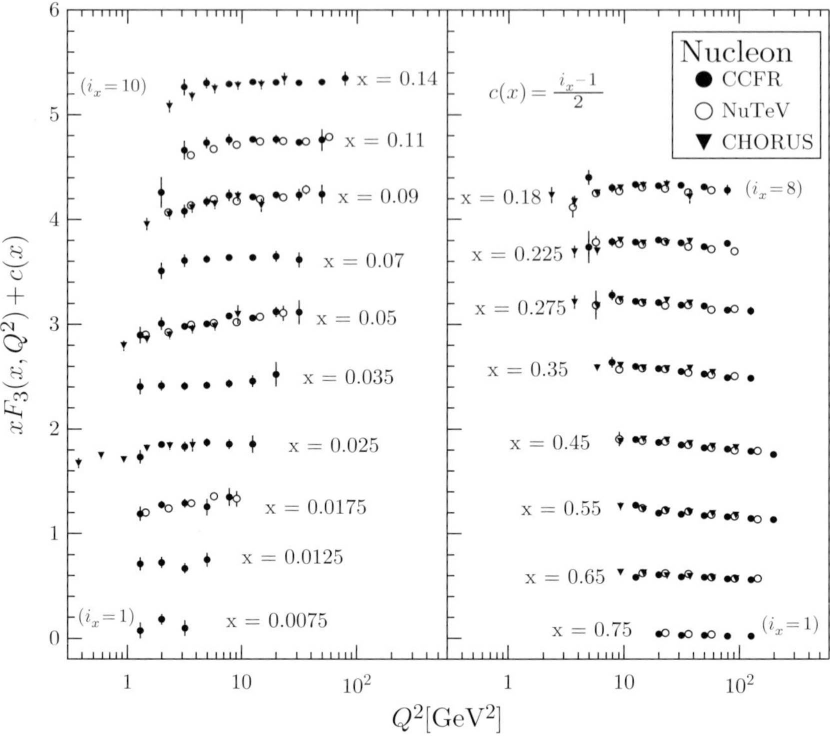

Fig. 18.16 The structure function $xF_3(x, Q^2)$ of the nucleon measured in ν-Fe scattering [PDG14]. The data are plotted as a function of Q^2 in bins of x. For the purpose of plotting, a constant $c(x) = 0.5(i_x - 1)$ is added to $xF_3(x, Q^2)$, where i_x is the number of the x-bin, as shown in the plot.

18.5 Outstanding Questions in Neutrino Physics

Despite considerable progress in understanding neutrino physics, nevertheless, there remain deep, fundamental questions to be addressed both experimentally and theoretically. These include the following:

Majorana or Dirac: Despite the remarkable success in determining that neutrinos must have mass, we still do not know if neutrinos are their own antiparticles. As discussed previously, the most sensitive means to reveal this particular feature of neutrinos comes from trying to observe neutrinoless double beta decay. A number of experimental searches using a range of isotopes and techniques all over the world are (or are soon to be) underway to determine the answer to this open question.

CP **violation:** Do neutrinos exhibit *CP* violation, as do their quark counterparts? If so, could they explain the observed matter/antimatter asymmetry in the universe? Now that it is known that *all* neutrino mixing angles are finite, one can look at the asymmetry between neutrinos and antineutrinos to determine if *CP*-violation is present. Employing neutrino oscillation accelerator experiments that have extremely long baselines, such as the deep underground neutrino experiment (DUNE) in the US or the Jiangmen Underground Neutrino Observatory (JUNO) in China, or use high-intensity cyclotron sources (Daeδalus) are potential contenders in the near future.

Neutrino mass and hierarchy: What is the neutrino mass scale and ordering of mass eigenstates? What fundamental mechanism gives rise to their mass? As the next generation of nuclear β-decay experiments and cosmological measurements continue to improve, much of the scale and structure of neutrino masses will become experimentally accessible.

Neutrino number: Are there only three families of neutrinos, or could there be more (often referred to as "sterile" neutrinos)? There are dedicated experiments looking for an additional oscillation scale, or even constraints from the relic neutrino density of the universe.

Providing answers to these questions represents a topic of experimental pursuit for the following decade, and perhaps even longer.

Exercises

18.1 Neutrino Propagation Through Matter

When neutrinos pass through matter, the propagation can be described in terms of the usual mixing matrix modified by an index of refraction term which describes the interaction of the neutrinos with the medium. This exercise studies how this comes about.

a) Show that when a beam travels through a medium the associated index of refraction n is given by

$$n - 1 = \frac{2\pi N}{k^2} f_k(0) \,,$$

where $f_k(0)$ is the forward scattering amplitude, N is the number density of scatterers, and k is the wavenumber of the incident beam.

Suggestion: Consider a slab of matter extending from $z = 0$ to $z = L$ and assume an incident beam e^{ikz} which travels along the z-direction. Calculate the wavefunction at location $z = Z$, where $z > L$, as a superposition of the incident wave and the scattered wave in the slab. Taking the distribution of scatterers to be uniform with number density N and using cylindrical coordinates ρ, ϕ, z this is given by

$$\psi(Z) = e^{ikZ} + N \int_0^L dz e^{ikz} \int_0^\infty d\rho \rho \frac{1}{r} e^{ikr} f_k(\theta) \,.$$

Introducing a factor which guarantees convergence at $\rho = \infty$, integrate by parts and show that the result

$$\psi(Z) = \left[1 + \frac{2\pi i NL}{k} f_k(0) \right] e^{ikZ} \approx e^{ikZ} e^{i \frac{2\pi NL}{k} f_k(0)}$$

implies the relation given above between the index of refraction and the forward scattering amplitude.

b) Show that the forward scattering amplitude of muon- or tau-neutrinos from electrons arises from Z^0-boson exchange and is given by

$$f_{\nu_\mu e}(0) = f_{\nu_\tau e}(0) = \sqrt{2}G_F\left(-\frac{1}{2} + 2\sin^2\theta_w\right) .$$

c) Show that the forward scattering amplitude of electron-neutrinos from electrons arises from both Z^0- and $W^\pm$-boson exchange and is given by

$$f_{\nu_e e}(0) = \sqrt{2}G_F\left(\frac{1}{2} + 2\sin^2\theta_w\right) .$$

Suggestion: Fierz has noted that any charged current-charged interaction $\bar{e}\mathcal{O}_+\nu_e\bar{\nu}_e\mathcal{O}_-e$ can always, be completeness, be written as a product of neutral current operators $\sum_i c_{ij}\bar{e}\mathcal{O}_i e\bar{\nu}_e\mathcal{O}_j\nu_e$, where $\mathcal{O}_i, \mathcal{O}_j$ range over the various possibilities VV, AA, VA, AV, SS, PS, PP, TT. Show that in the special case of the product of $V-A$ currents we have

$$\bar{e}\gamma_\mu(1-\gamma_5)\nu_e\bar{\nu}_e\gamma^\mu(1-\gamma_5)e = \bar{e}\gamma_\mu(1-\gamma_5)e\bar{\nu}_e\gamma^\mu(1-\gamma_5)\nu_e$$

and then note that one can drop the axial-vector electron current in looking at coherent scattering.

d) Neglecting the mixing between ν_e and ν_τ ($\theta_{13} = 0$) so only two-channel mixing is assumed, demonstrate that the resulting mixing matrix can be described via

$$i\frac{d}{dt}\begin{pmatrix} a_{\nu_e} \\ a_{\nu_\mu} \end{pmatrix} = \begin{pmatrix} -\frac{\Delta m_{12}^2}{4E}\cos 2\theta_{12} + \sqrt{\frac{1}{2}}G_F N_e & \frac{\Delta m_{12}^2}{4E}\sin 2\theta_{12} \\ \frac{\Delta m_{12}^2}{4E}\sin 2\theta_{12} & \frac{\Delta m_{12}^2}{4E}\cos 2\theta_{12} - \sqrt{\frac{1}{2}}G_F N_e \end{pmatrix}\begin{pmatrix} a_{\nu_e} \\ a_{\nu_\mu} \end{pmatrix},$$

where N_e is the electron number density.

18.2   Neutrino Mixing: Theoretical

The weak flavor (e, ν, τ) and mass $(1,2,3)$ eigenstates of the neutrino generations are related via a unitary mixing matrix U

$$\begin{pmatrix} e \\ \mu \\ \tau \end{pmatrix} = \begin{pmatrix} U_{e1} & U_{e2} & U_{e3} \\ U_{\mu 1} & U_{\mu 2} & U_{\mu 3} \\ U_{\tau 1} & U_{\tau 2} & U_{\tau 3} \end{pmatrix}\begin{pmatrix} 1 \\ 2 \\ 3 \end{pmatrix} .$$

The matrix U can be written in various forms, but one way is that given in the text

$$U = \begin{pmatrix} c_{12}c_{13} & s_{12}c_{13} & s_{13}e^{-i\delta} \\ -s_{12}c_{23} - c_{12}s_{23}s_{13}e^{i\delta} & c_{12}c_{23} - s_{12}s_{23}s_{13}e^{i\delta} & s_{23}c_{13} \\ s_{12}s_{23} - c_{12}c_{23}s_{13}e^{i\delta} & -c_{12}s_{23} - s_{12}c_{23}s_{13}e^{i\delta} & c_{23}c_{13} \end{pmatrix} .$$

a) The meaning of this rather complicated form can be seen by decomposing it into several pieces. Demonstrate that the above form for U_ν follows from the matrix multiplication of the simpler forms

$$
U_\nu =
\begin{pmatrix}
1 & 0 & 0 \\
0 & c_{23} & s_{23} \\
0 & -s_{23} & c_{23}
\end{pmatrix}
\times
\begin{pmatrix}
c_{13} & 0 & s_{13}e^{-i\delta} \\
0 & 1 & 0 \\
-s_{13}e^{i\delta} & 0 & c_{13}
\end{pmatrix}
\times
\begin{pmatrix}
c_{12} & s_{12} & 0 \\
-s_{12} & c_{12} & 0 \\
0 & 0 & 1
\end{pmatrix}.
$$

b) Suppose that at time $t = 0$ the neutrino (or antineutrino) is produced in weak flavor eigenstate $|\nu_a\rangle$ (or $|\bar{\nu}_a\rangle$). Demonstrate that the probability to be observed in state $|\nu_b\rangle$ (or $|\bar{\nu}_b\rangle$) at later time t is given by

$$
P_{\nu_a \to \nu_b}(t) = \delta_{ab} - 4 \sum_{i>j=1}^{3} \mathrm{Re}(U^*_{ai} U_{bi} U_{aj} U^*_{bj}) \sin^2\left(\Delta m^2_{ij} \frac{L}{4E}\right)
$$

$$
+ 2 \sum_{i>j=1}^{3} \mathrm{Im}(U^*_{ai} U_{bi} U_{aj} U^*_{bj}) \sin\left(\Delta m^2_{ij} \frac{L}{2E}\right)
$$

or, for antineutrinos,

$$
P_{\bar{\nu}_a \to \bar{\nu}_b}(t) = \delta_{ab} - 4 \sum_{i>j=1}^{3} \mathrm{Re}(U^*_{ai} U_{bi} U_{aj} U^*_{bj}) \sin^2\left(\Delta m^2_{ij} \frac{L}{4E}\right)
$$

$$
- 2 \sum_{i>j=1}^{3} \mathrm{Im}(U^*_{ai} U_{bi} U_{aj} U^*_{bj}) \sin\left(\Delta m^2_{ij} \frac{L}{2E}\right),
$$

where E is the neutrino energy and L is the distance between production and observation.

c) Show that in the case of two flavors with mixing angle θ the survival probability of state $|\nu_a >$ in the presence of mixing with state $|\nu_b >$ can be written in the familiar form

$$
P_{\nu_a \to \nu_a}(t) = 1 - \sin^2 2\theta_{ab} \sin^2\left(\Delta m^2_{ab} \frac{L}{4E}\right).
$$

d) In the three-flavor case suppose that we measure the probabilities for both $\nu_a \to \nu_b$ and $\bar{\nu}_a \to \bar{\nu}_b$ as functions of time. Show that a nonzero value for the difference

$$
P_{\nu_a \to \nu_b}(t) - P_{\bar{\nu}_a \to \bar{\nu}_b}(t)
$$

is a measure of *CP* violation, and is proportional to $s_{12}s_{23}s_{13}$, so that it must vanish if any of the mixing angles $\theta_{12}, \theta_{23}, \theta_{13}$ vanishes.

18.3   Neutrino Mixing: Experimental

In the previous exercise, the formalism to study neutrino oscillations was set up. In the present exercise, we examine various experimental aspects of mixing.

a) Atmospheric neutrinos: At Super-Kamiokande a giant water detector was used to detect electron and muon neutrinos which are emitted when muons are created in cosmic ray collisions at the top of the atmosphere. and travel to the Earth. Since the average energy of such neutrinos is ~ 1 GeV, their interactions with water can produce electrons or muons, but *not* taus. By looking at the Cherenkov radiation that the charged particles make as they pass through the water, the electrons can be distinguished from the muons and the direction of the neutrino determined.

If there were no mixing, one would expect roughly two muon neutrinos for each electron neutrino from the decay chain

$$\pi^+, K^+ \rightarrow \mu^+ + \nu_\mu \rightarrow e^+ + \nu_e + \bar{\nu}_\mu + \nu_\mu \,.$$

However, it was found that there was a severe reduction in the population of ν_μ ($\bar{\nu}_\mu$) which were produced on the opposite side of the globe and traveled through the Earth before interacting, presumably because they oscillated into ν_τ ($\bar{\nu}_\tau$). Use the two-channel mixing formula to show that this observation implies that $\theta_{23} \approx 45°$ and estimate the mass difference Δm_{23}^2. Compare both numbers with those found experimentally.

b) $\nu_e - \nu_\tau$ Mixing: A recent experiment at Daya Bay in China involved the use of eight 20 ton $\bar{\nu}_e$ detectors at distances of about 0.5 km and 1.8 km from a powerful nuclear reactor. The nearby detector was used to normalize the $\bar{\nu}_e$ flux and then the $\bar{\nu}_e$ rates measured at the more distant detector provided a measure of possible oscillation. Presuming that both $\bar{\nu}_e - \bar{\nu}_\mu$ and $\bar{\nu}_e - \bar{\nu}_\tau$ oscillations are present and that θ_{13} is small, show that the survival probability at distance L from the reactor is given by

$$P_{ee}(L) \approx 1 - \sin^2 2\theta_{13} \sin^2 \frac{\Delta m_{13}^2}{4E} L - \cos^4 \theta_{13} \sin^2 2\theta_{12} \sin^2 \frac{\Delta m_{12}^2}{4E} L \,.$$

A 6% rate reduction was found at the 1.8 km detector compared with that expected from normalizing to the nearby detector and assuming no oscillation. Using a typical reactor antineutrino energy of $\sim 4\,\mathrm{MeV}$ and experimental values of Δm_{12}^2 and Δm_{23}^2 determine $\sin^2 2\theta_{13}$. Compare with the experimental value.

c) Solar Neutrinos: In the standard pp-chain, three types of neutrinos are produced: low-energy $E_\nu < 0.42\,\mathrm{MeV}$ neutrinos from the reaction

$$p + p \rightarrow d + e^+ + \nu_e \,,$$

medium-energy 0.86 MeV neutrinos from the electron capture reaction

$$e^- + {}^7\mathrm{Be} \rightarrow \nu_e + {}^7\mathrm{Li} \,,$$

and high-energy $E < 14\,\mathrm{MeV}$ neutrinos from the beta-decay

$$^8\mathrm{B} \rightarrow {}^8\mathrm{Be} + e^+ + \nu_e \,.$$

In the standard solar model and assuming no mixing, the predicted capture rates for these reactions in gallium radiochemical experiments, which are primarily sensitive to the low-energy pp neutrinos, are $\sim$130–140 SNU. The two such experiments SAGE and GALLEX, however, yielded measurements about 70 SNU. (Here a SNU – solar neutrino unit – is a capture rate of 10^{-36} per second per atom.) Similarly Davis's Homestake experiment, which is primarily sensitive to the high-energy $^8\mathrm{B}$ neutrinos, found a flux only about 1/3 of the standard solar model prediction. Using the experimental values of mixing angles and mass differences, show that theory and experiment can only be resolved with the inclusion of MSW mixing, which is small for low- and medium-energy neutrinos ($E_\nu \ll 2\,MeV$), but is significant for high-energy neutrinos ($E_\nu \gg 2\,MeV$).

18.4   The RFG Model for CCν Reactions

In Chapter 16, the RFG model was developed for electron scattering in the quasielastic and Δ regions of excitation; the former was extended to discussions of CCν reactions from nuclei in the present chapter.

a) Verify the hadronic responses given in Eqs. (18.91).

b) Show that the "Rosenbluth" factors V_K, $K = L, T, \ldots$ in Eqs. (18.92) are obtained when the mass of the outgoing charged lepton, m' is kept nonzero.

c) Generate an extension of the electron scattering RFG Δ-region model of Chapter 16 which is appropriate to CCν reactions.

19.1 Introduction

The unique features of quantum chromodynamics (QCD), i.e. the quantum field theory of strong interactions, have been extensively discussed in Chapter 5. QCD is in striking contrast to quantum electrodynamics (QED) in that strong interactions are weak at short distances well below the proton radius of roughly 10^{-15} m, a central characteristic of the former known as asymptotic freedom. In contrast, strong interactions are nonperturbative at scales of the proton radius and beyond. This phenomenon leads to the well-known confinement of quarks and gluons inside hadrons. As discussed in Chapter 10, the region of small distances is routinely used in hadron collider programs to take advantage of the small-coupling regime where perturbative QCD calculations can be performed. Such a region can be realized not only in violent high-energy hadron collisions, but also at high temperatures and pressures where QCD predicts that quarks and gluons form a deconfined state, which is called the *quark–gluon plasma* (QGP). Such a deconfined state is expected to have existed at the very early stages of our universe with temperatures of about 2×10^{12} K, equivalent to about 150 MeV; QCD predicts that these conditions lead to the formation of a quark–gluon plasma.

The study of the transition from a hadron gas to a QGP and thus the evolution of the early universe requires a deep understanding of the QGP and its underlying dynamics. The thermodynamic properties of QCD have been the subject of theoretical work over the last few decades. The evolution from a QGP into a hadron gas has been successfully described using hydrodynamic theoretical tools, although understanding the dynamical properties of the QGP remains a formidable challenge.

Figure 19.1 shows the phase diagram of QCD with temperature plotted as a function of baryon density and indicating the phase change of a hadron-dominated phase to a deconfined state of quarks and gluons. A QGP is expected to play an important role in the characteristics of neutron stars. Experimental studies under laboratory conditions of the transition from a hadron gas to a QGP requiring very high temperatures could be realized by colliding heavy nuclei at very high energies.

The highest temperatures under laboratory conditions can be achieved in the relativistic heavy-ion program at the LHC which is focused on the collision of Pb–Pb ions of several TeV per beam, namely, at $\sqrt{s} = 2760$ GeV, as shown in Fig. 19.1. On the other hand, the RHIC relativistic heavy-ion program can reach higher net baryon densities at lower beam energies, giving that program a unique role to play. The location of a critical point is

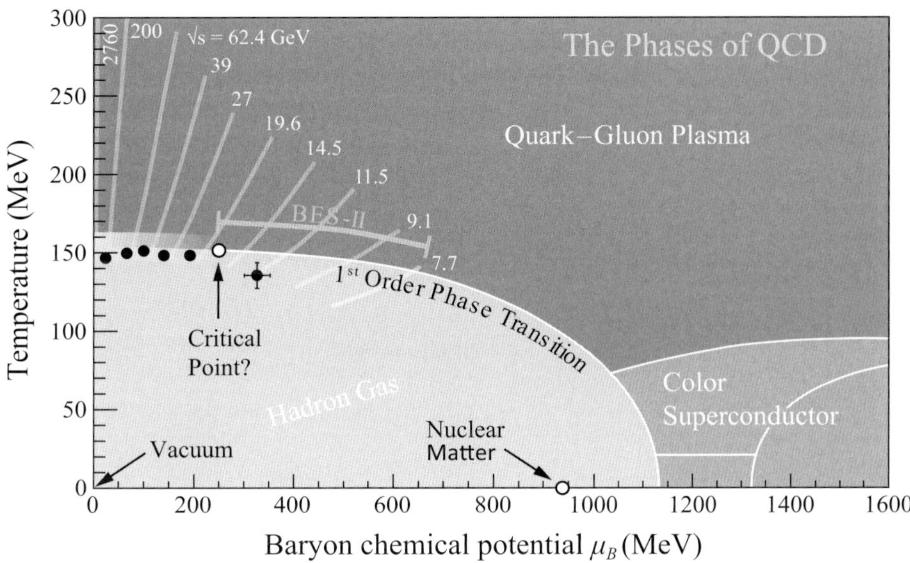

Fig. 19.1 Illustration of the QCD phase diagram of temperature as a function of the baryon density together with experimental and theoretical aspects highlighting where various experimental programs cover the expected first-order phase transition along with the approximate location of a critical point [Hei15]. Figure adapted from [LRP15].

indicated in Fig. 19.1 and this is under active theoretical and experimental study at present. The search for the critical point is being carried out in a beam-energy scan at RHIC over a range of $\sqrt{s}$ values running from 7.7 GeV to 200 GeV.

Prior to the RHIC and LHC collider programs reaching much higher center-of-mass energies, there have been two experimental programs, at the AGS (BNL) and the SPS (CERN). The AGS program suggested the creation of resonance matter with the creation of excited hadrons cascading and decaying into stable hadrons. Furthermore, the AGS program revealed the enhancement of high-p_T strangeness production. The creation of a QGP could not be ruled out, but clear experimental signatures were missing. On the other hand, the SPS program at CERN reached much larger center-of-mass energies compared with the AGS program at BNL. Results at the SPS of enhancement of multi-strange hadrons and the suppression of charm states led to intense discussions suggesting the formation of a QGP phase. A detailed study of a deconfined state of quarks and gluons is being carried out at RHIC and the LHC taking into account multiple, theoretically motivated signatures for the transition from a hadron gas to a QGP. Several review papers provide detailed discussions of these programs, including both the theoretical background and experimental achievements [Hei15, Jac12, Mul06, Mul12].

Table 19.1 provides an overview of the two major relativistic heavy-ion programs at LHC and RHIC in terms of their beam species, beam energy, center-of-mass energy, and luminosity. The main theoretical tools used to understand the QGP are lattice QCD [Bor12] and transport theory [Pet08]. In lattice QCD (see Section 5.3), the partition function is simulated on a spacetime lattice. Recent advances in computing have led to

Table 19.1 Main performance parameters of the RHIC and LHC relativistic heavy-ion programs

	RHIC	LHC
Date	Start in 2000	Start in 2010
Beam species	p+p / d+Au / Cu+Cu Cu+Au / Au+Au / U+U	p+p / Pb+Pb / p+Pb
Energy range	$\sqrt{s_{NN}} = 7.7 - 200\,\mathrm{GeV}$	$\sqrt{s_{NN}} = 2.76\,\mathrm{TeV}\ /\ 5.5\,\mathrm{TeV}$
Luminosity $(10^{27}\mathrm{cm}^{-2}\mathrm{s}^{-1})$	~ 8 (Au-Au)	~ 1 (Pb–Pb)
Experiments	BRAHMS / PHENIX PHOBOS / STAR	ALICE / ATLAS / CMS

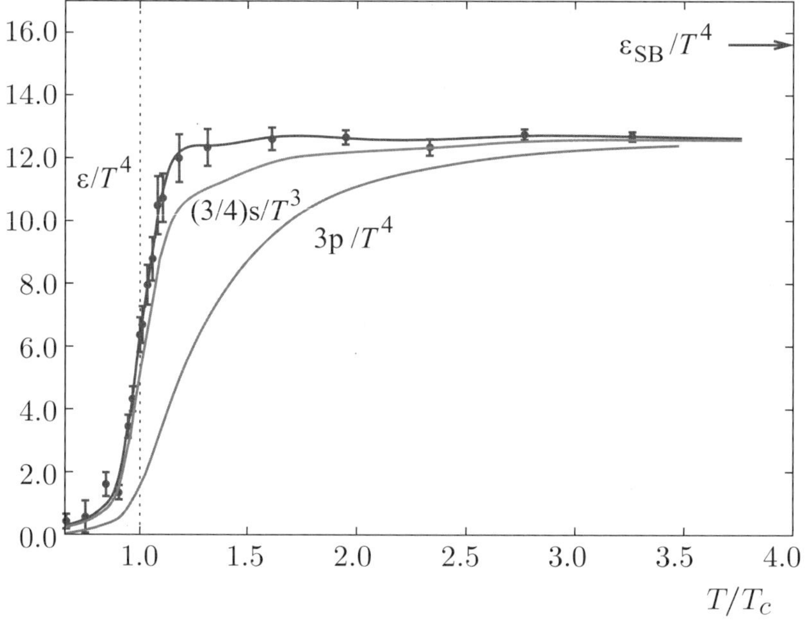

Fig. 19.2 Lattice QCD calculations for three quark flavors shown for energy density $\epsilon(T)/T^4$, pressure $P(T)/T^4$, and entropy $s(T)/T^3$ as functions of temperature scaled by the transition temperature T/T_c showing a drastic change around T_c in the range of 145–163 MeV. *SB* refers to the case of a massless Bose field [Mul06].

an enormous boost in the ability to undertake such studies, providing quantitative input for the equation-of-state and correlation functions in QCD, although, generally, lattice QCD calculations are still restricted to static properties. Figure 19.2 shows the results of lattice QCD computations for the energy density and three times the pressure of strongly-interacting matter as a function of T/T_c, with T_c being the transition temperature from a hadron gas to a QGP [Mul06]. The transition temperature is predicted to be in the range of 145–163 MeV, and over this temperature range, lattice QCD predicts that various thermodynamic properties change dramatically. Lattice QCD calculations include the equation-of-state, energy density, and pressure along with simulations concerning color confinement and chiral-symmetry breaking.

Transport theory is the main theoretical framework used to account for the dynamical properties of the hadron gas and QGP transition and formation. One typically distinguishes three distinct regions: the formation of the QGP itself initiated by the violent collisions of heavy ions at high temperature, the expansion of the QGP, and the subsequent transition to a gas of hadrons.

- Stage 1: Collision of gluons originating from the violent collision of heavy ions forming a nonlinear region of gluon fields, known as a *glasma* [Gel12].
- Stage 2: Rapid expansion of the QGP formulated on the basis of relativistic hydrodynamics.
- Stage 3: Cool-down and formation of the hadron gas characterizing the final expansion and freeze-out formulated using kinetic theory of hadrons.

The role of nonlinear dynamics has been presented in Chapter 9 in the context of low-x physics and is extensively discussed in [McL11] in particular for the color glass condensate. Over the last decade much theoretical work has been devoted to exploring the connection between QCD in the strong-coupling region and weakly coupled gravitation theories. It turns out that the duality of string theory in anti-de Sitter (AdS) space and conformal quantum field theory (CFT) provides an exact description of strongly-coupled systems [Kov05]. This correspondence provides a way to understand why a strongly coupled plasma of gauge fields can rapidly thermalize and why relativistic hydrodynamics provides a proper framework for the description of the rapid expansion of the QGP.

A critical measurement is the comparison of relativistic heavy-ion ($A + A$) collisions with control experiments of proton–proton ($p + p$) and $p + A$ collisions. Figure 19.3 shows an overview of these three experimental configurations. Various well-known probes are being employed at RHIC and the LHC, as illustrated in Fig. 19.4, which are sensitive to different aspects of the unknown, new medium under study. Photons and heavy bosons provide a direct probe of the very early stages of the QGP formation, since those are not affected by the presence of a QGP. Quark and gluon energy loss inside the QGP are used as diagnostics of the properties of the QGP. The dissociation of heavy quarks and heavy-quark bound states or quarkonia, including charm–quark and bottom-quark bound states,

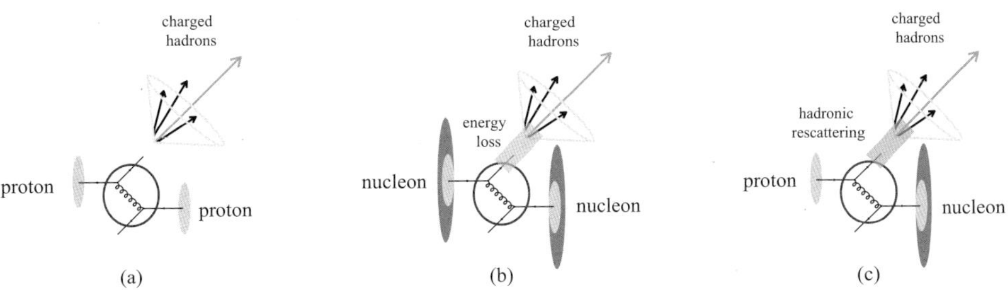

Fig. 19.3   Stages (a) of $p + p$, (b) of $A - A$, and (c) of $p/d - A$ collisions from the partonic level to the hadronization level focusing on final-state charged hadron production.

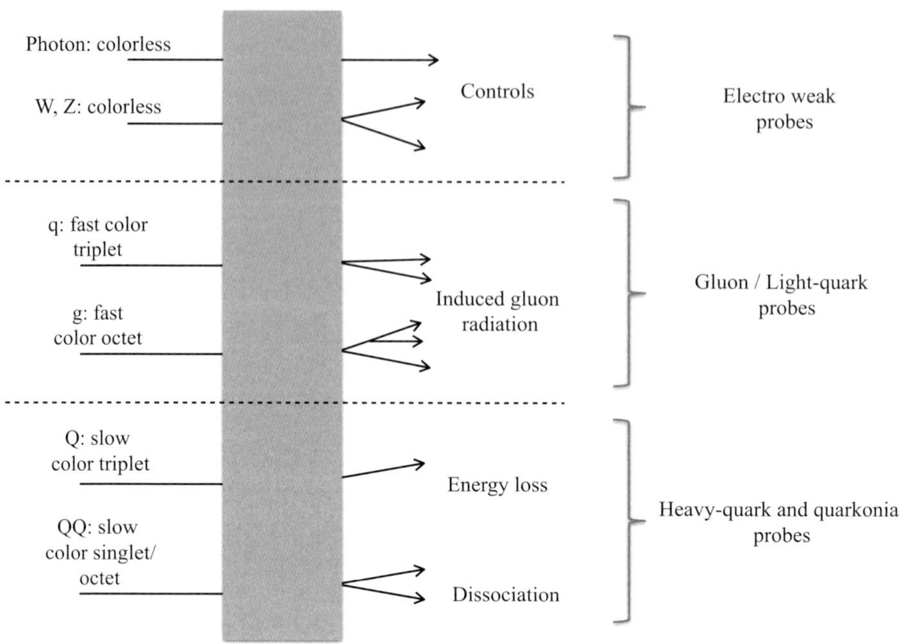

Fig. 19.4   Illustration of different probes (left) and their roles (right) to characterize the presence of an "unknown medium" of deconfined quarks and gluons shown in the center.

have long been discussed as sensitive probes for the formation of a QGP. The RHIC and LHC collider programs have established three main avenues in the study of relativistic heavy-ion collisions, all of which will be discussed in the following sections and illustrated with selected experimental results [Hei15, Jac12, Mul06, Mul12]. The main focus lies on various aspects of the expected transition from a hadron gas to a QGP:

- Global event characterization: State and dynamical evolution.
- Correlation measurements: Volume, lifetime and flow.
- Hard processes: Energy loss.

19.2 Global Event Characterization

Global event characterization includes the measurement of a large spectrum of identified hadrons in terms of their multiplicity, yield and momentum spectra, and correlations.

Multiplicity distributions have typically been the first results produced by the various relativistic heavy-ion programs discussed above. They are expressed in terms of the number of charged hadrons per unit pseudo-rapidity, i.e., $dN/d\eta$. The measured multiplicity produces a rough estimate of the energy density, first derived by Bjorken [Bjo83], relating the transverse energy distribution to the energy density of a system:

$$\epsilon \geq \frac{dE_T/d\eta}{\tau_0 \pi R^2} = \frac{3}{2}\left(\frac{E_T}{N}\right)\frac{dN_{ch}/d\eta}{\tau_0 \pi R^2}, \tag{19.1}$$

where R refers to the nuclear radius and τ_0 to the thermalization time. The ratio E_T/N is approximately $1\,\text{GeV}$ and denotes the transverse energy per emitted particle. It is interesting to compare the level of energy densities reached at RHIC and at LHC. At RHIC, an energy density of about $5\,\text{GeV fm}^{-3}$ is reached whereas at LHC a factor of three larger energy density of $15\,\text{GeV fm}^{-3}$ has been achieved. Multiplicity distributions can be used to characterize the actual collision geometry by the degree of centrality of a given relativistic heavy-ion collision ranging from central head-on collisions to grazing, peripheral collisions. The centrality of a given collision increases with the multiplicity. The degree of centrality can be interpreted using Glauber Monte-Carlo simulations [Mil07] to determine the number of participating nucleons N_{parts} in the overlap region of two colliding relativistic heavy nuclei as well as the number of nucleon–nucleon collisions.

Let us now focus on selected experimental results from the LHC and RHIC programs. Figure 19.5 shows the charged-particle pseudo-rapidity distribution, $dN_{ch}/d\eta$, per colliding nucleon pair as a function of the center-of-mass energy for $p + p$ and $A + A$ collisions (a) along with the charged-particle pseudo-rapidity distribution, $dN_{ch}/d\eta$, per colliding nucleon pair as a function of the number of participating nucleons (b). The behavior of $dN_{ch}/d\eta$ is well accounted for by a power-law behavior rather than a milder logarithmic behavior. It came as a surprise when the first multiplicity results from LHC experiments indicated that the behavior of $dN_{ch}/d\eta$ per colliding nucleon pair is rather flat as a function of the number of participating nucleons. Two effects can limit the growth with the number of participating nucleons and thus the centrality itself. Shadowing of the nuclear parton distribution functions besides QCD saturation phenomena are expected to limit the growth due to strong nuclear modifications.

The production of hadrons from elementary partonic collisions can be accounted for in relativistic heavy-ion collisions by statistical and hydrodynamical models. Figure 19.6 shows the transverse momentum spectra of identified particles for central $Pb + Pb$ collisions in comparison with hydrodynamical model calculations. Examples from the RHIC

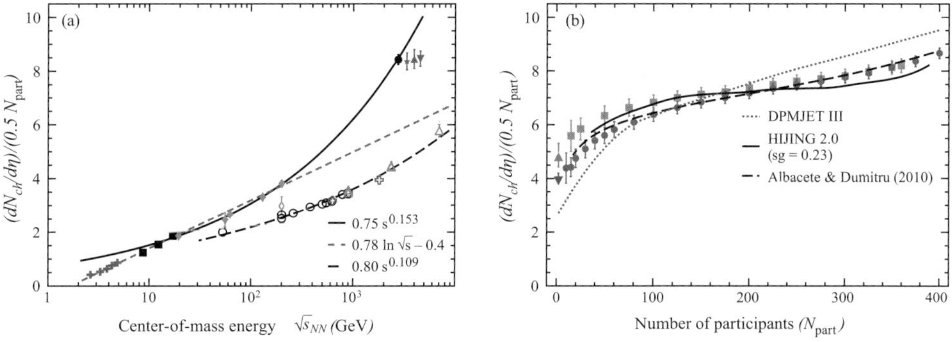

Fig. 19.5 (a) Charged-particle pseudo-rapidity density, $dN/d\eta$, per nucleon pair as a function of the pp and AA center-of-mass energy. (b) Charged-particle pseudo-rapidity density, $dN/d\eta$, per nucleon pair as a function of the number of participating nucleons; figure adapted from [Mul12].

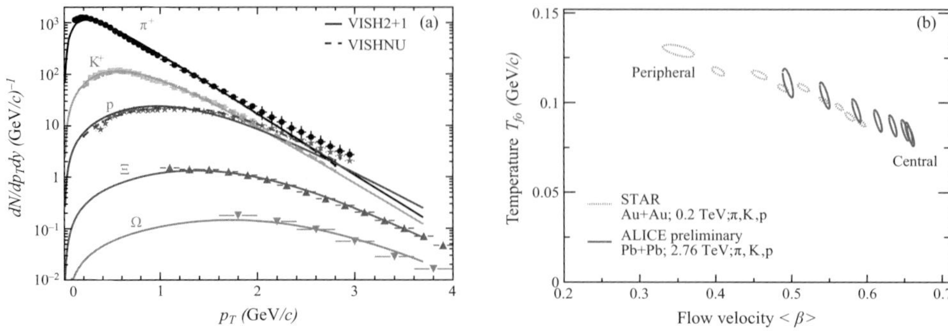

Fig. 19.6   Transverse momentum distribution for identified particles; figure adapted from [Mul12].

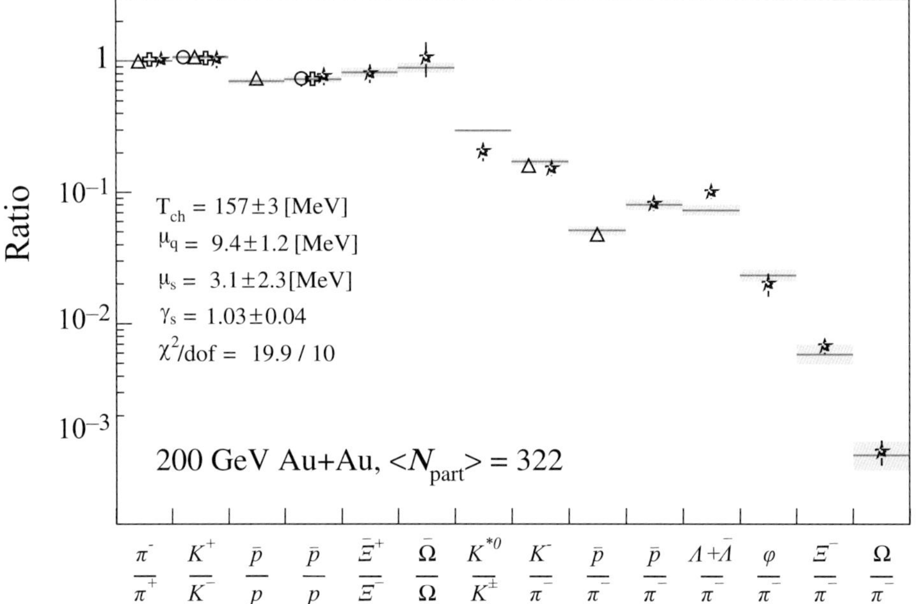

Fig. 19.7   Comparison of various particle-ratio measurements from RHIC experiments in comparison to thermal model calculations; figure adapted from [Mul06].

experiments featuring particle ratios are shown in Fig. 19.7 in comparison with thermal models, showing that the particle yields are consistent with a fully thermalized system.

19.3 Correlation Measurements

Particle correlation measurements have attracted considerable interest in characterizing the initial conditions in the formation of a QGP. Flow measurements have, in particular, created a lot of attention in the RHIC and LHC programs. The collision of two relativistic heavy ions with nonzero impact parameter has an almond shaped overlapping zone which

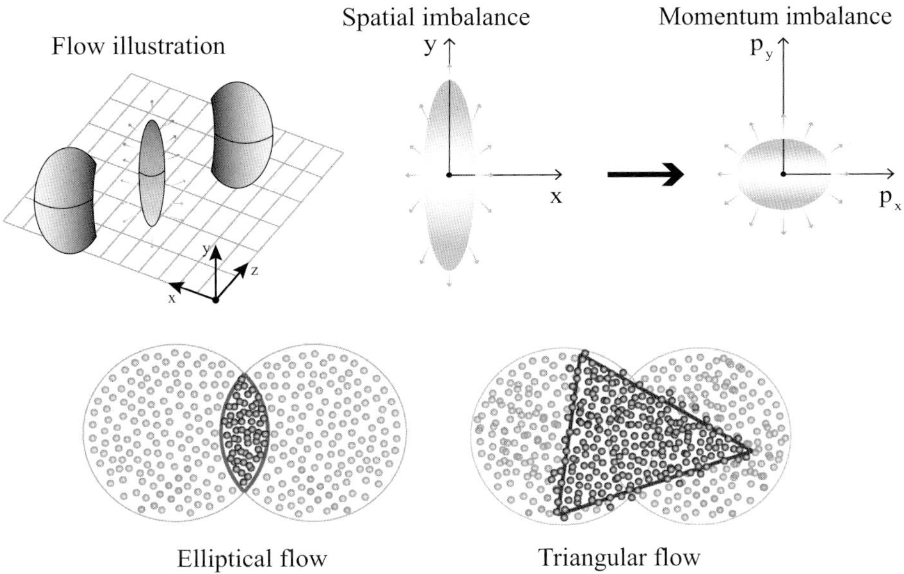

Fig. 19.8 Illustration of flow phenomena (top) and elliptical (bottom left) and triangular flow (bottom right).

changes as a function of centrality (Fig. 19.8). The pressure gradient is dependent on the azimuthal angle and is largest in the direction of the reaction plane angle ψ_{RP}, which coincides with the direction of the minor axis of the overlapping almond region. Collective flow of particles develops predominantly in the direction of the pressure gradient and thus along the reaction plane. This spatial anisotropy translates into the momentum anisotropy of produces particles, which is a direct measure of the interaction strength among the produced hadrons. With the system expanding in time, the initial spatial anisotropy becomes more uniform, while the momentum anisotropy is affected with the transition of partons to observed hadrons. The observed anisotropic distribution or flow of particles $dN/d\phi$ is quantified by a Fourier expansion as follows,

$$E\frac{d^3N}{d^3p} = \frac{1}{2\pi}\frac{d^2N}{p_T dp_T d\eta}\left(1 + 2\sum_{n=1}^{\infty} v_n \cos\left[n(\phi - \psi_n)\right]\right), \tag{19.2}$$

where the the Fourier coefficients v_n are functions of p_T, n is the order of the Fourier expansion and ϕ is the azimuthal angle. ψ_n is the angle of the spatial plane of symmetry for each order n of the Fourier expansion. Let us now provide some meaning to the Fourier coefficients: v_1 is also known as the directed flow, while v_2 is known as the elliptic flow and is largest for non-central collisions. Higher-order Fourier contributions v_n with $n > 2$ have also been observed; these are mainly due to fluctuations in the initial location of nucleons inside a given nucleus. Figure 19.8 provides an illustration of the elliptic flow (left), which increases with increasing impact parameter, and of higher-order contributions (right) also referred to as triangular flow. It turns out that hydrodynamical models successfully describe elliptic flow measurements. This then opens the possibility to characterize the behavior of a new phase in terms of hydrodynamical properties such as the ratio of the shear viscosity

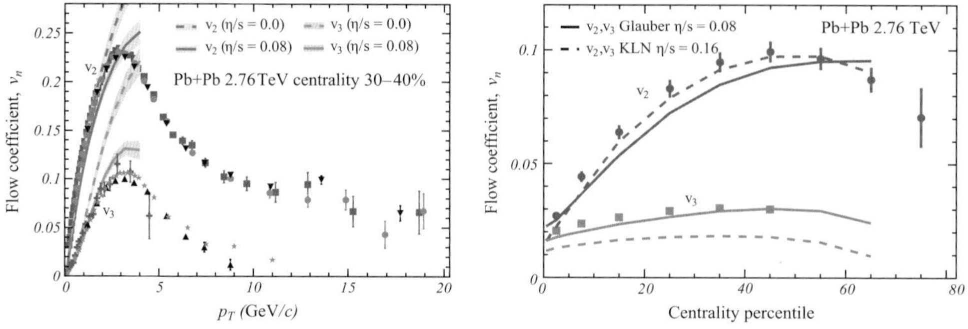

Fig. 19.9 The LHC results on flow coefficients as a function of transverse momentum (left) and centrality (right) in comparison with hydrodynamical calculations; figure adapted from [Mul12].

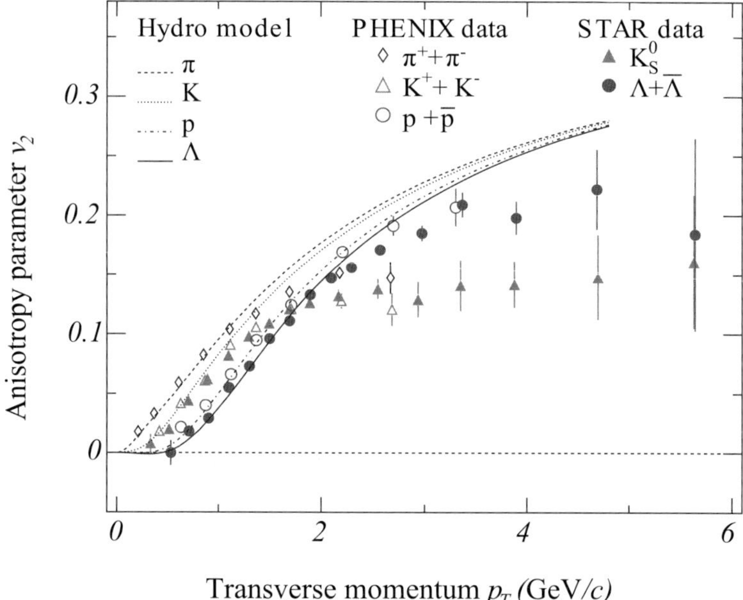

Fig. 19.10 The RHIC results on v_2 measurements as a function of transverse momentum in comparison to hydrodynamical calculations; figure adapted from [Mul06].

(η), i.e., the resistance to flow or the inability of matter to transport momentum, to the entropy density (s), i.e., η/s. Figure 19.9 shows the v_2 and v_3 dependence as a function of p_T and centrality in comparison with hydrodynamical models, and agreement is found in the region of overlap of measurements and hydrodynamical calculations. The Fourier coefficients v_2 and v_3 are found to increase with increasing centrality. Results from RHIC experiments are shown in Fig. 19.10.

The hydrodynamical interpretation of these results led to the conclusion that η/s is very small, suggesting that the QGP behaves as an almost perfect fluid. These findings were first seen at RHIC and confirmed later at the LHC. What are the expectations for this ratio? Earlier calculations on the dissipative phenomena of a QGP have been carried out

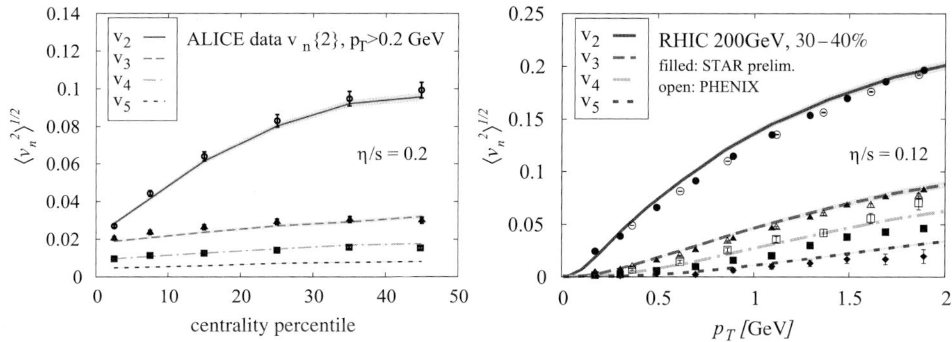

Fig. 19.11   Compilation of RHIC and LHC results of Fourier harmonics v_n as a function of centrality (left) and p_T (right) with extracted η/s values of 0.2 in the case of LHC and 0.12 in the case of RHIC; figure adapted from [Hei15].

in [Dan85] suggesting a value of η/s of approximately 0.1. A lower limit for η/s has been suggested in the context of the AdS/CFT correspondence and is given by $\eta/s \geq 1/4\pi \sim 0.08$, the so-called KSS bound [Kov05]. Figure 19.11 shows a side-by-side comparison of LHC and RHIC results for various measured Fourier harmonics v_n up to $n = 5$ as a function of p_T; the extracted values for η/s ranges from roughly 0.12 at $\sqrt{s} = 0.2\,\text{TeV}$ at RHIC to 0.2 at $\sqrt{s} = 2.76\,\text{TeV}$ at LHC, corresponding to roughly 1.5 and 2.5 of the KSS bound. It should be emphasized that all extracted Fourier harmonics including those for $n > 2$ have been taken into account in the extraction of η/s.

19.4 Hard Processes

Hard processes have long been suggested as a sensitive probe for investigating the occurrence of a new phase of matter in relativistic heavy-ion collisions in comparison with proton–proton collisions. The main idea is motivated by referring back to Fig. 19.3. Three main techniques are employed at RHIC and in particular at the LHC program in such comparisons:

- Single-particle spectra.
- Jet spectra.
- Quarkonium production.

The suppression of single-particle spectra was one of the early RHIC results that suggested the occurrence of a new phase of matter, followed by the study of jet production. The LHC program profits enormously from the larger partonic cross sections that occur with the increased center-of-mass energy. In addition, the LHC program allowed one to employ new probes such as W/Z production, in addition to γ production, which are not affected by a medium consisting of quarks and gluons. The main questions here are the roles that the color charge and partonic mass have on the partonic energy loss. It is expected that gluons should lose energy at about twice the rate that quarks do, and heavy quarks are expected to lose less energy than light quarks.

Each mechanism will be introduced below and discussed in the context of experimental data. Single-particle spectra resulting from the fragmentation of partons into observable hadrons have shown a strong suppression at RHIC in comparison with $p + p$ collisions. A larger kinematic phase space is accessible at the LHC in addition to new probes such as W/Z bosons. The suppression is quantified by the nuclear suppression factor defined as:

$$R_{AA}(p_T) = \frac{d^2 N_{AA}/dp_T d\eta}{\langle T_{AA} \rangle d^2 \sigma_{NN}/dp_T d\eta} \,. \tag{19.3}$$

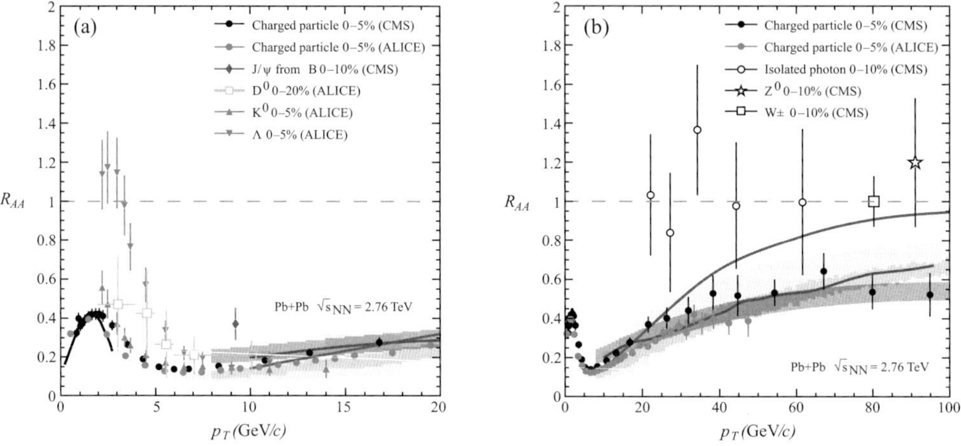

Fig. 19.12   Measurement of the nuclear modification factor R_{AA} from the LHC program as a function of p_T for different identified particles in comparison with theoretical calculations; figure adapted from [Mul12].

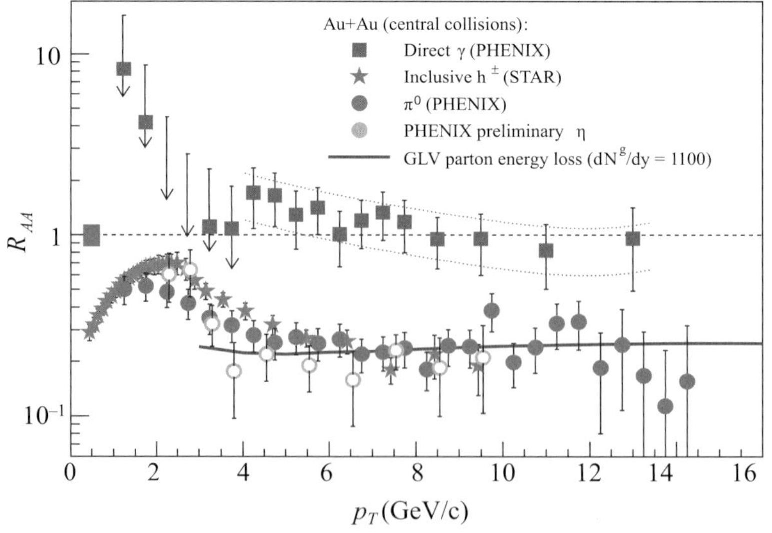

Fig. 19.13   Measurement of the nuclear modification factor R_{AA} from the RHIC program as a function of p_T for different identified particles in comparison with theoretical calculations; figure adapted from [Mul06].

Here N_{AA} and σ_{NN} refer to the number of particles and the cross section in $A+A$ collisions. And $\langle T_{AA}\rangle$ is the nuclear overlap function, defined as the ratio of the number of binary nucleon–nucleon collisions to the inelastic nucleon–nucleon cross section, where the number of binary collisions is determined from a Glauber model. The above ratio $R_{AA}(p_T)$ should be unity without any nuclear modifications; however, strong suppression of this ratio has been observed in several experimental results. Figures 19.12 and 19.13 display $R_{AA}(p_T)$ as a function of p_T where clear suppression is observed for various hadronic particles, whereas γ and W/Z bosons measured by the LHC program are not affected. Figure 19.13 shows results from the RHIC program with a similar observation.

The modification of jet production has been a prime focus of both the RHIC and LHC programs. The modification of reconstructed jets provides a powerful probe with which to search for a new phase of matter. The measurement of jet energy allows one to account for

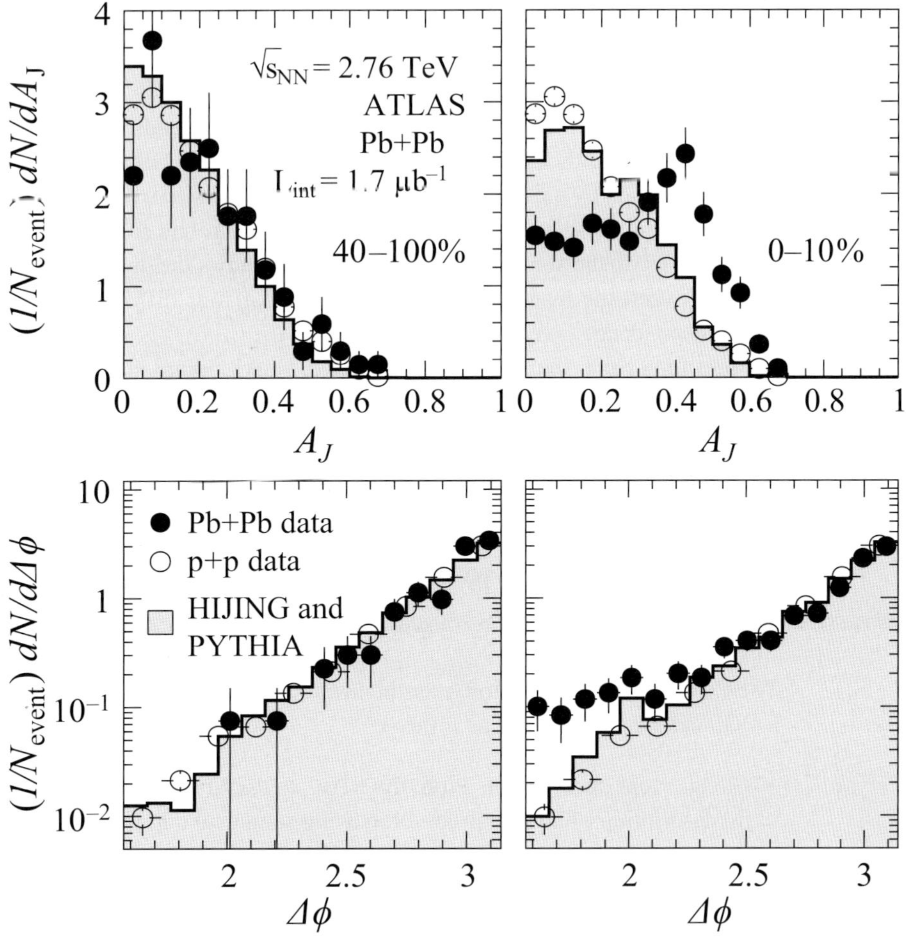

Fig. 19.14 LHC measurements (ATLAS) on (a) the calorimetric jet imbalance A_J in di-jet events and (b) the azimuthal angle $\Delta\phi$ between the leading and subleading jets, as functions of collision centrality for pp and $Pb+Pb$ collisions; figure adapted from [Mul12].

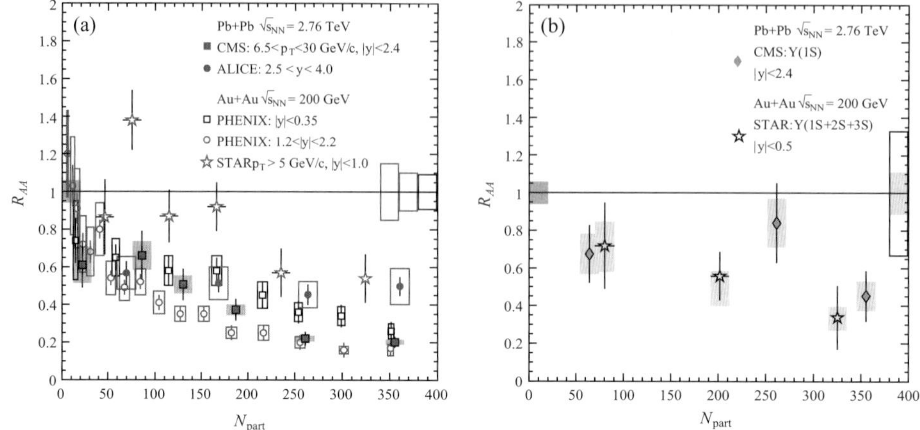

Fig. 19.15   Measurement of the nuclear modification factor R_{AA} for RHIC and LHC experiments as a function of centrality (a) for J/ψ and (b) for Υ production; figure adapted from [Mul12].

the energy redistribution from the leading parton to subleading partons and the dissipation of energy into the thermal medium. Such studies are, in particular, useful if one employs di-jet production by measuring the so-called di-jet asymmetry, $A_J \equiv (p_{T_1} - p_{T_2})/(p_{T_1} + p_{T_2})$, and comparing with $p + p$ collisions. The index 1 refers to the leading jet, whereas the index 2 refers to the subleading jet. Figure 19.14 shows the di-jet asymmetry and azimuthal angle between the leading and subleading jet for four different centrality bins, where a strong dependence on centrality is observed.

For a long time, the suppression of heavy quark bound states has been suggested as a sensitive probe due to the modification of a given bound state by the presence of a QGP. Figure 19.15 shows the nuclear modification factor as a function of centrality for J/ψ and Υ production measured at RHIC and the LHC.

19.5   Summary and Outlook

The experimental exploration over the last few decades of the transition of a hadron gas to a QGP can be summarized as follows:

- Global event observables: From the RHIC to the LHC energy regime one has a successful description of hadron production using statistical and hydrodynamical models.
- Hard probes: One sees a clear suppression of hadronic probes which are affected by the presence of a deconfined medium of quarks and gluons, in contrast to probes such as photons and electroweak boson which are not.
- Flow measurements: Precise measurement of flow coefficients v_n up to $n = 6$ provide a basis to determine the shear viscosity of a new phase of matter resembling the properties of a perfect liquid.

The deep connection in QCD concerning the interplay of QGP matter, chirality imbalance and strong magnetic fields has attracted a lot of attention, both theoretically, as well as on the experimental side, with the observation of charge-dependent hadron azimuthal correlations in heavy-ion collisions at LHC [Abe13a] and RHIC [Moh11]. This effect is known as chiral magnetic effect (CME) [Hir14]. Future studies will continue to focus on these aspects with higher precision, using the variability of beam species and energy of the RHIC and LHC facilities. This will include precision jet and quarkonium measurements, mapping of the QCD phase diagram using lower-energy data (e.g., the RHIC beam-energy scan), and precision flow measurements to characterize the transverse and longitudinal spatial dependence of the initial gluon fluctuation spectrum. A future electron–ion collider is in part motivated by the desire to understand the hot, dense matter made in heavy-ion collisions. It would provide an understanding of the initial partonic state of a high-energy nucleus and yield an understanding in cold hadronic matter of many of the processes discussed above in probing the hot, dense matter associated with the QGP.

While studies of the QGP will continue using heavy ions collisions, it is clear that a complete understanding of the new phenomena discovered will require experiments where the initial state in the high energy QCD collision is simpler and is changeable in a controlled way. In this book, we have seen consistently how lepton beams have these powerful characteristics to probe hadronic matter. Studying the nucleus at high energies from a fundamental quark–gluon perspective is a major scientific motivation for the future electron–ion collider introduced in Chapter 10 when nuclear beams are used.

Exercises

19.1 Heavy-Ion Hydrodynamics

A great deal of the phenomenology of heavy-ion collisions has been shown to follow from hydrodynamic considerations.

a) Consider an ultra–relativistic heavy-ion collision in the center-of-mass frame. The nuclei become two Lorentz-contracted pancakes. A short time (say 3 fm c^{-1}) after the collision, the pancakes recede from the collision point at the speed of light ($\gamma \gg 1$). Assuming an initial thermalization time of about 1 fm c^{-1}, show that the energy density is in the range 1-10 GeV fm^{-3}.

b) A basic feature of Bjorken's description [Bjo83] is that there is a central plateau structure for the rapidity distribution in nucleon–nucleon collisions, where rapidity is defined as

$$ y = \frac{1}{2} \ln \frac{t+z}{t-z} , $$

and proper time τ is defined as $\tau = \sqrt{t^2 - z^2}$. The system evolves so that throughout the central plateau region, the initial conditions, imposed at a proper time ≈ 1 fm c^{-1} after the collision time, are invariant with respect to

Lorentz transformations. By applying Landau hydrodynamics to the longitudinal evolution of the plasma, show that

$$\frac{(dN_\pi/dy)_{A-A}}{(dN_\pi/dy)_{pp}} \approx \left[\frac{2\ \text{fm}}{d_0}\right]^2 A^{2/3} \,,$$

where $1/d_0^2$ is the number of independent nucleon–nucleon collisions per unit area.

19.2   Scaling and Multiplicity

For a variety of nuclear collisions at low energies up through RHIC energies (200 GeV/nucleon NN center-of-mass), it was observed that there was approximate $N_{\text{participant}}$ scaling [Bus75] of the total charged-particle multiplicity, i.e., $N_{ch}/N_{ch}^{pp} = \frac{1}{2}N_{\text{part}}$, provided one scales heavy-ion data to pp data at twice the center-of-mass energy, as shown in the figure from [Bac05]. Make a plot that extends the range in energy also using available data from the LHC program to determine if this scaling is preserved at higher energies. Can one understand the origin of any deviations from scaling?

19.3   Viscosity

The concept of viscosity has proven to be useful in describing heavy-ion collisions.

a) Provide a definition for viscosity from a classical perspective and find the viscosity for water, oil, and mercury.

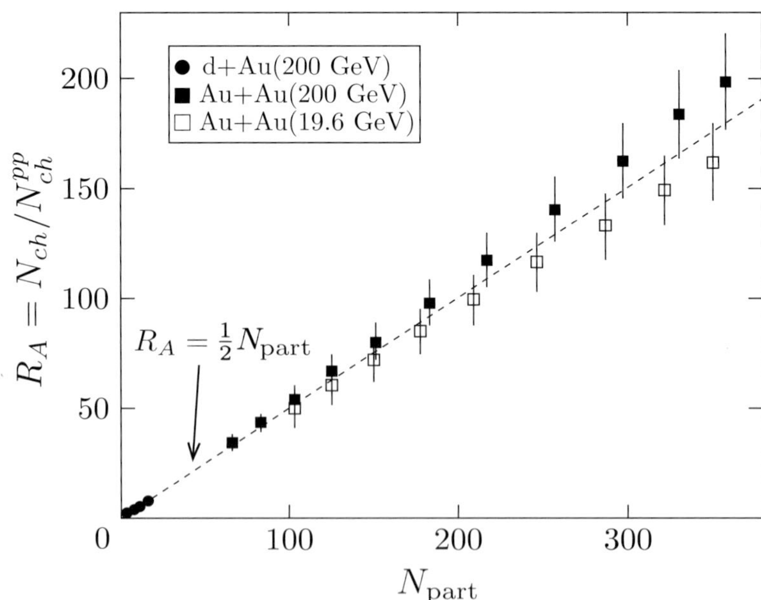

b) Discuss your result for a) from the perspective of the Navier–Stokes equation

$$\frac{\partial \rho v_i}{\partial t} = -\nabla_j \left(\rho v_j v_i + \delta_{ij} P - \sigma_{ij}\right) \,,$$

where ρ, $\boldsymbol{v}$ are the fluid density, velocity respectively. Here

$$\sigma_{ij} = \eta \left(\frac{\partial v_j}{\partial x_i} + \frac{\partial v_i}{\partial x_j} - \frac{2}{3} \delta_{ij} \nabla \cdot \boldsymbol{v} \right) + \zeta \delta_{ij} \nabla \cdot \boldsymbol{v} \,,$$

where η is the shear viscosity and ζ is the bulk viscosity.

19.4 Stokes' Law for Viscous Flow

In order to gain familiarity with the effects of viscosity, it is useful to examine a classic problem: Stokes' law for the flow of a viscous fluid. During the 1850s, G. G. Stokes began investigations of highly viscous flow. One of his results is that, for creeping viscous flow of an incompressible fluid past a sphere, the drag force is given by

$$\boldsymbol{F}_{\mathrm{drag}} = 6\pi \eta a v_0 \,,$$

where v_0 is the asymptotic flow velocity, a is the radius of the sphere, and η is the viscosity. In this problem we derive this result.

Assume that the most important term for large η is the diffusion term and that the inertial term $\rho \frac{\partial \boldsymbol{v}}{\partial t}$ can be neglected. Then, neglecting gravity, we have

$$-\nabla P + \eta \nabla^2 \boldsymbol{v} = 0 \,.$$

a) Show that the pressure satisfies Laplace's equation

$$\nabla^2 P = 0 \,.$$

Define the stream potential ψ in terms of axisymmetric spherical co-ordinates r, θ:

$$v_r = \frac{1}{r^2 \sin \theta} \frac{\partial \psi}{\partial \theta}, \qquad v_\theta = \frac{1}{r \sin \theta} \frac{\partial \psi}{\partial r} \,.$$

b) Show that for axially-symmetric flow

$$\nabla \times \boldsymbol{v} = 0 \,,$$

is required, and that ψ obeys the equation

$$\left(\frac{\partial^2}{\partial r^2} + \frac{1}{r^2} \frac{\partial^2}{\partial \theta^2} - \frac{\cot \theta}{r^2} \frac{\partial}{\partial \theta} \right)^2 \psi = 0 \,.$$

Now insert the boundary conditions:

i) no slip at the sphere surface

$$\frac{\partial \psi}{\partial r} = \frac{\partial \psi}{\partial \theta} = 0 \,,$$

ii) asymptotic flow $v_0 \hat{e}_z$

$$v_r \sim v_0 \cos \theta, \qquad v_\theta \sim -v_0 \sin \theta \quad \text{as} \quad r \to \infty \,.$$

c) Show that condition ii) is satisfied by the asymptotic form

$$\psi(r, \theta) \xrightarrow{r \to \infty} \frac{1}{2 v_0 r^2 \sin^2 \theta} \,,$$

which suggests the use of

$$\psi(r,\theta) = \frac{1}{2} v_0 f(r) \sin^2 \theta$$

as a trial solution.

d) Find the differential equation obeyed by $f(r)$ and solve to show that

$$\psi(r,\theta) = \frac{1}{2} v_0 \left(\frac{1}{2} \frac{a^3}{r} - \frac{3}{2} ar + r^2 \right) \sin^2 \theta .$$

e) Show that the pressure is given by

$$P(r,\theta) = P_\infty - \eta v_0 \frac{3a}{2r^2} \cos\theta .$$

f) Using $\nabla P = \eta \nabla^2 \mathbf{v}$, calculate the total drag force on the sphere due to the pressure differential.

g) Using $\hat{n} \cdot F = \int ds_i T_{ij} \hat{e}_j$ and the viscous friction component of the stress tensor $T_{ij}^2 = \eta \left(\frac{\partial v_i}{\partial x_j} + \frac{\partial v_j}{\partial x_i} \right)$, calculate the total drag force on the sphere due to viscous friction.

h) Add your results for f) and g) to yield Stokes' Law

$$\mathbf{F}_{\text{drag}} = 6\pi \eta a v_0 .$$

20

Astrophysics

With the term "experimental cosmology" becoming less of an oxymoron, the subject of nuclear and particle astrophysics is playing an increasingly important role. In this chapter we consider several aspects of this subject: Big Bang nucleosynthesis, stellar evolution, and cosmic rays. (Note that relevant material has also been presented in Chapters 4, 6, 12, and 17.)

20.1 Big Bang Nucleosynthesis

We begin by noting the experimental abundances by mass of various isotopes, which are shown in Table 20.1. It is clear from this listing that the Universe consists of 75% hydrogen, 25% helium plus trace amounts of other elements. Since helium is produced as a byproduct of stellar fusion, it is tempting to speculate that this may be the source of the observed helium abundance, as discussed in Chapter 12. However, a simple estimate quickly shows that this assumption must be incorrect. Indeed it was first shown by Bethe and Critchfield [Bet38], at the end of the 1930s, that the basic process which fuels solar burning is

$$4p \rightarrow {}^4\text{He} + 2e^+ + 2\nu_e + 25.7 \text{ MeV} . \tag{20.1}$$

Since the neutrinos carry off about 0.4 MeV, this leaves about 6.25 MeV for each proton consumed in the process and we leave it as an exercise to show that Eq. (20.1), combined with the solar constant 1.36 kW m^{-2} means that no more than 5% of its mass has been transformed into helium during the Sun's 4.6 Myr lifetime. If our Sun is a typical star, then it is apparent that solar burning is *not* responsible for the lion's share of the observed helium abundance, especially when it is realized that many (if not most) protons do not lie within stars. Thus we must look elsewhere in order to resolve the mystery.

The answer was provided by Gamow in the 1940s, who proposed what we now call "Big Bang cosmology." Essentially the idea is that the Universe began $\sim$14 Gyr ago as a highly compressed hot soup of electrons, protons, quarks, photons, etc. As the Universe expanded it cooled and, once the average energy dropped below the binding energy of deuterium, the neutrons and protons combined to form deuterium via

$$n + p \rightarrow d + \gamma .$$

In fact, since a free neutron decays via the weak interaction

$$n \rightarrow p + e^- + \bar{\nu}_e ,$$

Table 20.1 Isotopic abundances by mass of various isotopes in the universe

Element	Abundance
^{1}H	75%
^{4}He	25%
^{6}Li	7.75×10^{-10}
^{7}Li	1.13×10^{-8}
^{9}Be	3.13×10^{-10}
^{10}Be	5.22×10^{-10}
^{11}B	2.30×10^{-9}
^{12}C	3.87×10^{-3}
^{14}N	0.94×10^{-3}
^{16}O	8.55×10^{-3}
^{20}Ne	1.34×10^{-3}
^{24}Mg	0.58×10^{-3}
^{28}Si	0.75×10^{-3}
$A \geq 100$	many rare isotopes

with a lifetime $\tau_n \simeq 880\,\mathrm{sec}$ ($<<< 14\,\mathrm{Gyr}!$), all neutrons would have long since decayed away unless bound in nuclei, wherein such decay can be kinematically forbidden. At the present time, from Table 20.1, we have about 12 protons for each ^{4}He atom. Thus

$$\frac{\#n}{\#p} \simeq \frac{2}{2+12} = \frac{1}{7},$$

and relatively simple arguments allow an understanding of this number, as we show below.

Cosmology

Before proceeding, it is necessary to introduce a few basic ideas about cosmology. One is that the Universe is homogeneous and isotropic on cosmological length scales, which implies the Robertson–Walker metric [Wei72]

$$d\tau^2 = dt^2 - R^2(t)\left[\frac{dr^2}{1 - kr^2} + r^2 d\Omega^2\right] \tag{20.2}$$

with

$$k = \begin{cases} -1 & \text{negative curvature, infinite universe} \\ 0 & \text{flat, infinite universe} \\ +1 & \text{positive curvature, finite universe .} \end{cases} \tag{20.3}$$

Hence, an observer at fixed coordinates r, θ, ϕ (a so-called co-moving observer) would find the distance between him/herself and another such fixed observer at $r = 0$ changing, due to expansion, via

$$\Delta s = R(t) \int_0^r \frac{dr'}{\sqrt{1 - kr'^2}} . \tag{20.4}$$

This is akin, for the case of positive curvature, to sitting at a fixed location on the surface of a balloon which is being inflated. The details of how the distance evolves is contained in the function $R(t)$, whose time development follows (under simplifying assumptions) from the Einstein equation

$$\frac{\dot{R}}{R} \equiv H = \sqrt{\frac{8\pi G \rho(t)}{3}} \, , \tag{20.5}$$

where $\rho(t)$ is the total energy density of the Universe, while H is the Hubble constant and has the experimental value

$$H = (75 \pm 25) \, \frac{\text{m/s}}{\text{kpc}} \, .$$

(Here and below, a superscript dot signifies a time derivative.)

In the early Universe the temperature was much higher than the rest mass of the elementary particles and thus they were highly relativistic. In this case their mass becomes irrelevant and they act like photons for which the energy density satisfies

$$\rho = 3P \sim \frac{1}{R^4(t)} \, . \tag{20.6}$$

Then

$$\frac{\dot{\rho}}{\rho} = -4\frac{\dot{R}}{R} = -4\sqrt{\frac{8\pi G \rho}{3}} \tag{20.7}$$

$$\text{i.e.,} \quad dt = -\frac{1}{4}\sqrt{\frac{3}{8\pi G}}\frac{d\rho}{\rho^{\frac{3}{2}}} = \sqrt{\frac{3}{32\pi G}}d\left(\frac{1}{\rho^{\frac{1}{2}}}\right) , \tag{20.8}$$

whose solution is

$$\rho(t) = \frac{3}{32\pi G(t + t_0)^2} \, , \tag{20.9}$$

implying that the density monotonically falls as the expansion continues. For highly relativistic particles we also have the Stefan–Boltzmann Law, which relates the density to the temperature

$$\rho(t) = N\frac{\pi^2}{30}(kT)^4 \, , \tag{20.10}$$

where N is the number of degrees of freedom.[1] We see then that

$$R \sim \rho^{-\frac{1}{4}} \sim \frac{1}{T} \sim \sqrt{t} \, . \tag{20.11}$$

Thus as the Universe evolves, i.e., as t increases, the temperature and density fall, while the scale parameter increases.

Now let us analyze various epochs in the evolution process: We begin at $t \simeq 0.01$ sec. At this time the temperature is $\sim 10^{11}$ K, so $k_B T \sim 10\,$MeV. Since $k_B T >> 2m_e$, the neutrinos and electrons/positrons are kept in thermal equilibrium via the weak interactions

---

[1] This result is easily understood since the density of states goes like k_F^3 with each momentum state filled up to $k_F \sim T$. Since each particle carries energy $\sim T$, we find $\rho \sim T \cdot T^3$.

$$\nu_e + e^- \leftrightarrow \nu_e + e^-$$

$$e^+ + e^- \leftrightarrow \nu_e + \bar{\nu}_e .$$

(Note that for heavier leptons only the neutral current reactions are operative, since $k_B T << m_\mu, m_\tau$.) Similarly, since $k_B T >> |B_d| = 2.23$ MeV, the neutrons and protons are kept in equilibirium via

$$n + e^+ \leftrightarrow p + \bar{\nu}_e$$

$$p \leftrightarrow n + e^+ + \nu_e$$

$$n \leftrightarrow p + e^- + \bar{\nu}_e .$$

However, the relative neutron/proton number has begun to drop, since

$$\frac{n}{p} = \exp\left(-\frac{m_n - m_p}{k_B T}\right) \simeq 0.86 . \tag{20.12}$$

At the later time $t \simeq 0.1$ sec, the temperature has dropped to 3×10^{10} K so $k_B T \sim 3$ MeV. At this point the $\bar{\nu}_e$ "decouple" in that the rate at which reactions $\bar{\nu}_e + p \rightarrow n + e^+$ occur, which can destroy such particles, can no longer keep up with the expansion rate of the universe. Likewise, around this time, nucleons fall out of equilibrium. In order to understand this phenomenon more quantitatively, consider the reaction $n + \nu_e \leftrightarrow p + e^-$. The rate at which neutrons transform into protons is given by

$$\Lambda_{n \rightarrow p}(T) \sim n_\nu(T) <\sigma v> , \tag{20.13}$$

where $n_\nu(T)$ is the electron neutrino density, $v \simeq 1$ is the relative neutrino–neutron velocity, and

$$\sigma \sim G_F^2 E_\nu^2 \sim G_F^2 (k_B T)^2$$

is the weak interaction cross section. Since $n_\nu(T) \sim T^3$ we expect

$$\Lambda_{n \rightarrow p}(T) \sim G_F^2 (k_B T)^5 \sim 10^{-10} \frac{(k_B T)^5}{m_N^4} \sim \frac{0.15}{\text{sec}} \left(\frac{k_B T}{\text{MeV}}\right)^5 . \tag{20.14}$$

A more careful calculation yields

$$\Lambda_{n \rightarrow p}(T) \sim \frac{0.76}{\text{sec}} \left(\frac{k_B T}{\text{MeV}}\right)^5 , \tag{20.15}$$

which should be compared to the Hubble parameter, which characterizes the expansion rate of the universe

$$H(T) \sim \sqrt{\frac{8\pi G}{3} N \frac{\pi^2}{30} (k_B T)^4} \sim \frac{0.67}{\text{sec}} \left(\frac{k_B T}{\text{MeV}}\right)^2 . \tag{20.16}$$

Equating the neutron disappearance and expansion rates, we find

$$0.76 \left(\frac{k_B T}{\text{MeV}}\right)^5 = 0.67 \left(\frac{k_B T}{\text{MeV}}\right)^2 , \quad \text{i.e.,} \quad T \simeq 1 \text{ MeV} . \tag{20.17}$$

At this point in the evolution of the universe the n/p ratio is

$$\frac{n}{p} \sim \exp -\frac{1.3\,\text{MeV}}{k_B T} \sim 0.25 \quad \text{at} \quad t \sim 1\,\text{sec}, \tag{20.18}$$

which is the state of affairs at about $t \sim 1$ sec. During the next 10 seconds, the neutrino reactions continue to push things (albeit at a slower rate) toward the proton side, yielding

$$\frac{n}{p} \sim \frac{1}{6} \quad \text{at} \quad t \sim 10\,\text{sec}. \tag{20.19}$$

Neutron beta decay then becomes the dominant process. If nothing else were to occur, after 10 minutes or so there would be no neutrons left, only protons, and the Universe would be a dull place indeed.

Instead, around this time the reaction

$$n + p \rightarrow d + \gamma$$

becomes operative and essentially all of the remaining neutrons condense into deuterons. Since the deuteron binding energy is 2.225 MeV, one might have expected this condensation to have occurred at an earlier time, with a correspondingly larger n/p ratio. However, we have argued above that thermal equilibrium remains until about $k_B T \sim 1$ MeV. This is because, through processes we do not fully understand, the universe was created with nearly but not quite equal amounts of matter and antimatter. The resulting antimatter–matter collisions produced the universe we know, with only matter (from this tiny initial excess) plus *many* photons (from the annihilation),

$$\frac{\rho_N}{\rho_\gamma} \sim 10^{-9}\;!$$

The rate between reactions $n + p \rightarrow d + \gamma$ driving condensation and $\gamma + d \rightarrow n + p$ describing breakup is then

$$\frac{\Lambda_{\text{breakup}}}{\Lambda_{\text{condensation}}} \sim \frac{\rho_d \rho_\gamma}{\rho_p \rho_n} \exp -\frac{|B_d|}{k_B T}. \tag{20.20}$$

At high temperatures, $\exp(-|B_d|/k_B T) \sim 1$ and, since $\rho_\gamma/\rho_p \ggg 1$, we have $\Lambda_{\text{breakup}} \gg \Lambda_{\text{condensation}}$. However, deuterium begins to form when $\rho_d \sim \rho_n$, i.e., when

$$\exp\left(-\frac{|B_d|}{k_B T}\right) \sim \frac{\rho_p}{\rho_\gamma} \sim 10^{-9}. \tag{20.21}$$

Solving, we find

$$k_B T_d \sim (1.25 \times 10^9\,\text{K})k_B \sim 100\,\text{keV} \quad \text{at} \quad t \sim 100\,\text{sec}. \tag{20.22}$$

Numerical integration from the weak freezeout at $t \sim 1$ sec until this time of nucleosynthesis, $t_d \sim 100$ sec, yields $n/p \sim 1/7$. Thus, the n/p ratio is a sensitive indicator of the physics occurring in the early universe.

Once the bottleneck allowing neutron condensation into deuterium is broken, the remaining reactions proceed quickly

$$n + d \to {}^3\text{He} + \gamma \quad \text{and} \quad d + d \to {}^3\text{H} + p$$

$$p + d \to {}^3\text{He} + \gamma \quad \text{and} \quad d + d \to {}^3\text{He} + n$$

$$n + {}^3\text{He} \to {}^3\text{H} + p \,,$$

$$p + {}^3\text{H} \to {}^4\text{He} + \gamma \quad \text{and} \quad p + {}^3\text{H} \to {}^4\text{He} + n$$

$$n + {}^3\text{He} \to {}^4\text{He} + \gamma \quad \text{and} \quad d + {}^3\text{He} \to {}^4\text{He} + p$$

$${}^3\text{He} + {}^3\text{He} \to {}^4\text{He} + 2p \,.$$

The result is that essentially all the deuterium which has been produced ends up as ${}^4\text{He}$! Note that there will also be trace amounts of ${}^7\text{Li}$ and ${}^7\text{Be}$ produced via

$${}^4\text{He} + {}^3\text{H} \to {}^7\text{Li} + \gamma \quad \text{and} \quad {}^4\text{He} + {}^3\text{He} \to {}^7\text{Be} + \gamma \,.$$

One might also expect to be able to produce elements with $A = 5$ via $p + {}^4\text{He}$ and $A = 8$ with ${}^4\text{He} + {}^4\text{He}$, but no such species are stable. (There is also a significant Coulomb barrier here.) Thus, we have in this very simple picture explained the basic features of light element abundances in the Universe. This is very powerful evidence indeed for the essential correctness of the big bang scenario.

20.2 Nuclear Reaction Rates

Before proceeding to a study of stellar evolution, it will be useful to present a brief section on nuclear reaction rates in order that we can be a bit quantitative in our analysis, and that is our goal in this section [Hax14]. Consider then the generic reaction

$$a(p_1) + b(p_2) \longrightarrow c(p_1') + d(p_2').$$

The event rate per unit time per unit volume is given by

$$r_{ab} = \rho_a \rho_b |v_a - v_b| \sigma_{ab}(|v_a - v_b|) \frac{1}{1 + \delta_{ab}} \,, \tag{20.23}$$

where $\rho_a(\rho_b)$ is the number density of particles of type $a(b)$ respectively while $|v_a - v_b|$ is the relative velocity and can be expressed relativistically as[2]

$$|v_a - v_b| = \frac{1}{E_a E_b} \sqrt{(p_a \cdot p_b)^2 - m_a^2 m_b^2} \,. \tag{20.25}$$

The factor $1 + \delta_{ab}$ takes into account the feature that if the particles a, b are identical, there exists an overall factor of $1/2!$ asssociated with the statistics.

---

[2] This can be seen in the nonrelativistic limit via

$$\frac{1}{E_a E_b} \sqrt{(E_a E_b - p_a \cdot p_b)^2 - m_a^2 m_b^2} = \sqrt{(1 - v_a \cdot v_b)^2 - (1 - v_a^2)(1 - v_b^2)}$$

$$\simeq \sqrt{v_a^2 + v_b^2 - 2 v_a \cdot v_b} = |v_a - v_b| \,. \tag{20.24}$$

In the stellar medium we can assume that the reactions take place in thermal equilibrium at a temperature T. When fermions are involved, the momentum distribution of particles is given by

$$n_F(p) = \frac{g_F}{\exp\left[(E - E_F)/k_B T\right] + 1} , \tag{20.26}$$

where $E = \sqrt{p^2 + m^2}$ is the particle energy, E_F is the Fermi energy, and $g_F = 2s_F + 1$ is the degeneracy factor associated with the spin degrees of freedom. We see then that in the limit as $T \to 0$

$$n_F = \begin{cases} 0 & E > E_F \\ g_F & E < E_F , \end{cases} \tag{20.27}$$

as expected. The total number of occupied states is then given by

$$N = V \int \frac{d\mathbf{k}}{(2\pi)^3} \frac{g_F}{\exp\left[(\sqrt{k^2 + m^2} - E_F)/k_B T\right] + 1}$$

$$= V \frac{g_F}{2\pi^2} \int_m^\infty \frac{dE\,E\sqrt{E^2 - m^2}}{\exp\left[(E - E_F)/k_B T\right] + 1} . \tag{20.28}$$

In the nonrelativistic limit

$$E\sqrt{E^2 - m^2} \approx \sqrt{2m^3(E - m)} \tag{20.29}$$

and, redefining E as the kinetic energy, we have

$$N = \frac{V g_F}{2\pi^2}(2m^3)^{\frac{1}{2}} \int_0^\infty dE \sqrt{E} \frac{1}{\exp\frac{E - E_F}{k_B T} + 1} \xrightarrow{T \to 0} \frac{V g_F}{2\pi^2}(2m^3)^{\frac{1}{2}} \frac{2}{3} E_F^{\frac{3}{2}} , \tag{20.30}$$

which defines the Fermi energy in terms of the number density

$$E_F = \frac{1}{m} \left(\frac{3N}{V}\right)^{\frac{2}{3}} \left(\frac{\pi}{2g^2}\right)^{\frac{1}{3}} . \tag{20.31}$$

In general, however, the equation relating E_F and N must be solved numerically. The shape of the distribution function near the Fermi surface becomes smoothed at finite temperature compared to its rectangular $T = 0$ shape.

A typical astrophysical application involves low density, so that $E_F \to 0$. Then $(E - E_F)/k_B T >> 1$ and the Fermi–Dirac distribution goes over to its classical limit: the Maxwell–Boltzmann distribution

$$\frac{1}{\exp\left[(E - E_F)/k_B T\right] + 1} \longrightarrow \exp\left(-\frac{E}{k_B T}\right) . \tag{20.32}$$

Since in the nonrelativistic limit we have $E = \frac{1}{2}mv^2$ we can write

$$P(v) = N \left(\frac{m}{2\pi k_B T}\right)^{\frac{3}{2}} \exp\left(\frac{mv^2}{2k_B T}\right) v , \tag{20.33}$$

where N is the number density of particles.

For application to the solar interior we need to evaluate the reactions of nonrelativistic charged nuclei in a plasma, wherein said nuclei are distributed according to Eq. (20.33). The reaction rate formula Eq. (20.23) then generalizes to

$$r = \frac{N_a N_b}{1 + \delta_{ab}} v\sigma_{ab}(v) \rightarrow \frac{N_a N_b}{1 + \delta_{ab}} < v\sigma_{ab}(v) > , \tag{20.34}$$

where $<>$ designates a thermal average, i.e.,

$$< v\sigma_{ab}(v) > = \int d\mathbf{v}_a d\mathbf{v}_b \left(\frac{m_a}{2\pi k_B T}\right)^{\frac{3}{2}} \left(\frac{m_b}{2\pi k_B T}\right)^{\frac{3}{2}}$$

$$\exp\left[-\frac{(m_a v_a^2 + m_b v_b^2)}{2k_B T}\right] |\mathbf{v}_a - \mathbf{v}_b| \sigma_{ab}(|\mathbf{v}_a - \mathbf{v}_b|) . \tag{20.35}$$

In terms of center-of-mass coordinates

$$\mathbf{V} = \frac{m_a \mathbf{v}_a + m_b \mathbf{v}_b}{m_a + m_b}, \quad \mathbf{v} = \mathbf{v}_a - \mathbf{v}_b \tag{20.36}$$

we have

$$\mathbf{v}_a = \mathbf{V} + \frac{m_b \mathbf{v}}{m_a + m_b}, \quad \mathbf{v}_b = \mathbf{V} - \frac{m_a \mathbf{v}}{m_a + m_b} . \tag{20.37}$$

Then the exponential in Eq. (20.35) can be written as

$$\exp\left[\frac{-(m_a v_a^2 + m_b v_b^2)}{2k_B T}\right] = \exp\left[-((m_a + m_b)V^2 + \mu_r v^2)/2k_B T\right] , \tag{20.38}$$

where $\mu_r = m_a m_b/(m_a + m_b)$ is the reduced mass. Since

$$\int d\mathbf{v}_a d\mathbf{v}_b = \int d\mathbf{V} d\mathbf{v} , \tag{20.39}$$

Eq. (20.35) then becomes

$$< v\sigma_{ab}(v) > = \left(\frac{m_a}{2\pi k_B T}\right)^{\frac{3}{2}} \left(\frac{m_b}{2\pi k_B T}\right)^{\frac{3}{2}} \int d\mathbf{V} e^{-(m_a+m_b)V^2/2kT}$$

$$\int d\mathbf{v} v\sigma_{ab}(v) e^{-\mu_r v^2/2k_B T}$$

$$= \left(\frac{\mu_r}{2\pi k_B T}\right)^{\frac{3}{2}} \int d\mathbf{v} v\sigma_{ab}(v) e^{-\mu_r v^2/2k_B T} \tag{20.40}$$

and the reaction rate assumes the simple form

$$r = \frac{\rho_a \rho_b}{1 + \delta_{ab}} \left(\frac{\mu_r}{2\pi k_B T}\right)^{\frac{3}{2}} \int d\mathbf{v} v\sigma_{ab}(v) e^{-\mu_r v^2/2k_B T}$$

$$= \frac{4\pi \rho_a \rho_b}{1 + \delta_{ab}} \left(\frac{\mu_r}{2\pi k_B T}\right)^{\frac{3}{2}} \int dv\, v^3 \sigma_{ab}(v) e^{-\mu_r v^2/2k_B T} . \tag{20.41}$$

In terms of the center-of-mass energy $E = \mu_r v^2/2$ this becomes

$$r = \frac{\rho_a \rho_b}{1 + \delta_{ab}} \sqrt{\frac{8}{\pi \mu_r}} \left(\frac{1}{k_B T}\right)^{\frac{3}{2}} \int_0^\infty E dE \sigma_{ab}(E) e^{-E/k_B T} . \tag{20.42}$$

Nonresonant Reactions

With this background we can now attempt to calculate the reaction rate for processes taking place within the solar plasma. The primary problem here is that these are reactions between charged systems and therefore there exists a strong Coulomb repulsion which must be taken into account. We consider the generic reaction $a + b \rightarrow c \rightarrow a' + b'$ in terms of the scattering of particles a and b to form a compound nucleus c which subsequently decays via the channel $a' + b'$. For example, we could be considering the reaction

$$^3\text{He} + {}^4\text{He} \rightarrow {}^7\text{Be}^* \rightarrow {}^7\text{Be} + \gamma.$$

If one envisions the left-hand side of this process as the inverse of the alpha decay of the nucleus $^7\text{Be}^*$ then we know from simple quantum mechanics that there are two very different energy regimes. Considering $^7\text{Be}^*$ in terms of the interaction of ^3He and ^4He, there will exist certain energies at which quasi-bound states exist. Such states can be analyzed in terms of complex energy eigenstates of the system. They are nearly bound in the sense that they are quite long-lived. If the incoming energy corresponds to one of these quasi-bound state values, this is resonant scattering and will be analyzed shortly. For the present time we want to imagine that the incident energy is not near one of these resonant values, which is much the more common case.

We envision the generic reaction

$$A + B \rightarrow C \rightarrow D + \gamma$$

occurring as a combination of the process $C \rightarrow D + \gamma$ and the time-reversed decay $C \rightarrow A + B$. From the familiar alpha decay example in quantum mechanics, we can write the decay rate in the form

$$\Gamma_C = \text{Flux through a sphere of radius } r \rightarrow \infty$$
$$= v \int r^2 d\Omega |\psi(\mathbf{r})|^2 . \tag{20.43}$$

If we take the state C to have orbital angular momentum ℓ then we can write Eq. (20.43) in the form

$$\Gamma_C = \lim_{r \rightarrow \infty} v \int r^2 d\Omega \left|\frac{u_\ell(r)}{r}\right|^2 |Y_\ell^m(\Omega)|^2 = v|u_\ell(\infty)|^2 . \tag{20.44}$$

Often this result is written

$$\Gamma_C = v P_\ell |u_\ell(R)|^2 , \tag{20.45}$$

where R is the nuclear radius and

$$P_\ell = \frac{|u_\ell(\infty)|^2}{|u_\ell(R)|^2} \tag{20.46}$$

is called the penetration factor. Note that $u_\ell(R)$ is a strong-interaction quantity, which depends upon the dynamics in the interior of the nucleus. The suppression due to the Coulomb barrier arises from the region $R < r < \infty$ and is included in the penetration factor. Evaluation of P_ℓ involves solution of the partial-wave equation

$$\left(-\frac{1}{2\mu_r}\frac{d^2}{dr^2} + \frac{\ell(\ell+1)}{2\mu_r r^2} + \frac{Z_A Z_B \alpha}{r} - E \right) u_\ell(r) = 0 \,, \tag{20.47}$$

where r is the relative coordinate between A and B. Defining

$$p = \sqrt{2\mu_r E}, \quad \text{and} \quad \eta = \frac{Z_A Z_B \alpha}{v} = \alpha Z_A Z_B \sqrt{\frac{\mu_r}{2E}} \,, \tag{20.48}$$

the solution corresponding to outgoing waves involves the combination

$$G_\ell(pr) + i F_\ell(pr) \xrightarrow{r\to\infty} \exp i \left(pr - \ell\frac{\pi}{2} - \eta \log 2p + \sigma_\ell \right) \tag{20.49}$$

of the usual Coulomb solutions [Abr68]. Thus, the penetration factor is

$$P_\ell = \frac{|u_\ell(\infty)|^2}{|u_\ell(R)|^2} = \frac{1}{|F_\ell(pR)|^2 + |G_\ell(pR)|^2} \tag{20.50}$$

and can be found from tables of Coulomb functions. However, this form does not allow us to gain much analytic understanding of such processes. Thus, instead we shall use a simple analytic form which follows from application of the WKB approximation to such systems [Mer98]

$$P_\ell^{\text{WKB}}(E) \sim \sqrt{\frac{E_C}{E}} \exp\left[-\frac{2\pi Z_A Z_B}{v} \right], \tag{20.51}$$

where $E_C = Z_A Z_B \alpha / R$ is the Coulomb potential at the nuclear radius R. The term here is simply the usual nonrelativistic expression for the Coulomb wave function at the origin that was seen in our discussion of allowed beta decay in Chapter 17.

As a model for the (α, γ) reaction we take a complex pole picture

$$\text{Amp} \sim \frac{ge}{(E - E_0) - i\frac{\Gamma}{2}} \tag{20.52}$$

that would arise in a simple quantum mechanical treatment. The resulting cross section is then

$$\begin{aligned}
\sigma &= \frac{M_\alpha}{k_\alpha} \int \frac{ds_\gamma}{(2\pi)^3 2s_\gamma} 2\pi \delta(E_\alpha + M_1 - s_0 - M_2) \frac{g^2 e^2}{(E-E_0)^2 + \frac{1}{4}\Gamma^2} \\
&= \frac{M_\alpha}{k_\alpha} \frac{4\pi^2 s_\gamma}{(2\pi)^3} \frac{g^2 e^2}{(E-E_0)^2 + \frac{1}{4}\Gamma^2}.
\end{aligned} \tag{20.53}$$

Since for alpha decay

$$\Gamma_\alpha = g^2 \int \frac{dk_\alpha}{(2\pi)^3} 2\pi \delta(\Delta - E_\alpha) = g^2 k_\alpha M_\alpha / \pi \,, \tag{20.54}$$

while for γ-decay

$$\Gamma_\gamma = e^2 \int \frac{ds_\gamma}{s_\gamma} 2\pi\,\delta(\Delta - E_\gamma) = \frac{e^2 s_\gamma}{2\pi} \, , \tag{20.55}$$

the cross section assumes the form

$$\sigma = \frac{1}{k_\alpha^2} \frac{\pi\,\Gamma_\alpha \Gamma_\gamma}{(E_0 - E)^2 + \frac{1}{4}\Gamma^2} \, . \tag{20.56}$$

Of course, if there are multiple resonances which contribute, the total width is the sum of partial widths: $\Gamma = \Gamma_1 + \Gamma_2 + \cdots$. However, for a nonresonant reaction such as we are considering, $E - E_0$ is much larger than the width and so the width can be dropped. In this case, and especially if there are a number of contributing resonances, the denominator is rather smooth. Then, since

$$\Gamma_\alpha = v P_\ell |u_\ell(R)|^2 \propto \sqrt{E} \times \frac{1}{\sqrt{E}} \exp\left[-2\pi Z_A Z_B \alpha / v\right] \, , \tag{20.57}$$

we find that

$$\sigma \propto \frac{1}{E} \exp\left[-2\pi Z_A Z_B \alpha / v\right] \, . \tag{20.58}$$

This result motivates the definition of the astrophysical S-factor

$$\sigma \equiv \frac{1}{E} \exp\left[-2\pi Z_A Z_B \alpha / v\right] S(E) \tag{20.59}$$

in terms of which cross sections are usually quoted. The point here is that because of the large penetration factor suppression, the cross section itself is not a good measure of the dynamics of a given reaction. By factoring out the leading piece of the penetration factor, we have a better measure of the dynamical component of the cross section.

In fact another purpose is served here. Since energies at the stellar core are much lower than those involved in typical nuclear experiments, then, provided that $S(E)$ is a relatively slowly varying function, Eq. (20.59) can be used to extrapolate experimental laboratory cross sections down to the energies of interest for stellar evolution.

We could continue at this point by performing the requisite numerical integrations. However, it is useful to gain insight by makng a few approximations. We are after the velocity-averaged cross section $< v\sigma(v) >$ which can be written in terms of the S-factor as

$$< v\sigma(v) > = \sqrt{\frac{8}{\pi\mu_r}} \left(\frac{1}{k_B T}\right)^{\frac{3}{2}} \int_0^\infty dE\, E\, e^{-\frac{E}{k_B T}} \frac{1}{E} e^{-\frac{2\pi Z_A Z_B \alpha}{v}} S(E) \, . \tag{20.60}$$

Defining

$$b = 2\pi Z_A Z_B \sqrt{\frac{\mu_r}{2}} \, , \tag{20.61}$$

Equation (20.60) becomes

$$< v\sigma(v) > = \sqrt{\frac{8}{\pi\mu_r}} \left(\frac{1}{k_B T}\right)^{\frac{3}{2}} \int_0^\infty dE\, S(E)\, e^{-\left(\frac{E}{k_B T} + \frac{b}{\sqrt{E}}\right)} \, . \tag{20.62}$$

Obviously the velocity averaged rate involves a competition between the effects of the Coulomb barrier, which suppresses the low-energy events, and the Gamow factor, which suppresses those at high energy. Since $S(E)$ is assumed to be slowly varying, we can use the steepest descent approximation in order to evaluate the integral. In this procedure, if we have an integral of the form

$$I = \int_0^\infty dx f(x) e^{-g(x)} , \qquad (20.63)$$

then if x_0 is the point defined by $g'(x_0) = 0$ and $f(x)$ is assumed to be slowly varying, we can approximate

$$I \simeq f(x_0) e^{-g(x_0)} \int_{-\infty}^\infty dx\, e^{-\frac{1}{2} g''(x_0)(x-x_0)^2} = \sqrt{\frac{2\pi}{g''x_0)}} f(x_0) e^{-g(x_0)} . \qquad (20.64)$$

Applied to our case we require

$$\frac{d}{dE}\left(\frac{E}{k_B T} + \frac{b}{\sqrt{E}} \right) = 0, \quad \text{i.e.,} \quad E_0 = \left(\frac{b k_B T}{2} \right)^{\frac{2}{3}} , \qquad (20.65)$$

and since

$$\frac{d^2}{dE^2}\left(\frac{E}{k_B T} + \frac{b}{\sqrt{E}} \right)\Big|_{E=E_0} = \frac{3}{2 E_0 k_B T} , \qquad (20.66)$$

we identify

$$< v\sigma(v) > = \sqrt{\frac{32 E_0}{3\mu_r}} \frac{1}{k_B T} S(E_0) e^{-3E_0/k_B T} \qquad (20.67)$$

$$= \frac{16}{\sqrt{3\mu_r}} \frac{1}{2\pi \alpha Z_A Z_B} S(E_0) e^{-\frac{E_0}{k_B T}} \left(\frac{E_0}{k_B T} \right)^2 . \qquad (20.68)$$

Defining $\kappa \equiv \mu_r/M_N$ and putting in numbers we find

$$r_{AB} = \frac{N_A N_B}{1 + \delta_{AB}} (6.48 \times 10^{-18} \text{ cm}^3 \text{ sec}^{-1}) \frac{1}{\kappa Z_A Z_B} \frac{S(E_0)}{\text{keV} \cdot \text{b}} \left(\frac{E_0}{k_B T} \right)^2 . \qquad (20.69)$$

In order to get a feel for a typical case, let's return to the reaction

$$^3He + {}^4He \to {}^7Be + \gamma$$

and we find, using $k_B T = 1.5 \times 10^7 \text{ K} = 1.3 \text{ keV}$ as the temperature at the solar core,

$$\frac{E_0}{k_B T} = \left(\frac{\pi \alpha Z_H Z_{He}}{\sqrt{2}} \right)^{\frac{2}{3}} \left(\frac{\mu_r}{k_B T} \right)^{\frac{1}{3}} \simeq 16 . \qquad (20.70)$$

Thus most reactions take place far out on the tail of the Maxwell–Boltzmann distribution, wherein the effects of the Coulomb barrier are much smaller.

Resonant Reactions

If the relevant energy E_0 sits near a resonance, the analysis is different, of course. In this case, assuming that the resonance is narrow compared to the typical spread of energies of the colliding nuclei, we find

$$
\begin{aligned}
< v\sigma(v) > &= \sqrt{\frac{8}{\pi\mu_r}}\left(\frac{1}{k_BT}\right)^{\frac{3}{2}}\int_0^\infty dE E e^{-\frac{E}{k_BT}}\frac{\pi}{2\mu_r E}\frac{\Gamma_\alpha\Gamma_\gamma}{(E-E_0)^2+\frac{1}{2}\Gamma^2}\\
&\approx \sqrt{\frac{8}{\pi\mu_r}}\left(\frac{1}{k_BT}\right)^{\frac{3}{2}}\Gamma_\alpha\Gamma_\gamma e^{-\frac{E}{k_BT}}\int_{-\infty}^\infty\frac{dE}{(E-E_0)^2+\frac{1}{4}\Gamma^2}\\
&= \left(\frac{\Gamma_\alpha\Gamma_\gamma}{\Gamma}\right)^{\frac{3}{2}}e^{-\frac{E_0}{k_BT}}.
\end{aligned}
\tag{20.71}
$$

Now that we have the tools, we can proceed to our goal of understanding stellar evolution.

20.3   Stellar Evolution

Having understood the origin of the primary (H,^{4}He) component of the Universe as well as the light trace elements ($d,{}^3$He,^{3}H), we now move on to consider the much larger concentrations of heavier elements such as ^{12}C,^{16}O,^{20}Ne, etc., whose origin is associated with stellar evolution. We first consider hydrogen-burning stars, such as our Sun. Almost 80% of stars are believed to be hydrogen-burning and are said to be "main sequence," in that they lie along a common track on a Hertzsprung–Russell (HR) diagram. The Hertzsprung–Russell diagram, shown in Fig. 20.1, is a two-dimensional plot, with luminosity on the vertical axis and surface temperature (as gauged from the color of the star) on the horizontal. Typical luminosities range from $10^4 - 10^6$ times that of the Sun while surface temperatures vary from 2000 to 50000 K. Our Sun, with a surface temperature of about 6500 K, is then relatively cool on this scale.

That one might expect a relationship between temperature and luminosity can be understood by positing that a star radiates as a black body. In this case, the Stefan–Boltzmann law yields a total luminosity

$$
L = 4\pi R^2\sigma T^4
\tag{20.72}
$$

in terms of the stellar radius R. Scaling to solar values, we find

$$
\frac{L}{L_0} = \left(\frac{R}{R_0}\right)^2\left(\frac{T}{T_0}\right)^4.
\tag{20.73}
$$

Then to the extent that stars truly do radiate as black bodies, one expects a relation to exist between the luminosity and temperature in terms of the radius.

The evolution of such stars is governed by the so-called standard solar model (SSM), wherein the star is assumed to be in hydrostatic equilibrium, meaning the

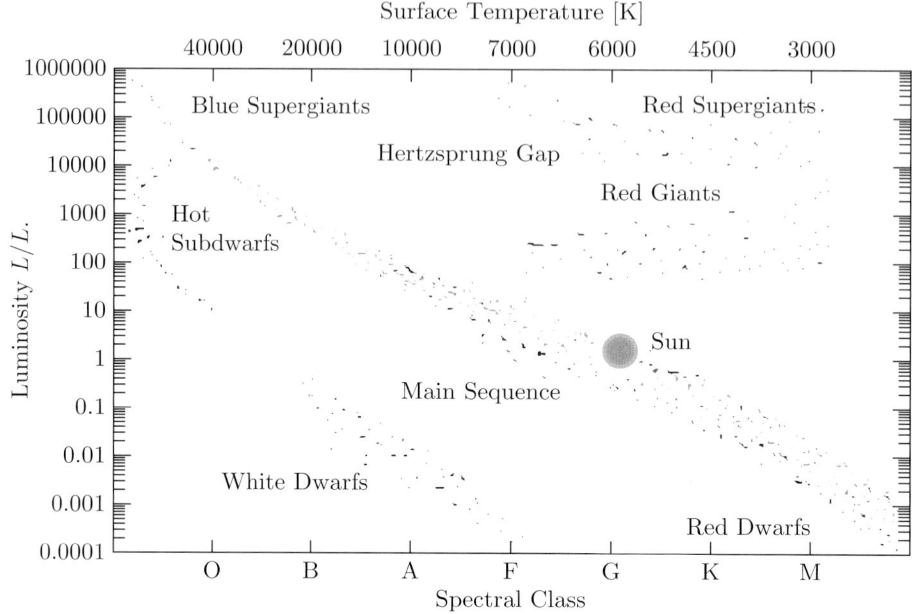

Fig. 20.1 The Hertzsprung–Russell diagram showing the relationship between the stars' absolute magnitude or luminosities versus their effective temperatures or spectral classification. Each star's brightness is plotted against its temperature (color). The location of the Sun is indicated.

gravitational attraction is balanced by the outward pressure due the combustion of the nuclear fuel. Thus

$$\frac{dP}{dr} = -\frac{GM(r)\rho(r)}{r^2} \, , \tag{20.74}$$

where G is the gravitational constant, $P(r)$ and $\rho(r)$ are the pressure and density at radial distance r from the star's center and

$$M(r) = \int_0^r dr' \rho(r') 4\pi r'^2 \tag{20.75}$$

is the mass contained within this radius. Also, if $L(r)$ is the luminosity and $\epsilon(r)$ is the energy production per unit mass, then energy balance requires

$$\frac{dL(r)}{dr} = 4\pi r^2 \rho(r)\epsilon(r) \, . \tag{20.76}$$

The flow of energy to the surface is controlled by the energy-transport equations

$$\frac{dT}{dt} = -\frac{3}{4\sigma}\frac{\bar{\kappa}}{T^3}\frac{L(r)}{4\pi r^2} \quad \text{radiative}$$

$$\frac{dT}{dt} = \left(1 - \frac{1}{\gamma}\right)\frac{T}{P}\frac{dP}{dt} \quad \text{convective} \, , \tag{20.77}$$

where T is the temperature, σ is the Stefan–Boltzmann constant, and $\gamma = C_P/C_V$ is the adiabatic index. The symbol $\bar{\kappa}$ stands for the opacity, which is calculated by huge computer codes that have been developed at the weapons laboratories for other reasons.

Finally, we require the equation of state

$$P = P(\rho,\, T,\, f_1,\, f_2,\, f_3)\,, \tag{20.78}$$

where f_1, f_2, f_3 are the fractional abundances of hydrogen, helium, and heavier elements. (Of course, the energy production ϵ and opacity $\bar{\kappa}$ are also dependent on these quantities.) Using the boundary conditions

$$r = 0, \quad M(r) = 0, \quad L(r) = 0$$
$$r = R_0, \quad M(r) = M_0 = 1.99 \times 10^{30} \text{ kg}$$
$$L(r) = L_0 = 3.90 \times 10^{33} \text{ erg sec}\,, \tag{20.79}$$

these equations can be numerically integrated in order to yield the present rate of nuclear reactions within the Sun. Actually the helium abundance f_2 is not well known, since no helium lines are in the visible spectrum. So generally one uses f_2 as a variable, with the value being fixed by demanding that the present luminosity L_0 be obtained at the age of the solar system $t \simeq 4.7 \times 10^9$ yr.

In this way we believe that we understand the basic process by which our Sun and other stars generate their prodigious amounts of energy. However, in order to be more quantitative, we must understand the basic reactions by which stellar fusion occurs.

The pp Chain

Since the strong $p + p$, $p + {}^4\text{He}$, ${}^4\text{He} + {}^4\text{He}$ reactions do not produce energy nor produce bound states, it is left to the weak interaction process

$$p + p \rightarrow d + e^+ + \nu_e$$

to start the entire chain going. In this reaction, once the Coulomb barrier is breached, the 0.931 MeV energy difference between the 1.294 MeV $n - p$ mass difference and the 2.225 MeV binding energy of deuterium is provided to the leptons. Since the deuteron has $S = 1$, while the low energy pp state must have $S = 0$ because of Fermi statistics, we see that the weak process must involve a Gamow–Teller reaction, so that the cross section can be written as

$$\sigma = \frac{G_F^2 |V_{ud}|^2 m_e^5}{2\pi^3 v_{rel}} f(E) \frac{1}{4} \sum_{spin} | < d|A^-|pp > |^2\,, \tag{20.80}$$

where

$$f(E) = \frac{1}{m_e^5} \int_{m_e}^{E_0} dE_e dE_\nu p_e E_e E_\nu^2 \delta(E + \Delta m - E_e - E - \nu) = 0.145 \tag{20.81}$$

is the weak phase-space factor defined in our discussion of nuclear beta decay. The challenge here is the calculation of the weak axial-vector matrix element connecting the pp and d states. Since the energy is so low, this is an ideal opportunity for the comparison of the use of effective field theory with traditional methods. We begin with the latter. In a lowest order (contact interaction) approach, the deuteron wavefunction is given by a simple exponential

$$\psi_1^M(\mathbf{r}) = \sqrt{\frac{\gamma}{2\pi}}\frac{1}{r}e^{-\gamma r}\eta_1^M\zeta_0^0 \equiv \sqrt{\frac{1}{4\pi}}\frac{1}{r}u(r)\eta_1^M\zeta_0^0 \, , \tag{20.82}$$

where $\gamma^2/m_N = B_d = 2.225\,\mathrm{MeV}$ is the deuteron binding energy (i.e., $\gamma = 45\,\mathrm{MeV}$) and η, ζ represent normalized spin, isospin states constructed from n, p spinors. In the case of the initial state, we write the even parity component as

$$\psi_{\mathbf{k}}^{(+)}(\mathbf{r}) = 4\pi\sqrt{2}\sum_{\ell=0,2,4..}\sum_{M_\ell} i^\ell Y_\ell^{M_\ell *}(\hat{\mathbf{k}})Y_\ell^{M_\ell}(\hat{r})e^{i\delta_\ell(k)}\frac{1}{kr}\chi_\ell(k,r)\eta_0^0\zeta_1^1 \, . \tag{20.83}$$

Here the radial wave function has the asymptotic normalization

$$\chi_\ell(k,r) \xrightarrow{r\to\infty} \cos\delta_\ell(k)F_\ell(kr) + \sin\delta_\ell(k)G_\ell(kr) \, , \tag{20.84}$$

where F_ℓ, G_ℓ are the usual Coulomb functions [Abr68] and $\delta_\ell(k)$ is the scattering phase shift. In impulse approximation we have

$$A^- = -g_A\sum_i \sigma_i\tau_i^- \, , \tag{20.85}$$

and, using the spherical vector notation

$$A_{\pm 1} = \mp\sqrt{\frac{1}{2}}(A_x \pm iA_y), \quad A_0 = A_z \, , \tag{20.86}$$

we find that

$$< d_1^M|A_n^-|pp> = g_A\delta_{n,M}\frac{\sqrt{16\pi}}{k}e^{i\delta_0(k)}\int_0^\infty dru(r)\chi_0(k,r) \, . \tag{20.87}$$

As a measure of this quantity and noting that

$$\int d\mathbf{r}u(r) = \int d\mathbf{r}\sqrt{2\gamma}e^{-\gamma r} = \sqrt{\frac{32\pi}{\gamma^3}}, \tag{20.88}$$

it is conventional to define [Sal52]

$$< d_1^M|A_n^-|pp> \equiv \delta_{n,M}g_A\sqrt{\frac{32\pi}{\gamma^3}}C_\eta\Lambda(E) \tag{20.89}$$

in terms of which

$$\begin{aligned}
S(E) &= 3\frac{E}{v_{rel}}\frac{G_F^2|V_{ud}|^2}{(2\pi)^3}m_e^5 f(E)g_A^2 C_\eta^2\frac{32\pi}{\gamma^3}\Lambda^2(E)\exp\frac{2\pi\alpha}{v_{rel}}\\
&= G_F^2|V_{ud}|^2\frac{12\alpha}{\pi}\frac{m_r m_e^5}{\gamma^3}g_A^2 f(E)\Lambda^2(E) \, .
\end{aligned} \tag{20.90}$$

Putting in numbers and noting that $1\,\mathrm{keV\,b} = 2.5\times 10^{-6}\,\mathrm{MeV^{-1}}$, we find

$$S(E) = \Lambda^2(E)\cdot 5.5\times 10^{-23}\,\mathrm{keV\,b} \, . \tag{20.91}$$

In order to proceed, we must evaluate the radial integral. Of course, this could be accomplished numerically. However, at least in the limit as $k \to 0$ one can do the job analytically [Rav01]. Specifically, since

$$
F_0(\rho) \xrightarrow{\rho \to 0} \frac{C_\eta}{\sqrt{2\eta(k)}} \sqrt{\rho} I_1\left(2\sqrt{2\eta(k)\rho}\right)
$$

$$
G_0(\rho) \xrightarrow{\rho \to 0} \frac{2}{C_\eta} \sqrt{2\eta(k)\rho} K_1\left(2\sqrt{2\eta(k)\rho}\right)
$$

$$
\sin\delta_0 \xrightarrow{k \to 0} ka_0 C_\eta^2 , \tag{20.92}
$$

we find

$$
\lim_{E \to 0} \Lambda(E) = \frac{1}{C_\eta}\sqrt{\frac{\gamma^3}{32\pi}} \sqrt{16\pi} \frac{1}{k} e^{i\delta_0} \int_0^\infty dr \sqrt{2\gamma} e^{-\gamma r} \left(\cos\delta_0 F_0(kr) + \sin\delta_0 G_0(kr)\right)
$$

$$
\approx \frac{\gamma}{k} \int_0^\infty d\rho \sqrt{\rho} e^{-\frac{\gamma}{k}\rho} \left(\frac{1}{\sqrt{2\eta(k)}} I_1(2\sqrt{2\eta(k)\rho}) - 2ka_0\sqrt{2\eta(k)} K_1(2\sqrt{2\eta(k)\rho}) \right) . \tag{20.93}
$$

The integrals can both be performed analytically, yielding

$$
\Lambda(0) = \frac{1}{\chi} e^{\frac{\chi}{2}} \left(M_{-1,\frac{1}{2}}(\chi) - 2m_r\alpha a_0 W_{-1,\frac{1}{2}}(\chi) \right), \tag{20.94}
$$

where $\chi = 2k\eta(k)/\gamma = 2m_r\alpha/\gamma = 0.15$. Using the experimental pp scattering length $a_0 = 7.85$ fm and

$$
M_{-1,\frac{1}{2}}(\chi) = -\chi e^{\frac{\chi}{2}}, \quad W_{-1,\frac{1}{2}} = \chi e^{-\frac{\chi}{2}} \left(\frac{1}{\chi} - e^\chi E_1(\chi) \right), \tag{20.95}
$$

where

$$
E_1 = \int_\chi^\infty \frac{ds}{s} e^{-s} \tag{20.96}
$$

is the exponential integral function, we determine

$$
\Lambda(0) = e^\chi - 2\alpha m_r a_0 \left(\frac{1}{\chi} - e^\chi E_1(\chi) \right) = 2.51 . \tag{20.97}
$$

Including effective range corrections, $\Lambda(0)$ becomes 2.66. Inclusion of meson-exchange corrections and the best modern wavefunctions yields the value 2.65. The corresponding S-factor is then

$$
S(0) = 3.94 \times 10^{-22} \text{ keV b} . \tag{20.98}
$$

Now let us consider the same calculation in an effective field theory approach [Rav01, Par01]. In this case we begin by evaluating the wavefunction renormalization constant of the deuteron, which is obtained, in lowest order, from the self-energy diagram, cf. Fig. 20.2, via [Kap99]

$$
Z_0 = \sqrt{\frac{1}{|d\Sigma_0/dE|_{E=-B}}} . \tag{20.99}
$$

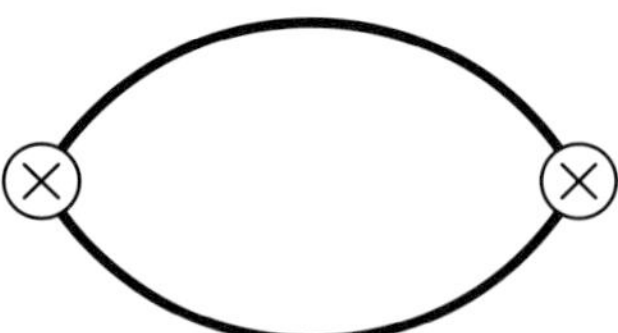

Diagram giving the leading contribution to the deuteron wavefunction renormalization. Here the solid lines designate nucleons and the cross-circle represents the deuteron wavefunction.

Here

$$\Sigma_0(E) = \int \frac{d\mathbf{k}}{(2\pi)^3} \frac{1}{E - \frac{k^2}{2m_r} + i\epsilon} \tag{20.100}$$

is the lowest-order contribution to the self energy. Using the PDS-regularized result (see Chapter 11)

$$\Sigma_0(E) = -\frac{m_r}{2\pi}\left(-\mu - \sqrt{-2m_r E}\right) , \tag{20.101}$$

we find the result

$$Z_0 = \sqrt{2\pi\gamma}/m_r . \tag{20.102}$$

We also need to review the connection between the four-nucleon coupling constant C_0^p in EFT and the scattering length a_0 in the case of pp scattering [Rav99]. In the absence of strong interactions, the wavefunction for the incoming pp state with center-of-mass momentum $\boldsymbol{p}$ is given by

$$\psi_{\boldsymbol{p}}(\boldsymbol{r}) = \frac{1}{pr}\sum_{\ell=0}^{\infty}(2\ell + 1)i^\ell e^{i\sigma_\ell} F_\ell(pr)P_\ell(\cos\theta) , \tag{20.103}$$

where $\sigma_\ell = \arg\Gamma(1 + \ell + i\eta)$ is the Coulomb phase shift and

$$F_0(pr) = C_\eta pr e^{-ipr} M(1 - i\eta, 2; 2ipr) . \tag{20.104}$$

The strong interactions can now be included via the diagrams shown in Fig. 20.3. In order to perform this sum we need to evaluate the Coulomb propagator bubble

$$\begin{aligned}
J_0(p) &= \int \frac{d\boldsymbol{q}}{(2\pi)^3} \frac{\psi_{\boldsymbol{q}}(\boldsymbol{r}' = 0)\psi_{\boldsymbol{q}}(\boldsymbol{r} = 0)}{\frac{p^2}{2m_r} - \frac{q^2}{2m_r}} \\
&= 2m_r \int \frac{d\boldsymbol{q}}{(2\pi)^3} \frac{2\pi\eta(q)}{e^{2\pi\eta(q)} - 1} \frac{1}{p^2 - q^2 + i\epsilon} ,
\end{aligned} \tag{20.105}$$

where we have defined $E = p^2/2m_r$. This integral can be evaluated exactly in PDS in $d = 3 - \epsilon$ dimensions, yielding the result [Gra65]

$$J_0(p) = \frac{\alpha m_r^2}{\pi}\left[\frac{1}{\epsilon} + \log\frac{\mu\sqrt{\pi}}{2\alpha m_r} + 1 - \frac{3}{2}\gamma_E - H(\eta)\right] - \frac{\mu m_r}{2\pi} , \tag{20.106}$$

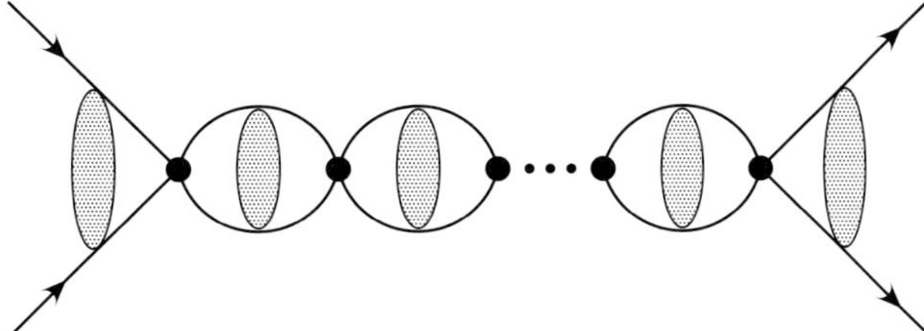

Fig. 20.3 Diagram giving the strong interaction corrections to pp scattering. Here the solid lines represent protons, the solid circle designates a strong interaction vertex, and the gray blob represents the static Coulomb interaction.

where $\gamma_E = 0.5772\ldots$ is Euler's constant and

$$H(x) = \psi(ix) + \frac{1}{2ix} - \log(ix) . \tag{20.107}$$

The scattering series can be summed exactly via

$$T(p) = |\psi_{\boldsymbol{p}}(\boldsymbol{r}) = 0)|^2 C_0^0(\mu) \sum_{n=0}^{\infty} (C_0^0(\mu) J_0(p))^n = \frac{C_\eta^2 e^{2i\sigma_0} C_0^0(\mu)}{1 - C_0^0(\mu) J_0(p)} . \tag{20.108}$$

In the limit as the momentum vanishes, we can relate $T(p)$ to the experimental scattering length $a_0 = -7.82$ fm via

$$T(p) \xrightarrow{p \to 0} \frac{C_\eta^2 e^{2i\sigma_0}}{\frac{1}{C_0^0(\mu)} - J_0(0)} \equiv C_\eta^2 e^{2i\sigma_0} \frac{2\pi a_0}{m_r} , \tag{20.109}$$

which provides the relation

$$\frac{1}{C_0^0(\mu)} = \frac{m_r}{2\pi a_0} + J_0(0) , \tag{20.110}$$

whereby the the divergence $1/\epsilon$ is absorbed into the renormalized coupling $C_0^p(\mu)$. Since $H(\eta)$ vanishes as $p \to 0$, it is also useful to define the scale-dependent scattering length

$$\frac{1}{a(\mu)} = \frac{1}{a_0} - 2\alpha m_r \left(\log \frac{\mu \sqrt{\pi}}{2\alpha m_r} + 1 - \frac{3}{2}\gamma_E \right) , \tag{20.111}$$

in terms of which

$$C_0^0(\mu) = \frac{2\pi}{m_r} \frac{1}{\frac{1}{a(\mu)} - \mu} . \tag{20.112}$$

With this as background, we can now evaluate the desired $pp \to de^+ \nu_e$ amplitude in terms of the lowest-order effective axial-vector couplings, which are shown in Fig. 20.4. The simple one-body term, Fig. 20.4(a), is the usual impulse approximation operator and leads to the diagrams shown in Fig. 11.7. Summing the series, we find then from these terms

$$Amp_{fi}(p) = Z_0 \left(A_0(p) + B_0(p) \frac{C_0^0(\mu)\psi_{\boldsymbol{p}}(\boldsymbol{r} = 0)}{1 - C_0^0(\mu) J_0(p)} \right) , \tag{20.113}$$

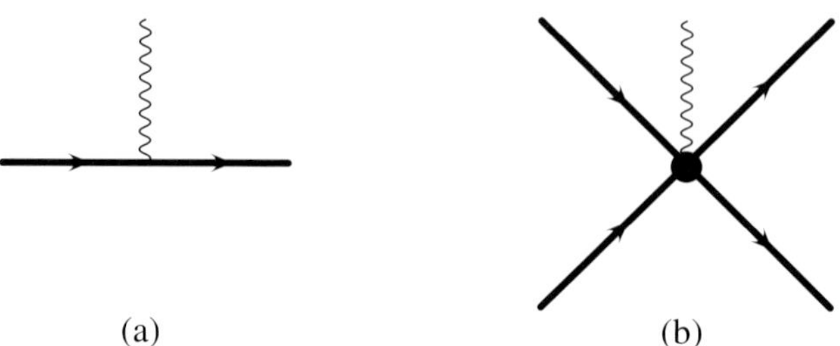

(a)   (b)

Fig. 20.4 Diagrams representing effective axial-current coupling to nucleons. Here the vertex (b) designates the isoscalar four-nucleon counterterm L_{1A}.

with

$$A_0(p) = 2m_r \int \frac{dk}{(2\pi)^3} \frac{\psi_p(k)}{k^2 + \gamma^2}, \quad B_0(p) = 2m_r \int \frac{dk}{(2\pi)^3} \int \frac{dk'}{(2\pi)^3} \frac{G_C(E; k, k')}{k^2 + \gamma^2}.$$
(20.114)

The integrations can be performed directly. In the case of $A_0(p)$ we have

$$A_0(p) = 2m_r C_\eta e^{i\sigma_0} \int_0^\infty dr r e^{-(\gamma + ip)r} M(1 - i\eta, 2; 2ipr)$$

$$= \frac{2m_r C_\eta e^{i\sigma_0}}{(\gamma + ip)^2} {}_2F_1\left(1 - i\eta, 2; 2; \frac{2ip}{\gamma - ip}\right) = \frac{2m_r C_\eta e^{i\sigma_0}}{(\gamma + ip)^2} \left(\frac{\gamma + ip}{\gamma - ip}\right)^{1-i\eta}$$

$$= C_\eta e^{i\sigma_0} \frac{2m_r}{p^2 + \gamma^2} e^{2\eta \arctan(p/\gamma)}.$$
(20.115)

In the case of the integral $B_0(p)$ we find

$$B_0(p) = 4m_r^2 \int \frac{dk}{(2\pi)^3} \int \frac{dq}{(2\pi)^3} \frac{\psi_q(k) \psi_q^*(r = 0)}{(k^2 + \gamma^2)(p^2 - q^2 + i\epsilon)}$$

$$= 2m_r \int \frac{dq}{(2\pi)^3} C_\eta(q) e^{-i\sigma_0(q)} \frac{A_0(q)}{p^2 - q^2 + i\epsilon}$$

$$= 4m_r^2 \int \frac{dq}{(2\pi)^3} \frac{e^{2\eta(q)\arctan(q/\gamma)}}{(q^2 + \gamma^2)(p^2 - q^2 + i\epsilon)} \frac{2\pi \eta(q)}{e^{2\pi \eta(q)} - 1}.$$
(20.116)

In the limit as $p \to 0$ we see that

$$A_0(p) \xrightarrow{p \to 0} C_\eta \frac{2m_r}{\gamma^2} e^{\chi + i\sigma_0},$$
(20.117)

where we have defined $\chi = 2\alpha m_r/\gamma$ as before, in which we find for case $B_0(p)$

$$B_0(p) \xrightarrow{p \to 0} -\frac{m_r^2}{\gamma^2 \pi^2} I(\chi),$$
(20.118)

where

$$I(\chi) = \int_0^\infty \frac{2x dx}{e^x - 1} \frac{\exp(\frac{x}{\pi} \arctan \frac{\pi \chi}{x})}{x^2 + \pi^2 \chi^2} = \frac{1}{\chi} - E_1(\chi).$$
(20.119)

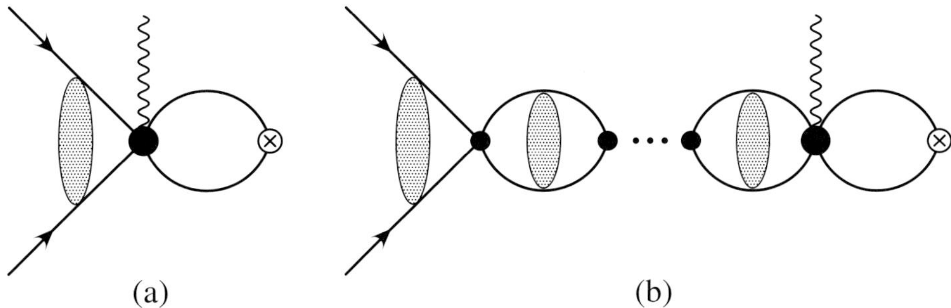

Fig. 20.5 Two-body operator diagrams contributing to $pp \rightarrow de^+ \nu_e$. Here the solid line represents a proton, the wiggly line designates emission the the lepton pair $e^+ \nu_e$, the cross-circle is the deuteron wavefunction, and the gray blobs represent the static Coulomb interaction. The solid circle indicates the four-nucleon axial-vector counterterm L_{1A}.

Combining, we have then

$$Amp_{fi} = \sqrt{\frac{8\pi}{\gamma^3}} C_\eta e^{i\sigma_0} \left(e^\chi - 2\alpha m_r I(\chi) \right) , \tag{20.120}$$

in complete agreement with the conventional wavefunction result.

However, what is different about the EFT calculation is that we have a systematic way to deal with higher-order contributions. Writing the higher-order contact interaction as [But01]

$$A_-^{(2)} = L_{1A} \left(N^T \frac{\sigma_2 \boldsymbol{\sigma} \, \tau_2}{\sqrt{8}} N \right)^\dagger \left(N^T \frac{\sigma_2 \tau_2 \tau_-}{\sqrt{8}} N \right) , \tag{20.121}$$

we can evaluate the diagrams shown in Fig. 20.5 to yield

$$Amp_{fi}^{ct} = Z_0 L_{1A}(\mu) C_\eta I_0(\gamma) \left(1 + \frac{C_0^p J_0(p)}{1 - C_0^p J_0(p)} \right) , \tag{20.122}$$

where

$$C_\eta I_0(\gamma) = \int \frac{d\boldsymbol{k}}{(2\pi)^3} \int \frac{d\boldsymbol{q}}{(2\pi)^3} \frac{2m_r \psi_p(\boldsymbol{q})}{\boldsymbol{k}^2 + \gamma^2} . \tag{20.123}$$

We have then

$$Amp_{fi}^{ct} = -Z_0 L_{1A}(\mu) C_\eta \frac{m_r}{2\pi} a_0(\mu - \gamma) \left(\mu - \frac{1}{a(\mu)} \right) . \tag{20.124}$$

Including this new contribution, we find then for the reduced matrix element

$$\Lambda_2(0) = \Lambda_0(0) - a_0(\mu - \gamma) \left(\mu - \frac{1}{a(\mu)} \right) \frac{\gamma^2}{4\pi} L_{1A}(\mu) . \tag{20.125}$$

Of course, the constant L_{1A} must have a dependence on the scale μ such that the total result is scale-independent. Thus, we define

$$L_{1A}(\mu)_{\mu >> \gamma} = \frac{2\pi \ell_{1A}}{m_r \mu^2} , \tag{20.126}$$

where ℓ_{1A} is a dimensionless constant which should be of order unity. In fact by comparing with the various calculations using meson exchange and state-of-the-art wavefunctions, we find the values

$$\ell_{1A} = \begin{cases} 5.6 \pm 2 & \text{Nakamura } et\ al. \text{ [Nak01]} \\ 6.94 \pm 2 & \text{Ying, Haxton, and Henley [Yin92]} \\ 6.5 \pm 2.4 & \text{Schiavilla } et\ al. \text{ [Sch98]} , \end{cases} \tag{20.127}$$

whose contribution can be characterized in term of

$$\Lambda(0) = 2.58 + 0.011\ell_{1A} + \cdots . \tag{20.128}$$

Thus, the predicted S-factor is known to $\sim 5\%$. However, it is possible to do even better by measuring the phenomenological constant ℓ_{1A} experimentally. This can be done in different ways. One is to confront the known $ft_{1/2}$ value for tritium decay with an effective field theory calculation. A second is to use muon capture on deuterium, $\mu^- + d \to \nu_\mu + n + n$, which is being performed at PSI (the MUSUN experiment). In either case, one should be able to pin down the S-factor to a percent or so.

Having the S-factor for the $pp \to de^+\nu_e$ reaction which initiates solar burning, we can determine the rates for the remaining solar reactions. Since (defining T_7 as the temperature in units of 10^7 K)

$$\frac{E_0}{k_B T} \sim \left(\frac{\pi\alpha Z_A Z_B}{\sqrt{2}} \right)^{\frac{2}{3}} \left(\frac{\mu_r}{k_B T} \right)^{\frac{1}{3}} \sim \frac{5.2}{T_7^{\frac{1}{3}}} , \tag{20.129}$$

and $T_7 \simeq 1.5$ in the solar core, we find $E_0 \sim 4.57\, k_B T = 6\,\text{keV}$ as the steepest descent energy for the Sun. The corresponding reaction rate is found to be

$$r_{pp} = 27.7 \times 10^{-38} \text{ cm}^3/\text{sec}\, \frac{\rho_p^2}{T_7^{\frac{2}{3}}} e^{-15.7/T_7^{\frac{1}{3}}} \sim 6.5 \times 10^{-44} \text{ cm}^3 \text{ sec}^{-1}. \tag{20.130}$$

Since in the center of the Sun $N_p \sim 6 \times 10^{25}$ cm^{-3}, we find

$$r_{pp} \sim 6 \times 10^7 \, /(\text{cm}^3 \text{ sec}) . \tag{20.131}$$

Since two protons are consumed for each such reaction, we have a simple estimate of the solar lifetime, $\tau \sim N_p/2r_{pp} \sim 5 \times 10^{17}$ sec $= 8$ Gyr. From this estimate then, our Sun, with its current age of 4.6 Gyr, is in its middle age.

With the initiating reaction now calibrated, we can work out the remaining components of the solar reaction cycle. At the very simplest level we have

$$\begin{aligned} p + p &\to d + e^+ + \nu_e & r_{pp} &\sim\, <v\sigma(v)>_{pp} \tfrac{1}{2}N_p^2 \\ d + p &\to {}^3\text{He} + \gamma & r_{pd} &\sim\, <v\sigma(v)>_{pd} N_p N_d \\ {}^3\text{He} + {}^3\text{He} &\to {}^4\text{He} + p + p & r_{\text{HeHe}} &\sim\, <v\sigma(v)>_{\text{HeHe}} \tfrac{1}{2}N_{He}^2. \end{aligned} \tag{20.132}$$

It is interesting here that both d and ^{3}He act as catalysts in that they are produced and then are totally consumed in the burning process. That means that once the Sun has reached equilibrium, the rates of production and of destruction of both elements must be equal. In the case of deuterium this means that

$$\frac{d\rho_d}{dt} = <v\sigma(v)>_{pp} \frac{1}{2}\rho_p^2 - <v\sigma(v)>_{pd} \rho_p\rho_d = 0 . \tag{20.133}$$

From the measured S-factors

$$S_{pp}(0) = 4.07 \times 10^{-22} \,\text{keV b}, \quad S_{pd}(0) = 2.5 \times 10^{-4} \,\text{keV b}, \tag{20.134}$$

and the rate formula

$$< v\sigma(v) >_{ab} = (2.9 \times 10^{-16} \,\text{cm}^3 \,\text{sec}^{-1})$$

$$\times \frac{1}{\kappa Z_a Z_b} \left[\frac{S(E_0)}{\text{keV b}} \frac{Z_a^2 Z_b^2 \kappa}{T_7}^{\frac{2}{3}} \exp\left(-19.7(Z_a^2 Z_b^2 \kappa/T_7)^{\frac{1}{3}}\right) \right], \tag{20.135}$$

we can put in numbers with $Z_a = Z_b = 1$, $\kappa_{pp} = 1/2$, $\kappa_{pd} = 2/3$, to find

$$\left(\frac{\rho_d}{\rho_p}\right) = 1.1 \times 10^{-18} \exp\left[-1.574/T_7^{\frac{1}{3}}\right]. \tag{20.136}$$

Thus, the higher the temperature, the less deuterium. The lowest concentration is then at the solar center, where $T_7 \simeq 1.5$ and $\rho_d/\rho_p \sim 4 \times 10^{-18}$. Since $\rho_p \sim 3 \times 10^{25} \,\text{cm}^3 \,\text{sec}^{-1}$, it follows that $\rho_d \sim 10^8 \,\text{cm}^3 \,\text{sec}^{-1}$. Since $r_{pp} \sim 0.6 \times 10^8 \,\text{cm}^3 \,\text{sec}^{-1}$, we see that the typical lifetime of a deuterium nucleus is only about a second or so, i.e., equilibrium is reached almost instantaneously.

Because of this feature, we can write the corresponding equation for ^{3}He as

$$\frac{d\rho_{\text{He}}}{dt} = < v\sigma(v) >_{pp} \frac{1}{2}\rho_p^2 - 2 < v\sigma(v) >_{\text{HeHe}} \frac{1}{2}\rho_{He}^2, \tag{20.137}$$

where the factor of two comes from the feature that two ^{3}He nuclei are destroyed in each reaction. At equilibrium then we have

$$\left(\frac{\rho_{\text{He}}}{\rho_p}\right) = \sqrt{\frac{< v\sigma(v) >_{pp}}{2 < v\sigma(v) >_{\text{HeHe}}}}. \tag{20.138}$$

In this case the experimental S-factor is found to be

$$S_{\text{HeHe}}(0) = 5.2 \times 10^3 \,\text{keV b}, \tag{20.139}$$

and

$$\left(\frac{\rho_{\text{He}}}{\rho_p}\right)_{\text{equil}} = 1.3 \times 10^{-13} \exp(20.65/T_7^{\frac{1}{3}}). \tag{20.140}$$

In this case, the concentration decreases the nearer one is to the core and, putting in numbers, one finds that

$$\rho_{\text{He}}/\rho_p \sim 9.1 \times 10^{-6} \quad \text{for} \quad T_7 = 1.5$$

$$\rho_{\text{He}}/\rho_p \sim 1.2 \times 10^{-4} \quad \text{for} \quad T_7 = 1.0,$$

so that a rather steep gradient of the ^{3}He concentration is established. The time to reach equilibrium is also a sharp function of the radius. This can be seen from the fact that, since the depletion rate is quadratic in the abundance, production will dominate until times when the abundance is very near its equilibrium value. Thus a reasonable estimate of the equilibration time is the time to produce $\sim 75\%$ of this number of ^{3}He nuclei. At the center of the Sun, the necessary abundance is $4.1 \times 10^{20} \,\text{cm}^{-3}$ and the corresponding estimated equilibration time is 2×10^5 years. On the other hand, the same calculation performed for a temperature $T_7 = 1.0$ finds a time which is a factor of 100 larger, while that for $T_7 \sim 0.65$ yields a time about the age of the Sun. This corresponds to a radius about 30% of that of the Sun and is right at the edge of the energy-producing core. Beyond this distance, equilibrium has still not been reached!

Of course, the above discussion is only the simplest one and, apropos of Chapter 18 on neutrinos, determines only the rate of the dominant low-energy neutrinos which arise in the original $p + p \rightarrow d + e^+ + \nu_e$ reaction. In order to understand how the higher-energy neutrinos which are detected by Kamiokande and/or SNO are produced, it is necessary to realize that there exist two alternative chains for ^{4}He production. The previous chain is called the ppI cycle. However, it is also possible to have the links

$$
\begin{aligned}
p + p &\rightarrow d + e^+ + \nu_e \\
d + p &\rightarrow {}^3\text{He} + \gamma \\
{}^3\text{He} + {}^4\text{He} &\rightarrow {}^7\text{Be} + \gamma \\
{}^7\text{Be} + e^- &\rightarrow {}^7\text{Li} + \nu_e \\
{}^7\text{Li} + p &\rightarrow {}^4\text{He} + {}^4\text{He} ,
\end{aligned}
\tag{20.141}
$$

which is called the ppII chain or the process

$$
\begin{aligned}
p + p &\rightarrow d + e^+ + \nu_e \\
d + p &\rightarrow {}^3\text{He} + \gamma \\
{}^3\text{He} + {}^4\text{He} &\rightarrow {}^7\text{Be} + \gamma \\
{}^7\text{Be} + p &\rightarrow {}^8\text{B} + \gamma \\
{}^8\text{B} &\rightarrow {}^8\text{Be} + e^+ + \nu_e \\
{}^8\text{Be} &\rightarrow {}^4\text{He} + {}^4\text{He} ,
\end{aligned}
\tag{20.142}
$$

which is called the ppIII chain. Obviously the ppII chain leads to the intermediate energy neutrinos arising from the ^{7}Li electron capture, while it is the ppIII chain which is responsible for the relatively high-energy neutrinos resulting from the decay of ^{8}B and to which both SNO and Kamiokande are primarily sensitive. While we now have the calculational tools to estimate the relative rates for such neutrino production, a detailed estimate is beyond the scope of our discussion. Figure 20.6 summarizes the three pp chains discussed here.

Heavy Element Production

Above we have outlined the basic processes by which the universe has evolved, from nucleosynthesis to stellar evolution. In both processes we have the transition from simple hydrogen to helium, with a few other elements in between. On the other hand, in the world that we live in, there also exist lots of heavier elements. How did they come about? The answer is an interesting one and has to do with what occurs when a star has used up most of its hydrogenic fuel. As the hydrogen concentration that powers the nuclear reactions in the star begins to wane, the corresponding reaction rates begin to fall and the star contracts under gravity, increasing the temperature and therefore the reaction rate at which the remaining hydrogen is burned. This increased rate of energy production leads to an expansion of the outer layers of the star, resulting in a decreased surface temperature.

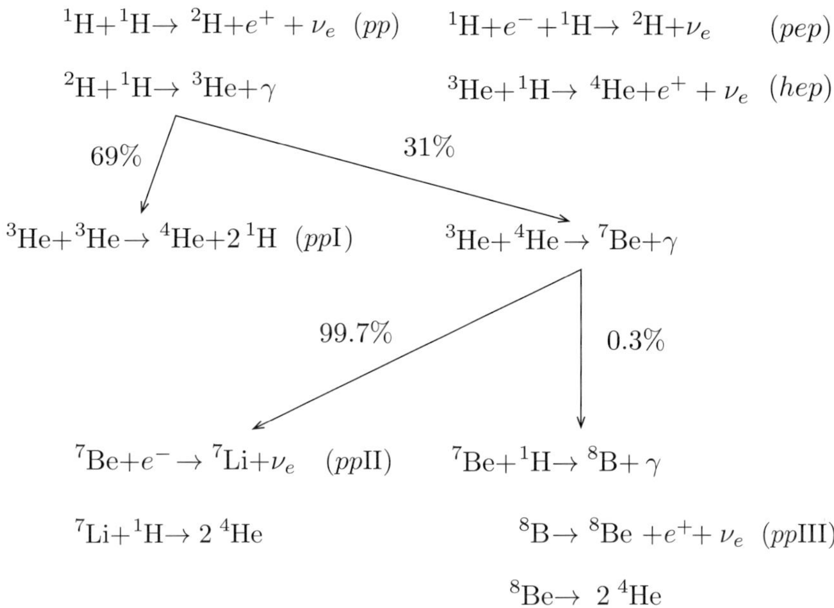

$$^1\mathrm{H}+^1\mathrm{H}\rightarrow\ ^2\mathrm{H}+e^+ + \nu_e \ \ (pp) \qquad\qquad ^1\mathrm{H}+e^-+^1\mathrm{H}\rightarrow\ ^2\mathrm{H}+\nu_e \qquad (pep)$$

$$^2\mathrm{H}+^1\mathrm{H}\rightarrow\ ^3\mathrm{He}+\gamma \qquad\qquad\qquad ^3\mathrm{He}+^1\mathrm{H}\rightarrow\ ^4\mathrm{He}+e^+ + \nu_e \ \ (hep)$$

$$69\% \qquad\qquad\qquad 31\%$$

$$^3\mathrm{He}+^3\mathrm{He}\rightarrow\ ^4\mathrm{He}+2\,^1\mathrm{H}\ \ (pp\mathrm{I}) \qquad ^3\mathrm{He}+^4\mathrm{He}\rightarrow\ ^7\mathrm{Be}+\gamma$$

$$99.7\% \qquad\qquad\qquad 0.3\%$$

$$^7\mathrm{Be}+e^-\rightarrow\ ^7\mathrm{Li}+\nu_e \ \ (pp\mathrm{II}) \qquad ^7\mathrm{Be}+^1\mathrm{H}\rightarrow\ ^8\mathrm{B}+\gamma$$

$$^7\mathrm{Li}+^1\mathrm{H}\rightarrow 2\,^4\mathrm{He} \qquad\qquad\qquad ^8\mathrm{B}\rightarrow\ ^8\mathrm{Be}+e^+ + \nu_e \ \ (pp\mathrm{III})$$

$$^8\mathrm{Be}\rightarrow\ 2\,^4\mathrm{He}$$

Fig. 20.6 Summary diagram of the pp chains (see also Fig. 18.3).

The star then becomes a so-called "red giant," having a much larger radius than before, but a cooler surface temperature (see Fig. 20.1). On the other hand, the core continues to heat up as the proton supply runs low and the contraction continues. When the temperature reaches about 10^8 K, the densities are high enough to allow for helium burning to commence. This does not occur directly ($^4\mathrm{He}+^4\mathrm{He}\rightarrow\ ^8\mathrm{Be}+\gamma$ cannot occur since $^8\mathrm{Be}$ is unstable and after about 10^{-16} sec. decays back to a pair of $^4\mathrm{He}$ nuclei) but rather via a two step process whereby, during the short existence of the $^8\mathrm{Be}$ nucleus, it collides with third $^4\mathrm{He}$ nucleus, producing $^{12}\mathrm{C}$ via

$$^8\mathrm{Be} + ^4\mathrm{He} \rightarrow\ ^{12}\mathrm{C} + \gamma$$

and the gravitational contraction is temporarily halted. The triple-alpha process proceeds via the $7.66\,\mathrm{MeV}$ resonance state of $^{12}\mathrm{C}$, the *Hoyle state,* predicted by Hoyle and subsequently discovered at Caltech in an experiment led by Whaling. The existence of the Hoyle state is essential to produce carbon and subsequently oxygen, two of the building blocks of life, in our universe. The prediction and discovery of the Hoyle state is cited as an example of the anthropic principle, a philosophical consideration that observations of the universe must be compatible with the sapient life that observes it.

The triple-alpha reaction converts $^4\mathrm{He}$ into $^{12}\mathrm{C}$ and eventually the $^{12}\mathrm{C}$ becomes so abundant that it can capture $^4\mathrm{He}$ to become $^{16}\mathrm{O}$. This happens via the $^{12}\mathrm{C}(^4\mathrm{He},\gamma)^{16}\mathrm{O}$ process, which strongly enriches $^{16}\mathrm{O}$ relative to the other oxygen isotopes. The $^{12}\mathrm{C}/^{16}\mathrm{O}$ ratio resulting from helium burning determines the nature of the subsequent carbon burning, which, in turn, determines the whole subsequent shell structure of the star. In particular, the carbon-to-oxygen ratio is a critical parameter in determining whether a supernova ends as a neutron star or a black hole. For this reason, the reaction $^{12}\mathrm{C}(^4\mathrm{He},\gamma)^{16}\mathrm{O}$, which is poorly known experimentally in the relevant stellar energy region, is the focus of intense experimental study at low energies. It is notable that the reaction $^{16}O + {}^4He \rightarrow {}^{20}Ne + \gamma$ does not occur at helium burning temperatures because of the lack of an appropriate resonance. Oxygen and carbon are the next abundant elements in the Universe after hydrogen and helium.

The higher temperature is required in order to overcome the Coulomb repulsion between two ^4He nuclei and ^8Be nuclei, which is considerably stronger than that between a pair of protons. Because the core is much hotter and the reaction rate much higher, this helium burning phase is considerably shorter than its life on the main sequence, and what happens next then depends upon the mass of the star. For a relatively light star, less than three or four solar masses, as the helium supply runs low and the core continues to collapse under the influence of gravity, the Pauli exclusion principle associated with the electron gas within the core wins out. The collapse is halted and the gravitational attraction is stabilized by the outward pressure of this electron gas. Through a mechanism still not fully understood, the envelope of the gaseous material surrounding the core is shed, producing a planetary nebula and leaving behind a white dwarf star, containing basically carbon nuclei and the electron gas. For stability the mass of this white dwarf must be less than 1.4 solar masses, the so-called Chandrasekhar limit [Cha35]. After this, the star cools by radiating energy into space and eventually becomes an invisible black dwarf.

More interesting and relevant for our discussion is what happens if the red giant star is much heavier than three or four solar masses. In this case the degeneracy pressure of electrons due to the exclusion principle is insufficient to halt further gravitational contraction, and, as the core compresses and continues to heat up, additional nuclear reactions are generated. Thus, at about 6×10^8 K, carbon itself begins to be consumed via

$$^{12}C + {}^{12}C \rightarrow \begin{cases} {}^{20}Ne + {}^4He \\ {}^{24}Mg + \gamma \\ {}^{16}O + {}^4He + {}^4He \\ \text{etc.} \end{cases}$$

The length of time spent in this phase, $\sim 10^3$ years, is even shorter than that spent in the helium burning mode and, as the carbon supply begins to dwindle, gravitational contraction heats the core even further, allowing neon burning

$$^{20}Ne + \gamma \rightarrow {}^{16}O + {}^4He$$
$$^{20}Ne + {}^4He \rightarrow {}^{24}Mg + \gamma$$
$$\text{etc.}$$

to commence at about 1×10^9 K. As the evolution continues and the core temperature reaches about 1.5×10^9 K, oxygen burning begins, producing a range of final products

$$^{16}O + {}^{16}O \rightarrow \begin{cases} {}^{32}S + \gamma \\ {}^{28}Si + {}^4He + \gamma \\ {}^{31}S + n + \gamma \\ {}^{24}Mg + {}^4He + {}^4He \\ \text{etc.} \end{cases}$$

Finally, at a temperature of about 3×10^9 K, silicon burning is started, again producing a range of heavier elements, among which the most prevalent and important is Fe. As is well known (cf. Chapter 13), nuclei in the vicinity of iron have the maximum binding energy per nucleon. Thus, as such elements accumulate in the core, no further thermonuclear reactions are possible. The configuration consists of a very hot and dense core containing primarily Fe, surrounded by a mantle consisting mainly of Si at somewhat lower temperature and pressure. Further out from the center exist additional shells with Mg, Ne, O, etc., until we reach the surface, where an envelope of hydrogen and helium still exists. Thus we have the so-called "onion skin model" of the star.

An obvious question here is that if no further thermonuclear reactions are taking place in the core what keeps the star from collapsing? The answer is that stability is provided by the electron degeneracy pressure, just as for the white dwarf. However, there is an important difference. As the Si burning continues, more and more Fe continues to pile up in the core and eventually the Chandrasekhar limit is exceeded. In this case, unable to sustain the tremendous inward crunch that is due to gravity, the core region collapses and begins rapidly to contract, resulting in a sharp rise in the temperature. Photonuclear reactions dissociate the remaining nuclei into their fundamental proton and neutron constituents, and the protons and electrons themselves combine to form even more neutrons, emitting an electron neutrino in the process: $p + e^- \rightarrow n + \nu_e$. This "neutronization" results in a core consisting primarily of neutrons, and eventually one reaches a stage wherein the central region consists of about a solar mass of neutrons contained within a radius of about 10 km. At this point the core is at solar density and the Pauli exclusion effect for these neutrons abruptly halts the collapse, as the core becomes virtually incompressible. Of course, stellar matter from outside the core region is still hurtling toward the center. When this material comes in contact with the now rigid neutron core, a rise in temperature and pressure develops, causing the inflowing material to "bounce" and begin to rush away toward the surface. In this fashion a shock wave develops, which sheds much of the outer gaseous stellar medium and leaves behind a neutron star at the center. Neutron stars are the densest and smallest stars known to exist in the Universe with a radius of only 8 to 15 km and a mass up to 2.5 times the mass of the Sun. Figure 20.7 shows a plot of neutron star mass versus radius for both hadronic and strange quark matter equations-of-state [Lat12].

What we have just described is a supernova explosion, of course, and the dominant mechanism for carrying away energy from the star is the emission of neutrinos. The first burst, as described above, is associated with the neutronization process and occurs during the very brief (fraction of a second) period of core collapse. A second flux of neutrinos is associated with the fact that the collapsing core involves very high temperatures ($T \sim 6 \times 10^{10}$ K) so that neutrinos are thermalized and produced by many means, e.g.,

   i) pair annihilation: $e^+ + e^- \rightarrow \nu_e + \bar{\nu}_e$,
   ii) photonuclear production: $\gamma + A \rightarrow A + \nu_e + \bar{\nu}_e$,
   iii) the Urca reaction: $e^- + (Z, N) \rightarrow \nu_e + (Z - 1, N + 1)$, and
   iv) etc.

The core densities are so high that the diffusion process by which neutrinos wend their way out of the core region takes several seconds and this is just the sort of time period observed

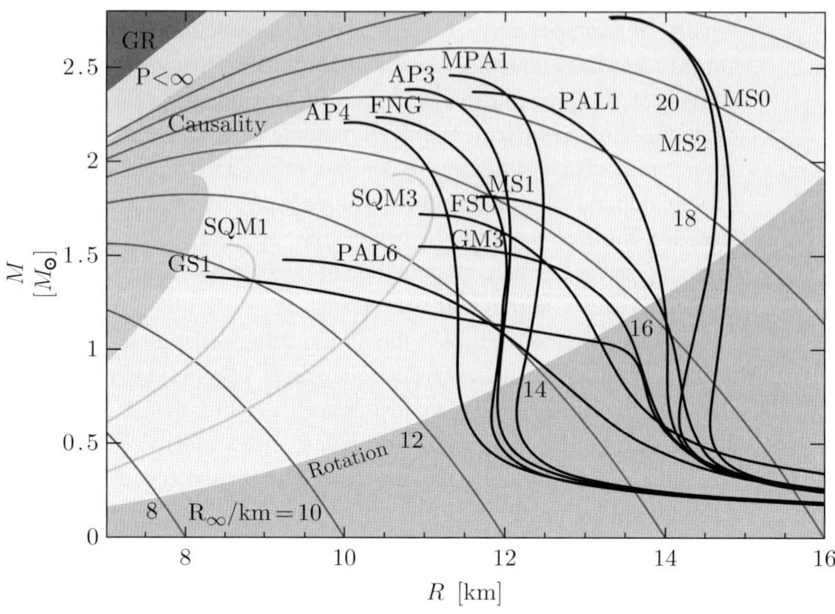

Fig. 20.7   A plot of neutron star mass versus radius curves for hadronic (heavy curves) and strange quark matter (light curves) equations-of-state from [Lat12]. Regions of the mass-radius plane excluded by general relativity, finite pressure, and causality are indicated. The region marked "rotation" is bounded by the realistic mass-shedding limit for the highest known pulsar frequency (715 Hz).

from the neutrinos generated from the explosion of SN1987A. The precise process by which such a supernova explosion occurs is a very complex one, and detailed numerical studies are required. Such work is underway at Oak Ridge National Laboratory, Caltech, and elsewhere.

In the evolution of a star from the main sequence to its ultimate fate we have thus generated a means for the production of elements in the region between ^{4}He and Fe. What about those elements which are even heavier? Such nuclei must be produced in the explosion of the supernova itself. Since the Coulomb barrier for the interaction of heavy nuclei is so large, a new mechanism is required and this is found in the flurry of (n, γ) reactions which take place involving the existing nuclei and the large number of neutrons which are present. This is called the *r-process*. The process entails a succession of rapid neutron captures by heavy seed nuclei, typically ^{56}Fe or other more neutron-rich heavy isotopes. As a result, the nuclei become neutron rich and decay quickly toward the axis of stability by β^- decay. The resultant nuclei are themselves inundated by a neutron flux and the process continues toward heavier and heavier nuclei. The other predominant mechanism for the production of heavy elements in the universe is the *s-process*, which is nucleosynthesis by means of slow capture of neutrons, primarily occurring in stars located in the Hertzsprung–Russell diagram (see Fig. 20.1) populated by evolving low- to medium-mass stars. All such heavy elements that make up our world then are the results of supernova explosions of stars that lived and died long before our own Sun began its own evolutionary cycle nearly 5 billion years ago. Indeed, as said most succinctly by Sagan, "We are made of star stuff!"

The recent observation of gravitational waves by LIGO [Abb16] resulting from inspiral and merger of two black holes opens a new window on the universe. The source of the signal is estimated to be located at a luminosity distance of about 400 Mpc. In particular, the inspiral and merger of two neutron stars is expected to occur relatively frequently and with a detectable signal. If this is realized, then the equation of state can be constrained in a completely new way. Further, comparison with theoretical expectations can shed light on whether such mergers are a significant r-process source. Thus, the details of an observed neutron star inspired gravitational-wave signal may provide insights into the nature and behavior of ultra-dense neutron matter.

We note that a major scientific motivation for studying the limits of stability via the production of rare isotopes (discussed in Chapter 13) is to understand the r-process. Highly radioactive nuclei are essential participants in this mechanism and a knowledge of their structure and properties is key to understanding the r-process. We direct the reader to the interactive chart of the nuclides at http://www.nndc.bnl.gov/chart/.

20.4 Cosmic Rays

Cosmic rays have been the subject of study ever since their discovery by Hess at the beginning of the past century, who observed the discharge of electrometers in balloon flights. The present status of measurements is shown in Fig. 20.8. These are energetic charged particles which impinge on the Earth and their energy spectrum has been studied from GeV energies up to 10^{13} GeV, which is ten million times the energy of the LHC! Over this energy range, the flux falls exponentially with energy of a slope $\sim E^{-2.5}$. However, there is interesting structure in this falloff. Specifically, at very low energy any cosmic rays are affected by the solar wind and the Earth's and Sun's magnetic fields. However, in the range from about 100 GeV to 10^6 GeV the steep exponential falloff, $\propto E^{-2.6}$, is operative and the differential flux of primary nucleons drops by *ten* orders of magnitude. At this point, called the "knee," the slope becomes a bit steeper, with an exponential falloff proportional to $\sim E^{-3.1}$ and the differential flux has dropped even further. This falloff continues to about 10^{10} GeV. At this point, called the "ankle," the slope becomes even steeper with a falloff proportional to $\sim E^{-3.7}$ and the event rate becomes even smaller and is barely measurable.

The charged particle makeup is primarily protons (about 80%), followed by alpha particles (about 16%), along with a substantial component of heavier nuclei (about 4%). All the atoms are completely ionized, meaning the acceleration mechanism completely strips the atoms. The relative abundance of these particles is very roughly comparable to that seen in the Sun (with important differences) and this feature suggests that the cosmic accelerator is associated with a highly evolved star, such as a core-collapse supernova. Unfortunately, as the charged particles traverse the $\sim 10^{-7}$ gauss galactic magnetic field on their way to their detection on the Earth, any directionality associated with such a supernova origin of cosmic rays is lost. The existence of the knee in the observed flux is attributed to some sort of cutoff of the supernova accelerator at an energy of 10^6 GeV, but the source of such a cutoff is unclear.

The general view of these particles is that they are primarily confined within the galactic disk, but can leak out slowly with a lifetime of order 2×10^7 yr. The confinement is due to

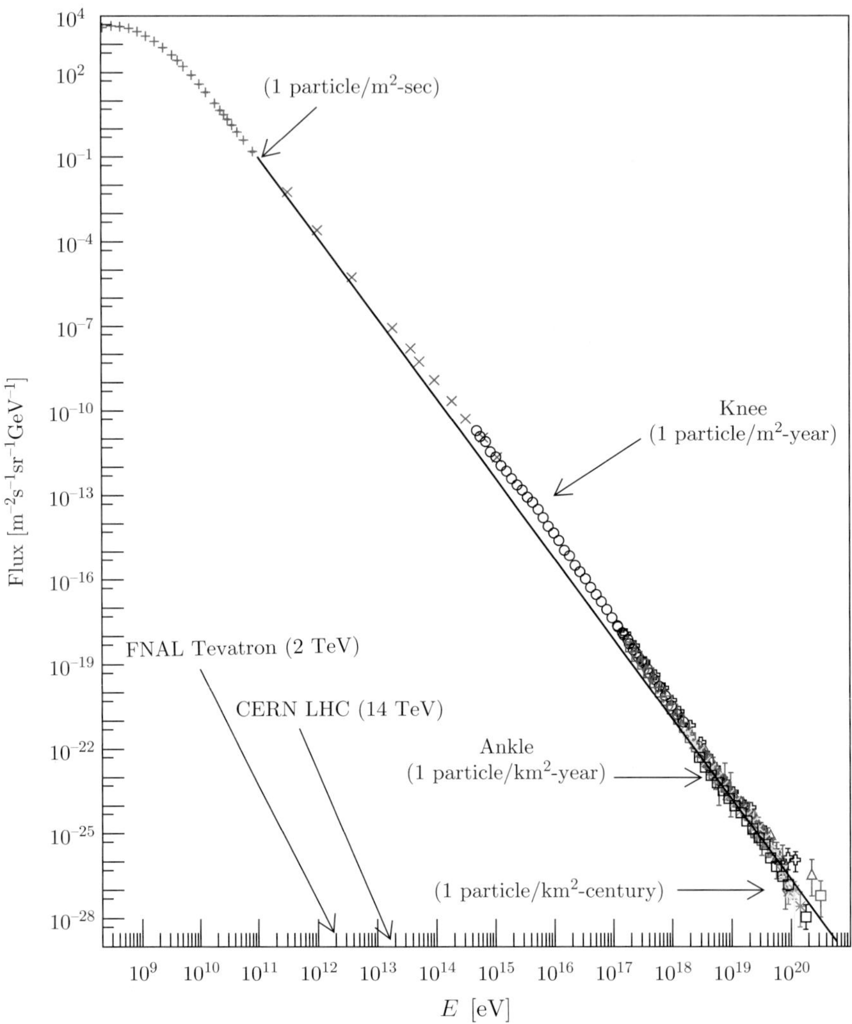

Fig. 20.8 The measured flux of cosmic rays versus energy; figure adapted from [Ama14].

the galactic magnetic field. Using $p \sim qBR$ from the Lorentz force law we find $R(\text{cm}) \sim p(\text{GeV})/3 \times 10^{-7} B(\text{gauss})$. Then for a 10^5 GeV proton we find a radius $R \sim 3 \times 10^{18}$ cm $\sim 10\,\text{ly} \sim 3$ parsec in units generally used by astronomers. Since the galactic radius is about 10^4 parsec, while the distance to a typical supernova remnant such as the Crab Nebula is about 2×10^3 parsec, we see that any directionality is indeed completely eliminated by dispersive transit through the galactic magnetic field. Of course, for cosmic rays with energy $\sim 10^9$ GeV the magnetic radius of curvature is *equal* to the galactic radius, so that any incoming particles with an energy higher than this value must be extragalactic and presumably do point back to their source.

An argument for a supernova genesis of cosmic rays can be made on energy grounds. Since the flux of primary cosmic rays with energy greater than 1 GeV is measured to be

$$I(E) \sim 2 \times 10^4 \left(\frac{E}{\text{GeV}} \right)^{-2.5} \quad \text{nucleons} / (\text{m}^2\text{sec sr GeV}) , \qquad (20.143)$$

the energy density of cosmic rays is approximately given by

$$\rho \sim \frac{4\pi m_N}{c} \int_1^\infty dx I_N(x) \sim 1 \, \text{eV cm}^{-3} \, , \qquad (20.144)$$

which can be compared to the $0.3 \, \text{eV cm}^{-3}$ energy density associated with starlight. Since the galactic volume is about 10^{12} parsec^3 or $\sim 10^{68}$ cm^3, the energy contained in galactic cosmic rays is $\sim 10^{68}$ eV. With a 20 million year cosmic ray lifetime, the associated energy production must be $10^{68} \text{eV}/2 \times 10^7 \, \text{sec} = 8 \times 10^{41} \, \text{erg sec}^{-1}$. Since a core collapse supernova releases $\sim 10^{52}$ erg, while the rate for such explosions in the galaxy is about 1/50 years, the energy which is put into cosmic rays in this scenario is about $2 \times 10^{49} \, \text{erg yr}^{-1} = 6 \times 10^{41} \, \text{erg sec}^{-1}$, which agrees with the number above coming from the cosmic ray energy density. This agreement depends, of course, on the cosmic ray lifetime given above, which comes from arguments dealing with the observed concentration of ^{10}Be in cosmic rays. The lifetime of this isotope is about 1.5 million years, so any observed ^{10}Be must come from cosmic rays. If one compares the observed ^{10}Be abundance with those of nearby stable isotopes, one finds a suppression of about 25% or so. Attributing this suppression to decay on the way to detection one identifies a lifetime of order 2×10^7 seconds, as claimed.

We have a strong argument then that, at energies below the knee, the origin of cosmic rays is from supernova explosions. Since the various charged particles are produced at relatively low energies, appropriate to nuclear reactions, we require an acceleration mechanism in order to explain how the high-energy components of the observed cosmic ray spectrum are produced. A good deal of study has been given to this question and it is now believed that this acceleration comes from diffusive shock amplification,[3] a mechanism first proposed by Fermi [Fer49, Fer54]. The basic idea is that energetic particles are scattered by magnetic irregularities. In an interaction with a magnetized cloud, the particle changes its direction of motion and gains or loses energy depending on whether the collision is head-on or tail-on. After averaging over many particle-cloud interactions, thanks to the fact that the head-on collisions turn out to be slightly more numerous, the process leads to a net energy gain ($\Delta E/E \approx 4v^2/c^2$), where v is the velocity of the cloud and c is the speed of light. Since the velocity of the cloud, or of the magnetic perturbation, is generally v/c, in this formulation the acceleration turns out to be very slow.

Things change dramatically, however, when the same process is considered in the context of a shock: the velocity of magnetic irregularities, which serve as scattering centers for particles, is negligible with respect to the fluid velocity on both sides of the shock. This implies that the scattering centers can be considered to be at rest with the fluid both upstream and downstream. In the shock system, whether a particle moves from upstream to downstream or vice-versa, it always sees the fluid on the other side of the shock as approaching. Therefore, each time an energetic particle crosses the shock, it always suffers head-on collisions with scatterers on the other side of the shock, thus gaining energy much faster: $\Delta E/E \propto v_S/c$, where v_S is now the shock velocity (which is also much higher than the typical velocity of magnetic perturbations in the galaxy). This linearity in velocity is the mechanism by which the energy is rapidly increased. The procedure also has the advantage

---

[3] The description of the acceleration mechanism here is due to Amato [Ama14].

that it produces a power-law particle spectrum

$$N(p) \propto p^{-\gamma_p} \, , \tag{20.145}$$

where

$$\gamma_p = \frac{4M_S^2}{M_S^2 - 1} \, , \tag{20.146}$$

with M_S being the shock Mach number. For a strong shock $M_S >> 1$ and $\gamma_p \approx 4$. Converting to energy we have then $N(E) \propto E^{-\gamma_p/2}$ which is also roughly observed experimentally.

For the cosmic rays between the knee and the ankle, if there is a mechanism yielding a maximum proton energy $E_{\max}$, then above this value for a system having charge Ze one would expect a maximum energy $ZE_{\max}$, and so the makeup of the cosmic rays would expected to be increasingly weighted toward heavier systems, and this phenomenon is roughly observed experimentally. However, above about 10^9 GeV, the magnetic curvature radius becomes larger than the radius of the galaxy, so that cosmic rays with these energies must be extragalactic and their origin is not really understood. There exists, however, a maximum energy of 10^{20} GeV which should be expected, according to arguments presented by Greisen, Zatsepin, and Kuzmin (GZK) [Gre66] and is explored in Exercise 20.5. The origin of the GZK limit arises from the fact that the universe is filled with relic photons, which are left over from the era of the big bang and no longer interact, since the electrons and nuclei have combined into atoms. The typical energy associated with these relic photons is of order 10^{-3} eV. A high-energy proton passing through this photon medium can photoproduce a pion via $\gamma + p \to \pi^+ + n$. The cross section for this process is known and leads to a mean free path of $\sim 3 \times 10^7$ parsecs for a 10^{20} GeV proton passing through this medium, which is a distance far shorter than the horizon. Thus, there should be a rather sharp cutoff for the proton flux at this energy. Data from AGASA do not show any evidence for such a cutoff. However, recent data from the Auger array do *not* reveal events above the GZK limit, as would be expected from these arguments.

Note that the cosmic ray spectrum described above is *not* what is directly observed by a ground-based detector, since when a high-energy hadron enters the atmosphere, strong interaction processes lead to scattering and the production of a shower of additional hadrons. The atmosphere acts then as a calorimeter, and the energy of the initiating hadron is inferred from the observed shower spectrum. Of course, when the shower begins at the top of the atmosphere it includes not only nucleons and nuclei, but also numerous charged pions. The pions themselves scatter and also decay, primarily to muons and neutrinos due to the weak interaction. The muons themselves subsequently decay to electrons and neutrinos and these pion and muon decays are the source of the atmospheric neutrinos discussed in Chapter 18. However, since the lifetime of a muon is about $2\,\mu$s, while the time for a relativistic pion to travel the $30\,$km from the upper atmosphere to the Earth's surface is about $100\,\mu$s, naively one might expect the muons to have decayed to electrons. However, due to time dilation, multi-GeV muons have ground frame lifetimes which are many times the restframe lifetime, so that muons are the primary component of cosmic ray backgrounds which affect terrestrial experiments, and the need to suppress such backgrounds in precision experiments is the reason for the existence of the many underground physics laboratories, such as SNO, Gran Sasso, Homestake/SURF, Soudan, WIPP, etc., around the world.

Exercises

20.1 Solar Fusion Reaction Rate

Show that the reaction rate near the most effective temperature $T = E_{max}/k_B$, given by the value $E_{max} = (\frac{1}{2}bk_BT)^{\frac{2}{3}}$ for solar nuclear fusion reactions, may be approximated by a normal distribution centered around E_{max} with a width $\sigma = \sqrt{2E_{max}k_BT/3}$. The parameter b is defined by the relation

$$e^{-2\pi Z_1 Z_2 \alpha c/v} = e^{-b/\sqrt{E}} ,$$

where Z_1 and Z_2 are the charge numbers of the projectile and target nucleus, α is the fine structure constant, and v is the relative velocity.

20.2 Building a Star: the Lane–Emden Equation

As discussed in the text, stellar dynamics is based on a few simple principles. When given an equation-of-state, which relates the pressure to the density, we can construct a model star, and that is the goal of this problem.

We assume a spherically symmetric star, which is governed by the equations

$$\frac{dp(r)}{dr} = -\frac{Gm(r)}{r^2}\rho(r) ,$$

which is the equation of hydrostatic equilibrium that relates the pressure gradient $dp(r)/dr$ to the density $\rho(r)$, and

$$\frac{dm(r)}{dr} = 4\pi r^2 \rho(r) ,$$

which relates the density to the mass gradient $dm(r)/dr$. Now assume a simple (polytropic) equation-of-state

$$p = \rho^\gamma ,$$

where $\gamma = c_P/c_V$ is the ratio of the specific heat at constant pressure to the specific heat at constant volume. For the Sun, a reasonable value is $\gamma = 4/3$. Now define the function

$$\Phi(r) = \left(\frac{\rho(r)}{\rho_0}\right)^{\gamma-1} = \left(\frac{p(r)}{p_0}\right)^{\frac{\gamma-1}{\gamma}} ,$$

where ρ_0 (p_0) is the density (pressure) at the star's center.

a) Demonstrate that

$$\frac{d\Phi(r)}{dr} = -\frac{\gamma-1}{\gamma}\frac{\rho_0}{p_0}\frac{Gm(r)}{r^2}$$

and

$$\frac{d^2\Phi(r)}{dr^2} + \frac{2}{r}\frac{d\Phi(r)}{dr} = -4\pi G\frac{\gamma-1}{\gamma}\frac{\rho_0^2}{p_0}\Phi(r)^{\frac{1}{\gamma-1}} .$$

b) Defining the dimensionless quantity $x = r/r_0$ where

$$r_0 = \left[\frac{p_0 \gamma}{4\pi G \rho_0^2 (\gamma - 1)} \right]^{\frac{1}{2}} ,$$

show that we obtain the Lane–Emden equation

$$\frac{d^2 \Phi(x)}{dx^2} + \frac{2}{x} \frac{d\Phi(x)}{dx} + \Phi(x)^{\frac{1}{\gamma-1}} = 0 ,$$

which must be solved subject to the boundary conditions $\rho(0) = \rho_0$ and $\rho'(0) = 0$, i.e., $\Phi(0) = 1$ and $\Phi'(0) = 0$.

c) Solve this equation numerically in the case that $\gamma = 4/3$ and show that the first zero is at $x_0 = 6.897$.

d) Perform the integral

$$J = \int_0^{x_0} dx x^2 \Phi(x)^{\frac{1}{\gamma-1}}$$

numerically and show that $J = 2.018$.

e) Using the central density $\rho_0 = 148 \, \text{g cm}^{-3}$, use your solution to determine the solar radius and compare to the known value of $R_{\text{sun}} = 6.96 \times 10^5$ km.

f) Compare the calculated value of p_0 with the known solar value of $2.29 \times 10^{17} \, \text{erg cm}^{-3}$.

g) Plot the predicted solar density and pressure profiles.

20.3 White Dwarfs

In the text we described the evolution of a star from its birth to death as a supernova. However, what happens to a star that runs out of fuel, but is not massive enough to explode as a supernova? Collapse still occurs, but ends when the electron degeneracy pressure halts the implosion. That is, one has a core with a mass of the order of the solar mass which supports itself from the degeneracy pressure of the electron gas, leading to densities which are of the order of 10^4 times the solar density. The purpose of this problem is to estimate the degeneracy pressure.

a) At low temperature the electrons will occupy the low-lying levels with two electrons in each. Show that this leads to a Fermi momentum k_F given by

$$n = 2 \int_0^{k_F} \frac{dk}{(2\pi)^3} = \frac{k_F^3}{3\pi^2} ,$$

where n is the electron number per unit volume.

b) Since nearly all the mass of the star is due to nucleons, calculate the Fermi energy in terms of the star density as

$$k_F = \left(\frac{3\pi^2 \rho}{m_N r} \right)^{\frac{1}{3}} ,$$

where r is the ratio of nucleons to electrons.

c) For electrons having momentum k show that the pressure on the walls is given by

$$p = \frac{nk^2}{3E} \, ,$$

where $E = \sqrt{k^2 + m_e^2}$ is the electron energy.

d) Integrate over the spectrum and show that

$$p = \frac{2}{3} \int_0^{k_F} \frac{dk}{(2\pi)^3} \frac{k^2}{\sqrt{k^2 + m_e^2}}$$

$$= \frac{m_e^4}{24\pi^2} \left[\left(2 \left(\frac{k_F^3}{m_e^3} \right) - 3 \frac{k_F}{m_e} \right) \left(1 + \frac{k_F^2}{m_e^2} \right)^{\frac{1}{2}} + 3\sinh^{-1} \frac{k_F}{m_e} \right] ,$$

which represents the equation-of-state, i.e., the pressure in terms of the density.

e) If $k_F << m_e$ show that the equation-of-state becomes

$$p = \frac{1}{15\pi^2 m_e} \left(\frac{3\pi^2}{rm_N} \rho \right)^{\frac{5}{3}} .$$

Given the equation-of-state, the Lane–Emden equation, derived in Exercise 20.2 above, allows the construction of the star and yields

$$M = \frac{2.72}{r^2} \left(\frac{\rho(0)}{\rho_c} \right)^{\frac{1}{2}} M_{\text{Sun}}$$

$$R = \frac{2.0 \times 10^4}{r} \left(\frac{\rho(0)}{\rho_c} \right)^{-\frac{1}{6}} \text{km} ,$$

where

$$\rho_c = \frac{m_N r}{3\pi^2} m_e^3 \simeq 1 \times 10^6 \, \text{g cm}^{-3}$$

is the critical density where $k_F = m_e$. Thus, the white dwarf has a mass of order the solar mass but is confined to a radius of order that of the Earth!

20.4 Neutron Stars

In the previous exercise we looked at what happens when a star collapses and is stabilized by the electron degeneracy pressure. However, Chandrasekhar showed that if the mass of the star is greater than about 1.4 solar masses, the electron degeneracy pressure is insufficient to stop the implosion. What happens if this is the case? The answer is that the collapse continues, but it can be halted by the neutron degeneracy pressure, i.e., we have a neutron star. We can understand the physics here by arguments similar to those given for the white dwarf, but with important differences.

a) At low temperature the neutrons will occupy the low-lying levels, with two neutrons in each. Show that this leads to a Fermi momentum k_F given by

$$n = 2 \int_0^{k_F} \frac{dk}{(2\pi)^3} = \frac{k_F^3}{3\pi^2} ,$$

where n is the neutron number per unit volume.

b) Since nearly all the mass of the star is due to nucleons, calculate the Fermi energy in terms of the star density as

$$k_F = \left(\frac{3\pi^2 \rho}{m_N}\right)^{\frac{1}{3}} .$$

c) For neutrons having momentum k show that the pressure on the walls is given by

$$p = \frac{nk^2}{3E} ,$$

where $E = \sqrt{k^2 + m_N^2} \simeq m_N$ is the neutron energy.

d) Integrate over the spectrum and show that

$$p = \frac{2}{3} \int_0^{k_F} \frac{dk}{(2\pi)^3} \frac{k^2}{\sqrt{k^2 + m_N^2}}$$

$$= \frac{m_N^4}{24\pi^2} \left[\left(2\left(\frac{k_F^3}{m_N^3}\right) - 3\frac{k_F}{m_N}\right)\left(1 + \frac{k_F^2}{m_N^2}\right)^{\frac{1}{2}} + 3\sinh^{-1}\frac{k_F}{m_N} \right] ,$$

which represents the equation of state, i.e., the pressure in terms of the density.

e) Calculate the neutron star energy density and show that

$$\rho = 8\pi \int_0^{k_F} dk k^2 \sqrt{k^2 + m_N^2}$$

$$= \frac{m_N^4}{8\pi^2} \left[\left(2\left(\frac{k_F^3}{m_N^3}\right) + \frac{k_F}{m_N}\right)\left(1 + \frac{k_F^2}{m_N^2}\right)^{\frac{1}{2}} - \sinh^{-1}\frac{k_F}{m_N} \right] .$$

f) Relating p to ρ yields a complicated equation of motion. However, in the nonrelativistic limit, show that we have a simple relation

$$p = \frac{1}{5}\rho_c \left(\frac{\rho}{\rho_c}\right)^{\frac{5}{3}} ,$$

where

$$\rho_c = \frac{m_N^4}{3\pi^2} \simeq 6 \times 10^{15}\, \mathrm{g\,cm^{-3}}$$

is the critical density where $k_F = m_N$.

Again, given the equation of state, the Lane–Emden equation allows the construction of the neutron star and yields

$$M = \frac{2.72}{r^2}\left(\frac{\rho(0)}{\rho_c}\right)^{\frac{1}{2}} M_{\mathrm{Sun}}$$

$$R = 10.9 \left(\frac{\rho(0)}{\rho_c}\right)^{-\frac{1}{6}}\, \mathrm{km} .$$

Thus, a neutron star has a mass of order the solar mass, but is now confined to a radius of order 10 km!

20.5   The GZK Limit

This problem explains why it is expected that no cosmic ray protons with energies greater than 10^{20} GeV should be observed.

a) The universe is filled with cosmic microwave background (CMB) radiation consisting of photons at an average temperature 2.72 K. Calculate the energy of these photons in eV.

b) Consider an ultrarelativistic proton of energy E_p passing through space filled only with CMB radiation. Calculate the threshold value of $E_p = E_{\text{GZK}}$ such that the reaction

$$p + \gamma \rightarrow \Delta^+(1232) \rightarrow n + \pi^+$$

can occur. One would not expect cosmic rays from deep space having $E_p > E_{\text{GZK}}$ because any such cosmic ray particles would scatter off the CMB photons and be lost. This value is known as the Greisen–Zatsepin–Kuzmin (GZK) cutoff [Gre66].

20.6   Lifetime of the Sun

The solar constant – 1.36 kW m^{-2} – measures the solar flux incident on the Earth's surface from the Sun.

a) As discussed in the text, the major source of energy production in the Sun is the "pp cycle" which may be summarized as

$$4p \rightarrow {}^4\text{He} + 2e^+ + 2\nu_e + 25.7\,\text{MeV} .$$

Assuming that the Sun has been burning fuel at the same rate since its birth 4.6 billion years ago, estimate the fraction of its proton "fuel" which has been used during this time.

b) Assuming that the solar burn rate stays the same, how many years does the Sun have remaining before running out of fuel?

c) Explain why the number calculated in b) is too large and why the Sun (and life on the Earth) will die long before this time.

21

Beyond the Standard Model Physics

21.1 Introduction

In the previous chapters we have seen how the Standard Model is able to explain all laboratory data to date in particle and nuclear physics [Don14]. In this picture the strong interactions are described in terms of an $SU(3)_{\text{color}}$ gauge model involving the exchange of massless colored gluons, while the electromagnetic and weak interactions are incorporated into a unified electroweak interaction based on the symmetries $SU(2)_L \times U(1)_Y$, involving the exchange of massive $W^{\pm}, Z^0$ bosons and the massless photon. On the experimental side, there have been numerous tests of this picture and in each case, there has been found to exist agreement between the experimental signal and the theoretical prediction. A concise summary of some of these tests is shown in Table 21.1, so that the Standard Model could be said to represent a triumph of modern physics. Nevertheless, there remain problems and there is good reason to think that there must exist "beyond the Standard Model" (BSM) physics and in this closing chapter we briefly review some of the BSM ideas which have gained traction.

There exist at least two quite different approaches to seeking out such effects. One is phenomenological, where one parameterizes possible BSM effects and designs experiments which will search for signals of such effects. An example dealing with possible modifications of the weak interaction will be outlined in the next section. A second approach is theoretical. Here one postulates the existence of a grand new idea which will try to answer some of the burning questions which remain in understanding the Standard Model. We shall outline some of these in Section 21.3. Finally, we close with a brief summary in Section 21.4.

21.2 BSM Physics: Phenomenological Approach

The Standard Model of the semileptonic weak interaction was developed over the years, first in 1933 when Fermi incorporated Pauli's neutrinos into a phenomenological theory of weak decay [Fer34] and then in 1957 when Feynman and Gell-Mann postulated the V–A theory for the weak current [Fey58]. Since that time, the theory has been subjected to and passed every test, but there are good reasons to believe that there exist effects from new (BSM) physics which will be manifested at higher energy scales. The development of innovative experimental techniques, such as atomic trapping, has meant that a new

Table 21.1 Experimental values versus Standard Model fits for various selected quantities

Quantity	Experimental Value	Standard Model Prediction
m_t [GeV]	173.4 ± 1.0	173.5 ± 1.0
M_W [GeV]	80.420 ± 0.031	80.381 ± 0.014
g_V^{ve}	-0.040 ± 0.015	-0.0398 ± 0.0003
g_A^{ve}	-0.507 ± 0.014	-0.5064 ± 0.0001
$Q_W(e)$	-0.0403 ± 0.0053	-0.0474 ± 0.0005
$Q_W(Cs)$	-73.20 ± 0.35	-73.23 ± 0.02
$Q_W(Tl)$	-116.4 ± 3.6	-116.88 ± 0.03
τ_τ [fs]	291.13 ± 0.43	290.75 ± 2.51
M_Z [GeV]	91.1876 ± 0.0021	91.1874 ± 0.0021
Γ_Z [GeV]	2.4952 ± 0.0023	2.4961 ± 0.0010
$\Gamma(had)$ [GeV]	1.7444 ± 0.0020	1.7426 ± 0.0010
$\Gamma(\ell^+\ell^-)$ [MeV]	83.984 ± 0.086	84.005 ± 0.015
R_e	20.804 ± 0.050	20.744 ± 0.011
R_μ	20.785 ± 0.033	20.744 ± 0.011
R_τ	20.764 ± 0.045	20.789 ± 0.011
R_b	0.21629 ± 0.00066	0.21576 ± 0.00004
R_c	0.1721 ± 0.0030	0.17227 ± 0.00004

precision low-energy experimental frontier has been reached. This class of experiments can probe Nature at much higher energies via an effective interaction of order v^2/M^2, where $v \sim (2\sqrt{2}G_F)^{-\frac{1}{2}} \sim 176\,\text{GeV}$ is the electroweak scale and $M \geq 1\,\text{TeV}$ is the scale of the new physics. This is commonly called the high-intensity frontier, in contrast to the high-energy frontier. At the present time, this approach is being applied at laboratories around the world. Defining the general effective Hamiltonian

$$\begin{aligned}
\mathcal{H}_{eff} = &\ \bar{p}n\bar{e}(C_S + C_S'\gamma_5)\nu_e \\
&+ \bar{p}\gamma^\mu n\bar{e}\gamma_\mu(C_V + C_V'\gamma_5)\nu_e \\
&+ \bar{p}\sigma^{\mu\nu}n\bar{e}(C_T + C_T'\gamma_5)\nu_e \\
&+ \bar{p}\gamma^\mu\gamma_5 n(C_A + C_A'\gamma_5)\nu_e \\
&- \bar{p}\gamma_5 n\bar{e}(C_P + C_P'\gamma_5)\nu_e + h.c.
\end{aligned} \tag{21.1}$$

as done by Lee and Yang [Lee56], the SM predicts the values $C_S = C_S' = C_T = C_T' = C_P = C_P' = 0$ and $C_V = C_V' = \frac{G_F}{\sqrt{2}}V_{ud} = \frac{1}{\lambda}C_A = \frac{1}{\lambda}C_A'$, where $\lambda = g_A/g_V \simeq 1.27$, any deviation from which would signal BSM physics. Presently, a series of such experiments on nuclear and neutron beta decay have provided limits $|C_S + C_S'| < 2 \times 10^{-3}|C_V|$, primarily from the analysis of $0^+ - 0^+$ analog Fermi decays and $|C_T + C_T'| < 6 \times 10^{-3}|C_A|$, primarily from the analysis of $\pi^+ \to e^+\nu_e\gamma$ decay, as discussed in detail in Chapter 17. The corresponding restrictions on the scale of new physics are $M_S > 3.8\,\text{TeV}$ and $M_T > 2.3\,\text{TeV}$, which are impressive limits from low-energy experiments [Cir13]. The improvement in the precision of these tests as well as direct probes at the LHC should allow significant improvements in these limits and/or the detection of new physics in the next few years.

21.3 BSM Physics: Theoretical Approaches

In this approach, one envisions a new theoretical paradigm describing particle interactions and examines the possible implications of the existence of such a mechanism. Below we examine a few of these ideas.

Grand Unified Theories (GUT)

One of the mysteries of the Standard Model is why there exist interactions having such different strengths – electromagnetic and weak sectors having strengths which are factors of $\sim 10^{-2}$ and $\sim 10^{-5}$ smaller than the strong interaction couplings. Before development of the electroweak theory, one could have said the same about the weak and electromagnetic interactions. However, we now realize that this disparity in strength is simply a low-energy phenomenon, associated with measurements at $Q^2 <<< M_W^2, M_Z^2$. It is only at energy scales much higher than the weak gauge boson masses that the unification can be seen clearly. In 1974 Georgi and Glashow proposed an exciting new idea – the existence of a symmetry which unified strong, electromagnetic, and weak interactions [Geo74]. The grand unified theory (GUT) which they proposed was based on the symmetry $SU(5)$, though since that time there have been many variations proposed. The symmetry $SU(5)$ is the simplest group that can contain the Standard Model via

$$SU(5) \longrightarrow SU(3) \otimes SU(2) \otimes U(1) .$$

In the simple $SU(5)$ picture, there exist for each generation 15 basic building blocks, left- and right-handed u- and d-type quarks and antiquarks carrying three colors each, together with left- and right-handed charged leptons and left-handed neutrinos.[1] These building blocks were placed into a quintet plus decuplet $SU(5)$ representation via (here $i = r, g, b$ represent color indices)

$$\begin{aligned} 5 \quad &: e_L, \, \nu_e, \, \bar{d}_R^i \\ 10 \quad &: e_R, \, u_R^i, \, \bar{u}_L^i, \, \bar{d}_L^i . \end{aligned} \tag{21.2}$$

Recall, of course, that the various gauge coupling "constants" are not really constant. They run with Q^2 in a well-defined pattern. It is well known that the strong and weak couplings decrease with increasing Q^2, while the electromagnetic coupling increases. It was suggested (hoped) by Glashow and Georgi that in a successful GUT, there would be a large energy scale at which all three couplings would become commensurate. If the known scaling is taken into account, this energy would occur in the vicinity of $10^{14} - 10^{16}$ GeV, the GUT scale. Another consequence of this running is that the $SU(5)$ prediction for the Weinberg angle, $\sin^2 \theta_W = 3/8$ at the GUT scale, runs to the value $\sin^2 \theta_W \simeq 0.21$ at low values of Q^2, which is in remarkable agreement with the experimental value $0.2223(21)$.

---

[1] In 1974 there was no evidence for a nonzero neutrino mass. Now we would have to include also a right-handed neutrino degree of freedom which would have to be placed in an $SU(5)$ singlet representation.

However, the hoped for matching of all three couplings does not quite work (more about that later, when we discuss supersymmetry).

In this GUT there are, of course, also $5^2 - 1 = 24$ gauge bosons, which correspond to the 8 gluons, the charged $W^\pm$, the neutral Z^0, and the photon, accompanied by 12 new such bosons. Six of these (X) have charge $\pm 4/3$ and mass of order the GUT scale and six (Y) have charge $\pm 1/3$ and mass of order the GUT scale. They couple leptons and quark/antiquark degrees of freedom, $\bar{d} \to e + X, \bar{u} \to e + Y$, and in this context are often called *leptoquarks*. Such quantities also couple quark and antiquark degrees of freedom, $u \to \bar{u} + X, d \to \bar{u} + Y$, and in this context are often called *diquarks*. Of course, this $SU(5)$ symmetry must be badly broken, as it would require the degeneracy of quark and lepton masses. Nevertheless, it makes one very strong and unique prediction, that the proton must be unstable, because of the leptoquark-exchange process $uud \to \bar{e} + ud$, which would be observed as $p \to e^+ + \pi^0$. Because of this leptoquark exchange and $M^Y \sim M_{\mathrm{GUT}} \sim 10^{16}$ GeV, we would expect a proton lifetime of order $\tau_p \sim M_Y^4/m_p^5 \sim 10^{39}$ sec $\sim 10^{32}$ yr. Even though this is more than 20 orders of magnitude larger than the $\sim 10^{10}$ yr lifetime of the universe, it can be tested, because of the many protons which are part of practical detectors. Thirty years of experimentation have yielded a limit $\tau_p > 8.2 \times 10^{33}$ yr [PDG14], so the simple $SU(5)$ model seems to be ruled out. Nevertheless, there have been other such GUT models proposed to take its place. Many have the proton lifetime larger than the current limit and physicists are reluctant to abandon this very beautiful idea. Indeed, in looking at a general theory, chiral anomalies often arise, due to the presence of triangle diagrams involving fermions and one must be sure that such effects do not spoil the gauge invariance. In order to guarantee anomaly cancelation, the sum of the charges in each multiplet must vanish. Thus, the condition $Q_e - 3Q_d$ must obtain, which requires the equality of the fundamental lepton and quark charges. Since this equality has been confirmed to better than a part in 10^{41} this is a very attractive prediction.

Although the simple $SU(5)$ model is ruled out by the absence of proton decay at the required $\sim 10^{32}$ year level, many other symmetry groups have been proposed as a basis for a GUT. The simplest of these groups is $SO(10)$, which is an attractive possibility because it has a 16-dimensional representation, which can contain the 15 basic fields contained in the 10+5 dimensional representations of $SU(5)$ plus a right-handed neutrino field. The symmetry breaking pattern is

$$SO(10) \longrightarrow SU(5) \longrightarrow SU(3) \otimes SU(2) \otimes U(1).$$

Various symplectic and quaternion groups have also been considered, even octonians, but none of the various possibilities appears to be consistent with known phenomenology [Ros85].

Supersymmetry: SUSY

There exist many symmetries which have been shown to be important in particle and nuclear physics and most have a proposed degeneracy of two or more particles which are connected by a symmetry transformation. Thus, for example, $SU(2)$ symmetry requires the equality of proton and neutron masses in the symmetry limit, cf. Chapter 2, while $SU(3)$

symmetry suggests the equality of m_N, m_Λ, m_Σ, m_Ξ, cf. Chapter 3. Of course, symmetry breaking effects remove this degeneracy in the real world, but the remaining symmetry predictions are often well satisfied and provide relations between various processes which are well satisfied experimentally. An example here are the baryon magnetic moments, whose basic values are in rough agreement with the simple $SU(3)$ predictions; see Chapter 3. In the symmetries cited above, the particles which are related by the symmetry have identical spin and parity. However, this need not be the case and indeed in the case of chiral symmetry, one has a relation in the symmetry limit between a particle and the particle plus a massless Goldstone boson, cf. Chapter 6, which again leads to very strong predictive power and successful predictions such as the Goldberger–Treiman relation $M_N g_A = F_\pi g_{\pi NN}$ which connect the strong and weak sectors of the theory [Gol58]. In 1974, however, Wess and Zumino introduced a new type of symmetry which related bosons and fermions [Wes74]. That is, in a conventional symmetry transformation, one looks for invariance under the replacement of a spinor ψ by its rotated version $U(\chi)\psi$, where the angles χ characterize the rotation. That is, writing $U(\chi) = \exp(-i\chi \cdot \lambda)$ one requires the Lagrangian to be unchanged by replacement of ψ by $(1 - i\chi \cdot \lambda)\psi$. Wess and Zumino imagined a transformation which mixed boson and fermion degrees of freedom. For example, a scalar field ϕ could mix with a spin-1/2 field ψ via

$$\delta\phi = 2\bar{\epsilon}\psi, \qquad \delta\psi = -i\gamma^\mu \epsilon \partial_\mu \phi, \tag{21.3}$$

where ϵ is an infinitesimal spinor describing the transformation. Indeed it is easy to see that the simple sum of free Lagrangians for ϕ and ψ

$$\mathcal{L} = \frac{1}{2}\left(\partial^\mu \phi^\dagger \partial_\mu \phi - m^2 \phi^\dagger \phi\right) + \bar{\psi}\left(i\partial^\mu \gamma_\mu - m\right)\psi \tag{21.4}$$

is invariant provided that the scalar and spinor masses are identical. There has been a tremendous amount of work done on SUSY since the original Wess–Zumino suggestion 30 years ago even though there has been no confirmation of its existence. That is because a supersymmetric theory solves many outstanding issues in particle and nuclear physics:

i) The existence of additional supersymmetric degrees of freedom modifies the running of the various gauge coupling constants such that they indeed converge, as first proposed by Georgi and Glashow, at an energy of order the GUT scale 10^{16} GeV.

ii) SUSY offers a "natural" solution to what is called the Higgs hierarchy problem. Now that the Higgs boson has been discovered to have a mass of 126 GeV, one must worry about loop corrections to the mass which arise in the Standard Model. For example, at one loop one finds the relation

$$\delta m_H^2 = \frac{3\zeta}{16\pi^2}\Lambda_H^2 \left[m_H^2 + 2m_W^2 + m_Z^2 - 4m_t^2\right], \tag{21.5}$$

where ζ is the quadratic parameter in the Higgs potential and Λ_H is a quadratic cutoff. In order that this term not completely overwhelm the Higgs mass itself, one must require its vanishing and this is called fine tuning. The trouble is that it requires a Higgs mass of 314 GeV. At the present time there exist no known Standard Model mechanisms to cure this sickness. However, the problem does not exist in SUSY models.

iii) A second hierarchy problem has to do with the cosmological constant. Recent measurements which show that the expansion of the universe is actually accelerating can be fit by adding a cosmological constant λ to the Einstein action [Wei72], whose value is

$$\lambda_{\text{exp}} \sim 10^{-47}\,\text{GeV}^4\,. \tag{21.6}$$

However, this presents a problem in that each quantum field should have a vacuum energy which is far greater than this value. In fact, the estimated vacuum energy is a factor of 122 orders of magnitude larger than the value in Eq. (21.6). There is no known mechanism in particle physics to solve this problem. In SUSY models, however, there exists a cancelation between the vacuum energy of the fermions, which is positive, and that of the bosons, which is negative. Thus there is a "natural" hierarchy solution.

iv) Most SUSY models have a lightest particle which is colorless, neutral, and stable, which makes it a candidate for dark matter, to be discussed below.

v) Aside from the goal of grand unification which is engendered by GUT models, many physicists have the further goal of inclusion of gravity into the unification, producing a "theory of everything" (TOE). So far, this seems only possible by use of supersymmetry.

Obviously then SUSY theories have a sort of siren's call in that they solve simultaneously many of the problems pointed out above. However, they also possess associated drawbacks. One is that each normal baryon in the particle data book must possess a supersymmetric bosonic analog (obtained by putting an *s* in front if the normal particle name: electron–selectron, neutrino–sneutrino, quark–squark, etc.) and each normal boson must have a supersymmetric fermion analog (obtained by putting an *ino* after the normal particle name: photon–photino, higgs–higgsino, *W*–Wino, etc.). If supersymmetry were unbroken these pairs would have to be degenerate. However, years of experimentation has yielded no sign of their existence, so supersymmetry must be strongly broken. In addition, even minimal SUSY models have scores of independent parameters, so even though many physicists believe in SUSY, it has proven extremely hard to verify [Kan00].

Dark Matter

The earliest suggestion for the existence of dark matter came in 1933 when Zwicky measured the velocity of galaxies in the Coma cluster [Zwi33]. This was done by looking at the Doppler shift of their atomic spectra and was used to determine the mass of the cluster. To his surprise, the number turned out to be 400 times larger than the visible mass, as observed from stars, indicating the presence of mass which was nonradiating, i.e., dark mass. More recently, Rubin and others have studied the tangential velocities of various galaxies as a function of the distance from the center [Rub70]. In a simple picture in which the stars orbit the central region the centripetal force should be provided by the gravitational attraction, meaning that $v^2/r \sim GM/r^2$. Thus we expect that $v \sim 1/\sqrt{r}$. Observationally, however, the velocity is asymptotically constant or even increases with large r, indicating the presence of a large amount of "dark matter" that has gravitational consequences but is not

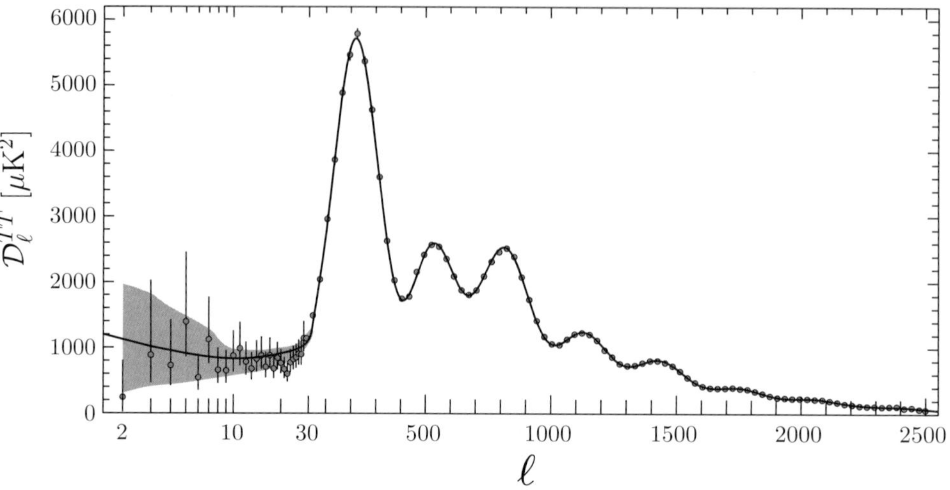

Fig. 21.1   This graph shows the temperature fluctuations in the cosmic microwave background detected by the Planck satellite at different angular scales on the sky, starting at 90° on the left-hand side of the graph, through to the smallest scales on the right-hand side [Pla15]. The multipole moments corresponding to the various angular scales are labeled ℓ. The dots are measurements made with the Planck satellite; these are shown with error bars that account for measurement errors as well as for an estimate of the uncertainty that is due to the limited number of points in the sky at which it is possible to perform measurements. This so-called cosmic variance is an unavoidable effect that becomes most significant at larger angular scales. The curve represents the best fit of the "standard model of cosmology" – currently the most widely accepted scenario for the origin and evolution of the universe – to the Planck data. The shading around the curve at large angles (small ℓ) illustrates the predictions of various modifications of the Standard Model.

seen in the form of stars. The presence of this dark matter is also suggested by gravitational lensing, as light from distant galaxies is bent by its presence on its way to Earth.[2]

The existence of dark matter then seems not to be in doubt. More recently, angular fluctuations in the cosmic microwave background radiation as measured by satellites such as WMAP and Planck have not only confirmed the existence of dark matter but have determined its contribution to the mass of the universe as being 27%, with ordinary matter contributing only 5% and dark energy, discussed later, contributing 68%. Figure 21.1 shows the recent precision measurements of temperature fluctuations as a function of anglar scale from the Planck experiment [Pla15]. Note that these experiments also verify that neutrinos come in only three types and their data can be interpreted to determine that the total mass of these three neutrino types is less than 0.2 eV. The burning question then is what is dark matter made of? There have been a number of suggestions in this regard, but nothing definitive can yet be said. Certainly, whatever constitutes the dark matter cannot interact very strongly with conventional matter or we would have observed it. Also, whatever it is must presumably be very heavy in order to overwhelm conventional matter in its gravitational effects. For this reason, Bahcall coined the term WIMPS (weakly interacting

---

[2] The correctness of general relativity is assumed at cosmic scales.

massive particles) for the substance that makes up dark matter. As to what these particles might be, some have suggested neutrinos, since there certainly are huge numbers left over from the Big Bang. However, the lightness of the standard neutrinos makes this possibility very unlikely. More plausible is the lightest supersymmetric particle, the higgsino or Zino or photino, or a mixture of same. Being neutral and the lightest supersymmetric particle, it should be stable and there should exist copious amounts left over from the Big Bang. But how can this be confirmed? At the present time there are a number of experimental dark matter searches underway. The basic idea is that the solar system has an orbital velocity of $\sim$220 km sec^{-1} with respect to the galactic center, while the Earth itself has a $\sim$30 km sec^{-1} orbital velocity around the Sun. Thus the Earth sits in a dark matter wind which is higher in the (northern hemisphere) in the summer and smaller in the winter. Even if such particles do not interact conventionally with ordinary matter, their presence should be observable by the conventional matter recoil which results from a collision of the dark and conventional matter. Thus far, no definitive signal has been seen, though there are a number of intriguing hints. Figure 21.2 summarizes the experimental status on a plot of WIMP–nucleon spin-independent cross section versus WIMP mass in 2014.

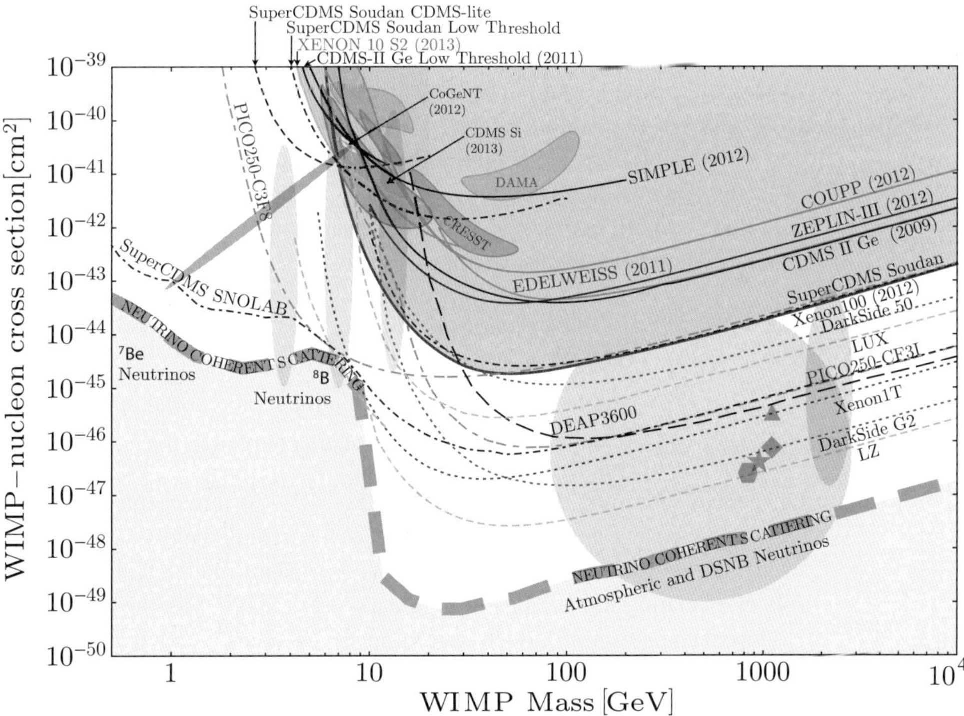

Fig. 21.2   A compilation of WIMP–nucleon spin-independent cross section limits (solid curves), hints for WIMP signals (shaded closed contours) and projections (dot and dot-dashed curves) for a selection of experiments that are expected to take data over the next decade (from [Sno13]). Also shown is the approximate band where coherent scattering of ^{8}B solar neutrinos (discussed in Chapter 18), atmospheric neutrinos, and diffuse supernova neutrinos with nuclei will begin to limit the sensitivity of direct detection experiments to WIMPs. Finally, a suite of theoretical model predictions is indicated by the shaded regions. See [Sno13] for references.

Dark Photons, Hidden Sectors, and Light Dark Matter

Dark matter and neutrino mass provide strong empirical evidence for BSM physics. Arguably, rather than suggesting any specific mass scale for new physics, they point to a hidden (or dark) sector, weakly coupled to the SM. These sectors also arise in many top-down models of beyond the SM [Ess13]. Dark sectors containing light stable degrees of freedom, with mass in the MeV–GeV range, are of particular interest as dark matter candidates, since this regime is poorly explored in comparison to the weak scale. Low-to-medium energy experiments with high precision and/or luminosity are ideally suited to explore this light dark sector landscape.

From a theoretical perspective, note that it is useful to consider [DeY14] a general parametrization of the interactions between the SM and a hidden or dark sector. A natural assumption is that any light dark sector states are SM gauge singlets. This automatically ensures weak coupling to the visible (SM) sector, while the impact of heavier charged states is incorporated in an effective field theory expansion at or below the weak scale,

$$\mathcal{L} \sim \sum_n \frac{c_n}{\Lambda^n} \mathcal{O}_{\text{SM}}^{(k)} \mathcal{O}_{\text{hidden}}^{(l)} \,, \tag{21.7}$$

where k and l denote operator dimensions and $n = k + l - 4$. The generic production cross section for hidden sector particles then scales as $\sigma \sim E^{2n-2}/\Lambda^{2n}$. Thus, lower-dimension interactions, unsuppressed by the heavy scale Λ, are preferentially probed at lower energy, and such interactions are natural targets for the intensity frontier more generally. The set of lowest-dimension interactions, or *portals*, which generalizes the right-handed neutrino coupling, is quite compact. Up to dimension five ($n \leq 1$), assuming SM electroweak symmetry breaking, the list of portals includes [Ess13]:

- $\mathcal{L} = -\frac{\epsilon}{2} B^{\mu\nu} F'_{\mu\nu}$: dark photons kinetically mixed with hypercharge;
- $\mathcal{L} = (AS + \lambda S^2) H^\dagger H$: dark scalars coupled to the Higgs;
- $\mathcal{L} = y_N LHN$: sterile neutrinos coupled via the lepton portal; and
- $\mathcal{L} = \frac{\partial_\mu a}{f_a} \overline{\psi} \gamma^\mu \gamma^5 \psi$: axion-like pseudoscalars coupled to the axial-vector current.

On general grounds, the couplings of these lowest-dimension operators are minimally suppressed by any heavy scale, and new weakly coupled physics would naturally manifest itself first via these portals in any generic top-down model. Thus, portals play a primary role in mediating interactions of light dark sector states with the SM.

One of the simplest hidden sectors involves states charged under a new $U(1)'$ gauge group. The corresponding gauge boson, dubbed the dark photon, A', is kinetically mixed with hypercharge via the vector portal above, which induces a coupling to the electromagnetic current $\mathcal{L} = -\epsilon e A'_\mu J^\mu_{\text{EM}}$. This scenario, and variations involving other nonanomalous currents such as $B-L$, provide some of the few relatively unconstrained UV-complete extensions of the SM. Models of this type, with a massive vector, have been the focus of considerable attention in recent years due, for example, to the role the dark vector mediator can play in models of dark matter. Initial interest arose from the utility of light vector mediators in building viable dark matter models to explain the enhanced positron

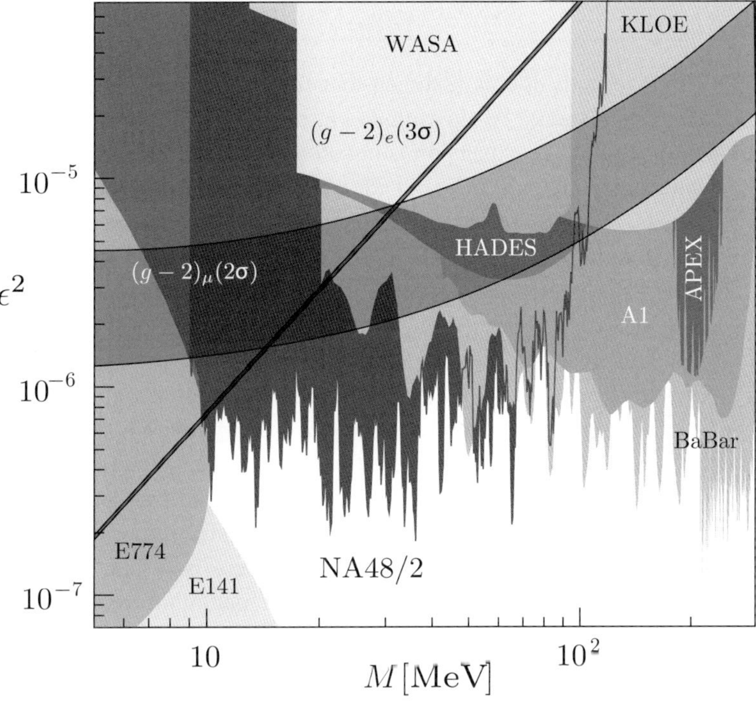

Fig. 21.3   Upper limits at 90% CL on the mixing parameter ϵ^2 versus the dark photon mass $m_{A'}$ obtained by the NA48 experiment [Bat15].

fraction observed in cosmic rays by PAMELA, FERMI and most recently AMS-02. WIMP dark matter annihilating via light mediators is Coulomb-enhanced at the low velocities $v \sim 10^{-3}$ relevant for the galactic halo. Subsequently, these scenarios have been explored within the context of light sub-GeV dark matter, in which dark matter lies in the hidden sector charged under $U(1)'$, and as the generic interaction channels accessible at low energy according to the operator analysis above. Further, the dark sector has been invoked to explain both the muon $g-2$ result as well as to understand the proton radius puzzle.

The A' might be produced in a number of ways: radiative production in electron scattering from a nucleus; π^0 decay to $e^+e^-\gamma$; $\Upsilon(2S, 3S) \rightarrow \gamma A'$, followed by $A' \rightarrow \mu^+\mu^-$; and in $e^+e^- \rightarrow \gamma A'$, followed by $A' \rightarrow e^+e^-$, $\mu^+\mu^-$. In addition, the A' can decay invisibly. Over the last several years, there has been an intensive worldwide effort to search for evidence of the A'. Ingenious searches have been conducted using data from past experiments completed within the last three decades. In addition, existing experiments built for other scientific purposes have been used to search for the A'.

To date, a large area of the coupling $\alpha' \equiv \epsilon^2 \alpha_{EM}$ vs. mass ($m_{A'}$) space for A' decay to e^+e^- has been excluded at the 2σ level. Figure 21.3 shows a recent review of experimental exclusion limits. To date, no evidence for the A' has been found. Further searches will employ innovative techniques that aim for higher precision in dedicated experiments. A number of nuclear physics facilities are well positioned to mount next-generation searches.

Dark Energy

Until recently it was assumed that the expansion of the universe which began with the Big Bang should be slowing down, due to gravitational effects. However, in 1998 it was realized that the expansion is not slowing down, but is actually speeding up [Per99]. This phenomenon seems to be outside of ordinary Einstein gravity and can be accommodated by inclusion of a so-called cosmological constant λ into the Einstein action (see, eg. [Pee03]). Einstein himself did this in order to produce a static universe. However, he discarded it when the expansion was observed by Hubble, calling his inclusion of a cosmological constant his "greatest blunder." However, that is exactly what is needed at the present time in order to explain the acceleration of the expansion. Just as λ provided the negative pressure which counteracted the gravitational contraction in Einstein's version, its inclusion in the present day theory (though with a smaller value, $\lambda \sim 10^{-29}\,\mathrm{g\,cm^{-3}}$), is just what is needed in order to explain the acceleration. As for the origin of this term, it could arise from the vacuum energy which fills all of space and is associated with the presence of quantum fields, although the value expected from field theoretic arguments is of order $10^{90}\,\mathrm{g\,cm^{-3}}$, meaning that there is a discrepancy with the measured value in this picture by about 120 orders of magnitude. An alternative to the cosmological constant is the idea of quintessence, which postulates the existence of a very light dynamical scalar field which fills the universe, but there is no evidence for its existence at this time.

String Theory

One of the problems with conventional quantum field theory is a large number of infinities which result when loop corrections are taken into account. These are tamed by inclusion of cutoffs or regularizing factors such as working in $n \neq 4$ dimensions (see Appendix A), which allow finite values to be obtained for observables when all diagrams are added together (see the renormalization discussion). Nevertheless it is a bit unnerving to note that the one-loop corrections to the electron mass due to quantum electrodynamics loops are formally infinite, implying that the bare electron mass must itself be infinite and negative in order that the sum of bare and one-loop corrected masses be finite and equal to the observed $511\,\mathrm{keV}$ mass. One solution to this problem comes from assuming that we are dealing not with point particles but rather with excitations of extended objects, i.e., one-dimensional strings or higher-dimensional branes. Divergences are indeed tamed in such a picture though the early versions required the existence of as many as 25 spatial dimensions. Connection with the real world was made by supposing that most of these are compacted (curled up) into currently unobservable scales. By inclusion of supersymmetry, it was possible to reduce the number of spatial dimensions to ten and to include a graviton, so that the theory could in principle be the long sought TOE (theory of everything). Though early researchers hoped that there might exist only one or two such theories with the right properties to explain our world, the development of M-theory has meant that, far from this hope, the number of allowed models occupies an entire "landscape" of possibilities, with perhaps as many as 10^{500} possible versions. Thus superstring theory is certainly promising as a route to TOE, but there is currently no real connection between the theoretical calculations in superstring theory and particle/nuclear phenomenology [Gre99].

Extra Dimensions

The theorist Kaluza was one of the first to propose a theory of extra dimensions when in 1921 he noted that by extending spacetime to five dimensions, electromagnetism could be unified with general relativity [Kal21]. Five years later Klein, proposed that the fourth spatial dimension could curl up on itself after a very small distance, the idea that is now called compactification [Kle26]. A modern application of compactification attempts to explain why the strength of the gravitational interaction is anomalously small when compared to its strong, weak, and electromagnetic analogs. Note that the ratio of the electric to gravitational forces between a pair of electrons is of order 10^{40}. This is an enormous number and an obvious question is how to explain this value. One suggestion, due to Dirac, was that the only way to conjure up such a large number was to compare the age of the universe, $\sim 10^{17}$ seconds, with the time it takes for light to traverse the $\sim 10^{-15}$ m size of an elementary particle, $\sim 10^{-23}$ sec. The ratio of these numbers is indeed 10^{40}, and if there is anything to this idea, then the size of this ratio should change as the universe evolves, meaning that, for example, the strength of the electromagnetic interaction may have been smaller in the remote past. This hypothesis can be checked by looking at distant galaxies which were formed early in the evolution of the universe and there does not seem to be any change in such strength. In the meantime a new paradigm, based on superstrings, has developed. The basic idea here is that while the strong, weak, and electromagnetic interactions propagate on the four-dimensional brane on which we exist, the gravitational interaction exists throughout spacetime even in the curled up dimensions which are required for the correctness of the supergravity idea. For this reason the gravitational force is strongly reduced with respect to the three others and this is a possible reason for its weakness [Ran05].

Matter–Antimatter Asymmetry

The origin of matter and antimatter is believed to have occured in early epochs of the Big Bang, but almost certainly in equal amounts. It is therefore a mystery why the Universe is found to consist almost entirely of matter, with very little evidence of antimatter content. In 1967, Sakharov considered this problem and identified three conditions which must obtain in order to this to occur [Sak67]:

  i) Violation of lepton and baryon number conservation laws.
  ii) Lack of thermal equilibrium.
  iii) Violation of CP invariance.

The reason for item i) is clear, since we must end up with a strong preponderance of baryons and leptons, while beginning with equal numbers of baryons/leptons and antibaryons/antileptons. Such violation is part of many grand unified theories. There must be an era of thermal nonequilibrium for the same reason, since in thermal equilibrium all reactions would proceed equally in both directions, meaning that there could be no net change of baryon or lepton number. Also, in thermal equilibrium the only thing that reaction rates can depend on is the temperature and the mass of the system. By *CPT* invariance, the baryon and antibaryon masses must be identical, so no asymmetry can be generated without thermal nonequilibrium. More interesting is the third requirement, which is needed in order that there be net particle production. Without it the rates for particle and antiparticle reactions would be identical, so that no net baryon or lepton number production could result. Of course, *CP* violation is part of the Standard Model in the nonzero phase δ of the CKM matrix. However, this phase is associated only with the tiny two-generation-change element V_{td} and is far too small to explain the matter/antimatter asymmetry of the universe. It is not yet known whether there is a similar nonzero phase in the corresponding PMNS neutrino mixing matrix, but the fact that this is associated with the weak interactions indicates that, again, this cannot be the B/L asymmetry source. We need then an additional source of *CP* violation. One possibility comes from the anomalous interaction

$$\mathcal{L}_{\text{anom}} = \frac{g_3^2}{32\pi^2}\theta G_{\mu\nu}^a \tilde{G}_a^{\mu\nu} \,, \tag{21.8}$$

where $G_{\mu\nu}^a$ is the color gluon field tensor, g_3 is the color gluon coupling constant, and θ is an arbitrary parameter introduced in Chapter 5. Although θ is not constrained by any principle of the Standard Model, a nonzero value would contribute to a neutron electric dipole moment, whose size has been experimentally constrained to the incredibly small value $d_n < 2.9 \times 10^{-26}$ e-cm [Bak07]. In order to explain the smallness of θ, Peccei and Quinn proposed the existence of a global symmetry which is spontaneously broken [Pec77] – see Chapter 5. This spontaneous breaking leads, via Goldstone's theorem, to the existence of a light particle called the axion which itself could be a candidate for dark matter. However, as yet, there is no evidence for the existence of the axion, and the origin of the CP violation required to explain the matter/antimatter asymmetry of the universe remains unclear.

21.4 Summary

We have seen in this book the incredible success which the Standard Model has had in explaining the entire range of phenomena in particle and nuclear physics. Nevertheless, there remain many outstanding issues to be explained, some small, some large. Certainly in order to understand the size of the cosmological constant or the Higgs mass without fine tuning, a major new idea, such as supersymmetry, must be introduced. Alternatively, a trenchant understanding of the relative size of the strong, weak, and electromagnetic interactions needs an idea such as grand unification. In any case, it appears that in order to solve these deep issues, there must exist some sort of new physics between the energy scale which characterizes electroweak physics, $v \sim 246$ GeV, and the Planck scale, $M_{Pl} \sim 10^{19}$ GeV, where gravitational effects become important. The LHC offers our best current hope to directly access some of this BSM physics, although precision low-energy tests may also offer some hint of what is to come. In any case, there remains much to be discovered.

Appendix A Useful Information

Conversion Factors ($\hbar = c = k_B = 1$):

$$1\,\text{GeV}^{-1} = 6.582122 \times 10^{-25}\,\text{s} = 0.197327\,\text{fm}$$
$$1\,\text{GeV} = 1.16 \times 10^{13}\,\text{K} = 1.78 \times 10^{-24}\,\text{g} \tag{A.1}$$

Physical Constants [PDG14]:

$$G_\mu = 1.1663787(6) \times 10^{-5}\,\text{GeV}^{-2} \quad \alpha^{-1} = 137.035999074(44)$$
$$m_W = 80.385(15)\,\text{GeV} \quad m_Z = 91.1876(21)\,\text{GeV}$$
$$F_\pi = 92.2(2)\,\text{MeV} \quad F_K = 110.4(8)\,\text{MeV}$$
$$G_N^{-1/2} = M_{\text{Pl}} = 1.2 \times 10^{19}\,\text{GeV} \quad \sin^2\theta_W^{\bar{MS}}(m_Z) = 0.23125(16) \tag{A.2}$$

Masses

Gauge bosons:

$$m_W = 80.385(15)\,\text{GeV} \quad m_Z = 91.1876(21)\,\text{GeV}$$

Higgs boson:

$$m_H = 125.09(24)\,\text{GeV}$$

Charged leptons:

$$m_e = 0.510998928(11)\,\text{MeV} \quad m_\mu = 105.6583715(35)\,\text{MeV}$$
$$m_\tau = 1776.82(16)\,\text{GeV}$$

Baryons:

$$m_p = 938.272046(21)\,\text{MeV} \quad m_n = 939.56539(21)\,\text{MeV}$$
$$m_\Lambda = 1115.683(6)\,\text{MeV} \quad m_{\Sigma^\pm} = 1189.37(7)\,\text{MeV}$$
$$m_{\sigma^0} = 1192.642(24)\,\text{MeV} \quad m_{\Xi^\pm} = 1321.71(7)\,\text{MeV}$$
$$m_{\Xi^0} = 1314.86(20)\,\text{MeV}$$

Mesons:

$$m_{\pi^\pm} = 139.57018(35)\,\text{MeV} \quad m_{\pi^0} = 134.9766(6)\,\text{MeV}$$
$$m_{K^\pm} = 493.677(16)\,\text{MeV} \quad m_{K^0} = 497.614(24)\,\text{MeV}$$
$$m_{\eta^0} = 547.862(18)\,\text{MeV}$$

Quarks:

$$m_u = 2.3^{+0.7}_{-0.5}\,\text{MeV} \qquad m_d = 4.8^{+0.5}_{-0.3}\,\text{MeV}$$

$$m_c = 1.275(25)\,\text{GeV} \quad m_s = 95(5)\,\text{MeV}$$

$$m_t = 175(6)\,\text{GeV} \qquad m_b = 4.66(3)\,\text{GeV}$$

CKM Matrix Elements

$$|V_{ud}| = 0.97427(15) \quad |V_{us}| = 0.22534(65) \quad |V_{ub}| = 0.00351(15)$$

$$|V_{cd}| = 0.22520(65) \quad |V_{cs}| = 0.97344(16) \quad |V_{cb}| = 0.0412(10)$$

$$|V_{td}| = 0.00867(30) \quad |V_{ts}| = 0.0404(10) \quad |V_{tb}| = 0.99915(4)$$

$$\theta_{12} = 13^o \qquad \theta_{23} = 2.4^o \qquad \theta_{13} = 0.20^o \qquad \delta = 69^o \tag{A.3}$$

Wolfenstein parameter representation is given by

$$\lambda = 0.22535(5) \quad A = 0.811^{+0.022}_{-0.012}$$

$$\bar{\rho} = 0.131^{+0.026}_{-0.013} \quad \bar{\eta} = 0.345^{+0.013}_{-0.014}$$

PNMS Matrix Elements

$$|V_{1e}| = 0.827 \quad |V_{2e}| = 0.543 \quad |V_{3e}| = 0.156$$

$$|V_{1\mu}| = 0.480 \quad |V_{2\mu}| = 0.531 \quad |V_{3\mu}| = 0.699$$

$$|V_{1\tau}| = 0.296 \quad |V_{2\tau}| = 0.651 \quad |V_{3\tau}| = 0.699$$

$$\theta_{12} = 33.3^o \qquad \theta_{23} = 45.0^o \qquad \theta_{13} = 9.0^o \qquad \delta = ? \tag{A.4}$$

Note that values of $|V_{ij}|$ are evaluated using $\delta = 0$.

A.1 Notations and Identities

$$\text{Metric tensor:} \quad \eta_{\mu\nu} = (1, -1, -1, -1)_{\text{diag}} \quad \eta^{\mu}{}_{\mu} = 4 \tag{A.5}$$

$$\text{Levi-Civita antisymmetric tensor:} \quad \epsilon_{0123} = +1 \tag{A.6}$$

$$\text{Pauli matrices:} \quad \sigma_1 = \begin{pmatrix} 0 & 1 \\ 1 & 0 \end{pmatrix} \quad \sigma_2 = \begin{pmatrix} 0 & -i \\ i & 0 \end{pmatrix} \quad \sigma_3 = \begin{pmatrix} 1 & 0 \\ 0 & -1 \end{pmatrix}$$

$$\sigma^j \sigma^k = \delta^{jk} I + i\epsilon^{jkl}\sigma^l \qquad (j, k, l = 1, 2, 3)$$

$$\sigma^j_{ab}\sigma^j_{cd} = 2\delta_{ad}\delta_{bc} - \delta_{ab}\delta_{cd} \qquad (a, b, c, d = 1, 2) \tag{A.7}$$

$$\text{Dirac matrices:} \quad \gamma^0 = \begin{pmatrix} 1 & 0 \\ 0 & -1 \end{pmatrix} \quad \gamma^i = \begin{pmatrix} 0 & \sigma_i \\ -\sigma_i & 0 \end{pmatrix}$$

$$\gamma_5 = i\gamma^0\gamma^1\gamma^2\gamma^3 \quad \sigma^{\mu\nu} = \frac{i}{2}[\gamma^\mu, \gamma^\nu]$$

$$\gamma^\mu\gamma^\nu\gamma^\alpha = \eta^{\mu\nu}\gamma^\alpha + \eta^{\nu\alpha}\gamma^\mu - \eta^{\alpha\mu}\gamma^\nu + i\epsilon^{\mu\nu\alpha\beta}\gamma_\beta\gamma_5$$

$$\gamma^0\Gamma_i^\dagger\gamma^0 = \Gamma_i \qquad (\Gamma_i = 1, \gamma^\mu, \gamma^\mu\gamma_5, \sigma^{\mu\nu})$$

$$\gamma^0\Gamma_i^\dagger\gamma^0 = -\Gamma_i \qquad (\Gamma_i = \gamma_5) \tag{A.8}$$

$$\text{Trace relations:} \quad \operatorname{Tr}\left(\gamma^\mu\right) = 0 \quad \operatorname{Tr}\left(\gamma_5\right) = 0$$

$$\operatorname{Tr}\left(\gamma^\mu\gamma^\nu\right) = 4\eta^{\mu\nu} \quad \operatorname{Tr}\left(\gamma^\mu\gamma^\nu\gamma_5\right) = 0$$

$$\operatorname{Tr}\left(\gamma^\mu\gamma^\nu\gamma^\alpha\gamma^\beta\right) = 4(\eta^{\mu\nu}\eta^{\alpha\beta} - \eta^{\mu\alpha}\eta^{\nu\beta} + \eta^{\mu\beta}\eta^{\nu\alpha})$$

$$\operatorname{Tr}\left(\gamma_5\gamma^\mu\gamma^\nu\gamma^\alpha\gamma^\beta\right) = 4i\epsilon^{\mu\nu\alpha\beta}$$

$$\operatorname{Tr}\left(\displaystyle{\not}a_1 \cdots \displaystyle{\not}a_{2n+1}\right) = 0$$

$$\operatorname{Tr}\left(\displaystyle{\not}a_1 \cdots \displaystyle{\not}a_{2n}\right) = \operatorname{Tr}\left(\displaystyle{\not}a_{2n} \cdots \displaystyle{\not}a_1\right) \tag{A.9}$$

Plane-wave Solutions

The Dirac spinor $u(\mathbf{p}, s)$ is a positive-energy eigenstate of the momentum $\mathbf{p}$ and energy $E = \sqrt{\mathbf{p}^2 + m^2}$. Antifermions are described in terms of the Dirac spinor $v(\mathbf{p}, s)$. The adjoint solutions are denoted by $\bar{u} \equiv u^\dagger\gamma^0$ and $\bar{v} \equiv v^\dagger\gamma^0$. Note that our normalization of Dirac spinors behaves smoothly in the massless limit.

$$(\not{p} - m)u(\mathbf{p}, s) = 0 \quad \bar{u}(\mathbf{p}, s)(\not{p} - m) = 0$$

$$(\not{p} + m)v(\mathbf{p}, s) = 0 \quad \bar{v}(\mathbf{p}, s)(\not{p} + m) = 0$$

$$\bar{u}(\mathbf{p}, r)u(\mathbf{p}, s) = 2m\delta_{rs} \quad \bar{v}(\mathbf{p}, r)v(\mathbf{p}, s) = -2m\delta_{rs}$$

$$u^\dagger(\mathbf{p}, r)u(\mathbf{p}, s) = 2E\delta_{rs} \quad v^\dagger(\mathbf{p}, r)v(\mathbf{p}, s) = 2E\delta_{rs}$$

$$\sum_s u(\mathbf{p}, s)\bar{u}(\mathbf{p}, s) = \not{p} + m \quad \sum_s v(\mathbf{p}, s)\bar{v}(\mathbf{p}, s) = \not{p} - m. \tag{A.10}$$

Gordon decomposition for a fermion of mass m:

$$\bar{u}(\mathbf{p}', r)\gamma^\mu u(\mathbf{p}, s) = \bar{u}(\mathbf{p}', r)\left(\frac{(p' + p)^\mu}{2m} + \frac{i\sigma^{\mu\nu}(p' - p)_\nu}{2m}\right)u(\mathbf{p}, s) \tag{A.11}$$

$$u(\mathbf{p}, s) = \sqrt{\frac{E + m}{2m}}\begin{pmatrix} \chi_s \\ \frac{\sigma \cdot \mathbf{p}\chi_s}{E+m} \end{pmatrix} \qquad v(\mathbf{p}, s) = \sqrt{\frac{E + m}{2m}}\begin{pmatrix} \frac{\sigma \cdot \mathbf{p}\chi_s}{E+m} \\ \chi_s \end{pmatrix}. \tag{A.12}$$

Propagators

The propagators associated with fields $\phi(x)$, $\psi(x)$, $W_\lambda(x)$ having spins $0, 1/2, 1$, and masses μ, m, M are respectively

$$i\Delta_F(x) = {<}0|T(\phi(x)\phi^\dagger(0))|0{>} = \int \frac{d^4p}{(2\pi)^4}\, e^{-ip\cdot x}\, \frac{i}{p^2 - \mu^2 + i\epsilon}$$

$$iS_{F\beta\alpha}(x) = \langle 0|T(\psi_\beta(x)\bar\psi_\alpha(0))|0 \rangle = \int \frac{d^4p}{(2\pi)^4}\, e^{-ip\cdot x}\, \frac{i\,(\slashed{p}+m)_{\beta\alpha}}{p^2 - m^2 + i\epsilon}$$

$$iD_F^{\lambda\nu}(x) = \langle 0|T(W^\lambda(x)W^{\dagger\nu}(0))|0 \rangle$$

$$= \int \frac{d^4p}{(2\pi)^4}\, e^{-ip\cdot x}\, \frac{i\left(-\eta_{\lambda\nu} + \frac{(1-\xi)p_\lambda p_\nu}{p^2 - \xi M^2 + i\epsilon}\right)}{p^2 - M^2 + i\epsilon}, \tag{A.13}$$

where ξ is a gauge-dependent parameter.

A.2 Decay Lifetimes and Cross Sections

The normalization of charged spin zero and spin one-half fields is

$$S = 0: \quad \phi(x) = \int \frac{d^3k}{(2\pi)^3 2E_\mathbf{k}}\, (a(\mathbf{k})e^{-ik\cdot x} + a^\dagger(\mathbf{k})e^{ik\cdot x})$$

$$S = \frac{1}{2}: \quad \psi(x) = \sum_s \int \frac{d^3p}{(2\pi)^3 2E_\mathbf{p}}(b(\mathbf{p},s)u(\mathbf{p},s)e^{-ip\cdot x} + d^\dagger(\mathbf{p},s)v(\mathbf{p},s)e^{ip\cdot x}) \tag{A.14}$$

with momentum-space algebraic relations

$$[a(\mathbf{k}), a^\dagger(\mathbf{k}')] = (2\pi)^3 2E_\mathbf{k}\delta^3(\mathbf{k} - \mathbf{k}')\,,$$

$$\{b(\mathbf{p},r), b^\dagger(\mathbf{p}',s)\} = (2\pi)^3 2E_\mathbf{p}\delta_{rs}\delta^3(\mathbf{p} - \mathbf{p}') \tag{A.15}$$

and single-particle states

$$|\mathbf{k}\rangle_B = a^\dagger(\mathbf{k})|0\rangle\,, \qquad |\mathbf{p},s\rangle_F = b^\dagger(\mathbf{p},s)|0\rangle. \tag{A.16}$$

Using the above, one can express the single-particle expectation value of the quantum mechanical probability density as

$$\rho_i = 2E_i. \tag{A.17}$$

Lifetimes

From the decay law $N(t) = N(0)e^{-t/\tau}$, the mean life τ is seen to be the transition rate per decaying particle, $\Gamma = \tau^{-1} = -\dot{N}/N$. For decay of a particle of energy E_1 into a total of $n-1$ bosons and/or fermions, the S-matrix amplitude can be written in terms of a reduced (or invariant) amplitude $\mathcal{M}_{\mathrm{fi}}$ as

$$\langle f|S - 1|i\rangle = -i(2\pi)^4 \delta^{(4)}(p_1 - p_2 \cdots - p_n)\, \mathcal{M}_{\mathrm{fi}}. \tag{A.18}$$

The inverse lifetime is computed from the squared S-matrix amplitude per spacetime volume VT and incident particle density ρ_1, integrated over final state phase space. Using

$$\text{Phase space per particle} \equiv \int \frac{d^3\mathbf{k}}{(2\pi)^3)2E_\mathbf{k}}, \tag{A.19}$$

we have

$$\langle \mathbf{p}'|\mathbf{p}\rangle = \int \frac{d^3k}{(2\pi)^3 2E_{\mathbf{k}}} \langle \mathbf{p}'|\mathbf{k}\rangle \langle \mathbf{k}|\mathbf{p}\rangle. \tag{A.20}$$

The inverse lifetime (or decay width) is then given by

$$\tau^{-1} = \Gamma = \frac{1}{2E_1} \frac{1}{\mathcal{Z}} \int \left(\prod_{k=2}^{n} \frac{d^3 p_k}{(2\pi)^3 2E_{\mathbf{p}_k}}\right) \frac{|S-1|^2_{\mathrm{fi}}}{VT}$$

$$= \frac{1}{2E_1} \frac{1}{\mathcal{Z}} \int \left(\prod_{k=2}^{n} \frac{d^3 p_k}{(2\pi)^3 2E_{\mathbf{p}_k}}\right) (2\pi)^4 \delta^4(p_1 - \cdots - p_n) \sum_{\mathrm{int}} |\mathcal{M}_{\mathrm{fi}}|^2, \tag{A.21}$$

where $\mathcal{Z} = \prod_j n_j!$ is a statistical factor accounting for the presence of n_j identical particles of type j in the final state, and the sum 'int' is over internal degrees of freedom such as spin and color.

Scattering Cross Sections

For the reaction $1 + 2 \to 3 + \cdots n$, the cross section σ is the transition rate per incident flux. The incident flux f_{inc} can be represented as

$$f_{\mathrm{inc}} = 4E_1 E_2 |\mathbf{v}_1 - \mathbf{v}_2| = 4\sqrt{(p_1 \cdot p_2)^2 - m_1^2 m_2^2}, \tag{A.22}$$

and the cross section becomes

$$\sigma = \frac{1}{\mathcal{Z}} \frac{1}{4\sqrt{(p_1 \cdot p_2)^2 - m_1^2 m_2^2}}$$

$$\times \int \left(\prod_{k=3}^{n} \frac{d^3 p_k}{(2\pi)^3 2E_{\mathbf{p}_k}}\right) (2\pi)^4 \delta^4(p_1 + p_2 - \cdots - p_n) \sum_{\mathrm{int}} |\mathcal{M}_{\mathrm{fi}}|^2. \tag{A.23}$$

Gamma, Psi, and Beta Functions

$$\Gamma(z) = \int_0^\infty dt\, e^{-t}\, t^{z-1} \qquad (\mathrm{Re}\, z > 0),$$

$$\Gamma(z+1) = z\Gamma(z) = z(z-1)\Gamma(z-1) = \cdots = z! \qquad (z\ \text{integer}),$$

$$\Gamma\left(-n + \frac{\epsilon}{2}\right) = \frac{(-)^n}{n!} \left[\frac{2}{\epsilon} + \psi(n+1) + \mathcal{O}(\epsilon)\right] \qquad (n\ \text{integer}),$$

$$d\Gamma(z)/dz = \Gamma(z)\psi(z) \quad \text{where} \quad \psi(z+1) = \psi(z) + 1/z,$$

$$\psi(1) = -\gamma = -\lim_{n\to\infty}\left(1 + \frac{1}{2} + \cdots + \frac{1}{n} - \ln n\right) \simeq -0.5772,$$

$$d\psi(z+1)/dz \equiv \psi'(z+1) = \psi'(z) - 1/z^2 \quad \text{with} \quad \psi'(1) = \pi^2/6$$

$$B(z,w) = \frac{\Gamma(n)\Gamma(m)}{\Gamma(n+m)} = 2\int_0^1 dt\, \frac{t^{2z-1}}{(t^2+1)^{z+w}} \qquad (\mathrm{Re}\, z,\ \mathrm{Re}\, w > 0). \tag{A.24}$$

A.3 Mathematics in d Dimensions

Dirac Algebra

In dimensional regularization one employs a metric $\eta_{\mu\nu}$ corresponding to a spacetime of continuous dimension d, but to maintain certain $d = 4$ properties of the Dirac matrices.[1] In the following, I_d is a diagonal d-dimensional matrix with $\mathrm{Tr}\, I_d = 4$ and $\epsilon \equiv 4 - d$.

$$\eta_\mu{}^\mu = d$$
$$\{\gamma_\mu, \gamma_\nu\} = 2\eta_{\mu\nu} I_d$$
$$\gamma_\mu \gamma^\mu = d\, I_d$$
$$\gamma_\mu \not{p} \gamma^\mu = (\epsilon - 2)\, \not{p}$$
$$\gamma_\mu \not{p} \not{q} \gamma^\mu = 4p \cdot q I_d - \epsilon \not{p} \not{q}$$
$$\gamma_\mu \not{p} \not{q} \not{r} \gamma^\mu = -2 \not{r} \not{q} \not{p} + \epsilon \not{p} \not{q} \not{r}$$
$$\not{p} \not{q} \not{r} + \not{r} \not{q} \not{p} = +2p \cdot q \not{r} + 2q \cdot r \not{p} - 2p \cdot r \not{q}. \tag{A.25}$$

Feynman Parameterization

$$\frac{1}{a^n b^m} = \frac{\Gamma(n+m)}{\Gamma(n)\Gamma(m)} \int_0^1 dx \, \frac{x^{n-1}(1-x)^{m-1}}{[ax + b(1-x)]^{n+m}} \qquad (n, m > 0)$$
$$\frac{1}{abc} = 2 \int_0^1 x \, dx \int_0^1 dy \, \frac{1}{[a(1-x) + bxy + cx(1-y)]^3}. \tag{A.26}$$

Feynman Integrals

For the integrals below we constrain $n \geq 1$ and denote $i\epsilon$ as the infinitesimal Feynman parameter.

$$\int \frac{d^d p}{(2\pi)^d} \frac{1}{((p-q)^2 - m^2 + i\epsilon)^n} = (-)^n \frac{i}{(4\pi)^{\frac{d}{2}}} \frac{\Gamma(n - \frac{d}{2})}{\Gamma(n)} \frac{1}{(m^2)^{n - \frac{d}{2}}} \tag{A.27}$$

$$\int \frac{d^d p}{(2\pi)^d} \frac{p^\mu}{((p-q)^2 - m^2 + i\epsilon)^n} = (-)^n \frac{i}{(4\pi)^{\frac{d}{2}}} \frac{\Gamma(n - \frac{d}{2})}{\Gamma(n)} \frac{q^\mu}{(m^2)^{n - \frac{d}{2}}} \tag{A.28}$$

$$\int \frac{d^d p}{(2\pi)^d} \frac{p^\mu p^\nu}{((p-q)^2 - m^2 + i\epsilon)^n} = (-)^n \frac{i}{(4\pi)^{\frac{d}{2}}} \frac{1}{\Gamma(n)} \left[\Gamma\left(n - \frac{d}{2}\right) \frac{q^\mu q^\nu}{(m^2)^{n - \frac{d}{2}}} \right.$$
$$\left. - \frac{1}{2} \eta^{\mu\nu} \Gamma\left(n - \frac{d}{2} - 1\right) \frac{1}{(m^2)^{n - \frac{d}{2} - 1}} \right]$$
$$\tag{A.29}$$

---

[1] Since there exists no well-defined continuation of the γ_5 matrix to d-dimensions, we restrict all γ_5 relations to $d = 4$.

$$\int \frac{d^d p}{(2\pi)^d} \frac{p^\mu p^\nu p^\lambda}{((p-q)^2 - m^2 + i\epsilon)^n}$$

$$= (-)^n \frac{i}{(4\pi)^{\frac{d}{2}}} \frac{1}{\Gamma(n)} \left[\Gamma\left(n - \frac{d}{2}\right) \frac{q^\mu q^\nu q^\lambda}{(m^2)^{n-\frac{d}{2}}} \right.$$

$$\left. - \Gamma\left(n - \frac{d}{2} - 1\right) \frac{\frac{1}{2}(\eta^{\mu\nu} q^\lambda + \eta^{\mu\lambda} q^\nu + \eta^{\nu\lambda} q^\mu)}{(m^2)^{n-\frac{d}{2}-1}} \right] \tag{A.30}$$

$$\int \frac{d^d p}{(2\pi)^d} \frac{p^\mu p^\nu p^\lambda p^\sigma}{((p-q)^2 - m^2 + i\epsilon)^n}$$

$$= (-)^n \frac{i}{(4\pi)^{\frac{d}{2}}} \frac{1}{\Gamma(n)} \left[\Gamma\left(n - \frac{d}{2}\right) \frac{q^\mu q^\nu q^\lambda q^\sigma}{(m^2)^{n-\frac{d}{2}}} \right.$$

$$- \Gamma\left(n - \frac{d}{2} - 1\right) \frac{\frac{1}{2}(\eta^{\mu\nu} q^\lambda q^\sigma + \eta^{\lambda\sigma} q^\mu q^\nu + 4\,\text{perm.})}{(m^2)^{n-\frac{d}{2}-1}}$$

$$\left. + \Gamma\left(n - \frac{d}{2} - 2\right) \frac{\frac{1}{4}(\eta^{\mu\nu}\eta^{\lambda\sigma} + \eta^{\mu\sigma}\eta^{\nu\lambda} + \eta^{\nu\sigma}\eta^{\mu\lambda})}{(m^2)^{n-\frac{d}{2}-2}} \right]. \tag{A.31}$$

For the next integrals, we define the denominator function

$$\mathcal{D} \equiv m_1^2 x + m_2^2 (1 - x) - q^2 x(1 - x) - i\epsilon \ ,$$

and constrain $n_1, n_2 \geq 1$.

$$\int \frac{d^d p}{(2\pi)^d} \frac{1}{\left[(p-q)^2 - m_1^2 + i\epsilon\right]^{n_1} \left[p^2 - m_2^2 + i\epsilon\right]^{n_2}}$$

$$= (-1)^{n_1+n_2} \frac{i}{(4\pi)^{d/2}} \frac{\Gamma(n_1 + n_2 - d/2)}{\Gamma(n_1)\Gamma(n_2)} \int_0^1 dx \, \frac{x^{n_1-1}(1-x)^{n_2-1}}{\mathcal{D}^{n_1+n_2-d/2}} \tag{A.32}$$

$$\int \frac{d^d p}{(2\pi)^d} \frac{p^\mu}{\left[(p-q)^2 - m_1^2 + i\epsilon\right]^{n_1} \left[p^2 - m_2^2 + i\epsilon\right]^{n_2}}$$

$$= (-1)^{n_1+n_2} q^\mu \frac{i}{(4\pi)^{d/2}} \frac{\Gamma(n_1 + n_2 - d/2)}{\Gamma(n_1)\Gamma(n_2)} \int_0^1 dx \, \frac{x^{n_1}(1-x)^{n_2-1}}{\mathcal{D}^{n_1+n_2-d/2}} \tag{A.33}$$

$$
\int \frac{d^d p}{(2\pi)^d} \frac{p^\mu p^\nu}{\left[(p-q)^2 - m_1^2 + i\epsilon\right]^{n_1} \left[p^2 - m_2^2 + i\epsilon\right]^{n_2}}
$$

$$
= \frac{i}{(4\pi)^{d/2}} \frac{(-1)^{n_1+n_2}}{\Gamma(n_1)\Gamma(n_2)} \left[q^\mu q^\nu \Gamma(n_1+n_2-d/2) \int_0^1 dx \, \frac{x^{n_1+1}(1-x)^{n_2-1}}{\mathcal{D}^{n_1+n_2-d/2}} \right.
$$

$$
\left. - \frac{\eta^{\mu\nu}}{2} \Gamma(n_1+n_2-1-d/2) \int_0^1 dx \, \frac{x^{n_1-1}(1-x)^{n_2-1}}{\mathcal{D}^{n_1+n_2-1-d/2}} \right]
\tag{A.34}
$$

$$
\int \frac{d^d p}{(2\pi)^d} \frac{p^\mu p^\nu p^\lambda}{\left[(p-q)^2 - m_1^2 + i\epsilon\right]^{n_1} \left[p^2 - m_2^2 + i\epsilon\right]^{n_2}}
$$

$$
= \frac{i}{(4\pi)^{d/2}} \frac{(-1)^{n_1+n_2}}{\Gamma(n_1)\Gamma(n_2)} \left[q^\mu q^\nu q^\lambda \Gamma(n_1+n_2-d/2) \int_0^1 dx \, \frac{x^{n_1+2}(1-x)^{n_2-1}}{\mathcal{D}^{n_1+n_2-d/2}} \right.
$$

$$
\left. - \frac{1}{2}(\eta^{\mu\nu}q^\lambda + \eta^{\mu\lambda}q^\nu + \eta^{\nu\lambda}q^\mu)\Gamma(n_1+n_2-1-d/2) \int_0^1 dx \, \frac{x^{n_1}(1-x)^{n_2-1}}{\mathcal{D}^{n_1+n_2-1-d/2}} \right]
$$

$$
\tag{A.35}
$$

$$
\int \frac{d^d p}{(2\pi)^d} \frac{p^\mu p^\nu p^\lambda p^\sigma}{\left[(p-q)^2 - m_1^2 + i\epsilon\right]^{n_1} \left[p^2 - m_2^2 + i\epsilon\right]^{n_2}}
$$

$$
= \frac{i}{(4\pi)^{d/2}} \frac{(-1)^{n_1+n_2}}{\Gamma(n_1)\Gamma(n_2)} \left[q^\mu q^\nu q^\lambda q^\sigma \, \Gamma(n_1+n_2-d/2) \int_0^1 dx \, \frac{x^{n_1+3}(1-x)^{n_2-1}}{\mathcal{D}^{n_1+n_2-d/2}} \right.
$$

$$
- \frac{1}{2}(\eta^{\mu\nu}q^\lambda q^\sigma + \eta^{\mu\lambda}q^\nu q^\sigma + 4\,\text{perm.})\Gamma(n_1+n_2-1-d/2)\int_0^1 dx \frac{x^{n_1+1}(1-x)^{n_2-1}}{\mathcal{D}^{n_1+n_2-1-d/2}}
$$

$$
\left. + \frac{1}{4}(\eta^{\mu\nu}\eta^{\lambda\sigma} + \eta^{\mu\lambda}\eta^{\nu\sigma} + \eta^{\mu\sigma}\eta^{\nu\lambda})\Gamma(n_1+n_2-2-d/2)\int_0^1 dx \frac{x^{n_1-1}(1-x)^{n_2-1}}{\mathcal{D}^{n_1+n_2-2-d/2}} \right].
$$

$$
\tag{A.36}
$$

Appendix B Quantum Theory

In this appendix we review some of the basics of nonrelativistic and relativistic quantum mechanics, as well as a brief summary of quantum field theory.[1]

B.1 Nonrelativistic Quantum Mechanics

We here present some of the basic ideas from nonrelativistic quantum mechanics which are employed in the text.

Path Integral Propagator

The path integral is a technique to evaluate the quantum mechanical time evolution operator. Specifically, it offers an alternative way to evaluate the quantity

$$D(x_f, t_f; x_i, t_i) = \langle x_f \left| e^{-i\hat{H}(t_f - t_i)} \right| x_i \rangle \theta(t_2 - t_1). \tag{B.1}$$

This matrix element, generally called the *propagator*, is the amplitude for a particle located at position x_i at time t_i to be found at position x_f at subsequent time t_f. The propagator can also be written as a functional integral

$$D(x_f, t_f; x_i, t_i) = \int \mathcal{D}[x(t)]\, e^{iS[x(t)]} \tag{B.2}$$

where the integration is over all histories (i.e. paths) of the system which begin at spacetime point x_i, t_i and end at x_f, t_f. The paths are identified by specifying the coordinate x at each intermediate time t, so that the symbol $\mathcal{D}[x(t)]$ represents a sum over all such trajectories. The contribution of each path to the integral is weighted by the exponential involving the classical action

$$S[x(t)] = \int_{t_i}^{t_f} dt \left(\frac{m}{2}\dot{x}^2(t) - V(x(t)) \right) \tag{B.3}$$

which, since it depends on the detailed shape of $x(t)$, is a functional of the trajectory.[2] Although the validity of the path integral representation, Eq. (B.2), may not be obvious,

---

[1] Note that much of this appendix is based on similar material in [Don14].

[2] It is important to understand the difference between the concept of a function and that of a functional. A real-valued function involves the mapping from the space of real numbers onto themselves

$$\text{Reals} \longleftarrow [f : \text{Reals}] \ .$$

its correctness can be verified by beginning with Eq. (B.1) and breaking the time interval $t_f - t_i$ into N discrete steps of size $\epsilon = (t_f - t_i)/N$. Using the completeness relation

$$1 = \int_{-\infty}^{\infty} dx_n \, |x_n\rangle\langle x_n| \, ,$$

one can write Eq. (B.1) as

$$D(x_f, t_f; x_i, t_i) = \int_{-\infty}^{\infty} dx_{N-1} \cdots \int_{-\infty}^{\infty} dx_1$$
$$< x_N|e^{-i\epsilon H}|x_{N-1}><x_{N-1}|e^{-i\epsilon H}|x_{N-2}>$$
$$\cdots < x_1|e^{-i\epsilon H}|x_0 >, \tag{B.4}$$

where $x_0 \equiv x_i$, $x_N \equiv x_f$. In the limit of large N the time slices become infinitesimal, implying

$$< x_\ell \left|e^{-iH\epsilon}\right| x_{\ell-1} > \; = \; < x_\ell \left| e^{-i\epsilon\left(\frac{p^2}{2m}+V(x)\right)} \right| x_{\ell-1} >$$

$$= e^{-i\epsilon V(x_\ell)} \left\langle x_\ell \left| e^{-i\epsilon \frac{p^2}{2m}} \right| x_{\ell-1} \right\rangle + \mathcal{O}(\epsilon^2). \tag{B.5}$$

Inserting a complete set of momentum states and introducing a convergence factor $e^{-\kappa p^2}$ for the resulting integral over momentum, we have

$$\left\langle x_\ell \left| e^{-i\epsilon \frac{p^2}{2m}} \right| x_{\ell-1} \right\rangle = \lim_{\kappa \to 0} \int_{-\infty}^{\infty} \frac{dp}{2\pi} \, e^{ip\cdot(x_\ell - x_{\ell-1}) - i\epsilon \frac{p^2}{2m} - \kappa p^2}$$

$$= \left(\frac{m}{2\pi i\epsilon}\right)^{\frac{3}{2}} e^{i\frac{m}{2\epsilon}(x_\ell - x_{\ell-1})^2}. \tag{B.6}$$

Taking the continuum limit we obtain

$$D\left(x_f, t_f; x_i, t_i\right) = \lim_{N\to\infty} \left(\frac{m}{2\pi i\epsilon}\right)^{\frac{3N}{2}} \left[\prod_{n=1}^{N-1} \int_{-\infty}^{\infty} dx_n\right] e^{i\sum_{\ell=1}^{N}\left(\frac{m}{2}\frac{(x_\ell - x_{\ell-1})^2}{\epsilon} - \epsilon V(x_\ell)\right)}. \tag{B.7}$$

It is clear then that we can make connection with Eq. (B.2) by identifying each path with the sequence of locations $(x_1, \ldots, x_{N-1})$ at times $\epsilon, 2\epsilon, \ldots, (N-1)\epsilon$. Integration over these intermediate positions is what is meant by the symbol $\int \mathcal{D}\,[x(t)]$, viz.

$$\int \mathcal{D}\,[x(t)] \equiv \lim_{N\to\infty} \left(\frac{m}{2\pi i\epsilon}\right)^{3N/2} \prod_{n=1}^{N-1} \int_{-\infty}^{\infty} dx_n \, . \tag{B.8}$$

Each trajectory has an associated exponential factor $\exp iS\,[x(t)]$, where the quantity

$$S\,[x(t)] = \sum_{\ell=1}^{N} \epsilon \left(\frac{m\,(x_\ell - x_{\ell-1})^2}{2\,\epsilon^2} - V(x_\ell)\right) \tag{B.9}$$

becomes the classical action in the limit $N \to \infty$. We have thus demonstrated the equivalence of the operator (Eq. (B.1)) and path integral (Eq. B.2)) representations of the

---

On the other hand, a real-valued functional such as $S\,[x(t)]$ is a mapping from the space of functions $x(t)$ onto real numbers

$$\text{Reals} \longleftarrow [S : x(t)] \; .$$

propagator.[3] It is important to realize that in the latter all quantities are *classical* – no operators are involved.

The path integral propagator contains a great deal of information, and there are a variety of techniques for extracting it. For example, the spatial wavefunctions and energies are all present, as can be seen by inserting a complete set of energy eigenstates $\{|\, n\rangle\}$ satisfying

$$\left(\frac{-\nabla_2^2}{2m} + V(x_2) \right) \psi_n(x_2) = E_n \psi_n(x_2) \tag{B.11}$$

into the definition of the propagator given in Eq. (B.1)

$$D(x_f, t_f; x_i, t_i) = \begin{cases} \sum_{n=0}^{\infty} \psi_n(x_f) \psi_n^*(x_i) e^{-iE_n(t_f - t_i)} & t_f > t_i \\ 0 & t_f < t_i. \end{cases} \tag{B.12}$$

Note that the propagator is the Green's function for the Schrödinger equation, satisfying

$$\left(\frac{-\nabla_2^2}{2m} + V(x_2) - i\frac{\partial}{\partial t_2} \right) D(x_2, t_2; x_1, t_1) = -i\delta(x_2 - x_1)\delta(t_2 - t_1). \tag{B.13}$$

This is clear since both for $t_2 > t_1$ and $t_2 < t_1$ we have

$$\left(\frac{-\nabla_2^2}{2m} + V(x_2) - i\frac{\partial}{\partial t_2} \right) D(x_2, t_2; x_1, t_1) = 0 \tag{B.14}$$

while, integrating both sides of Eq. (B.12) over t_2 from $t_1 - \epsilon$ to $t_1 + \epsilon$ we have

$$\lim_{\epsilon \to 0} \int_{t_1 - \epsilon}^{t_1 + \epsilon} dt_2 - i\delta(x_2 - x_1)\delta(t_2 - t_1) = -i\delta(x_2 - x_1) \tag{B.15}$$

on the RHS and

$$\lim_{\epsilon \to 0} \int_{t_1 - \epsilon}^{t_1 + \epsilon} dt_2 \left(\frac{-\nabla_2^2}{2m} + V(x_2) - i\frac{\partial}{\partial t_2} \right) D(x_2, t_2; x_1, t_1)$$

$$= \left[2\epsilon \left(\frac{-\nabla_2^2}{2m} + V(x_2) \right) D(x_2, t_1; x_1, t_1) - i(D_F(x_2, t_1 + \epsilon; x_1, t_1) \right.$$

$$\left. - D(x_2, t_1 - \epsilon; x_1, t_1)) \right] = -iD(x_2, t_1; x_1, t_1)$$

$$= -i\sum_n \psi_n(x_2)\psi_n(x_1) = -i\delta(x_2 - x_1), \tag{B.16}$$

---

[3] For completeness, we note that by combining Eqs. (B.5) – (B.8), one can also write the propagator in a corresponding Hamiltonian path integral representation

$$D(x_f, t_f; x_i, t_i) = \lim_{N \to \infty} \int \frac{dp_0}{(2\pi)^3} dx_1 \frac{dp_1}{(2\pi)^3} dx_2 \cdots dx_{N-1} \frac{dp_{N-1}}{(2\pi)^3}$$

$$\times e^{i\sum_{\ell=1}^{N} \left(p_\ell \cdot (x_\ell - x_{\ell-1}) - \left(\frac{p_\ell^2}{2m} + V(x_\ell) \right)\epsilon \right)}$$

$$\equiv \int \mathcal{D}[x(t)]\, \mathcal{D}[p(t)]\, e^{i\int dt\, (p\cdot\dot{x} - H(p,x))}. \tag{B.10}$$

This form is useful when one is dealing with non-cartesian variables or with constrained systems.

where we have used Completeness.

Note that for a free particle we have, from Eq. (B.12)

$$D^{(0)}(x_f, t_f; x_i, t_i) = \theta(t_f - t_f) \int \frac{dp}{(2\pi)^3} \exp(i(p \cdot (x_f - x_i) - i\frac{p^2}{2m}(t_f - t_i))$$

$$= \theta(t_f - t_i) \left(\frac{m}{2\pi i(t_f - t_f)}\right)^{\frac{3}{2}} \exp\left(i\frac{m(x_f - x_i)^2}{2(t_f - t_i)}\right). \qquad (B.17)$$

Alternatively, we can evaluate the free propagator by use of the path integral. In this case we have

$$D^{(0)}(x_f, t_f; x_i, t_i) = \lim_{N \to \infty} \left(\frac{m}{2\pi i\epsilon}\right)^{\frac{3N}{2}} \prod_{i=1}^{N-1} \left(\int dx_i\right) \exp i\sum_{j=1}^{N} \frac{m}{2}(x_i - x_{i-1})^2. \qquad (B.18)$$

The integrations can be performed sequentially

$$\int_{-\infty}^{\infty} dx_1 \exp i\frac{m}{2\epsilon}\left((x_1 - x_0)^2 + (x_2 - x_1)^2\right) = \left(\frac{2\pi i\epsilon}{2m}\right)^{\frac{3}{2}} \exp i\frac{m}{2 \cdot 2\epsilon}(x_2 - x_0)^2$$

$$\int_{-\infty}^{\infty} dx_2 \exp i\frac{m}{2\epsilon}\left(\frac{1}{2}(x_2 - x_0)^2 + (x_3 - x_2)^2\right) = \left(\frac{2\pi i\epsilon \cdot 2}{3m}\right)^{\frac{3}{2}} \exp i\frac{m}{3 \cdot 2\epsilon}(x_3 - x_0)^2$$

$$\vdots$$

$$\int_{-\infty}^{\infty} dx_{N-1} \exp i\frac{m}{2\epsilon}\left(\frac{1}{N-1}(x_{N-1} - x_0)^2 + (x_N - x_{N-1})^2\right)$$

$$= \left(\frac{2\pi i\epsilon(N-1)}{Nm}\right)^{\frac{3}{2}} \exp i\frac{m}{N \cdot 2\epsilon}(x_N - x_0)^2. \qquad (B.19)$$

We have then

$$D^{(0)}(x_f, t_f; x_i, t_i) = \lim_{N \to \infty} \left(\frac{m}{2\pi iN\epsilon}\right)^{\frac{3}{2}} \exp \frac{im(x_N - x_0)^2}{2N\epsilon}$$

$$= \left(\frac{m}{2\pi i(t_f - t_i)}\right)^{\frac{3}{2}} \exp \frac{im(x_f - x_i)^2}{2(t_f - t_i))} \qquad (B.20)$$

in agreement with the result calculated in Eq. (B.17).

Time-Dependent Perturbation Theory

One can derive the time-dependent perturbation expansion within the path integral formalism. We begin with

$$<x'\left|\hat{U}(t, 0)\right|x> = \int \mathcal{D}[x(t)] \exp i\int_0^t dt' \left(\frac{1}{2}m\dot{x}^2(t') - V(x(t'))\right)$$

$$= \int \mathcal{D}[x(t)] \exp i\int_0^t dt' \frac{1}{2}m\dot{x}^2(t')$$

$$\times \left(1 - i\int_0^t dt_1 V(x(t_1)) + \frac{1}{2!}\left(-i\int_0^t dt_1 V(x(t_1))\right)^2 + \cdots\right).$$

$$(B.21)$$

Note that here since we are dealing only with *classical* quantities – no operators – we do not have to worry about lack of commutativity and can expand the exponential involving the potential straightforwardly. The first term in the expansion is recognized to be the free particle propagator

$$\left\langle x' \left| \hat{U}^{(0)}(t,0) \right| x \right\rangle = \int \mathcal{D}\left[x(t)\right] \exp i \int_0^t dt' \frac{1}{2} m\dot{x}^2(t'). \tag{B.22}$$

For the term linear in the potential, we interchange orders of integration over time t_1 and paths $x(t)$ yielding

$$- i \int_0^t dt_1 \int \mathcal{D}\left[x(t)\right] V\left(x(t_1)\right) \exp i \int_0^t \frac{1}{2} m\dot{x}^2(t') dt' \tag{B.23}$$

and in the path integration we separate the paths into two pieces:

i) a path which begins at x at time $t = 0$ and connects in all possible ways with point $x'' = x(t_1)$ at time t_1; and

ii) a path which begins at $x'' = x(t_1)$ at time t_1 and connects in all possible ways with point x' at time t.

Of course, x'' is not fixed but must take on all possible values, so that the linear term becomes

$$- i \int_0^t dt_1 \int_{-\infty}^{\infty} dx'' \left\langle x' \left| \hat{U}^{(0)}(t,t_1) \right| x'' \right\rangle V(x'') \left\langle x'' \left| \hat{U}^{(0)}(t_1,0) \right| x \right\rangle$$

$$= (\text{by completeness}) \ - i \int_0^t dt_1 \left\langle x' \left| \hat{U}^{(0)}(t,t_1) \hat{V} \, \hat{U}^{(0)}(t_1,0) \right| x \right\rangle. \tag{B.24}$$

Likewise we can analyze the quadratic term. In this case, however, we divide the time integration into two regions depending on whether $t_1 > t_2$ or $t_1 < t_2$:

$$\frac{(-i)^2}{2!} \int_0^t dt_2 \int_0^{t_2} dt_1 \int_{-\infty}^{\infty} dx'' \int_{-\infty}^{\infty} dx''' \left\langle x' \left| \hat{U}^{(0)}(t,t_2) \right| x''' \right\rangle V(x''')$$

$$\times \left\langle x''' \left| \hat{U}^{(0)}(t_2,t_1) \right| x'' \right\rangle V(x'') \left\langle x'' \left| \hat{U}^{(0)}(t_1,0) \right| x \right\rangle$$

$$+ \frac{(-i)^2}{2!} \int_0^t dt_2 \int_{t_2}^{t} dt_1 \int_{-\infty}^{\infty} dx'' \int_{-\infty}^{\infty} dx''' \left\langle x' \left| \hat{U}^{(0)}(t,t_1) \right| x'' \right\rangle V(x'')$$

$$\times \left\langle x'' \left| \hat{U}^{(0)}(t_1,t_2) \right| x''' \right\rangle V(x''') \left\langle x''' \left| \hat{U}^{(0)}(t_2,0) \right| x \right\rangle. \tag{B.25}$$

In the second term we can change the order of integration and if we now interchange the identities of the variables $t_2 \leftrightarrow t_1$ the second term in Eq. (B.25) is seen to be identical to the first, canceling the 2!. The quadratic piece of the expansion becomes then

$$(-i)^2 \int_0^t dt_2 \int_0^{t_2} dt_1 \left\langle x' \left| \hat{U}^{(0)}(t,t_2) \hat{V}(t_2) \, \hat{U}^{(0)}(t_2,t_1) \hat{V}(t_1) \, \hat{U}^{(0)}(t_1,0) \right| x \right\rangle \tag{B.26}$$

and we recognize the general form of the propagator to be

$$\left\langle x' \left| \hat{U}(t,0) \right| x \right\rangle = \left\langle x' \left| \hat{U}^{(0)}(t,0) - i \int_0^t dt_1 \, U^{(0)}(t,t_1) \hat{V}(t_1) \hat{U}^{(0)}(t_1,0) \right. \right.$$
$$\left. \left. + (-i)^2 \int_0^t dt_2 \int_0^{t_2} dt_1 \, \hat{U}^{(0)}(t,t_2) \hat{V}(t_2) \hat{U}^{(0)}(t_2,t_1) \hat{V}(t_1) \hat{U}^{(0)}(t_1,0) + \cdots \right| x \right\rangle \quad (B.27)$$

which is the usual result.

Time-Independent Perturbation Theory

In the case that the potential is time-independent, it is useful to work in frequency space. We write the time development operator (assuming $\hat{V}$ is time-independent) as

$$\hat{U}(t,0)\theta(t) = e^{-i(\hat{H}_0 + \hat{V})t} \theta(t)$$
$$= \int_{-\infty}^{\infty} \frac{d\omega}{2\pi} e^{-i\omega t} \frac{i}{\omega - \hat{H}_0 - \hat{V} + i\epsilon} \equiv \int_{-\infty}^{\infty} \frac{d\omega}{2\pi} e^{-i\omega t} \hat{\mathcal{K}}(\omega) \quad (B.28)$$

where we have defined the "full propagator" in frequency space as

$$\hat{\mathcal{K}}(\omega) = \frac{i}{\omega - \hat{H}_0 - \hat{V} + i\epsilon} \quad (B.29)$$

We may generate a perturbative expansion by use of the operator identity

$$\frac{1}{\hat{A} - \hat{B}} = \frac{1}{\hat{A}} + \frac{1}{\hat{A}} \hat{B} \frac{1}{\hat{A}} + \frac{1}{\hat{A}} \hat{B} \frac{1}{\hat{A}} \hat{B} \frac{1}{\hat{A}} + \cdots \quad (B.30)$$

[Note: The proof of Eq. (B.30) is provided by multiplication by $\hat{A} - \hat{B}$

$$(\hat{A} - \hat{B}) \left[\frac{1}{\hat{A}} + \frac{1}{\hat{A}} \hat{B} \frac{1}{\hat{A}} + \frac{1}{\hat{A}} \hat{B} \frac{1}{\hat{A}} \hat{B} \frac{1}{\hat{A}} + \cdots \right]$$
$$= 1 - \hat{B} \frac{1}{\hat{A}} + \hat{B} \frac{1}{\hat{A}} - \hat{B} \frac{1}{\hat{A}} \hat{B} \frac{1}{\hat{A}} + \hat{B} \frac{1}{\hat{A}} \hat{B} \frac{1}{\hat{A}} - \cdots = 1]. \quad (B.31)$$

The full propagator can then be written as

$$-i\hat{\mathcal{K}}(\omega) = \frac{1}{\omega - \hat{H}_0 - \hat{V} + i\epsilon} = \frac{1}{\omega - \hat{H}_0 + i\epsilon} + \frac{1}{\omega - \hat{H}_0 + i\epsilon} \hat{V} \frac{1}{\omega - \hat{H}_0 + i\epsilon}$$
$$+ \frac{1}{\omega - \hat{H}_0 + i\epsilon} \hat{V} \frac{1}{\omega - \hat{H}_0 + i\epsilon} \hat{V} \frac{1}{\omega - \hat{H}_0 + i\epsilon} + \cdots$$
$$= -i \left(\hat{K}^{(0)}(\omega) + \hat{K}^{(0)}(\omega)(-i\hat{V})\hat{K}^{(0)}(\omega) \right.$$
$$\left. + \hat{K}^{(0)}(\omega)(-i\hat{V})\hat{K}^{(0)}(\omega)(-i\hat{V})\hat{K}^{(0)}(\omega) + \cdots \right) \quad (B.32)$$

where

$$\hat{K}^{(0)}(\omega) = \frac{i}{\omega - \hat{H}_0 + i\epsilon} \quad (B.33)$$

is the free propagator. Returning to a real time rather than frequency representation we have

$$\hat{U}(t,0) = \int_{-\infty}^{\infty} \frac{d\omega}{2\pi} e^{-i\omega t} \hat{\mathcal{K}}(\omega)$$

$$= \hat{U}^{(0)}(t,0) - i \int_0^t dt_1 \hat{U}^{(0)}(t,t_1) \hat{V} \hat{U}^{(0)}(t_1,0)$$

$$+ (-i)^2 \int_0^t dt_2 \int_0^{t_2} dt_1 \hat{U}^{(0)}(t,t_2) \hat{V} \hat{U}^{(0)}(t_2,t_1) \hat{V} \hat{U}^{(0)}(t_1,0) + \cdots \tag{B.34}$$

where

$$\hat{U}^{(0)}(t,0) = \int_{-\infty}^{\infty} \frac{d\omega}{2\pi} e^{-i\omega t} \frac{i}{\omega - \hat{H}_0 + i\epsilon} = \theta(t) e^{-i\hat{H}_0 t} \tag{B.35}$$

is the free propagator. (Eq. (B.35) follows via, e.g.

$$\int_{-\infty}^{\infty} \frac{d\omega}{2\pi} e^{-i\omega t} \frac{1}{\omega - \hat{H}_0 + i\epsilon} \hat{V} \frac{1}{\omega - \hat{H}_0 + i\epsilon}$$

$$= \int_{-\infty}^{\infty} \frac{d\omega}{2\pi} \int_{-\infty}^{\infty} \frac{d\omega'}{2\pi} \int_{-\infty}^{\infty} dt_1 e^{-i\omega(t-t_1)} e^{-i\omega' t_1} \frac{1}{\omega - \hat{H}_0 + i\epsilon} \hat{V} \frac{1}{\omega' - \hat{H}_0 + i\epsilon}$$

$$= \int_{-\infty}^{\infty} dt_1 \hat{U}^{(0)}(t,t_1) \hat{V} \hat{U}^{(0)}(t_1,0) = \int_0^t dt_1 e^{-i\hat{H}_0(t-t_1)} \hat{V} e^{-i\hat{H}_0 t_1} \tag{B.36}$$

and similarly for the other terms.) Then for a transition from state $|i\rangle$ at time $t = 0$ to state $|f\rangle$ at (later) time t we have

$$\mathrm{Amp}_{fi}(t) = \left\langle f \left| \hat{U}(t,0) \right| i \right\rangle = i \int_{-\infty}^{\infty} \frac{d\omega}{2\pi} e^{-i\omega t} \left\{ \delta_{fi} \frac{1}{\omega - E_i + i\epsilon} \right.$$

$$+ \frac{1}{\omega - E_f + i\epsilon} \left\langle f|\hat{V}|i \right\rangle \frac{1}{\omega - E_i + i\epsilon} + \sum_a \frac{1}{\omega - E_f + i\epsilon} \left\langle f|\hat{V}|a \right\rangle$$

$$\times \frac{1}{\omega - E_a + i\epsilon} \left\langle a|\hat{V}|i0 \right\rangle \frac{1}{\omega - E_i + i\epsilon} + \cdots \left. \right\}$$

$$= \delta_{fi} e^{-iE_i t} + \frac{\left\langle f|\hat{V}|i \right\rangle}{E_f - E_i + i\epsilon} \left(e^{-iE_f t} - e^{-iE_i t} \right) + \cdots \tag{B.37}$$

which is the form which follows from Eq. (B.27) in the case that $\hat{V}$ is time-independent, which can be seen from the result

$$\int \frac{d\omega}{2\pi} e^{-i\omega(t_2 - t_1)} \left\langle \boldsymbol{x}_2 \left| \frac{i}{\omega - \hat{H}_0 + i\epsilon} \right| \boldsymbol{x}_1 \right\rangle$$

$$= \theta(t_2 - t_1) \left\langle \boldsymbol{x}_2 | e^{-i\hat{H}_0(t_2 - t_1)} | \boldsymbol{x}_1 \right\rangle = D^{(0)}(\boldsymbol{x}_2, t_2; \boldsymbol{x}_1, t_1). \tag{B.38}$$

B.2 Relativistic Quantum Mechanics

We now present some of the basics of relativistic quantum mechanics used in the text.

Green's Function Methods

Propagator methods are also useful in relativistic quantum mechanics. As in the nonrelativistic case the propagator is the Green's function. For a free spinless system we use the Klein–Gordon equation so that

$$(\Box_2 + m^2)D^{(0)}(x_2, x_1) = -i\delta^4(x_2 - x_1). \tag{B.39}$$

Then the free propagator is

$$D^{(0)}(x_2, x_1) = \left\langle x_2 \left| \frac{-i}{\Box + m^2} \right| x_1 \right\rangle = \int \frac{d^4q}{(2\pi)^4} e^{iq\cdot(x_2-x_1)} \frac{i}{q^2 - m^2 + i\epsilon} \tag{B.40}$$

If the spinless particle has charge e then we must make the "minimal substitution" replacement

$$\partial_\mu \rightarrow \partial_\mu + ieA_\mu$$

whereby the Green's function satisfies

$$(\Box_2 + ie\{\partial_\mu, A^\mu\} - e^2 A_\mu A^\mu) + m^2)D(x_2, x_1) = -i\delta^4(x_2 - x_1) \tag{B.41}$$

and the propagator becomes

$$D(x_2, x_1) = \left\langle x_2 \left| \frac{-i}{\Box_2 + ie\{\partial_\mu, A^\mu\} - e^2 A_\mu A^\mu) + m^2} \right| x_1 \right\rangle \tag{B.42}$$

which can be evaluated perturbatively, yielding

$$D(x_2, x_1) = D^{0)}(x_2, x_1) + \int d^4z_1 D^{(0)}(x_2, z)(ie\{A_\mu(z), \partial_z^\mu\} - e^2 A^2(z))D^{(0)}(z, x_1)$$

$$+ \int d^4z \int d^4y D^{(0)}(x_2, z)(ie\{A_\mu(z), \partial_z^\mu\})D^{(0)}(z, y)(ie\{A_\nu(y), \partial_y^\nu\})D^{(0)}(y, x_1) + \cdots . \tag{B.43}$$

In momentum space this result yields the usual Feynman rules:

a) propagator for particle with four-momentum q:

$$\frac{i}{q^2 - m^2 + i\epsilon};$$

b) one-photon vertex between particles with four-momenta p_1, p_2:

$$-ie\epsilon \cdot (p_1 + p_2); \text{ and}$$

c) two-photon vertex:

$$2ie^2 \epsilon_1 \cdot \epsilon_2.$$

In the spin 1/2 case we have the free particle Green's function equation

$$(i\,\slashed{\partial}_2 - m)S^{(0)}(x_2,x_1) = i\delta^4(x_2 - x_1) \tag{B.44}$$

yielding the free propagator

$$S^{(0)}(x_2,x_1) = \int \frac{d^4q}{(2\pi)^4} e^{iq\cdot(x_2-x_1)} \frac{i}{\slashed{q} - m + i\epsilon} = \int \frac{d^4q}{(2\pi)^4} e^{iq\cdot(x_2-x_1)} \frac{i(\slashed{q} + m)}{q^2 - m^2 + i\epsilon}. \tag{B.45}$$

If the particle has charge e the Green's function satisfies

$$(i\,\slashed{\partial}_2 - e\,\slashed{A}(x_2) - m)S^{(0)}(x_2,x_1) = i\delta^4(x_2 - x_1) \tag{B.46}$$

so that the propagator satisfies

$$S(x_2,x_1) = \left\langle x_2 \left| \frac{i}{i\,\slashed{\partial} - m - e\,\slashed{A}} \right| x_1 \right\rangle \tag{B.47}$$

which has the perturbative solution

$$S(x_2,x_1) = S^{(0)}(x_2,x_1) - ie \int d^4z\, S^{(0}(x_2,z)\, \slashed{A}(z)S^{(0)}(z,x_1) + \cdots \tag{B.48}$$

and yields, in momentum space, the usual Feynman rules

a) propagator for particle with four-momentum q:

$$\frac{i(\slashed{q} + m)}{q^2 - m^2 + i\epsilon}$$

b) one-photon vertex between particles with four-momenta p_1, p_2:

$$ie\,\slashed{\epsilon}.$$

Quantum Field Theory

The relativistic results obtained via Green's function methods can also be obtained using quantum field theoretic techniques. In the case of nonrelativistic quantum mechanics we found that the theory could be formulated equally well via traditional methods and path integral techniques. In the relativistic analog we have seen how a perturbative expansion can be generated within the context of a Green's function approach. It is also possible to derive such results from a path integral framework, and we shall end our formal presentation by outlining how this is accomplished. The field theoretic methods which we describe are now widespread and underlie much of contemporary work in particle, nuclear, and condensed matter physics. Consequently there are a number of texts which survey this material in depth and it is not our purpose to attempt a detailed development of field theory via path integral methods. Nevertheless, it is interesting and important to see how the simple quantum mechanical techniques can be generalized and we thus present here a brief summary.

An advantage of the path integral to quantum mechanics is that it can be taken over rather directly to quantum field theory. An important difference is that instead of trajectories $x(t)$ which pick out a particular point in space at a given time, one must deal with fields $\phi(\boldsymbol{x}, t)$

which are defined at *all* points in space at a given time t. Instead of a sum $\int \mathcal{D}[x(t)]$ over trajectories one has a sum $\int [d\phi(x)]$ over all possible field configurations. Nevertheless, the analogy is rather direct. The formal transition from quantum mechanics to field theory can be accomplished by partitioning spacetime – both space *and* time – into a set of tiny four-dimensional cubes of volume $\delta t\, \delta x\, \delta y\, \delta z$. Within each cube one takes the field

$$\phi\left(x_i, y_j, z_k, t_\ell\right) \tag{B.49}$$

as a constant. Derivatives are defined in terms of differences between fields in neighboring blocks, e.g.

$$\partial_t \phi \Big|_{x_i, y_j, z_k, t_\ell} \simeq \frac{1}{\delta t}\left[\phi\left(x_i, y_j, z_k, t_\ell + \delta t\right) - \phi\left(x_i, y_j, z_k, t_\ell\right)\right]. \tag{B.50}$$

The Lagrangian density is easily found

$$\mathcal{L}\left(\phi, \partial_\mu \phi\right)\Big|_{x_i, y_j, z_k, t_\ell} \simeq \mathcal{L}\left[\phi\left(x_i, y_j, z_k, t_\ell\right), \partial_\mu \phi\left(x_i, y_j, z_k, t_\ell\right)\right] \tag{B.51}$$

and the corresponding action can be written as

$$S \simeq \sum_{ijk\ell} \delta x\, \delta y\, \delta z\, \delta t\, \mathcal{L}\left[\phi\left(x_i, x_j, z_k, t_\ell\right) \partial_\mu \phi\left(x_i, y_j, z_k, t_\ell\right)\right]. \tag{B.52}$$

The field theoretic analog of the path integral can then be constructed by summing over all possible field values in each cell

$$F \sim \prod_{ikj\ell} \int_{-\infty}^{\infty} d\phi\left(x_i, y_j, z_k, t_\ell\right) \exp iS\left[\phi\left(x_i, y_j, z_k, t_\ell\right), \partial_\mu \phi\left(x_i, y_j, z_k, t_\ell\right)\right]. \tag{B.53}$$

Formally, in the limit in which the cell size is taken to zero Eq. (B.53) is written as

$$F \sim \int [d\phi(x)] \exp iS\left[\phi, \partial_\mu \phi\right]. \tag{B.54}$$

By analogy with the quantum mechanical situation, since the time integration in the action, Eq. (B.52), is from $-\infty$ to $+\infty$, it is suggestive that this amplitude is to be identified with the vacuum-to-vacuum amplitude

$$< 0|0 > \sim N \int [d\phi(x)] \exp iS\left[\phi, \partial_\mu \phi\right]. \tag{B.55}$$

Generally quantum field theory is formulated in terms of vacuum expectation values of time ordered products of fields – the Green's functions of the theory – and it is conventional to fix the normalization constant N by dividing out the vacuum-to-vacuum amplitude. Thus we have

$$G^{(n)}\left(x_1, x_2 \ldots x_n\right) = \frac{< 0\,|T\left(\phi(x_1) \ldots \phi(x_n)\right)| 0 >}{< 0|0 >} \tag{B.56}$$

and by analogy to the quantum mechanical case one is led to the path integral definition

$$G^{(n)}\left(x_1, x_2, \ldots x_n\right) = \frac{\int [d\phi(x)] \exp iS\left[\phi, \partial_\mu \phi\right] \phi(x_1) \ldots \phi(x_n)}{\int [d\phi(x)] \exp iS\left[\phi, \partial_\mu \phi\right]}. \tag{B.57}$$

These Green's functions can most straightforwardly be evaluated by use of a generating functional

$$W[j] = N \int [d\phi(x)] \exp\left(iS[\phi, \partial_\mu\phi] + i \int d^4x\, j(x)\phi(x)\right) \tag{B.58}$$

in terms of which[4]

$$G^{(n)}(x_1, x_2, \ldots x_n) = (-i)^n \frac{1}{W[0]} \frac{\delta^n}{\delta j(x_1)\ldots\delta j(x_n)} W[j]\Big|_{j=0}. \tag{B.60}$$

For our purpose we shall deal only with the propagator or two-point function

$$G^{(2)}(x_1, x_2) = (-i)^2 \frac{1}{W[0]} \frac{\delta^2}{\delta j(x_1)\delta j(x_2)} W[j]\Big|_{j=0}. \tag{B.61}$$

Consider the application of this formalism to free scalar field theory. For simplicity we consider a neutral particle so that the field ϕ may be taken to be Hermitian. The Lagrangian density then is given by

$$\mathcal{L}^{(0)}(x) = \frac{1}{2}\partial_\mu\phi(x)\partial^\mu\phi(x) - \frac{1}{2}m^2\phi^2(x) . \tag{B.62}$$

That this is the appropriate form can be verified by use of the Euler–Lagrange relation

$$\partial_\mu \frac{\delta\mathcal{L}^{(0)}}{\delta\partial_\mu\phi(x)} - \frac{\delta\mathcal{L}^{(0)}}{\delta\phi(x)} = 0, \tag{B.63}$$

which yields the Klein–Gordon equation

$$(\Box + m^2)\phi(x) = 0. \tag{B.64}$$

The generating functional $W^{(0)}[j]$ is given by

$$W^{(0)}[j] = N \int [d\phi(x)] \exp i \int d^4x \left(\frac{1}{2}\partial_\mu\phi(x)\partial^\mu\phi(x) - \frac{1}{2}m^2\phi^2(x) + j(x)\phi(x)\right) , \tag{B.65}$$

where, in order to make the integral convergent for large ϕ^2 it is necessary to give the mass a negative imaginary part

$$m^2 \to m^2 - i\epsilon . \tag{B.66}$$

Integrating by parts, Eq. (B.65) becomes

$$W^{(0)}[j] = \int [d\phi(x)] \exp\left(-\frac{i}{2}\int d^4x \int d^4y\, \phi(x)\mathcal{O}(x-y)\phi(y) + \int d^4x\, j(x)\phi(x)\right) \tag{B.67}$$

where

$$\mathcal{O}(x-y) = \left(\Box_x + m^2 - i\epsilon\right)\delta^4(x-y) \equiv \langle x|\Box + m^2 - i\epsilon|y\rangle . \tag{B.68}$$

---

[4] Here functional differentiation is defined via

$$\frac{\delta j(y)}{\delta j(x)} = \frac{\delta}{\delta j(x)} \int d^4x\, \delta^4(y-x)j(x) = \delta^4(y-x). \tag{B.59}$$

Finally, defining a shifted field[5]

$$\phi'(x) = \phi(x) - \int d^4y\, \mathcal{O}^{-1}(x-y)j(y) \tag{B.71}$$

we obtain

$$W^{(0)}[j] = N \int [d\phi(x)] \exp\left(-\frac{i}{2}\int d^4x \int d^4y \left(\phi'(x)\mathcal{O}(x-y)\phi'(y)\right.\right.$$

$$\left.\left. - j(x)\mathcal{O}^{-1}(x-y)j(y)\right)\right)$$

$$= \left[N \int [d\phi'(x)] \exp\left(-\frac{i}{2}\int d^4x \int d^4y\, \phi'(x)\mathcal{O}(x-y)\phi'(y)\right)\right]$$

$$\times \exp\frac{i}{2}\int d^4x \int d^4y\, j(x)\mathcal{O}^{-1}(x-y)j(y) \tag{B.72}$$

where we have used

$$[d\phi(x)] = [d\phi'(x)] \ . \tag{B.73}$$

In Eq. (B.72) we recognize the factor in brackets as $W^{(0)}[0]$. Thus

$$W^{(0)}[j] = W^{(0)}[0] \exp\frac{i}{2}\int d^4x \int d^4y\, j(x)\mathcal{O}^{-1}(x-y)j(y) \ , \tag{B.74}$$

and we determine the scalar propagator as[6]

$$G^{(2)}(x_1,x_2) = (-i)^2 \frac{\delta^2}{\delta j(x_1)\delta j(x_2)} \frac{1}{W^{(0)}[0]} W^{(0)}[j]\bigg|_{j=0}$$

$$= -i\mathcal{O}^{-1}(x_1 - x_2)$$

$$= i\int \frac{d^4k}{(2\pi)^4} \frac{e^{-ik\cdot(x_1-x_2)}}{k^2 - m^2 + i\epsilon} \equiv D^{(0)}(x_1 - x_2) \tag{B.76}$$

in agreement with the result found using the Green's function.

If the electromagnetic interaction is added via the minimal substitution then Eq. (B.65) becomes

---

[5] Here the inverse operator is defined via the relation

$$\int d^4x\, \mathcal{O}^{-1}(z-x)\mathcal{O}(x-y) = \delta^4(z-y) \tag{B.69}$$

and is given by

$$\mathcal{O}^{-1}(z-x) = \frac{1}{\Box_z + m^2 - i\epsilon}\delta^4(z-x) \equiv \left\langle z\left|\frac{1}{\Box + m^2 - i\epsilon}\right|x\right\rangle \ . \tag{B.70}$$

[6] The explicit form for the operator $\mathcal{O}^{-1}(x_1 - x_2)$ is obtained by insertion of a complete set of momentum states

$$\frac{1}{\Box_{x_1} + m^2 - i\epsilon}\delta^4(x_1 - x_2) = -\int \frac{d^4k}{(2\pi)^4} \frac{e^{-ik\cdot(x_1-x_2)}}{k^2 - m^2 + i\epsilon} \ . \tag{B.75}$$

$$W[j,j^\dagger] = N \int [d\phi(x)] \left[d\phi^\dagger(x) \right]$$

$$\times \exp i \int d^4x \left((D_\mu \phi(x))^\dagger D^\mu \phi(x) - m^2 \phi^\dagger(x)\phi(x) + j^\dagger(x)\phi(x) + \phi^\dagger(x)j(x) \right), \quad \text{(B.77)}$$

and the Green's function is

$$G^{(2)}(x_1, x_2) = (-i)^2 \frac{\delta^2}{\delta j(x_1)\delta j^\dagger(x_2)} \frac{1}{W[0]} W[j] \Big|_{j=j^\dagger=0}$$

$$= \left\langle x_2 \left| \frac{-i}{\mathcal{O} + ie\{\partial^\mu, A_\mu\} - e^2 A^2} \right| x_1 \right\rangle \quad \text{(B.78)}$$

which may be solved perturbatively by use of the identity

$$[\mathcal{O} - \Delta]^{-1} = \mathcal{O}^{-1} + \mathcal{O}^{-1}\Delta\mathcal{O}^{-1} + \cdots \quad \text{(B.79)}$$

and yields the Feynman rules given earlier.

Thus far, our development of quantum field theory has been based upon the simple example of scalar fields. For completeness it is important to treat also the case of fermion fields wherein the requirements of antisymmetry impose important modifications to the functional integration techniques. The key to the treatment of anticommuting fields is the use of so-called Grassmann variables. Thus, while ordinary c-number quantities (hereafter denoted by roman letters) $a, b, \ldots$ commute with one another

$$[a, a] = [a, b] = [a, c] = \cdots = 0, \quad \text{(B.80)}$$

Grassmann numbers (hereafter denoted by Greek letters) $\alpha, \beta, \ldots$ anticommute even though they are c-number quantities

$$\{\alpha, \alpha\} = \{\alpha, \beta\}, = \{\alpha, \gamma\} = \cdots = 0 \quad \text{(B.81)}$$

This means that the square of a Grassmann quantity must vanish

$$\alpha^2 = \beta^2 = \gamma^2 = \cdots = 0 \quad \text{(B.82)}$$

and that any function of Grassmann variables must have a very simple expansion

$$f(\alpha) = f_0 + f_1 \alpha$$
$$g(\alpha, \beta) = g_0 + g_1\alpha + g_2\beta + g_3\alpha\beta . \quad \text{(B.83)}$$

Differentiation is defined correspondingly via

$$\left\{ \frac{d}{d\alpha}, \alpha \right\} = \left\{ \frac{d}{d\beta}, \beta \right\} = \cdots = 1$$

$$\left\{ \frac{d}{d\alpha}, \beta \right\} = \left\{ \frac{d}{d\alpha}, \gamma \right\} = \cdots = 0 . \quad \text{(B.84)}$$

Thus

$$\frac{d}{d\alpha} f(\alpha) = f_1$$

$$\frac{d}{d\beta} g(\alpha, \beta) = g_2 - g_3\alpha , \quad \text{(B.85)}$$

and second derivatives have the property

$$\frac{d^2}{d\alpha\,d\alpha} = 0 \ . \tag{B.86}$$

We must also define the concept of Grassmann integration. Since we demand that integration be translation invariant

$$\int d\alpha\, f(\alpha) = \int d\alpha\, f(\alpha + \beta) \tag{B.87}$$

we require

$$\int d\alpha\, f_1 \beta = 0 \quad \text{i.e.,} \quad \int d\alpha = 0 \ . \tag{B.88}$$

We normalize the diagonal integral via

$$\int d\alpha\, \alpha = 1 \ , \tag{B.89}$$

so that

$$\int d\alpha\, f(\alpha) = f_1 \ . \tag{B.90}$$

The formalism for treating Fermi fields can now be developed in parallel to that for the scalar field case. Using the free field Lagrangian density

$$\mathcal{L}_0\left(\bar\psi(x), \psi(x)\right) = \bar\psi(x)\left(i\,\slashed\nabla_x - m\right)\psi(x) \tag{B.91}$$

the generating functional for the free spin 1/2 field becomes

$$W[\eta, \bar\eta] = \int [d\psi(x)]\,[d\bar\psi(x)]\exp\left(i\int d^4x \int d^4y\, \bar\psi(x)\mathcal{O}(x-y)\psi(y) \right.$$
$$\left. + i\int d^4x\, \bar\eta(x)\psi(x) + i\int d^4x\,\bar\psi(x)\eta(x)\right) \tag{B.92}$$

where

$$\mathcal{O}(x-y) = \left(i\,\slashed\nabla_x - m + i\epsilon\right)\delta^4(x-y) \tag{B.93}$$

and $\bar\eta(x), \eta(x)$ are Grassmann fields. Changing variables to

$$\psi'(x) = \psi(x) + \int d^4y\,\mathcal{O}^{-1}(x-y)\eta(y)$$
$$\bar\psi'(x) = \bar\psi(x) + \int d^4y\,\bar\eta(y)\mathcal{O}^{-1}(y-x) \tag{B.94}$$

we find that an alternative form for the generating functional is

$$W[\eta, \bar\eta] = \int [d\psi'(x)]\,[d\bar\psi'(x)]\exp\left(i\int d^4x \int d^4y\,\left[\bar\psi'(x)\mathcal{O}(x-y)\psi'(y)\right.\right.$$
$$\left.\left. -\ \bar\eta(x)\mathcal{O}^{-1}(x-y)\eta(y)\right]\right)$$
$$= W[0,0]\exp\left(-i\int d^4x \int d^4y\,\bar\eta(x)\mathcal{O}^{-1}(x-y)\eta(y)\right) . \tag{B.95}$$

The two particle Green's function is given by

$$G^{(2)}(x_1, x_2) = (-i)^2 \frac{1}{W[0,0]} \frac{\delta^2 W[\eta, \bar{\eta}]}{\delta \eta(x_2) \delta \bar{\eta}(x_1)} \Big|_{\eta=\bar{\eta}=0}$$

$$= i\mathcal{O}^{-1}(x_1 - x_2) = \int \frac{d^4 k}{(2\pi)^4} e^{-ik \cdot (x_1 - x_2)} \frac{i}{\slashed{k} - m + i\epsilon} \tag{B.96}$$

which is the usual Feynman propagator.

Electromagnetic effects may be included via the minimal substitution

$$\nabla_\mu \to \nabla_\mu + ieA_\mu \tag{B.97}$$

whereby the propagator becomes

$$iS(x_1, x_2) = \left\langle x_1 \Big| \frac{i}{i \slashed{\nabla} - e \slashed{A}(\hat{x}) - m + i\epsilon} \Big| x_1 \right\rangle \tag{B.98}$$

which is exact but no longer soluble. As in the bosonic case using completeness we can develop a perturbation series

$$iS(x_1, x_2) = iS^{(0)}(x_1, x_2) - ie \int d^4 y\, iS^{(0)}(x_1, y)\, \slashed{A}(y) iS^{(0)}(y, x_2)$$

$$+ (-ie)^2 \int d^4 y \int d^4 z\, iS^{(0)}(x_1, y)\, \slashed{A}(y) iS^{(0)}(y, z)\, \slashed{A}(z) iS^{(0)}(z, x_2)$$

$$+ \mathcal{O}(e^3) \tag{B.99}$$

which is equivalent to the Feynman rules found earlier.

B.3 Elastic Scattering Theory

Consider the case of low-energy (nonrelativistic) elastic scattering. The time independent Schrödinger equation for a particle of mass m moving under the influence of a potential $\hat{V}$ is

$$(E_i - \hat{H}_0 - \hat{V})|\psi_i >= 0, \tag{B.100}$$

where $\hat{H}_0 = -\nabla^2/2m$ is the kinetic energy, and a scattering solution involving an incoming plane wave having momentum p and an outgoing spherical wave is

$$|\psi_{p_i}^+ >= |\phi_{p_i} > + \frac{1}{E_i - \hat{H}_0 + i\epsilon} \hat{V}|\psi_{p_i}^+ >, \tag{B.101}$$

where $|\phi_{p_i} >$ is a plane wave solution and satisfies the Schrödinger equation

$$(E_i - \hat{H}_0)|\phi_{p_i} >= 0. \tag{B.102}$$

Projecting onto coordinate space we have the scattering wavefunction

$$\psi_{p_i}^+(r) =< r|\psi_{p_i}^+ >,$$

where

$$\psi_{p_i}^{+}(r) = e^{ip_i \cdot r} + \int dr' G_0(r,r')V(r')\psi_{p_i}^{+}(r') \tag{B.103}$$

and

$$G_0(r,r') = \left\langle r \left| \frac{1}{E_i - \hat{H}_0 + i\epsilon} \right| r' \right\rangle = -\frac{e^{ip_i|r-r'|}}{|r-r'|} \tag{B.104}$$

is the Green's function. In the large r limit

$$\psi_{p_i}^{+}(r) \xrightarrow{r\to\infty} e^{ip_i \cdot r} + \frac{e^{ip_i r}}{r} f_{p_i}(\theta), \tag{B.105}$$

where

$$f_{p_i}(\theta) = -\frac{m}{2\pi} < \phi_f|\hat{V}|\psi_i^{+} > \equiv -\frac{m}{2\pi} < \phi_f|\hat{T}|\phi_i > \tag{B.106}$$

is the scattering amplitude. If the incident velocity is v_i the scattering cross section is defined by

$$d\sigma = \frac{1}{v_i}\int \frac{dp_f}{(2\pi)^3}2\pi\,\delta(E_f - E_i)| < \phi_f|\hat{T}|\phi_i > |^2 = d\Omega_f|f_{p_i}(\theta)|^2. \tag{B.107}$$

If I_0 = number of particles per unit time per unit area is the incident flux then the differential cross section $d\sigma/d\Omega_f$ is defined such that

$$I_0\frac{d\sigma}{d\Omega} = \text{number of particles per second scattered into solid angle } d\Omega.$$

Making an expansion into partial-wave amplitudes $a_\ell(p)$ via

$$f_{p_i}(\theta) = \sum_{\ell=0}^{\infty}(2\ell + 1)a_\ell(p_i)P_\ell(\cos\theta), \tag{B.108}$$

where $P_\ell(\cos\theta)$ are the Legendre polynomials, we have for plane waves

$$e^{ip_i z} = \sum_{\ell=0}^{\infty}(2\ell+1)i^\ell j_\ell(p_i r)P_\ell(\cos\theta) \xrightarrow{r\to\infty} \frac{1}{2ip_i r}\sum_{\ell=0}^{\infty}(2\ell+1)(e^{ip_i r} - e^{-i(p_i r-\ell\pi)})P_\ell(\cos\theta),$$
$$\tag{B.109}$$

where $j_\ell(p_i r)$ is a spherical Bessel function. Defining the scattering phase shift $\delta_\ell(p_i)$ via

$$\psi_{p_i}(r) \xrightarrow{r\to\infty} \frac{1}{2ip_i r}\sum_{\ell=0}^{\infty}(2\ell + 1)(e^{i(p_i r+\delta_\ell(p_i))} - e^{-i(p_i r-\ell\pi)})P_\ell(\cos\theta), \tag{B.110}$$

the scattering amplitude is given by

$$f_{p_i}(\theta) = \sum_{\ell=0}^{\infty}(2\ell + 1)\left(\frac{e^{2i\delta_\ell(p_i)} - 1}{2ip_i}\right)P_\ell(\cos\theta). \tag{B.111}$$

That is,

$$a_\ell(p_i) = \frac{e^{2i\delta_\ell(p_i)} - 1}{2ip_i} = \frac{1}{p_i}e^{i\delta_\ell(p_i)}\sin\delta_\ell(p_i). \tag{B.112}$$

Conservation of probability yields the optical theorem

$$\mathrm{Im} f_{p_i}(\theta = 0) = \frac{p_i}{4\pi}\sigma_{\mathrm{tot}}, \tag{B.113}$$

where

$$\sigma_{\mathrm{tot}} = \int d\Omega \frac{d\sigma}{d\Omega} = \frac{4\pi}{p_i^2}\sum_{\ell=0}^{\infty}(2\ell+1)\sin^2\delta_\ell(p_i) \tag{B.114}$$

is the total scattering cross section.
An alternate form for the partial wave amplitude is

$$a_\ell(p_i) = \frac{1}{p_i\mathrm{ctn}\delta_\ell(p_i) - ip_i} \tag{B.115}$$

and at low energy the quantity $p_i\mathrm{ctn}\delta_0(p_i)$ is often represented via the effective range expansion

$$p_i\mathrm{ctn}\delta_0(p_i) = -\frac{1}{a_0} + \frac{1}{2}p_i^2 r_e + (p_i^4) \tag{B.116}$$

where a_0 is the scattering length and r_e is the effective range.
The T-matrix has the perturbative form

$$< \phi_f|\hat{T}|\phi_i > = < \phi_f|\hat{V}|\phi_i > + \left\langle \phi_f \left| \hat{V}\frac{1}{E - \hat{H}_0 + i\epsilon}\hat{V} \right| \phi_i \right\rangle + \cdots , \tag{B.117}$$

so at lowest order the scattering amplitude is given by the Born approximation

$$f_{p_i}^{\mathrm{Born}}(\theta) = -\frac{m}{2\pi} < \phi_f|\hat{V}|\phi_i > = -\frac{m}{2\pi}\int dr e^{iq\cdot r}V(q), \tag{B.118}$$

where $q = p_i - p_f$ is the momentum transfer and $V(q) = < \phi_f|\hat{V}|\phi_i >$. The coordinate-space representation of the potential can then be given via the Fourier transform

$$V(r) = \int \frac{dq}{(2\pi)^3}e^{-iq\cdot r}V(q). \tag{B.119}$$

B.4 Fermi–Watson Theorem

The unitarity property, which is equivalent to conservation of probability, can be expressed via the relation

$$S^\dagger S = SS^\dagger = 1 , \tag{B.120}$$

where $S = 1 + iT$ is the S-matrix. The condition on transition amplitudes is then

$$2\mathrm{Im} < f|T|i > = \sum_n < f|T^\dagger|n >< n|T|i > . \tag{B.121}$$

If we consider a case where elastic scattering is the only kinematically allowed reaction and consider a particular partial wave channel carrying angular momentum ℓ, then the S-matrix is written in the form

$$S_\ell(E) = \exp(i2\delta_\ell(E)) , \tag{B.122}$$

where $\delta_\ell(E)$ is the scattering phase shift at the energy E. Unitarity is then obvious since

$$S_\ell(E)S^\dagger(E) = \exp(i2\delta_\ell(E))\exp(-i2\delta_\ell(E)) = 1 . \tag{B.123}$$

The corresponding T-matrix element is then

$$T_\ell(E) = -i(S_\ell(E) - 1) = 2\exp(i\delta_\ell(E))\sin\delta_\ell(E) , \tag{B.124}$$

which satisfies the unitarity relation Eq. (B.121) since, using $i = f$

$$2\mathrm{Im}T_\ell(E) = 4\sin^2\delta_\ell(E) = T_\ell^*(E)T_\ell(E) . \tag{B.125}$$

Now consider the case of a decay amplitude from state $|i>$ to state $|f>$, in which case the unitarity condition reads

$$2\mathrm{Im} <f|T|i> = 2\sin\delta_\ell^f(E)\exp(-i\delta_\ell^f(E)) <f|T|i> \tag{B.126}$$

whose solution is

$$<f|T|i> = | <f|T|i> | \exp(i\delta_\ell^f(E_i)) , \tag{B.127}$$

i.e., the phase of the decay amplitude is identical to the final state scattering phase shift at the initial state energy. This identity is called the Fermi–Watson theorem. Note that it is violated once we are in the inelastic region, since the sum over intermediate states $|n>$ contains more than a single channel.

Exercises

B.1   The Optical Theorem

One can use the result that, under an integral,

$$\frac{1}{E - E_0 + i\epsilon} = P\frac{1}{E - E_0} - i\pi\delta(E - E_0)$$

where P signifies the principle value integral, to prove both Fermi's golden rule *and* the optical theorem.

The counter in a scattering experiment can be considered as a device for measuring the probability that a particle described by wavefunction $\psi_a^{(+)}(r)$ will be in state $|\phi_b>$ which takes it into the counter. Here $|\phi_b\rangle$ is an eigenstate of the free Hamiltonian

$$\hat{H}_0|\phi_b\rangle = E_b|\phi_b\rangle$$

while $|\psi_a^{(+)}\rangle$ is an eigenstate of the full Hamiltonian $\hat{H} = \hat{H}_0 + \hat{V}$

$$\hat{H}|\psi_a^{(+)}\rangle = E_a|\psi_a^{(+)}\rangle .$$

a) Show that the state

$$|\psi_a^{(+)}\rangle = |\phi_a\rangle + \frac{1}{E_a - \hat{H}_0 + i\epsilon}\hat{V}|\psi_a^{(+)}\rangle$$

has this property. The overlap of state $|\psi_a^{(+)}\rangle$ with $|\phi_b\rangle$ is then found to be

$$< \phi_b|\psi_a^{(+)} > \; = \langle \phi_b| \left(|\phi_a > + \frac{1}{E_a - E_b + i\epsilon}\hat{V}|\psi_a^{(+)} > \right)$$

$$= \delta_{ab} + \frac{1}{E_a - E_b + i\epsilon} < \phi_b|\hat{T}|\phi_a >$$

where $< \phi_b|\hat{T}|\phi_a > = < \phi_b|\hat{V}|\psi_a^{(+)} >$ is the transition matrix element.

b) If $w_{ba} = \left| < \phi_b|\psi_a^{(+)} > \right|^2$ then the transition rate is

$$\frac{dw_{ba}}{dt} = \left(< \dot{\phi}_b|\psi_a^{(+)} > + < \phi_b|\dot{\psi}_a^{(+)} > \right) < \phi_b|\psi_a^{(+)} >^* + \text{c.c.}$$

Use the equations satisfied by $|\psi_a^{(+)}\rangle$ and $|\phi_b\rangle$ to show

$$\frac{dw_{ba}}{dt} = -i\left[< \phi_b|\hat{V}|\psi_a^{(+)} >< \phi_b|\psi_a^{(+)} >^* - \text{c.c.} \right].$$

c) Show that if $b \neq a$

$$\frac{dw_{ba}}{dt} = 2\pi\,\delta(E_a - E_b)\,|\mathcal{F}(E_a)|^2$$

which is Fermi's golden rule.

d) Use the fact that $\sum_b \frac{dw_{ba}}{dt} = 0$ to derive

$$\text{Im} < \phi_a|\hat{V}|\psi_a^{(+)} > \cong -\frac{p_i}{2m}\sigma_a$$

where σ_a is the total scattering cross section. Since

$$f_k(\theta) = -\frac{m}{2\pi} < \phi_b|\hat{V}|\psi_a^{(+)} >$$

is the scattering amplitude, we have

$$\text{Im}\, f_k(0) = \frac{p_i}{4\pi}\sigma_a$$

which is the standard optical theorem.

B.2   Effective Lagrangian for a Constant B or E Field

Consider a charged scalar field interacting with an external magnetic field $\boldsymbol{B} = B\hat{e}_z$ for which the corresponding Klein–Gordon equation reads

$$(D^2 + m^2)\phi(x) = 0$$

where $D_\mu = \partial_\mu + ieA_\mu$ is the covariant derivative and $A_\mu(x)$ is the vector potential associated with the magnetic field. The effective action is then given by

$$e^{iS_{\text{eff}}(B)} = \frac{\int [d\phi(x)][d\phi^*(x)]e^{i\int d^4x\,\phi^*(x)(D^2+m^2)\phi(x)}}{\int [d\phi(x)][d\phi^*(x)]e^{i\int d^4x\,\phi^*(x)(\Box+m^2)\phi(x)}}$$

or

$$S_{\text{eff}}(B) = i\text{Tr}\left(\log \frac{D^2 + m^2}{\Box + m^2}\right) \ .$$

Of course, the operation "Tr log" applied to a differential operator is not a trivial one, and the purpose of this problem is to evaluate this quantity for the case at hand.

a) Demonstrate that

$$\log\frac{a}{b} = \int_0^\infty \frac{ds}{s}(e^{-bs} - e^{-as})$$

so that

$$S_{\text{eff}}(B) = i\text{Tr}\int_0^\infty \frac{ds}{s}e^{-m^2 s}(e^{-\Box s} - e^{-D^2 s}) \ .$$

In order to evaluate the trace we require a complete set of solutions to the equations

$$D^2 \phi_n(x, y, z, t) = \lambda_n \phi_n(x, y, z, t)$$
$$\Box \phi_n(x, y, z, t) = \kappa_n \phi_n(x, y, z, t).$$

Then we may write

$$S_{\text{eff}}(B) = i\sum_n \int_0^\infty \frac{ds}{s}e^{-m^2 s}(e^{-\kappa_n s} - e^{-\lambda_n s}) \ .$$

b) Show that when $B = 0$ the eigenstates are given by

$$\phi(x, y, z, t) = \exp i(k_x x + k_y y + k_z z - k_t t)$$

with eigenvalues

$$\kappa_n = -k_t^2 + k_x^2 + k_y^2 + k_z^2 \ .$$

c) For the gauge choice

$$A_\mu = (0, Bx\hat{e}_y)$$

show that the eigenstates in the presence of the magnetic field become

$$\phi(x, y, z, t) = \exp i(k_z z + k_y y - k_t t)\psi_n\left(x - \frac{k_y}{eB}\right),$$

where $\psi_n(x)$ is an eigenstate of the harmonic oscillator Hamiltonian with frequency

$$\omega = eB$$

and that the corresponding eigenvalues are given by

$$\lambda_n = -k_t^2 + k_z^2 + eB(2n + 1) \ .$$

Now rotate to Euclidean space

$$k_t \rightarrow ik_0$$

so that

$$\kappa_n^2 = k_0^2 + k_x^2 + k_y^2 + k_z^2$$

$$\lambda_n^2 = k_0^2 + k_z^2 + eB(2n+1)$$

and evaluate the trace using box quantization. Taking a box with sides L_1, L_2, L_3 and a time interval T, we have

$$\kappa : \quad \sum_n \rightarrow L_1 L_2 L_3 T \int_{-\infty}^{\infty} \frac{d^4 k}{(2\pi)^4}$$

$$\lambda : \quad \sum_n \rightarrow L_2 L_3 T \int_0^{eBL_1} dk_y \int_{-\infty}^{\infty} \frac{dk_0 dk_z}{(2\pi)^2} \sum_{n=0}^{\infty}$$

where the integration on k_y is over all values with $x' = x - k_y/eB$ kept positive.

d) Now evaluate the effective action

$$S_{\text{eff}}(B) = L_1 L_2 L_3 T \int_0^{\infty} \frac{ds}{s} \int_{-\infty}^{\infty} \frac{dk_0 dk_z}{(2\pi)^2} e^{-(m^2 + k_0^2 + k_z^2)s}$$

$$\times \left[\frac{eB}{2\pi} \sum_{n=0}^{\infty} e^{-eB(2n+1)s} - \int_{-\infty}^{\infty} \frac{dk_x dk_y}{(2\pi)^2} e^{-(k_x^2 + k_y^2)s} \right]$$

and show that

$$S_{\text{eff}}(B) = L_1 L_2 L_3 T \frac{1}{16\pi^2} \int_0^{\infty} \frac{ds}{s^3} e^{-m^2 s} \left(\frac{eBs}{\sinh eBs} - 1 \right) .$$

The "physics" of this result can be seen via an alternative derivation – the effective action is simply the shift in the vacuum energy due to the presence of the magnetic field times the interaction time. Since the zero point energy is given by $\sum_n \omega_n$ (i.e., $\frac{1}{2}\omega_n$ associated with the positively charged states and $\frac{1}{2}\omega_n$ with the negatively charged states) we have

$$S_{\text{eff}}(B) = T \sum_n (\omega_n(B) - \omega_n(0)) .$$

e) Show that

$$\omega_n(B) = \sqrt{k_z^2 + eB(2n+1) + m^2}$$

$$\omega_n(0) = \sqrt{k_x^2 + k_y^2 + k_z^2 + m^2}$$

so that

$$S_{\text{eff}}(B) = L_2 L_3 T \int_{-\infty}^{\infty} \frac{dk_z}{2\pi}$$

$$\times \left[\int_0^{eBL_1} \frac{dk_x}{2\pi} \sum_n \sqrt{k_z^2 + eB(2n+1) + m^2} \right.$$

$$\left. - L_1 \int_{-\infty}^{\infty} \frac{dk_y dk_x}{(2\pi)^2} \sqrt{k_x^2 + k_y^2 + k_z^2 + m^2} \right].$$

f) Use the representation

$$\sqrt{a} = \frac{1}{2\sqrt{\pi}} \int_0^\infty \frac{ds}{s^{\frac{3}{2}}} e^{-as}$$

to prove the identity of the two expressions for $S_{\text{eff}}(B)$.

Now suppose we have a constant electric field $\boldsymbol{E} = E\hat{\boldsymbol{e}}_z$.

g) Working in the gauge $\phi = -Ez$, show that the Euclidean Klein–Gordon equation becomes

$$[(\hat{p}_0 - e\phi)^2 + \hat{\boldsymbol{p}}^2 + m^2]\chi_n(\boldsymbol{x}, t) = \lambda_n \chi_n(\boldsymbol{x}, t)$$

with

$$\chi_n(\boldsymbol{x}, t) = e^{ip_0 t + ip_x x + ip_y y} \psi_n\left(z - \frac{p_0}{eE}\right).$$

h) Then proceeding as before and making the substitution $E \to iE$ to rotate back to Minkowski coordinates, show that

$$S_{\text{eff}}(E) = -TL_1 L_2 L_3 \frac{1}{16\pi^2} \int_0^\infty \frac{ds}{s^3} \left(\frac{eEs}{\sin eEs} - 1\right) e^{-m^2 s}.$$

Although the forms of $S_{\text{eff}}(E)$ and $S_{\text{eff}}(B)$ appear similar, there is an important difference in that $\sin eEs$ has a series of zeroes at the points $s = n\pi/eE$ along the real axis so that contour integration must be used.

i) Show that the effective action becomes complex, with

$$\text{Im}(S_{\text{eff}}(E)) = TL_1 L_2 L_3 \frac{e^2 E^2}{16\pi^2} \sum_{n=1}^\infty \frac{(-)^{n+1}}{n^2} \exp\left(-\frac{n\pi m^2}{eE}\right)$$

indicating that in the presence of sufficiently strong electric fields, the vacuum breaks down and $e^+ e^-$ pairs are produced. Sometimes this is called "sparking the vacuum" and this vacuum breakdown is the analog of the Hawking effect in gravity.

B.3 Euler–Heisenberg Lagrangian and Photon–Photon Scattering

Now repeat the previous problem but with parallel electric and magnetic fields.

a) Show that the effective action becomes

$$\mathcal{L}_{\text{eff}}[E, B] = -\frac{1}{16\pi^2} \int_0^\infty \frac{ds}{s^3} e^{-m^2 s} \left(\frac{e^2 EB}{\sinh(eBs) \sin(eEs)} - 1\right).$$

b) Expanding in powers of E, B show that

$$\mathcal{L} = \frac{\alpha}{24\pi}(E^2 - B^2) \int_0^\infty \frac{ds}{s} e^{-m^2 s} + \frac{\alpha^2}{360}\left[7(E^2 - B^2) + 4E^2 B^2\right] \int_0^\infty dss e^{-m^2 s} + \cdots.$$

Here the first piece, which is divergent and proportional to $F_{\mu\nu} F^{\mu\nu} = 2(B^2 - E^2)$, is absorbed into renormalization of the field tensor, while the second piece is finite.

c) Write this finite component in terms of the vector potential A_μ and show that it has the general form

$$\mathcal{L}_{\text{eff}} = \frac{\alpha^2}{1440m^4} \left[7(F_{\mu\nu}F^{\mu\nu})^2 + 4(F_{\mu\nu}\tilde{F}^{\mu\nu})^2 \right] + \cdots$$

where

$$\tilde{F}^{\mu\nu} = \frac{1}{2}\epsilon^{\mu\nu\alpha\beta}F_{\alpha\beta}$$

is the dual tensor. This result was first derived by W. Heisenberg and H. Euler and is called the Euler–Heisenberg Lagrangian.

We can use this result to derive the cross section for low energy ($\omega << m$) photon–photon scattering.

d) Summing over final polarizations and averaging over initial polarizations show that

$$\frac{1}{4}\sum_{\text{pol.}} |\text{Amp}_{\gamma\gamma}|^2 = \frac{556\alpha^4}{8100m^8}(s^2t^2 + s^2u^2 + t^2u^2).$$

e) Use the result of d) to calculate the total photon-photon scattering cross section and show that

$$\sigma_{\text{tot}} = \frac{973\alpha^4}{10125\pi}\frac{\omega^6}{m^8}.$$

References

[Abb16] B.P. Abbott *et al.* (LIGO Scientific Collaboration and Virgo Collaboration), *Phys. Rev. Lett.*, **116**, 061102 (2016).

[Abe02] H. Abele et al., *Phys. Rev. Lett.* **88**, 211801 (2002).

[Abe08] S. Abe et al., *Phys. Rev. Lett.* **100**, 221803 (2008).

[Abe13] K. Abe et al., (The T2K Collaboration), *Phys. Rev.* **D87**, 092003 (2013).

[Abe13a] B. Abelev et al., (The ALICE Collaboration), *Phys. Rev. Lett.* **110**, 012301 (2013).

[Abe14] K. Abe et al., *Phys. Rev. Lett.* **112**, 061802 (2014).

[Abh05] A. Deshpande et al., *Ann. Rev. Nucl Part. Sci.* **55**, 165 (2005).

[Abr68] M. Abramowitz and I. A. Stegun, *Handbook of Mathematical Functions,* National Burean Standards, Washington (1968), Ch. 14.

[Ach07] A. Acha et al., *Phys. Rev. Lett.* **98**, 032301 (2007).

[Ack98a] K. Ackerstaff et al., (The HERMES Collaboration), *Phys. Rev. Lett.* **81**, 5519 (1988).

[Ada66] M. J. Adamovitch et al., *Sov. J. Nucl. Phys.* **2**, 95 (1966).

[Ada09] T. Adams et al., *Int. J. Mod. Phys.* **A24**, 671 (2009).

[Ada12] L. Adamczyk et al., (The STAR Collaboration), *Phys. Rev.* **D85**, 092010 (2012).

[Ada13] P. Adamson et al., *Phys. Rev. Lett.* **110**, 251801 (2013).

[Ada14] L. Adamczyk et al., (The STAR Collaboration), *Phys. Rev. Lett.* **113**, 072301 (2014).

[Ada15] C. Adams et al., *Proceedings of the Workshop on the intermediate Neutrino Program*, Brookhaven National Laboratory, Upton, New York, February 4–6, FERMILAB-CONF-15-120-ND; arXiv 1503.06637, unpublished (2015).

[Ada15a] L. Adamczyk et al., (The STAR Collaboration), *Phys. Rev. Lett.* **115**, 092002 (2015).

[Ade99] E. Adelberger et al., *Phys. Rev. Lett.* **83**, 1299 (1999).

[Adi15] D. Adikaram et al., *Phys. Rev. Lett.* **114**, 062003 (2015).

[Aga14] N. Agofonova et al., *Phys. Rev.* **D89**, 051102 (2014).

[Agu10] A. A. Aguilar-Arevalo et al., *Phys. Rev.* **D81**, 092005 (2010).

[Agu13] A. A. Aguilar-Arevalo et al., *Phys. Rev.* **D88**, 032001 (2013).

[Aha05] B. Aharmim et al., *Phys. Rev.* **C72**, 055502 (2005).

[Ahm01] Q. R. Ahmad et al., *Phys. Rev. Lett.* **87**, 071301 (2001).

[Ahm02a] Q. R. Ahmad et al., *Phys. Rev. Lett.* **89**, 011301 (2002).

[Ahm02b] Q. R. Ahmad et al., *Phys. Rev. Lett.* **89**, 011302 (2002).

[Ahn12] J. K. Ahn et al., *Phys. Rev. Lett.* **108**, 191802 (2012).

[Ahr05] J. Ahrens et al., *Eur. Phys. J.* **A23**, 113 (2005).

[Aid13] C. A. Aidala S. D. Bass, D. Hasch, and G. K. Mallot, *Rev. Mod. Phys.* **85**, 655 (2013).

[Air07] A. Airapetian et al., (The HERMES Collaboration), *Nucl. Phys.* **B780**, 1 (2007).

[Alb88] W. M. Alberico, A. Molinari, T. W. Donnelly, L. Kronenberg, and J. W. Van Orden, *Phys. Rev.* **C38**, 1801 (1988).

[Ale99] L. C. Alexa et al., (The Jefferson Lab Hall A Collaboration), *Phys. Rev. Lett.* **82**, 1374 (1999).

[Ali02] G. Alimonte et al., *Astropart. Phys.* **16**, 203 (2002).

[Ali10] A. Ali and G. Kramer, *Eur. Phys. J.* **H36**, 245 (2011).

[Alp14] B. Alpert et al., *Eur. Phys. J.* **C75** 112 (2015).

[Alt77] G. Altarelli and G. Parisi, *Nucl. Phys.* **B126**, 298 (1977).

[Ama99] J. E. Amaro, M. B. Barbaro, J.A. Caballero, T. W. Donnelly, and A. Molinari, *Nucl. Phys.* **A657**, 161 (1999).

[Ama05] J. E. Amaro, M. B. Barbaro, J. A. Caballero, T. W. Donnelly, A. Molinari, and I. Sick, *Phys. Rev.* **C71**, 015501 (2005).

[Ama14] E. Amato, *Int. J. Mod. Phys.* **D23**, 1430013 (2014).

[Ama14a] J. E. Amaro, M. B. Barbaro, J. A. Caballero, T. W. Donnelly, and C. F. Williamson, *Phys. Lett.* **B696**, 151 (2014).

[An12] F. P. An et al., *Phys. Rev. Lett.* **108**, 171803 (2012).

[And83] B. Andersson, G. Gustafson, G. Ingelman, and T. Sjöstrand, *Phys. Rep.* **97**, 31 (1983).

[Ang05] J. Angrik et al., *KATRIN Design Report 2004*, preprint FZKA-7090 (2005).

[Ani06] K. A. Aniol et al., *Phys. Rev. Lett.* **96**, 022003 (2006).

[Ant83] Yu. M. Antipov et al., *Phys. Lett.* **B121**, 445 (1983).

[Ant85] Yu. M. Antipov et al., *Z. Phys.* **C26**, 495 (1985).

[Ant10] M. Antonelli et al., *Eur. Phys. J.* **C69**, 399 (2010).

[Arg88] P. Argan et al., *Phys. Lett.* **B206**, 4 (1988).

[Arm12] D. S. Armstrong and R. D. McKeown, *Ann. Rev. Nucl. and Part. Sci.* **62**, 337 (2012).

[Arp08] C. Arpesella et al., *Phys. Rev. Lett.* **101**, 091302 (2008).

[Ase11] V. N. Aseev et al., *Phys. Rev.* **D84**, 112003 (2011).

[Ask94a] A. J. Askew et al., *Phys. Rev.* **D49**, 4402 (1994).

[Ask94b] A. J. Askew et al., *Phys. Lett.* **B325**, 212 (1994).

[Asn14] D. M. Asner et al., (Project 8 Collaboration), *Phys. Rev. Lett.* **16**, 162501 (2015).

[Atl12] ATLAS Collaboration, *Phys. Lett.* **B716**, 1 (2012).

[Att91] C. Ciofi degli Atti et al., *Phys. Rev.* **C43**, 1155 (1991).

[Aub85] J. J. Aubert et al., (The EMC Collaboration), *Phys. Lett.* **B160**, 417 (1985).

[Auf85] S. Auffret et al., *Phys. Rev.* **55**, 1362 (1985).

[Aul11] K. Aulenbacher *Hyperfine Interactions* **200**, 3 (2011).

[Bab92] D. Babusci et al., *Phys. Lett.* **B277**, 158 (1992).

[Bac05] B.B. Back et al., (The PHOBOS Collaboration), *Nucl. Phys.* **A757**, 28 (2005).

[Bah97] J. N. Bahcall *Phys. Rev.* **C56**, 3391 (1997).

[Bah05] J. N. Bahcall, A. M. Serenelli, and S. Basu, *Ap. J.* **621**, L85 (2005).

[Bak07] C. A. Baker et al., *Phys. Rev. Lett.* **98**, 149102 (2007). For a review, see S. K. Lamoreaux and R. Golub, *J. Phys.* **G36**, 104002 (2009).

[Bal77] R. M. Baltrusaitis and F. M. Calaprice, *Phys. Rev. Lett.* **38**, 464 (1977).

[Bal78] Y. Balitki and L. N. Lipatov, *Phys. Rev. Lett.* **28**, 822 (1978); E. A. Kuraev et al., *Phys. Rev. Lett.* **44**, 443 (1976); E. A. Kuraev et al., *Phys. Rev. Lett.* **45**, 199 (1977).

[Bar81a] V. Barger, K. Whisnant, and R. J. N. Phillips, *Phys. Rev.* **D24**, 538 (1981); S. L. Glashow and L. M. Krauss, *Phys. Lett.* **B190**, 199 (1987).

[Bat15] J. R. Batley et al., (The NA48/2 Collaboration), *Phys. Lett.* **B746**, 178 (2015).

[Bea99] J. Beacom and P. Vogel, *Phys. Rev.* **D60**, 053003 (1999).

[Bea01] S. R. Beane and M. J. Savage, *Nucl. Phys.* **A694**, 511 (2001).

[Bea02] J. Beacom, J. W. Farr, and P. Vogel, *Phys. Rev.* **D66**, 033001 (2002).

[Bea14] S. R. Beane et al., *Phys. Rev. Lett.* **113**, 252001 (2014).

[Bea15] S. R. Beane et al., arXiv:1505.02422 (2015).

[Beg77] M. A. B. Beg, R. Budney, R. Mohapatra, and A. Sirlin, *Phys. Rev. Lett.* **38**, 1252 (1977); R. Rizzo and G. Senjanovic, *Phys. Rev. Lett.* **46**, 1315 (1978).

[Bel14] G. Bellini et al., *Phys. Rev.* **D89**, 112007 (2014).

[Ben91] S. Bentvelsen et al., in *Proceedings of the Workshop on Physics at HERA*, eds. W. Buchmüller and G. Ingelman (Hamburg, Germany, 1991).

[Ben10] W. Bentz et al., *Phys. Lett.* **B693**, 462 (2010).

[Ber74] V. B. Beretetskii, E. M. Litshitz, and J. L. Pitaevski, *Relativistic Quantum Theory I*, Pergamon Press, New York (1974).

[Ber81] M. Bernheim et al., *Phys. Rev. Lett.* **46**, 402 (1981).

[Ber91a] V. Bernard, N. Kaiser, and U-G. Meißner, *Phys. Rev. Lett.* **67**, 1515 (1991); *Nucl. Phys.* **B373**, 364 (1992).

[Ber91b] V. Bernard, J. Gasser, N. Kaiser, and U-G. Meißner, *Phys. Lett.* **B268**, 291 (1991).

[Ber92] V. Bernard, N. Kaiser, J. Kambor, and U-G. Meißner, *Nucl. Phys.* **B388**, 315 (1992).

[Ber95] V. Bernard, N. Kaiser, and U-G. Meissner, *Int. J. Mod. Phys.* **E4**, 193 (1995).

[Ber96] J. C. Bergstrom et al., *Phys. Rev.* **C53**, R1052 (1996).

[Ber98] J. Bergstrom, private communication.

[Ber14] J. C. Bernauer et al., (The A1 Collaboration), *Phys. Rev.* **C90**, 015206 (2014).

[Bet34] H. A. Bethe and R. E. Peirls, *Nature* **133**, 532 (1934).

[Bet38] H. A. Bethe and C. L. Chritchfield, *Phys. Rev.* **54**, 248 (1938).

[Bet47] H. Bethe, *Phys. Rev.* **72**, 339(L) (1947).

[Bet86] H. Bethe, *Phys. Rev. Lett.* **56**, 1305 (1986).

[Bij14] J. Bijnens and G. Ecker, to be published in *Ann. Rev. Nucl. Part. Sci.*

[Bjo64] J. D. Bjorken and S. D. Drell, *Relativistic Quantum Mechanics*, McGraw-Hill, New York (1964).

[Bjo66] J. D. Bjorken, *Phys. Rev.* **148**, 1467 (1966).

[Bjo67] J. D. Bjorken, *Proc. 1967 International Symposium on Electron and Photon Interactions at High Energies*, Stanford, CA, September 5–9, 1967, p. 109.

[Bjo83] J. D. Bjorken, *Phys. Rev.* **D27**, 140 (1983).

[Bla84] B. Blankleider and R. M. Woloshyn, *Phys. Rev.* **C29**, 538 (1984).

[Bla13] K. Blaum et al., arXiv:1306.2655 (2013).

[Blu05] P. G. Blunden, W. Melnitchouk, and J. A. Tjon, *Phys. Rev.* **C72**, 034612 (2005).

[Blu08] J. Blümlein and H. Böttcher, *Phys. Lett.* **B662**, 336 (2008).

[Bob94] I. Bobeldijk et al., *Phys. Rev. Lett.* **73**, 2684 (1994).

[Bop86] P. Bopp et al., *Phys. Rev. Lett.* **56**, 919 (1986).

[Bor12] S. Borsányi et al., (The Wuppertal-Budapest Collaboration), *Nucl. Phys.* **A904**, 270 (2013).

[Bro68] L. S. Brown and R. L. Goble, *Phys. Rev. Lett.* **20**, 346 (1968).

[Bro78] L. M. Brown, *Physics Today*, September 1978, 23–28; see also *Los Alamos Science* **25**, 1 (1997).

[Bud95] B. Budliner et al., *Phys. Rev. Lett.* **74**, 4396 (1995).

[Bur65] J. P. Burg, *Ann. De Phys. (Paris)* **10**, 363 (1965).

[Bur98] C. P. Burgess, *Phys. Rep.* **330**, 193 (2000).

[Bus75] W. Busza et al., (The Fermilab E178 Experiment), *Phys. Rev. Lett.* **34**, 836 (1975).

[But01] M. Butler and J.-W. Chen, *Phys. Lett.* **B520**, 87 (2001).

[Byr02] J. Byrne et al., *J. Phys.* **G28**, 1325 (2002).

[Cab63] N. Cabibbo, *Phys. Rev. Lett.* **10**, 531 (1963).

[Cac08] M. Cacciari, G. P. Salam and G. Soyez, *JHEP* **0804**, 063 (2008).

[Cal67] C. G. Callen and S. B. Treiman, *Phys. Rev.* **162**, 1494 (1967).

[Cal75] F. P. Calaprice et al., *Phys. Rev. Lett.* **35**, 1566 (1975); D. Schreiber, Princeton University PhD dissertation (1982).

[Cam14] A. Camsonne et al., *Phys. Rev. Lett.* **112**, 132503 (2014).

[Car88] J. Carlson, *Phys. Rev.* **C38**, 1879 (1988).

[Car91] A. S. Carnoy et al., *Phys. Rev.* **C43**, 2825 (1991).

[Cha35] S. Chandrasekhar, *Mon. Not. Roy. Ast. Soc.* **95**, 207 (1935); L. D. Landau, *Phys. Z. Sowjet Union* **1**, 285 (1932).

[Cha14] S. Chatrchyan et al., (The CMS Collaboration), *Phys. Rev.* **D89**, 092007 (2014).

[Chu12] T. E. Chupp et al., *Phys. Rev.* **C86**, 035505 (2012).

[Cir11] V. Cirigliano and H. Neufeld, *Phys. Lett.* **B700**, 7 (2011).

[Cir12] V. Cirigliano et al., *Rev. Mod. Phys.* **84**, 399 (2012).

[Cir13] See, e.g., V. Cirigliano, S. Gardner, and B. R. Holstein, *Prog. Part. Nucl. Phys.* **71**, 93 (2013).

[Cle98] B. T. Cleveland et al., *Ap. J.* **496**, 505 (1998).

[Clo79] F. E. Close, *An Introduction to Quarks and Partons*, Academic Press, London and New York, (1979).

[Cms12] CMS Collaboration, *Phys. Lett.* **B716**, 30 (2012).

[Col84] P. D. B. Collins and A. D. Martin, *Hadron Interactions*, Adam Hilger, Bristol (1984).

[Col89] J. C. Collins, D. E. Soper, and G. F. Sterman, *Adv. Ser. Direct. High Energy Phys.* **5**, 1 (1988).

[Con35] E. U. Condon and G. H. Shortley, *The Theory of Atomic Spectra*, Cambridge University Press, Cambridge (1935).

[Cow56] C. L. Cowan, Jr., et al., *Science* **124**, 103 (1956).

[Cra61] H. Crannell, R. Helm, H. Kendall, J. Oeser, and M. Yearian, *Phys. Rev.* **123**, 923 (1961).

[Cra10] C. Crawford et al., *Phys. Rev.* **C82**, 045211 (2010).

[Dal82] E. B. Dally et al., *Phys. Rev. Lett.* **48**, 375 (1982).

[Dan85] P. Danielewicz and M. Gyulassy, *Phys. Rev.* **D31**, 53 (1985).

[Das69] R. Dashen and M. Weinstein, *Phys. Rev.* **188**, 2330 (1969); J. L. Goity, R. Lewis, and M. Schvelinger, *Phys. Lett.* **B454**, 115 (1999).

[Day90] D. B. Day, J. S. McCarthy, T. W. Donnelly, and I. Sick, *Ann. Rev. Nucl. Part. Sci.* **40** 357 (1990).

[Day15] D. B. Day et al., online database of quasielastic electron scattering data at http://faculty.virginia.edu/qes-archive/maintained by D. B. Day at the University of Virginia.

[Deb70] P. deBaenst, *Nucl. Phys.* **B24**, 613 (1970).

[Dec80] J. Dechargé and D. Gogny, *Phys. Rev.* **C21**, 1568 (1980).

[Def66] T. deForest, Jr. and J. D. Walecka, *Adv. in Phys.* **15**, 1 (1966).

[Def69] T. de Forest, Jr., *Nucl. Phys.* **A132**, 305 (1969).

[deF09] D. de Florian, R. Sassot, M. Stratmann, and W. Vogelsang, *Phys. Rev.* **D80**, 034030 (2009).

[deF14] D. de Florian, R. Sassot, M. Stratmann, and W. Vogelsang, *Phys. Rev. Lett.* **113**, 012001 (2014).

[DeG10] A. DeGrush, *Single and Double Polarization Observables in the Electrodisintegration of the Deuteron from BLAST*, MIT PhD thesis, unpublished (2010).

[Der58] G. Derrick and J. M. Blatt, *Nucl. Phys.* **8**, 310 (1958).

[Des54] S. Deser, M. L. Goldberger, K. Baumann, and W. Thirring, *Phys. Rev.* **96**, 774 (1954).

[Des74] A. deShalit and H. Feshbach, *Theoretical Nuclear Physics*, Wiley, New York (1974).

[DeT83] C. E. DeTar and J. F. Donoghue, *Ann. Rev. Nucl. Part. Sci.* **33**, 235 (1983).

[DeY14] T. DeYoung et al., *Whitepaper on New Phenomena and Inititaives* (unpublished), November 2014.

[Doi12] T. Doi et al., *Prog. of Theor. Phys.* **127**, 723 (2012).

[Dok77] Yu. L. Dokshitzer, *Sov. Phys. JETP* **46**, 641 (1977).

[Don68] T. W. Donnelly, J. D. Walecka, I. Sick, and E. B. Hughes, *Phys. Rev. Lett.* **21**, 1196 (1968).

[Don72] T. W. Donnelly and J. D. Walecka, *Phys. Lett.* **41B**, 275 (1972).

[Don72a] T. W. Donnelly, in *Dynamic Structure of Nuclear States*, eds. D. J. Rowe, L. E. H. Trainor, S. S. M. Wong, and T. W. Donnelly, University of Toronto Press, Toronto (1972).

[Don73] T. W. Donnelly, *Phys. Lett.* **43B**, 93 (1973).

[Don73a] T. W. Donnelly and J. D. Walecka, *Phys. Lett.* **44B**, 330 (1973).

[Don74] T. W. Donnelly, D. Hitlin, M. Schwartz, J. D. Walecka, and S. J. Wiesner, *Phys. Lett.* **49B**, 8 (1974).

[Don75] T. W. Donnelly and J. D. Walecka, *Ann. Rev. Nucl. Sci.* **15**, 329 (1975).

[Don76] T. W. Donnelly and J. D. Walecka, *Nucl. Phys.* **A274**, 368 (1976).

[Don78] T. W. Donnelly, S. J. Freedman, R. S. Lytel, R. D. Peccei, and M. Schwartz, *Phys. Rev.* **D18**, 1607 (1978).

[Don79] T. W. Donnelly and R. D. Peccei, *Phys. Rep.* **50**, 1 (1979).

[Don79a] T. W. Donnelly and W. C. Haxton, *At. Data and Nucl. Tables* **23**, 103 (1979) (note the typographical error in Eq. (6) where the factor in the numerator should read $(2z)^{\ell}$).

[Don82] J. F. Donoghue and B. R. Holstein, *Phys. Lett.* **B113**, 382 (1982).

[Don82a] T. W. Donnelly, in *Proc. of Los Alamos Neutrino Workshop*, eds. F. Boehm and G. J. Stephenson, LA-9358-C (1982).

[Don84] T. W. Donnelly and I. Sick, *Rev. Mod. Phys.* **56**, 461 (1984).

[Don85] T. W. Donnelly, *Prog. Part. Nucl. Phys.* **13**, 183 (1985).

[Don86] T. W. Donnelly and A. S. Raskin, *Ann. Phys.* **169**, 247 (1986).

[Don86a] T. W. Donnelly, in *New Vistas in Electro-Nuclear Physics*, eds. E. L. Tomusiak, H. S. Caplan, and E. T. Dressler, Plenum, New York (1986), p. 151.

[Don89] T. W. Donnelly, J. Dubach and I. Sick, *Nucl. Phys.* **A503**, 589 (1989).

[Don89a] J. F. Donoghue and B. R. Holstein, *Phys. Rev.* **D40**, 2378 (1989).

[Don97] J. F. Donoghue and E. S. Na, *Phys. Rev.* **D56**, 7073 (1997).

[Don99] T. W. Donnelly and I. Sick, *Phys. Rev. Lett.* **82**, 3212 (1999).

[Don99a] T. W. Donnelly and I. Sick, *Phys. Rev.* **C60**, 065502 (1999).

[Don14] J. F. Donoghue, E. Golowich, and B. R. Holstein, *Dynamics of the Standard Model*, 2nd Ed., Cambridge University Press, New York (2014).

[Dow88] K. Dow et al., *Phys. Rev. Lett.* **61**, 1706 (1988).

[Dre66] S. Drell and A. C. Hearn, *Phys. Rev. Lett.* **16**, 908 (1966); S. Gerasimov, *Sov. J. Nucl. Phys.* **2**, 430 (1966).

[Dre70] S. Drell and T.-N. Yan, *Phys. Rev. Lett.* **25**, 316 (1970).

[Edm74] A. R. Edmonds, *Angular Momentum in Quantum Mechanics* (third printing with corrections), Princeton University Press, Princeton, NJ (1974).

[EIC12] EIC White Paper, arXiv: 1212.1701, December 2012.

[Ell74] J. R. Ellis and R. L. Jaffe, *Phys. Rev.* **D9**, 1444 (1974).

[Ell93] S. D. Ellis and D. E. Soper, *Phys. Rev.* **D48**, 3160 (1993).

[Eng64] F. Englert and R. Brout, *Phys. Rev. Lett.* **13**, 321 (1964).

[Epe12] E. Epelbaum et al., *Phys. Rev. Lett.* **109**, 252501 (2012).

[Epe13] E. Epelbaum et al., *Phys. Rev. Lett.* **110**, 112502 (2013).

[Ero97] B. G. Erozolimsky et al., *Phys. Lett.* **B412**, 240 (1997).

[Ess13] R. Essig et al., Report of the Community Summer Study 2013 (Snowmass), arXiv:1311.0029 (2013).

[Fed91] F. J. Federspiel et al., *Phys. Rev. Lett.* **67**, 1511 (1991); E. L. Hallin et al., *Phys. Rev.* **C48**, 1497 (1993); A. Zeiger et al., *Phys. Lett.* **B278**, 34 (1992); B. E. MacGibben et al., *Phys. Rev.* **C52**, 2097 (1995); G. Blanpied et al., *Phys. Rev.* **C64**, 025203 (2001); V. Olmos de Leon et al., *Eur. Phys. J.* **A10**, 207 (2001).

[Fei75] G. Feinberg, *Phys. Rev.* **D12**, 3575 (1975).

[Fer34] E. Fermi, *Z. Phys.* **88**, 161 (1934). See also E. Fermi, *Nuovo Cim.* **11**, 1 (1934).

[Fer49] E. Fermi, *Phys. Rev.* **75**, 1169 (1949).

[Fer54] E. Fermi, *Ap. J.* **119**, 1 (1954).

[Fet71] A. L. Fetter and J. D. Walecka, *Quantum Theory of Many-Particle Systems*, McGraw-Hill, San Francisco (1971).

[Fey58] R. P. Feynman and M. Gell-Mann, *Phys. Rev.* **109**, 193 (1958).

[Fey63] R. P. Feynman, R. B. Leighton, and M. Sands, *The Feynman Lectures on Physics*, Addison-Wesley, Reading, MA (1963).

[Fie15] L. Fields et al., (MINERνA Collaboration), *Phys. Rev. Lett.* **111**, 022501 (2013).

[Fio13] G. A. Fiorentini et al., (MINERνA Collaboration), *Phys. Rev. Lett.* **111**, 022502 (2013).

[Flo07a] D. de Florian, R. Sassort, and M. Stratmann, *Phys. Rev.* **D75**, 114010 (2007).

[Flo07b] D. de Florian, R. Sassort, and M. Stratmann, *Phys. Rev.* **D76**, 074033 (2007).

[Fod12] Z. Fodor and C. Hoelbling, *Rev. Mod. Phys.* **84**, 449 (2012).

[Fom12] N. Fomin et al., *Phys. Rev. Lett.* **108**, 092502 (2012).

[For12] J. A. Formaggio, E. Figueroa-Feliciano, and A. J. Anderson, *Phys. Rev.* **D85**, 013009 (2012).

[Fre74] D. Z. Freedman, *Phys. Rev.* **D9**, 1389 (1974).

[Fri78] R. Friedberg and T. D. Lee, *Phys. Rev.* **D18**, 2623 (1978).

[Fri84] J. L. Friar et al., *Ann. Rev. Nucl. Part. Sci.* **34**, 403 (1984).

[Fri90] J. L. Friar et al., *Phys. Rev.* **C42**, 2310 (1990).

[Fri12] J. Friedrich, *Proc. of Science (Confinement)* X, 120 (2012).

[Fro61] M. Froissart, *Phys. Rev.* **123**, 1053 (1961).

[Fro87] B. Frois and C. N. Papanicolas, *Ann. Rev. Nucl. Part. Sci.* **37**, 133 (1987).

[Fuc96] M. Fuchs et al., *Phys. Lett.* **B368**, 20 (1996).

[Fuk98] Y. Fukuda et al., *Phys. Rev. Lett.* **81**, 1158 (1998).

[Fuk02] Y. Fukuda et al., *Phys. Lett.* **B539**, 179 (2002).

[Fur80] W. Furmanski and R. Petronzio, *Phys. Lett.* **B97**, 437 (1980).

[Gan11] A. Gando et al., *Phys. Rev.* **D83**, 052002 (2011).

[Gan13] A. Gando et al., (KamLAND Collaboration), *Phys. Rev.* **D88**, 033001 (2013).

[Gar86] M. Gari and W. Krumpelmann, *Phys. Lett.* **B173**, 10 (1986).

[Gar01] G. T. Garvey and J.-C. Peng, *Prog. Part. Nucl. Phys.* **47**, 203 (2001).

[Gar14] G. T. Garvey et al., *Phys. Rev. Lett.* **112**, 202501 (2014).

[Gas69] S. Gasiorowicz and D. A. Geffen, *Rev. Mod. Phys.* **41**, 531 (1969).

[Gas84] J. Gasser and H. Leutwyler, *Ann. Phys. (NY)* **158**, 142 (1984); *Nucl. Phys.* **B250**, 465 (1985).

[Gas88] J. Gasser, M. Sainio, and A. Svarc, *Nucl. Phys.* **B307**, 779 (1988).

[Geg98] J. Gegelia, *Phys. Lett.* **B249**, 227 (1998).

[Gei08] E. Geis et al., *Phys. Rev. Lett.* **101**, 042501 (2008).

[Gel61] M. Gell-Mann, Caltech Rept. CTSL-20 (1961); S. Okubo, *Prog. Theo. Phys.* **27**, 949 (1962).

[Gel64a] M. Gell-Mann, *Phys. Lett.* **8**, 214 (1964).

[Gel64b] M. Gell-Mann and Y. Ne'eman, eds., *The Eightfold Way*, W.A. Benjamin, New York. LCCN 65013009, (1964).

[Gel10] F. Gelis, E. Iancu, J. Jalilian-Marian, and R. Venugopalan, *Ann. Rev. Nucl. Part. Sci.* **60**, 463 (2010).

[Gel12] F. Gelis, *Int. J. Mod. Phys.* **A28**, 1330001 (2013).

[Geo74] H. Georgi and S. L. Glashow, *Phys. Rev. Lett.* **32**, 438 (1974).

[Gil02] R. Gilman and F. Gross, *J. Phys.* **G28**, R37 (2002).

[Giu07] C. Giunti and C. W. Kim, *Fundamentals of Neutrino Physics and Astrophysics*, Oxford University Press, New York (2007).

[Giu12] A. Giuliani and A. Poves, *Adv. High. En. Phys.* **2102**, 857016 (2012).

[Gla61] S. L. Glashow, *Nucl. Phys.* **22**, 579 (1961).

[Gla70] S. L. Glashow, J. Iliopoulos, and L. Maiani, *Phys. Rev.* **D2**, 1285 (1970).

[Gla79] S. L. Glashow, *Physica* **96A**, 27 (1979).

[Glu98] F. Glück, *Nucl. Phys.* **A628**, 493 (1998).

[Goc06] M. Göckler et al., *Phys. Rev.* **D73**, 014513 (2006).

[Gol48] M. Goldhaber and E. Teller, *Phys. Rev.* **74**, 1046 (1948).

[Gol58] M. L. Goldberger and S. B. Treiman, *Phys. Rev.* **110**, 1478 (1958).

[Gol61] J. Goldstone, *Nuovo Cim.* **19**, 154 (1961); J. Goldstone, A. Salam, and S. Weinberg, *Phys. Rev.* **127**, 965 (1962).

[Gol66] E. L. Goldwasser et al., *Proc. XII Int. Conf. on High Energy Physics, Dubna, 1964*, ed. Ya.-A Smorodinsky, Atomizdat, Moscow (1966).

[Gom94] J. Gomez et al., *Phys. Rev.* **D49**, 4348 (1994).

[Gon13] R. Gonzalez-Jimenez, J. A. Caballero, and T. W. Donnelly, *Phys. Rep.* **524**, 1 (2013).

[Gon14] R. Gonzalez-Jimenez, J. A. Caballero, and T. W. Donnelly, *Phys. Rev.* **D90**, 033002 (2014).

[Gor05] A. Gorelov et al., *Phys. Rev. Lett.* **92**, 142501 (2005).

[Gra65] I. S. Gradshteyn and I. M. Ryzhik, *Table of Integrals Series and Products*, Academic Press, New York (1965).

[Gre53] A. E. S. Green and D. F. Edwards, *Phys. Rev.* **91**, 46 (1953).

[Gre54] A. E. S. Green, *Phys. Rev.* **95**, 1006 (1954).

[Gre66] K. Greisen, *Phys. Rev. Lett.* **16**, 748 (1966); G. T. Zatsepin and V. A. Kuz'min, *JETP Lett.* **4**, 78 (1966).

[Gre99] B. Greene, *The Elegant Universe*, Norton, New York (1999).

[Gre12] J. R. Green et al., *Phys. Rev.* **D86**, 114509 (2012); T. Bhattacharya et al., *Phys. Rev.* **D85**, 054512 (2012).

[Gri72] V. N. Gribov, L. N. Lipatov, *Sov. J. Nucl. Phys.* **15**, 438 (1972).

[Gri08] D. Griffiths, *Introduction to Elementary Particles*, (Second, Revised Edition) Wiley-VCH, Weinheim (2008).

[Gro73] D. J. Gross and F. Wilczek, *Phys. Rev. Lett.* **30**, 1343 (1973).

[Hal84] F. Halzen and A. D. Martin, *Quarks and Leptons*, John Wiley and Sons, New York (1984).

[Ham62] M. Hamermesh, *Group Theory*, Addison-Wesley Reading, MA (1962).

[Han65] M. Y. Han and Y. Nambu, *Phys. Rev.* **139**, B1006 (1965).

[Han99] T. Hannah, *Phys. Rev.* **D59**, 057502 (1999).

[Haw98] E. A. Hawker et al., (The E866 Collaboration), *Phys. Rev. Lett.* **80**, 3715 (1998).

[Hax86] W. C. Haxton, *Phys. Rev. Lett.* **57**, 1271 (1986); S. J. Parke, *Phys. Rev. Lett.* **57**, 1275 (1986).

[Hax14] Much of this material is taken from lecture notes by W. C. Haxton, (Institute for Nuclear Throey, University of Washington, Seattle). Available at www.int.washington.edu [accessed 12 August 2016].

[Hei15] U. Heinz, P. Sorensen, A. Deshpande, C. Gagliardi, F. Karsch, T. Lappi, Z. E. Meziani, and R. Milner, et al., arXiv:1501.06477 [nucl-th], (2015).

[Hen13] O. Hen, et al. *Int. J. Mod. Phys.* **E22**, 1330017 (2013).

[Hen16] B. S. Henderson et al., (The OLYMPUS Collaboration), arXiv:1611.04685, Nov. 15 2016.

[Hig64] P. Higgs, *Phys. Rev. Lett.* **13**, 508 (1964).

[Hir14] Y. Hirono, T. Hirano, and D. E. Kharzeev, *Phys. Rev.* **C91**, 054915 (2015).

[Hoh76] G. Hohler et al., *Nucl. Phys.* **B114**, 505 (1976).

[Hol71] B. R. Holstein and S. B. Treiman, *Phys. Rev.* **C3**, 1921 (1971); S. P. Rosen, *Phys. Rev.* **D5**, 760 (1972).

[Hol72] B. R. Holstein, *Phys. Rev.* **C3**, 1964 (1972).

[Hol74] B. R. Holstein, *Rev. Mod. Phys.* **46**, 789 (1974).

[Hol76] B. R. Holstein and S. B. Treiman, *Phys. Rev.* **D13**, 3059 (1976).

[Hol77] B. R. Holstein and S. B. Treiman, *Phys. Rev.* **D16**, 2369 (1977).

[Hol90a] B. R. Holstein, *Phys. Lett.* **B244**, 83 (1990).

[Hol90b] B. R. Holstein, *Comm. Nucl. Part. Phys.* **19**, 221 (1990).

[Hol99a] B. R. Holstein, *Phys. Rev.* **D60**, 114030 (1999).

[Hol10] R. J. Holt and C. D. Roberts, *Rev. Mod. Phys.* **82**, 2991 (2010).

[Hol14] B. R. Holstein, *J. Phys* **G41**, 114001 (2014).

[Hol14a] B. R. Holstein, *Topics in Advanced Quantum Mechanics*, Dover, New York (2014).

[Hoo71] G. 't Hooft, *Phys. Lett.* **B37**, 195 (1971).

[Hub03] R. Huber et al., *Phys. Rev. Lett.* **90**, 202301 (2003).

[Hug99] E. W. Hughes and R. Voss, *Ann. Rev. Nucl. Part. Sci.* **49**, 303 (1999).

[Hum90] E. Hummel and J. A. Tjon, *Phys. Rev.* **C42**, 423 (1990).

[Iac73] F. Iachello, A. D. Jackson, and A. Lande, *Phys. Lett.* **B43**, 191 (1973).

[Iac87] F. Iachello and A. Arima, *The Interacting Boson Model*, Cambridge University Press, Cambridge, (1987).

[Iac91] F. Iachello and P. Van Isacker, *The Interacting Boson-Fermion Model*, Cambridge University Press, Cambridge, (1991).

[Iac03] F. Iachello, in *From Nuclei and their Constitutents to Stars*, Società Italiana di Fisica, (IOS Press), 2003, p. 1.

[Isg99] N. Isgur, *Phys. Rev. Lett.* **83**, 272 (1999).

[Ish07] N. Ishii, S. Aoki, and T. Hatsuda, *Phys. Rev. Lett.* **99**, 022001 (2007).

[Ish12] N. Ishii et al., *Phys. Lett.* **B712**, 437 (2012).

[Iva15] M. V. Ivanov et al., arXiv:1506.00801v1 [nucl-th] (2015).

[Jac50] J. D. Jackson and J. M. Blatt, *Rev. Mod. Phys.* **22**, 77 (1950).

[Jac57] J. D. Jackson, S. B. Treiman, and H. W. Wyld, *Nucl. Phys.* **4**, 206 (1957) and *Phys. Rev.* **106**, 517 (1957).

[Jac75] J. D. Jackson, *Classical Electrodynamics*, Wiley, New York (1975).

[Jac12] B. V. Jacak and B. Müller, *Science* **337**, 310 (2012).

[Jaf90] R. L. Jaffe and A. Manohar, *Nucl. Phys.* **B337**, 509 (1990).

[Jaq79] F. Jaquet, A. Blondel, *Proceedings of the study of an ep facility for Europe* ed. U. Amaldi DESY 79/48 (1979).

[Jar85] C. Jarlskog, *Phys. Rev. Lett.* **55**, 1039 (1985).

[Jen92] E. Jenkins and A. V. Manohar, in *Effective Theories of the Standard Model*, ed. U-G. Meißner, World Scientific, Singapore (1992).

[Ji95] X. Ji, *Phys. Rev. Lett.* **74**, 1072 (1995).

[Ji97] X. Ji, *Phys. Rev. Lett.* **78**, 610 (1997).

[Joh63] C. H. Johnson, F. Pleasonton, and T. A. Carlson, *Phys. Rev.* **132**, 1149 (1963).

[Kal21] T. Kaluza, *Sitzungber. Pruess. Akad. Wiss., Berlin*, 966 (1921).

[Kan00] G. Kane, *Supersymmetry: Unveiling the Ultimate Laws of Nature*, Perseus, Reading, MA (2000).

[Kap98] D. B. Kaplan, M. J. Savage, and M. B. Wise, *Phys. Lett.* **B424**, 390 (1998) and *Nucl. Phys.* **B534**, 329 (1998).

[Kap99] D. B. Kaplan, M. J. Savage, and M. B. Wise, *Phys. Rev.* **C59**, 617 (1999).

[Kat12] U. F. Katz et al., *Prog. Part. Nucl. Phys.* **67**, 651 (2012).

[Kaw75] Y. Kawazoe, *Prog. Theor. Phys.* **54**, 1394 (1975) 1394.

[Kay05] *Tables of Physical and Chemical Constants*, Kaye and Laby Online, Version 1.0 (2005) at http://www.kayelaby.npl.co.uk

[Kle26] O. Klein, *Zeit. für Phys.* **A37**, 895 (1926).

[Kob73] M. Kobayashi and T. Maskawa, *Prog. Theo. Phys.* **49**, 652 (1973).

[Koc80] R. Koch and E. Pieterinin, *Nucl. Phys.* **A336**, 331 (1980).

[Kop95] S. Kopecky et al., *Phys. Rev. Lett.* **74**, 2427 (1995).

[Kov97] M. Kovash, πN *Newsletter* **12**, 51 (1997).

[Kov05] P. Kovtun, D. T. Son, and A. O. Starinets, *Phys. Rev. Lett.* **94**, 111601 (2005).

[Koz09] A. Kozela et al., *Phys. Rev. Lett.* **102**, 172301 (2009); *Phys. Rev.* **C85**, 045501 (2012).

[Kra05] Ch. Kraus et al., *Eur. Phys. J.* **C40**, 447 (2005).

[Kro54] N. Kroll and M. A. Ruderman, *Phys. Rev.* **93**, 233 (1954).

[Kro12] A. Kronfeld, *Ann. Rev. Nucl. Part. Sci.* **62**, 265 (2012).

[Kwi94] J. Kwiecinski et al., *Phys. Rev.* **D50**, 217 (1994).

[Kwi97] J. Kwiecinski et al., in *Proceedings of the Workshop on Deep Inelastic Scattering and QCD (DIS 97)*, ed. J. Repond, AIP, Chicago, (1997).

[Lam47] W. E. Lamb and R. C. Retherford, *Phys. Rev.* **72**, 241 (1947).

[Lan77] L. D. Landau and E. M. Lifshitz, *Quantum Mechanics: Nonrelativistic Theory*, Pergamon Press, Toronto (1977).

[Lap93] L. Lapikas, *Nucl. Phys.* **A553**, 297 (1993).

[Lat12] J. M. Lattimer, *Ann. Rev. Nucl. Part. Sci.* **62**, 485 (2012).

[Lee56] T. D. Lee and C. N. Yang, *Phys. Rev.* **104**, 254 (1956).

[Lee63] Y. K. Lee, L. W. Mo, and C. S. Wu, *Phys. Rev Lett.* **10**, 253 (1963).

[Lev97] A. Levy, *DESY report 97-013* (1997).

[Lia97] P. Liaud et al., *Nucl. Phys.* **A612**, 53 (1997).

[Lle72] C. H. Llewellyn-Smith, *Phys. Rep.* **3** (1972) 261 .

[Low54] F. Low, *Phys. Rev.* **96**, 1428 (1954); M. Gell-Mann and M. L. Goldberger, *Phys. Rev.* **96**, 1433 (1954).

[LRP15] *Reaching for the Horizon*, the 2015 Long Range Plan for Nuclear Science prepared for the US Department of Energy and the National Science Foundation by the Nuclear Science Advisory Committee, October 2015.

[Mac01] R. Machleidt and I. Slaus, *J. Phys. G* **27**, R69 (2001).

[Mac86] R. Machleidt, *The Meson Theory of Nuclear Forces and Nuclear Matter*, In *Relativistic Dynamics and Quark-Nuclear Physics*, eds. M. B. Johnson and A. Picklesimer, Wiley, New York (1986).

[Mai02] C. Maieron, T. W. Donnelly, and I. Sick, *Phys. Rev.* **C65**, 025502 (2002).

[Maj37] E. Majorana, *Nuovo Cim.* **14**, 171 (1937).

[Mak62] Z. Maki, M. Nakagawa, and S. Sakata, *Prog. Theo. Phys.* **28**, 870 (1962).

[Mam89] W. Mampe et al., *Phys. Rev. Lett.* **63**, 593 (1989).

[MaM93] B. Mampe et al., *JETP Lett.* **57**, 82 (1993).

[Man84] A. Manohar and H. Georgi, *Nucl. Phys.* **B234**, 189 (1984); J. F. Donoghue, E. Golowich, and B. R. Holstein, *Phys. Rev.* **D30**, 587 (1984).

[Man96] A. Manohar, arXiv:hep-ph/9606222 (1996); D. Kaplan, arXiv:nucl-th/9506035 (1995); H. Georgi, *Ann. Rev. Nucl. Part. Sci.* **43**, 209 (1995).

[Mar03] W. J. Marciano and Z. Parsa, *J. Phys.* **G29**, 2629 (2003).

[Mar06] W. J. Marciano and A. Sirlin, *Phys. Rev. Lett.* **96**, 032002 (2006).

[Mat07] V. Mateu and J. Portoles, *Eur. Phys. J.* **C52**, 325 (2007).

[McG99] P. L. McGaughey, J. M. Moss, and J. C. Peng, *Ann. Rev. Nucl. Part. Sci.* **49**, 217 (1999).

[McL11] L. McLerran, J. Dunlop, D. Morrison, and R. Venugopalan, *Nucl. Phys.* **A854**, 1 (2011).

[Meg14] G. D. Megias et al., *Phys. Rev.* **D89**, 093002 (2014).

[Meg15] G. D. Megias et al., *Phys. Rev.* **D91**, 073004 (2015).

[Meh99] T. Mehen and I. W. Stewart, *Phys Lett.* **B445**, 37 (1999) and *Phys. Rev.* **C59**, 2365 (1999).

[Mer98] E. Merzbacher, *Quantum Mechanics*, Wiley, New York (1998), Ch. 7.

[Mik85] S. P. Mikheyev and A. Yu. Smirnov, *Sov. J. Nucl. Phys.* **42**, 913 (1985); *Nuovo Cim.* **9C**, 17–26 (1986).

[Mil07] M. L. Miller, K. Reygers, S. J. Sanders, and P. Steinberg, *Ann. Rev. Nucl. Part. Sci.* **57**, 205 (2007).

[Mil07a] G. A. Miller, *Phys. Rev. Lett.* **99**, 112001 (2007).

[Mil14] R. Milner et al., (The OLYMPUS Collaboration), *Nucl. Instr. and Meth.* **A471**, 1, (2014).

[Min01] K. Minamisono, et al., *Phys. Rev.* **C65**, 015501 (2001); T. Minimisono et al., *Phys. Rev. Lett.* **80**, 4132 (1998).

[Min11] K. Minamisono, et al., *Phys. Rev.* **C84**, 055501 (2011); N. Dupuis-Rolin, *et al.*, *Phys. Lett.* **B79**, 359 (1978); R. E. Tribble and D. P. May, *Phys. Rev.* **C18**, 2704 (1978); R. E. Tribble, D. P. May, and D. M. Tanner, *Phys. Rev.* **C23**, 2245 (1981); R. D. Rosa et al., *Phys. Rev.* **C37**, 2722 (1988).

[Moh11] B. Mohanty et al., [STAR Collaboration], *J. Phys.* **G38**, 124023 (2011).

[Mon71] E. J. Moniz et al., *Phys. Rev. Lett.* **26**, 445 (1971).

[Mon09] B. Monreal and J. A. Formaggio, *Phys. Rev.* **D80**, 051301 (2009).

[Mor09] O. Moreno et al., *Nucl. Phys.* **A828**, 306 (2009).

[Mor14] O. Moreno and T. W. Donnelly, *Phys. Rev.* **C89**, 01550 (2014).

[Mor14a] O. Moreno, T. W. Donnelly, J. W. Van Orden, and W. P. Ford, *Phys. Rev.* **D90**, 013014 (2014).

[Mor15] O. Moreno et al., *J. Phys.* **G42**, 034006 (2015).

[Mor15a] O. Moreno and T. W. Donnelly, arXiv:1506.04733 [nucl-th] (2015).

[Mor15b] O. Moreno, T. W. Donnelly, J. W. Van Orden, and W. P. Ford, *Phys. Rev.* **D92** (2015) 0253006 .

[Mue90] A. H. Mueller, *Nucl. Phys. B (Proc. Suppl.)* **18C**, 125 (1990).

[Mul06] B. Müller and J. L. Nagle, *Ann. Rev. Nucl. Part. Sci.* **56**, 93 (2006).

[Mul12] B. Müller, J. Schukraft, and B. Wyslouch, *Ann. Rev. Nucl. Part. Sci.* **62**, 361 (2012).

[Mus92] M. J. Musolf and T. W. Donnelly, *Nucl. Phys.* **A546**, 509 (1992) and **A550**, 564 (1992).

[Mus94] M. J. Musolf et al., *Phys. Rep.* **239**, 1 (1994).

[Nak01] S. Nakamura, T. Sato, V. Gudkov, and K. Kubodera, *Phys. Rev.* **C63**, 034617 (2001); Erratum *Phys. Rev.* **C73**, 049904 (2006).

[Nak01a] I. Nakagawa et al., *Phys. Rev. Lett.* **86**, 5446 (2001).

[Nav01] F. S. Navarra, M. S. Robilatta, and G. Krein (eds.), *Proc. 7th Int. Workshop on Hadron Physics*, World Scientific, Singapore (2001).

[Nav09] O. Naviliat-Cuncic and N. Severijns, *Phys. Rev. Lett.* **102**, 142302 (2009).

[Nes92] V. V. Nesvizhevsky et al., *JETP* **75**, 405 (1992).

[Nic05] J. Nico et al., *Phys. Rev.* **C71**, 055502 (2005).

[Noc14] E. R. Nocera et al., (NNPDF Collaboration), *Nucl. Phys.* **B887**, 276 (2014).

[OCo72] J. S. O'Connell, T. W. Donnelly, and J. D. Walecka, *Phys. Rev.* **C6**, 719 (1972).

[Oly16] Olympus data (2016).

[Omn58] R. Omnes, *Nuovo Cim.* **8**, 1244 (1958).

[Pap03] C. N. Papanicolas, *Eur. Phys. J.* **A18**, 141 (2003).

[Par01] T.-S. Park, K. Kubodera, D.-P. Min, and M. Rho, arXiv:nucl-th/0108050 (2001).

[Pau33] W. Pauli, *Proc. VII Sovay Congress, Brussels*, 324 (1933); a translation can be found in [Bro78].

[PDG14] K. A. Olive et al., (Particle Data Group), *Chin. Phys.* **C38**, 090001 (2014); also at http://pdg.lbl.gov.

[Pec77] R. D. Peccei and H. R. Quinn, *Phys. Rev. Lett.* **38**, 1440 (1977); *Phys. Rev.* **D16**, 1791 (1977).

[Pee03] P. J. E. Peebles and B. Ratra, *Rev. Mod. Phys.* **75**, 559 (2003).

[Per99] S. Perlmutter et al., *Ap. J.* **517**, 565 (1999).

[Pes95] M. E. Peskin and D. V. Schroeder, *An Introduction to Quantum Field Theory*, Perseus, Reading, MA (1995).

[Pet08] H. Petersen, J. Steinheimer, G. Burau, M. Bleicher, and H. Stocker, *Phys. Rev.* **C78**, 044901 (2008).

[Pet00] G. G. Petratos et al., (The Hall A Collaboration), *Nucl. Phys.* **A663**, 357c (2000).

[Pia15] M. Piarulli et al., *Phys. Rev.* **C91**, 024003 (2015).

[Pla15] Copyright ESA and the Planck Collaboration.

[Poc04] D. Pocanic et al., *Phys. Rev. Lett.* **93**, 181803 (2004).

[Poc14] D. Pocanic, E. Frlez, and A. van der Shaef, *J. Phys.* **G41**, 114002 (2014).

[Pol73] H. D. Politzer, *Phys. Rev. Lett.* **30**, 1346 (1973).

[Pon57] B. Pontecorvo, *Sov. Phys. JETP* **6**, 429 (1957).

[Pon68] B. Pontecorvo, *Sov. Phys. JETP* **26**, 984-88 (1968).

[Pov08] B. Povh et al., *Particles and Nuclei: an Introduction to the Physical Concepts*, Springer, New York (2008).

[Pre62] M. A. Preston, *Physics of the Nucleus*, Addison-Wesley, Reading, MA (1962).

[Pry93] K. Prytz, *Phys. Lett.* **311**, 2861 (1993).

[Qui88] E. N. M. Quint, *Limitations of the Mean-Field Description for Nuclei in the Pb-Region, Observed with the $(e, e'p)$ Reaction*, PhD thesis, University of Amsterdam, 1988 (unpublished).

[Rad15] I. A. Radchek et al., *Phys. Rev. Lett.* **114**, 062005 (2015).

[Ran05] See, e.g., L. Randall, *Warped Passages: Unraveling the Universe's Hidden Dimensions*, Ecco Press, New York (2005).

[Rar41] W. Rarita and J. Schwinger, *Phys. Rev.* **59**, 436 (1941).

[Ras89] A. S. Raskin and T. W. Donnelly, *Ann. Phys.* **191**, 78 (1989).

[Rav99] F. Ravndal and X. Kong, *Nucl. Phys.* **A665**, 137 (2000); *Phys. Lett.* **B450**, 320 (1999).

[Rav01] F. Ravndal and X. Kong, *Phys. Rev.* **C64**, 044002 (2001).

[Rei76] F. Reines, H. S. Gurr, and H. W. Sobel, *Phys. Rev. Lett.* **37**, 315 (1976).

[Ris72] D. O. Riska and G. E. Brown, *Phys. Lett.* **B38**, 193 (1972).

[Rom64] P. Roman, *Theory of Elementary Particles*, North-Holland, Amstersam (1964).

[Ros85] For reviews see, e.g., G. Ross, *Grand Unified Theories*, Perseus, Reading, MA (1985); P. Langacker, *Phys. Rep.* **72**, 185 (1981).

[Row72] D. J. Rowe, in *Dynamic Structure of Nuclear States*, eds. D. J. Rowe, L. E. H. Trainor, S. S. M. Wong, and T. W. Donnelly, University of Toronto Press, Toronto (1972).

[Row10] D. J. Rowe, *Nuclear Collective Motion: Models and Theory*, World Scientific, Singapore (2010).

[Rub70] V. C. Rubin and W. K. Ford, Jr., *Ap. J.* **159**, 379 (1970).

[Sak67] A. D. Sakharov, *JETP Lett.* **5**, 24 (1967).

[Sak94] J. J. Sakurai, *Modern Quantum Mechanics*, (Revised Edition) Addison-Wesley, Reading, MA (1994).

[Sak14] H. Sakurai, plenary talk at joint meeting of JPS and DNP nuclear physics societies, Hawaii, October 2014.

[Sal52] E. E. Salpeter, *Phys. Rev.* **88**, 547 (1952).

[Sal69] A. Salam, *Nobel Symposium No. 8*, ed. N. Svartholm, Almqvist and Wiksell, Stockholm (1969).

[Sch55] L. I. Schiff, *Quantum Mechanics*, McGraw-Hill (1955).

[Sch98] R. Schiavilla et al., *Phys. Rev.* **C58**, 1263 (1998).

[Sch04] J. P. Schiffer et al., *Phys. Rev. Lett.* **92**, 162501 (2004).

[Sch06] K. Scholberg, *Phys. Rev.* **D73**, 033005 (1906).

[Sch07] R. Schiavilla et al., *Phys. Rev. Lett.* **98**, 132501 (2007).

[Sch14] M. D. Schwartz, *Quantum Field Theory and the Standard Model*, Cambridge University Press, New York (2014).

[Ser86] B. D. Serot and J. D. Walecka, *Adv. in Nucl. Phys.* **16**, 1 (1986), eds. J. W. Negele and E. Vogt, Plenum Press, New York. (1986).

[Ser05] A. Serebrov et al., *Phys. Lett.* **605**, 72 (2005).

[Sis04] M. Sisti et al., *Nucl. Inst. Meth.* **A520**, 125 (2004).

[Smi72] R. A. Smith and E. J. Moniz, *Nucl. Phys.* **B43**, 605 (1972).

[Sno13] Snowmass CF1 Summary: WIMP Dark Matter Direct Detection, Convenors: P. Cushman, C. Galbiati, D. N. McKinsey, H. Robertson, and T. M. P. Tait, arXiv: 1310.8327, November 2013.

[Spi88] P. E. Spivak et al., *JETP* **67**, 1735 (1988).

[Ste78] G. Steigman, D. Schramm, and J. Gunn, *Phys. Lett.* **B66**, 202 (1978); J. Yang, et al., *Ap. J.* **281**, 493 (1984).

[Ste95] G. Sterman et al., *Rev. Mod. Phys.* **67**, 157 (1995).

[Ste13] D. Steppenbeck et al., *Nature* **502**, 207 (2013).

[Sub08] R. Subedi et al., *Science* **320**, 1476 (2008).

[Sum11] K. Sumikama et al., *Phys. Rev.* **C83**, 065501 (2011).

[Sur99] B. Surrow *Eur. Phys. J.* **C1**, 2 (1999).

[Tay91] See the published versions of the Physics Nobel Prize lectures: R. E. Taylor, *Rev. Mod. Phys.* **63**, 573 (1991); H. W. Kendall, *Rev. Mod. Phys.* **63**, 597 (1991); J. I. Friedman, *Rev. Mod. Phys.* **63**, 615 (1991).

[Tow09] I. S. Towner and J. C. Hardy, *Phys. Rev.* **C79**, 055502 (2009).

[Tow10] I. S. Towner and J. C. Hardy, *Rept. Prog. Phys.* **73**, 046301 (2010).

[Tow14] I. S. Towner and J. C. Hardy, *J. Phys.* **G41**, 114004 (2014).

[Tow15] I. S. Towner and J. C. Hardy, *Phys. Rev.* **C91**, 025501 (2015).

[Tru61] T. L. Trueman, *Nucl. Phys.* **26**, 57 (1961).

[Tru88] T. N. Truong, *Phys. Rev. Lett.* **61**, 2526 (1988).

[Van78] J. W. Van Orden, PhD thesis Stanford University (unpublished, 1978).

[Van95] J. W. Van Orden, N. Devine, and F. Gross, *Phys. Rev. Lett.* **75**, 4369 (1995).

[Vet08] P. A. Vetter et al., *Phys. Rev.* **C77**, 035502 (2008).

[Von35] C. F. Von Weizsäcker, *Z. Phys.* **96**, 431 (1935).

[Von60] R. von Gehlen, *Phys. Rev.* **118**, 1445 (1960); M. Gourdin, *Nuovo Cim.* **21**, 1094 (1961).

[von90] K. von Reden et al., *Phys. Rev.* **C41**, 1084 (1990).

[Wal75] J. D. Walecka, in *Muon Physics*, Vol. 2, eds. V. W. Hughes and C. S. Wu, Academic Press, New York (1975), p. 113.

[Wal77] J. D. Walecka, *Nucl. Phys.* **A285**, 349 (1977).

[Wal95] J. D. Walecka, *Theoretical Nuclear and Subnuclear Physics*, Oxford University Press, Oxford (1995).

[Wau09] F. Wauters et al., *Phys. Rev.* **C80**, 062501(R) (2009).

[Wau10] F. Wauters et al., *Phys. Rev.* **C82**, 055502 (2010).

[Wei58] S. Weinberg, *Phys. Rev.* **112**, 1375 (1958).

[Wei67] S. Weinberg, *Phys. Rev. Lett.* **19**, 1264 (1967).

[Wei72] S. Weinberg, *Gravitation and Cosmology,* Wiley, New York (1972).

[Wei79] S. Weinberg, *Physica* **A96**, 327 (1979).

[Wei05] S. Weinberg, *The Quantum Theory of Fields*, Vols. I, II, III, Cambridge University Press, New York (2005).

[Wei11] L. Weinstein et al., *Phys. Rev. Lett.* **106**, 052301 (2011).

[Wen10] R. Wendell et al., *Phys. Rev.* **D81**, 092004 (2010).

[Wes74] J. Wess and B. Zumino, *Phys. Lett.* **B49**, 58 (1974).

[Wes75] G. B. West, *Phys. Rep.* **18**, 263 (1975).

[Wic87] V. A. Wichers et al., *Phys. Rev. Lett.* **58**, 1821 (1987).

[Wil82] F. Wilczek, *Ann. Rev. Nucl. Part. Sci.* **32**, 177 (1982).

[Wol78] L. Wolfenstein, *Phys. Rev.* **D17**, 2369 (1978).

[Wol83] L. Wolfenstein, *Phys. Rev Lett.* **51**, 1945 (1983).

[Won98] S. S. M. Wong, *Introductory Nuclear Physics*, Wiley-Interscience New York (1998).

[Yam71] A. Yamaguchi, T. Terasawa, K. Nakahara, and Y. Torizuka, *Phys. Rev.* **C3**, 1750 (1971).

[Yin92] S. Ying, W. C. Haxton, and E. M. Henley, *Phys. Rev.* **C45**, 1982 (1992); *Phys. Rev.* **D40**, 3211 (1989).

[Zha11] C. Zhang et al., (The BLAST Collaboration), *Phys. Rev. Lett.* **107**, 252501 (2011).

[Zwe64] G. Zweig, *An SU(3) Model for Strong Interaction Symmetry and its Breaking, I and II*, CERN Reports No. 8182/TH.401 and No. 8419/TH.412 (1964).

[Zwi33] F. Zwicky, *Helv. Phys. Acta* **6**, 110 (1933).